*I often used to be in doubt,
but now I'm not so sure anymore.*

Tor G Syvertsen

Kolbein Bell

An engineering approach to

FINITE ELEMENT ANALYSIS

of linear structural mechanics problems

FAGBOKFORLAGET

Copyright © 2013 by
Vigmostad & Bjørke AS
All Rights Reserved

First Edition 2013 / Printing 4, 2022

ISBN: 978-82-321-0268-6

Graphic production: John Grieg, Bergen
Cover design: The author
Typeset by: The author

Enquiries about this text can be directed to:

Fagbokforlaget
Kanalveien 51
5068 Bergen, Norway

Tel.: +47 55 38 88 00

e-mail: fagbokforlaget@fagbokforlaget.no

www.fagbokforlaget.no

All rights reserved. No part of this publication may be reproduced, stored in a retrieval system, or transmitted, in any form or by any means, electronic, mechanical, photocopying, recording, or otherwise, without the prior written permission of the publisher.

Preface

Is there really a need for yet another book on basic finite element analysis? Probably not. However, it has proved quite difficult to find a suitable text for our basic finite element course, in spite of the large number of books on the market. Such a text should preferably represent a smooth transition from courses leading up to the FEM course, not just in terms of content, but also in terms of terminology and notation. Other authors have probably been driven by similar arguments, thus explaining the many titles available.

This particular book would have been written in Norwegian, as was its predecessor on matrix theory of structures, but for a recent decision made at our university which requires all texts used in the last two years (the M.Sc part) of our 5 year engineering programme, to be in English.

The book is written for engineers and engineering students by an engineer. I have been active in and around FEM for almost 50 years, but it is some time since I did research in the field. From the very beginning my main interest has been in the programming of the method, with the purpose of making tools for solving practical engineering problems. Such FEM based tools are now embedded in sophisticated graphical user interfaces, and they may seem easy to use. However they are not, and never will be, "push-button" systems for just anyone to use. FEM is an approximate method, and its successful use requires knowledgeable users. I believe that the teaching aspect of mechanics and FEM is important, and the incredible developments in hardware technology have now given us a computational capability that I am fairly convinced will bring about a change in the way in which we teach FEM analysis to engineering students who are primarily concerned with the practical use of the method. Some chapters reflect this conviction, but much of the content is still fairly traditional. However I have included some personal comments and preferences in a short chapter at the end of the book, where I also speculate on the impact of computational power on the teaching and practical use of FEM.

The book is based on lecture notes I first prepared a long time ago; at the basic level the core material has not changed very much, but interaction with students and colleagues has clearly influenced the presentation. I therefore take the opportunity to thank many of my previous students as well as friends and colleagues at the Department of Structural Engineering at NTNU, in particular Professors Kjell Magne Mathisen and Kjell Holthe, for valuable input and discussions. I would also like to thank Professors Ray Clough and Ed Wilson of UC Berkeley who set me on my course during a one year stay at UC Berkeley in the mid 1960s, and with whom I have been fortunate enough to have had a life-long friendship. Thanks are also due to the late Professor Ivar Holand, my mentor, whose encouragement and support in the early years of my professional life were very important, as were his example and friendship until his untimely death in 2000. If Professor Greg Fenves ever gets to see this book he might recognize some passages; I sat in on his excellent FEM course during a sabbatical at UC Berkeley in 1989-90, so thanks are also due to him. Last, but by no means least, my thanks go to my wife Janet for her patience and encouragement and above all for her many corrections and improvements of my English.

Trondheim, April 2013,

Kolbein Bell

kolbein.bell@ntnu.no

Contents

Preface   v

Symbols and abbreviations   xv

Software agreement   xix

1. **Introduction**   1
    1.1 **Approach**   1
        The engineering approach   1
        The mathematical approach   3
    1.2 **A brief historical note**   5
    1.3 **Notation and layout**   7
    1.4 **Organization**   9

2. **Summary of matrix structural analysis**   15
    2.1 **Conventions, assupmtions and simple beam theory**   15
        Reference coordinates and kinematic degrees of freedom   15
        Basic assumptions   17
        EULER-BERNOULLI beam theory   17
    2.2 **The fundamental requirements in matrix notation**   21
        Computational model   21
        Force-displacement   23
        Static equilibrium   25
        Kinematic compatibility   27
        Discussion   29
    2.3 **Principle of virtual displacements (PVD)**   29
    2.4 **The system stiffness relation**   33

	2.5	Distributed loading – element load vector	37
	2.6	Transformations	39
		Summary of chapter 2	43
3.		**Summary of linear elasticity theory**	**53**
	3.1	Three-dimensional stress analysis	53
		Stress and equilibrium	53
		Strains and kinematic compatibility	61
		Stress-strain relationship	65
		Initial strain	69
	3.2	Axisymmetric stress and strain	71
	3.3	Stress and strain in two dimensions	73
		Plane stress	73
		Plane strain	75
	3.4	Stress and strain in beams	77
	3.5	Plate bending	81
		Plate kinematics	81
		Stress-strain	85
		Equilibrium	87
		Strain energy	87
4.		**Mathematical basics of FEM**	**91**
	4.1	The approximate nature of FEM – the basic assumption	91
	4.2	Example problem	97
	4.3	Strong and weak form	97
	4.4	Principle of minimum potential energy (PMPE)	103
	4.5	The RAYLEIGH-RITZ method	103
		Equilibrium in the R-R process	109
		Accuracy and convergence of the R-R process	111
		Problems with classical R-R solutions	111
		RAYLEIGH-RITZ as a finite element method	113
	4.6	Weighted residual methods – GALERKIN	119
	4.7	General formulation of FEM using PVD	123
5.		**Element analysis I** Natural coordinates and interpolation	**137**

5.1	**Types and classifications of elements**	137
	Characteristics of an individual element	137
	Classification of elements	139
5.2	**Natural coordinates**	139
	One-dimensional element	141
	Plane (2D) rectangle	143
	The cuboid	143
	Plane triangle – area coordinates	145
	Tetrahedron	151
5.3	**Polynomials**	153
5.4	**Nodal points and degrees of freedom**	155
5.5	**The shape functions**	157
5.6	**Indirect interpolation – generalized displacements**	163
5.7	**Direct interpolation**	169
	C^0 elements	169
	C^1 elements	179
	Hierarchic C^0 elements (2D)	181
	Summary of chapter 5	185

6. Element analysis II
Mapping and numerical integration — 191

6.1	**Mapping – isoparametric formulation**	191
	4-node C^0 quadrilateral	193
	Higher order C^0 quadrilaterals – curved edges	203
	Triangular, isoparametric C^0 elements	205
6.2	**Numerical integration**	207
	One-dimensional element	209
	Two- and three-dimensional elements	217
6.3	**Integration schemes**	219
	Full integration	219
	Reduced integration	221
	Selective reduced integration	221
6.4	**Integration and convergence**	221
6.5	**Element instabilities – mechanisms**	225

Summary of chapter 6 — 229

7. Element analysis III
Element loads and stresses — 235

7.1 Static equivalent nodal point loads — 235
    Consistent element load vector — 235
    Load lumping — 237

7.2 Stress recovery and stress smoothing — 245
    Stresses from computed displacements — 247
    Interpolation / extrapolation — 249
    Nodal point averaging — 251
    Global smoothing — 251
    Stresses from nodal point forces — 255

8. Accuracy and convergence — 265

8.1 Energy bounds in R-R consistent FEM solutions — 269

8.2 Error and rate of convergence — 271

8.3 Error estimates — 275

8.4 Convergence criteria — 277

8.5 Element tests — 281
    Eigenvalue test — 283
    The patch test — 285

8.6 Exact solution at the nodal points — 291

Summary of chapter 8 — 293

9. System analysis — 295

9.1 Mesh generation — 295

9.2 Storage formats and node renumbering — 299
    Storage formats — 299
    Renumbering schemes — 301

9.3 Assembly of K and R — 303

9.4 Boundary conditions – an overview — 305
    Definitions — 305
    Implementation — 309
    Formal treatment — 309

9.5	Boundary conditions – elimination of *dofs*	311
9.6	Boundary conditions – LAGRANGE multipliers	317
9.7	Boundary conditions – penalty functions	321
9.8	Boundary conditions – rigid elements	327
9.9	Solution of **Kr** = **R**	329
9.10	Static condensation and substructure analysis	333
9.11	Numerical issues	339

10. Programming issues — 345

10.1	Programming paradigms and languages	347
10.2	Data structures and storage formats	349
10.3	Stiffness matrix for an isoparametric element	359
10.4	Two typical FEM programs	365
	Program **CrossX**	365
	Program **FEMplate**	367

11. Plane stress and plane strain — 369

11.1	Triangular elements	371
	The linear triangle	371
	The quadratic triangle	373
	The cubic triangle	373
	Formulation for computer implementation	375
11.2	Quadrilateral elements	385
	4-node element – the basic version	385
	4-node element – incompatibleincompatible version	391
	Stabilization – hourglass stiffness	399
	Higher order quadrilateral elements	401
	Discussion	403
11.3	Boundary conditions and singularities	405

12. Axisymmetric stress analysis — 409

12.1	Axisymmetric loading	409
	Isoparametric elements	415
	Load vectors	415

xi

Boundary conditions	417
Convergence	417
12.2 Non-symmetric loading	419

13. Three-dimensional stress analysis — 425

13.1 The basics — 425
- Theory of elasticity — 425
- Element shapes and natural coordinates — 427

13.2 Common solid elements — 429
- Hexahedral elements — 429
- Tetrahedral elements — 433

14. Bending of beams and plates — 435

14.1 The two-dimensional beam problem — 435
- Euler-Bernoulli beam theory — 437
- Timoshenko beam theory — 437
- Mindlin beam theory — 443
- A discrete Kirchhoff element — 451

14.2 Beam element in 3D space — 453
- Deformation stiffness for a double symmetric cross section — 453
- Including the rigid body modes — 455
- Arbitrary cross section — 455
- Transformation to global axes — 457
- Offset nodes – eccentricity — 459

14.3 Triangular (thin) plate bending elements — 461
- The Morley triangle – **T6** — 463
- Cubic triangle – **T10** — 471
- Quartic triangle – **T15** — 471
- The quintic triangle – **T21** and **T18** — 473

14.4 Mindlin plate bending elements — 479

14.5 Discrete Kirchhoff elements — 485

14.6 A hybrid 9-node triangular element — 489

14.7 A note on boundary conditions — 491

14.8 Comparison of some plate elements — 491

15. Arches and shells — 499

- 15.1 Curved beams and arches — 501
- 15.2 The shell problem — 507
    - Shells of revolution — 507
    - Flat shell elements — 511
    - Thick shell elements — 517
    - Solid elements in shell analysis — 517

16. St. Venant torsion — 519

- 16.1 St. Venant torsion – theoretical basis — 521
- 16.2 Finite element torsion analysis — 525
    - Shear stress distribution and torsional stiffness — 529
    - Position of shear centre — 531
- 16.3 Finite element shear analysis of prismatic beam — 533
    - Theoretical approach — 535
- 16.4 Numerical examples — 537
    - Rectangular, massive sections — 537
    - Massive circular sections — 537

17. Practical use of FEM — 541

- 17.1 The Sleipner accident — 541
- 17.2 Advice and guidelines — 547
    - Know your program and your own limitations — 547
    - From structure to FEM model — 551
    - Interpretation and assessment of results — 557

18. Personal comments — 563

- 18.1 The past — 563
- 18.2 The present — 569
- 18.2 The future — 573

19. References — 575

APPENDIX

A	**Matrix algebra**	581
	A.1 Definitions	581
	A.2 Addition and multiplication	587
	A.3 Matrix partitioning – submatrices	593
	A.4 Determinants	593
	A.5 Linear dependencies – rank	599
	A.6 Liear systems of equations	601
	A.7 Matrix inversion	601
	A.8 Quadratic forms and definiteness	605
	A.9 The eigenvalue problem	607
	A.10 Matrices and differentiation	609
	A.11 Least square approximation	613
B	**Coordinate transformation**	617
C	**Numerical integration**	625
	C.1 Quadrilaterals	625
	C.2 Triangles	627
D	**Source code**	629
	D.1 Introduction	629
	D.2 Subroutine MPQ61	630
	D.3 Subroutine MPQ62	634
	D.4 Subroutine MPQ63	637
	D.5 Subroutine MPQ64	641
	D.6 Subroutine SHPQ49	644
	D.7 Subroutine GAUSQ2	647
Index		649

Symbols and abbreviations

The following is a list, in alphabetical order, of the principal symbols used in the text. Some of the symbols have a page reference to where they are introduced or defined. It should be noted that some of the symbols in the list may, in some cases, also be used with a different meaning than that given in the list.

Scalar quantities

These are represented by italic Latin or ordinary Greek letters.

A	-	area
B	-	bulk modulus
D	-	flexural rigidity of a plate
E	-	modulus of elasticity
G	-	shear modulus
H	-	load potential
I	-	second moment of area (moment of inertia)
J	-	Jacobian (determinant of the Jacobian matrix), functional
L	-	length parameter
M	-	bending moment
N	-	axial force, shape function
P	-	concentrated (external) load
Q	-	generalized force
R	-	nodal point force at system level
S	-	nodal point force at element level
T	-	temperature, torque
U	-	strain energy
V	-	shear force
c	-	curvature
f	-	flexibility coefficient
h	-	thickness/depth parameter, characteristic element size
k	-	stiffness coefficient
p	-	load intensity (force per unit length), polynomial degree
q	-	generalized displacement
r	-	nodal point displacement at system level
u,v,w	-	displacement components (in x-, y- og z-direction, respectively); v also denotes a nodal point displacement at element level
x,y,z	-	cartesian coordinates; note that in case of some 2D analyses

	(frames and trusses) we operate in the *x-z* plane
Π	- total potential energy
α	- angle, coefficient of thermal expansion, shear parameter (p 439)
β	- angle
γ	- shear strain
ε	- normal strain
θ	- rotation of cross section, rate of twist
κ	- shear deformation factor (p 437)
λ	- LAMÉS constant
ν	- POISSON's ratio
ρ	- radius of curvature
ϕ	- slope of beam axis, angle of twist
σ	- normal stress
τ	- shear stress
ξ, η, ζ	- natural (dimensionless) coordinates
ζ_i	- area ($i = 1, 2, 3$) or volume ($i = 1, 2, 3, 4$) coordinates

Matrices and vectors

Matrices and vectors are identified by boldface type. An arbitrary element of a matrix is designated by the same letter as the matrix, in *italic* but with ordinary print (not bold), and with indices indicating row and column number. A_{ij}, for instance, designates the element in row *i* and column *j* og matrix **A**.

A	- designates the relationship between nodal point displacements **v** and generalized displacements **q** (p 165)
a	- compatibility matrix (p 27)
B	- strain-displacement matrix (p 125)
b	- equilibrium matrix (p 27)
C	- elasticity (or constitutive) matrix (relating **σ** and **ε**, p 67)
c	- curvature vector
D	- designates normally a diagonal matrix
F	- system flexibility matrix, volume forces
f	- element flexibility matrix
g	- equilibrium matrix (p 25)
H	- constraint matrix (p 309)
I	- unit or identity matrix
J	- Jacobian matrix
K	- system stiffness matrix (**Kr = R**)
k	- element stiffness matrix (**kv = S**)
L	- lower triangular matrix
m	- moment vector (p 85)
N	- shape function matrix
q	- generalized displacement vector (element level)
R	- system load vector (matrix)
r	- nodal point displacements at system level

S	- nodal point forces at element level
S^0	- element load vector ("fixed-end" forces)
T, t	- transformation matrices
U	- upper triangular matrix
u	- vector of displacement components
v	- nodal point displacements at element level
0	- the null matrix
Δ	- operator matrix
ε	- strain vector
σ	- stress vector
Φ	- traction vector

Indices

A superindex on a vector or matrix symbol normally designates an element number, while a subindex designates a nodal point number.

T as superindex on a vector or matrix symbol *always* designates the *transpose* of the vector or matrix.

An asterisk (*) as superindex on a vector or matrix symbol indicates that the symbol is associated with a state of pure deformation (no rigid body motion included)

A horizontal bar on top of a symbol, *e.g.* $\bar{\mathbf{k}}$, indicates that the element or elements associated with the symbol are referred to the global, reference coordinate axes.

A tilde (~) on top of a symbol desigantes normally a *virtual* quantity.

0 as superindex on a vector or matrix symbol, *e.g.* **S**^{0}, designates a "fixed end" situation (**v** = **0**).

Abbreviations

BVP	- boundary value problem
DKT	- discrete KIRCHHOFF triangle
CST	- constant strain triangle
dof	- degree of freedom
dofs	- degrees of freedom
FEM	- finite element method
GB	- giga byte
GBS	- gravity base structure
GUI	- graphical user interface
LST	- linear strain triangle
MTS	- matrix theory of structures
OOP	- object oriented programming
PC	- personal computer
PMPE	- principle of minimum potential energy

PVD	- principle of virtual displacements
QED	- quod erat demonstrandum (which was to be shown)
RAM	- random access memory
R-R	- Rayleigh-Ritz method
RCM	- Reverse Cuthill-McKee
SPR	- superconvergent patch recovery
SSD	- solid state disk

Software agreement

As the owner of this book you are entitled to download the two Windows-programs **FEMplate** and **CrossX** to your own personal PC from

http://www.fagbokforlaget.no/fem

These programs are used in connection with examples in the book, and a brief description of the them is given at the end of Chapter 10. Both programs were developed some ten years ago for earlier Windows platforms; they are no longer maintained, but they seem to function without problems on current platforms (*e.g.* Windows 7).

Permission to install and use **FEMplate** and **CrossX** is given subject to the following conditions:

- there is no guarantee that they will work,
- no support is available, and
- all use of the programs and of results obtained by the them, directly or indirectly, is the user's own responsibility.

And

It is a breach of this agreement to pass on the licence information below to anyone who does not own a copy of this book.

In order to install a full version of the programs you need the following *licence information*:

For **FEMplate**:

Licence code : e029 L4L6 79KU z50R
User : Book Owner
Organization : FEM

For **CrossX**:

Licence code : U061 n7n4 39vM R28w
User : Book Owner
Organization : FEM

Note that all three parts of the licence information must be typed into the registration dialog boxes exactly as they appear above.

Good luck!

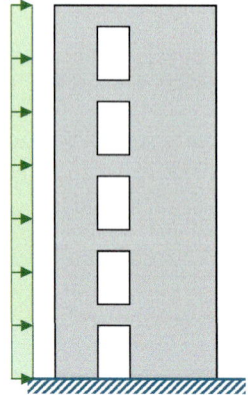
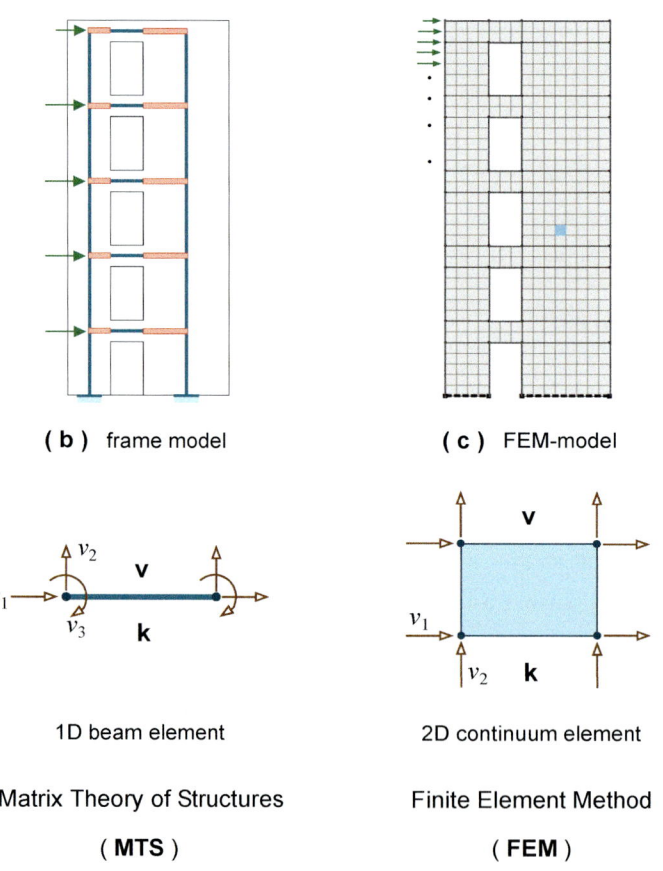

Figure 1.1 Modelling of a wall diaphragm – engineering approach

1

Introduction

Here we set the scene and define our purpose and scope. We introduce some basic terminology, explain our notation and layout, include a very brief historical note and describe the organisation of the material included in the book.

The problems we aim to solve belong to the rather wide area of *structural mechanics*, though we limit the scope by concentrating on *linear static* problems. The target parameters are stresses and displacements, and most of our presentation will be in a language familiar to the structural engineer. Having said this it should be pointed out that much of the methodology is problem independent and can be applied to any physical problem governed by differential equations.

1.1 Approach

The finite element method, or FEM as we shall use most of the time, can be explained from basically two points of view: the *physical* or *engineering view* and the *mathematical view*. There is of course an area of overlap between these two approaches; the engineering approach cannot do without some mathematics, and being a numerical method, the mathematical approach needs parameters and principles from the physical world.

The engineering approach. The basic idea is one of *divide and conquer*; a system with a complex behaviour is divided into subsystems with "known" behaviour. The wall diaphragm of Fig. 1.1 may serve as an example. The figure shows two discretized models of the diaphragm, one consisting of an assemblage of one-dimensional beam elements, Fig. 1.1b, and one in which the wall is modelled by rectangular "continuum" elements, Fig. 1.1c. Both models assume the structure to be replaced by *elements* that are connected at *nodal points* or *nodes*, and their behaviour is completely defined by certain displacement parameters (degrees of freedom) at the nodal points.

The behaviour of the individual beam elements is expressed by the relationship between the nodal displacements **v** and the corresponding nodal forces **S**, that is

$$\mathbf{S} = \mathbf{k}\mathbf{v} \tag{1-1}$$

where **k** is the element's *stiffness matrix*. This matrix is readily established by

Introduction

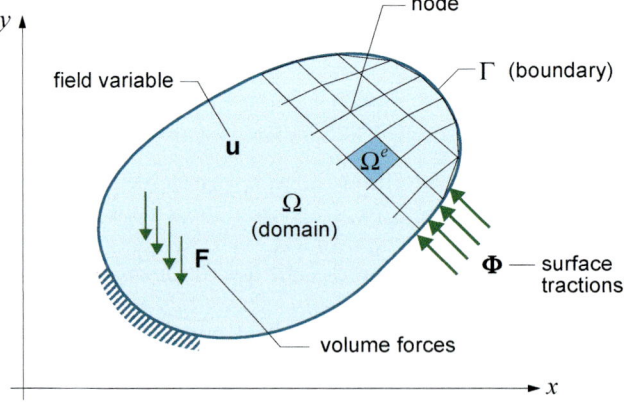

Figure 1.2 Elastic body in two dimensions

standard structural principles. We assume the reader is familiar with this procedure, dealt with in courses entitled *matrix theory of structures* (MTS) or something similar.

The engineering approach to FEM generalizes the basic idea of MTS, that is the notion of elements of a certain geometry interconnected at nodal points (with some unique position relative to the element geometry), and whose behaviour is described by Eq. (1-1). The key step is to determine **k**, a task which the engineering approach solves by using mainly physical arguments and the principle of *virtual displacements* (PVD).

Once the element behaviour is determined, the elements are *assembled* into a *system* by requiring *kinematic compatibility* and *static equilibrium* at all nodes. For the frame model in Fig. 1.1 compatibility is satisfied by simply requiring compatibility at the nodal points. For the 2- and 3-dimensional continuum finite elements this is obviously not enough; we have to make sure that certain continuity requirements along the element borders (edges or surfaces) are satisfied as a direct consequence of compatibility at the nodes. The assembly process leads to the *system stiffness relation*,

$$\mathbf{Kr} = \mathbf{R} \qquad (1\text{-}2)$$

where **K** is the *system stiffness matrix*, **r** is a vector of the nodal point displacements (which at the system level are denoted by the letter *r*) and **R** is a vector of (known) nodal forces (loading). The mathematical problem of the wall diaphragm which is governed by differential equations, has been transformed, through the discretization process, into a system of linear, algebraic equations. This system is solved with respect to the unknown displacement components **r**, with due regard to the problem's boundary conditions.

Basically the only difference between MTS and FEM is the element analysis. While we can always use the finite element techniques to establish the (stiffness) relations for one-dimensional beam and bar elements, we *cannot* use standard MTS methods to find the corresponding relations for 2- and 3-dimensional elements.

The mathematical approach. In this more abstract world the terminology is different. The finite element method is a procedure for finding an approximate solution to a *boundary value problem* (BVP) defined over a domain Ω with boundary Γ. The *field variables* (**u**), *e.g.* displacement components, are the unknown parameters, while the volume forces (**F**) and surface tractions (**Φ**) represent the known *actions*, see Fig. 1.2.

The field variables (**u**) are approximated by *assumed* local, piece-wise interpolations over sub-domains (Ω_e), expressed in terms of the values of the field variables at the nodes (and possibly their derivatives). The union of these *shape functions* represents test or *trial functions* that are used to determine the stationary value of a *functional* in a *variational formulation* or to minimize an *error residual* in a *weighted residual method*.

While FEM, as we know it, was conceived by engineers in the 1950s and made into a powerful technique by engineers in the 1960s, the mathematical approach that emerged in the late 1960s, was necessary to explain questions concerning *convergence* and *errors*. It also provides a more consistent frame-

Introduction

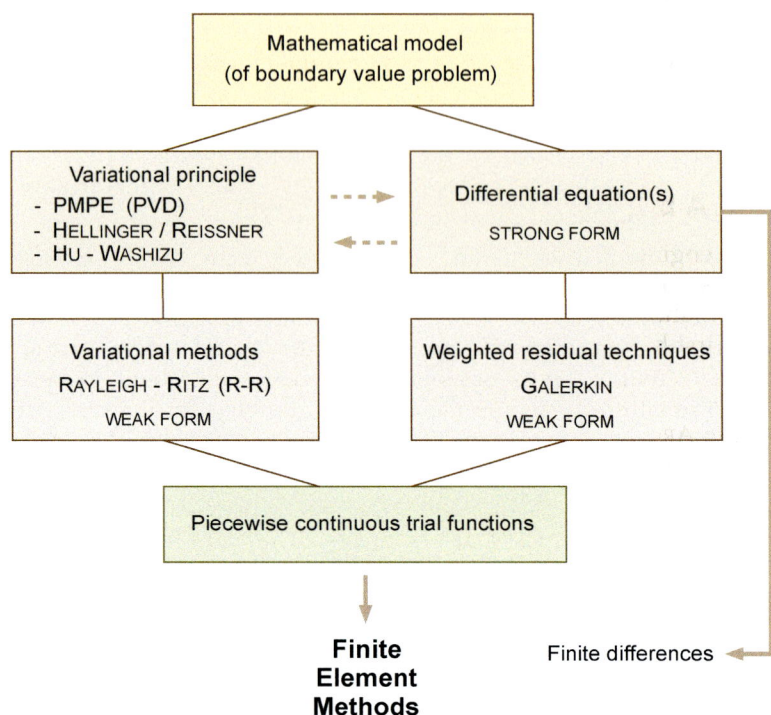

work for the development of "new" elements, although engineering need and intuition have played an important role in the development of the method.

This presentation is in the spirit of the engineering approach, and although we cannot make do without some mathematics and mathematical manipulations, we try to use physical arguments and a language familiar to engineers whenever we can.

1.2 A brief historical note

The engineering approach is based on classical structural mechanics principles which have their roots in the eighteenth century, and was properly established by MAXWELL and MOHR around 1870. Important contributions in the first half of the twentieth century were made by CROSS in 1932 (moment distribution), HRENIKOFF in 1941 (framework method) and McHENDRY in 1944 (lattice analogy). For the finite element method the fundamental references are to ARGYRIS [1], TURNER et al [2] and CLOUGH [3], the latter being the first to use the term **finite element**.

The mathematical approach can be traced back to giants like EULER and LAGRANGE (calculus of variation), and later to RAYLEIGH (1880), RITZ (1909), GALERKIN (1915) and COURANT (1941). The first comprehensive text, by STRANG and FIX [4], appeared in 1973.

It is too early to write the history of the finite element method, but it should come as no surprise if the three big names of FEM turn out to be, in alphabetical order, John ARGYRIS (London/Stuttgart), Ray CLOUGH (Berkeley) and Olgierd (Olek) ZIENKIEWICZ (Swansea). If we confine ourselves to the "pioneering" period, the late 1950s and the 1960s, names like Bruce IRONS (Swansea), FRAEIJS de VEUBEKE (Liège), Theodore PIAN (MIT), Richard GALLAGHER (Cornell), Robert TAYLOR (Berkeley), Robert MELOSH (University of Washington/Boeing), Ed WILSON (Berkeley) and John TINSLEY ODEN (Oklahoma State University) will also most certainly qualify for a place in the "FEM hall of fame", as will Carlos FELIPPA (Berkeley/Boeing).

Alf SAMUELSSON (Chalmers) and Ivar HOLAND (NTH) were of great importance for the development of FEM in the Nordic countries.

The development of the finite element method goes hand in hand with the development of the digital computer; without this device no FEM. The synthesis of FEM is the computer program that implements the series of numerical operations required to produce a usable result. Typical of any FEM based program is the need to solve large systems of linear, algebraic equations. From the crude computers of the late 1950s and early 1960s, when solving a system of a couple of hundred unknowns was quite a feat, to today's computers, which enable us to solve systems with several million unknowns, FEM has evolved to become an effective and powerful tool for most branches of engineering.

Perhaps the most remarkable development over the past 50 years is the incredible reduction in cost of computer computations. From being an expensive tool that only universities, research organizations and large companies could afford in the 1960s, affordable PCs, with computational

Introduction

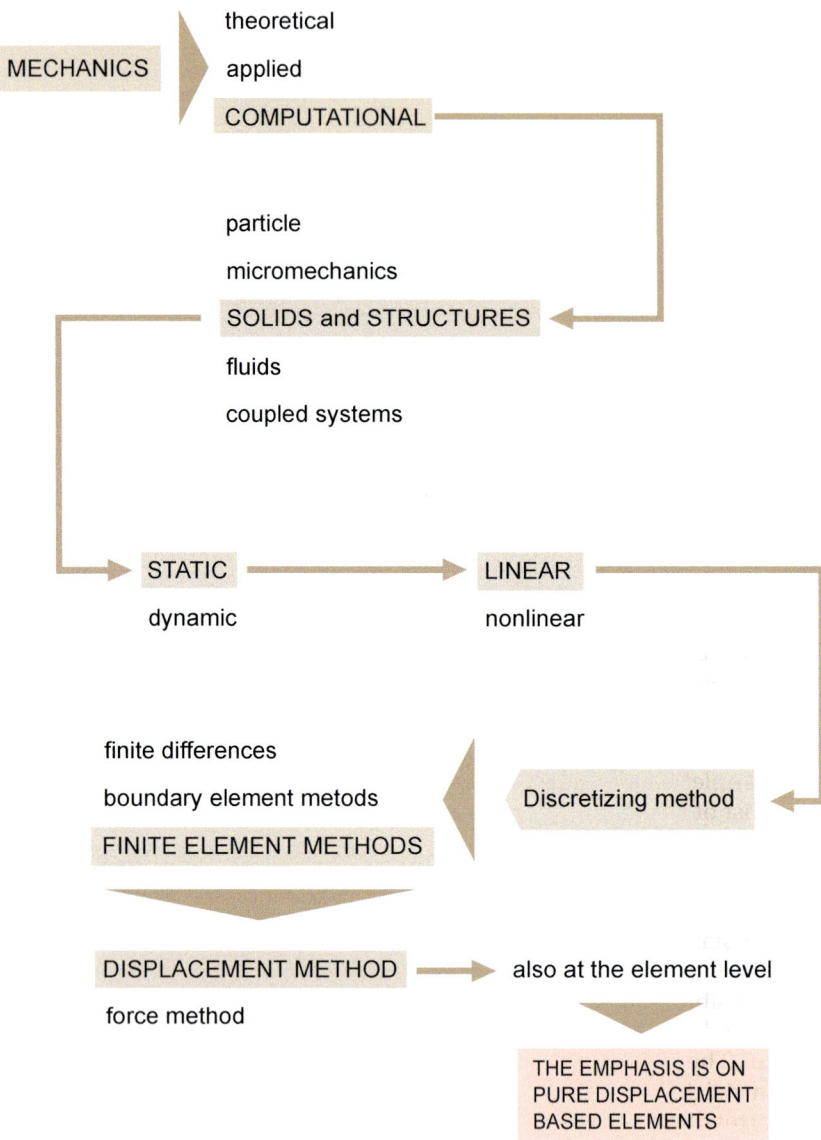

The content of the book in the bigger picture

capabilities beyond anyone's imagination in the early days of FEM, are now available, as desk-tops or laptops, to every engineer and engineering student. And MOORE's law, that postulates a doubling of the number of transistors on a given area of silicon (or chip performance) every second year (modified to every 18 months), seems to still hold, nearly 50 years after it was first proposed.

To sum it up in one sentence we can say that FEM, as an engineering tool, is a mix of *mechanics*, *numerical analysis* and *computer science* (programming).

1.3 Notation and layout

Both MTS and FEM use many symbols, and apart from the symbols used for some of the most basic concepts, no one set of symbols seems to have gained wide acceptance. If you pick up ten textbooks on these themes you will probably find that no two books use the same symbols. The symbols used in this text are quite similar to those we started out with at our university when it all started in the 1960s; they were at the time strongly influenced by ARGYRIS who came to Trondheim in 1963 and gave his Trondheim lectures (which he kept referring to, although few if any have seen a printed version of these lectures). Hence, our symbols are perhaps more historically than logically motivated, but they are now well established in our courses and we see no compelling reason to change to another set of symbols.

As for coordinate systems and sign conventions we believe we are well placed in mainstream, but the reader with a different background than our students will probably find "irregularities" also here. Hopefully we are reasonably consistent throughout the book, and that is perhaps the best a student of FEM can expect in terms of notation when she or he picks up a text on this topic.

A complete list of the symbols we use is included in the front of the book, as is a list of the abbreviations used.

The layout of this book is a bit unusual. You will find text only on the right-hand sides. The left-hand sides are reserved for figures, illustrations and other items (some of which are blocks of text) that can support and aid in understanding the text on the opposite page. This has both pros and cons. The obvious advantage is that figures and illustrations can be placed as close as possible to the text they are meant to clarify, and a figure can easily be repeated in order to avoid the reader having to turn back to its first appearance. The disadvantage is that this layout will inevitably lead to more pages than strictly necessary since it is not always possible to find enough useful information to fill the left-hand pages.

However, you will not find a totally blank left-hand side, although some have very little information on them. The space available on left-hand pages has enabled us to include short "biographies" of many of the giants we often make references to, such as HOOKE's law, GAUSS quadrature, EULER-BERNOULLI beam theory etc. Most of these multitalented individuals are impressive to say the least, and all engineering students ought to know something about them. We provide but a glimpse, taken from such sources as KURRER [5], TIMOSHENKO [6], and the internet, mainly Wikipedia, the free encyclopedia.

Introduction

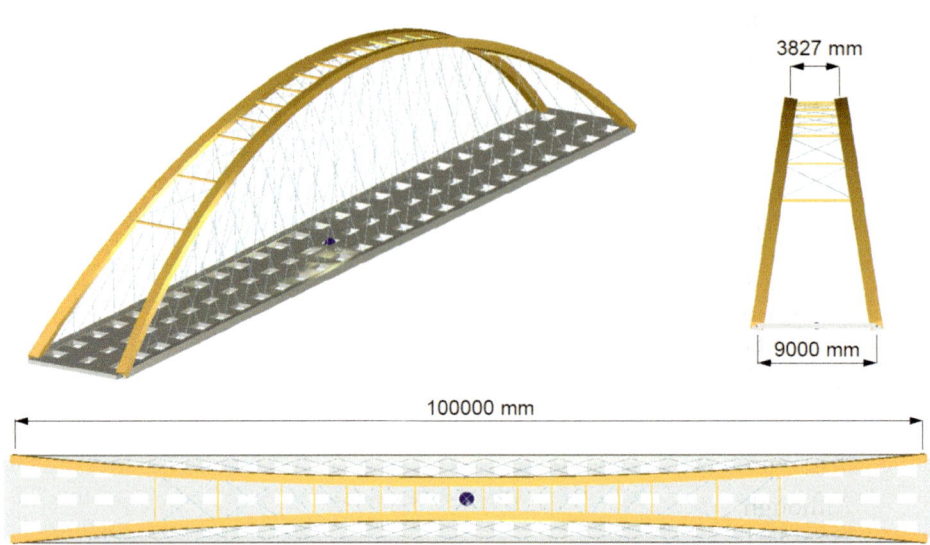

FEM model of a network arch bridge with timber (glulam) arches and concrete deck

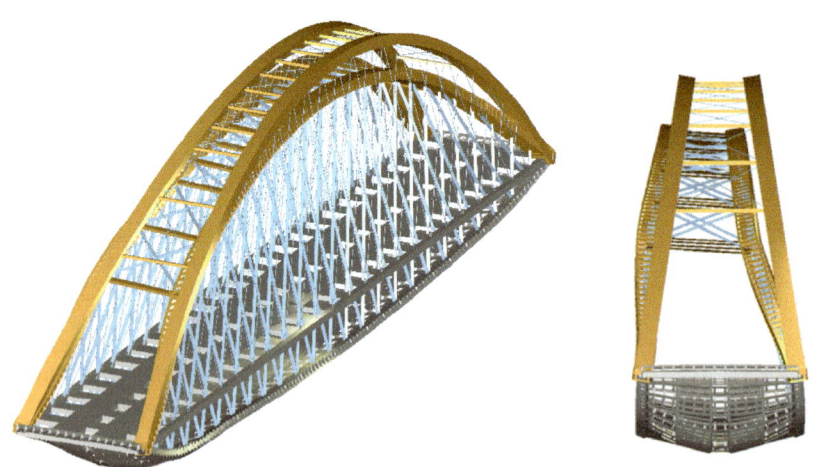

Deformation of bridge subjected to critical (ultimate limit state) loading – grossly exaggerated

Figure 1.3 Example of FEM analysis; bridge modelled by straight 1D bar and beam elements

The interested reader is encouraged to visit these and other sources for further information.

 This symbol in the left-hand margin signals an important topic or statement.

1.4 Organization

The material of a book like this can be organized in various ways. Since we emphasize the engineering approach it might have been most appropriate to present the method and its techniques in connection with specific applications. However, we have chosen almost the opposite line of action: first we explain the underlying concepts and basic techniques, more or less independent of the applications, and then we show how the method is applied to specific problems of structural mechanics. The advantage of this approach is that the problems can be addressed and fully explained in a logical sequence. The downside is that a lengthy and slightly "theoretical" prelude may stretch the engineering student's patience.

Although there is a certain progression from chapter to chapter, it is quite possible to teach certain chapters in a sequence different from that presented here.

We assume the reader is familiar with basic matrix theory of structures, but in order to refresh this knowledge we include in Chapter 2 a brief recap of the main relationships of linear (1st order) structural analysis in matrix notation. This chapter also introduces our coordinate systems, sign conventions and basic notation and terminology.

Chapter 3 contains a summary of some fundamental equations for two- and three-dimensional theory of elasticity, as well as the assumptions and simplifications made in some of the methods we seek FEM solutions for.

Chapter 4 presents the mathematical basics of FEM. We try to demonstrate the mathematical justification of the finite element method within a physical framework. A very simple example is used to show how the method can be derived using different principles and also how it ties up with the classical RAYLEIGH-RITZ method. We do not present a rigorous mathematical treatment, but hopefully the chapter will explain the most central concepts, connections and terminology. The chapter concludes that, for our purpose, the *principle of virtual displacements* (PVD) is the preferred vehicle for the detailed derivations.

The next three chapters are devoted to the analysis of the individual elements, that is the *element analysis*. This is where FEM differs from MTS, and we present the "tools of the trade" for the development of displacement based finite elements. Chapter 5 deals with *natural coordinates* and *interpolation*, while Chapter 6 is devoted to *mapping* and *numerical integration*; the *isoparametric* concept is introduced in this chapter. Finally in Chapter 7 our task is to convert distributed element *loading* into statically equivalent concentrated nodal forces corresponding to the nodal point displacements or *dofs* (degrees of freedom), and we also address the problem of *stress recovery*, that is, the determination of element stresses once the prime variables, the nodal displacements, have been determined.

Introduction

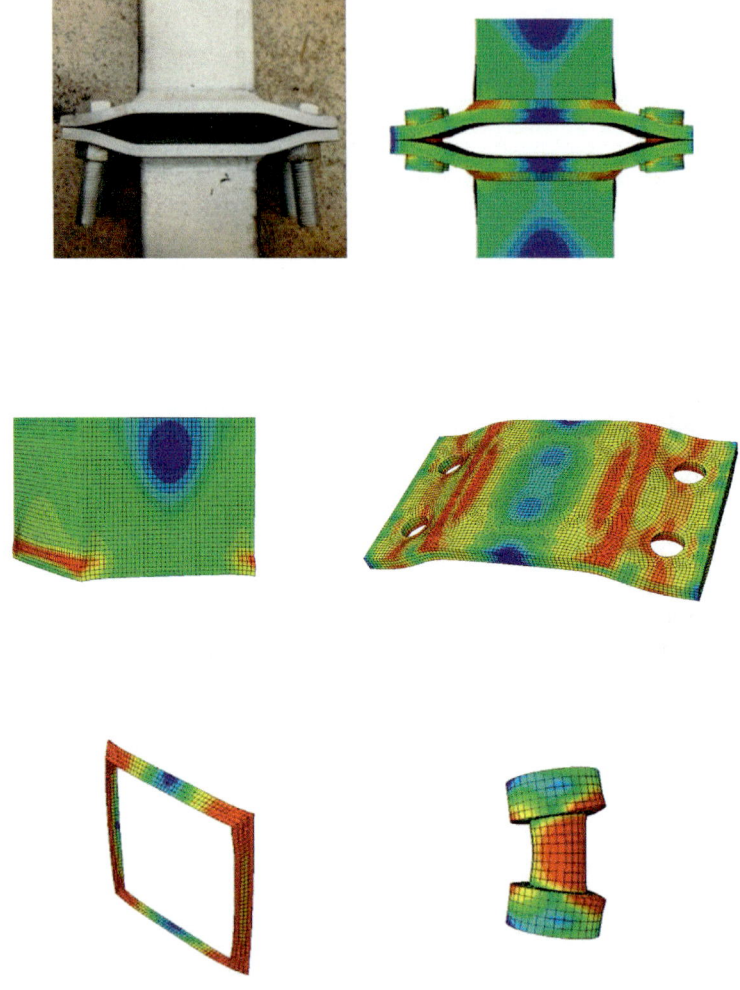

Figure 1.4 Example of FEM analysis; detail of a steel connection analysed by 3D solid elements

Organization

In Chapter 8 we take a closer look at *accuracy* and *convergence*; we formulate necessary and sufficient requirements for convergence to the correct answers. We also describe some useful element tests, in particular the all-important *patch* test.

Chapter 9 is devoted to the *system analysis*. This consists of the *assembly* process, the systematic formation of the system stiffness matrix **K** and load vector **R**, the handling of the *boundary conditions*, and the *solution* of the system stiffness relation, **Kr** = **R**, with respect to the unknown nodal point displacements **r**. This chapter is also concerned with automatic or semi-automatic *mesh generation*, storage formats for the system stiffness matrix **K** and automatic node renumbering schemes (in order to reduce the amount of storage space and solution time). A partial solution technique, referred to as *static condensation*, and related subjects like *substructure analysis* and *super-elements* are also discussed, as are potential numerical problems.

Some *programming issues* are dealt with in Chapter 10. This is a vast problem area, and we only scratch the surface by indicating how one (procedural) approach to the numerical manipulations of the problem (as opposed to the user interface) can be implemented. The chapter concludes with a brief description of a couple of locally developed FEM programs, which are believed to be fairly representative and which the owner of the book can download to her or his PC.

Finally we are ready for the application of FEM to specific problems of structural mechanics, and we start with *in-plane loading of flat plates* in Chapter 11. The emphasis is on the most common displacement-type elements of triangular and quadrilateral shape. We also look at some of the "tricks" applied to improve performance and/or avoid potential problems.

Two short chapters follow, one on *axisymmetric stress analysis*, Chapter 12, and one on *volume elements*, Chapter 13.

Plate bending is the main theme of Chapter 14, but we also pay a visit to the simpler but related problem of bending of straight beams. Most of this chapter deals with triangular shaped plate elements.

Chapter 15 looks at *curved structures*, such as *arches* and various forms of *shell structures*.

Chapter 16 deals with St. Venant torsion, a problem well suited for finite element analysis.

In Chapter 17 we address the important, but difficult problem of modelling practical engineering problems. Efficient and safe use of the method is not just a question of having access to powerful computer programs. Without knowledge and experience, FEM can easily become an exercise in "colourized mechanics". A large part of the chapter is a description of one of the most costly FEM errors ever made, the error that was in part responsible for sending a huge reinforced concrete offshore platform (Sleipner) to the bottom of the sea.

As an "old timer" in the FEM "game" the author rounds off the book with some personal comments.

Introduction

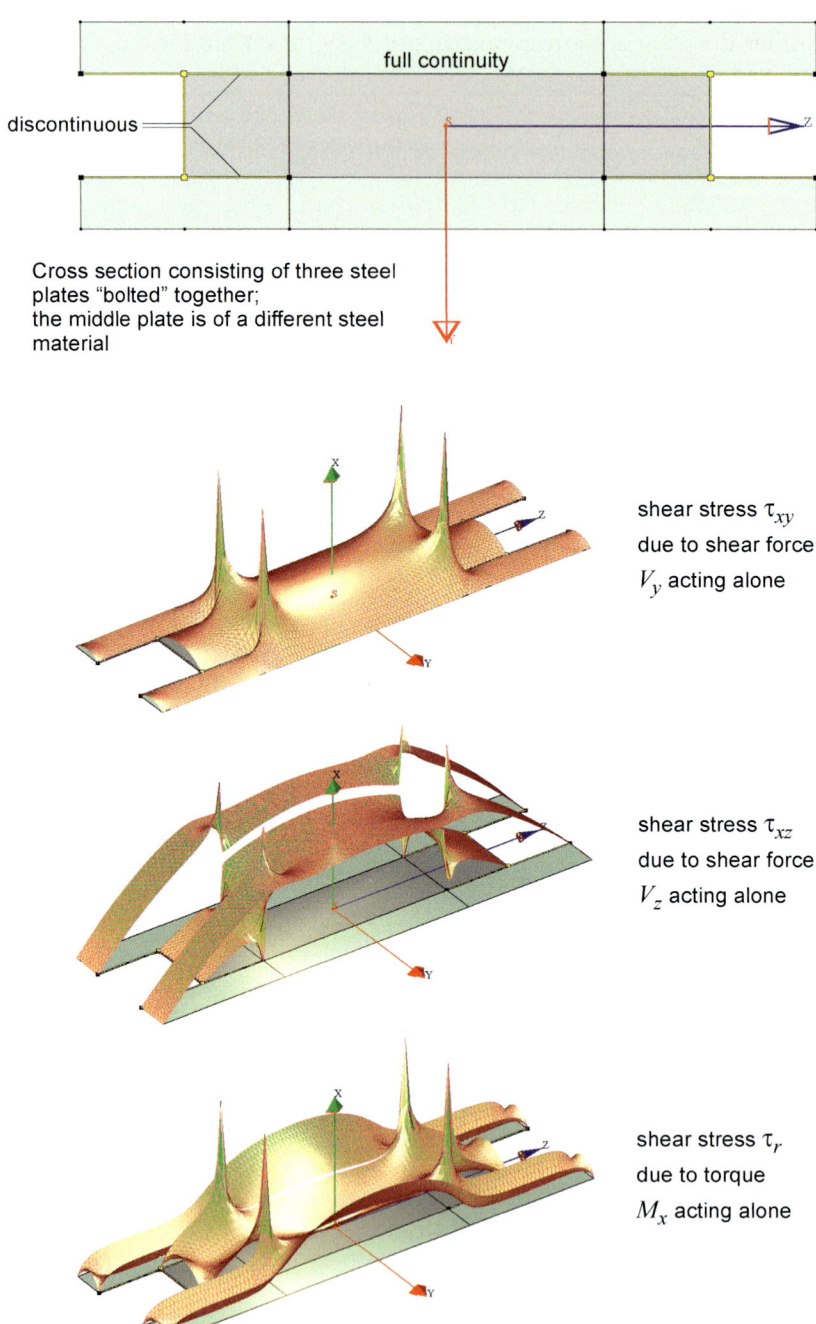

Figure 1.5 Example of FEM analysis; cross section analysed by approximately 6800 elements (6-node triangles)

Four appendices are included at the end. One summarizes, without much proof, definitions, terminology and some useful "rules" of matrix algebra. Another deals with coordinate transformation in 3D, while a third presents some useful information about integration. The last appendix contains some Fortran code for the determination of stiffness matrices of quadrilateral plane stress and plane strain finite elements.

The book includes a fair number of examples and a few problems can be found at the end of some, but not all chapters.

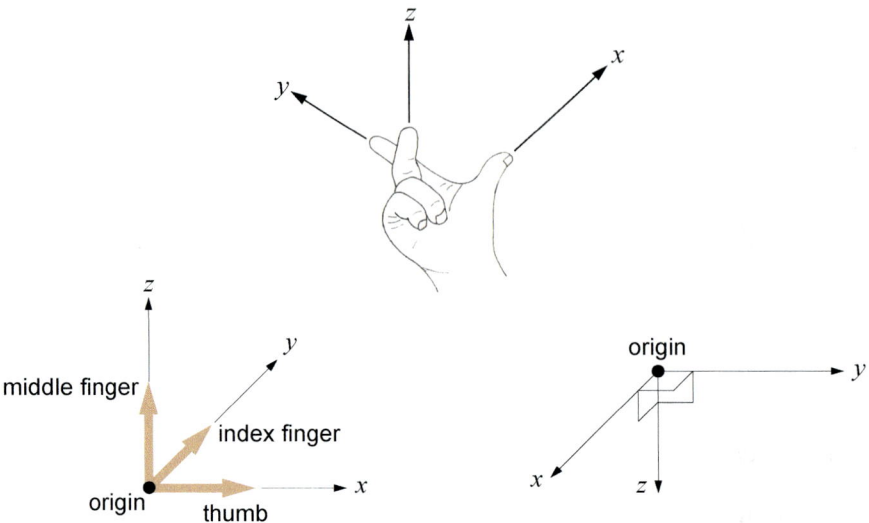

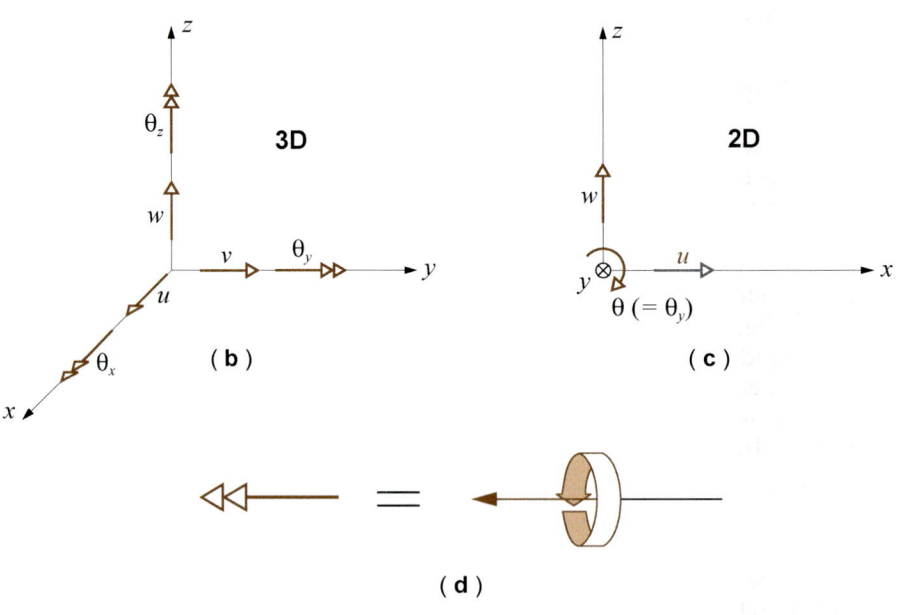

(a) right-handed cartesian coordinate systems; defined by three fingers of the right hand

Figure 2.1 Coordinate systems

2

Summary of matrix structural analysis

A brief recap of the main relationships of linear (1st order) structural analysis in matrix notation, including definitions of coordinate systems, sign rules and basic notation, all of which are taken from the predecessor [7] to this text.

In this chapter we limit ourselves to problems that can be modelled by one-dimensional (1D) elements, such as beams and bars. In matrix notation such (structural) problems are almost exclusively formulated by the *displacement or stiffness method* of analysis. It will be shown that the other classical method of structural analysis, the *force* or *flexibility method*, does not lend itself to automated computations. Since we will view the finite element method as a generalization of the matrix formulation of the structural displacement method, this chapter sets the scene for the rest of the book.

2.1 Conventions, assumptions and simple beam theory

Reference coordinates and kinematic degrees of freedom. All models are referred to a *global* reference system. For a three-dimensional (3D) problem this is always a *right-handed* cartesian coordinate system (x,y,z), see Fig. 2.1a. The relevant displacement parameters of a point, or rather a particle or small rigid body, in 3D space are the three *translations*, u, v and w, in the directions of the reference axes, and the three *rotations*, θ_x, θ_y and θ_z, about the same axes[1], Fig. 2.1b. These 6 parameters are the *kinematic degrees of freedom (dofs)* of a rigid body in 3D space. Translations, represented by single-headed arrows, are positive in the direction of positive axes. Rotations, represented by double-headed arrows, are positive in the direction of right-handed screws driven in positive axis directions, see Fig. 2.1b and d.

For two-dimensional (2D) problems the coordinate system will depend on the problem at hand. If we consider a 2D frame model as a special case of the 3D model, the x-z plane is the natural reference plane for the model, see Fig. 2.1c. The number of degrees of freedom of a particle in 2D space is reduced to 3, two translations (u and w) and one rotation (θ), see Fig. 2.1c. Note that positive rotation is clock-wise; this follows from the 3D definition, since the y-axis proceeds away from us and into the x-z plane.

---

1. It should be noted that the term displacement will normally include both translations and rotations.

Summary of matrix structural analysis

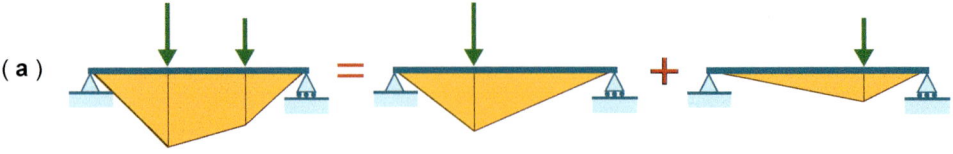

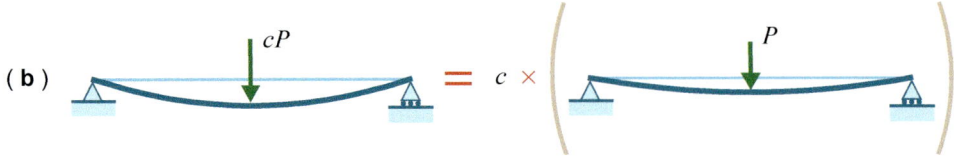

Figure 2.2 The principle of superposition

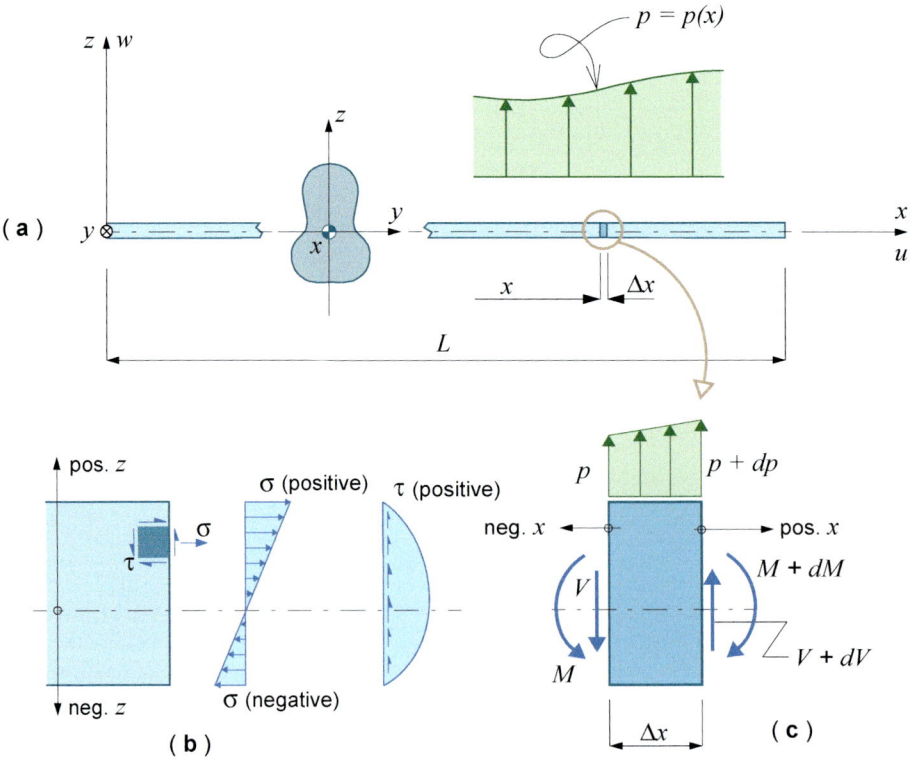

Figure 2.3 Beam sign conventions

A deviation from the coordinate system definitions of Fig. 2.1 will be found for 2D plates (membranes) with in-plane loading, *i.e.*, plane stress and plane strain problems. For such problems we will be using the *x-y* plane. That a plate in bending lies in the *x-y* plane is, however, in accordance with Fig. 2.1.

For simplicity, but without much loss of generality, we will exemplify the remaining discussions in this chapter with 2D problems only. The extension to 3D is fairly straightforward and is dealt with in Chapter 14.

Basic assumptions. Most of this book will be concerned with idealized models obeying the laws of a simplified world where all materials are elastic and homogeneous, and where the structural response to all actions can be determined with satisfactory accuracy by *linear* or so-called 1st order theory.

Linear theory is based on two basic assumptions:

1) The *displacements are so small* that we can, with sufficient accuracy, base both equilibrium and kinematic compatibility on the original, undeformed geometry.

2) All materials are *linear elastic, i.e.,* the relationship between stress and strain is linear and reversible.

An important consequence of these assumptions is that the *principle of superposition* has unlimited validity. A graphical representation of this principle is shown in Fig. 2.2.

EULER-BERNOULLI beam theory. We consider a prismatic beam of length L subjected to an arbitrary distributed transverse loading p, see Fig. 2.3. The beam is referred to a *local* coordinate system, *x-z*, where the origin is located at the area centre of the cross section at the left-hand end of the beam. The load p is positive in the direction of positive z. The lower part of Fig. 2.3 shows positive stresses, axial (σ) and shear (τ), as well as positive stress resultants, bending moment (M) and shear force (V). The axial force (N), which is not present in the figure, is positive as tension. The reasoning behind the sign convention for the bending moment is slightly involved, but the upshot is that the bending moment is positive when it stretches the fibres having positive *z*-coordinates.

The sign convention for the stress resultants shown in Fig. 2.3 is a relative one, a classical "engineering" convention based on equilibrium of the forces present on opposite sides of a small element of the beam. For a matrix formulation of the fundamental relationships, it is more convenient to use the absolute sign convention shown in Fig. 2.4. Here all forces, irrespective of which face they act upon, are positive in the direction of positive corresponding displacements. Furthermore, all forces are now identified by the letter S and a subscript indicating "direction". We are slightly ahead of ourselves here, but as will become apparent, S will be our preferred notation for *internal* or element forces.

EULER-BERNOULLI beam theory is based on the following assumptions:

1) Small displacements.
2) Linear elastic and homogeneous material.

Summary of matrix structural analysis

NAVIER's hypothesis:

Strong form: Plane sections perpendicular to the beam axis remain plane and perpendicular to the beam axis after deformation

Weak form: Plane sections perpendicular to the beam axis remain plane but not necessarily perpendicular to the beam axis after deformation

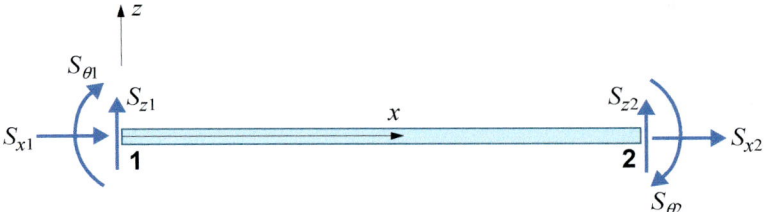

Figure 2.4 Force notation and sign convention for matrix formulation

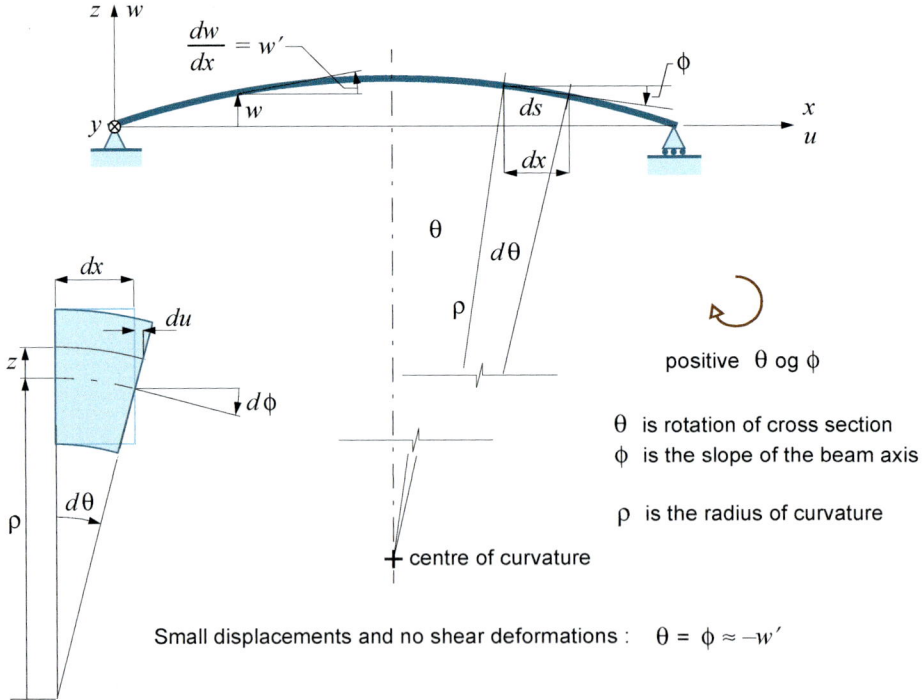

positive θ og ϕ

θ is rotation of cross section
ϕ is the slope of the beam axis
ρ is the radius of curvature

Small displacements and no shear deformations: $\theta = \phi \approx -w'$

Figure 2.5 Beam kinematics

Conventions, assumptions and simple beam theory

3) The strong form of NAVIER's hypothesis applies, that is, plane sections perpendicular to the beam axis remain plane and perpendicular to the beam axis after deformation.

4) The beam is prismatic; this means that parameters *E, A* and *I* are constant along the entire beam, where
   E is the modulus of elasticity (YOUNG's modulus),
   A is the cross section area, and
   I is the second moment of the cross section area (moment of inertia).

5) Normal stress in the *z*-direction (σ_z) is neglected.

It should be noted that Assumption 4 is not strictly necessary; the theory applies with good accuracy also for beams with moderate curvature and varying cross section.

Kinematics (see Fig. 2.5):

$$\varepsilon = \frac{du}{dx} = z\frac{d\theta}{dx} = -z\frac{d^2w}{dx^2} = -zw'' \tag{2-1}$$

Material law (HOOKE):

$$\sigma = E\varepsilon = -zEw'' \tag{2-2}$$

By definition:

$$M = \int_A \sigma z \, dA \quad \Rightarrow \quad M = -\int_A Ew''z^2 dA = -Ew'' \int_A z^2 dA = -EIw'' \tag{2-3}$$

Equilibrium (see Fig. 2.3):

Force: $\quad V + dV + p\,dx - V = 0 \quad \Rightarrow \quad \dfrac{dV}{dx} = V' = -p \tag{2-4}$

Moment: $\quad dM - V\,dx = 0 \quad \Rightarrow \quad \dfrac{dM}{dx} = M' = V \tag{2-5}$

Combining (2-4) and (2-5):

$$M'' = V' = -p \quad \Rightarrow \quad \frac{d^2M}{dx^2} + p = 0 \tag{2-6}$$

Differentiating (2-3) twice and substitution into (2-6) yields the *differential equation* for the beam

$$\frac{d^4w}{dx^4} \equiv w'''' = \frac{p}{EI} \tag{2-7}$$

By solving this equation for different loading and boundary conditions we can establish a set of useful *beam formulas*. *EI* is the *flexural rigidity* or bending stiffness of the beam.

Summary of matrix structural analysis

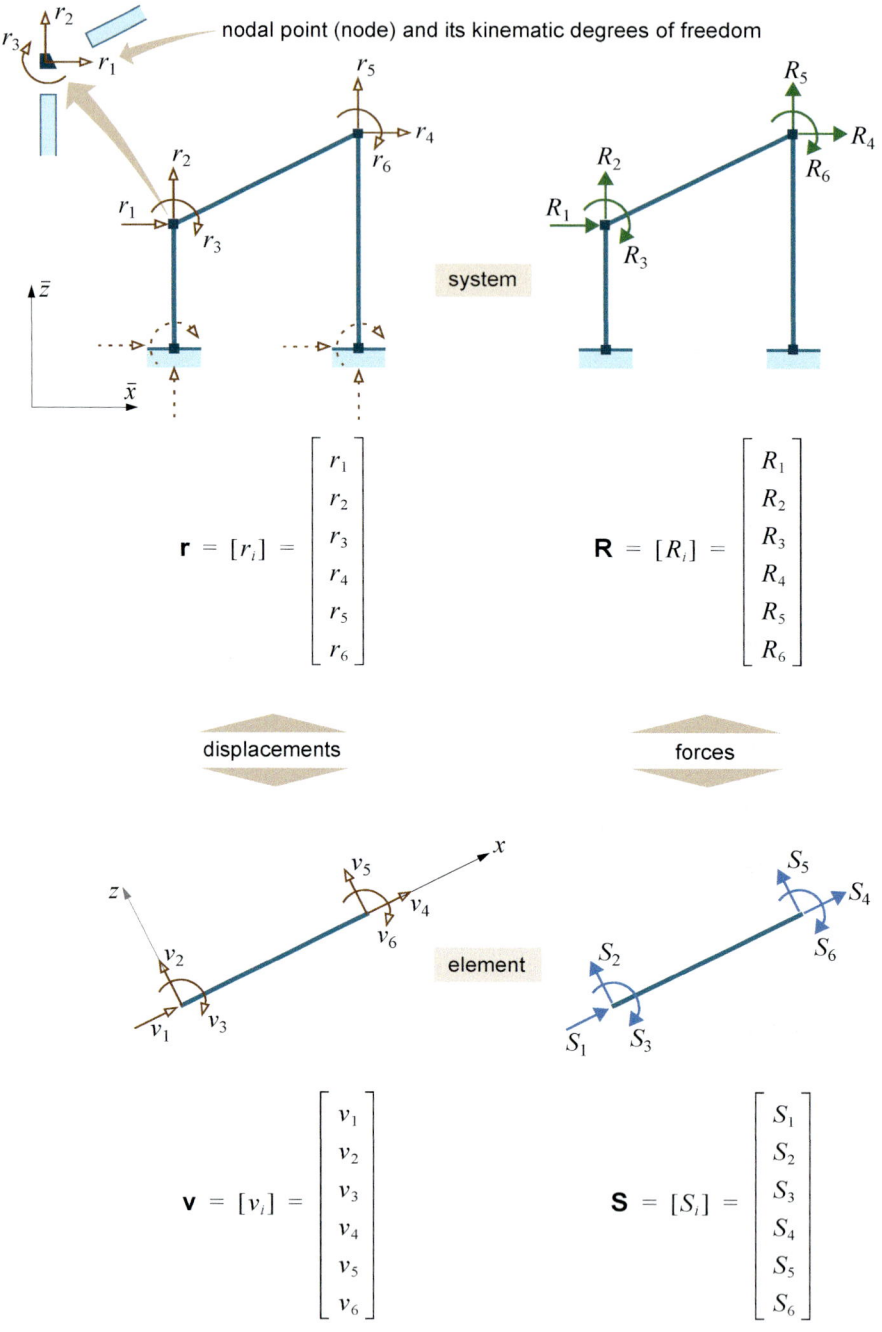

Figure 2.6 Corresponding displacements and forces at element and structure level

2.2 The fundamental requirements in matrix notation

A valid solution of a structural statics problem must satisfy the fundamental requirements of

- **static equilibrium**,
- **kinematic compatibility**, and
- the **force-displacement** (stress-strain) relationship(s) for the material(s) involved; material law for short.

If a solution satisfies all three requirements, the solution is correct, and if the problem is linear, as in our case, the solution is also unique.

Computational model. Our computational model consists of a number of beam *elements* interconnected at *nodal points* or simply *nodes*, as shown schematically in the upper part of Fig. 2.6. The model implies a *discretization* of the problem in that we postulate that the deformation of the structure is uniquely defined by the *finite* number of nodal point displacements or, more precisely, kinematic degrees of freedom (*dofs*), denoted r_i. This holds true if all actions (loading) can be represented by concentrated *nodal forces*, R_i, corresponding to the nodal displacements, r_i.

> Corresponding displacement and force (rotation and moment) act at the same point and in the same direction.

Our postulate is easily verified for a *truss*; if we know the displacements of all nodal points, we also know the deformation of each (bar) element. However, the same is also true for the beam element. If we know the displacements r_1 to r_6 for the frame in Fig. 2.6, we also know the end displacements v_1 to v_6 of the element connecting the two (top) nodal points of the frame (through simple transformations). It can be easily shown that if there is no external action between the ends of the beam element, the six end degrees of freedom (v_1 to v_6) will uniquely describe the deformation (change of form and length) of the element caused by any combination of corresponding *element forces*, S_1 to S_6.

At the structure or *system* level we now collect, in a systematic manner, all *dofs* in the *vector* **r**, the *displacement vector*, and the corresponding nodal forces or loads in the vector **R**, the *load vector*. In the *displacement method* of analysis, **r** contains the *unknown* parameters of our problem, whereas **R** contains the *known* action or loading on the structure. It should be noted that the *dofs* of the two base nodes of the frame in Fig. 2.6 are known, due to the boundary conditions at these nodes, and hence omitted from **r**.

As will become apparent below, it is convenient to distinguish between the *system* level and the *element* level. The latter is shown in the lower part of Fig. 2.6, and as already mentioned, at this level we denote displacements and corresponding forces by v_i/**v** and S_i/**S**, respectively. Furthermore, at the element level it is convenient to refer the parameters to a *local*, element coordinate system (x,z). Whenever it is necessary to distinguish this local coordinate system from the global reference system, the latter will be marked with a bar on top $(\bar{x}, \bar{z})$.

Summary of matrix structural analysis

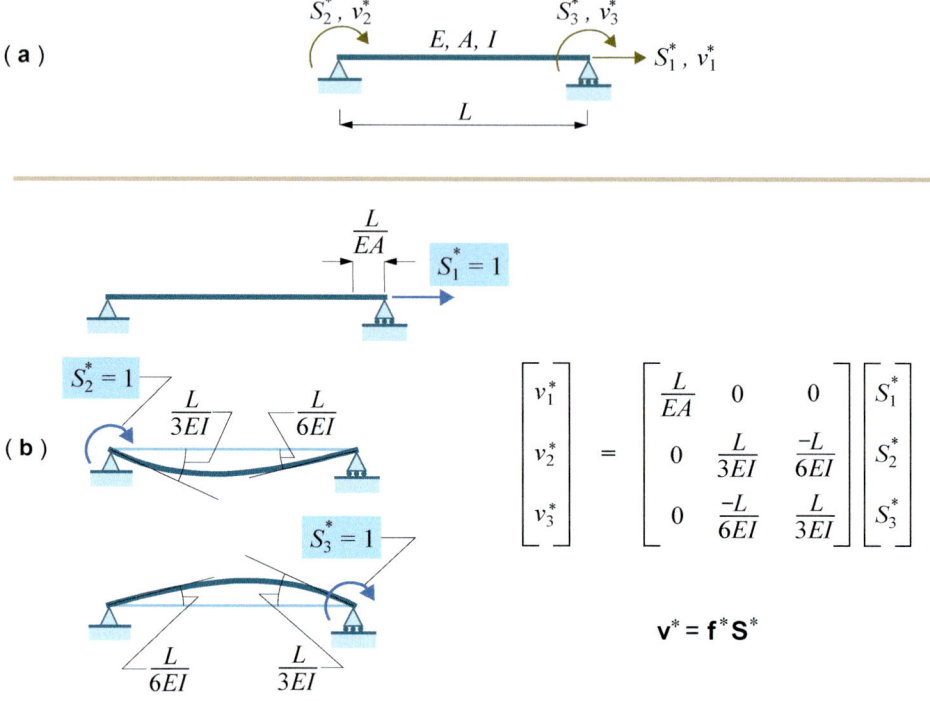

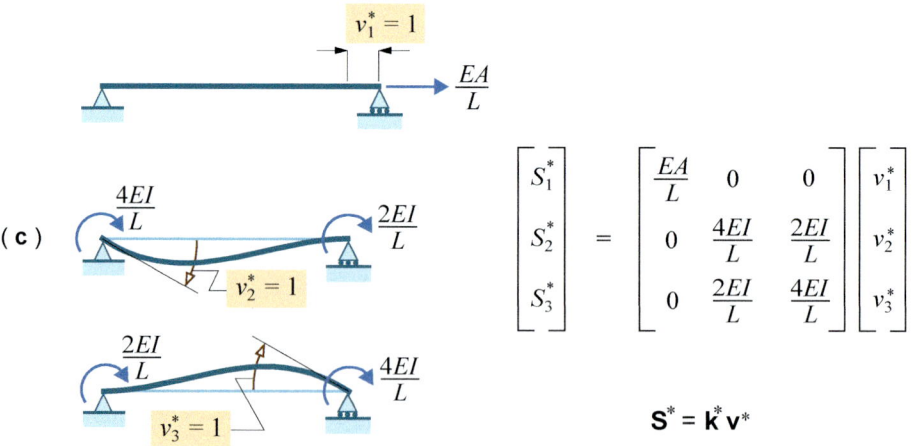

Figure 2.7 Displacement-force relationship for deformation *dofs* of a 2D beam element

The fundamental requirements in matrix notation

Our task is now to establish a relationship between the unknown parameters **r** and the known parameters **R**, in such a way that the three fundamental requirements are satisfied. For the time being we will continue to consider a model where all actions are in the form of concentrated nodal loads (including moments). This is an unacceptable constraint, but, as will be shown, we can easily circumvent it.

Force-displacement. We start with the *material law,* and we seek a relationship between **S** and **v**. The simplest 1D element is the axial element of a truss, a *bar* element. However, we continue with the beam element, but we will reduce its *dofs* and include only those that are necessary to describe the *deformation* of the element. The 6 element *dofs* of Fig. 2.6 can describe both the deformation of the element and its displacements as a *rigid body*. The element has 3 independent *rigid body modes* of displacement, two translations and one rotation. That leaves three degrees of freedom to describe the deformation of the element; these are shown in Fig. 2.7a, and the symbols are marked by an asterisk (*), designating deformation only.

In Figure 2.7b we have applied *unit* forces, one at a time, and from standard beam formulas we find the displacements caused by these forces. The net result of this is the matrix relation

$$\mathbf{v}^* = \mathbf{f}^*\mathbf{S}^* \tag{2-8}$$

The matrix $\mathbf{f}^*$ is the element *flexibility* matrix with respect to the deformation *dofs*. We note that since we have placed the nodal points at the centre of the cross section area there is no coupling of axial and bending effects.

In Figure 2.7c we have applied *unit* displacements, one at a time and such that the other *dofs* are zero. Again, from standard beam formulas, we find the forces necessary to maintain the deformation patterns compatible with the nodal displacements, and we can now express the nodal forces in terms of nodal displacements as

$$\mathbf{S}^* = \mathbf{k}^*\mathbf{v}^* \tag{2-9}$$

The matrix $\mathbf{k}^*$ is the element *stiffness* matrix with respect to the deformation *dofs*. If we multiply the flexibility matrix with the stiffness matrix, or vice versa, we find that

$$\mathbf{f}^*\mathbf{k}^* = \mathbf{k}^*\mathbf{f}^* = \mathbf{I} \tag{2-10}$$

where **I** is the unit or *identity* matrix. Hence, since both matrices are regular (non-singular),

$$\mathbf{k}^* = (\mathbf{f}^*)^{-1} \tag{2-11}$$

In other words, stiffness and flexibility are, as in the case of a simple elastic spring, *inverse* quantities. We observe another very important quality of the flexibility and stiffness matrices, they are both *symmetric* matrices.

The procedure used in Fig. 2.7 could equally well have been applied to the beam element with 6 *dofs*, in which case we would have ended up with

23

Summary of matrix structural analysis

Jacob (Jacques) BERNOULLI (1654-1705) was one of a number of well known mathematicians in the BERNOULLI family. He was born in Basel and studied theology (following his father's wish), but contrary to the desires of his parents he also studied mathematics and astronomy; BERNOULLI was the first to develop a technique for solving separable differential equations, and he also discovered the constant *e*, by studying a question of compound interest which required him to find the value of the limit of $(1 + 1/n)^n$ as *n* goes to infinity. His most notable achievement however, was in the field of *probability* – it has been said that he "discovered" the *theory of probability*.

GALILEI (1564-1642) has been credited as the first to attempt developing a beam theory, but recent studies suggest that da VINCI (1452-1519) was the first to make the most relevant observations; however, he lacked HOOKE's law and calculus to complete the theory. Jacob BERNOULLI is attributed to have determined the elastic curve (of a cantilever beam), but it was EULER and Jacob's nephew Daniel BERNOULLI who first put together a useful theory around 1750. So, when we talk about the EULER-BERNOULLI beam theory we actually acknowledge two members of the BERNOULLI family.

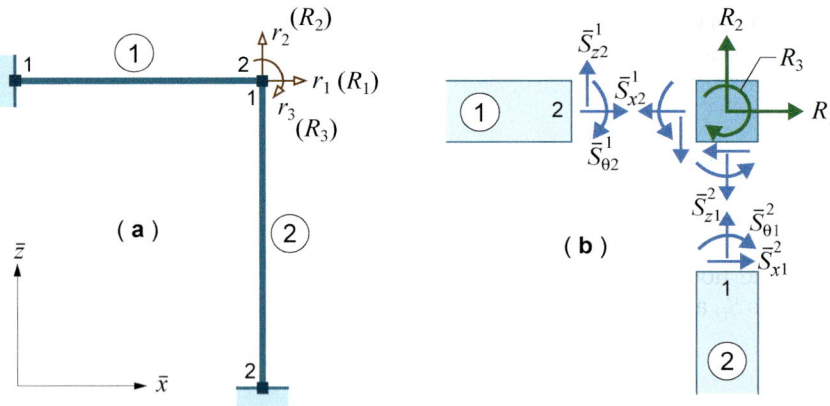

Figure 2.8 Example of external (R) and internal (S) nodal forces

A note on notation: element numbers are indicated by superindices, whereas nodal point numbers are subindices

The fundamental requirements in matrix notation

$$\mathbf{v} = \mathbf{fS} \quad \text{and} \quad \mathbf{S} = \mathbf{kv} \quad \text{(2-12a and b)}$$

where **f** and **k** are 6 by 6 symmetric matrices. However, they are no longer regular. In fact their *rank deficiency* is 3, since **v** now contains the 3 rigid body modes of displacement. Equation (2-10) still applies, *i.e.*, their product is still equal to the identity matrix, but we *cannot* obtain one of the matrices by inverting the other one.

With reference to Fig. 2.7c, a note on *unit displacements* is perhaps in order. Rotations are measured in radians, and the moments in Fig. 2.7c are in fact those caused by rotations of 1 (one) radian. But what happened to small displacements? In linear theory it is quite possible to compute a deflection of, say 10 m for a 5 m long beam by simply multiplying a reasonable load by a large number. We certainly maintain our assumption of small displacements, and in Fig. 2.7c the real moments caused by, for instance v_2 acting alone, are

$$v_2 \cdot (4EI/L) \quad \text{and} \quad v_2 \cdot (2EI/L),$$

respectively. Although it makes little sense physically, mathematically there is no problem with the notion of a unit rotation. What matters is the value we compute at the end; that will decide whether or not we are within the bounds of our theory. The force-displacement relationships in Fig. 2.7 suggest that the flexibility and stiffness coefficients are *influence* coefficients. A flexibility coefficient has dimension *displacement per unit force*, whereas a stiffness coefficient is *force per unit displacement*.

Static equilibrium. In order to satisfy this requirement we need to make sure that the internal forces (S) are in static equilibrium with the external forces (R). This we can do in two ways: (1) we can make sure that the known external forces are in equilibrium with the unknown internal forces, or (2) we can establish, at least in theory, how the internal forces must relate to the external forces in order for equilibrium to be maintained.

In our discretized model the nodal points represent the total structure, or system, and if all nodal points are in equilibrium, the system as such is also in equilibrium. The external loading acts as concentrated forces directly at the nodal points. For a plane frame model each beam element exerts forces on the nodal points to which it is connected; these forces are the element forces S_i, acting in opposite direction. In matrix notation we can write this as

$$\mathbf{R} - (\mathbf{g}^1\mathbf{S}^1 + \mathbf{g}^2\mathbf{S}^2 + .. + \mathbf{g}^i\mathbf{S}^i + .. + \mathbf{g}^m\mathbf{S}^m) = \mathbf{0}$$

or

$$\mathbf{R} = \mathbf{g}^1\mathbf{S}^1 + \mathbf{g}^2\mathbf{S}^2 + .. + \mathbf{g}^i\mathbf{S}^i + .. + \mathbf{g}^m\mathbf{S}^m = \sum_{i=1}^{m} \mathbf{g}^i\mathbf{S}^i \quad (2\text{-}13)$$

Here m is the total number of elements in the model, and the term $\mathbf{g}^i\mathbf{S}^i$ is an expression for the forces that element number i exerts on *all* nodal points. Since element i is connected to only two nodes, it is obvious that for a model of some size, the matrix $\mathbf{g}^i$ contains mostly zero elements. This "equilibrium matrix" becomes particularly simple if we express the element stiffness relation in the same (global) coordinate system as the system *dofs*, see Fig. 2.8. For this simple example the following equilibrium expression:

Summary of matrix structural analysis

Leonhard EULER (1707-1783) was a pioneering Swiss mathematician and physicist who made important discoveries in fields as diverse as infinitesimal calculus and graph theory. He introduced much of modern mathematical terminology and notation, and he is also well known for his work in mechanics, fluid dynamics, optics and astronomy.

EULER spent most of his life in St. Petersburg and Berlin, and he was one of the most prolific mathematicians ever. A statement attributed to another great mathematician of that time, Pierre-Simon LAPLACE (1749-1827) expresses EULER's influence on mathematics: "Read EULER, read EULER, he is our teacher in all things."

In structural mechanics EULER is perhaps best known for his beam theory (developed around 1750 with help from two members of the BERNOULLI family) and for determining the instability load of an ideal column some years later. However, EULER did not get it quite right in that he never attempted to discuss his constant C which he called "absolute elasticity"; he merely states that it depends on the elastic properties of the material and that, in the case of beams with rectangular cross sections, it is proportional to the width and to the *square* (!) of the depth. It was left to NAVIER to show (in 1825) that C is equal to EI.

EULER is credited for having invented the *calculus of variations*, including its most significant result, the EULER-LAGRANGE equation.

EULER's remarkable production is even more impressive in view of his poor eyesight; after a fever in 1735 he became nearly blind in his right eye, and when he suffered a cataract in his left eye in 1766, it left him practically blind for the last 17 years of his life. Nevertheless, he continued to produce high quality material at an incredible rate.

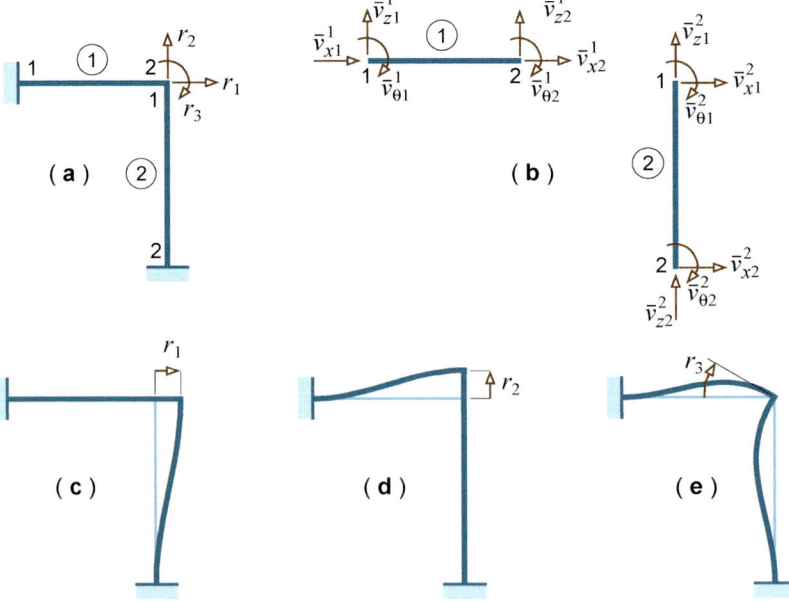

Figure 2.9 Example of deformation patterns and kinematic relationship between **v** og **r**

The fundamental requirements in matrix notation

$$\mathbf{R} = \begin{bmatrix} R_1 \\ R_2 \\ R_3 \end{bmatrix} = \begin{bmatrix} \overset{\text{end 1}}{0\ 0\ 0} & \overset{\text{end 2}}{1\ 0\ 0} \\ 0\ 0\ 0 & 0\ 1\ 0 \\ 0\ 0\ 0 & 0\ 0\ 1 \end{bmatrix} \begin{bmatrix} \bar{S}^1_{x1} \\ \vdots \\ \bar{S}^1_{\theta 2} \end{bmatrix} + \begin{bmatrix} \overset{\text{end 1}}{1\ 0\ 0} & \overset{\text{end 2}}{0\ 0\ 0} \\ 0\ 1\ 0 & 0\ 0\ 0 \\ 0\ 0\ 1 & 0\ 0\ 0 \end{bmatrix} \begin{bmatrix} \bar{S}^2_{x1} \\ \vdots \\ \bar{S}^2_{\theta 2} \end{bmatrix} = \bar{\mathbf{g}}^1 \bar{\mathbf{S}}^1 + \bar{\mathbf{g}}^2 \bar{\mathbf{S}}^2$$

is readily verified. In this case the $\bar{\mathbf{g}}^i$ matrices contain only zero and unit elements, and for an arbitrary model, large or small, no $\bar{\mathbf{g}}^i$ will contain more than 6 non-zero elements, all of which are equal to unity.

In principle we can write the equilibrium equations the other way around, that is, as a series of equations of the form

$$\mathbf{S}^i = \mathbf{b}^i \mathbf{R} \tag{2-14}$$

However, the elements of the **b** matrix can only be established directly for *statically determinate* structures, and even then the determination of the matrix is not easily automated. This is a severe limitation; furthermore, this matrix is, in most cases, far more densely populated than the **g** matrix.

Kinematic compatibility. This requirement demands that any element end cross section must be "glued" to the "end surface" of the nodal point it is connected to. In other words, the end cross section of an element must have the same displacements (including rotation) as the node it is connected to. In matrix notation we can express this as

$$\mathbf{v}^1 = \mathbf{a}^1 \mathbf{r}, \ldots, \mathbf{v}^i = \mathbf{a}^i \mathbf{r}, \ldots, \mathbf{v}^m = \mathbf{a}^m \mathbf{r} \tag{2-15}$$

Again, m is the total number of elements in the model. Each **a** matrix has as many rows as there are elements in **v**, and as many columns as there are elements in **r**. Since each element is connected to only two nodal points it follows that the **a**-matrices are, in general, very sparsely populated. As in the case of equilibrium, the "compatibility matrix" **a** becomes particularly simple if we express the element *dofs* in the same (global) coordinate system as we do the elements of **r**.

A simple example is shown in Fig. 2.9. This is the same frame as in Fig. 2.8, and this figure shows how the frame deforms when subjected to the system *dofs*, one at a time, and such that when one is different from zero the other two are equal to zero. For this simple example the following kinematic relationships

$$\bar{\mathbf{v}}^1 = \begin{bmatrix} \bar{v}^1_{x1} \\ \bar{v}^1_{z1} \\ \bar{v}^1_{\theta 1} \\ \bar{v}^1_{x2} \\ \bar{v}^1_{z2} \\ \bar{v}^1_{\theta 2} \end{bmatrix} = \begin{bmatrix} 0 & 0 & 0 \\ 0 & 0 & 0 \\ 0 & 0 & 0 \\ 1 & 0 & 0 \\ 0 & 1 & 0 \\ 0 & 0 & 1 \end{bmatrix} \mathbf{r} = \bar{\mathbf{a}}^1 \mathbf{r} \quad \text{and} \quad \bar{\mathbf{v}}^2 = \begin{bmatrix} \bar{v}^2_{x1} \\ \bar{v}^2_{z1} \\ \bar{v}^2_{\theta 1} \\ \bar{v}^2_{x2} \\ \bar{v}^2_{z2} \\ \bar{v}^2_{\theta 2} \end{bmatrix} = \begin{bmatrix} 1 & 0 & 0 \\ 0 & 1 & 0 \\ 0 & 0 & 1 \\ 0 & 0 & 0 \\ 0 & 0 & 0 \\ 0 & 0 & 0 \end{bmatrix} \mathbf{r} = \bar{\mathbf{a}}^2 \mathbf{r}$$

are readily established.

Summary of matrix structural analysis

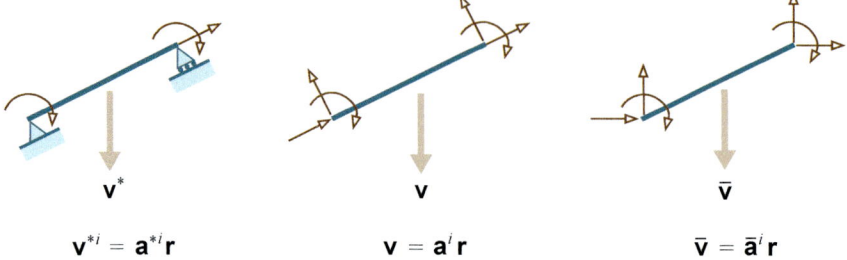

$$v^{*i} = a^{*i}r \qquad v = a^i r \qquad \bar{v} = \bar{a}^i r$$

Figure 2.10 Beam element reference systems

**Claude-Louis NAVIER** (1785-1836) was a French engineer and physicist. He studied at both the École Polytechnique and the École des Ponts et Chaussées where he graduated in 1806. He succeeded his uncle as *Inspecteur general* at the Corps des Ponts et Chaussées. In 1824 NAVIER was admitted into the French Academy of Science, and in 1830 he became professor at the École Nationale des Ponts et Chaussées. The following year he succeeded exiled Augustin Louis CAUCHY as professor of calculus and mechanics at the École Polytechnique.

Although occupied with theoretical work and with editing books, NAVIER always had some practical work at hand, usually in connection with bridge engineering. In 1826, the first printed edition of his book on strength of materials appeared, a milestone in structural mechanics. He redefines YOUNG's modulus of elasticity (E) in the form we use today, and assuming that *cross sections remain plane during bending* and using three equations of statics, he concludes that the neutral axis passes through the centroid of the cross section area and that the radius of curvature (ρ) is given by the equation $EI/\rho = M$.

NAVIER is often considered to be the founder of modern structural analysis, and yet, his major contribution, and what he is best remembered for, is the NAVIER-STOKES equations, central to fluid mechanics.

Principle of virtual displacements (PVD)

As for equilibrium, the kinematic requirement can also be expressed the other way around, that is, as

$$\mathbf{r} = \mathbf{h}^1\mathbf{v}^1 + \mathbf{h}^2\mathbf{v}^2 + .. + \mathbf{h}^i\mathbf{v}^i + .. + \mathbf{h}^m\mathbf{v}^m = \sum_{i=1}^{m}\mathbf{h}^i\mathbf{v}^i \qquad (2\text{-}16)$$

but as for the **b**-matrices, the **h**-matrices are not easily available. In fact they can only be established directly for very simple structures.

Discussion. First of all, Fig. 2.10 may help clarify some of the notation used above.

We have now expressed the three fundamental requirements of structural analysis in matrix form. This has been made possible through a discretization of the problem at hand. In this process we have made two important and, to some extent, limiting assumptions. First of all we have assumed that all actions (external loading) can be represented by concentrated forces (and moments) acting at the nodal points. Secondly, we have assumed full continuity of displacements (including rotations) at all nodes. Both assumptions put severe restrictions on our ability to model real problems, and we therefore need to relax them. It will be shown later how that can be done, but for the time being, and without much loss of generality, we proceed with these assumptions in place.

The duality of the presentation should be noted. Each requirement is presented in two versions. One set of relationships is used in the displacement method, the other forms the basis for the force method.
The *displacement method* makes use of Eqs. (2-12b), (2-13) and (2-15), that is of matrices **k**, **g** and **a**, while the *force method* makes use of matrices **f**, **b** and **h** of Eqs. (2-12a), (2-14) and (2-16), respectively. In other words, the two troublesome matrices, **b** and **h**, are associated with the force method, and without labouring the point, these matrices are the main reason why the force method does not lend itself to automatic computation.

The upshot of this is that, as of now, we retire the force method, and the matrices **b** and **h** will not feature in the remaining part of this book, at least not with the meaning they have in this section. However, the element flexibility matrix **f** may crop up again, but only as a means of determining the element stiffness matrix **k**.

2.3 Principle of virtual displacements (PVD)

Before we make use of the relationships derived in the previous section, we formulate, without formal proof[1], a very useful principle, namely that of *virtual displacements* which is a special version of the *principle of virtual work*. In words, the principle states that

> if a system of *real* forces, **R** and **S**^{i} ($i = 1, 2, 3, ... , m$) that are in *static equilibrium*, are subjected to a system of *virtual*, but *kinematically compatible* displacements and deformations $\tilde{\mathbf{r}}$ and $\tilde{\mathbf{v}}^i$ ($i = 1, 2, ... , m$), then the *virtual work* performed by the real forces **R** over the virtual

---
1. That will come later.

Summary of matrix structural analysis

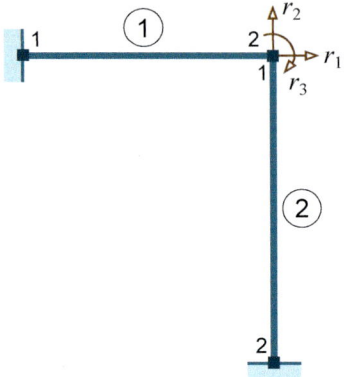

$$\bar{g}^1 = \begin{bmatrix} 0 & 0 & 0 & 1 & 0 & 0 \\ 0 & 0 & 0 & 0 & 1 & 0 \\ 0 & 0 & 0 & 0 & 0 & 1 \end{bmatrix} \qquad \bar{g}^2 = \begin{bmatrix} 1 & 0 & 0 & 0 & 0 & 0 \\ 0 & 1 & 0 & 0 & 0 & 0 \\ 0 & 0 & 1 & 0 & 0 & 0 \end{bmatrix}$$

$$\bar{a}^1 = \begin{bmatrix} 0 & 0 & 0 \\ 0 & 0 & 0 \\ 0 & 0 & 0 \\ 1 & 0 & 0 \\ 0 & 1 & 0 \\ 0 & 0 & 1 \end{bmatrix} = (\bar{g}^1)^T \qquad \bar{a}^2 = \begin{bmatrix} 1 & 0 & 0 \\ 0 & 1 & 0 \\ 0 & 0 & 1 \\ 0 & 0 & 0 \\ 0 & 0 & 0 \\ 0 & 0 & 0 \end{bmatrix} = (\bar{g}^2)^T$$

Principle of virtual displacements (PVD)

displacements $\tilde{\mathbf{r}}$ will be *equal* to the *internal virtual* work performed by the real internal forces $\mathbf{S}^i$ ($i = 1, 2, 3, \ldots, m$) over the *virtual deformations* $\tilde{\mathbf{v}}^i$ ($i = 1, 2, \ldots, m$),

or, in matrix notation

$$\underbrace{\mathbf{R}^T \tilde{\mathbf{r}}}_{\text{static equilibrium}} = \sum_{i=1}^{m} (\mathbf{S}^i)^T \tilde{\mathbf{v}}^i \qquad (2\text{-}17)$$

(kinematic compatibility)

A tilde (~) on top of a symbol designates a virtual quantity. Such a quantity is an imaginary quantity, and in our case it should be stressed that the virtual displacements and deformations, $\tilde{\mathbf{r}}$ and $\tilde{\mathbf{v}}^i$, are quite arbitrary and they have absolutely *nothing* to do with the real forces, $\mathbf{R}$ and $\mathbf{S}^i$. However, they must satisfy kinematic compatibility which is ensured by requiring that

$$\tilde{\mathbf{v}}^i = \mathbf{a}^i \tilde{\mathbf{r}} \qquad (2\text{-}18)$$

Substitution into Eq. (2-17):

$$\mathbf{R}^T \tilde{\mathbf{r}} = \sum_{i=1}^{m} (\mathbf{S}^i)^T \mathbf{a}^i \tilde{\mathbf{r}} = \left(\sum_{i=1}^{m} (\mathbf{S}^i)^T \mathbf{a}^i \right) \tilde{\mathbf{r}} \qquad (2\text{-}19)$$

This equation is valid for *any* virtual displacement ($\tilde{\mathbf{r}}$). Hence

$$\mathbf{R}^T = \sum_{i=1}^{m} (\mathbf{S}^i)^T \mathbf{a}^i \quad \text{or} \quad \mathbf{R} = \sum_{i=1}^{m} (\mathbf{a}^i)^T \mathbf{S}^i \qquad (2\text{-}20)$$

From before we have Eq. (2-13), that is,

$$\mathbf{R} = \sum_{i=1}^{m} \mathbf{g}^i \mathbf{S}^i$$

Since the last two equations must hold for any loading we must have

$$\mathbf{g}^i = (\mathbf{a}^i)^T \qquad (2\text{-}21)$$

This means that we can use the PVD to *replace* the equilibrium conditions with the kinematic conditions. Equation (2-21) is very important, and it is obviously independent of the coordinate system used. Hence

$$\bar{\mathbf{g}}^i = (\bar{\mathbf{a}}^i)^T$$

As shown on the opposite page, the $\bar{\mathbf{g}}$- and $\bar{\mathbf{a}}$-matrices associated with Figs. 2.8 and 2.9 satisfy Eq. (2-21).

Summary of matrix structural analysis

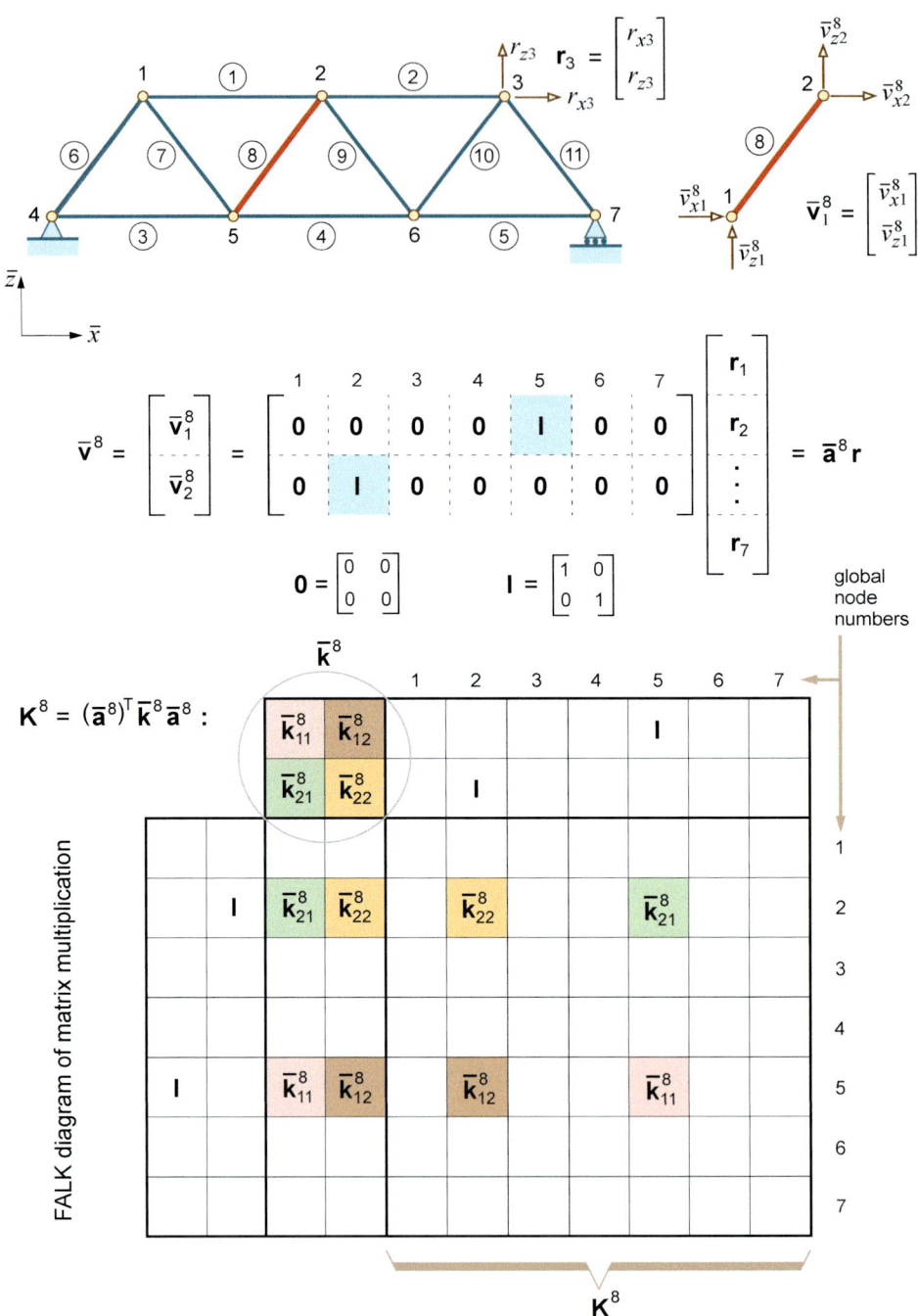

Figure 2.11 Example of stiffness assembly – the direct stiffness method

2.4 The system stiffness relation

At the system level we have one set of known parameters, the nodal loads **R**, and another corresponding set of unknown parameters, the nodal displacements **r**. We now seek a relationship between these two parameter sets that satisfies all the three fundamental requirements. We start with Eq. (2-13):

$$\text{static equilibrium:} \qquad \mathbf{R} = \sum_{i=1}^{m} \mathbf{g}^i \mathbf{S}^i$$

Next we substitute for $\mathbf{S}^i$ from Eq. (2-12b):

$$\text{force-displacement:} \qquad \mathbf{R} = \sum_{i=1}^{m} \mathbf{g}^i \mathbf{S}^i = \sum_{i=1}^{m} \mathbf{g}^i \mathbf{k}^i \mathbf{v}^i$$

Finally we invoke Eq. (2-15):

$$\text{kinematic compatibility:} \quad \mathbf{R} = \sum_{i=1}^{m} \mathbf{g}^i \mathbf{k}^i \mathbf{v}^i = \left(\sum_{i=1}^{m} \mathbf{g}^i \mathbf{k}^i \mathbf{a}^i \right) \mathbf{r} = \mathbf{K}\mathbf{r} \quad (2\text{-}22)$$

This is the governing equation of the displacement method of analysis, and if we turn it around, that is,

$$\mathbf{K}\mathbf{r} = \mathbf{R} \qquad (2\text{-}23)$$

we see clearly that it represents a system of algebraic equations. The matrix of coefficients, the *system stiffness matrix* **K**, is

$$\mathbf{K} = \sum_{i=1}^{m} \mathbf{g}^i \mathbf{k}^i \mathbf{a}^i \stackrel{\text{PVD}}{=} \sum_{i=1}^{m} (\mathbf{a}^i)^T \mathbf{k}^i \mathbf{a}^i = \sum_{i=1}^{m} (\mathbf{a}^{*i})^T \mathbf{k}^{*i} \mathbf{a}^{*i} = \sum_{i=1}^{m} (\bar{\mathbf{a}}^i)^T \bar{\mathbf{k}}^i \bar{\mathbf{a}}^i \qquad (2\text{-}24)$$

Here we have made use of the principle of virtual displacements (PVD) to replace the equilibrium matrices ($\mathbf{g}^i$) with the compatibility matrices ($\mathbf{a}^i$) This equation tells us that the system stiffness matrix **K** is established by adding together the stiffness contributions from each element of the model, and the equation is a *recipe* for how this *assembly* process is carried out.

Equation (2-24) is arguably the most important equation in this book. It makes no difference which reference system the element stiffness matrix is referred to, but **k** and **a** must be consistent. The assembly process becomes particularly simple if the element stiffness relation refers to the global reference system, as in the last right-hand side of Eq. (2-24). In this case we have a one-to-one relationship between the elements of $\bar{\mathbf{v}}$ and the appropriate elements of **r**, and Eq. (2-24) simply tells us where in **K** each individual element of $\bar{\mathbf{k}}$ is to be added (to the current content). This is exemplified by the simple truss problem in Fig. 2.11. Since the nodal displacements are ordered similarly at the element and system level, we can consider 2 by 2 submatrices of $\bar{\mathbf{k}}$ instead of individual elements. We see that the only information we need to know in order to place the submatrices of $\bar{\mathbf{k}}$ in the correct positions in **K**, is the number of the system nodes at which element nodes 1 and 2 are connected. Hence, for each beam element we need to know two numbers. In this version, Eq. (2-24) is the mathematical form of what is often referred to as *the direct stiffness method*.

Summary of matrix structural analysis

A linear elastic system with *dofs* **r** is subjected to forces $\mathbf{R}_a$ and $\mathbf{R}_b$.

$\mathbf{R}_a$ alone cause displacements $\mathbf{r}_a$, whereas $\mathbf{R}_b$ alone cause displacements $\mathbf{r}_b$

MAXWELL (2 *dofs*):

$$R_{1a}r_{1b} = R_{2b}r_{2a} \Rightarrow r_{1b} = r_{2a}$$

BETTI (any number of *dofs*):

$$\mathbf{R}_a^T \mathbf{r}_b = \mathbf{R}_b^T \mathbf{r}_a$$

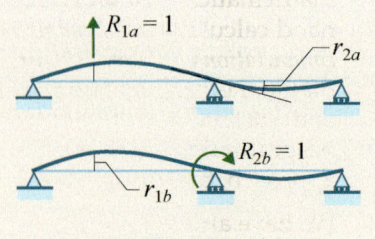

Reciprocity theorem

K_{ij} is the force (per unit displacement) at *dof i* due to a *unit* displacement at *dof j* while all other nodal displacements (r_k) are zero

F_{ij} is the displacement (per unit force) at *dof i* due to a *unit* force at *dof j* while all other nodal forces (R_k) are zero

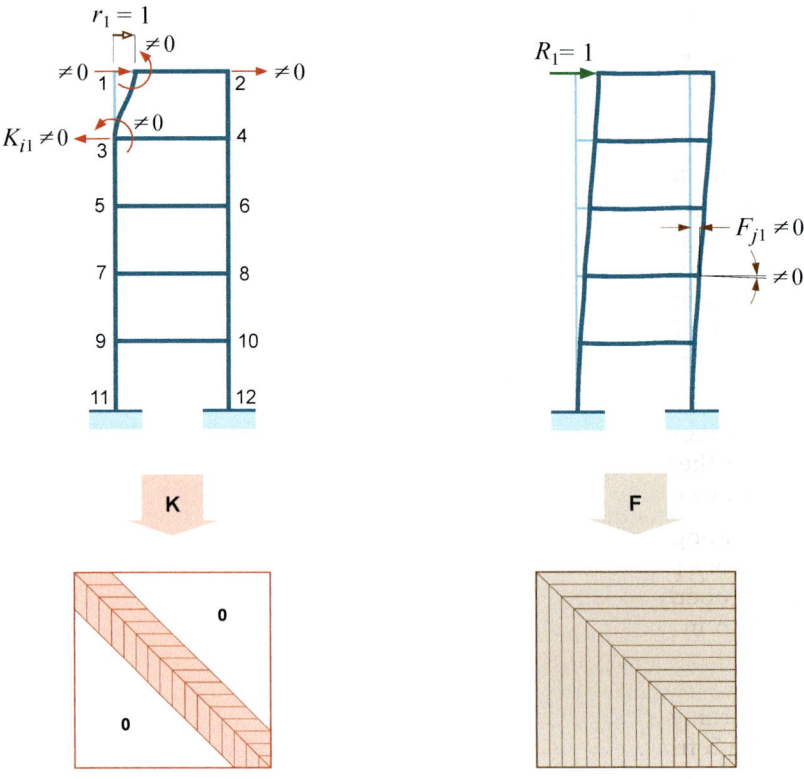

Figure 2.12 Schematic, but characteristic form of **K** and **F**

From this we can conclude that while the **a**-matrices play a central role in the mathematical formulation of the problem, and may even be useful for some hand calculations, *they are never formed in the case of automated, program driven computations.* From the programmer's point of view, the interesting information held by the **a**-matrix is the connectivity information between element and system nodes. How this information is recorded depends on which programming approach is used, the *procedural* approach or the *object oriented* approach, but for any 1D element it basically boils down to two numbers.

We have already noted that the element stiffness matrix **k** is symmetric. This property of **k** is a direct consequence of the *reciprocal theorem* of BETTI and MAXWELL. It follows then from Eq. (2-24) that **K** is also a symmetrical matrix, that is

$$\mathbf{K} = \mathbf{K}^T \qquad (2\text{-}25)$$

Multiplying Eq. (2-23) by $\frac{1}{2}\mathbf{r}^T$ we get

$$\tfrac{1}{2}\mathbf{r}^T\mathbf{K}\mathbf{r} = \tfrac{1}{2}\mathbf{r}^T\mathbf{R} = \text{strain energy} \ > \ 0 \quad \text{for all} \ \ \mathbf{r} \neq \mathbf{0} \qquad (2\text{-}26)$$

Thus, in linear theory, the stiffness matrix **K** also has the important property of being *positive definite*.

The last, but not least important property of **K** is its *sparseness* – it contains relatively few non-zero elements. Furthermore, by suitable numbering of the system nodes and thereby the problem *dofs*, it is possible to collect all non-zero elements in a narrow *band* along the main diagonal. This is indicated by Fig. 2.12, where the deformations are drawn on the basis of the definition of stiffness and flexibility coefficients, both of which are quoted on the opposite page. The shaded areas contain non-zero elements (and also some zero elements) whereas non-shaded areas contain only zero elements. Stiffness and flexibility are inverse quantities also at the system level. But while the inverse of a symmetric matrix is also symmetric, the inverse of a band matrix is, in general, not a band matrix, or if it is its bandwidth will be significantly larger than that of the matrix it came from.

The diagonal dominance that we can almost always obtain for **K** tends to render the matrix numerically well behaved; **K** is normally well *conditioned*. The same cannot always be said for the flexibility or force method.

Depending on how we have treated the *dofs* at the supports during the assembly process we may have to make sure that all boundary conditions are properly taken care of. This is an important aspect that we will come back to in some detail. For now we assume that the boundary conditions have been incorporated, in which case **K** is a regular (non-singular) matrix and we can proceed to solve Eq. (2-23) with respect to the unknown nodal displacements **r**. Knowing **r** we can recover **v** for each element and then obtain **S** from **S** = **kv**, and our problem is basically solved!

We end this section by emphasizing the fact that everything we have said here is independent of the element; all of it applies to other types of displacement based elements, whether 1D, 2D and 3D elements.

Summary of matrix structural analysis

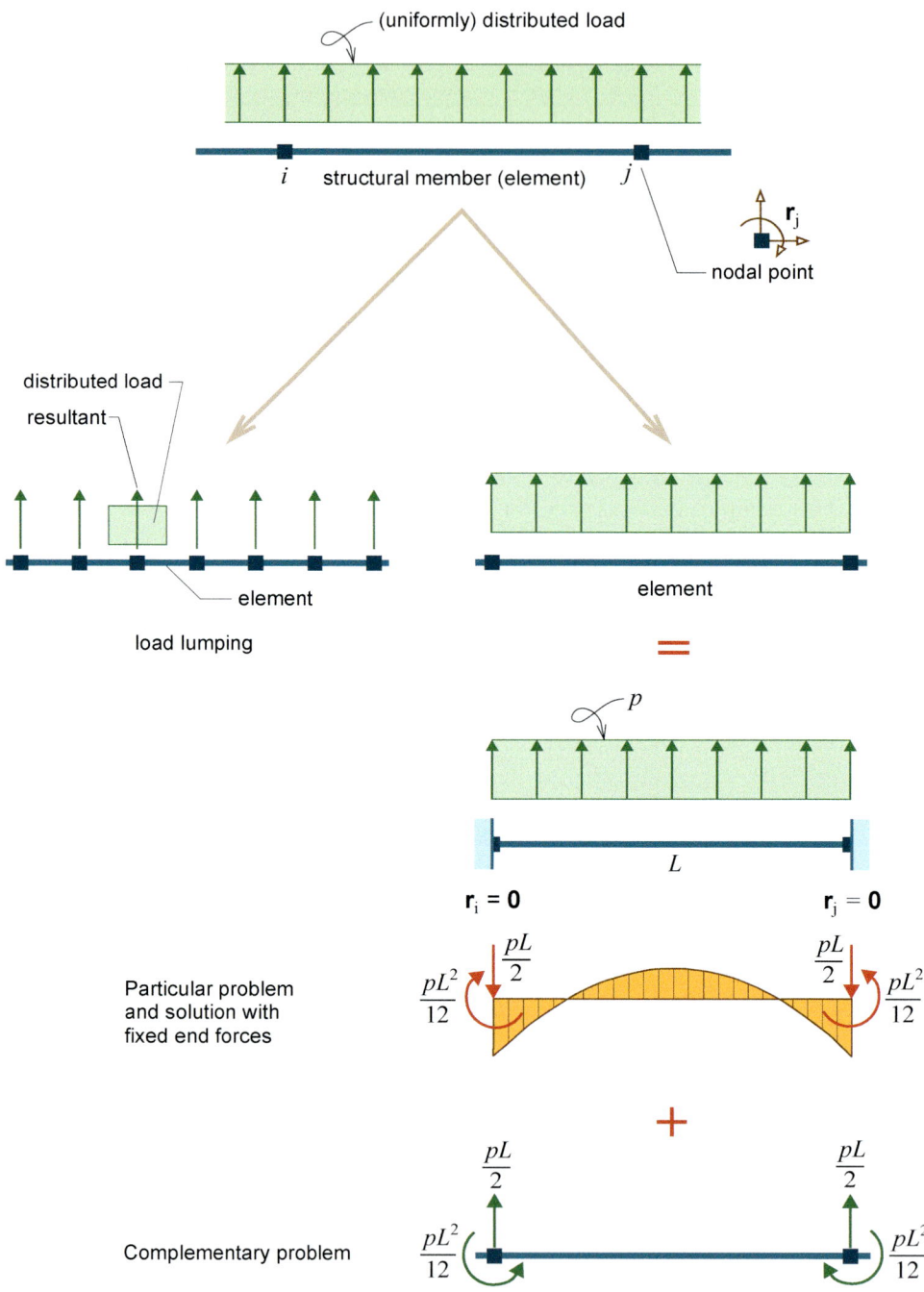

Figure 2.13 Distributed load on beam element

2.5 Distributed loading – element load vector

In dealing with distributed loading between nodal points we have basically two options, as shown in Fig. 2.13. The first and obvious solution is an approximate "sledgehammer" technique in which we "model" us out of the problem by simply dividing the structural members (beams and columns) into many short elements and *lump* the loading into statically equivalent concentrated forces. We are thus back to where we started this section, all loading acts as concentrated forces at the nodal points. The price is of course a significant increase in the number of degrees of freedom, and thus increased computational effort.

The second approach is to split the problem into two:
(1) the so-called *particular* problem, and (2) the *complementary* problem. The particular problem is one in which *all dofs* are suppressed by *fictitious nodal forces* (including moments) such that **r** = **0**. The determination of the internal forces (due to the distributed loading) of this "fixed end" problem is fairly straightforward (and is solved element by element), as is the determination of the fictitious nodal forces necessary to maintain **r** = **0**.

The complementary problem is one in which the distributed loading is replaced by the fictitious nodal forces, applied to the model with *opposite* sign. We are thus back to a familiar problem with only concentrated nodal loading.

If we now add the two problems together, the fictitious nodal forces cancel out and the loading of the combined model is the one we started with. The final solution is therefore the sum of the solutions of the two problems. If our original problem had some concentrated nodal forces in addition to the distributed loading, these are simply included in the loading of the complementary problem.

Let us formalize this and put it in matrix notation. We start with a single element subjected to some action between its ends (element nodes); this action may be transverse distributed load or temperature or some other form of *initial strain*. In order to maintain **v** = **0**, we need to apply (fixed end) forces (including moments) corresponding to the element *dofs* **v**. We denote these fixed end forces by $\mathbf{S}^0$. We can now rewrite the element stiffness relation of Eq. (2-12b), which applies to an element with no external action between its nodes, to one that also allows external action between the element nodes:

$$\mathbf{S} = \mathbf{k}\mathbf{v} + \mathbf{S}^0 \tag{2-27}$$

When an element interacts with the rest of the structure, which is subjected to concentrated nodal forces (**R**), it receives an elastic deformation defined by **v**; this causes *elastic* forces **kv**. If the element is also subjected to some external action between its (end) nodes, a set of forces $\mathbf{S}^0$ must be applied at the element ends in order to secure zero end displacements (**v** = **0**). The sum of these two "sources" gives the final element nodal forces **S**, as expressed by Eq. (2-27).

Just as **k** is the element stiffness matrix, $\mathbf{S}^0$ is the element *load vector*. In our definition $\mathbf{S}^0$ follows the same sign rules as **S**; it may represent all types of external action on the element, *e.g.* distributed loading and temperature.

37

Summary of matrix structural analysis

 **James Clerk MAXWELL** (1831-1879) was a Scottish physicist and mathematician, by many considered the greatest physicist between NEWTON and EINSTEIN. He made decisive contributions to electromagnetism, colour analysis, control theory and kinetic theory and thermodynamics (MAXWELL-BOLTZMANN).

MAXWELL also made many contributions to structural mechanics and analysis, but much of his work was published in an abstract form without many illustrating figures and therefore remained unnoticed by engineers of his time. He formulated a comprehensive theory of statically indeterminate frames (trusses) which contains the *principle of virtual forces* and the *reciprocity theorem* (which he formulated for the simple case of two forces and corresponding displacements of a truss). Ten years later, Otto MOHR rediscovered (without knowledge of MAXWELL's work) the essence of this work and made it available for practical application. MAXWELL's contribution to *graphical statics* was adopted and generalized by Luigi CREMONA. MAXWELL also showed how stresses can be measured *photoelastically*.

 **Enrico BETTI** (1823-1892) was an Italian mathematician, now remembered mostly for his 1871 paper on topology that led to the later naming of him of the BETTI numbers (used to distinguish topological spaces). He also worked on the theory of equations.

His main impact on structural mechanics is his generalization of MAXWELL's *reciprocity theorem* to include any number of degrees of freedom.

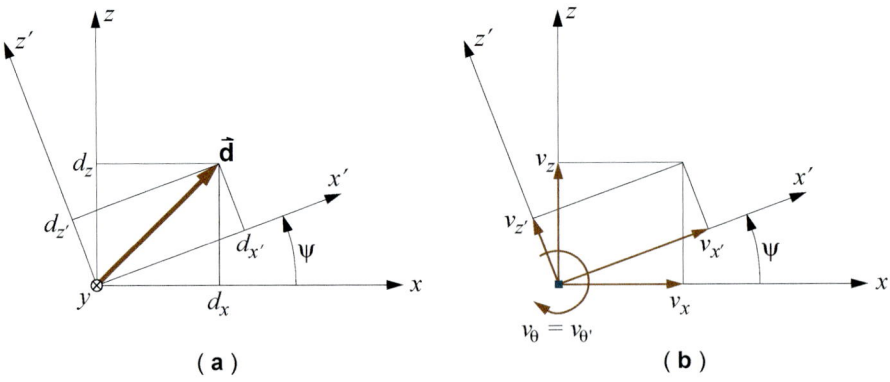

Figure 2.14 Rotation of coordinate axes

Denoting the part of the external loading that actually consists of concentrated nodal forces by $\mathbf{R}^c$, and using Eq. (2-27) as our force-displacement relation, we can rewrite Eq. (2-22) as

$$\mathbf{R}^c = \left(\sum_{i=1}^{m}(\mathbf{a}^i)^T\mathbf{k}^i\mathbf{a}^i\right)\mathbf{r} + \sum_{i=1}^{m}(\mathbf{a}^i)^T\mathbf{S}^{0i} = \mathbf{Kr} + \mathbf{R}^0 \quad \Rightarrow \quad \mathbf{Kr} = \mathbf{R} \qquad (2\text{-}28)$$

where

$$\mathbf{R} = \mathbf{R}^c - \mathbf{R}^0 \qquad (2\text{-}29)$$

is the "final" load vector. Equation (2-29) is consistent with our arguments above: the fictitious fixed end forces are applied with *opposite* sign, hence the minus sign.

For external distributed element loading we will recommend the simple load *lumping* technique; we will support this recommendation by an example at the end of the chapter. However, in order to handle loading best described as *initial strain* (*e.g.* temperature), we will also need to form the element load vector $\mathbf{S}^0$.

2.6 Transformations

Figure 2.14a shows a physical vector $\mathbf{d}$ in 2D space and its two components in two different coordinate systems, x-z and x'-z'. The latter is obtained by a rotation of the "basis" system (x-z) by an angle ψ about the y-axis. From the figure the following relations are readily obtained:

$$d_{x'} = d_x\cos\psi + d_z\sin\psi$$

$$d_{z'} = -d_x\sin\psi + d_z\cos\psi$$

or

$$\mathbf{d}' = \begin{bmatrix} d_{x'} \\ d_{z'} \end{bmatrix} = \begin{bmatrix} c & s \\ -s & c \end{bmatrix}\begin{bmatrix} d_x \\ d_z \end{bmatrix} = \mathbf{t}_2\mathbf{d} \qquad (2\text{-}30a)$$

where we have used the shorthand notation $c \equiv \cos\psi$ and $s \equiv \sin\psi$. The inverse relation is easily obtained:

$$\mathbf{d} = \begin{bmatrix} d_x \\ d_z \end{bmatrix} = \begin{bmatrix} c & -s \\ s & c \end{bmatrix}\begin{bmatrix} d_{x'} \\ d_{z'} \end{bmatrix} = \mathbf{t}_2^{-1}\mathbf{d}' \qquad (2\text{-}30b)$$

We see that

$$\mathbf{t}_2^{-1} = \mathbf{t}_2^T \qquad (2\text{-}31)$$

and the *rotation matrix* $\mathbf{t}_2$ is an *orthogonal* matrix. We take this a step further, and consider a nodal point with three *dofs*, two translations and one rotation, as shown in Fig. 2.14b. We can express the *dofs* in the two coordinate systems x-z and x'-z'. The rotation (v_θ) is the same in both systems, and by use of Eq. (2-30a) we can write

Summary of matrix structural analysis

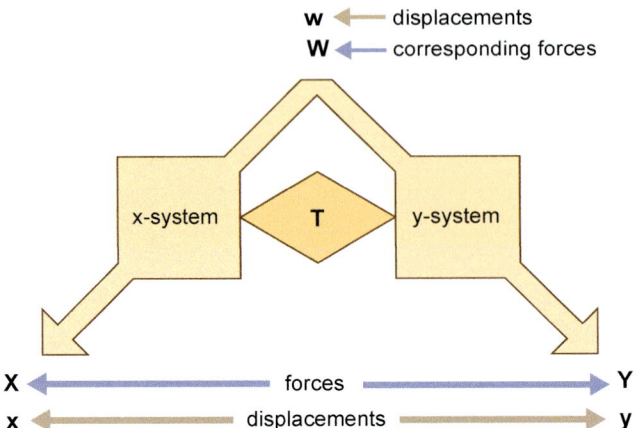

Figure 2.15 Different representations of displacements and corresponding forces

$$\mathbf{v'} = \begin{bmatrix} v_{x'} \\ v_{z'} \\ v_{\theta'} \end{bmatrix} = \begin{bmatrix} c & s & 0 \\ -s & c & 0 \\ 0 & 0 & 1 \end{bmatrix} \begin{bmatrix} v_x \\ v_z \\ v_\theta \end{bmatrix} = \mathbf{t}_3 \mathbf{v} \quad \text{and} \quad \mathbf{v} = \mathbf{t}_3^T \mathbf{v'} \tag{2-32}$$

The arrows in Fig. 2.14b could equally well have been force components. Hence

$$\mathbf{S'} = \mathbf{t}_3 \mathbf{S} \quad \text{and} \quad \mathbf{S} = \mathbf{t}_3^T \mathbf{S'} \tag{2-33}$$

We can generalize this to a much wider notion of coordinate or reference systems. Figure 2.15 indicates schematically how a system of displacements (**w**) and corresponding forces (**W**) can be represented in two different reference systems "x" and "y". The total work performed by a given set of forces over the corresponding displacements is a scalar, *independent* of the reference system. Hence

$$\mathbf{W}^T \mathbf{w} = \mathbf{X}^T \mathbf{x} = \mathbf{Y}^T \mathbf{y} \tag{2-34}$$

We assume that a unique linear relationship exists between the two reference systems, that is, the components in the one system can be expressed by a unique linear combination of the components in the other system, as

$$\mathbf{y} = \mathbf{T}\mathbf{x} \tag{2-35}$$

The transformation matrix **T** contains "constants" which are independent of the magnitude of the elements of **x** and **y**. Equation (2-35) substituted into (2-34) yields:

$$\mathbf{X}^T \mathbf{x} = \mathbf{Y}^T \mathbf{T} \mathbf{x}$$

This relation must hold for any set of displacements **x**. Hence

$$\mathbf{X}^T = \mathbf{Y}^T \mathbf{T} \quad \text{or} \quad \mathbf{X} = \mathbf{T}^T \mathbf{Y} \tag{2-36}$$

Equations (2-35) and (2-36) represent a very important transformation, often referred to as *contragredient* transformation. We see that the coordinate transformations of Eqs. (2-32) and (2-33) satisfy the general rule, and so does in effect also the PVD transformation expressed by Eq. (2-21), even if this association is more subtle. It should be emphasized that the transformation matrix **T** does not have to be a square matrix; the example below will show an example of a non-square transformation matrix.

Consider two representations of an element stiffness relation,

$$\mathbf{S}_a = \mathbf{k}_a \mathbf{v}_a \quad \text{and} \quad \mathbf{S}_b = \mathbf{k}_b \mathbf{v}_b \quad \text{where}$$

$$\mathbf{v}_a = \mathbf{T}_b \mathbf{v}_b \tag{2-37}$$

then, according to our transformation rule,

$$\mathbf{S}_b = \mathbf{T}_b^T \mathbf{S}_a = \mathbf{T}_b^T \mathbf{k}_a \mathbf{v}_a = \mathbf{T}_b^T \mathbf{k}_a \mathbf{T}_b \mathbf{v}_b = \mathbf{k}_b \mathbf{v}_b \quad \Rightarrow \quad \mathbf{k}_b = \mathbf{T}_b^T \mathbf{k}_a \mathbf{T}_b \tag{2-38}$$

Summary of matrix structural analysis

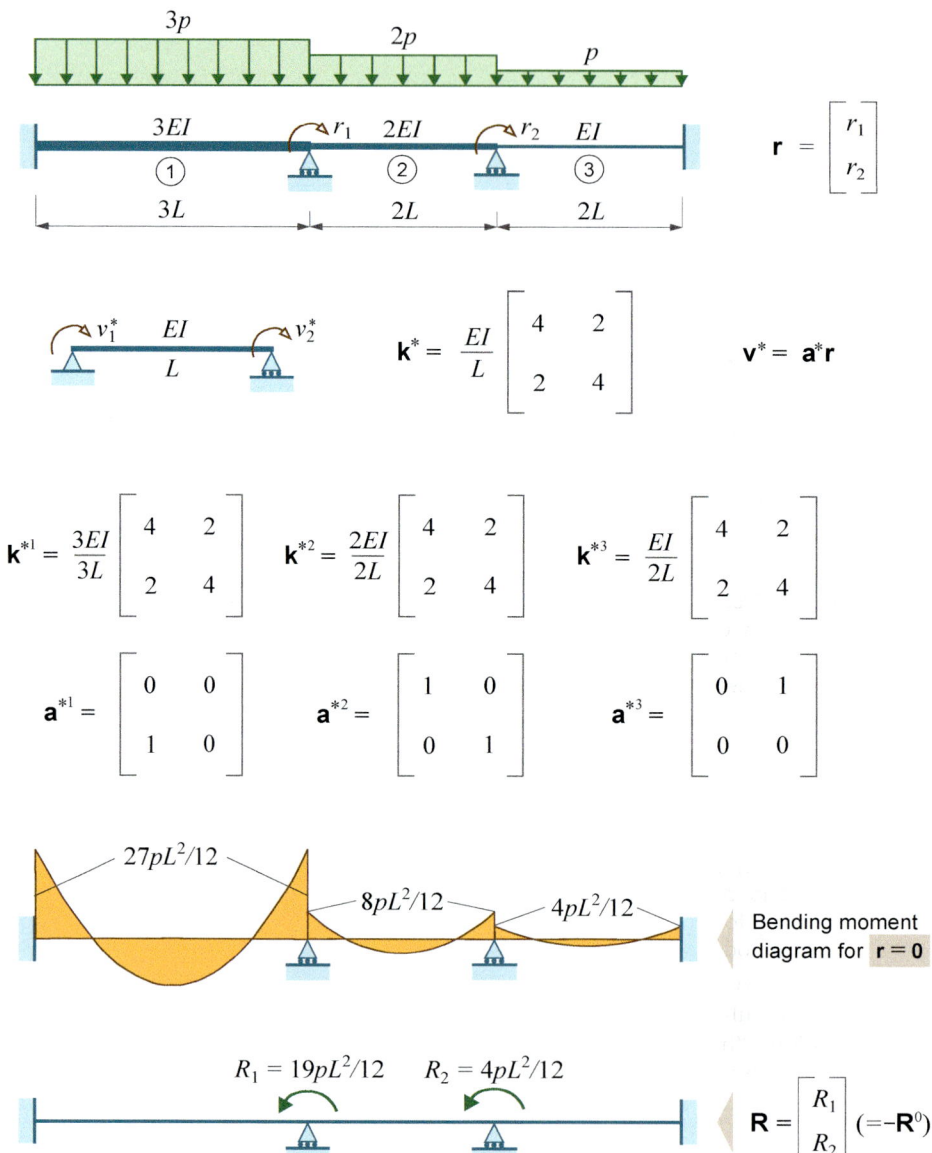

Figure 2.16 A simple beam problem

Summary of Chapter 2

This chapter gives a brief overview of the matrix formulation of the *displacement method* of structural analysis, exemplified by a 2D beam element. It also sets up the framework for the presentation to come, in terms of coordinate systems, terminology, notation and sign rules.

It should be emphasized that, within the assumptions of linear theory and simple beam theory, we have made no approximations. We have simply reformulated standard structural mechanics procedures. Nevertheless, a large part of what we have presented will also apply in the *finite element world*. In fact, the only thing that will be different is the *element analysis*, that is to say all computations related to the element themselves. We will also show later on that we can determine our beam element matrices (**k** and **S**⁰) by the procedures we will develop for finite element analysis.

We conclude this summary chapter with some examples.

Example 2-1 - System stiffness relation

At the top of Fig. 2.16 is shown a simple beam problem for which we seek the system stiffness matrix **K** and load vector **R**.

Solution

Due to the loading and support conditions we only need to consider two *dofs*, the rotation of the two internal support points, r_1 and r_2. Since we have no axial forces the simplest, 2-*dof*, element suffices, and the stiffness matrices for the three elements are readily obtained from Fig. 2.7. The corresponding **a***-matrices are very simple, and the matrix multiplications of Eq. (2-24) are trivial. Using the matrices of Fig. 2.16, the result of this multiplications is:

$$\mathbf{K} = \frac{EI}{L}\left(\begin{bmatrix} 4 & 0 \\ 0 & 0 \end{bmatrix} + \begin{bmatrix} 4 & 2 \\ 2 & 4 \end{bmatrix} + \begin{bmatrix} 0 & 0 \\ 0 & 2 \end{bmatrix} \right) = \frac{EI}{L}\begin{bmatrix} 8 & 2 \\ 2 & 6 \end{bmatrix}$$

This result could have been obtained more easily by applying the technique of Fig. 2.7 directly to the entire beam structure, that is, imposing unit rotations, one at a time, and determining (from beam formulas) the concentrated moments (corresponding to the *dofs*) required to produce the deformation patterns consistent with the unit nodal displacements.

In order to establish the load vector we consider the bending moment diagram of the particular solution (**r** = **0**). The unbalanced moments at internal supports are the fictitious "fixed end" forces (**R**⁰), and these forces, with opposite sign, are the loading in the complementary solution. Hence

$$\mathbf{R} = \frac{-pL^2}{12}\begin{bmatrix} 19 \\ 4 \end{bmatrix}$$

and our problem is solved. The rest, finding **r** and the moments it causes and adding these to the particular solution, is straightforward.

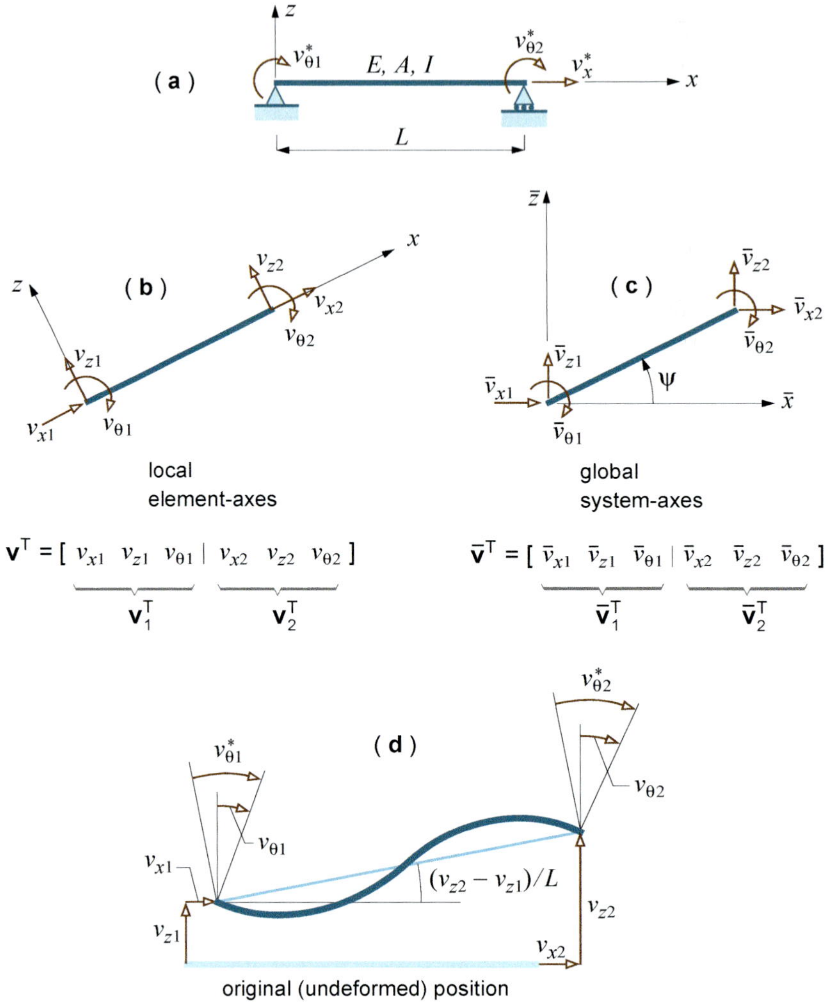

Summary of matrix structural analysis

$$\mathbf{v}^T = [\ v_{x1}\ v_{z1}\ v_{\theta 1}\ |\ v_{x2}\ v_{z2}\ v_{\theta 2}\ ]$$

$$\underbrace{\phantom{v_{x1}\ v_{z1}\ v_{\theta 1}}}_{\mathbf{v}_1^T} \quad \underbrace{\phantom{v_{x2}\ v_{z2}\ v_{\theta 2}}}_{\mathbf{v}_2^T}$$

$$\bar{\mathbf{v}}^T = [\ \bar{v}_{x1}\ \bar{v}_{z1}\ \bar{v}_{\theta 1}\ |\ \bar{v}_{x2}\ \bar{v}_{z2}\ \bar{v}_{\theta 2}\ ]$$

$$\underbrace{\phantom{\bar{v}_{x1}\ \bar{v}_{z1}\ \bar{v}_{\theta 1}}}_{\bar{\mathbf{v}}_1^T} \quad \underbrace{\phantom{\bar{v}_{x2}\ \bar{v}_{z2}\ \bar{v}_{\theta 2}}}_{\bar{\mathbf{v}}_2^T}$$

Figure 2.17 Different reference systems for a 2D beam element

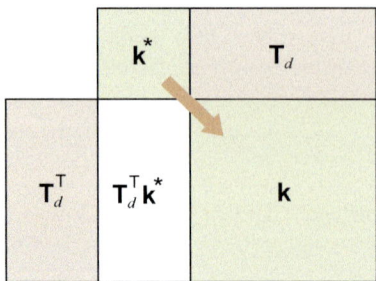

Example 2-2 - Transformations

Figure 2.17a, b and c show the same 2D beam element in three different reference systems. In Fig. 2.17a the rigid body modes are removed and only the deformation *dofs* are retained. The deformation stiffness (k^*) is derived in Fig. 2.7. The question we now pose is: can we obtain k and $\bar{k}$ for the elements of Fig. 2.17b and c, respectively, by transformation of k^*?

Solution

We need to establish a relationship between v^* and v. Figure 2.17d shows the element in its original, undeformed position and in its final deformed position. The transition from the original to the final position can be thought of as a pure rigid body movement followed by a pure deformation, and while v can describe both, v^* can only describe the latter. The components of both vectors are shown in the figure, and we can easily write down their relationship:

$$v^* = \begin{bmatrix} v_x^* \\ v_{\theta 1}^* \\ v_{\theta 2}^* \end{bmatrix} = \begin{bmatrix} -1 & 0 & 0 & 1 & 0 & 0 \\ 0 & -1/L & 1 & 0 & 1/L & 0 \\ 0 & -1/L & 0 & 0 & 1/L & 1 \end{bmatrix} \begin{bmatrix} v_{x1} \\ v_{z1} \\ v_{\theta 1} \\ v_{x2} \\ v_{z2} \\ v_{\theta 2} \end{bmatrix} = T_d v \qquad (2\text{-}39)$$

Note carefully how we "measure" the components of the two sets, and keep in mind that displacements in Fig. 2.17 d are exaggerated; displacements are still *small*. Since no work is performed over the rigid body displacements we have

$$(S^*)^T v^* = S^T v$$

and the transformation rule of Eqs. (2-35) and (2-36) apply. Hence

$$S = T_d^T S^* = T_d^T k^* v^* = T_d^T k^* T_d v \quad \Rightarrow \quad k = T_d^T k^* T_d \qquad (2\text{-}40)$$

The result of this matrix multiplication is

$$k = \begin{bmatrix} \frac{EA}{L} & 0 & 0 & \frac{-EA}{L} & 0 & 0 \\ 0 & \frac{12EI}{L^3} & \frac{-6EI}{L^2} & 0 & \frac{-12EI}{L^3} & \frac{-6EI}{L^2} \\ 0 & \frac{-6EI}{L^2} & \frac{4EI}{L} & 0 & \frac{6EI}{L^2} & \frac{2EI}{L} \\ \frac{-EA}{L} & 0 & 0 & \frac{EA}{L} & 0 & 0 \\ 0 & \frac{-12EI}{L^3} & \frac{6EI}{L^2} & 0 & \frac{12EI}{L^3} & \frac{6EI}{L^2} \\ 0 & \frac{-6EI}{L^2} & \frac{2EI}{L} & 0 & \frac{6EI}{L^2} & \frac{4EI}{L} \end{bmatrix} \qquad (2\text{-}41)$$

Next we seek a relation between v and $\bar{v}$. According to Eq. (2-32) we can write

Summary of matrix structural analysis

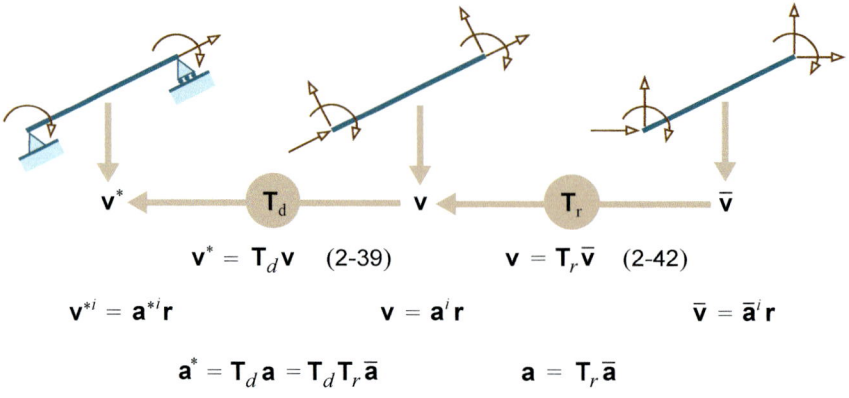

$$\mathbf{v}^* = \mathbf{T}_d \mathbf{v} \quad (2\text{-}39) \qquad \mathbf{v} = \mathbf{T}_r \bar{\mathbf{v}} \quad (2\text{-}42)$$

$$\mathbf{v}^{*i} = \mathbf{a}^{*i} \mathbf{r} \qquad \mathbf{v} = \mathbf{a}^i \mathbf{r} \qquad \bar{\mathbf{v}} = \bar{\mathbf{a}}^i \mathbf{r}$$

$$\mathbf{a}^* = \mathbf{T}_d \mathbf{a} = \mathbf{T}_d \mathbf{T}_r \bar{\mathbf{a}} \qquad \mathbf{a} = \mathbf{T}_r \bar{\mathbf{a}}$$

Figure 2.18 Some important relationships

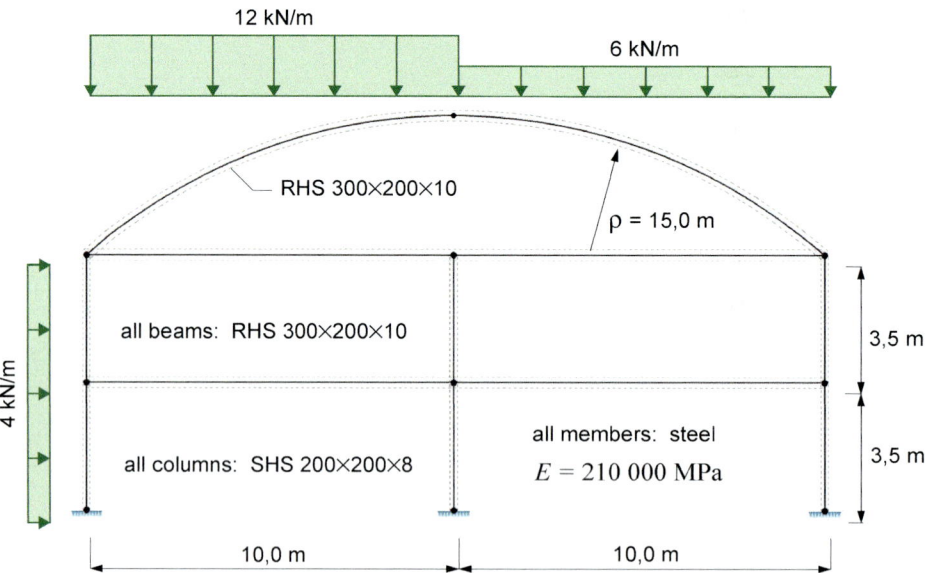

Figure 2.19 Test frame

$$\mathbf{v} = \begin{bmatrix} \mathbf{v}_1 \\ \mathbf{v}_2 \end{bmatrix} = \begin{bmatrix} \mathbf{t}_r & \mathbf{0} \\ \mathbf{0} & \mathbf{t}_r \end{bmatrix} \begin{bmatrix} \bar{\mathbf{v}}_1 \\ \bar{\mathbf{v}}_2 \end{bmatrix} = \mathbf{T}_r \bar{\mathbf{v}} \qquad (2\text{-}42)$$

where $\mathbf{t}_r$ is the rotation matrix $\mathbf{t}_3$ of Eq. (2-32). Then, applying our transformation rule

$$\bar{\mathbf{S}} = \mathbf{T}_r^T \mathbf{S} = \mathbf{T}_r^T \mathbf{k} \mathbf{v} = \mathbf{T}_r^T \mathbf{k} \mathbf{T}_r \bar{\mathbf{v}} = \bar{\mathbf{k}} \bar{\mathbf{v}}$$

and

$$\bar{\mathbf{k}} = \mathbf{T}_r^T \mathbf{k} \mathbf{T}_r = \mathbf{T}_r^T \mathbf{T}_d^T \mathbf{k}^* \mathbf{T}_d \mathbf{T}_r \qquad (2\text{-}43)$$

Figure 2.18 sums up some important relationships. We end this example with a word of warning. We cannot get more from a transformation than that which is contained in the matrix/vector we transform. For instance, we cannot transform $\mathbf{S}^{0*} = [0 \quad pL^2/12 \quad -pL^2/12]^T$, for uniformly distributed transverse loading, to $\mathbf{S}^0$, by use of $\mathbf{T}_d$. The three elements of $\mathbf{S}^{0*}$ have no knowledge of the parabolic shape of the moment diagram along the beam, and hence of the end shear forces. However, $\mathbf{T}_r$ will transform $\mathbf{S}^0$ into $\bar{\mathbf{S}}^0$.

Example 2-3 – A computer aided solution

Matrix theory of structures and computers go hand in hand. For manual or hand calculations we would probably not bother with matrices.

Perhaps one of the most significant developments in structural analysis in the first half of the 20th century was the *method of moment distribution* developed by Hardy CROSS [8]. This clever scheme, well suited for hand calculations, is an iterative solution based on the displacement method of analysis. As for all hand calculations the emphasis is on minimizing the amount of computational work. Hence, *clever* is a key word in hand calculations.

The computer, however, is not all that clever, but it is extremely fast and accurate at the few simple operations it can perform. The focus therefore is now on *simplicity* and *systematics*, both of which the matrix formulation of the displacement method offers. Today, the challenges of designing and developing a computer program for static analysis of frame type structures by linear theory, are not so much associated with the actual computations as they are with the graphical user interface (GUI).

In order to demonstrate the capabilities of a fairly typical plane frame analysis program, we consider the test frame of Fig. 2.19, and we make use of program **fap2D** [9], a homegrown program with considerable student participation, still in the development stages. The test frame serves merely as an example problem, it has nothing to do with real life. All members are made of hollow steel sections of standard dimensions, and the only loading considered consists of distributed line loads applied on the "outside" of the structure, as shown in the figure. The load on the arch is a projection load (*e.g.* snow). All joints are assumed to be completely rigid, and shear deformations are neglected.

fap2D is designed with the computational power of a typical PC (as of 2011) in mind. A model consisting of many simple elements and load lumping, is therefore a natural choice. Computationally this clearly means more work, but this is, as we will see, insignificant and more than outweighed by the advantages.

Summary of matrix structural analysis

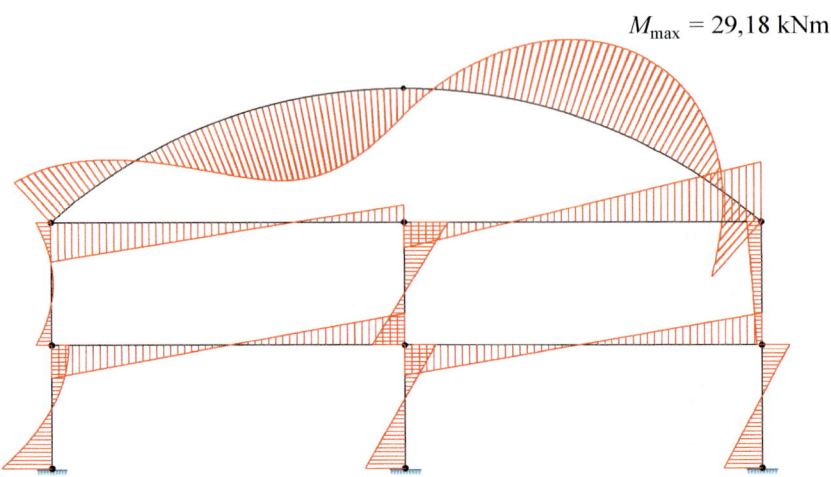

(a) bending moment diagram (drawn on the tensile side of the cross section)

$M_{max} = 29{,}18$ kNm

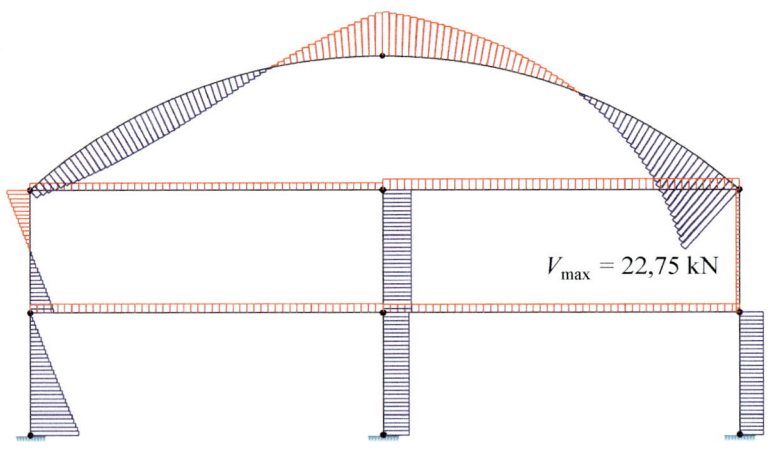

(b) shear force diagram

$V_{max} = 22{,}75$ kN

Figure 2.20 Computational results (obtained by fap2D)

Example 2-3 – A computer aided solution

Figure 2.20 shows the computed bending moments (a) and shear forces (b). Each line perpendicular to the member axis represents a nodal value. Thus the model contains

  500 short and straight beam elements,

    498 nodal points, and

    1498 simultaneous equations to be solved (dimension of **K**).

The total time spent in the Fortran "computational engine", on a standard PC, is (in 2011) about 10 milliseconds. This includes most of the number handling: the assembly process, solving the simultaneous equations and computing internal forces. If this is the downside of the approach, look at the upside. The modelling of curved members and members of varying cross section by a series of straight beam elements with constant cross section is a straightforward and easily automated process, and so is the handling of distributed loading.

The "many simple element" approach also lends itself to some of the graphical issues. For instance, *all* lines in Fig. 2.20 are *straight*, a fact welcomed by most programmers. Furthermore, if we go beyond simple linear analysis to for instance a geometrically nonlinear problem, the issue of geometrical *imperfection*, in the form of say a buckling or displacement mode shape is quite easily handled.

To some ("puritans") the shear force diagram in Fig. 2.20b may cause some concern. Compared to an "exact" solution of the problem posed by Fig. 2.19, the shear forces are, due to the load lumping, computed with considerably less accuracy than are the bending moments. However, from a practical point of view, this is hardly an issue at all. The maximum shear force of Fig. 2.20b is probably within a couple of per cent of the "exact" value, and considering our knowledge about the actual loading, and some of the other uncertainties of our problem, an engineer would not grumble at this. In fact, he or she would most likely accept results from a much coarser element "mesh".

Finally a word of warning; there is clearly a limit to how many (very short) elements a member can be divided into before numerical problems creep into the solution. A well designed commercial program should issue a warning if the user asks for very short elements. We shall come back to such problems in Chapters 17 and 18.

Summary of matrix structural analysis

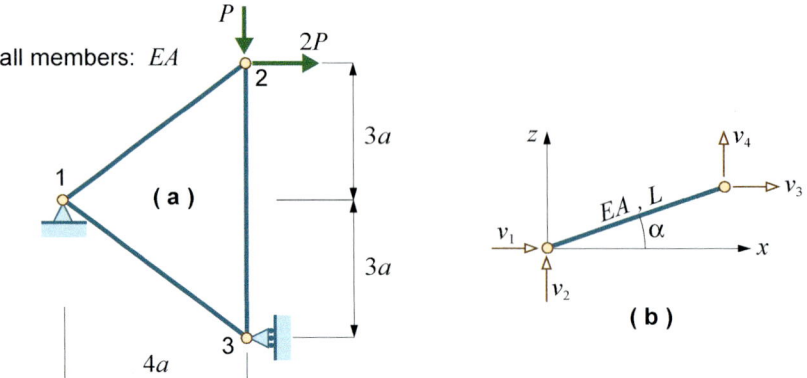

Figure 2.21

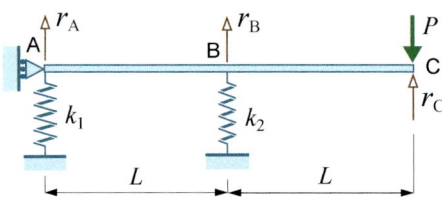

Figure 2.22

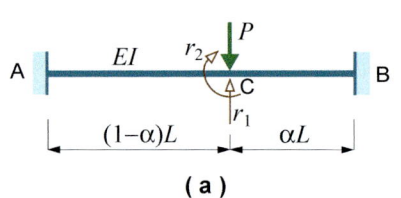

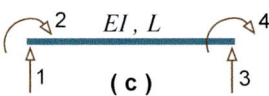

Figure 2.23

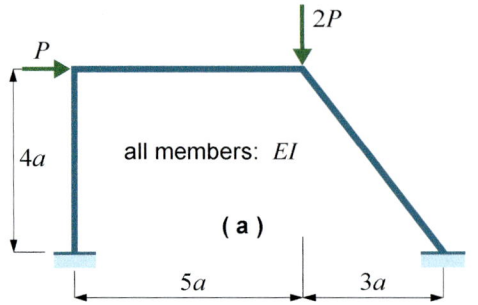

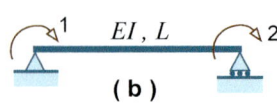

Figure 2.24

Problems

Problem 2.1

Figure 2.21a shows a very simple *truss*. All three bar members have the same axial stiffness EA. Figure 2.21b shows an arbitrary bar element with 4 *dofs*.

a) Determine the stiffness matrix **k** for the element in Fig. 2.21b in terms of EA, L and α.
b) Establish the element stiffness matrix for each element in the truss, in terms of EA and a.
c) The truss has 3 non-zero degrees of freedom. Establish **K** for the truss by assembling the element stiffness matrices found in **b)** with due regard to the boundary conditions.
d) Establish **R** and solve for the unknown displacements **r**. Determine the bar forces from the displacements and check equilibrium at node 2.

Problem 2.2

Figure 2.22 shows a completely *rigid* bar A-B-C supported by two springs in A and B and prevented from horizontal displacement. The spring stiffness is k_1 and k_2, respectively, as shown in the figure. The bar has two independent degrees of freedom, and it is loaded by a vertical force P at point C.

a) Establish, by a direct equilibrium approach, the stiffness relation $\mathbf{K}_1\mathbf{r}_1 = \mathbf{R}_1$ for the system when r_A and r_B are chosen as degrees of freedom, that is $\mathbf{r}_1^T = [r_A \ r_B]$.
b) Establish, by a direct equilibrium approach, the stiffness relation $\mathbf{K}_2\mathbf{r}_2 = \mathbf{R}_2$ for the system when r_B and r_C are chosen as degrees of freedom, that is $\mathbf{r}_2^T = [r_B \ r_C]$.
c) Establish a relationship between $\mathbf{r}_1$ and $\mathbf{r}_2$ and determine $\mathbf{K}_2$ and $\mathbf{R}_2$ by *transforming* $\mathbf{K}_1$ and $\mathbf{R}_1$, respectively.

Problem 2.3

Figure 2.23a shows a uniform beam with bending stiffness EI, fixed at both ends and subjected to a vertical load P at point C. We define two degrees of freedom at point C as shown, and we consider only bending deformation.

a) Establish the stiffness matrix **K** for the problem, using first the simple 2-*dof* element in Fig. 2.23b, and then the 4-*dof* element in Fig. 2.23c
b) Solve the system stiffness relation **Kr** = **R** and use the result to establish formulas for the fixed end moments due to a concentrated force at an arbitrary point on the beam.

Problem 2.4

Figure 2.24a shows a simple plane frame subjected to two concentrated forces as shown. All members have bending stiffness EI.

a) Disregard axial and shear deformations and choose suitable (unknown) degrees of freedom for the frame.
b) Use the simple 2-*dof* element in Fig. 2.24b and establish **a**-matrices for all three elements of the frame. Use these matrices and the element stiffness matrices to establish the stiffness matrix **K** for the frame.
c) Determine the load vector **R**, solve the stiffness relation **Kr** = **R** and find (and draw) the bending moment diagram for the frame.

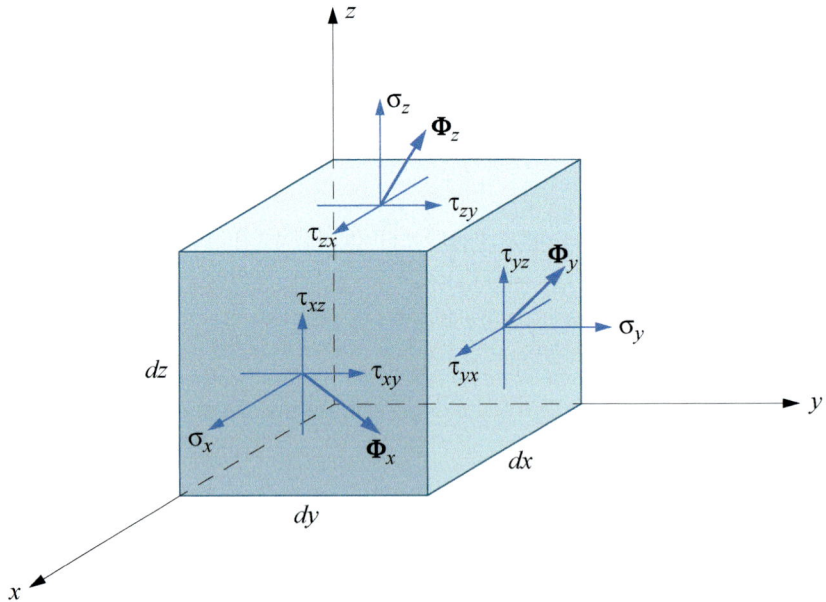

Figure 3.1 Stresses (σ and τ) and tractions (Φ) in 3D

3
Summary of linear elasticity theory

In this chapter we summarize some fundamental equations for two- and three-dimensional theory of linear elasticity as well as the assumptions and simplifications made in some of the methods we seek FEM solutions for.

Our focus is elastic stress analysis, for all types of structures. We will start with the complete, three-dimensional (3D) problem which is fairly straightforward since we need not make any simplifying assumptions.

Plate problems are next. If all loading is in the plane of the plate, we have a 2D (membrane) problem. Depending on the simplification made in moving from three to two dimensions, we distinguish between *plane stress* and *plane strain* problems. If the loading acts normal to the plane of the plate, we have a *plate bending* problem which is somewhere between a two- and a three-dimensional problem. As a prelude to the plate bending problem we include a section on the related problem of bending of beams. We include both medium thick and thin plates, but the emphasis is on thin plates.

A chapter on *shell* problems is included. However, shell theories will not be dealt with here. Our treatment of shells will be limited to models consisting of flat plate elements, combining bending and membrane action.

Another somewhat specialized problem, well suited for finite element analysis, is St.Venant torsion. A chapter devoted to this problem is included, but since it is slightly peripheral, the theoretical basis is treated in that chapter, along with the finite element formulation, and not here.

3.1 Three-dimensional stress analysis

Stress and equilibrium. Figure 3.1 shows an infinitesimal volume element *dxdydz* with the *tractions*, and their stress components, acting on the "positive" faces, that is faces with surface normals in the direction of positive axes. The three *normal* stress components are designated by σ and an index indicating its direction. The six *shear* components are designated by τ and two indices, the first of which indicates the direction of the surface normal and the second the direction of the stress component itself. The *traction* vectors on the three surfaces are denoted by Φ and an index indicating the direction of the surface normal; they are defined as:

Summary of linear elasticity theory

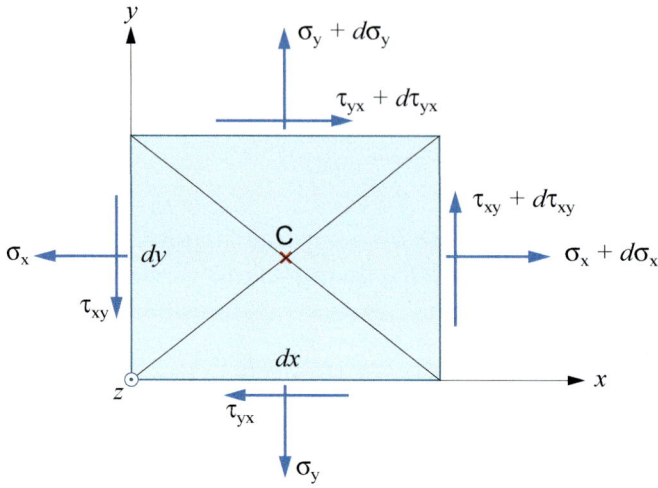

Figure 3.2 Stresses on faces parallel with z-axis

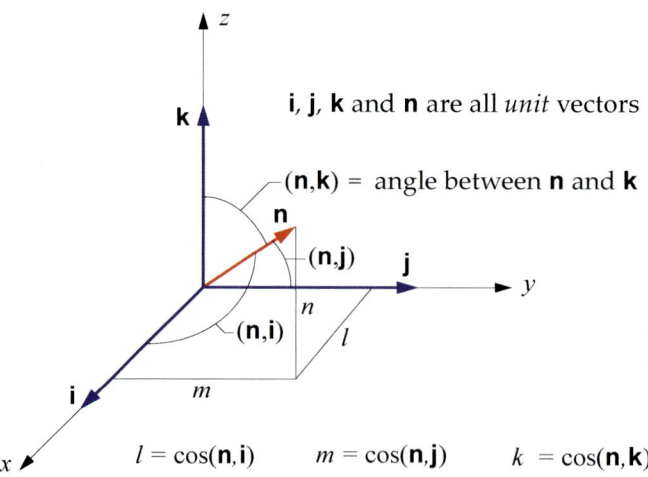

Figure 3.3 Direction cosines, l, m and n

Three-dimensional stress analysis

$$\boldsymbol{\Phi}_x = \begin{bmatrix} \sigma_x \\ \tau_{xy} \\ \tau_{xz} \end{bmatrix}, \quad \boldsymbol{\Phi}_y = \begin{bmatrix} \tau_{yx} \\ \sigma_y \\ \tau_{yz} \end{bmatrix} \quad \text{and} \quad \boldsymbol{\Phi}_z = \begin{bmatrix} \tau_{zx} \\ \tau_{zy} \\ \sigma_z \end{bmatrix} \tag{3-1}$$

The *stress matrix* is defined as

$$\boldsymbol{\sigma}_M = \begin{bmatrix} \boldsymbol{\Phi}_x^T \\ \boldsymbol{\Phi}_y^T \\ \boldsymbol{\Phi}_z^T \end{bmatrix} = \begin{bmatrix} \sigma_x & \tau_{xy} & \tau_{xz} \\ \tau_{yx} & \sigma_y & \tau_{yz} \\ \tau_{zx} & \tau_{zy} & \sigma_z \end{bmatrix} \tag{3-2}$$

Moment equilibrium about an axis parallel with the *z*-axis through point C in Fig. 3.2, gives (when we disregard higher order terms):

$$\tau_{xy} = \tau_{yx} \tag{3-3}$$

Similarly for axes parallel with *x* and *y*, i.e., $\tau_{yz} = \tau_{zy}$ and $\tau_{xz} = \tau_{zx}$. Hence, the stress matrix is *symmetric*, that is

$$\boldsymbol{\sigma}_M = \boldsymbol{\sigma}_M^T \tag{3-4}$$

In matrix notation it is more convenient to arrange the 6 independent stress components in a *stress vector*:

$$\boldsymbol{\sigma} = \begin{bmatrix} \sigma_x \\ \sigma_y \\ \sigma_z \\ \tau_{xy} \\ \tau_{yz} \\ \tau_{zx} \end{bmatrix} \tag{3-5}$$

Before we proceed we need to define the *direction cosines* of an arbitrary *unit vector* **n**. Referred to a cartesian coordinate system *x,y,z* as shown in Fig. 3.3, the direction cosines to the angles between **n** and the unit vectors **i**, **j**, and **k** along axes *x*, *y* and *z*, respectively, are defined as

$$l = \cos(\mathbf{n}, \mathbf{i}), \quad m = \cos(\mathbf{n}, \mathbf{j}) \quad \text{and} \quad n = \cos(\mathbf{n}, \mathbf{k}) \tag{3-6a}$$

Since **n** is a unit vector, the direction cosines are its components along the cartesian axes, that is

$$\mathbf{n} = \begin{bmatrix} l \\ m \\ n \end{bmatrix} \tag{3-6b}$$

Next we consider the equilibrium of a small tetrahedron OABC, see Fig. 3.4a. The triangle ABC is a boundary surface, with an area *S*, on which a surface traction **Φ** acts. Internal stresses act on the remaining three sides (surfaces)

Summary of linear elasticity theory

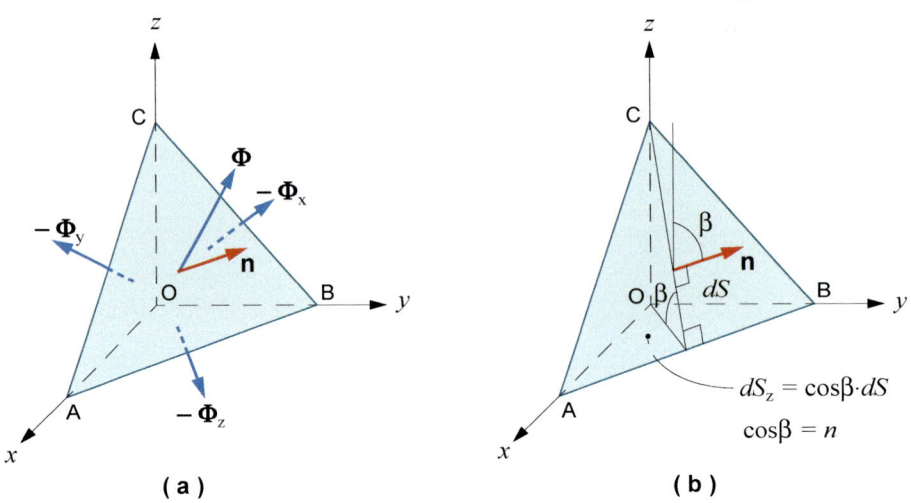

Figure 3.4 Boundary surface

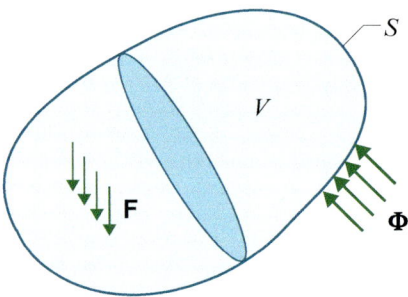

Figure 3.5 Elastic body in equilibrium

Divergence theorem of GAUSS:

$$\iiint_V \mathrm{div}\,\mathbf{u}\,dV = \iint_S \mathbf{u}^\mathsf{T}\mathbf{n}\,dS$$

where **n** is the outer unit normal vector of the surface S

of the tetrahedron, the resultants of which are $-\boldsymbol{\Phi}_x$, $-\boldsymbol{\Phi}_y$ and $-\boldsymbol{\Phi}_z$. Neglecting volume forces (**F**), equilibrium of the tetrahedron requires (keep in mind that traction is force per unit area)

$$\boldsymbol{\Phi} dS - \boldsymbol{\Phi}_x dS_x - \boldsymbol{\Phi}_y dS_y - \boldsymbol{\Phi}_z dS_z = \mathbf{0} \qquad (3\text{-}7)$$

The three side areas can be expressed in terms of S and the direction cosines, see Fig. 3.4b, as

$$dS_x = l \cdot dS, \qquad dS_y = m \cdot dS \quad \text{and} \quad dS_z = n \cdot dS$$

Hence

$$\boldsymbol{\Phi} = \begin{bmatrix} \boldsymbol{\Phi}_x & \boldsymbol{\Phi}_y & \boldsymbol{\Phi}_z \end{bmatrix} \begin{bmatrix} l \\ m \\ n \end{bmatrix} = \boldsymbol{\sigma}_M^T \mathbf{n} \qquad (3\text{-}8a)$$

This is referred to as CAUCHY's equation which, due to the symmetry property of the stress matrix, can be written as

$$\boldsymbol{\Phi} = \boldsymbol{\sigma}_M \mathbf{n} \qquad (3\text{-}8b)$$

or, in component form as,

$$\boldsymbol{\Phi}_x = \boldsymbol{\Phi}_x^T \mathbf{n}, \qquad \boldsymbol{\Phi}_y = \boldsymbol{\Phi}_y^T \mathbf{n} \quad \text{and} \quad \boldsymbol{\Phi}_z = \boldsymbol{\Phi}_z^T \mathbf{n} \qquad (3\text{-}8c)$$

If we consider an elastic body, CAUCHY's equation is valid for the outer surface as well as the surface of an internal section, and it states that we can find the traction vector $\boldsymbol{\Phi}$ in an arbitrary direction $\mathbf{n}$ providing the state of stress is known. On the body surface CAUCHY's equation represents a boundary condition since it relates surface loading to internal stresses.

We now turn to the elastic body of Fig. 3.5 (the well-known "potato") with a volume of V and a total surface of S. The body is subjected to *volume* forces **F** and surface forces $\boldsymbol{\Phi}$. Equilibrium of the body requires that

$$\int_S \boldsymbol{\Phi} dS + \int_V \mathbf{F} dV = \mathbf{0} \qquad (3\text{-}9)$$

or, in component form

$$\int_S \boldsymbol{\Phi}_x dS + \int_V F_x dV = 0, \quad \int_S \boldsymbol{\Phi}_y dS + \int_V F_y dV = 0 \quad \text{and} \quad \int_S \boldsymbol{\Phi}_z dS + \int_V F_z dV = 0 \quad (3\text{-}10)$$

Using the component form of CAUCHY's equation, Eq. (3.8c), the first of the equations in (3-10) can be expressed as

$$\int_S \boldsymbol{\Phi}_x^T \mathbf{n} dS + \int_V F_x dV = 0$$

Invoking the *divergence* theorem on the surface integral gives:

Summary of linear elasticity theory

By definition: $\quad \text{div } \mathbf{v} = \dfrac{\partial v_x}{\partial x} + \dfrac{\partial v_y}{\partial y} + \dfrac{\partial v_z}{\partial z}$

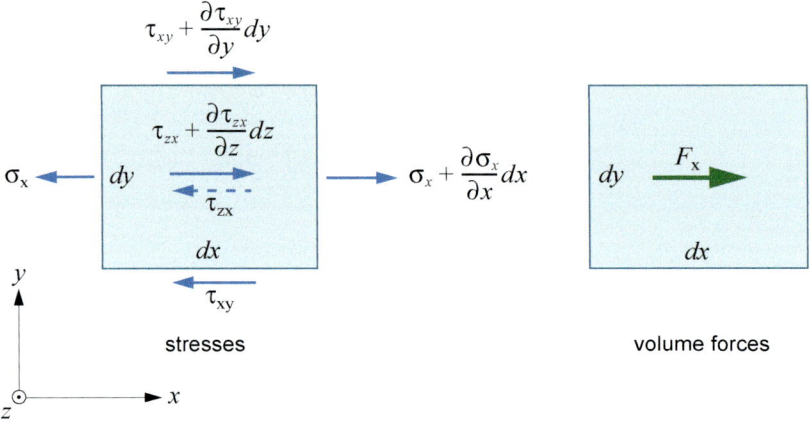

Figure 3.6 Stresses and forces in x-direction

 **Augustin-Louis Cauchy** (1789-1857) was a French mathematician who exercised a great influence over other mathematicians, both contemporaries and successors. He was a prolific writer, and his writings cover the entire range of mathematics and mathematical physics; he wrote five complete textbooks and some eight hundred research articles.

An early pioneer of analysis, Cauchy started the project of formulating and proving the theorems of infinitesimal calculus in a rigorous manner, discarding the heuristic principles exploited by earlier authors. He defined continuity in terms of infinitesimals and gave several important theorems in complex analysis.

Cauchy was admitted to the École Polytechnique where he finished at the age of 18 and went on to the École des Ponts et Chaussées where he graduated in civil engineering with the highest honours. He worked as an engineer for some years, but he was more and more attracted to the abstract beauty of mathematics. Helped by his reactionary religious and political beliefs he was made professor at École Polytechnique. However, the same beliefs and stubbornness made him go into exile – he was out of France for about 8 years.

Cauchy also contributed significantly to research in mechanics, substituting the notion of the continuity of geometrical displacements for the principle of the continuity of matter. He wrote on the equilibrium of rods and elastic membranes and on waves in elastic media. He introduced a 3×3 symmetric matrix of numbers that is known as the Cauchy stress tensor. In elasticity, he originated the theory of stress, and his results are nearly as valuable as those of Poisson.

Three-dimensional stress analysis

$$\int_V (\text{div}\boldsymbol{\Phi}_x + F_x)dV = 0 \qquad (3\text{-}11)$$

Since this equation must hold for any volume V we have:

$$\text{div}\boldsymbol{\Phi}_x + F_x = 0 \quad \text{or} \quad \frac{\partial \sigma_x}{\partial x} + \frac{\partial \tau_{xy}}{\partial y} + \frac{\partial \tau_{zx}}{\partial z} + F_x = 0$$

Applying the same procedure to the remaining two equations of Eqs. (3-10) we can now write down the *equilibrium equations* of our elastic body:

$$\frac{\partial \sigma_x}{\partial x} + \frac{\partial \tau_{xy}}{\partial y} + \frac{\partial \tau_{zx}}{\partial z} + F_x = 0 \qquad (3\text{-}12a)$$

$$\frac{\partial \tau_{xy}}{\partial x} + \frac{\partial \sigma_y}{\partial y} + \frac{\partial \tau_{yz}}{\partial z} + F_y = 0 \qquad (3\text{-}12b)$$

$$\frac{\partial \tau_{zx}}{\partial x} + \frac{\partial \tau_{yz}}{\partial y} + \frac{\partial \sigma_z}{\partial z} + F_z = 0 \qquad (3\text{-}12c)$$

Equilibrium of the body requires that these three equations are satisfied at *every* point in the body.

We could have arrived at these equations more directly. Figure 3.6 shows the stresses and forces acting in the x-direction on an infinitesimal element ($dxdydz$). Force equilibrium in the x-direction gives us Eq. (3-12a) directly.

The equilibrium equations, Eqs. (3-12), can also be expressed in *operator* form as

$$\boldsymbol{\Delta}^T \boldsymbol{\sigma} + \mathbf{F} = \mathbf{0} \qquad (3\text{-}13)$$

where the differential *operator matrix* $\boldsymbol{\Delta}$ is defined as

$$\boldsymbol{\Delta} = \begin{bmatrix} \frac{\partial}{\partial x} & 0 & 0 \\ 0 & \frac{\partial}{\partial y} & 0 \\ 0 & 0 & \frac{\partial}{\partial z} \\ \frac{\partial}{\partial y} & \frac{\partial}{\partial x} & 0 \\ 0 & \frac{\partial}{\partial z} & \frac{\partial}{\partial y} \\ \frac{\partial}{\partial z} & 0 & \frac{\partial}{\partial x} \end{bmatrix} \qquad (3\text{-}14)$$

It should be emphasized that $\boldsymbol{\sigma}$ is *not* a vector in the *physical* sense. The state of stress at a point depends on *both* the traction vector ($\boldsymbol{\Phi}$) and the orientation of the section considered; the latter is defined by **n**.

59

Summary of linear elasticity theory

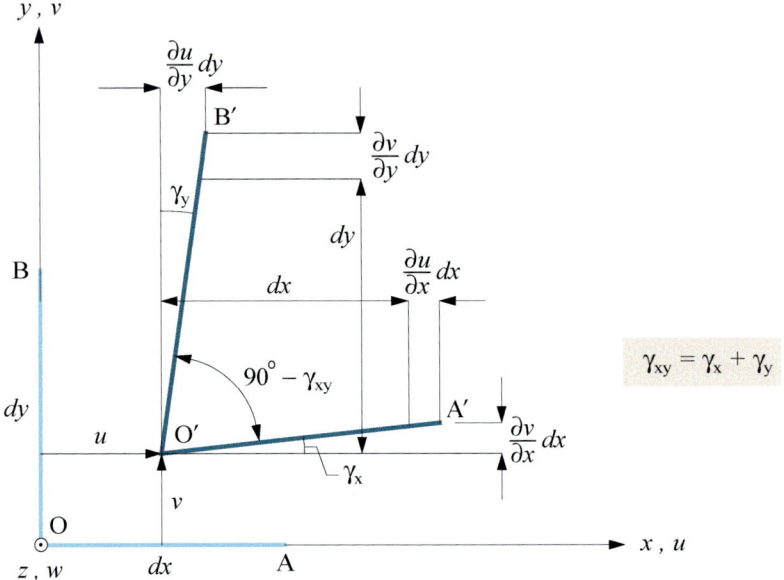

Figure 3.7 Displacements and deformations

Three-dimensional stress analysis

Strains and kinematic compatibility. An elastic body subjected to any kind of loading will deform. We consider two infinitesimal straight lines O-A and O-B within the body forming 90 degrees with each other before loading is applied. Line O-A is parallel with the *x*-axis and line O-B is parallel with the *y*-axis, as shown in Fig. 3.7. After loading we find the lines at positions O', A' and B'. The change in position consists of both a *rigid body* movement, defined by displacements *u* and *v* in the *x*- and *y*-direction, respectively, and a *deformation* (change of length).

Here we are mainly interested in the deformation, and keeping in mind that displacements are *small* we find that the change of length of O-A is *du* which we can write as

$$du = \frac{\partial u}{\partial x} dx \quad \text{(since both } dy \text{ and } dz \text{ are zero for line O-A)}$$

Hence, the relative change of length of O-A, defined as the *normal strain* in the *x*-direction at point O, is

$$\varepsilon_x = \frac{|O\text{-}A|}{|O'\text{-}A'|} = \frac{\partial u}{\partial x} \tag{3-15a}$$

Similarly for the *y*- and *z*-direction:

$$\varepsilon_y = \frac{\partial v}{\partial y} \tag{3-15b}$$

$$\varepsilon_z = \frac{\partial w}{\partial z} \tag{3-15c}$$

Figure 3.7 shows that the deformation of the body has also changed the right angle between the lines O-A and O-B; this change is the angle γ_{xy} which defines the *shear strain* in the *x-y* plane. For small displacements we have

$$\tan \gamma_x = \frac{\frac{\partial v}{\partial x} dx}{dx} = \frac{\partial v}{\partial x} \approx \gamma_x$$

Similarly

$$\gamma_y = \frac{\partial u}{\partial y}$$

and

$$\gamma_{xy} = \gamma_y + \gamma_x = \frac{\partial u}{\partial y} + \frac{\partial v}{\partial x} \tag{3-16a}$$

By changing axes *x* and *y* in Fig. 3.7 to *y* and *z* and then to *z* and *x* we find:

$$\gamma_{yz} = \frac{\partial v}{\partial z} + \frac{\partial w}{\partial y} \tag{3-16b}$$

and

$$\gamma_{zx} = \frac{\partial u}{\partial z} + \frac{\partial w}{\partial x} \tag{3-16c}$$

Summary of linear elasticity theory

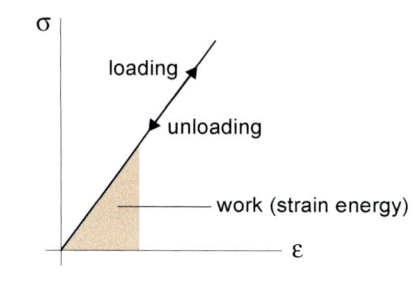

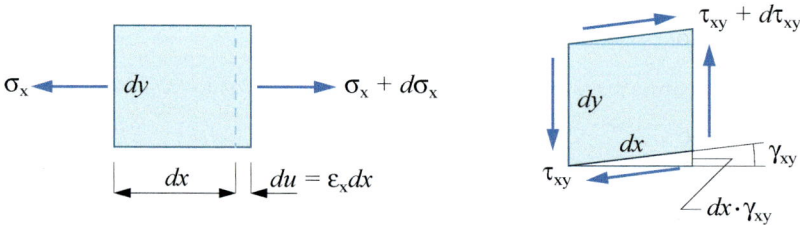

Figure 3.8 Corresponding stress and strain

The relationship between strain (which is a measure of deformation) and displacements can now be expressed in matrix notation as

$$\begin{bmatrix} \varepsilon_x \\ \varepsilon_y \\ \varepsilon_z \\ \gamma_{xy} \\ \gamma_{yz} \\ \gamma_{zx} \end{bmatrix} = \begin{bmatrix} \frac{\partial}{\partial x} & 0 & 0 \\ 0 & \frac{\partial}{\partial y} & 0 \\ 0 & 0 & \frac{\partial}{\partial z} \\ \frac{\partial}{\partial y} & \frac{\partial}{\partial x} & 0 \\ 0 & \frac{\partial}{\partial z} & \frac{\partial}{\partial y} \\ \frac{\partial}{\partial z} & 0 & \frac{\partial}{\partial x} \end{bmatrix} \begin{bmatrix} u \\ v \\ w \end{bmatrix} \quad \text{or} \quad \boldsymbol{\varepsilon} = \boldsymbol{\Delta} \mathbf{u} \qquad (3\text{-}17)$$

where $\boldsymbol{\varepsilon}$ is the *strain vector* and $\mathbf{u} = [\ u\ \ v\ \ w\ ]^T$ is the *displacement vector*.

Equation (3-17) expresses the *kinematic compatibility* requirement of small displacement 3D theory of elasticity. The strain vector corresponds to the stress vector in the sense that

$\boldsymbol{\sigma}^T \boldsymbol{\varepsilon}$ = the work performed by the stresses $\boldsymbol{\sigma}$ over the deformations $\boldsymbol{\varepsilon}$.

This can be verified by considering the two simplified stress states in Fig. 3.8. For the case of uniaxial stress (σ_x) the work performed on the volume element *dxdydz* by the stress over the displacements it causes is, when we neglect higher order terms,

$$\frac{1}{2}(\sigma_x dydz)du = \frac{1}{2}(\sigma_x dydz)\varepsilon_x dx = \frac{1}{2}\sigma_x \varepsilon_x dV$$

For a state of pure shear deformation, due to shear stress τ_{xy}, the work performed is (again neglecting higher order terms):

$$\frac{1}{2}(\tau_{xy}dydz)dx\gamma_{xy} = \frac{1}{2}\tau_{xy}\gamma_{xy}dV,$$

This last expression is easily verified if all displacement is assumed to take place on the right-hand face of the element.

Summing up we can now express the total *elastic work* (W) which is stored as *strain energy* (U) as

$$dW_{elastic} = dU = \frac{1}{2}\boldsymbol{\sigma}^T\boldsymbol{\varepsilon}dV \qquad (3\text{-}18)$$

With reference to Eqs. (3-13) and (3-17) we see that the operator matrix $\boldsymbol{\Delta}$ plays a role in both equilibrium and compatibility, in much the same way as the **a**-matrix at the element level, which is one (integrated) step up from the infinitesimal volume element.

Summary of linear elasticity theory

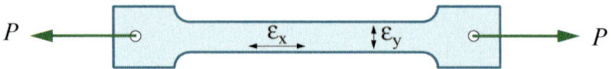

Figure 3.9 Uniaxial stress test

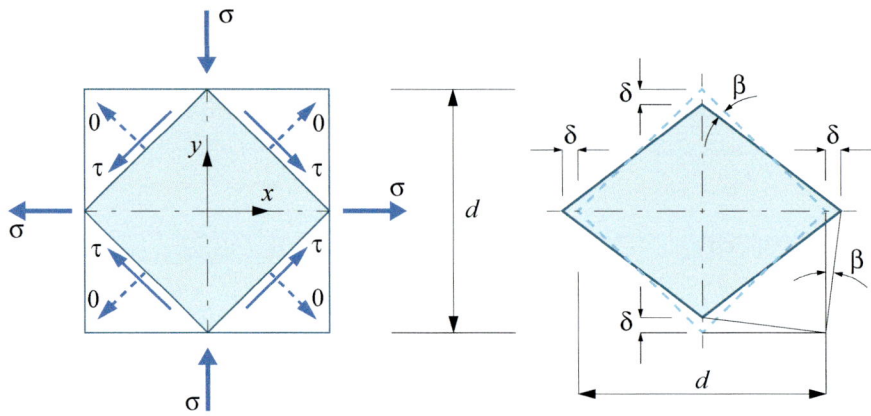

Figure 3.10 Relationship between axial and shear stress and deformation

Some typical values:

Material	E [MPa]	ν
Steel	210 000	0,3
Concrete	30 000	0,2
Aluminum	70 000	0,3
Timber [1]	11 000	—

[1] Structural timber (including glulaminated timber) is not an isotropic material; it is basically an *orthotropic* material. The value for E in the table is the mean value in fibre direction for a medium timber quality (softwood).
Since the ratio E/G is approximately 16 (for most timber qualities), Eq. (3-25) obviously does not apply to timber. Hence, for 1D timber components, the relevant elastic constants are E and G.

Three-dimensional stress analysis

Stress-strain relationship. If we carry out a uniaxial stress test as shown in Fig. 3.9 we will find that the strain ε_x can be expressed in terms of the stress σ_x as

$$\varepsilon_x = \frac{\sigma_x}{E_x} \qquad (3\text{-}19)$$

This is commonly known as HOOKE's law. For real materials E_x is not a constant, but in linear theory we assume it to be constant, an assumption that is quite good for some materials (such as steel), but not all that good for others (like reinforced concrete). The "constant" E_x is the *modulus of elasticity,* also known as YOUNG's modulus.

The test in Fig. 3.9 also reveals a *lateral contraction* resulting in a *lateral strain* ε_y, called the POISSON effect. The ratio of the lateral strain to the axial strain is known as POISSON's *ratio* and denoted by ν; thus

$$\nu_y = -\frac{\text{lateral strain}}{\text{axial strain}} = -\frac{\varepsilon_y}{\varepsilon_x} \qquad (3\text{-}20)$$

from which

$$\varepsilon_y = -\nu_y \varepsilon_x = -\nu_y \frac{\sigma_x}{E_x} \qquad (3\text{-}21)$$

Similarly, if we cut our test specimen in the *y*-direction and repeated our uniaxial stress test we would find

$$\varepsilon_y = \frac{\sigma_y}{E_y} \quad \text{and} \quad \varepsilon_x = -\nu_x \frac{\sigma_y}{E_y} \qquad (3\text{-}22a, b)$$

For an *isotropic* material we have:

$$\nu_x = \nu_y = \nu \quad \text{and} \quad E_x = E_y = E \qquad (3\text{-}23)$$

For the problem in Fig. 3.10 we readily find the following kinematic relations:

$$\gamma_{xy} = 2\beta \quad \text{and} \quad \beta \approx \tan\beta = \frac{\delta}{d/2} = \frac{2\delta}{d} \quad \Rightarrow \quad \varepsilon_x = \frac{2\delta}{d} = \beta = \frac{\gamma_{xy}}{2}$$

Equilibrium requires: $\quad \tau = \tau_{xy} = \sigma$

HOOKE's law, assuming isotropic material, states

$$\varepsilon_x = \frac{1}{E}[\sigma - \nu(-\sigma)] = \frac{\sigma}{E}(1+\nu)$$

Combining these equations we can now write

$$\frac{\gamma_{xy}}{2} = \varepsilon_x = \frac{\sigma}{E}(1+\nu) = \frac{\tau_{xy}}{E}(1+\nu)$$

Hence

$$\tau_{xy} = \frac{E}{2(1+\nu)}\gamma_{xy} \quad \text{or} \quad \tau_{xy} = G\gamma_{xy} \qquad (3\text{-}24)$$

65

Summary of linear elasticity theory

Thomas YOUNG (1773-1829) was an English polymath. He is perhaps best known for having partly deciphered Egyptian hieroglyphs (specifically the Rosetta Stone), but he also made notable scientific contributions to the fields of vision, light, solid mechanics, energy, physiology, language and musical harmony.

YOUNG was born the eldest of ten children to a Quaker family. At the age of fourteen he had learned Greek and Latin and was acquainted with French, Italian, Hebrew, German, Chaldean, Syriac, Samaritan, Arabic, Persian, Turkish and Amharic.

YOUNG began to study medicine in London, moved on to Edinburgh and obtained the degree of doctor of physics in Göttingen. In 1799 he established himself as physician; he published many of his first academic articles anonymously to protect his reputation as a physician. He was appointed professor of natural philosophy at the Royal Institution (1801), but resigned two years later, fearing that its duties would interfere with his medical practice.

YOUNG described the characterization of elasticity that came to be known as YOUNG's modulus in 1807, and further described it in his *Course of Lectures on Natural Philosophy and the Mathematical Arts*. However, the first use of the YOUNG's modulus in experiments was by RICCATI in 1782, predating YOUNG by 25 years, and the idea can be traced back to a paper by EULER published as early as in 1727.

Robert HOOKE (1635-1703) was an English natural philosopher, architect and polymath. He was at one time simultaneously the curator of experiments of the Royal Society, Gresham Professor of Geometry and a Surveyor to the City of London after the Great Fire of London (1666). He was also an important architect of his time, and is by a historian characterized as "England's Leonardo".

HOOKE's pioneering achievements in many areas of natural science and technology (he invented the microscope) were founded methodically on the idea of formulating a hypothesis which was then proved right or wrong through experiments. He came close to explain that gravity follows an inverse square law, an idea which was subsequently developed by NEWTON, with whom he had several disputes. Much has been written about the unpleasant side of HOOKE's personality, and one biographer described him as "despicable, mistrustful and jealous". Another states that "he was a difficult man in an age of difficult men".

In 1678 HOOKE published the paper *The Potentiâ Restitutiva* ("Of Spring"). It contains the results of his experiments with elastic bodies, and it is the first published paper in which the elasitc properties of materials are discussed; it is the basis of HOOKE's law.

Siméon Denis POISSON (1781-1840) was a French mathematician, geometer and physicist. Born into a poor family he had little formal schooling before he was 15 years old, at which age he was sent to his uncle and was able to visit mathematics classes. In 1798 he passed the entrance examinations of the École Polytechnique in Paris, and soon began to attract the notice of the professors of the school, amongst them LAGRANGE and LAPLACE. Less than two years after his entry, he published two memoirs. After finishing his studies at the École Polytechnique, he was appointed teaching assistant there and a couple of years later professor.

As a teacher of mathematics POISSON is said to have been extraordinarily successful, and as a scientific worker, his productivity has rarely if ever been equalled. In spite of his many official duties he found time to publish more than three hundred works, several of them extensive treatises. He is quoted to have said that "Life is good for only two things: doing mathematics and teaching it".

POISSON is known for, and has given name to many aspects of mathematics and physics, such as POISSON's *process, equation, kernel, distribution, bracket, regression, spot* and *ratio*.

A slightly negative experience was that as the final leading opponent of the *wave theory of light*, as a member of the elite *l'Académie française*, he was proven wrong.

where G is the *shear modulus*, which according to Eq. (3-24) can be expressed in terms of E and v as

$$G = \frac{E}{2(1+v)} \qquad (3\text{-}25)$$

for a linear elastic, *isotropic* material.

Extending the arguments above to three dimensions for an isotropic material is straightforward. For the normal components we have:

$$\varepsilon_x = \frac{1}{E}(\sigma_x - v\sigma_y - v\sigma_z) = \frac{1+v}{E}\sigma_x - \frac{3v}{E}\bar{\sigma}$$

$$\varepsilon_y = \frac{1}{E}(-v\sigma_x + \sigma_y - v\sigma_z) = \frac{1+v}{E}\sigma_y - \frac{3v}{E}\bar{\sigma} \qquad (3\text{-}26)$$

$$\varepsilon_z = \frac{1}{E}(-v\sigma_x - v\sigma_y + \sigma_z) = \frac{1+v}{E}\sigma_z - \frac{3v}{E}\bar{\sigma}$$

where

$$\bar{\sigma} = (\sigma_x + \sigma_y + \sigma_z)/3 \qquad (3\text{-}27)$$

is the *hydrostatic* or *mean* stress. Similarly for the shear components:

$$\gamma_{xy} = \frac{2(1+v)}{E}\tau_{xy} = \frac{1}{G}\tau_{xy}$$

$$\gamma_{yz} = \frac{2(1+v)}{E}\tau_{yz} = \frac{1}{G}\tau_{yz} \qquad (3\text{-}28)$$

$$\gamma_{zx} = \frac{2(1+v)}{E}\tau_{zx} = \frac{1}{G}\tau_{zx}$$

In matrix notation:

$$\begin{bmatrix} \varepsilon_x \\ \varepsilon_y \\ \varepsilon_z \\ \gamma_{xy} \\ \gamma_{yz} \\ \gamma_{zx} \end{bmatrix} = \frac{1}{E} \begin{bmatrix} 1 & -v & -v & 0 & 0 & 0 \\ -v & 1 & -v & 0 & 0 & 0 \\ -v & -v & 1 & 0 & 0 & 0 \\ 0 & 0 & 0 & 2(1+v) & 0 & 0 \\ 0 & 0 & 0 & 0 & 2(1+v) & 0 \\ 0 & 0 & 0 & 0 & 0 & 2(1+v) \end{bmatrix} \begin{bmatrix} \sigma_x \\ \sigma_y \\ \sigma_z \\ \tau_{xy} \\ \tau_{yz} \\ \tau_{zx} \end{bmatrix} \quad \text{or} \quad \boldsymbol{\varepsilon} = \mathbf{C}^{-1}\boldsymbol{\sigma} \quad (3\text{-}29)$$

$\mathbf{C}^{-1}$ is the *compliance* or *flexibility* matrix. The inverse relation is

$$\begin{bmatrix} \sigma_x \\ \sigma_y \\ \sigma_z \\ \tau_{xy} \\ \tau_{yz} \\ \tau_{zx} \end{bmatrix} = \begin{bmatrix} \lambda+2G & \lambda & \lambda & 0 & 0 & 0 \\ \lambda & \lambda+2G & \lambda & 0 & 0 & 0 \\ \lambda & \lambda & \lambda+2G & 0 & 0 & 0 \\ 0 & 0 & 0 & G & 0 & 0 \\ 0 & 0 & 0 & 0 & G & 0 \\ 0 & 0 & 0 & 0 & 0 & G \end{bmatrix} \begin{bmatrix} \varepsilon_x \\ \varepsilon_y \\ \varepsilon_z \\ \gamma_{xy} \\ \gamma_{yz} \\ \gamma_{zx} \end{bmatrix} \quad \text{or} \quad \boldsymbol{\sigma} = \mathbf{C}\boldsymbol{\varepsilon} \quad (3\text{-}30)$$

where $\mathbf{C}$ is the *elasticity matrix*, also called the constitutive matrix, and

Summary of linear elasticity theory

Table 3.1 Relationships between the elastic constants for isotropic materials

	Young's mod. E	Poisson's ratio ν	Shear mod. G	Lamé's const. λ	Bulk mod. B
E, ν			$\dfrac{E}{2(1+\nu)}$	$\dfrac{\nu E}{(1+\nu)(1-2\nu)}$	$\dfrac{E}{3(1-2\nu)}$
E, G		$\dfrac{E-2G}{2G}$		$\dfrac{(2G-E)G}{E-3G}$	$\dfrac{GE}{3(3G-E)}$
E, λ		$\dfrac{\sqrt{(E+\lambda)^2+8\lambda^2}}{4\lambda} - \dfrac{E+\lambda}{4\lambda}$	$\dfrac{\sqrt{(E-3\lambda)^2+8\lambda E}}{4} + \dfrac{E-3\lambda}{4}$		$\dfrac{\sqrt{(3\lambda+E)^2-4\lambda E}}{6} + \dfrac{3\lambda+E}{6}$
E, B		$\dfrac{3B-E}{6B}$	$\dfrac{3EB}{9B-E}$	$\dfrac{3B(3B-E)}{9B-E}$	
ν, G	$2G(1+\nu)$			$\dfrac{2G\nu}{1-2\nu}$	$\dfrac{2G(1+\nu)}{3(1-2\nu)}$
ν, λ	$\dfrac{\lambda(1+\nu)(1-2\nu)}{\nu}$		$\dfrac{\lambda(1-2\nu)}{2\nu}$		$\dfrac{\lambda(1+\nu)}{3\nu}$
ν, B	$3B(1-2\nu)$		$\dfrac{3B(1-2\nu)}{2(1+\nu)}$	$\dfrac{3B\nu}{1+\nu}$	
G, λ	$\dfrac{G(3\lambda+2G)}{\lambda+G}$	$\dfrac{\lambda}{2(\lambda+G)}$			$\dfrac{3\lambda+2G}{3}$
G, B	$\dfrac{9BG}{3B+G}$	$\dfrac{3B-2G}{6B+2G}$		$\dfrac{3B-2G}{3}$	
λ, B	$\dfrac{9B(B-\lambda)}{3B-\lambda}$	$\dfrac{\lambda}{3B-\lambda}$	$\dfrac{3(B-\lambda)}{2}$		

$$\lambda = \frac{\nu E}{(1+\nu)(1-2\nu)} \qquad (3\text{-}31)$$

is LAMÉ's *constant*. It should be noted that **C** is basically a stiffness matrix. Adding the three equations in Eq. (3-26) gives

$$\bar{\varepsilon} = \varepsilon_x + \varepsilon_y + \varepsilon_z = \frac{3(1+\nu)}{E}\bar{\sigma} - \frac{9\nu}{E}\bar{\sigma} = \frac{3(1-2\nu)}{E}\bar{\sigma} = \frac{\bar{\sigma}}{B} \qquad (3\text{-}32)$$

where

$$B = \frac{E}{3(1-2\nu)} \qquad (3\text{-}33)$$

is the *bulk modulus*. The physical interpretation of $\bar{\varepsilon}$ is a change of volume:

Before deformation: $\quad \Delta V_0 = dxdydz$

After deformation: $\quad \Delta V_1 = dx(1+\varepsilon_x)dy(1+\varepsilon_y)dz(1+\varepsilon_z) \approx dxdydz(1+\bar{\varepsilon})$
$\qquad\qquad\qquad\qquad = \Delta V_0(1+\bar{\varepsilon}) = \Delta V_0 + \Delta(\Delta V_0)$

Hence

$$\bar{\varepsilon} = \frac{\Delta(\Delta V_0)}{\Delta V_0} \qquad (3\text{-}34)$$

is the relative volume change, the *volumetric strain*, also called *dilatation*. It follows from Eq. (3-32) that for a given state of normal stress, $\bar{\varepsilon}$ will diminish with increasing B. We also see that B increases with increasing ν, and $\nu = 0{,}5$ represents a limit value for which B becomes infinitely large and $\bar{\varepsilon} = 0$.

Materials with POISSON's ratio close to 0,5, such as rubber, are said to be *incompressible* in elastic deformation.

We have now introduced 5 elastic material constants: E, ν, G, λ and B. An isotropic material is, however, uniquely defined by only two of these constants which therefore must be interrelated. We have already seen how G, λ and B can be expressed by E and ν. In Table 3.1 on the opposite page all relationships are given.

Initial strain. Some actions, such as temperature change, shrinkage and swelling, that will produce stresses in (statically indeterminate) structures are conveniently treated as *initial strain*, denoted by $\boldsymbol{\varepsilon}_0$. The total strain in an elastic body is now expressed as

$$\boldsymbol{\varepsilon} = \mathbf{C}^{-1}\boldsymbol{\sigma} + \boldsymbol{\varepsilon}_0 \qquad (3\text{-}35)$$

Multiplying this equation by **C** gives

$$\boldsymbol{\sigma} = \mathbf{C}(\boldsymbol{\varepsilon} - \boldsymbol{\varepsilon}_0) \qquad (3\text{-}36)$$

A physical interpretation of the initial strain is as follows: we disassemble the structural components and subject them to the initial strain (whatever the cause, *e.g.* temperature increase/decrease) *without* any constraints. In other words, the components can deform freely, and this deformation is the initial strain $\boldsymbol{\varepsilon}_0$. The components are then "forced" back in place (by some fictitious forces which we shall come back to), and the net result is stress

Summary of linear elasticity theory

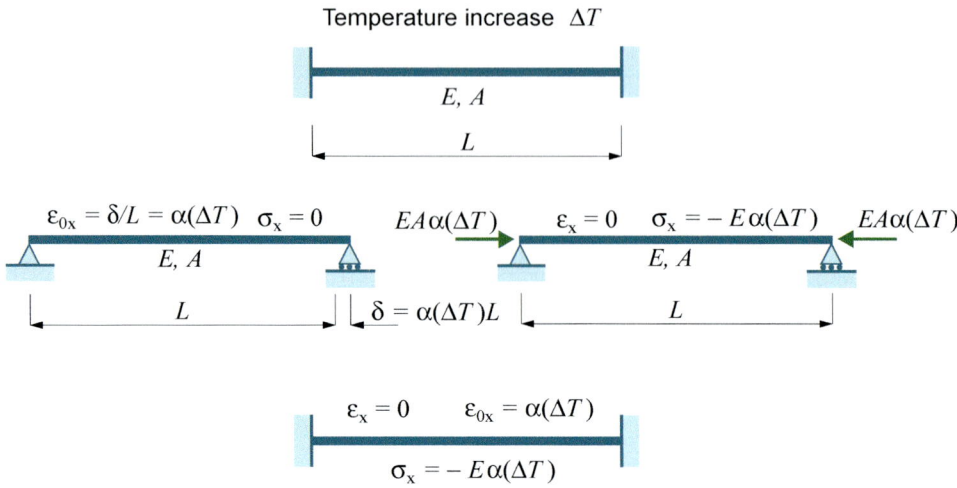

Figure 3.11 Stress and strain of a fixed rod subjected to temperature increase

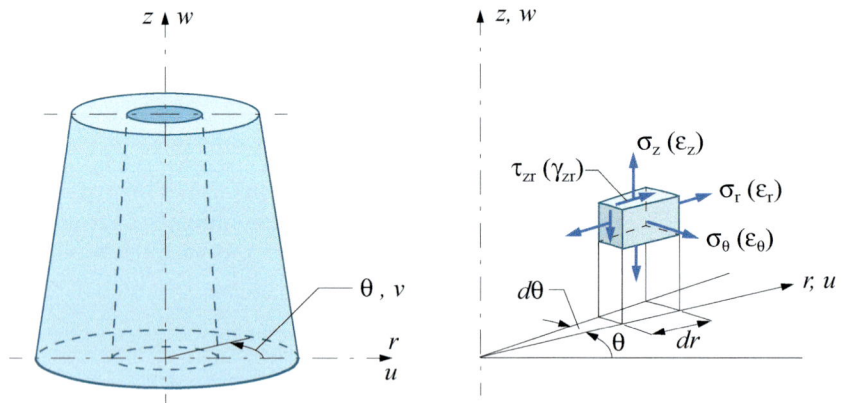

Figure 3.12 Axisymmetric stress and strain

computed by Eq. (3-36), where **ε** is the final (total) strain at points in the assembled structure. This abstract model is illustrated in Fig. 3.11 for a simple fixed rod subjected to a temperature change of ΔT.

For an isotropic material the thermal strain is:

$$\boldsymbol{\varepsilon}_0 = \alpha \cdot \Delta T [\,1 \quad 1 \quad 1 \quad 0 \quad 0 \quad 0\,]^T \tag{3-37}$$

where α is the material's coefficient of thermal expansion, and ΔT is the change in temperature relative to some reference value. The coefficient of thermal expansion is a material constant for moderate temperature changes; it is sometimes designated by λ, but in order to avoid confusion with LAMÉ's constant we have here used α.

3.2 Axisymmetric stress and strain

With respect to Fig. 3.12 we make the following assumptions:

Geometry, material properties, loading and boundary conditions are independent of the angle θ.

This implies

$$v = 0 \quad \text{and} \quad \tau_{r\theta} = \tau_{\theta z} = \gamma_{r\theta} = \gamma_{\theta z} = 0 \tag{3-38}$$

For an arbitrary value of θ the problem is thus reduced to a 2D problem. The kinematics of the problem are:

$$\varepsilon_r = \frac{\partial u}{\partial r}, \quad \varepsilon_z = \frac{\partial w}{\partial z}, \quad \varepsilon_\theta = \frac{2\pi(r+u) - 2\pi r}{2\pi r} = \frac{u}{r} \quad \text{and} \quad \gamma_{zr} = \frac{\partial u}{\partial z} + \frac{\partial w}{\partial r}$$

In matrix notation this becomes

$$\boldsymbol{\varepsilon} = \begin{bmatrix} \varepsilon_r \\ \varepsilon_\theta \\ \varepsilon_z \\ \gamma_{zr} \end{bmatrix} = \begin{bmatrix} \frac{\partial}{\partial r} & 0 \\ \frac{1}{r} & 0 \\ 0 & \frac{\partial}{\partial z} \\ \frac{\partial}{\partial z} & \frac{\partial}{\partial r} \end{bmatrix} \begin{bmatrix} u \\ w \end{bmatrix} = \boldsymbol{\Delta}\mathbf{u} \tag{3-39}$$

By use of Eqs. (3-30), (3-31) and (3-25) we can, for an isotropic material, write down the following stress-strain relation for an axisymmetric solid problem:

$$\boldsymbol{\sigma} = \begin{bmatrix} \sigma_r \\ \sigma_\theta \\ \sigma_z \\ \tau_{zr} \end{bmatrix} = \frac{E}{(1+\nu)(1-2\nu)} \begin{bmatrix} 1-\nu & \nu & \nu & 0 \\ \nu & 1-\nu & \nu & 0 \\ \nu & \nu & 1-\nu & 0 \\ 0 & 0 & 0 & \frac{1-2\nu}{2} \end{bmatrix} \begin{bmatrix} \varepsilon_r \\ \varepsilon_\theta \\ \varepsilon_z \\ \gamma_{zr} \end{bmatrix} \tag{3-40}$$

Summary of linear elasticity theory

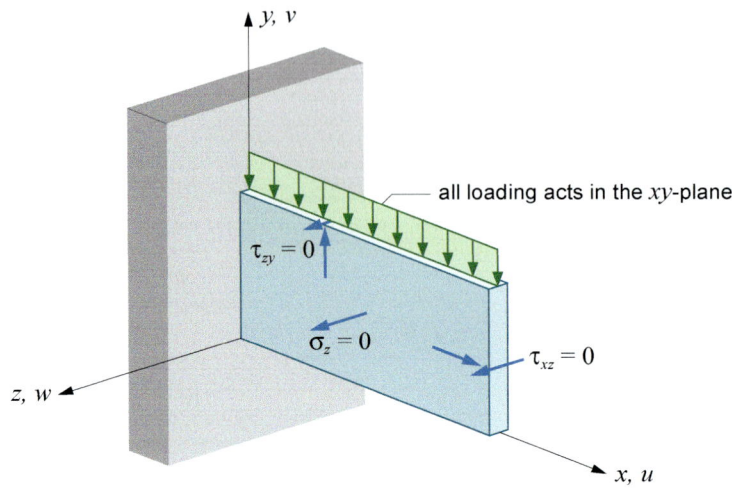

Figure 3.13 Plane stress

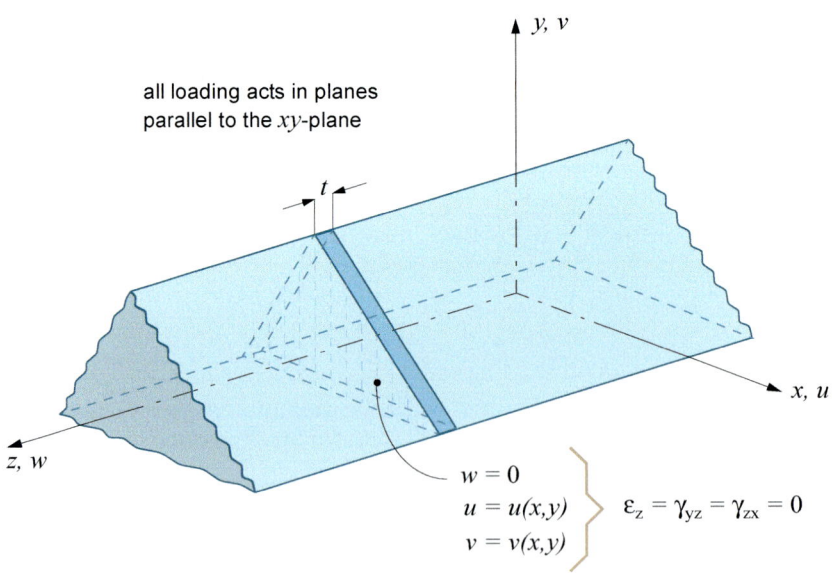

Figure 3.14 Plane strain

3.3 Stress and strain in two dimensions

We now turn to problems for which geometry, material properties, loading and boundary conditions depend on only *two* axes, x and y; in other words, these properties are independent of the third, z-axis. For such a problem the strain-displacement relation is found from Eq. (3-17) by simply deleting three rows and a column:

$$\boldsymbol{\varepsilon} = \begin{bmatrix} \varepsilon_x \\ \varepsilon_y \\ \gamma_{xy} \end{bmatrix} = \begin{bmatrix} \dfrac{\partial}{\partial x} & 0 \\ 0 & \dfrac{\partial}{\partial y} \\ \dfrac{\partial}{\partial y} & \dfrac{\partial}{\partial x} \end{bmatrix} \begin{bmatrix} u \\ v \end{bmatrix} = \boldsymbol{\Delta} \mathbf{u} \qquad (3\text{-}41)$$

However, for the stress-strain relation we now have to distinguish between two cases: so-called *plane stress* and *plane strain*.

Plane stress is characterized by a state of stress where the components σ_x, τ_{yz} and τ_{zx} vanish; in other words:

for plane stress: $\sigma_z = \tau_{yz} = \tau_{zx} = 0$

The cantilever plate in Fig. 3.13 exemplifies this type of problem.

Plane strain on the other hand is characterized by a state of strain for which the components ε_z, γ_{yz} and γ_{zx} vanish; in other words:

for plane strain: $\varepsilon_z = \gamma_{yz} = \gamma_{zx} = 0$

An example of this type of problem is indicated in Fig. 3.14, where a thin slice of a long "solid", for which the basic assumptions made at the start of this section apply, represents the actual problem.

Both models are approximations to real world problems.

Plane stress. For an isotropic material we find the relationship between strain and stress from Eq. (3-29) by deleting rows and columns corresponding to the vanishing stress components:

$$\boldsymbol{\varepsilon} = \begin{bmatrix} \varepsilon_x \\ \varepsilon_y \\ \gamma_{xy} \end{bmatrix} = \frac{1}{E} \begin{bmatrix} 1 & -\nu & 0 \\ -\nu & 1 & 0 \\ 0 & 0 & 2(1+\nu) \end{bmatrix} \begin{bmatrix} \sigma_x \\ \sigma_y \\ \tau_{xy} \end{bmatrix} = \mathbf{C}^{-1} \boldsymbol{\sigma} \qquad (3\text{-}42)$$

The inverse relation is

$$\boldsymbol{\sigma} = \begin{bmatrix} \sigma_x \\ \sigma_y \\ \tau_{xy} \end{bmatrix} = \frac{E}{1-\nu^2} \begin{bmatrix} 1 & \nu & 0 \\ \nu & 1 & 0 \\ 0 & 0 & \dfrac{1-\nu}{2} \end{bmatrix} \begin{bmatrix} \varepsilon_x \\ \varepsilon_y \\ \gamma_{xy} \end{bmatrix} = \mathbf{C} \boldsymbol{\varepsilon} \qquad (3\text{-}43)$$

While σ_z is zero we see from Eq. (3-29) that $\varepsilon_z = -\nu(\sigma_x + \sigma_y)/E$ is normally *not* zero, whereas γ_{yz} and γ_{zx} are zero.

Summary of linear elasticity theory

Gabriel Léon Jean Baptiste LAMÉ (1795-1870) was a French mathematician, who became well known for his general theory of curvilinear coordinates and his notation and study of ellipse-like curves, now known as LAMÉ curves.

LAMÉ had his undergraduate training at École Polytechnique and he went on to study engineering at École des Mines in Paris where he graduated in 1820. The same year he went to Russia where he was soon appointed professor in St Petersburg. At first things were difficult for him there, but later his visit proved highly productive; he went back to France in 1832 and soon became professor of physics at the École Polytechnique.

LAMÉ worked on a variety of different topics. Problems he undertook in the engineering tasks often led him to study mathematical questions. For example his work on the stability of vaults and on the design of suspension bridges led him to work on elasticity theory. In fact this was not a passing interest, for LAMÉ made substantial contributions to this topic. In linear elasticity, the LAMÉ parameters, λ and G, are named after him.

LAMÉ was considered the leading French mathematician of his time by many, in particular GAUSS, who was never one to give praise easily, held this opinion. It is interesting to note that he was more highly thought of outside France than inside; the French seemed to feel that he was too practical for a mathematician and yet too theoretical for an engineer.

Example: $\nu = 0{,}25$

Plane stress :
$$\mathbf{C} = E \begin{bmatrix} 1{,}067 & 0{,}267 & 0 \\ 0{,}267 & 1{,}067 & 0 \\ 0 & 0 & 0{,}4 \end{bmatrix}$$

Plane strain :
$$\mathbf{C} = E \begin{bmatrix} 1{,}2 & 0{,}4 & 0 \\ 0{,}4 & 1{,}2 & 0 \\ 0 & 0 & 0{,}4 \end{bmatrix}$$

For an *orthotropic* material, that is a material with different properties in two orthogonal directions (x and y), we have according to Eqs. (3-21) and (3-22)

$$\begin{bmatrix} \varepsilon_x \\ \varepsilon_y \end{bmatrix} = \begin{bmatrix} \dfrac{1}{E_x} & \dfrac{-v_x}{E_y} \\ \dfrac{-v_y}{E_x} & \dfrac{1}{E_x} \end{bmatrix} \begin{bmatrix} \sigma_x \\ \sigma_y \end{bmatrix} \quad (3\text{-}44)$$

The inverse relation is

$$\begin{bmatrix} \sigma_x \\ \sigma_y \end{bmatrix} = \frac{1}{1 - v_x v_y} \begin{bmatrix} E_x & v_x E_x \\ v_y E_y & E_y \end{bmatrix} \begin{bmatrix} \varepsilon_x \\ \varepsilon_y \end{bmatrix} \quad (3\text{-}45)$$

We assume that Eq. (3-24) still applies; hence the constitutive relation for a state of orthotropic, plane stress can be expressed as

$$\begin{bmatrix} \sigma_x \\ \sigma_y \\ \tau_{xy} \end{bmatrix} = \begin{bmatrix} E_x' & E'' & 0 \\ E'' & E_y' & 0 \\ 0 & 0 & G \end{bmatrix} \begin{bmatrix} \varepsilon_x \\ \varepsilon_y \\ \gamma_{xy} \end{bmatrix} \quad (3\text{-}46)$$

where

$$E_x' = E_x/(1 - v_x v_y)$$
$$E_y' = E_y/(1 - v_x v_y) \quad (3\text{-}47)$$
$$E'' = \frac{v_x E_x}{1 - v_x v_y} = \frac{v_y E_y}{1 - v_x v_y}$$

Note that we need to have $v_x E_x = v_y E_y$ in order to maintain symmetry.

In two-dimensional analysis an orthotropic material is characterized by four elastic parameters (E_x', E_y', E'' and G) while two (E and v) are sufficient for an isotropic material. For a general, *anisotropic* material we may have six independent parameters (the elements on and above the main diagonal of matrix **C**).

Plane strain. For an *isotropic* material the constitutive relation in the case of plane strain is obtained by deleting rows and columns corresponding to vanishing strain components in Eq. (3-30):

$$\begin{bmatrix} \sigma_x \\ \sigma_y \\ \tau_{xy} \end{bmatrix} = \begin{bmatrix} \lambda + 2G & \lambda & 0 \\ \lambda & \lambda + 2G & 0 \\ 0 & 0 & G \end{bmatrix} \begin{bmatrix} \varepsilon_x \\ \varepsilon_y \\ \gamma_{xy} \end{bmatrix} \quad (3\text{-}48a)$$

or, with the usual elastic constants,

$$\sigma = \begin{bmatrix} \sigma_x \\ \sigma_y \\ \tau_{xy} \end{bmatrix} = \frac{E}{(1+v)(1-2v)} \begin{bmatrix} (1-v) & v & 0 \\ v & (1-v) & 0 \\ 0 & 0 & \dfrac{1-2v}{2} \end{bmatrix} \begin{bmatrix} \varepsilon_x \\ \varepsilon_y \\ \gamma_{xy} \end{bmatrix} = \mathbf{C}\varepsilon \quad (3\text{-}48b)$$

Summary of linear elasticity theory

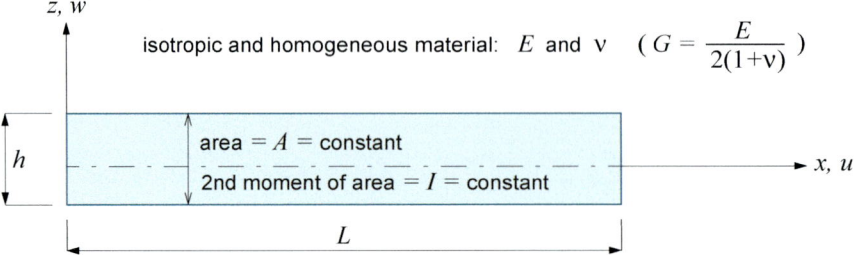

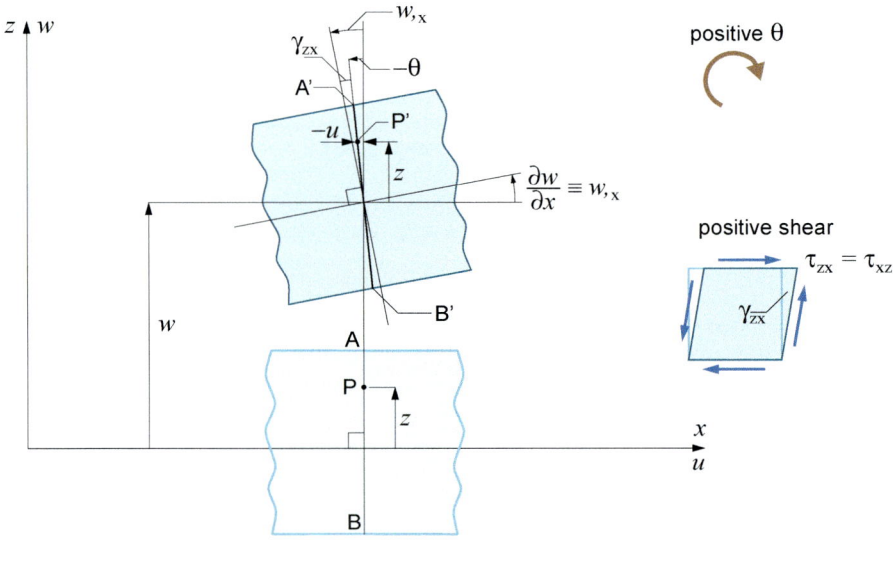

Assumptions: $w = w(x)$ and $\theta = \theta(x)$

Figure 3.15 Bending of prismatic 2D beam

3.4 Stress and strain in beams

We have already visited the beam in Section 2.1, but it will be useful to expand somewhat on the brief presentation given there. This section also serves as a prelude to the following one. Again, we limit the discussion to a homogeneous, prismatic plane (2D) beam, as shown in Fig. 3.15.

The basic assumption is the weak form of NAVIER's hypothesis which says that plane sections remain plane, that is

$$-u = z(-\theta) \quad \Rightarrow \quad u = z\theta \tag{3-49}$$

The relationships between strain and displacements are, see Eqs. (3-15a) and (3-16c):

$$\varepsilon_x = \frac{\partial u}{\partial x} = z\theta_{,x} \tag{3-50}$$

$$\gamma_{zx} = \frac{\partial u}{\partial z} + \frac{\partial w}{\partial x} = \theta + w_{,x} = \gamma_{xz} \tag{3-51}$$

A consequence of NAVIER's hypothesis is that the shear strain is constant over the height h (independent of z); it must therefore be viewed as an average strain. For the stresses we assume that

$$\sigma_z = 0 \tag{3-52}$$

Hence, the only non-zero stress components are the axial stress σ_x and the shear stress γ_{xz} ($=\gamma_{zx}$). Stress-strain relations are simply

$$\varepsilon_x = \frac{\sigma_x}{E} \quad \Rightarrow \quad \sigma_x = E\varepsilon_x \tag{3-53}$$

$$\gamma_{xz} = \frac{\tau_{xz}}{G} \quad \Rightarrow \quad \tau_{xz} = G\gamma_{xz} \tag{3-54}$$

Consistent with the constant (average) shear strain over the height, we assume a corresponding (average) shear stress

$$\tau_{xz}^{av} = \frac{V}{A_s} = \frac{\kappa V}{A} \tag{3-55}$$

where V is the section shear force and

$$A_s = \frac{A}{\kappa} \tag{3-56}$$

is the effective shear area. The parameter κ, which depends on the geometric shape of the cross section, can be found by a virtual work procedure for different cross sections. We have, for instance, $\kappa = 1{,}2$ for a rectangle and $\kappa = 2{,}0$ for a thin-walled circular tube cross section.

The strain energy for the beam in Fig. 3.15 can be expressed as

$$U = U_b + U_s \tag{3-57}$$

where

Summary of linear elasticity theory

Curvature c

by definition: $c = \dfrac{1}{\rho}$

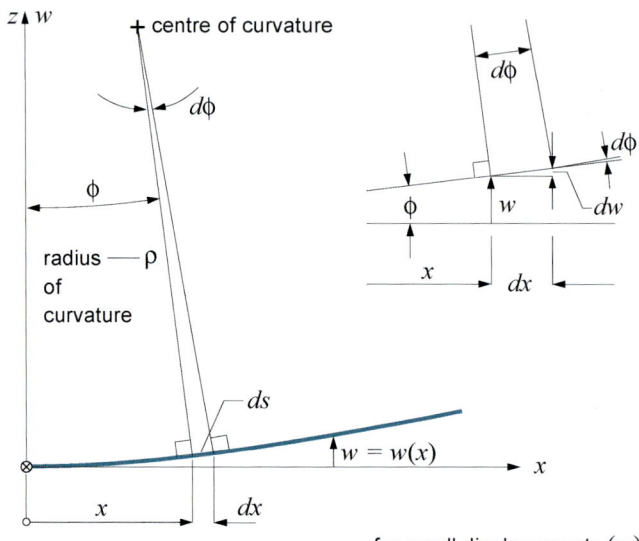

From the figure:
$$c = \dfrac{1}{\rho} \approx \dfrac{d\theta}{dx} = \dfrac{d}{dx}\dfrac{dw}{dx} = w''$$
(for small displacements (w))

Mathematically correct:
$$c = \dfrac{1}{\rho} = \dfrac{|w''|}{[1+(w')^2]^{3/2}}$$

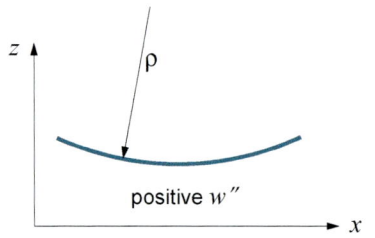

positive w''

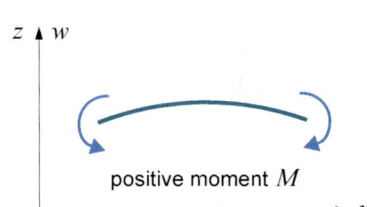

positive moment M

$$U_b = \int_V \frac{1}{2}\sigma_x\varepsilon_x dV = \frac{1}{2}\int_0^L\int_A E\varepsilon_x^2 dA\,dx = \frac{E}{2}\int_0^L\int_A z^2 dA(\theta_{,x})^2 dx = \frac{EI}{2}\int_0^L (\theta_{,x})^2 dx \quad (3\text{-}58)$$

is the strain energy due to bending deformation and

$$U_s = \int_0^L \frac{1}{2}V\gamma_{xz}dx = \int_0^L \frac{A}{2\kappa}\tau_{xz}^{av}\gamma_{xz}dx = \frac{A}{2\kappa}\int_0^L G\gamma_{xz}^2 dx = \frac{GA}{2\kappa}\int_0^L (w_{,x}+\theta)^2 dx \quad (3\text{-}59)$$

is the strain energy due to shear deformation.

Depending on how we now treat the shear deformation, we can formulate different beam theories.

1) The most common assumption is to *neglect* shear deformations altogether, that is

$$\gamma_{xz} = 0 \quad \Rightarrow \quad \theta = -w_{,x} \quad (3\text{-}60)$$

and

$$U = U_b = \int_V \frac{1}{2}\sigma_x\varepsilon_x dV = \frac{EI}{2}\int_0^L (w_{,xx})^2 dx \quad (3\text{-}61)$$

This represents the classical EULER-BERNOULLI beam theory, for which the normal strain, according to Eqs. (3-50) and (3-60), can be expressed as

$$\varepsilon_x = -zw_{,xx} \quad (3\text{-}62)$$

where $|w_{,xx}|$ is (an approximate expression of) the *curvature*, see opposite page.

2) If we include shear deformations, based on average shear strain, which, according to Eqs. (3-54) and (3-55), can be expressed as

$$\gamma_{xz} = \frac{\kappa V}{GA} \quad (3\text{-}63)$$

the cross section rotation, θ, is no longer equal to the slope of the deflection curve ($-w_{,x}$). According to Eq. (3-51) it now becomes

$$\theta = \gamma_{xz} - w_{,x} \quad (3\text{-}64)$$

The inclusion of shear deformations defined by Eq. (3-63) was first suggested by TIMOSHENKO, and modification of EULER-BERNOULLI theory by replacing Eq. (3-60) by Eqs. (3-64) and (3-63), leads to what is commonly known as TIMOSHENKO beam theory.

In a finite element context the TIMOSHENKO beam element can be derived in various ways. The most common method is to first derive the flexibility matrix, which is fairly straightforward, and then obtain the stiffness matrix through inversion. However, the element stiffness can also be derived directly by an assumed displacement field for the lateral displacement w, using indirect interpolation with due regard to Eq. (3-64), see Section 14.1.

Summary of linear elasticity theory

Stephen P. TIMOSHENKO (1878-1972) was a Russian-American engineer who is said to be the father of modern engineering mechanics. He wrote many influential works in the areas of engineering mechanics, elasticity and strength of materials, some of which are still widely used today.

TIMOSHENKO was born in a part of the Russian Empire that is now in Ukraine where he also had his first training. He continued his education towards a university degree in St Petersburg. After graduating in 1901 he stayed on teaching in St Petersburg until he was appointed to the Chair of Strengths of Materials at the Kiev Polytechnic Institute in 1906. The next 5 years he carried out pioneering work on buckling and also published the first version of his famous *Strength of Materials* textbook. In 1911 he signed a protest against the Minister of Education and was dismissed from the Kiev Polytechnic Institute. He went back to St Petersburg, and for the next 7 years he lectured and did research on theory of elasticity and developed the theory of beam deflection, and continued to study buckling. In 1918 he returned to Kiev, but due to the upheaval of the revolution he ended up i Zagreb for a while where he got a professorship at the Zagreb Polytechnic Institute. In 1920, during the brief takeover of Kiev by the Polish army, he went to Kiev, reunited with his family and returned with his family to Zagreb.

In 1922 TIMOSHENKO moved to the United States where he worked for the Westinghouse Electric Corporation from 1923 to 1927, after which he became a faculty professor at the University of Michigan. His textbooks have been published in almost 40 languages. From 1936 onward he was a professor at Stanford University. In 1960 he moved to Wuppenthal in West Germany to live with his daughter.

In addition to his textbooks TIMOSHENKO also wrote an excellent historical survey, *History of Strength of Materials,* first published in 1953.

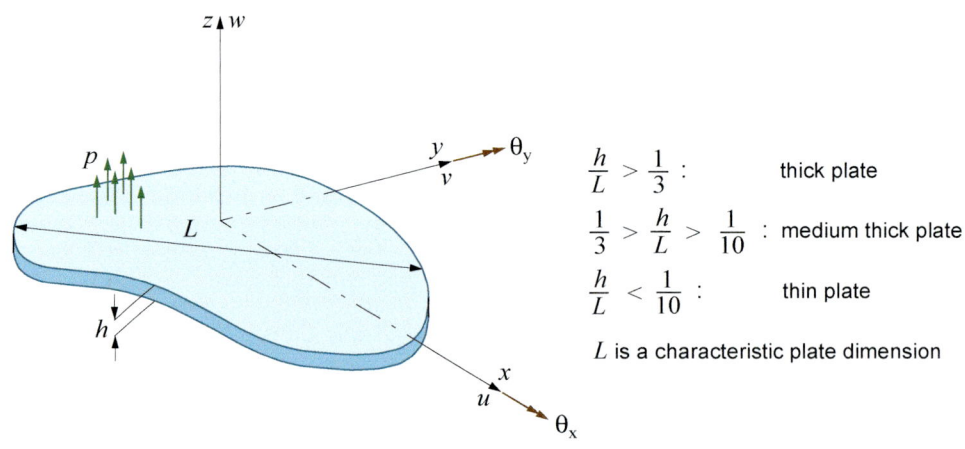

$\dfrac{h}{L} > \dfrac{1}{3}$: thick plate

$\dfrac{1}{3} > \dfrac{h}{L} > \dfrac{1}{10}$: medium thick plate

$\dfrac{h}{L} < \dfrac{1}{10}$: thin plate

L is a characteristic plate dimension

Figure 3.16 Plate bending parameters

3) Shear can also be accounted for by treating θ and *w* as independent variables, for which *independent* displacement fields are assumed. This two-field solution was inspired by similar approaches suggested independently by MINDLIN and REISSNER for plate bending problems.

This beam theory, by some authors called MINDLIN theory, will, if applied by standard finite element displacement procedures, lead to an element that will be far too stiff for elements with a low *h* to *L* ratio. However, through clever manipulations, the theory may be made to produce the same robust element as TIMOSHENKO theory, see for instance [10].

There is some confusion in the literature as to which theory the various beam elements are attributed to. We shall return to this issue later, in Chapter 14, when we deal more specifically with the problems associated with the finite element formulation.

3.5 Plate bending

Figure 3.16 indicates the nature of a *plate bending* problem. All loading is perpendicular to the plane (*x-y*) of the plate. Depending on the ratio of the plate thickness (*h*) and a characteristic length dimension *L* (measured in the plane of the plate), we distinguish between *thick*, *medium thick* and *thin* plates, as shown in the figure. Thick plates are basically 3D solids and should be treated as such. Hence this section is concerned with thin and medium thick plates.

The following assumptions are made:

$$u_0 = u(x, y, 0) = 0 \tag{3-65a}$$

$$v_0 = v(x, y, 0) = 0 \tag{3-65b}$$

and

a straight line (of material) that is perpendicular to the plane of the undeformed plate (*x-y*) remains straight, but not necessarily perpendicular to the deformed middle plane of the plate – this is equivalent to the weak form of NAVIER's hypothesis for beams.

Plate kinematics. With reference to Fig. 3.15 and the notation and sign conventions of Fig. 3.16 we now have

$$u = z\theta_y(x, y) \qquad v = -z\theta_x(x, y) \qquad w = w(x, y) \tag{3-66}$$

and

$$\theta_x = \theta_x(x, y) \qquad \theta_y = \theta_y(x, y) \tag{3-67}$$

θ_x and θ_y are the rotations of the plate normal about the *x*- and *y*-axis, respectively. Equations (3-17) and (3-66) give the strain-displacement relations:

$$\varepsilon_x = \frac{\partial u}{\partial x} = z\theta_{y,x} \tag{3-68a}$$

$$\varepsilon_y = \frac{\partial v}{\partial y} = -z\theta_{x,y} \tag{3-68b}$$

$$\varepsilon_z = \frac{\partial w}{\partial z} = 0 \tag{3-68c}$$

Summary of linear elasticity theory

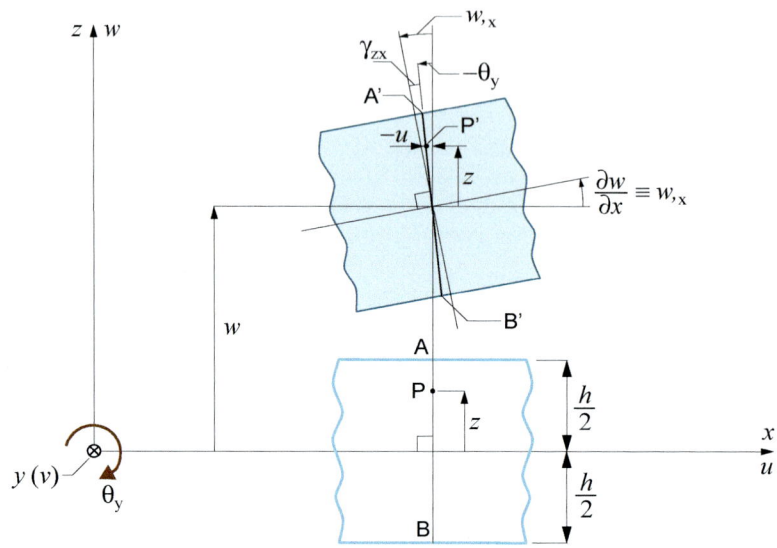

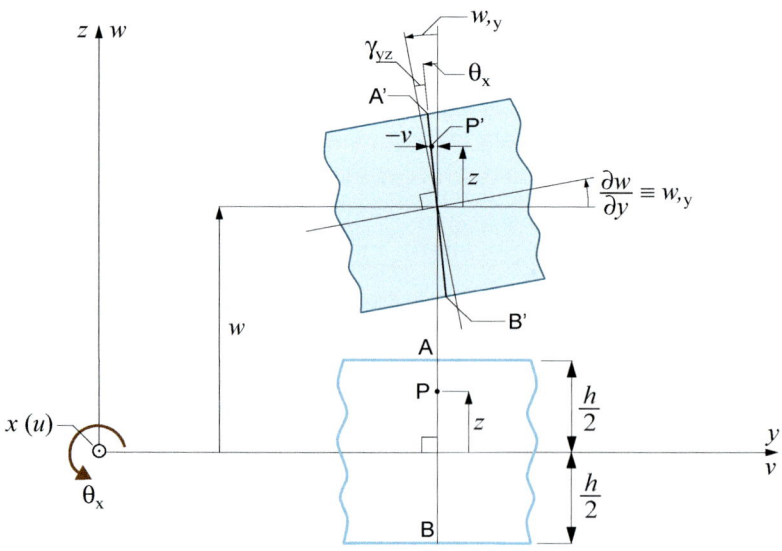

Plate bending

$$\gamma_{xy} = \frac{\partial u}{\partial y} + \frac{\partial v}{\partial x} = z(\theta_{y,y} - \theta_{x,x}) \qquad (3\text{-}68d)$$

$$\gamma_{yz} = \frac{\partial v}{\partial z} + \frac{\partial w}{\partial y} = -\theta_x + w_{,y} \qquad (3\text{-}68e)$$

$$\gamma_{zx} = \frac{\partial w}{\partial x} + \frac{\partial u}{\partial z} = \theta_y + w_{,x} \qquad (3\text{-}68f)$$

or

$$\boldsymbol{\varepsilon} = \begin{bmatrix} \boldsymbol{\varepsilon}_b \\ \boldsymbol{\varepsilon}_s \end{bmatrix} \qquad (3\text{-}69)$$

where

$$\boldsymbol{\varepsilon}_b = \begin{bmatrix} \varepsilon_x \\ \varepsilon_y \\ \gamma_{xy} \end{bmatrix} = -z \begin{bmatrix} -\theta_{y,x} \\ \theta_{x,y} \\ \theta_{x,x} - \theta_{y,y} \end{bmatrix} = -z\mathbf{c} \quad \text{and} \quad \mathbf{c} = \begin{bmatrix} -\theta_{y,x} \\ \theta_{x,y} \\ \theta_{x,x} - \theta_{y,y} \end{bmatrix} \qquad (3\text{-}70)$$

and

$$\boldsymbol{\varepsilon}_s = \begin{bmatrix} \gamma_{yz} \\ \gamma_{zx} \end{bmatrix} = \begin{bmatrix} -\theta_x + w_{,y} \\ \theta_y + w_{,x} \end{bmatrix} \qquad (3\text{-}71)$$

The last two expressions apply to medium thick plates (with shear deformations $\boldsymbol{\varepsilon}_s$).

 For *thin plates* we *neglect shear deformations* normal to the plate plane, that is

$$\boldsymbol{\varepsilon}_s = \mathbf{0} \qquad (3\text{-}72)$$

This is known as KIRCHHOFF's hypothesis, and thin plate theory is often referred to as KIRCHHOFF theory. With this assumption Eqs. (3-68e and f) give

$$\theta_x = w_{,y} \quad \text{and} \quad \theta_y = -w_{,x} \qquad (3\text{-}73)$$

and we can write the *thin plate kinematics* as

$$\boldsymbol{\varepsilon} = \boldsymbol{\varepsilon}_b = \begin{bmatrix} \varepsilon_x \\ \varepsilon_y \\ \gamma_{xy} \end{bmatrix} = -z \begin{bmatrix} w_{,xx} \\ w_{,yy} \\ 2w_{,xy} \end{bmatrix} = -z\mathbf{c}_K \qquad (3\text{-}74)$$

where

$$\mathbf{c}_K = \begin{bmatrix} w_{,xx} \\ w_{,yy} \\ 2w_{,xy} \end{bmatrix} = \begin{bmatrix} \dfrac{\partial^2}{\partial x^2} \\ \dfrac{\partial^2}{\partial y^2} \\ 2\dfrac{\partial^2}{\partial x \partial y} \end{bmatrix} w = \boldsymbol{\Delta}_K w \qquad (3\text{-}75)$$

is the *curvature* vector for thin (KIRCHHOFF) plates.

Summary of linear elasticity theory

> **Gustav Robert Kirchhoff** (1824-1887) was a German physicist who is known for his contribution to the fundamental understanding of electrical circuits, spectroscopy, and the emission of black-body radiation by heated objects. He coined the term "black body" radiation, and his name is associated with two sets of independent concepts in both circuit theory and thermal emission.
>
> In 1845, while still a student at the University of Köningsberg, Kirchhoff formulated his circuit laws, which are now everywhere in electrical engineering. He completed this study as a seminar exercise; it later became the theme of his doctoral dissertation. In a short biography of Kirchhoff his contribution to our field is hardly mentioned. However, being a pupil of Franz Neumann, he soon became interested in theory of elasticity. In 1850 he published his important paper on the theory of plates in which we find the first satisfactory theory of bending of plates. He based his theory on assumptions which are now generally accepted, for thin plates. In his lectures, Kirchhoff subsequently extended his theory of plates to cover the case where the deflections are not very small. Another important contribution he made to the theory of elasticity was his theory of deformation of thin bars, and he also wrote a paper on the vibration of bars of variable cross section.

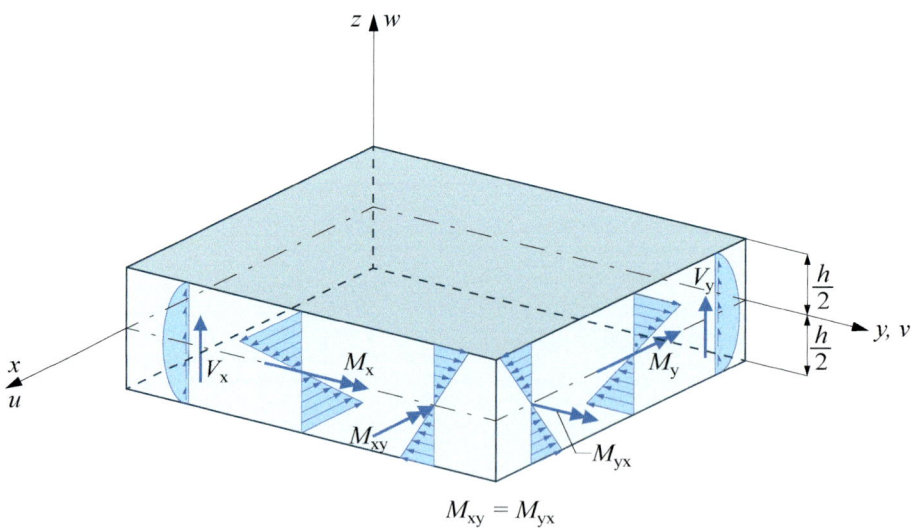

Figure 3.17 Stresses and stress resultants – plate bending

Plate bending

Stress-strain. We assume that $\sigma_z = 0$ and write, consistent with the separation of the strain components,

$$\boldsymbol{\sigma} = \begin{bmatrix} \boldsymbol{\sigma}_b \\ \boldsymbol{\sigma}_s \end{bmatrix} = \begin{bmatrix} \mathbf{C}_b & 0 \\ 0 & \mathbf{C}_s \end{bmatrix} \begin{bmatrix} \boldsymbol{\varepsilon}_b \\ \boldsymbol{\varepsilon}_s \end{bmatrix} = \mathbf{C}\boldsymbol{\varepsilon} \qquad (3\text{-}76)$$

Hence

$$\boldsymbol{\sigma}_b = \begin{bmatrix} \sigma_x \\ \sigma_y \\ \tau_{xy} \end{bmatrix} = \mathbf{C}_b \boldsymbol{\varepsilon}_b = -z\mathbf{C}_b \mathbf{c} \quad \text{and} \quad \boldsymbol{\sigma}_s = \begin{bmatrix} \tau_{yz} \\ \tau_{zx} \end{bmatrix} = \mathbf{C}_s \boldsymbol{\varepsilon}_s \qquad (3\text{-}77\text{a and b})$$

For a linear elastic, isotropic material matrix $\mathbf{C}_b$ is the same as matrix $\mathbf{C}$ of Eq. (3-43), and matrix $\mathbf{C}_s$ is simply a 2 by 2 *diagonal* matrix with the shear modulus G on the diagonal.

It should be noted that the assumption $\sigma_z = 0$ is basically inconsistent with the assumption that $\varepsilon_z = 0$. However, this is an insignificant discrepancy (which we also have in simple beam theory).

The stress resultants corresponding to $\boldsymbol{\sigma}_b$ are, see Fig. 3.17,

$$\mathbf{m} = \begin{bmatrix} M_x \\ M_y \\ M_{xy} \end{bmatrix} = \int_{-h/2}^{h/2} \boldsymbol{\sigma}_b z\,dz = -\mathbf{C}_b \int_{-h/2}^{h/2} z^2 dz\,\mathbf{c} = -\frac{h^3}{12}\mathbf{C}_b \mathbf{c} = -\mathbf{D}\mathbf{c} \qquad (3\text{-}78)$$

where

$$\mathbf{D} = \frac{h^3}{12}\mathbf{C}_b = D\begin{bmatrix} 1 & \nu & 0 \\ \nu & 1 & 0 \\ 0 & 0 & \frac{1-\nu}{2} \end{bmatrix} \quad \text{and} \quad D = \frac{Eh^3}{12(1-\nu^2)} \qquad (3\text{-}79)$$

D is the *flexural rigidity* of the plate. Similarly for the shear forces,

$$\mathbf{V} = \begin{bmatrix} V_y \\ V_x \end{bmatrix} = \int_{-h/2}^{h/2} \boldsymbol{\sigma}_s dz = \int_{-h/2}^{h/2} \mathbf{C}_s \boldsymbol{\varepsilon}_s dz = h\mathbf{C}_s \boldsymbol{\varepsilon}_s = Gh\begin{bmatrix} \gamma_{yz} \\ \gamma_{zx} \end{bmatrix} \qquad (3\text{-}80)$$

Equations (3-78) and (3-80) apply to medium thick plates including shear deformations.

For *thin* plates we replace $\mathbf{c}$ by $\mathbf{c}_K$ in Eq. (3-78), that is

$$\mathbf{m} = \begin{bmatrix} M_x \\ M_y \\ M_{xy} \end{bmatrix} = -\mathbf{D}\mathbf{c}_K = -\mathbf{D}\begin{bmatrix} w_{,xx} \\ w_{,yy} \\ 2w_{,xy} \end{bmatrix} \qquad (3\text{-}81)$$

Equation (3-80) on the other hand cannot be used for thin plates where the shear strains on vertical sections vanish (that is $\gamma_{yz} = \gamma_{zx} = 0$). In KIRCHHOFF

85

Summary of linear elasticity theory

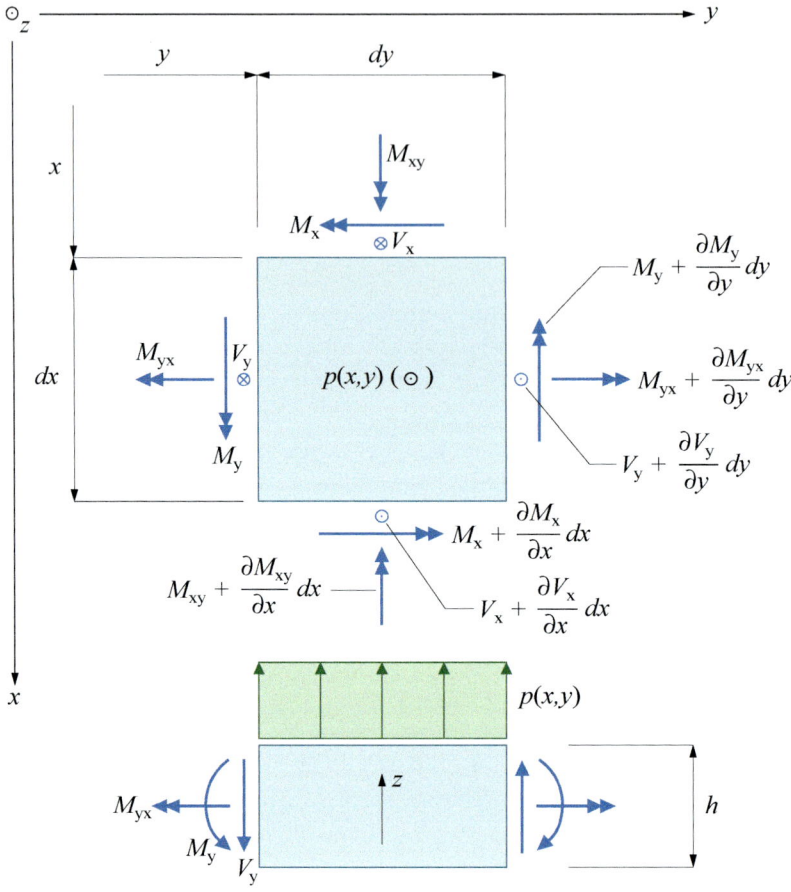

Figure 3.18 Forces and moments acting on an infinitesimal plate element

Plate bending

theory the shear forces can only be determined indirectly, from equilibrium considerations, see below.

Equilibrium. Consider the infinitesimal element $dxdy$ of a plate shown in Fig. 3.18; all forces and moments acting on the element are shown. Moment equilibrium about an axis parallel to the x-axis through the centre of the element requires:

$$\frac{\partial M_x}{\partial x}dxdy + \frac{\partial M_{xy}}{\partial y}dydx - V_x dydx = 0$$

or

$$V_x = M_{x,x} + M_{xy,y} \qquad (3\text{-}82\text{a})$$

Similarly for moment equilibrium about the y-axis:

$$V_y = M_{y,y} + M_{xy,x} \qquad (3\text{-}82\text{b})$$

Force equilibrium in the z-direction:

$$V_{x,x} + V_{y,y} + p = 0 \qquad (3\text{-}82\text{c})$$

Substituting the moments of Eq. (3-81) into (3-82a) gives:

$$V_x = -D(w_{,xxx} + \nu w_{,yyx} + \{1-\nu\}w_{,xyy}) = -D(w_{,xxx} + w_{,yyx})$$

or

$$V_x = -D\frac{\partial}{\partial x}(w_{,xx} + w_{,yy}) = -D\frac{\partial}{\partial x}\nabla^2 w \qquad (3\text{-}83\text{a})$$

Similarly

$$V_y = -D\frac{\partial}{\partial y}\nabla^2 w \qquad (3\text{-}84)$$

Combining the three equilibrium equations (3-82a, b and c) gives:

$$M_{x,xx} + 2M_{xy,xy} + M_{y,yy} = -p \qquad (3\text{-}84\text{b})$$

This equation applies to both medium thick and thin plates. If we substitute the thin plate relation (3-81) into (3-84) we get the well known differential equation for thin (KIRCHHOFF) plates:

$$w_{,xxxx} + 2w_{,xxyy} + w_{,yyyy} = \frac{p}{D}$$

or

$$\nabla^4 w = \frac{p}{D} \qquad (3\text{-}85)$$

Strain energy. The strain energy of the plate is

$$U = \int_V \frac{1}{2}\boldsymbol{\sigma}^T \boldsymbol{\varepsilon} dV = \int_V \frac{1}{2}\boldsymbol{\varepsilon}^T \mathbf{C}\boldsymbol{\varepsilon} dV \qquad (3\text{-}86)$$

By splitting the strain into two parts, one due to bending and the other due to shear and using the notation of Eqs. (3-77) we can write this as

Summary of linear elasticity theory

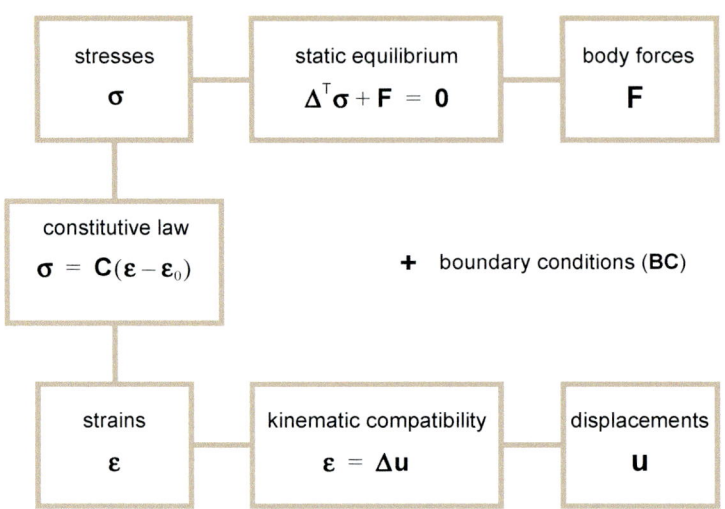

SUMMARY

strain energy

$$U = \int_V \frac{1}{2}\boldsymbol{\sigma}^T\boldsymbol{\varepsilon}\,dV = \int_V \frac{1}{2}\boldsymbol{\varepsilon}^T\mathbf{C}\boldsymbol{\varepsilon}\,dV$$

$$U = \frac{1}{2}\int_V \boldsymbol{\varepsilon}_b^\mathsf{T} \mathbf{C}_b \boldsymbol{\varepsilon}_b dV + \frac{1}{2}\int_V \boldsymbol{\varepsilon}_s^\mathsf{T} \mathbf{C}_s \boldsymbol{\varepsilon}_s dV = \frac{1}{2}\int_A \int_{-h/2}^{h/2} \mathbf{c}^\mathsf{T} z^2 \mathbf{C}_b \mathbf{c}\, dz\, dA + \frac{1}{2}\int_A \int_{-h/2}^{h/2} \boldsymbol{\varepsilon}_s^\mathsf{T} \mathbf{C}_s \boldsymbol{\varepsilon}_s dz\, dA$$

or

$$U = \frac{1}{2}\int_A \mathbf{c}^\mathsf{T}\mathbf{D}\mathbf{c}\, dA + \frac{1}{2}\int_A h\boldsymbol{\varepsilon}_s^\mathsf{T}\mathbf{C}_s\boldsymbol{\varepsilon}_s dA \qquad (3\text{-}87)$$

For *thin* plates we have $\boldsymbol{\varepsilon}_s = \mathbf{0}$ and $\mathbf{c} = \mathbf{c}_K$. Hence

$$U = \frac{1}{2}\int_A \mathbf{c}_K^\mathsf{T}\mathbf{D}\mathbf{c}_K\, dA \qquad (3\text{-}88)$$

This concludes our summary of linear elasticity theory. Even though we shall also, in Chapter 15, say something about *shell* elements and shell analysis, we leave the rather complex *shell theories* alone. One reason is that they are beyond a text like this, another is that our dealing with shells will mainly take the view that shells are curved plates with both in-plane (membrane) action and bending action, and that they can therefore be adequately modelled as an assemblage of flat plate facets.

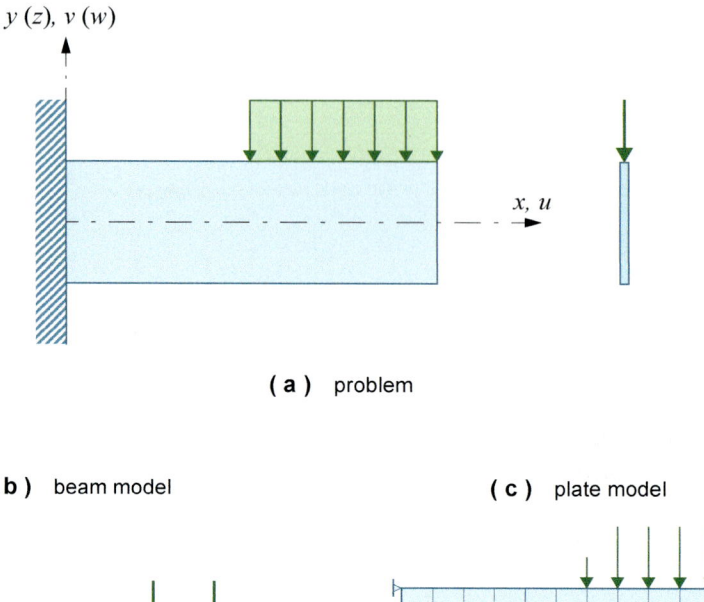
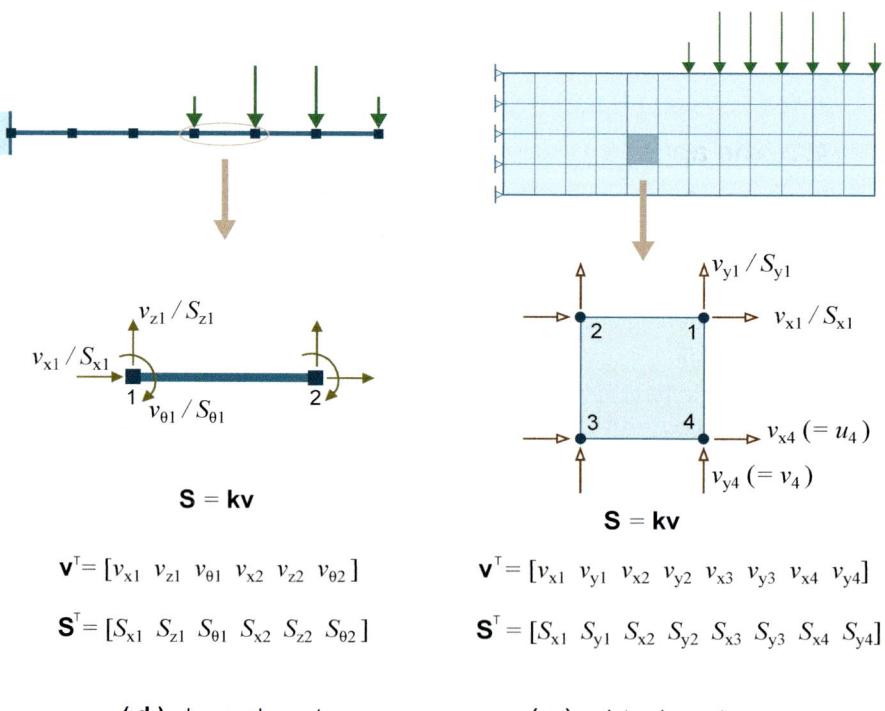

Figure 4.1 Different finite element models of a plate problem

4

Mathematical basics of FEM

In this chapter we try to demonstrate the mathematical justification of the finite element method within a physical framework. By using a very simple example we will show how the method can be derived using different principles, and also how it ties up with the classical RAYLEIGH-RITZ method. We do not present a rigorous mathematical treatment, but hopefully the chapter will explain the most central concepts, connections and terminology. We shall conclude that, for our purposes, the principle of virtual displacements (PVD) is the preferred vehicle for the derivations.

4.1 The approximate nature of FEM – the basic assumption

In Chapter 2 we outlined a discrete computational model of a *frame* or *truss* structure, consisting of a series of 1D *elements* interconnected at *nodal points*. This *physical* approach or interpretation can be generalized to a continuum type problem. However, there are some very distinct differences that we need to recognize and make allowances for.

Let us consider the simple problem of a short cantilever beam as shown in Fig. 4.1a. In order to determine the displacement of the beam we might make a simple beam model (and account for shear deformations) as shown in Fig. 4.1b, or we might consider the problem as a plane stress plate problem and make a model of, say, rectangular elements, as shown in Fig. 4.1c.

For the beam model the subdivision into beam elements is fairly straightforward and determined only by how the loading is handled, by lumping or by a consistent approach. The plate model on the other hand is far from straightforward; here we are faced with some hard questions, none of which has a single correct answer:

- which geometrical shape should the elements have?
- where in the element do we place the nodal points?
- which kinematic degrees of freedom do we define at the nodal points?
- how many elements do we need to use?

For our particular problem, the rectangular element is an obvious choice, but if the problem domain has an irregular geometry, a triangular or quadrilateral

Mathematical basics of FEM

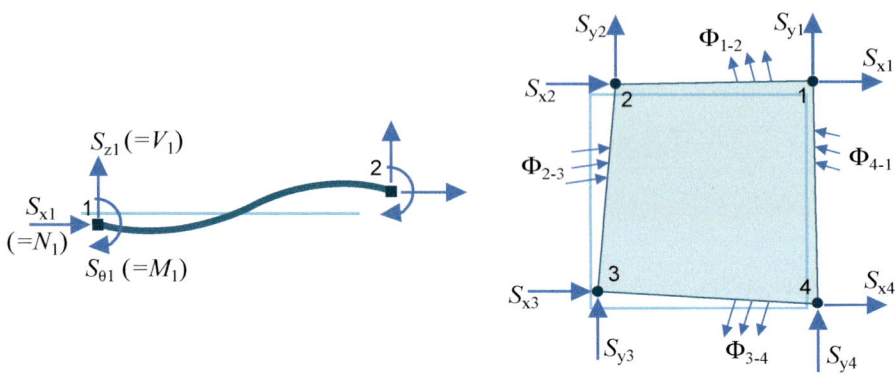

nodal forces S_{xi} and S_{yi} are *resultants* of the tractions (stresses) $\Phi_{j\text{-}k}$

Figure 4.2 Elements as free-bodies – interpretation of the nodal forces S

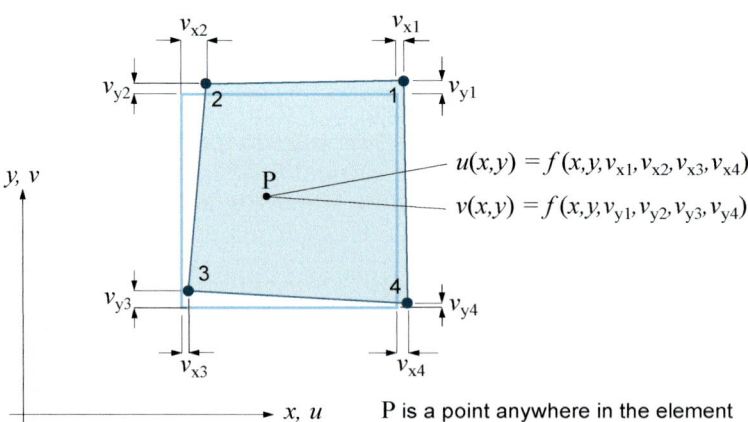

$u(x,y) = f(x,y,v_{x1},v_{x2},v_{x3},v_{x4})$
$v(x,y) = f(x,y,v_{y1},v_{y2},v_{y3},v_{y4})$

P is a point anywhere in the element

Figure 4.3 Assumed displacements are functions of x, y and **v**

92

shape would be more appropriate. Placing nodal points or nodes at the element corners also seems quite reasonable, but as we shall see later on there are other options. As kinematic degrees of freedom we have indicated the nodal values of the displacement components u and v in the global coordinate directions x and y, respectively. Again this is a fairly obvious choice, but not the only one.

Corresponding to the (eight) nodal degrees of freedom (*dofs*), **v**, we have a set of concentrated nodal forces **S**. What are these forces? For the beam element they are the section forces (N, V and M) at the element's end sections (or nodes). In other words, the nodal forces (S) are the stress resultants on the sections where the beam element is "cut out" from the structure. The plate element is more complicated in that it has common boundaries with other (neighbouring) elements *other than* at the nodal points. The concentrated nodal forces S should therefore be interpreted as resultants of the stresses or tractions on the *entire* element boundary (the section along which we "cut" the element out as a free-body), see Fig. 4.2. Alas, we do not know exactly how the nodal forces are related to the edge stresses, only that they represent the resultants of the stresses and therefore can replace the stresses in some crucial work equations.

How can we now establish a relationship between **S** and **v**, that is, how do we find **k** in the relation **S** = **kv** for the plate element, or any other 2 and 3D element? This question is what a large part of this book is all about. Here it suffices to say that we have no simple way of establishing an "exact" relation in the way we can for the beam element; not even for the simple rectangular element in Fig. 4.1.

The standard FEM way of determining the element stiffness relation, and the only approach we will pursue in detail in our treatment, is to *assume* a displacement field within the element. This assumption, which is the key step in the procedure, expresses the displacement within the element in terms of the nodal degrees of freedom, such that a displacement component at a particular point inside the element is a *unique* function of the nodal displacements **v**. In other words, the element displacement components are *interpolated* between the nodal degrees of freedom. For our simple rectangular element, the displacement u in the x-direction is expressed in terms of v_{x1} to v_{x4}, and v in terms of v_{y1} to v_{y4}, see Fig. 4.3.

In order to answer the last question in our list above, we need to recognize the *approximate* nature of the FEM solution. We cannot expect to be able to assume the correct displacement field, not even for very simple problems. We can only hope for a good approximation of the displacement field, and thus a reasonable solution based on this approximation. Intuitively we expect the approximation to be better the more elements we employ; this quality, however, is not automatically achieved, but it is obviously a feature of our element that we need to achieve. But even if we do obtain convergence towards the correct result when the element mesh is made finer, we would also like to know the rate of convergence and better still, how large is the error for a given element and mesh? These are basic questions that we shall try to answer satisfactorily, if not conclusively, in chapters to come.

Mathematical basics of FEM

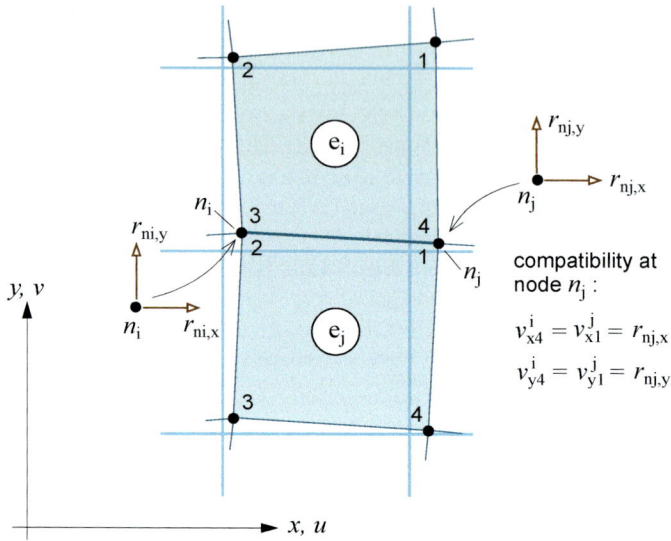

Figure 4.4 Continuity of displacements between neighbouring elements

The approximate nature of FEM – the basic assumption

For beam elements we satisfy kinematic compatibility by simply requiring that the elements are properly joined at the nodes. A model consisting of 2- or 3-dimensional continuum elements that are joined only at the nodes and not necessarily along the boundaries between the nodes, seems more than a little suspicious. As it turns out this suspicion is well founded, and for most continuum elements we need to make sure that our assumed displacement fields are such that continuity at the nodal points also implies some degree of continuity along the common boundaries between elements.

If we take our simple rectangular plate element as an example we need to make sure that the assumed displacement functions, or *shape functions* as we shall be calling them, are such that displacements u and v are the same in both neighbouring elements along their common boundary. If so we have displacement continuity between the elements of the model, which is, as will be shown, a necessary requirement for plane stress/plane strain elements. From Fig. 4.4, which shows two neighbouring elements, e_i and e_j, with a common edge (between system nodes n_i and n_j), we see that this requirement is satisfied if the displacement components along the common edge are unique functions of the relevant *dofs* at the two end nodes of the edge (here n_i and n_j), and nothing else. In our example this will be a linear function. More about this later.

It should be pointed out that while the direct approach used to derive the element matrices for 1D elements in Chapter 2 cannot be used for continuum elements, the opposite presents no problem. The stiffness matrix for a simple beam element can readily be derived on the basis of appropriately assumed displacement functions. What makes the beam element special is that we can (for a straight element with constant cross section properties) "assume" the (mathematically) correct displacements. Otherwise the 1D elements are finite elements as good as any, and in the rest of this chapter we shall use the very simplest 1D bar (axial) problem to exemplify and justify some basic aspects of the method.

In order to shed some light on the important questions of convergence we need a somewhat more mathematical approach than that of Chapter 2, but we will not abandon the physical or engineering perspective.

We conclude this section by stating the basic assumption of the finite element method in matrix notation:

$$\mathbf{u} = \mathbf{Nv} \qquad (4\text{-}1)$$

where $\mathbf{u}$ is a vector containing the relevant displacement components (field variables) of the problem, $\mathbf{N}$ is a matrix containing the *shape* or *interpolation* functions and $\mathbf{v}$ is the nodal displacement vector or the vector of element degrees of freedom. For our simple rectangular plane stress element of Fig. 4.1, Eq. (4-1) may read:

$$\mathbf{u} = \begin{bmatrix} u \\ v \end{bmatrix} = \begin{bmatrix} N_1(x,y) & 0 & \cdots & N_4(x,y) & 0 \\ 0 & N_1(x,y) & \cdots & 0 & N_4(x,y) \end{bmatrix} \begin{bmatrix} v_{x1} \\ v_{y1} \\ \vdots \\ v_{y4} \end{bmatrix} = \mathbf{Nv}$$

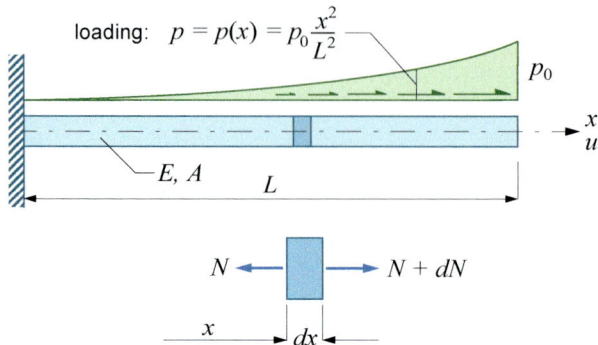

Figure 4.5 Cantilever bar subjected to a parabolic varying tangential load

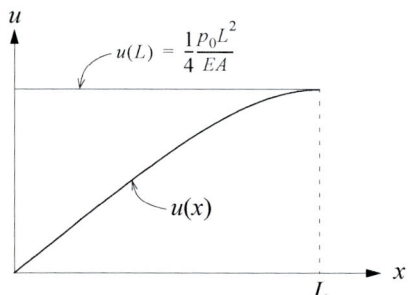
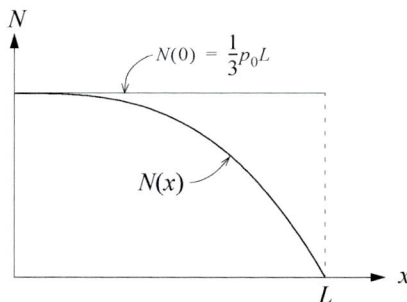

4.2 Example problem

Figure 4.5 shows a very simple problem, a cantilever prismatic bar subjected to a tangential distributed load varying parabolically along the bar. We assume the load is applied at the area centre of the bar cross section; hence no bending.

Equilibrium of an infinitesimal element of the bar:

$$dN + pdx = 0 \quad \Rightarrow \quad \frac{dN}{dx} = -p(x)$$

Stress-strain:

$$\sigma = \frac{N}{A} = E\varepsilon = E\frac{du}{dx} \quad \Rightarrow \quad N = EA\frac{du}{dx}$$

Boundary conditions: $u(0) = 0$ and $N(L) = EA\left(\frac{du}{dx}\right)_{x=L}$

Combining equilibrium and stress-strain:

$$\frac{d}{dx}\left(EA\frac{du}{dx}\right) = -p(x) = -p_0\frac{x^2}{L^2} \qquad (4\text{-}2)$$

Since EA is constant over the entire length of the bar the solution to this differential equation, with due regard to the boundary conditions, is readily found to be

$$u(x) = \frac{p_0 L^2}{EA}\left[\frac{1}{3}\left(\frac{x}{L}\right) - \frac{1}{12}\left(\frac{x}{L}\right)^4\right]$$

From this we get

$$u(L) = \frac{1}{4}\frac{p_0 L^2}{EA} \quad \text{and} \quad N(x) = EA\frac{du}{dx} = \frac{p_0 L}{3}\left[1 - \left(\frac{x}{L}\right)^3\right]$$

which is the "exact" solution to the problem. In the next sections we will use this example to demonstrate some important concepts, priciples and methods.

4.3 Strong and weak form

The 2nd order differential equation (4-2) is, together with the boundary conditions, the so-called *strong form* of the problem. It specifies the requirements that have to be met at every point of the problem domain.

The strong form can be expressed in *operator form* as

$$\Lambda u = f \qquad (4\text{-}3)$$

where $\Lambda = \frac{d}{dx}\left(EA\frac{d}{dx}\right)$ is a differential *operator*, u is the independent *variable*, and $f = -p(x)$, which is a function that is independent of u, is the *data* of the equation. The *residual r* is formed by subtracting the data *f* from the *image* Λu

97

Mathematical basics of FEM

Integration by parts:

$$\int_a^b \frac{dy}{dx} dx = \int_a^b dy = y(x)\Big|_a^b$$

Let $y(x) = \alpha(x)\beta(x) \quad \Rightarrow \quad \frac{dy}{dx} = \frac{d\alpha}{dx}\beta + \alpha\frac{d\beta}{dx}$

Then $\quad \int_a^b \left(\frac{d\alpha}{dx}\beta + \alpha\frac{d\beta}{dx}\right) dx = [\alpha(x)\beta(x)]_a^b$

and $\quad \int_a^b \alpha\beta' dx = [\alpha\beta]_a^b - \int_a^b \alpha'\beta dx$

$$r = \Lambda u - f \quad (4\text{-}4)$$

The residual $r = r(x)$ is a function of x over the problem domain, and the *residual form* of our problem is

$$r = 0 \quad (4\text{-}5)$$

We now dabble in a branch of mathematics called *calculus of variation* and seek a *functional J* whose so-called EULER-LAGRANGE equation is a particular differential equation, namely the strong form of the problem. To this end we start with the residual form of our problem, that is

$$r = \frac{d}{dx}\left(EA\frac{d}{dx}\right)u + p = 0 \quad (4\text{-}6)$$

and multiplies both sides by the function $\delta u(x)$,

$$\left[\frac{d}{dx}\left(EA\frac{du}{dx}\right) + p\right]\delta u = 0 \quad (4\text{-}7)$$

At this stage we put no other restrictions on δu, the *variation* of u, than that it is first order continuous in the range $<0, L>$.

The integral of Eq. (4-7) over the problem domain is called the *variation of J* and termed δJ, which is the *inner product* of r and δu,

$$\delta J = \int_0^L \left[\frac{d}{dx}\left(EA\frac{du}{dx}\right) + p\right]\delta u\, dx = \int_0^L r\delta u\, dx = (r, \delta u) = 0 \quad (4\text{-}8)$$

Let $\quad \alpha = \delta u \quad \Rightarrow \quad \alpha' \equiv \frac{d\alpha}{dx} = \frac{d(\delta u)}{dx} = \delta\frac{du}{dx}$

and $\quad \beta = EAu'$

then Eq. (4-8) can be written as

$$\delta J = \int_0^L \alpha\beta'\, dx + \int_0^L p\delta u\, dx \quad (4\text{-}9)$$

Using *integration by parts* on the first term on the right-hand side yields:

$$\delta J = \left[EA\frac{du}{dx}\delta u\right]_0^L - \int_0^L EAu'\delta u'\, dx + \int_0^L p\delta u\, dx = 0$$

or

$$\delta J = \int_0^L (p\delta u - EAu'\delta u')\, dx + \left[EA\frac{du}{dx}\delta u\right]_0^L = 0 \quad (4\text{-}10)$$

The last term is the boundary term for which we have

Mathematical basics of FEM

Some rules from the calculus of variation:

Let $J = I(f(x), g(x))$ where I is an integral over a domain in x.
Then

$$\delta J = \frac{\partial J}{\partial f}\delta f + \frac{\partial J}{\partial g}\delta g \quad \text{(confer total differential)}$$

$$\frac{d}{dt}(\delta J) = \delta\left(\frac{dJ}{dt}\right)$$

$$\int_t \delta J\, dt = \delta \int_t J\, dt$$

$$\delta(J^2) = 2J\delta J$$

In his book *The Mathematical Foundation of Structural Mechanics* [11], FRIEDEL HARTMANN makes the following observation (page 3):

The principle of "conservation of energy" and the δ-symbol are, loosely spoken, the jolly jokers of structural mechanics. A talented engineer proves all and everything with these two wild cards.

The reader should perhaps keep this in mind; our presentation may not stand up to a strict mathematical scrutiny. However, the essence and the end results are believed to be in good agreement with current knowledge.

$u(0) = 0 \quad \Rightarrow \quad \delta u|_0 = 0$ is an *essential* boundary condition (BC) and

$\left(EA\dfrac{du}{dx}\right)_{x=L} = 0 \quad \Rightarrow \quad \dfrac{du}{dx}\bigg|^L = 0$ is a *natural* (mechanical) BC.

Thus (see rules on opposite page)

$$\delta J = \int_0^L (p\delta u - EAu'\delta u')dx = \int_0^L \delta(pu - \tfrac{1}{2}EA(u')^2)dx = 0$$

or

$$\delta J = \delta \int_0^L (pu - \tfrac{1}{2}EA(u')^2)dx = 0 \qquad (4\text{-}11)$$

This equation, together with the boundary conditions $u(0) = 0$ and $N(L) = 0$, represent the *weak form* of the problem. From this equation we find the functional we set out to uncover to be

$$J = \tfrac{1}{2}\int_0^L EA\left(\dfrac{du}{dx}\right)^2 dx - \int_0^L pu\,dx \qquad (4\text{-}12)$$

We have changed the signs (which clearly makes no difference) in order to simplify the physical interpretation we make in the next section. In our context, a functional is a *function of functions* ($J = J(u,u')$).

It should be emphasized that the weak form of the problem is mathematically equivalent to the strong form. The transition from the strong to the weak form was achieved through mathematically correct manipulations. So why bother? Why not just settle for the strong form? Closed form solutions of differential equations are hard to obtain, even for very simple problems. For most practical problems governed by differential equations we have to use approximate methods of solution, and this is where the advantage of the weak form becomes evident.

Without labouring the point, this advantage has to do with the fact that the order of differentiation is lower in the weak form – in our example we have u'' in the strong form, but only u' in the weak form. We have "replaced" differentiation with integration which, in an "error sense" is advantageous. An error in u is amplified by differentiation while it is damped or "smeared" by integration. This lower order of differentiation also enables the use of simpler shape functions (**N**).

Another advantage of the weak form is that it makes it easy to handle discontinuities (in geometry or material properties).

We conclude this section by noting that here we have followed the opposite route of that which is commonly used in variational calculus. It is usual to start with a stationary principle, like Eq. (4-11), and either solve the problem directly by some approximate method starting with the weak form, or, via the principles and techniques of variational calculus, establish the strong form of the problem, that is, the differential equation (the EULER equation) and its boundary conditions. This route is *always* open, which means that

Mathematical basics of FEM

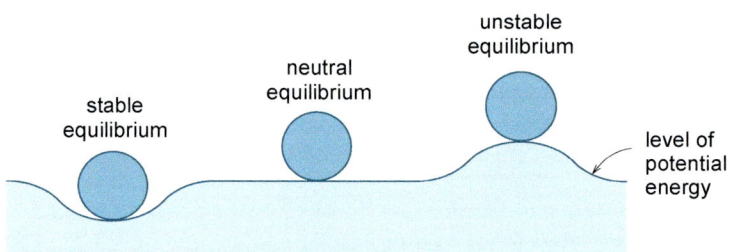

every problem that has a functional (potential) with a stationary requirement also has an equivalent differential equation. However, the opposite route is not always possible. In other words, a physical boundary value problem that has its differential equation(s) and boundary conditions, may not have a corresponding functional. That does not mean that the problem cannot be solved by a finite element formulation, but in such a case (which we will not encounter in our problem world) other approaches, such as GALERKIN's weighted residual method, may be used.

4.4 Principle of minimum potential energy (PMPE)

In Equation (4-12) we recognize the first term as the strain energy (U) of the rod, while the second term, including the sign, is the potential (H) of the external loading. In other words, the functional J in Eq. (4-12) is identical with the *total potential energy* (Π) of the problem, *i.e.*,

$$J = \Pi = U + H \qquad (4\text{-}13)$$

The weak form of the problem, Eq. (4-11), therefore expresses the *principle of minimum potential energy* (PMPE) which states that

> *for all admissible displacement configurations of a conservative system, the one or those that satisfy equilibrium will give the total potential energy of the system a stationary value with respect to a small admissible variation of its displacements.*

It should be noted that while the displacements are being varied, the loads are kept constant.

This principle is valid whether the load-displacement relation is linear or not. This is indicated by the suggestion that there may be more than one equilibrium position rendering the potential energy a stationary value, as would be the case for neutral equilibrium (bifurcation).

If the stationary condition is a relative *minimum*, the state of equilibrium is a *stable* one.

4.5 The RAYLEIGH-RITZ method

If the problem has a variational form, *i.e.* has a functional, we can establish an approximate solution for the independent variable(s) that will minimalize the functional (or give it a stationary value).

Classical RAYLEIGH-RITZ (R-R) discretizes the problem by assuming the displacement(s) as a sum of *trial functions* in the form,

$$u \approx \hat{u} = \sum_{k} a_k \varphi_k(x) = \boldsymbol{\varphi}\mathbf{a} \qquad (4\text{-}14)$$

where each trial function $\varphi_k(x)$, in order to be *admissible*, must satisfy compatibility (be sufficiently continuous and smooth) and also the essential (geometric) boundary conditions.

Mathematical basics of FEM

> NOTE: The parameters or amplitudes, **a**, are also referred to as *generalized coordinates* or *generalized degrees of freedom*.

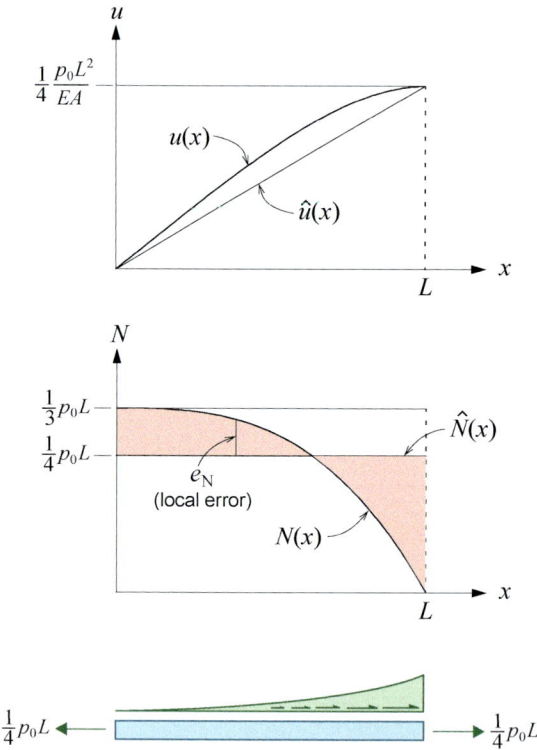

Figure 4.6 R-R with only one trial function ($= ax$)

The unknown parameters (amplitudes) a_k are determined so as to give the functional, with u replaced by $\hat{u}$, a stationary (minimum) value.

The functional for our example problem, the total potential energy, is given by Eq. (4.12),

$$\Pi = \frac{1}{2}\int_0^L EA\left(\frac{du}{dx}\right)^2 dx - \int_0^L pu\,dx = \frac{1}{2}\int_0^L EA\left(\frac{du}{dx}\right)^2 dx - \int_0^L p_0\frac{x^2}{L^2}u\,dx$$

We now assume only one trial function, and a very simple one at that, namely

$$\hat{u} = ax \qquad (4\text{-}15)$$

In spite of its simplicity it is admissible in that it is continuous and smooth and it satisfies the essential boundary condition, $u(0) = 0$.
We now proceed as follows

$$\hat{\varepsilon} = \frac{d\hat{u}}{dx} = a$$

$$\hat{\Pi} = \frac{1}{2}\int_0^L EAa^2\,dx - \int_0^L ap_0\frac{x^3}{L^2}dx$$

For equilibrium: $\delta\hat{\Pi} = 0 \quad \Rightarrow \quad \delta\hat{\Pi} = \frac{d\hat{\Pi}}{da}\delta a = 0 \quad \Rightarrow \quad \frac{d\hat{\Pi}}{da} = 0$

$$\frac{d\hat{\Pi}}{da} = \int_0^L EAa\,dx - \int_0^L p_0\frac{x^3}{L^2}dx = EALa - \frac{1}{4}p_0L^2 = 0 \quad \Rightarrow \quad a = \frac{1}{4}\frac{p_0L}{EA}$$

Hence $\hat{u}(x) = \frac{1}{4}\frac{p_0L}{EA}x \quad \Rightarrow \quad \hat{u}(L) = \frac{1}{4}\frac{p_0L^2}{EA}$ which is "exact" (!)

and $\quad \hat{N}(x) = EA\frac{d\hat{u}}{dx} = \frac{1}{4}p_0L$ which is constant.

These results are shown graphically in Fig. 4.6. Displacements are not too bad, but we certainly do not have equilibrium in a "strong" (point-wise) sense.

Since we have only used *one* trial function this is an example of RAYLEIGH's method. RITZ extended the method to include more than one trial function. This more general form of the method we can express in matrix notation:

$$\hat{\Pi} = \frac{1}{2}\int_0^L EA(\boldsymbol{\varphi}_{,x}\mathbf{a})(\boldsymbol{\varphi}_{,x}\mathbf{a})dx - \int_0^L \boldsymbol{\varphi}\mathbf{a}p\,dx \qquad (4\text{-}16)$$

Since $\boldsymbol{\varphi}_{,x}\mathbf{a}$ is a scalar ($\boldsymbol{\varphi}$ is a row matrix) we have $\boldsymbol{\varphi}_{,x}\mathbf{a} = \mathbf{a}^T(\boldsymbol{\varphi}_{,x})^T$ and

Mathematical basics of FEM

> A complete 2nd order polynomial reads
> $$\hat{u} = a_0 + a_1 x + a_2 x_2$$
> In order to satisfy the essential BC, $u(0) = 0$, we must have $a_0 = 0$.

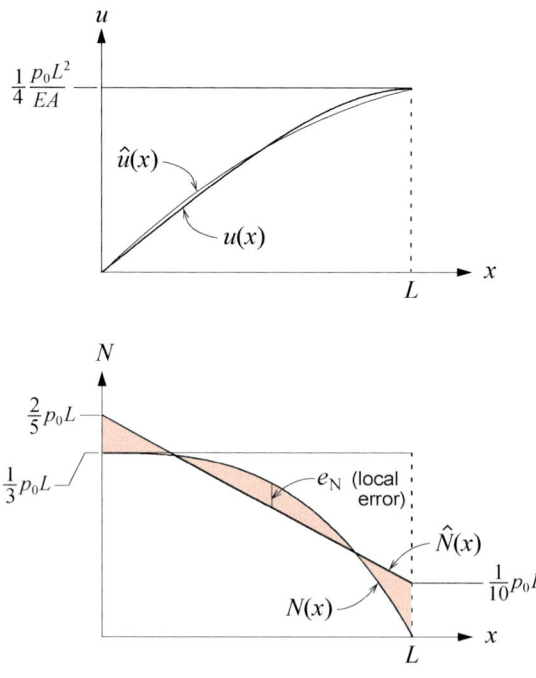

Figure 4.7 R-R with two trial functions

The RAYLEIGH-RITZ method

$$\hat{\Pi} = \frac{1}{2}\int_0^L \mathbf{a}^T(\boldsymbol{\varphi}_{,x})^T EA\boldsymbol{\varphi}_{,x}\mathbf{a}\,dx - \int_0^L \mathbf{a}^T\boldsymbol{\varphi}^T p\,dx \qquad (4\text{-}17)$$

Since $\frac{\partial}{\partial \mathbf{x}}(\mathbf{x}^T \mathbf{A} \mathbf{x}) = \mathbf{A}\mathbf{x} + \mathbf{A}^T\mathbf{x}$, see APPENDIX A, the variation of $\hat{\Pi}$ becomes

$$\frac{\partial \hat{\Pi}}{\partial \mathbf{a}} = \underbrace{\int_0^L (\boldsymbol{\varphi}_{,x})^T EA\boldsymbol{\varphi}_{,x}\,dx}_{\mathbf{K}}\mathbf{a} - \underbrace{\int_0^L \boldsymbol{\varphi}^T p\,dx}_{\mathbf{R}} = \mathbf{0} \qquad (4\text{-}18)$$

or

$$\mathbf{K}\mathbf{a} = \mathbf{R} \qquad (4\text{-}19)$$

where **K** can be regarded as a form of stiffness and **R** is effective load. This equation, which expresses average or integrated equilibrium, has a solution only if **K** is regular (non-singular). This puts yet another requirement on the trial functions, *they must be linearly independent*.

We go back to our example and assume two trial functions,

$$\varphi_1 = x \quad \text{and} \quad \varphi_2 = x^2$$

which are both admissible. With $\varphi_{1,x} = 1$ and $\varphi_{2,x} = 2x$ we find

$(\boldsymbol{\varphi}_{,x})^T\boldsymbol{\varphi}_{,x}$:

	1	2x
1	1	2x
2x	2x	$4x^2$

$$K_{11} = \int_0^L EA\,dx = EAL$$

$$K_{12} = K_{21} = \int_0^L 2xEA\,dx = EAL^2$$

$$K_{22} = \int_0^L 4x^2 EA\,dx = \frac{4}{3}EAL^3$$

Also $R_1 = \int_0^L p_0\frac{x^3}{L^2}dx = \frac{1}{4}p_0 L^2$ and $R_2 = \int_0^L p_0\frac{x^4}{L^2}dx = \frac{1}{5}p_0 L^3$

Substituting into (4-19) and solving yields,

$$a_1 = \frac{2p_0 L}{5 EA} \quad \text{and} \quad a_2 = -\frac{3}{20}\frac{p_0}{EA}$$

Hence,

$$\hat{u} = a_1\varphi_1 + a_2\varphi_2 = \frac{p_0 L^2}{EA}\left[\frac{2x}{5L} - \frac{3}{20}\left(\frac{x}{L}\right)^2\right]$$

and

$$\hat{N}(x) = EA\frac{d\hat{u}}{dx} = \frac{p_0 L}{10}\left[4 - 3\left(\frac{x}{L}\right)\right]$$

These results are shown graphically in Fig. 4.7.

107

Mathematical basics of FEM

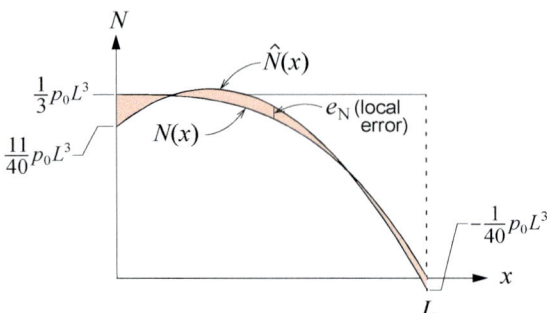

Figure 4.8 R-R with three trial functions

If we include the "next" admissible term in $\boldsymbol{\varphi}$, that is $\varphi_3 = x^3$ we will find:

$$\hat{u} = a_1\varphi_1 + a_2\varphi_2 + a_3\varphi_3 = \frac{p_0 L^2}{40EA}\left[11\frac{x}{L} + 9\left(\frac{x}{L}\right)^2 - 10\left(\frac{x}{L}\right)^3\right]$$

and

$$\hat{N}(x) = EA\frac{d\hat{u}}{dx} = \frac{p_0 L}{40}\left[11 + 18\left(\frac{x}{L}\right) - 30\left(\frac{x}{L}\right)^2\right]$$

The latter is shown in Fig. 4.8. If we also include the 4th term, $\varphi_4 = x^4$, $\hat{u}$ is able to express the "exact" displacement, and R-R *will give the correct solution*. We leave it to the reader to demonstrate this.

Equilibrium in the R-R process. Let us rewrite Eq. (4.18) slightly,

$$\frac{\partial \hat{\Pi}}{\partial \mathbf{a}} = \int_0^L (\boldsymbol{\varphi}_{,x})^T EA \underbrace{\boldsymbol{\varphi}_{,x}\mathbf{a}}_{\frac{d\hat{u}}{dx}} dx - \int_0^L \boldsymbol{\varphi}^T p\, dx = \mathbf{0} \qquad (4\text{-}20)$$

We do not know the exact solution, but we do know that it must satisfy the differential equation

$$EA\frac{d^2 u}{dx^2} + p = 0 \quad \text{with boundary conditions} \quad u(0) = 0 \quad \text{and} \quad EA\left(\frac{du}{dx}\right)_{x=L} = 0.$$

We multiply the differential equation by $\boldsymbol{\varphi}^T$ and integrate over the length,

$$\int_0^L \boldsymbol{\varphi}^T\left(EA\frac{d^2 u}{dx^2} + p\right)dx = \mathbf{0}$$

Partial integration of the first term yields

$$\boldsymbol{\varphi}^T EA\frac{du}{dx}\bigg|_0^L - \int_0^L (\boldsymbol{\varphi}_{,x})^T EA\frac{du}{dx}dx + \int_0^L \boldsymbol{\varphi}^T p\, dx = \mathbf{0}$$

The first term is zero (= **0**), due to the boundary conditions (remember all terms of $\boldsymbol{\varphi}$ are zero at $x=0$). If we now add what remains of this equation to Eq. (4-20) and change sign, we get

$$\int_0^L (\boldsymbol{\varphi}_{,x})^T EA\frac{d}{dx}(u - \hat{u})dx = \int_0^L (\boldsymbol{\varphi}_{,x})^T \underbrace{(N - \hat{N})}_{e_N \text{ (error)}} dx = ((\boldsymbol{\varphi}_{,x})^T, e_N) = \mathbf{0} \qquad (4\text{-}21)$$

From this we can conclude that R-R satisfies equilibrium in an average sense in that the integral, that is the sum, of the error over the problem domain is zero. We also see that the error is orthogonal to the derivative of the trial functions.

109

Mathematical basics of FEM

> A note on strain energy (U) and load potential (H) of a linear elastic system

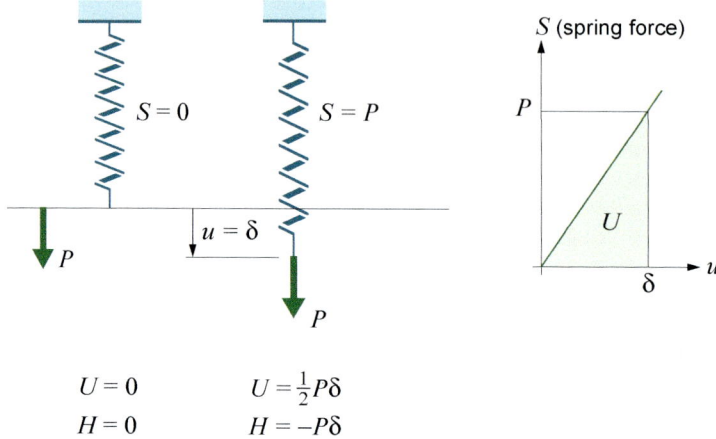

$U = 0$ $U = \frac{1}{2}P\delta$
$H = 0$ $H = -P\delta$

> There may seem to be a discrepancy here. We assume that the load P is applied gradually from zero to its final value and sufficiently slowly to avoid dynamic effects. This leads to the force-displacement diagram shown, and the work done by the force during this loading process is stored in the spring as strain energy (conservation of energy).
>
> So why is there not a factor 1/2 in the expression for the load potential H? It may well be argued that the load's *loss* of capacity to do work is equal to the work it has done, namely $P\delta/2$.
>
> However, and this is important, in assessing the potential energy of the external loading only the *final* or *current position* is of interest; how we got there is irrelevant. Hence no 1/2 factor.

Accuracy and convergence of the R-R process. The technique demonstrated above for a very simple 1D problem can readily be extended to 2 and 3D problems. As already stated the assumed *trial functions* need to

- be sufficiently continuous,
- satisfy the essential boundary conditions, and
- be linearly independent.

We shall come back to what is meant by the first point. In practical applications, polynomials and trigonometric functions are used almost exclusively as trial functions. What about accuracy and convergence? These are vital questions, but difficult to answer conclusively. However, we can say something, and this something is why the R-R process is important for the finite element method.

Suppose we are solving our problem several times, and each time we include a new trial function (as we did in the examples above). We thus form a sequence of trial solutions, and we would expect, or certainly hope, that this sequence converges towards the correct solution, not only for the functional Π (energy), but also for displacements and stresses. Our examples seem to support this expectation. A necessary requirement for convergence is that the set of trial functions is *complete*. This means that the series of trial functions must be such that by including the next function in the series we will *always* come closer to, or at least not further away from the correct values of the displacements (field variables) and their derivatives in the expression for Π. By including sufficiently many trial functions in such a complete set, we should be able to come as close to the correct displacements (and the associated derivatives) as we wish.

Subject to completeness being satisfied, it can be shown that the functional Π, that is the total potential energy, converges (monotonically) towards the correct value for a series of admissible trial functions. The proof is completely dependent on the way the series is formed: whenever a new function is added, previous terms must *not* be removed. In other words, *completeness* demands that *all* admissible functions up to a certain order are present (also the lowest).

Polynomials can form complete series and so can FOURIR series. It should be emphasized that convergence for R-R solutions can, strictly speaking, only be proved for the energy. It can also be shown that a "properly" conducted R-R solution is either correct or too "stiff".

Problems with classical R-R solutions. The method as we have described it above has some serious drawbacks in practical applications:

- For 2 and 3D problems it is quite difficult, even for simple geometries, to choose admissible trial functions, and the integrals in the expression for **K** become troublesome.
- The solution may be too smooth – it is difficult to pick up kinks and discontinuities.
- **K** is always a full matrix.
- Every trial function (φ_i) must satisfy the essential boundary conditions.

Mathematical basics of FEM

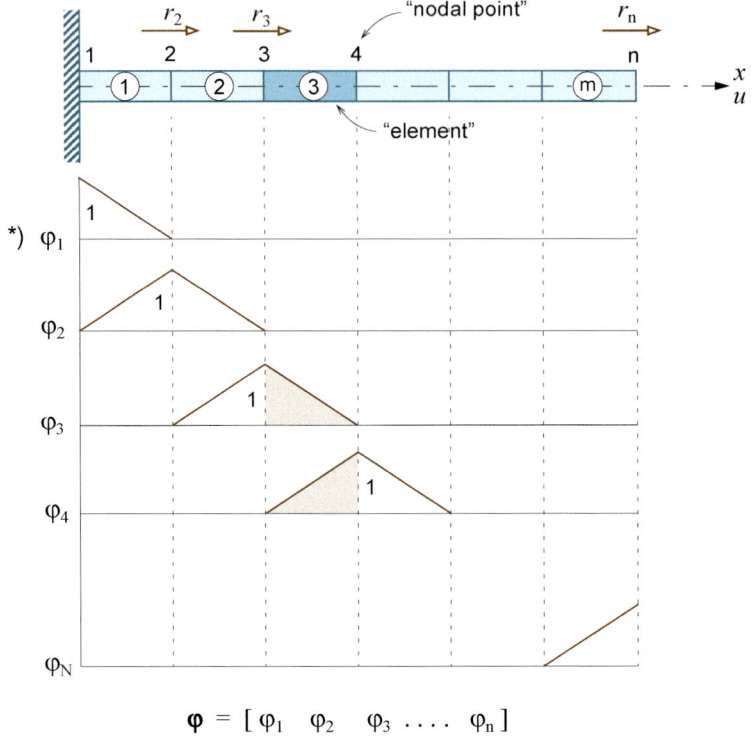

Figure 4.9 Trial functions over sub-domains

*) We have also included r_1, even if we know that it is equal to zero. This is our essential boundary condition which we will have to account for later.

The RAYLEIGH-RITZ method

Most of these problems can be overcome by the following simple remedy:

piece-wise assumption of the displacement field.

This amounts to expressing the displacement field as a sum of (simple) functions defined over sub-domains. With reference to Fig. 4.9 we now assume

$$u(x) \approx \hat{u}(x) = \boldsymbol{\varphi}\mathbf{a} = \boldsymbol{\varphi}\mathbf{r} \quad \text{where} \quad \mathbf{r} = [r_i] \tag{4-22}$$

and r_i is the "nodal point" value of the displacement $\hat{u}$, that is $r_i = \hat{u}(x_i)$. The amplitudes (generalized coordinates) **a** have thus been given a distinct physical meaning, in itself a clear advantage. Other advantages are:

It is quite simple and straightforward to select trial functions, which, in accordance with what has been said about completeness, must be as simple functions as possible – inside every "element" (or sub-domain) it must be possible to express $\hat{u}^e = c_0 + c_1 x$.

Minimum degree of smoothness; the total displacement field does not have a higher order of smoothness (continuity) than the mathematical model requires (with respect to the highest order of derivative present in the functional).

Most of the trial functions φ_i are decoupled, or orthogonal (in an inner product sense); this means that

$$\int \varphi_{i,x} \varphi_{j,x} dx = 0$$

for many combinations of *i* and *j*, which in turn will give us a sparse **K** matrix.

It is much easier to satisfy the essential boundary conditions. While the boundary conditions must be satisfied *a priori* in the classical R-R metod, in that every trial function needs to satisfy this requirement, the sub-domain approach can satisfy this requirement *a posteriori* by specifying "correct" values for those components in **r** that have prescribed essential (geometric) values.

RAYLEIGH-RITZ as a finite element method. We will now demonstrate that a piece-wise or sub-domain based R-R solution can be presented as a "finite element formulated" R-R method or, *vice versa*, as an R-R formulated finite element method.

We take the functional in Eq. (4-17) as our starting point, that is

$$\hat{\Pi} = \frac{1}{2}\mathbf{r}^T \int_0^L (\boldsymbol{\varphi}_{,x})^T EA \boldsymbol{\varphi}_{,x} dx \, \mathbf{r} - \mathbf{r}^T \int_0^L \boldsymbol{\varphi}^T p \, dx \tag{4-23}$$

where we have replaced **a** by **r**. We consider an arbitrary element *e* between nodal points *i* and *i*+1, whose coordinates are x_i and x_{i+1}, and define the quantity

$$\hat{\Pi}_e = \frac{1}{2}\mathbf{r}^T \int_{x_i}^{x_{i+1}} (\boldsymbol{\varphi}_{,x})^T EA \boldsymbol{\varphi}_{,x} dx \, \mathbf{r} - \mathbf{r}^T \int_{x_i}^{x_{i+1}} \boldsymbol{\varphi}^T p \, dx \tag{4-24}$$

Mathematical basics of FEM

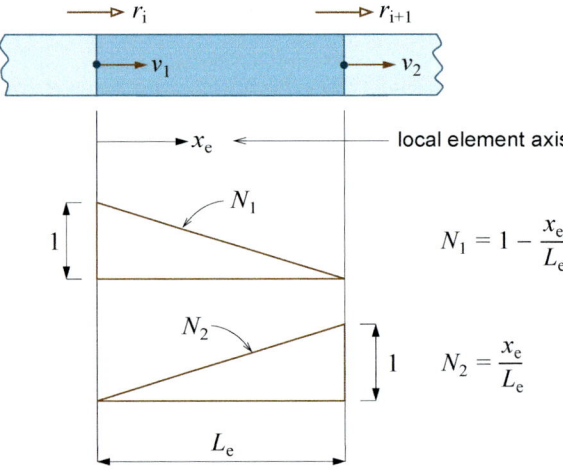

Figure 4.10 Shape functions for a simple 1D axial element

The RAYLEIGH-RITZ method

Can we now claim that $\hat{\Pi} = \sum_e \hat{\Pi}_e$, which is equivalent to stating that

$\int_L (\ )dx = \sum_e \int_{L_e} (\ )dx$? The answer is *yes*, provided the integrals in (4-24) are *finite*. The highest order of derivative in our functional is $m = 1$. Hence we can allow a "jump" (discontinuity) in this quantity, but the jump must be finite. It follows from this that the derivative of order $m - 1 = 1 - 1 = 0$ must be continuous, not only within the elements, but also between the elements. This is called C^0 continuity. For our problem the function ($\hat{u}$) itself must be continuous, whereas we can accept a discontinuous first derivative. Not only can we accept a discontinuous first derivative (which is proportional with the axial stress σ_x), it is in fact an advantage in some cases. Consider for instance a rod with a sudden jump in cross sectional height (or width); it is the axial force and not the stress that is continuous across the jump. Hence if we had a continuous first derivative across the jump we would force the stress to be continuous, which it should not be (a case of so-called over-conformity). This paragraph explains what *sufficiently continuous* means, which was one of the requirements we demanded of the trial functions.

We have here demonstrated a general and very important requirement that our trial functions, or *shape functions* as we shall call them in FEM analyses, need to satisfy in order for the convergence properties of classical R-R to be valid. In addition to the continuity requirement, the shape functions obviously need to be *m*-order differentiable within the elements. We shall come back to a formal and general definition of these requirements in the next chapter. Here it suffices to state that for our problem, for which C^0 continuity is required, the functions of Fig. 4.9 satisfy the continuity requirement.

We will now consider a typical (generic) element of our example problem, see Fig. 4.10, and we will use our "finite element symbols".

The basic assumption for the displacement we express as:

$$\hat{u}(x_e) = N_1 v_1 + N_2 v_2 = \mathbf{Nv} \tag{4-25}$$

where

$$\mathbf{N} = [\ N_1 \quad N_2\ ] \tag{4-26}$$

is the *shape function* matrix, and the vector

$$\mathbf{v} = [\ v_1 \quad v_2\ ] \tag{4-27}$$

contains the nodal displacements (or kinematic degrees of freedom) of the element. The strain is

$$\hat{\varepsilon} = \frac{d\hat{u}}{dx} = \frac{d\hat{u}}{dx_e} = \frac{d}{dx_e}\mathbf{Nv} = \mathbf{Bv} \tag{4-28}$$

where

$$\mathbf{B} = \frac{d}{dx_e}\mathbf{N} = \left[-\frac{1}{L_e} \quad \frac{1}{L_e}\right] \tag{4-29}$$

Mathematical basics of FEM

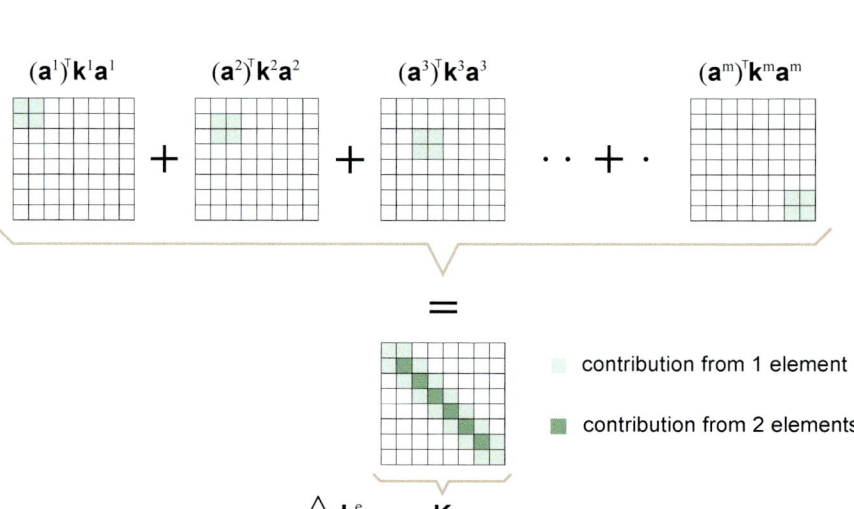

Figure 4.11 The assembly process

The RAYLEIGH-RITZ method

is the *strain-displacement* matrix.

The element functional, that is the element's contribution to the rod's total potential energy, can now be written as

$$\hat{\Pi}_e = \frac{1}{2}\int_0^{L_e} \hat{\sigma}\hat{\varepsilon}\,dx_e - \int_0^{L_e} \hat{u}p(x_e)\,dx_e = \frac{1}{2}\int_0^{L_e} EA\hat{\varepsilon}^2\,dx_e - \int_0^{L_e} \hat{u}p(x_e)\,dx_e$$

It should be emphasized that here we do *not* consider the element as a free body (the element is simply a sub-region of the rod). This equation can be expressed as

$$\hat{\Pi}_e = \frac{1}{2}\mathbf{v}^T\int_0^{L_e} \mathbf{B}^T EA\mathbf{B}\,dx_e\,\mathbf{v} - \mathbf{v}^T\int_0^{L_e} \mathbf{N}^T p(x_e)\,dx_e \qquad (4\text{-}30)$$

This we can write as

$$\hat{\Pi}_e = \frac{1}{2}\mathbf{v}^T\mathbf{k}\mathbf{v} + \mathbf{v}^T\mathbf{S}^0 \qquad (4\text{-}31)$$

where

$$\mathbf{k} = \int_0^{L_e} \mathbf{B}^T EA\mathbf{B}\,dx_e \qquad (4\text{-}32)$$

is the element's *stiffness matrix*, and

$$\mathbf{S}^0 = -\int_0^{L_e} \mathbf{N}^T p(x_e)\,dx_e \qquad (4\text{-}33)$$

is the *consistent* element *load vector*; the sign of $\mathbf{S}^0$ has been chosen in view of previous definitions (see Chapter 2).

What about the corresponding system matrices? If we introduce the connection between element displacements $\mathbf{v}$ and system displacements $\mathbf{r}$,

$$\mathbf{v} = \mathbf{a}^e \mathbf{r} \qquad (4\text{-}34)$$

where $\mathbf{a}^e$ is the *connectivity* matrix, with the same properties as in Chapter 2, then we can write for the complete rod (the system)

$$\hat{\Pi} = \sum_e \hat{\Pi}_e = \frac{1}{2}\mathbf{r}^T\left(\left(\sum_e (\mathbf{a}^e)^T \mathbf{k}\mathbf{a}^e\right)\right)\mathbf{r} + \mathbf{r}^T\left(\sum_e (\mathbf{a}^e)^T \mathbf{S}^0\right) \qquad (4\text{-}35)$$

where

$$\sum_e (\mathbf{a}^e)^T \mathbf{k}^e \mathbf{a}^e = \mathbf{K} = \underset{e}{\mathbf{A}}\,\mathbf{k}^e \qquad (4\text{-}36)$$

and

$$\sum_e (\mathbf{a}^e)^T \mathbf{S}^{0e} = \mathbf{R}^0 = \underset{e}{\mathbf{A}}\,\mathbf{S}^{0e} \qquad (4\text{-}37)$$

are the system stiffness matrix and load vector, respectively. These two equations are the mathematical expressions for the *assembly* process, illustrated in Fig. 4.11 for the stiffness matrix, which is no longer a full matrix. The functional in (4-35) becomes

Mathematical basics of FEM

Recall $(\mathbf{ABC})^T = \mathbf{C}^T\mathbf{B}^T\mathbf{A}^T$

If $\underset{n\times n}{\mathbf{A}} = \mathbf{A}^T$ (symmetric)

and $\underset{n\times m}{\mathbf{B}}$ is an arbitrary matrix

then $\underset{m\times m}{\mathbf{C}} = \mathbf{B}^T\mathbf{A}\mathbf{B}$ is also symmetric

since $(\mathbf{B}^T\mathbf{A}\mathbf{B})^T = \mathbf{B}^T\mathbf{A}^T\mathbf{B} = \mathbf{B}^T\mathbf{A}\mathbf{B}$

$$\hat{\Pi} = \tfrac{1}{2}\mathbf{r}^T\mathbf{K}\mathbf{r} + \mathbf{r}^T\mathbf{R}^0 \qquad (4\text{-}38)$$

A stationary value with respect to a variation of the displacement

$$\delta\hat{\Pi} = \frac{\partial\hat{\Pi}}{\partial\mathbf{r}}\delta\mathbf{r} = 0$$

yields

$$\tfrac{1}{2}(\mathbf{K}\mathbf{r} + \mathbf{K}^T\mathbf{r}) + \mathbf{R}^0 = 0 \quad \text{or} \quad \mathbf{K}\mathbf{r} = -\mathbf{R}^0 \qquad (4\text{-}39)$$

The last transition is valid only if $\mathbf{K}$ is symmetric, which it is, as a direct consequence of Eqs. (4-32) and (4-36). If the rod (system) is also subjected to concentrated axial forces applied at the nodal points (we can always adjust the element lengths to accommodate this), and these forces are contained in the vector $\mathbf{R}^k$, then we need to adjust the load potential in Eq. (4-38), that is

$$\hat{\Pi} = \tfrac{1}{2}\mathbf{r}^T\mathbf{K}\mathbf{r} + \mathbf{r}^T\mathbf{R}^0 - \mathbf{r}^T\mathbf{R}^k \quad \Rightarrow \quad \mathbf{K}\mathbf{r} = \mathbf{R}^k - \mathbf{R}^0 = \mathbf{R} \qquad (4\text{-}40)$$

Before we can solve this equation with respect to the unknown nodal displacements $\mathbf{r}$ we must make sure that the essential (geometric) boundary conditions (here $u = r_1 = 0$ at node 1 of Fig. 4.9) are satisfied (*a posteriori* satisfaction).

The latter part of this section is in fact the FEM version of PMPE (or *vice versa*). In summarizing the section we have, by means of a very simple example, demonstrated the close relationship between the finite element method and the RAYLEIGH-RITZ (R-R) method used in connection with PMPE. This is important when it comes to understanding questions concerning convergence of FEM solutions. We now have a better and more reliable basis for making statements about the key step in the FEM process, the selection of the shape functions $\mathbf{N}$.

4.6 Weighted residual methods – GALERKIN

This section describes very briefly another family of methods, one member of which is quite popular in the FEM world. Again we take as our starting point the one-dimensional differential equation (4-3) representing, together with some specified (essential) boundar conditions, the strong form of a problem defined in the range $g \leq x \leq h$, that is

$$\Lambda u(x) = f(x) \qquad g \leq x \leq h \qquad (4\text{-}41)$$

The residual is defined as

$$r = \Lambda u - f = 0 \qquad (4\text{-}42)$$

Again we assume an approximate solution to be of the form

$$\hat{u} = a_1\varphi_1 + a_2\varphi_2 + \ldots + a_n\varphi_n = \boldsymbol{\varphi}\mathbf{a} \qquad (4\text{-}43)$$

In general $\hat{u}$ does not satisfy (4-42) and we therefore have

Mathematical basics of FEM

Lord RAYLEIGH (John William STRUTT, 1842-1919) was one of the very few members of the English higher nobility (third Baron RAYLEIGH) who won fame as an outstanding scientist.

His younger years were marred with ill-health, but after years of private lessons he entered Trinity College, Cambridge, and graduated in the Mathematical Tripos in 1865.

Lord RAYLEIGH's first research was mainly mathematical, concerning optics and vibrating systems, but his later work ranged over almost the whole field of physics. His most famous work, *Theory of Sound*, was published in two volumes in 1877/78, and his other extensive studies are reported in his *Scientific Papers* – six volumes issued during 1889-1929. 446 papers are reprinted in his collected works.

He was awarded the *Nobel Prize* in Physics for 1904.

Walter RITZ (1878-1909) was a Swiss theoretical physicist.

As a very gifted student the young RITZ excelled academically in his home town of Sion, and in 1897 he entered ETH in Zurich where he began studies in engineering. He was not happy with the approximations and compromises involved with engineering so he switched to more mathematically exacting studies in physics. Albert EINSTEIN was one of his classmates with whom he later (in 1908-1909) battled a "war" (in *Physikalische Zeitschrift*) over, amongst other things, the theoretical origin of the second law of thermodynamics. The discussion ended with an "agreement to disagree".

He is perhaps best known for his work with Johannes RYDBERG on the RYDBERG-RITZ *combination principle* (to explain relationship of the spectral lines for all atoms!). In structural mechanics and structural dynamics he is best known for the RITZ method (also known as the RAYLEIGH-RITZ method), which is a direct method to find an approximate solution for boundary value problems.

Boris Grigoryevich GALERKIN (1871-1945) was a Russian mathematician and engineer. He was trained at Petersburg Technological Institute (PTI) where he graduated in1899, and then went to work as an engineer. In 1907 he was jailed for a year and a half for his (leftish) political activities. In prison he lost interest in politics and turned to science and engineering, and in 1909 he went back to PTI and began teaching. After some years, mostly in engineering, his academic work turned to the area for which he is today best known, approximate solution of differential equations. His well-known method was published in 1915.

In 1920 when he held two chairs, one in elasticity at the Leningrad Institute of Communications Engineers and one in structural mechanics at Leningrad University he was promoted to Head of Structural Mechanics at PTI. At this time BGG was not only a well known scientist, he was also considered an authority among design engineers. GALERKIN is also known for his work on thin elastic plates; his monograph *Thin Elastic Plates* was published in 1937.

From 1940 until his death he was head of the Institute of Mechanics of the Soviet Academy of Sciences.

$$\Lambda\hat{u} - f = \hat{r} \neq 0$$

where the residual $\hat{r}$ is a measure of the *error*.

Next we multiply (4-42) by an arbitrary *weight function* v and integrate the product over the problem domain, i.e.,

$$\int_g^h v(\Lambda u - f)dx = 0 \qquad (4\text{-}44)$$

This equation is mathematically equivalent to (4-42), but it is *not* the weak form of the problem. Using the approximate residual, Eq. (4-44) becomes

$$\int_g^h v(\Lambda\hat{u} - f)dx = \int_g^h v(x)\hat{r}(x)dx = 0 \qquad (4\text{-}45)$$

The interpretation of this equation, which will be used to determine the parameters **a**, is that the residual ($\hat{r}$) is given a certain weight (v) and that the integral of this *weighted residual* over the problem domain is required to be zero. The various weighted residual methods differ in how the weight function v is chosen.

It is convenient to express the weight function as a sum of functions ψ_i,

$$v = \psi_1(x)b_1 + \psi_2(x)b_2 + \ldots = \boldsymbol{\psi}\mathbf{b} \qquad (4\text{-}46)$$

This is similar to Eq. (4-14). ψ_i is an assumed (known) function of x, and b_i is a parameter independent of x. Since

$$v = \boldsymbol{\psi}\mathbf{b} = \mathbf{b}^T\boldsymbol{\psi}^T$$

we can write (4-45) as

$$\mathbf{b}^T\int_g^h \boldsymbol{\psi}^T\hat{r}\,dx = 0$$

This equation must hold for *any* parameters **b**; hence

$$\int_g^h \boldsymbol{\psi}^T\hat{r}\,dx = \int_g^h \boldsymbol{\psi}^T(\Lambda\boldsymbol{\varphi}\mathbf{a} - f)dx = 0$$

or

$$\int_g^h \boldsymbol{\psi}^T\Lambda\boldsymbol{\varphi}\mathbf{a}\,dx = \int_g^h \boldsymbol{\psi}^T f\,dx \qquad (4\text{-}47)$$

There are various methods of solving this equation with respect to the "amplitudes" **a** of the trial functions $\boldsymbol{\varphi}$, for instance *collocation methods* (point and sub-domain collocation) and *least squares methods*. In our context, however, the method of interest is due to GALERKIN. In this method the weight functions $\boldsymbol{\psi}$ are chosen to be the same as the trial functions $\boldsymbol{\varphi}$, i.e.,

Mathematical basics of FEM

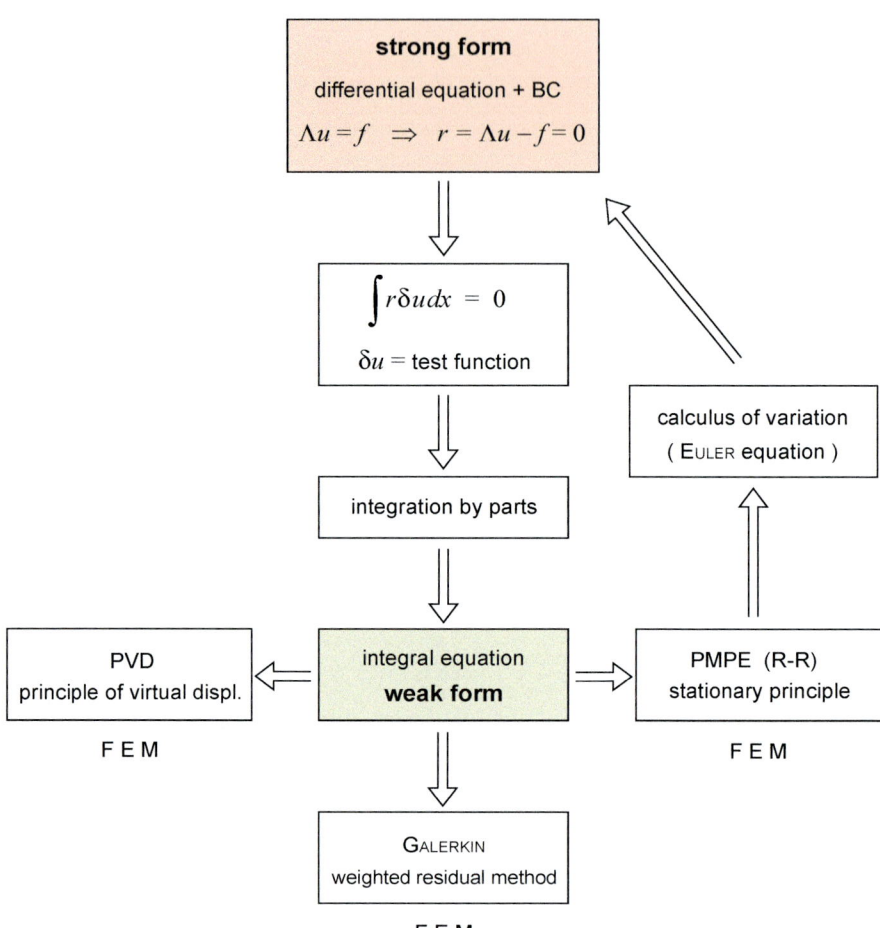

$$\text{GALERKIN:} \quad \psi = \varphi \qquad (4\text{-}48)$$

and

$$\int_g^h \varphi^T A \varphi \mathbf{a}\, dx = \int_g^h \varphi^T f\, dx \qquad (4\text{-}49)$$

Applied to this equation the method will produce a *non-symmetric* matrix of coefficients. This drawback is overcome by applying the method to the *weak form* of the problem. This form is obtained by partial integration of the left-hand side of Eq. (4-49), an operation that will produce the same system of equations for the solution of **a** as the one obtained by PMPE.

We notice that GALERKIN's method needs no functional, and the method is therefore applicable to most physical boundary value problems that can be described by differential equations.

We also notice that in this and the previous section we have not made use of an important concept in matrix structural analysis, namely that of *separation* (into individual elements), *element analysis* and finally *assembly* of the system matrices by satisfying compatibility and equilibrium at the system nodal points. Here the elements are more "mathematical" than they were in our physical approach in Chapter 2, in that they merely define sub-domains for the interpolation and integration processes. However, as will be demonstrated in the next section, the equations we have arrived at are the same as those which a more physical approach, based on the principle of virtual displacements (PVD), will produce.

4.7 General formulation of FEM using PVD

The principle of minimum potential energy (PMPE) and GALERKIN's method are commonly used approaches to finite element analysis in engineering. A third approach makes use of *virtual work*, and if we limit ourselves to the *displacement* method of analysis, as we shall, it is the *principle of virtual displacements*, VPD, that comes into play.

If we go back to Section 4.3 and replaces δu by a *virtual displacemet* $\tilde{u}$ and go through the same mathematical manipulations (integration by parts) we would end up with the weak form of the problem representing the principle of virtual displacements. Instead of repeating this process we go straight to the principle itself, PVD, which we can state as follows:

> If a system of *real* external and internal *forces* (stresses) that are in *static equilibrium* are subjected to *any* set of *virtual*, but *kinematically compatible*, displacements and deformations, then the *virtual work* performed by the real external forces over the virtual displacements is *equal* to the *virtual work* performed by the real internal forces (stresses) over the virtual deformations (strains).

We now apply the principle to a *single* element, not to the entire problem domain as we did for both PMPE and GALERKIN's method. Figure 4.12 shows an arbitrary, but general finite element that has been "cut out" from the element mesh; in other words, we consider the element as a free-body.

Mathematical basics of FEM

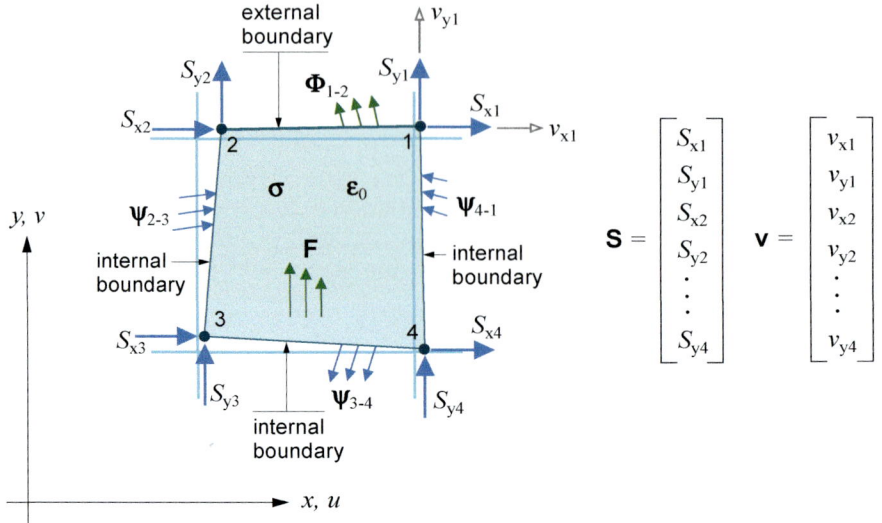

Figure 4.12 An arbitrary finite element as a free-body

Eq. (3-17): $\varepsilon = \Delta \mathbf{u}$

Eq. (3-36): $\sigma = \mathbf{C}(\varepsilon - \varepsilon_0)$

The element has three internal boundaries along which "cuts" have been made. The resultants of the stresses or tractions ($\psi_{i\text{-}j}$) along these internal boundaries are the nodal forces **S**, which, in this context, are *external* forces acting on the element. Other external forces are the volume forces **F** and the tractions ($\Phi_{1\text{-}2}$) on the external (free) boundary. The *internal* forces are the stresses σ which are in equilibrium with the external forces. The element is also subjected to initial strains, represented by ε_0. The displacement components, *u* and *v*, are contained in **u** and the nodal displacements in **v**; the real strains, compatible with **u**, are contained in ε.

For the *real* stresses and strains we recall the two basic equations, (3-17) and (3-36), and we also recall the basic *assumption*, Eq. (4-1) which reads

$$\mathbf{u} = \mathbf{N}\mathbf{v}$$

Substituting this into (3-17) gives

$$\varepsilon = \Delta\mathbf{u} = \Delta(\mathbf{N}\mathbf{v}) = \Delta\mathbf{N}\mathbf{v} = \mathbf{B}\mathbf{v} \qquad (4\text{-}50)$$

where

$$\mathbf{B} = \Delta\mathbf{N} \qquad (4\text{-}51)$$

is the general expression for the strain-displacement matrix.

We now subject the element to a set of virtual nodal displacements $\tilde{\mathbf{v}}$, and by choosing the virtual displacements inside the element as

$\tilde{\mathbf{u}} = \mathbf{N}\tilde{\mathbf{v}}$, the corresponding virtual strains become $\tilde{\varepsilon} = \mathbf{B}\tilde{\mathbf{v}}$,

and we have secured the kinematic compatibility requirement for the virtual displacements[1].

PVD states that the virtual work performed by the real external forces over the virtual displacements ($\tilde{\mathbf{u}}$ and $\tilde{\mathbf{v}}$) is equal to the virtual work performed by the real internal stresses over the virtual strains ($\tilde{\varepsilon}$), compatible with ($\tilde{\mathbf{u}}$ and $\tilde{\mathbf{v}}$), that is $\tilde{W}_e = \tilde{W}_i$.

For our element this work equation reads (S_T is the part of the element surface subjected to external distributed forces or tractions)

$$\mathbf{S}^T\tilde{\mathbf{v}} + \int_{V_e}\mathbf{F}^T\tilde{\mathbf{u}}\,dV + \int_{S_T}\Phi^T\tilde{\mathbf{u}}\,dS = \int_{V_e}\sigma^T\tilde{\varepsilon}\,dV \qquad (4\text{-}52)$$

or, more conveniently (remember that work is a scalar),

$$\tilde{\mathbf{v}}^T\mathbf{S} + \int_{V_e}\tilde{\mathbf{u}}^T\mathbf{F}\,dV + \int_{S_T}\tilde{\mathbf{u}}^T\Phi\,dS = \int_{V_e}\tilde{\varepsilon}^T\sigma\,dV$$

Substituting for $\tilde{\mathbf{u}}$, $\tilde{\varepsilon}$ and σ, and rearranging give

---

1. All virtual quantities are marked by a tilde (~) above their respective symbols.

Mathematical basics of FEM

> It should be emphasized that the virtual displacements $\tilde{\mathbf{v}}$, and hence $\tilde{\mathbf{u}}$ have absolutely nothing to do with the real forces ($\mathbf{R}$ and $\mathbf{S}$). This also explains the lack of a factor 1/2 in the work expressions; the final real forces do work over the complete virtual (imaginary) displacements.
>
> Our choice of expressing the virtual displacements as $\tilde{\mathbf{u}} = \mathbf{N}\tilde{\mathbf{v}}$, and hence $\tilde{\boldsymbol{\varepsilon}} = \mathbf{B}\tilde{\mathbf{v}}$, is motivated by the requirement that the virtual displacements must satisfy kinematic compatiblity (which is basically the only requirement they need to satisfy); this is a simple way to achieve this. An important consequence of this choice is that it leads to a symmetric stiffness matrix.

$$\tilde{\mathbf{v}}^T \mathbf{S} = \tilde{\mathbf{v}}^T \int_{V_e} \mathbf{B}^T \mathbf{C}(\mathbf{B}\mathbf{v} - \boldsymbol{\varepsilon}_0) dV - \tilde{\mathbf{v}}^T \int_{V_e} \mathbf{N}^T \mathbf{F} dV + \tilde{\mathbf{v}}^T \int_{S_T} \mathbf{N}^T \boldsymbol{\Phi} dS$$

This equation must be valid for any $\tilde{\mathbf{v}}$, hence

$$\mathbf{S} = \int_{V_e} \mathbf{B}^T \mathbf{C}\mathbf{B} dV \mathbf{v} - \int_{V_e} \mathbf{B}^T \mathbf{C}\boldsymbol{\varepsilon}_0 dV - \int_{V_e} \mathbf{N}^T \mathbf{F} dV + \int_{S_T} \mathbf{N}^T \boldsymbol{\Phi} dS = \mathbf{k}\mathbf{v} + \mathbf{S}^0 \qquad (4\text{-}53)$$

where

$$\mathbf{k} = \int_{V_e} \mathbf{B}^T \mathbf{C}\mathbf{B} dV \qquad (4\text{-}54)$$

is the element *stiffness matrix*, and

$$\mathbf{S}^0 = \mathbf{S}^0_{\varepsilon 0} + \mathbf{S}^0_F + \mathbf{S}^0_\Phi \qquad (4\text{-}55)$$

is the *consistent* element *load vector* where the contributions are

$$\mathbf{S}^0_{\varepsilon 0} = -\int_{V_e} \mathbf{B}^T \mathbf{C}\boldsymbol{\varepsilon}_0 dV \qquad (4\text{-}56a)$$

$$\mathbf{S}^0_F = -\int_{V_e} \mathbf{N}^T \mathbf{F} dV \qquad (4\text{-}56b)$$

$$\mathbf{S}^0_\Phi = -\int_{S_T} \mathbf{N}^T \boldsymbol{\Phi} dS \qquad (4\text{-}56c)$$

due to initial strain, volume forces and surface forces (tractions), respectively. In the element stiffness relation

$$\mathbf{S} = \mathbf{k}\mathbf{v} + \mathbf{S}^0 \qquad (4\text{-}57)$$

the first term $\mathbf{k}\mathbf{v}$ represents the elastic forces due to the deformations ($\mathbf{v}$), whereas the last term $\mathbf{S}^0$ represents the forces due to "external" actions with $\mathbf{v} = \mathbf{0}$; in other words, $\mathbf{S}^0$ are the "fixed end forces" (in line with the beam element terminology).

It is worth noting that we have satisfied equilibrium *indirectly*, through the virtual work equation (4-52); in other words, PVD "replaces" the equilibrium equations.

An important feature of PVD is that it is *independent* of the material law; the principle does not require linear elastic material the way PMPE does.

The derivation above is completely general and applies to any displacement based element, whether 1-, 2- or 3-dimensional. The remaining system analysis, that is assembly, handling of boundary conditions and solution of the system of equations, is exactly the same as described in Chapter 2 for 1D elements. However, the recovery of stresses, once the displacements have been determined, differs considerably from the simple procedure used for

Mathematical basics of FEM

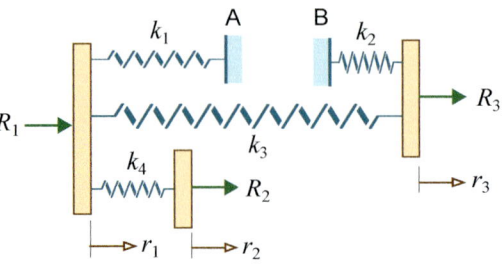

Figure 4.13 System of springs

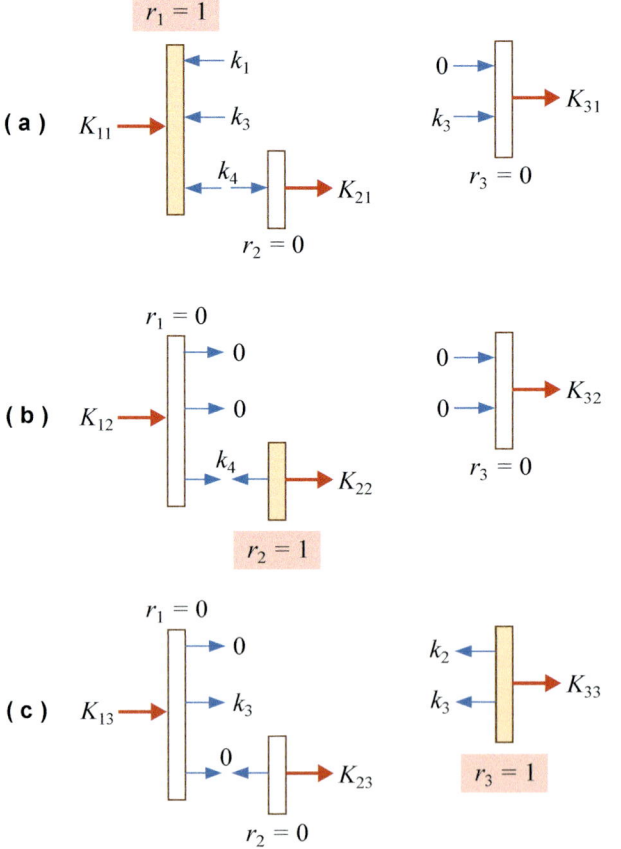

Figure 4.14 Forces (per unit length) due to unit displacements

Example 4-1 - A simple system of springs

beam and bar elements. We shall come back to this issue in some detail in Chapter 7.

For linear static analysis of structural mechanics problems by finite element methods, any one of the three methods or principles dealt with in this chapter will do. Which one to use is almost a question of taste. Since this text is primarily aimed at engineers we favour a "physical" approach and we would also like to keep as close as possible to the way of thinking described in Chapter 2. PVD is therefore our preferred choice.

Example 4-1 - A simple system of springs

Figure 4.13 shows a system of elastic springs. Three rigid bodies that can only move horizontally are interconnected and connected to two fixed points A and B, via 4 elastic springs with stiffness k_j. Each rigid body is subjected to an external force (R_i) in the direction of its degree of freedom (r_i).

Establish the stiffness matrix **K** in the relation **Kr = R**,

a) by simple equilibrium, and
b) by use of PMPE.

Solution

Equilibrium

We give the rigid bodies (representing the problem *dofs*) unit displacements in turn, as shown in Fig. 4.14. For these displacement patterns the elastic forces (per unit displacement) exerted by the springs on the rigid bodies are known, and by definition the stiffness coefficient K_{ij} is the force at *dof i*, due to a unit displacement (r_j) at *dof j*, that will have to act in order to maintain equilibrium of the rigid body corresponding to *dof i* when in displaced position.

Figure 4.14a provides the first column of **K**, 4.14b the second and 4.14c the third. Hence

$$\mathbf{K} = \begin{bmatrix} k_1 + k_3 + k_4 & -k_4 & -k_3 \\ -k_4 & k_4 & 0 \\ -k_3 & 0 & k_2 + k_3 \end{bmatrix}$$

PMPE

Strain energy: $\quad U = \frac{1}{2}k_1 r_1^2 + \frac{1}{2}k_2 r_3^2 + \frac{1}{2}k_3(r_3 - r_1)^2 + \frac{1}{2}k_4(r_2 - r_1)^2$

Load potential: $\quad H = -R_1 r_1 - R_2 r_2 - R_3 r_3$

Total potential energy: $\quad \Pi = U + H$

PMPE: $\quad \delta\Pi = \dfrac{\partial \Pi}{\partial \mathbf{r}} \delta\mathbf{r} = \mathbf{0} \quad \Rightarrow \quad \dfrac{\partial \Pi}{\partial r_i} = 0 \quad (i = 1, 2, 3)$

$\dfrac{\partial \Pi}{\partial r_1} = k_1 r_1 - k_3 r_3 + k_3 r_1 - k_4 r_2 + k_4 r_1 - R_1 = 0$

Mathematical basics of FEM

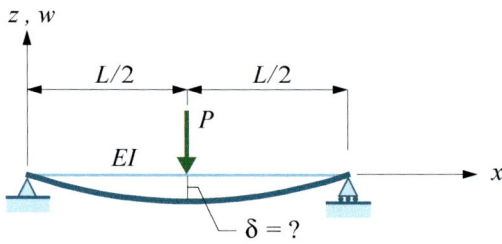

Figure 4.15 Simply supported beam problem

Strain energy (internal work) for a 2D beam:

Kinematics: $\varepsilon = \dfrac{du}{dx} = z\dfrac{d\theta}{dx} = -z\dfrac{d^2w}{dx^2} = -zw''$

Material law: $\sigma = E\varepsilon = -zEw''$

Hence:

$$U = \int_0^L \int_A \tfrac{1}{2}\sigma\varepsilon\, dA\, dx = \tfrac{1}{2}\int_0^L \int_A (-zEw'')(-zw'')\, dA\, dx = \tfrac{1}{2}\int_0^L EI(w'')^2\, dx$$

since

$$I = \int_A z^2\, dA$$

$$\frac{\partial \Pi}{\partial r_2} = k_4 r_2 - k_4 r_1 - R_2 = 0$$

$$\frac{\partial \Pi}{\partial r_3} = k_2 r_3 + k_3 r_3 - k_3 r_1 - R_3 = 0$$

In matrix notation this becomes

$$\begin{bmatrix} k_1 + k_3 + k_4 & -k_4 & -k_3 \\ -k_4 & k_4 & 0 \\ -k_3 & 0 & k_2 + k_3 \end{bmatrix} \begin{bmatrix} r_1 \\ r_2 \\ r_3 \end{bmatrix} = \begin{bmatrix} R_1 \\ R_2 \\ R_3 \end{bmatrix} \quad \text{or} \quad \mathbf{Kr = R}$$

and we have the same **K** as obtained by direct equilibrium.

Example 4-2 - Simply supported beam by R-R

Figure 4.15 shows a simply supported beam subjected to a concentrated force P at mid-span; we seek the maximum lateral displacement δ. This is of course a trivial problem, for which the correct solution is $PL^3/48EI$; it is, however, well suited for demonstrating some features of the classical R-R method.

Solution

First we recognize the essential boundary conditions which are: $w(0) = w(L) = 0$.

If we decide to use polynomial trial functions, what is the simplest admissible function we can use? A linear function cannot satisfy the essential boundary conditions, so our best bet is to assume a quadratic (parabolic) function:

$$\hat{w} = a_0 + a_1 x + a_2 x^2$$

BC: $w(0) = 0 \quad \Rightarrow \quad a_0 = 0$

$\quad\quad w(L) = 0 \quad \Rightarrow \quad a_1 L + a_2 L^2 = 0 \quad \Rightarrow \quad a_1 = -a_2 L$

Hence: $\hat{w} = a_2(x^2 - xL)$ is the lowest order trial function we can use.

$$\hat{w}'' = 2a_2$$

Strain energy: $\quad U = \frac{1}{2}\int_0^L EI(\hat{w}'')^2 dx = 2EIL a_2^2$

Load potential: $\quad H = -(-P)\hat{w}(L/2) = P(a_1 L/2 + a_2 L^2/4) = -P a_2 L^2/4$

Potential energy: $\quad \Pi = U + H = 2EIL a_2^2 - P a_2 L^2/4$

PMPE: $\quad \dfrac{\partial \Pi}{\partial a_2} = 4EIL a_2 - PL^2/4 = 0 \quad \Rightarrow \quad a_2 = \dfrac{PL}{16EI}$

This gives the following solution for the lateral displacement,

$$\hat{w} = \frac{PL}{16EI}(x^2 - xL) \quad \Rightarrow \quad \hat{w}_{max} = -\frac{PL^3}{64EI} < -\frac{PL^3}{48EI}$$

Mathematical basics of FEM

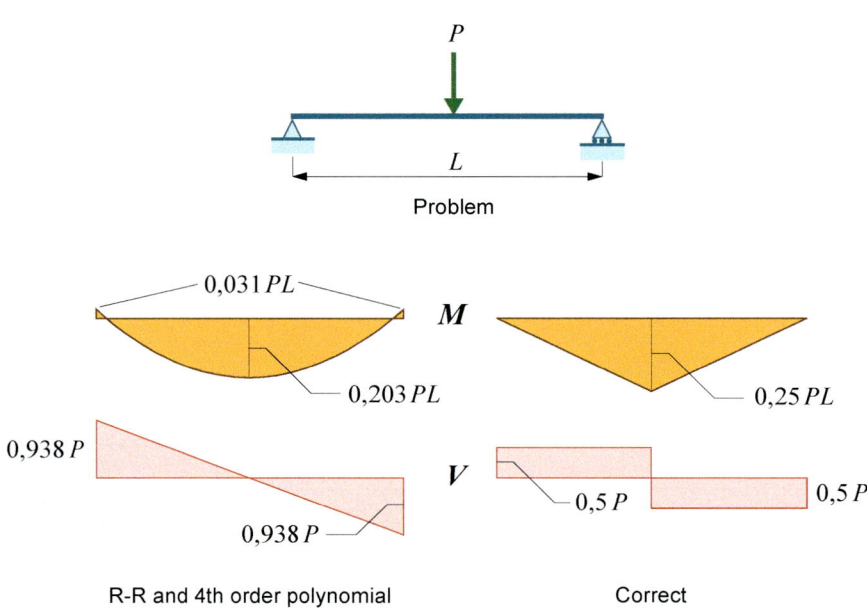

Figure 4.16 Bending moment (M) and shear force (V) diagrams

Example 4-2 - Simply supported beam by R-R

The assumed parabolic displacement is reasonable, although 25 per cent too stiff. If we were to use the moment curvature relation, $M = -EIw''$, with the approximate displacement $\hat{w}$, we would get the rather useless result of $M = -PL/8$ for all x!

In order to improve the solution, the obvious "next" choice will be to assume

$$\hat{w} = a_0 + a_1 x + a_2 x^2 + a_3 x^3$$

For this expression to satisfy the essential boundary conditions we need to have

$w(0) = 0 \quad \Rightarrow \quad a_0 = 0$

$w(L) = 0 \quad \Rightarrow \quad a_1 L + a_2 L^2 + a_3 L^3 = 0 \quad \Rightarrow \quad a_1 = -a_2 L - a_3 L^2$

If we proceed with this assumption for w we will find that $a_3 = 0$, and a_2 is the same as we obtained with the previous (quadratic) assumption. This does *not* contradict the alleged properties of the R-R method. We have not come out *worse* by including an extra term.

So what next? Again the obvious choice seems to be

$$\hat{w} = a_0 + a_1 x + a_2 x^2 + a_3 x^3 + a_4 x^4$$

It should be noted that we need to include the cubic term. The fact that the process cancelled this term in our previous assumption does not mean that it will also be inactive here. If we proceed with this assumption we will find (after some tedious, but straightforward operations) the solution (note the cubic term),

$$\hat{w} = \frac{PL^3}{64EI}\left(-4\left(\frac{x}{L}\right) - \left(\frac{x}{L}\right)^2 + 10\left(\frac{x}{L}\right)^3 - 5\left(\frac{x}{L}\right)^4\right) \text{ with a value at mid-span of}$$

$$\hat{w}_{x=L/2} = -\frac{21}{1024}\frac{PL^3}{EI} = -\frac{PL^3}{48,8EI} \text{ , which is only 1,6 \% off the correct value.}$$

The second derivative, which is proportional to the bending moment ($M = -EIw''$) is

$$\hat{w}'' = \frac{PL}{64EI}\left(-2 + 60\frac{x}{L} - 60\left(\frac{x}{L}\right)^2\right)$$

This gives a bending moment diagram as shown in Fig. 4.16. The shear force, which is the derivative of the bending moment (the third derivative of w), is also shown in Fig. 4.16 and it is, as expected, worse than the moment. All in all perhaps not very impressive for such a simple example. We should keep in mind, however, that concentrated loading as we have here is difficult for most approximate methods. We would find that if we replaced P with a uniformly distributed load over the length of the beam, our approximate solutions would improve quite noticeably.

As a last example we try the following simple trigonometric trial function

$$\hat{w} = a_1 \sin\frac{\pi x}{L} \text{ which satisfies the essential BC explicitly.}$$

If we carry out the differentiations and integration we find

$$a_1 = -\frac{2PL^3}{\pi^4 EI} \text{ and } \hat{w} = -\frac{2PL^3}{\pi^4 EI}\sin\frac{\pi x}{L}$$

and the displacement at mid-span is

$$\hat{w}_{x=L/2} = -\frac{PL^3}{48,7EI} \text{ , which is quite good, and as good as a quartic polynomial.}$$

Mathematical basics of FEM

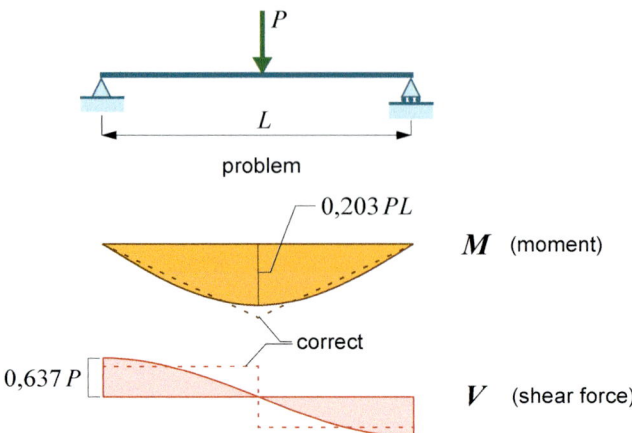

Figure 4.17 R-R solution with one trial function ($= a_1 \sin\frac{\pi x}{L}$)

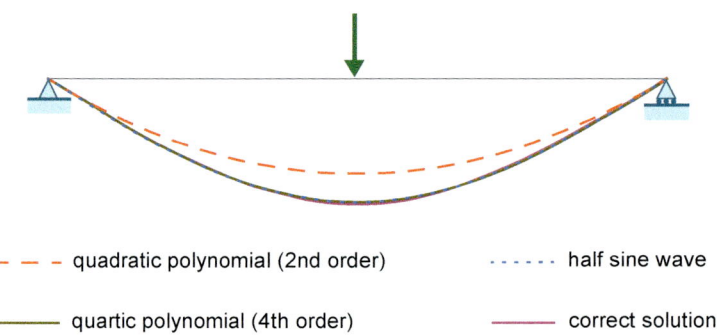

— — — quadratic polynomial (2nd order)   · · · · · · half sine wave

——— quartic polynomial (4th order)   ——— correct solution

Figure 4.18 Lateral displacements by different R-R solutions

Example 4-2 - Simply supported beam by R-R

The bending moment diagram is shown in Fig. 4.17 along with the shear force diagram. We see that the maximum value of M is the same as the one obtained by a quartic function for w (involving three generalized coordinates), but here the end values are correct. What is perhaps more interesting is that even the shear force is quite reasonable, and clearly better than that obtained by the quartic polynomial.

Figure 4.18 shows the lateral displacements obtained by the three different R-R solutions compared with the correct beam solution. Apart from the parabolic trial function, which gives far too stiff a solution, it is quite interesting to see how close the latter two are to each other and also to the correct solution. Even so, once we start to differentiate, the differences become very noticeable.

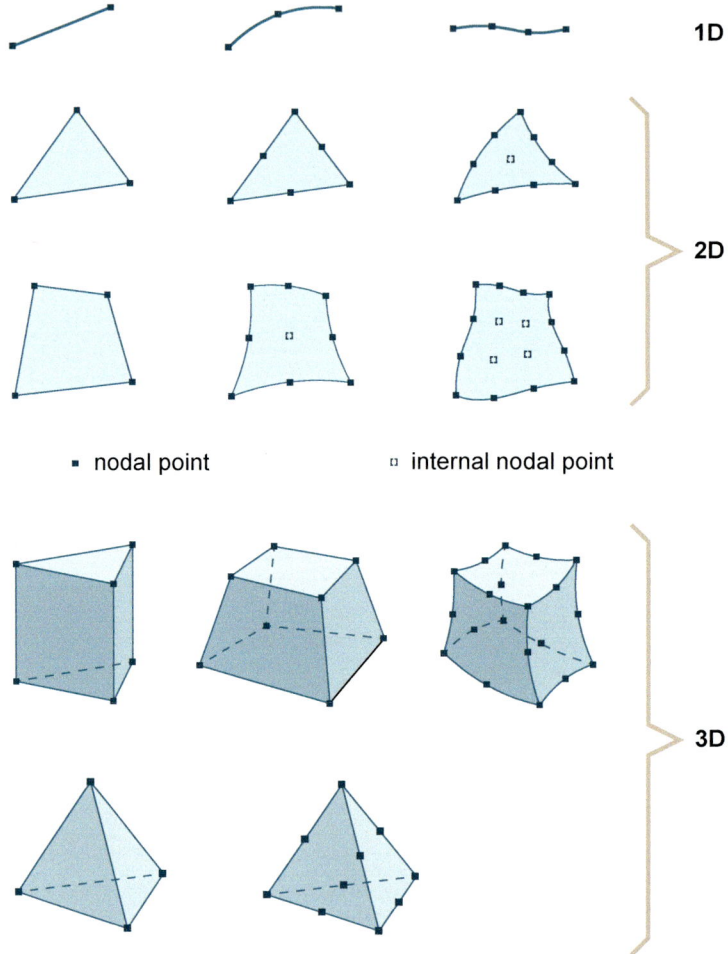

Figure 5.1 Typical element geometries and nodal point positions

5
Element analysis I
Natural coordinates and interpolation

In this and the next two chapters we focus on the "tools of the trade" for the development of displacement based finite elements. We start by reviewing types and classes of elements, and then we go on and try to establish general instruments and techniques applicable to most displacement type elements. Natural coordinates, shape functions and interpolation are key words for this chapter. Elements for specific applications will be dealt with in later chapters.

In this chapter we are concerned with problems associated with generation of the matrices **k** and $\mathbf{S}^0$ for displacement type elements in general. We attempt generality, but the emphasis will be on plane (2D) elements. The central issue here is the assumed *shape functions* (**N**) and thus principles for interpolation.

5.1 Types and classification of elements

Characteristics of an individual element:

Dimension – 1, 2 or 3 dimensions, see Fig. 5.1.

Nodal points or *nodes* – number and position. The nodes define the element geometry and, via their degrees of freedom, the coupling to neighbouring elements. They are usually placed at the end of 1D elements or at the corner points of 2 and 3D elements, but also on edges and surfaces (3D) and sometimes at the interior of the element.

Geometry – usually simple geometric shapes defined by the position of the nodal points. 1D elements, edges (2 and 3D elements) and surfaces (3D elements) may be curved.

Kinematic degrees of freedom (dofs) **v** – define, via shape functions, the state of the field variable(s) (displacements) of the entire element. Usually the *dofs* are the nodal values of the field variables and their derivatives that need to be continuous. However, higher derivatives than those dictated by continuity requirements are also used.

Dofs that are independent of nodal points (*nodeless dofs*) are sometimes used.

Element analysis I - Natural coordinates and interpolation

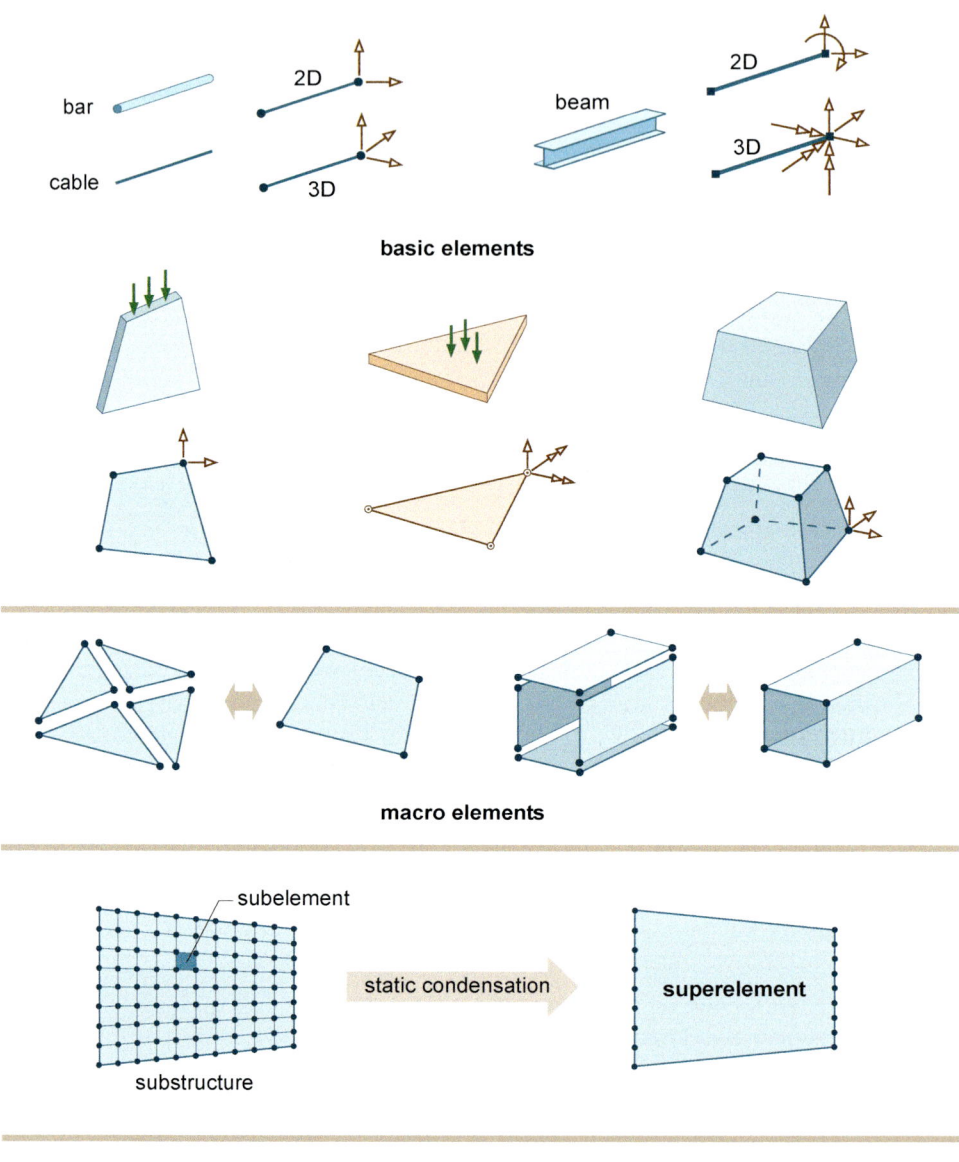

Figure 5.2 Classes of finite elements

Nodal forces **S** – one-to-one correspondence with the nodal *dofs* **v**.

Specific element properties are material properties (usually constant), expressed by the relationship between stress and strain (matrix **C**), and geometric data associated with the dimension of the element, *e.g.* cross section properties for a beam element or the thickness of a plate element.

Classification of elements. Figure 5.2 shows one way of classifying finite elements for structural analysis.

Basic elements are the "primitive" elements in the sense that they cannot be broken down into "smaller" entities. Their matrices are established on the basis that the element is *one* entity and the assumptions made apply equally to all points of this entity. Primitive does not reflect on the complexity or sophistication of the element – a primitive element can be very complex indeed.

Macro elements are similar to basic elements, and they are used in modelling the structural problem in exactly the same way as basic elements. The difference is in their "creation"; macro elements are formed through some special assembly of basic elements that often impose certain constraints.

Superelements are similar to macro elements in that they are also formed by assembling other elements and then *eliminating* (through a substructure analysis) *internal dofs*. However, there is a subtle, but important difference. The substructure analysis, which is basically a *static condensation* of certain *dofs* that are local to the substructure, is "exact" in the sense that it does not introduce new assumptions or constraints of any kind. If the subelements in the substructure in Fig. 5.2 are all basic elements, the resulting superelement is a *first level* superelement. Such an element can be used together with other superelements and possibly basic elements to form a 2nd level superelement, and so on.

Special elements are elements developed for particular problems as opposed to basic elements that are general purpose elements. Some special elements are indicated in Fig. 5.2; there may be other.

We shall for the most part concentrate on basic elements, but superelements and some special elements will also be mentioned.

5.2 Natural coordinates

Equations (4-54) and (4-56), in which **N** contains the shape functions and **B** is the strain-displacement matrix defined by Eq. (4-51), express the mathematical problem we have to solve in order to determine the element matrices. It is implied that the element degrees of freedom (**v**) refer to the global cartesian reference system (x, y, z), and so does the operator Δ. For the interpolation process, that is the determination of the shape functions, and for the integration process, the cartesian reference coordinates are not very suitable. These processes are more conveniently (and numerically robustly) handled in local, dimensionless or *natural*, coordinates associated with the element shape rather than its real size and orientation in space.

Element analysis I - Natural coordinates and interpolation

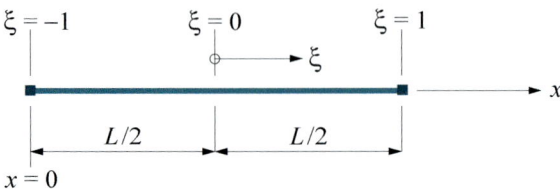

Figure 5.3 Natural coordinate (ξ) for a one-dimensional element

Chain rule:

Let $w = f(x,y)$ and let $x = x(t)$ and $y = y(t)$

Then $\quad \dfrac{dw}{dt} = \dfrac{\partial w}{\partial x}\dfrac{dx}{dt} + \dfrac{\partial w}{\partial y}\dfrac{dy}{dt}$

 **Carl Gustav Jacob JACOBI** (1804-1851) was a German mathematician. He had a reputation of being a very inspiring teacher, and he was considered to be one of the greatest mathematicians of his generation.
He has a matrix and its determinant and a crater on the moon called after him.

Natural coordinates

In this section we will present natural coordinate systems for the most common element shapes and we start with the simplest of them all, the straight one dimensional element.

One-dimensional element. A straight one-dimensional element of length L is shown in Fig. 5.3. Without loss of generality we assume a local cartesian coordinate system where the x-axis coincides with the element axis and the origin is placed at the left-hand end of the element. The most common *normalized* or *natural coordinate* (ξ) along the element has its origin at the mid-point and varies from -1 to $+1$, as shown. This gives the following relationship between ξ and x:

$$\xi = \frac{2x}{L} - 1 \tag{5-1}$$

The strain is the derivative with respect to x, and we therefore need to determine d/dx. The *chain rule* gives

$$\frac{d}{dx} = \frac{d}{d\xi}\frac{d\xi}{dx} = J^{-1}\frac{d}{d\xi} = \frac{2}{L}\frac{d}{d\xi} \tag{5-2}$$

where

$$J^{-1} = \frac{1}{J} = \frac{d\xi}{dx} = \frac{2}{L} \tag{5-3}$$

In multi-dimensional problems J is a matrix, but for a one-dimensional problem it is a scalar and it therefore represents both the *Jacobian matrix* and its determinant, the *Jacobian determinant* (for a scalar these two quantities coincide). The Jacobian determinant is often referred to as simply the *Jacobian*. In our case it is equal to $L/2$ and we have

$$dx = J d\xi = \frac{L}{2}d\xi \tag{5-4}$$

Consider the integral from 0 to L of an arbitrary function

$$f = f_x(x) = f_\xi(\xi)$$

that is

$$\int_0^L f\,dx = \int_0^L f_x(x)\,dx = \int_{-1}^1 f_\xi(\xi) J\,d\xi \tag{5-5}$$

Example 5-1

$$f_x(x) = x^2 \quad \Rightarrow \quad f_\xi(\xi) = \frac{L^2}{4}(\xi^2 + 2\xi + 1)$$

$$\int_0^L x^2\,dx = \frac{1}{3}L^3 \quad \text{and} \quad \int_{-1}^1 \frac{L^2}{4}(\xi^2 + 2\xi + 1)\overset{J}{\frac{L}{2}}d\xi = \frac{L^3}{8}\left[\frac{1}{3}\xi^3 + \xi^2 + \xi\right]_{-1}^1 = \frac{1}{3}L^3$$

We see that the Jacobian is a *scale factor* indicating the physical length (dimension) of dx on the dimensionless reference length $d\xi$.

141

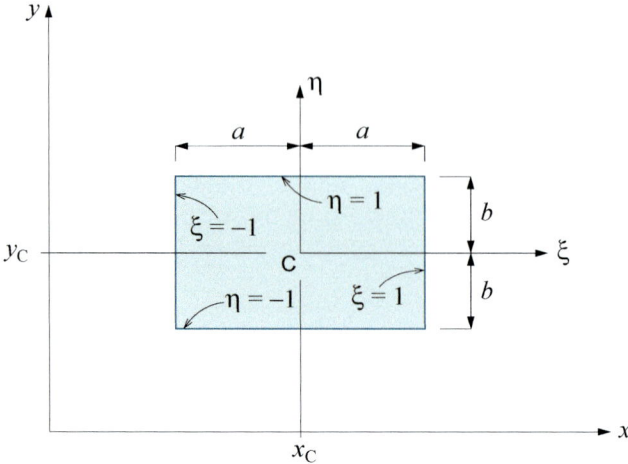

Figure 5.4 Natural coordinates (ξ, η) for a plane rectangle

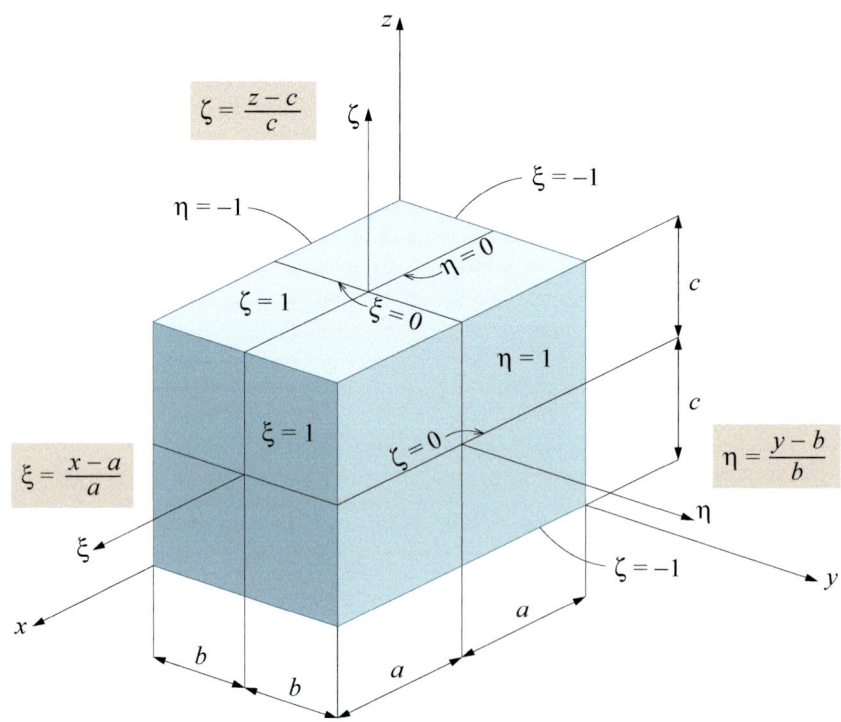

Figure 5.5 Natural coordinates (ξ, η, ζ) for a cuboid

Plane (2D) rectangle. In line with the definition of the natural coordinate (ξ) for a one-dimensional element, the natural coordinates for a plane rectangle, ξ and η, are defined as shown in Fig. 5.4:

$$\xi = \frac{x - x_C}{a} \quad \Rightarrow \quad d\xi = \frac{1}{a}dx \tag{5-6}$$

$$\eta = \frac{y - y_C}{b} \quad \Rightarrow \quad d\eta = \frac{1}{b}dy \tag{5-7}$$

Without loss of generality we may set $x_C = a$ and $y_C = b$. For

$$f = f_{xy}(x, y) = f_{\xi\eta}(\xi, \eta)$$

we have that the integral of f over the rectangle is

$$\int_A f dA = \int_0^{2a}\int_0^{2b} f_{xy} dx dy = \int_{-1}^{1}\int_{-1}^{1} f_{\xi\eta} \overbrace{ab}^{J} d\xi d\eta \tag{5-8}$$

Hence the scaling factor in this case is ab. In order to see that this is the determinant of a special matrix we consider the following relationships:

$$\frac{\partial}{\partial \xi} = \frac{\partial}{\partial x}\frac{\partial x}{\partial \xi} + \frac{\partial}{\partial y}\frac{\partial y}{\partial \xi} = a\frac{\partial}{\partial x} + 0\frac{\partial}{\partial y} \tag{5-9a}$$

$$\frac{\partial}{\partial \eta} = \frac{\partial}{\partial x}\frac{\partial x}{\partial \eta} + \frac{\partial}{\partial y}\frac{\partial y}{\partial \eta} = 0\frac{\partial}{\partial x} + b\frac{\partial}{\partial y} \tag{5-9b}$$

or

$$\begin{bmatrix} \frac{\partial}{\partial \xi} \\ \frac{\partial}{\partial \eta} \end{bmatrix} = \begin{bmatrix} \frac{\partial x}{\partial \xi} & \frac{\partial y}{\partial \xi} \\ \frac{\partial x}{\partial \eta} & \frac{\partial y}{\partial \eta} \end{bmatrix} \begin{bmatrix} \frac{\partial}{\partial x} \\ \frac{\partial}{\partial y} \end{bmatrix} = \mathbf{J} \begin{bmatrix} \frac{\partial}{\partial x} \\ \frac{\partial}{\partial y} \end{bmatrix} \tag{5-10}$$

where

$$\mathbf{J} = \begin{bmatrix} a & 0 \\ 0 & b \end{bmatrix} \tag{5-11}$$

is the *Jacobian* matrix and $J = \det(\mathbf{J}) = ab$ is the *Jacobian*.

The cuboid. This is a hexahedron consisting of three pairs of rectangles and is the 3D version of the plane rectangle, see Fig. 5.5.

The extension from the rectangle (2D) to the cuboid (3D) is straightforward, and from Fig. 5.5 we have

$$d\xi = \frac{1}{a}dx, \quad d\eta = \frac{1}{b}dy \quad \text{and} \quad d\zeta = \frac{1}{c}dz \tag{5-12}$$

and the Jacobian is

$$J = abc \tag{5-13}$$

Element analysis I - Natural coordinates and interpolation

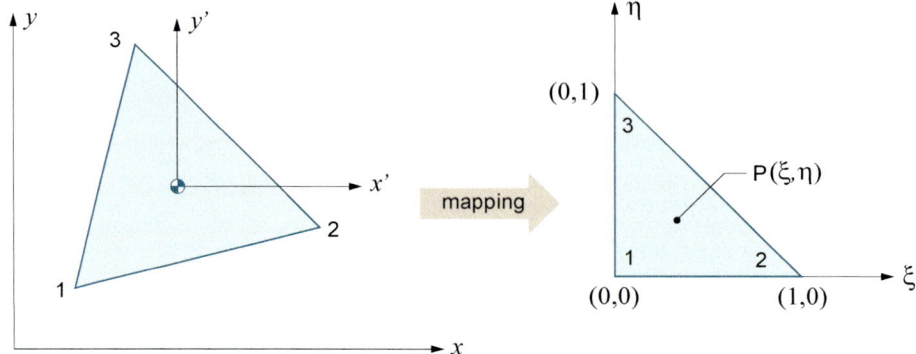

(a) global and local cartesian coordinates (b) normalized cartesian coordinates

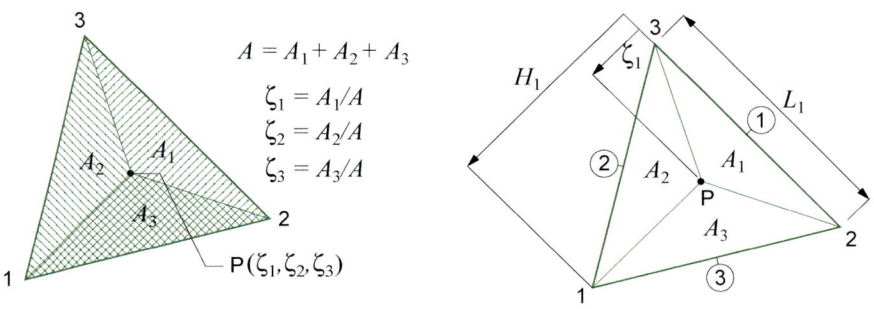

(c) area coordinates $(\zeta_1, \zeta_2, \zeta_3)$

Figure 5.6 Coordinate systems for a plane triangle

Plane triangle – area coordinates. Figure 5.6 shows different coordinate systems for a plane triangle. Figure 5.6a shows the global cartesian reference system (x,y) and a local cartesian system (x',y') with axes parallel to the global axes and with the origin placed at the area centre of the triangle. It is possible, but not recommended to work with cartesian coordinates; if we do, the local system (x',y') in Fig. 5.6a offers some advantages.

The normalized systems have clear advantages, and for the triangle we have two possible candidates, one in which the triangular element is *mapped* on to a right-angle, isosceles "unit" triangle in the $\xi\eta$-plane as indicated in Fig. 5.6b, and one making use of the so-called *area* or *triangular* coordinates ζ_1, ζ_2, ζ_3 as shown in Fig. 5.6c.

Using the notation

$$x_{ij} = x_i - x_j \quad \text{and} \quad y_{ij} = y_i - y_j \qquad (5\text{-}14)$$

the mapping in Fig. 5.6b is uniquely defined by

$$x = x_1 + x_{21}\xi + x_{31}\eta \qquad (5\text{-}15a)$$

$$y = y_1 + y_{21}\xi + y_{31}\eta \qquad (5\text{-}15b)$$

Conversely

$$\xi = \frac{1}{2A}[y_{31}x + x_{13}y + (x_3y_1 - x_1y_3)] \qquad (5\text{-}16a)$$

$$\eta = \frac{1}{2A}[y_{12}x + x_{21}y + (x_1y_2 - x_2y_1)] \qquad (5\text{-}16b)$$

Using the chain rule and Eqs. (5-15) we can readily establish the Jacobian matrix and its determinant. We can also derive a simple and useful formula for integrals of terms like $\xi^p\eta^q$ over the triangle area (for arbitrary values of p and q). That is basically all we need, and while this is a possible alternative, we much prefer the area coordinates; hence the rather superficial treatment of the ξ,η coordinates. Interpolation is more straightforward and unbiased in area coordinates. Another finding, although not conclusive, is that numerical results [12] seem to indicate that the normalized cartesian coordinates ξ and η yield results that are not quite symmetrical in cases where they should have been; area coordinates, on the other hand, satisfy symmetry for the same problem.

Area coordinates are, with reference to Fig. 5.6c, defined as

$$\zeta_i = \frac{A_i}{A} = \frac{\frac{1}{2}z_iL_i}{\frac{1}{2}H_iL_i} = \frac{z_i}{H_i} \quad i = 1, 2, 3 \quad \text{and} \quad A = \sum_{i=1}^{3} A_i \qquad (5\text{-}17)$$

Once the corners are numbered all the other numbering (of sub-areas, edges etc) follows. We see that ζ_i of an arbitrary point P within the triangle can be interpreted as a normalized (dimensionless) distance from edge i to the point. From the definition in Eq. (5-17) it follows that the three area coordinates of point P are coupled by the equation

Element analysis I - Natural coordinates and interpolation

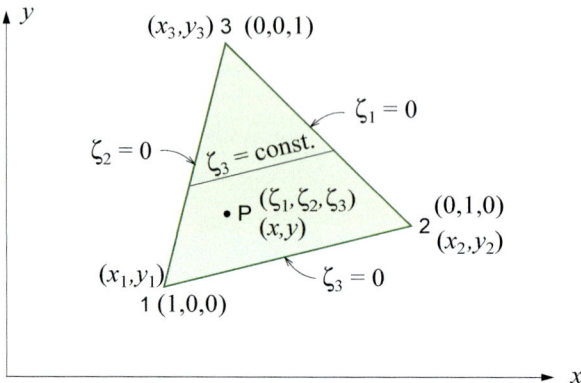

Figure 5.7 Area coordinates

Consider lines 1-2 and 1-3 of the triangle 1-2-3 (numbered in counter clockwise direction) to be (physical) vectors, and use the definition of the *vector* or *cross product* of these two vectors (see APPENDIX B) to show that the area A of the triangle is half the determinant of the matrix in Eq. (5-20).

With reference to APPENDIX A:

Use the formula $\quad \mathbf{A}^{-1} = \dfrac{\mathrm{adj}(\mathbf{A})}{\det(\mathbf{A})} = \dfrac{[C_{ij}]^T}{\det(\mathbf{A})} \quad$ to derive Eq. (5-21).

The *cofactor* $C_{ij} = (-1)^{i+j} M_{ij}$ where the *minor* M_{ij} of $\det(\mathbf{A})$ is the determinant left when row i and column j of $\det(\mathbf{A})$ are deleted.

Natural coordinates

$$\zeta_1 + \zeta_2 + \zeta_3 = 1 \qquad (5\text{-}18)$$

Hence one of the three coordinates (usually ζ_3) can be expressed in terms of the other two. We see that the equation for edge i is $\zeta_i = 0$, and $\zeta_i =$ constant between zero and one represents a line parallel to edge i, see Fig. 5.7. Also, the area coordinates of the area centre (centroid) of the triangle are

$$\zeta_1 = \zeta_2 = \zeta_3 = \frac{1}{3}$$

Next we need a relationship between the cartesian reference coordinates and the area coordinates. Consider the point P within the triangle in Fig. 5.7; its cartesian coordinates are x and y, while its area coordinates are ζ_1, ζ_2 and ζ_3. Denoting the corner coordinates x_i and y_i ($i = 1, 2, 3$) we see that

$$x = x_1\zeta_1 + x_2\zeta_2 + x_3\zeta_3 \qquad (5\text{-}19a)$$

$$y = y_1\zeta_1 + y_2\zeta_2 + y_3\zeta_3 \qquad (5\text{-}19b)$$

These relations, which obviously apply at the corners and along the edges and thus at any point within the triangle, are valid for any position of the origin of the cartesian reference system. In order to find the inverse relation we need to invoke Eq. (5-18) and write the following matrix equation

$$\begin{bmatrix} x \\ y \\ 1 \end{bmatrix} = \begin{bmatrix} x_1 & x_2 & x_3 \\ y_1 & y_2 & y_3 \\ 1 & 1 & 1 \end{bmatrix} \begin{bmatrix} \zeta_1 \\ \zeta_2 \\ \zeta_3 \end{bmatrix} \qquad (5\text{-}20)$$

It can be shown (see opposite) that the determinant of this 3 by 3 matrix is equal to twice the triangle area A, and for all valid triangles the inverse therefore exists. With a bit of effort (see opposite) we find

$$\begin{bmatrix} \zeta_1 \\ \zeta_2 \\ \zeta_3 \end{bmatrix} = \frac{1}{2A} \begin{bmatrix} y_{23} & x_{32} & (x_2y_3 - x_3y_2) \\ y_{31} & x_{13} & (x_3y_1 - x_1y_3) \\ y_{12} & x_{21} & (x_1y_2 - x_2y_1) \end{bmatrix} \begin{bmatrix} x \\ y \\ 1 \end{bmatrix} \qquad (5\text{-}21)$$

We have here made use of the notation of Eq. (5-14), and it should be emphasized that *the corners must be numbered 1, 2 and 3 in the counter clockwise direction*.

Differentiation

From (5-19) and (5-21) we find the following relations:

$$\frac{\partial x}{\partial \zeta_i} = x_i \quad \text{and} \quad \frac{\partial y}{\partial \zeta_i} = y_i \qquad (5\text{-}22)$$

and

$$\frac{\partial \zeta_i}{\partial x} = \frac{y_{jk}}{2A} \quad \text{and} \quad \frac{\partial \zeta_i}{\partial y} = \frac{x_{kj}}{2A} \qquad (5\text{-}23)$$

Element analysis I - Natural coordinates and interpolation

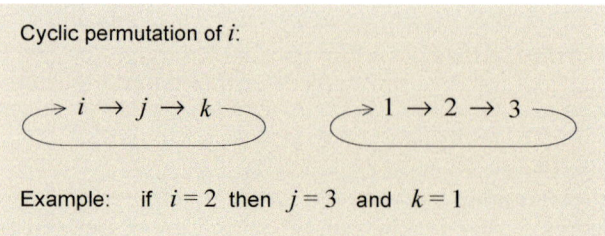

Cyclic permutation of i:

$\to i \to j \to k \to$ $\to 1 \to 2 \to 3 \to$

Example: if $i = 2$ then $j = 3$ and $k = 1$

where *j* and *k* are *cyclic permutations* of *i*. The derivatives of an arbitrary function $f(\zeta_1,\zeta_2,\zeta_3)$ with respect to *x* and *y* follow directly from the chain rule and the two equations above:

$$\frac{\partial f}{\partial x} = \frac{1}{2A}\left(\frac{\partial f}{\partial \zeta_1}y_{23} + \frac{\partial f}{\partial \zeta_2}y_{31} + \frac{\partial f}{\partial \zeta_3}y_{12}\right)$$

$$\frac{\partial f}{\partial y} = \frac{1}{2A}\left(\frac{\partial f}{\partial \zeta_1}x_{32} + \frac{\partial f}{\partial \zeta_2}x_{13} + \frac{\partial f}{\partial \zeta_3}x_{21}\right)$$

or, in matrix notation,

$$\begin{bmatrix}\dfrac{\partial}{\partial x}\\[6pt]\dfrac{\partial}{\partial y}\end{bmatrix} = \frac{1}{2A}\begin{bmatrix}y_{23} & y_{31} & y_{12}\\ x_{32} & x_{13} & x_{21}\end{bmatrix}\begin{bmatrix}\dfrac{\partial}{\partial \zeta_1}\\[6pt]\dfrac{\partial}{\partial \zeta_2}\\[6pt]\dfrac{\partial}{\partial \zeta_3}\end{bmatrix} \qquad (5\text{-}24)$$

For a triangle with *straight* edges, for which we can always express the area coordinates by *x* and *y* through Eq. (5-21), Eq. (5-24) enables us to find the derivatives with respect to *x* and *y* of any function of the area coordinates. This is precisely what we need in order to determine strains from shape functions in area coordinates.

However, if we want to take on triangular elements with curved edges we need to map the element on to a triangle with straight edges. In this process, as we shall see in the next chapter when we investigate the *isoparametric* concept, we will know *x* and *y* in terms of area coordinates, but not vice versa. We will therefore need the "inverse" of the relation in Eq. (5-24). But we cannot invert a rectangular matrix. The problem is Eq. (5-18), the three area coordinates are not independent.

In order to establish unambiguous and invertible rules for differentiation we now consider ζ_1 and ζ_2 to be our independent variables, while

$$\zeta_3 = 1 - \zeta_1 - \zeta_2 \qquad (5\text{-}25)$$

Similar to Eq. (5-10) we can now write

$$\begin{bmatrix}\dfrac{\partial}{\partial \zeta_1}\\[6pt]\dfrac{\partial}{\partial \zeta_2}\end{bmatrix} = \begin{bmatrix}\dfrac{\partial x}{\partial \zeta_1} & \dfrac{\partial y}{\partial \zeta_1}\\[6pt]\dfrac{\partial x}{\partial \zeta_2} & \dfrac{\partial y}{\partial \zeta_2}\end{bmatrix}\begin{bmatrix}\dfrac{\partial}{\partial x}\\[6pt]\dfrac{\partial}{\partial y}\end{bmatrix} = \mathbf{J}\begin{bmatrix}\dfrac{\partial}{\partial x}\\[6pt]\dfrac{\partial}{\partial y}\end{bmatrix} \qquad (5\text{-}26)$$

But now

$$\frac{\partial x}{\partial \zeta_1} = \frac{\partial x}{\partial \zeta_1}\frac{\partial \zeta_1}{\partial \zeta_1} + \frac{\partial x}{\partial \zeta_2}\frac{\partial \zeta_2}{\partial \zeta_1} + \frac{\partial x}{\partial \zeta_3}\frac{\partial \zeta_3}{\partial \zeta_1} = \frac{\partial x}{\partial \zeta_1} - \frac{\partial x}{\partial \zeta_3} \qquad (5\text{-}27\text{a})$$

$$\frac{\partial x}{\partial \zeta_2} = \frac{\partial x}{\partial \zeta_1}\frac{\partial \zeta_1}{\partial \zeta_2} + \frac{\partial x}{\partial \zeta_2}\frac{\partial \zeta_2}{\partial \zeta_2} + \frac{\partial x}{\partial \zeta_3}\frac{\partial \zeta_3}{\partial \zeta_2} = \frac{\partial x}{\partial \zeta_2} - \frac{\partial x}{\partial \zeta_3} \qquad (5\text{-}27\text{b})$$

Element analysis I - Natural coordinates and interpolation

FALK diagram for $\mathbf{JJ}^{-1}$:

		$(y_2 - y_3)$	$(y_3 - y_1)$	$\times \dfrac{1}{2A}$		
		$(x_3 - x_2)$	$(x_1 - x_3)$			
$(x_1 - x_3)$	$(y_1 - y_3)$	α	0	$\times \dfrac{1}{2A} =$	1	0
$(x_2 - x_3)$	$(y_2 - y_3)$	0	α		0	1

QED

$$\alpha = \det\{\text{matrix in Eq. (5-20)}\} = 2A$$

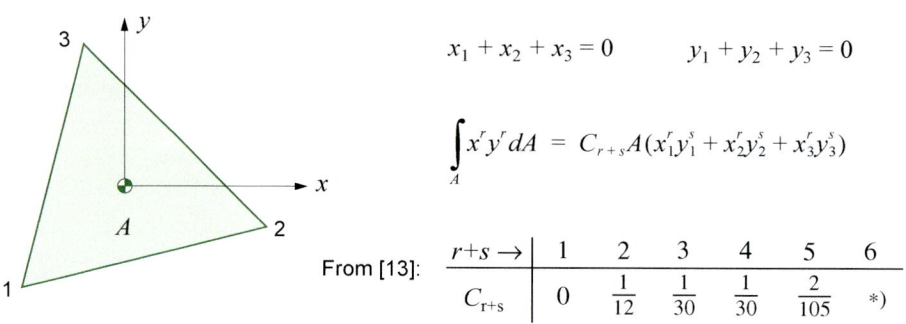

$$x_1 + x_2 + x_3 = 0 \qquad y_1 + y_2 + y_3 = 0$$

$$\int_A x^r y^r \, dA = C_{r+s} A (x_1^r y_1^s + x_2^r y_2^s + x_3^r y_3^s)$$

From [13]:

$r+s \rightarrow$	1	2	3	4	5	6
C_{r+s}	0	$\dfrac{1}{12}$	$\dfrac{1}{30}$	$\dfrac{1}{30}$	$\dfrac{2}{105}$	*)

*) This expression exists, but it is more complex (see [13])

$$V = V_1 + V_2 + V_3 + V_4$$

$\zeta_1 = V_1/V$
$\zeta_2 = V_2/V$
$\zeta_3 = V_3/V$
$\zeta_4 = V_4/V$

Figure 5.8 Tetrahedron – volume coordinates

and similar for $\partial y/\partial\zeta_1$ and $\partial y/\partial\zeta_2$. For a triangle with straight edges Eqs. (5-19) apply, and the Jacobian matrix is readily found to be

$$\mathbf{J} = \begin{bmatrix} x_{13} & y_{13} \\ x_{23} & y_{23} \end{bmatrix} \tag{5-28}$$

The inverse of Eq. (5-26) follows directly from Eq. (5-24), that is

$$\begin{bmatrix} \dfrac{\partial}{\partial x} \\ \dfrac{\partial}{\partial y} \end{bmatrix} = \frac{1}{2A}\begin{bmatrix} y_{23} & y_{31} \\ x_{32} & x_{13} \end{bmatrix}\begin{bmatrix} \dfrac{\partial}{\partial \zeta_1} \\ \dfrac{\partial}{\partial \zeta_2} \end{bmatrix} = \mathbf{J}^{-1}\begin{bmatrix} \dfrac{\partial}{\partial \zeta_1} \\ \dfrac{\partial}{\partial \zeta_2} \end{bmatrix} \tag{5-29}$$

It is shown on the opposite page that the matrices of (5-28) and (5-29) really are inverse matrices.

Integration

In most cases the integrals we need to evaluate in order to find $\mathbf{k}$ and $\mathbf{S}^0$ involve polynomial expressions. This favours the normalized coordinates since the following simple formula can be derived (see next page)

$$I = \int_A \zeta_1^m \zeta_2^n \zeta_3^p \, dA = 2A\frac{m!n!p!}{(m+n+p+2)!} \tag{5-30}$$

where m, n and p are non-negative integers. This formula can also be used to find the integral over the triangle area of polynomial expressions in cartesian coordinates up to order 6 [13], se the page opposite.

Tetrahedron. Just as a point inside a triangle splits the area into three sub-areas, a point inside a tetrahedron splits the volume into four sub-volumes, see Fig. 5-8. This can be used to define normalized tetrahedron or *volume* coordinates

$$\zeta_i = V_i/V \quad (i = 1, 2, 3, 4) \tag{5-31}$$

in the same way we defined the area coordinates for a plane triangle. We clearly have

$$\zeta_1 + \zeta_2 + \zeta_3 + \zeta_4 = 1 \tag{5-32}$$

and we can proceed to develop procedures and formulas as we did for the triangle. For instance, the integration of a polynomial term over the volume is

$$I = \int_V \zeta_1^m \zeta_2^n \zeta_3^p \zeta_4^q \, dV = 6V\frac{m!n!p!q!}{(m+n+p+q+3)!} \tag{5-33}$$

Element analysis I - Natural coordinates and interpolation

				1					linear (3 terms)
			x		y				
			x^2		xy		y^2		quadratic (6 terms)
		x^3		x^2y		xy^2		y^3	cubic (10 terms)
	x^4		x^3y		x^2y^2		xy^3		y^4
x^5		x^4y		x^3y^2		x^2y^3		xy^4	y^5
x^6	x^5y		x^4y^2		x^3y^3		x^2y^4		xy^5 y^6

(Rows shown: linear, quadratic, cubic, quartic (15 terms), quintic (21 terms), sextic (28 terms))

Figure 5.9 Complete polynomials in two variables

Integration in area coordinates

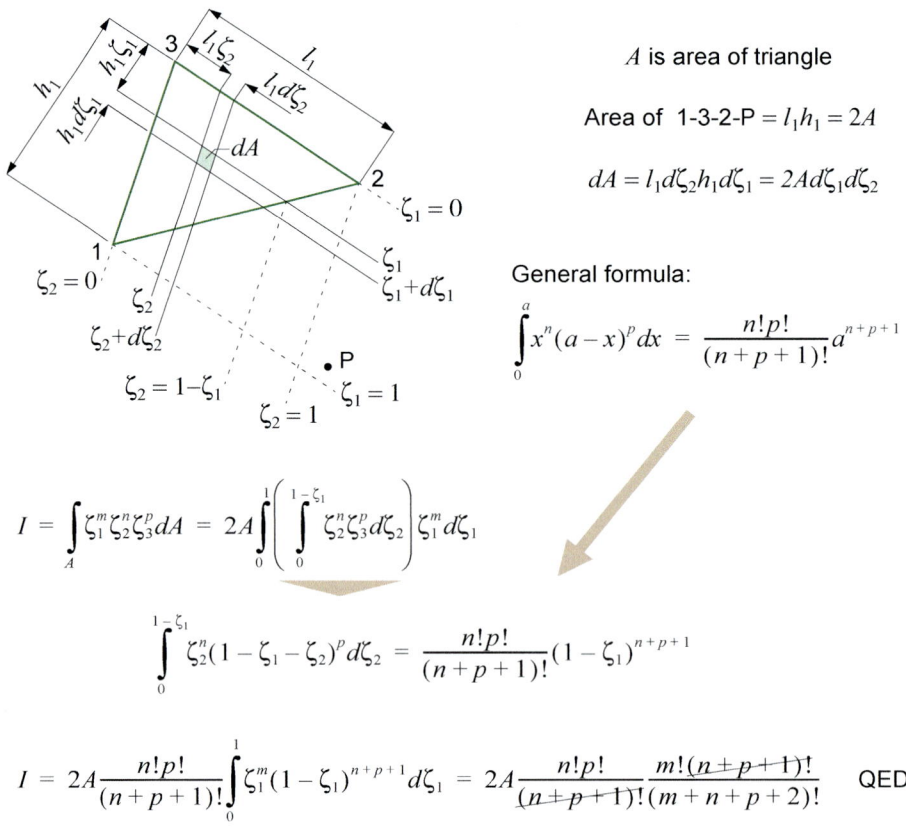

A is area of triangle

Area of 1-3-2-P = $l_1 h_1 = 2A$

$dA = l_1 d\zeta_2 h_1 d\zeta_1 = 2A d\zeta_1 d\zeta_2$

General formula:

$$\int_0^a x^n (a-x)^p dx = \frac{n!\,p!}{(n+p+1)!} a^{n+p+1}$$

$$I = \int_A \zeta_1^m \zeta_2^n \zeta_3^p dA = 2A \int_0^1 \left(\int_0^{1-\zeta_1} \zeta_2^n \zeta_3^p d\zeta_2 \right) \zeta_1^m d\zeta_1$$

$$\int_0^{1-\zeta_1} \zeta_2^n (1 - \zeta_1 - \zeta_2)^p d\zeta_2 = \frac{n!\,p!}{(n+p+1)!} (1-\zeta_1)^{n+p+1}$$

$$I = 2A \frac{n!\,p!}{(n+p+1)!} \int_0^1 \zeta_1^m (1-\zeta_1)^{n+p+1} d\zeta_1 = 2A \frac{n!\,p!}{(n+p+1)!} \cdot \frac{m!(n+p+1)!}{(m+n+p+2)!} \quad \text{QED}$$

5.3 Polynomials

Polynomials play an important role in the element analysis since they are the dominant provider of shape functions. We will deal more systematically with shape functions in the next section; here we will merely point out the importance of and connection between *complete* and *homogeneous polynomials*. We take the 2-dimensional case as an example and consider two coordinate systems, a cartesian system x and y (or ξ and η) and a system of area or triangular coordinates ζ_1, ζ_2 and ζ_3.

Complete polynomials of different order in x and y are conveniently demonstrated by arranging the terms in a triangle as shown in Fig. 5.9 (this is akin to PASCAL's triangle of the binomial coefficients). We see that a complete quadratic polynomial has 6 terms and a complete cubic polynomial has 10 terms. The use of complete polynomials guarantees symmetry in x and y as well as invariance under coordinate transformation.

A *homogeneous* polynomial in the triangular coordinates, ζ_1, ζ_2 and ζ_3, of a certain *order d*, is a polynomial that consists of all independent terms of the form

$$\zeta_1^r \zeta_2^s \zeta_3^t \quad \text{where } r + s + t = \text{constant} = d \tag{5-34}$$

For $d = 2$, for instance, we have the terms

$$\zeta_1^2, \ \zeta_2^2, \ \zeta_3^2, \ \zeta_1\zeta_2, \ \zeta_2\zeta_3 \ \text{and} \ \zeta_3\zeta_1$$

It can readily be shown that the number of independent terms of a homogeneous polynomial is

$$n = \frac{1}{2}(d+1)(d+2) \tag{5-35}$$

This is exactly the same number as the number of terms of a complete polynomial in x and y of the same order d.

According to Eq. (5-21) we can write

$$\zeta_i = a_i x + b_i y + c_i \quad (i = 1, 2, 3) \tag{5-36}$$

where a_i, b_i and c_i are constants (for a given triangle). If we substitute these expressions into (5-34) we get terms of the type $x^p y^q$ where $p+q \leq d$.
Conversely, by use of Eqs. (5-19) and (5-18) we can write the general term $x^p y^q$ ($p+q \leq d$) as

$$x^p y^q = (x_1\zeta_1 + x_2\zeta_2 + x_3\zeta_3)^p (y_1\zeta_1 + y_2\zeta_2 + y_3\zeta_3)^q (\zeta_1 + \zeta_2 + \zeta_3)^{d-p-q} \tag{5-37}$$

which is a homogeneous expression in triangular coordinates of order d, since every term on the right-hand side is of the type in Eq. (5-34). From this we conclude that

> a *complete* polynomial in x and y can be expressed as a *homogeneous* polynomial of the same order in triangular coordinates, and vice versa.

Element analysis I - Natural coordinates and interpolation

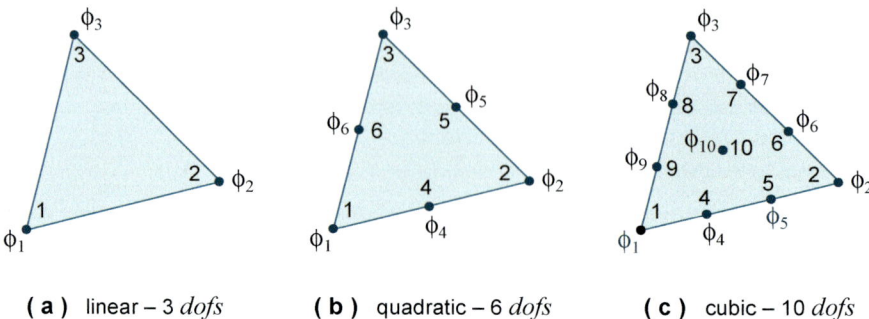

(a) linear – 3 *dofs* (b) quadratic – 6 *dofs* (c) cubic – 10 *dofs*

Figure 5.10 Possible arrangements of nodes and *dofs* of a triangular element

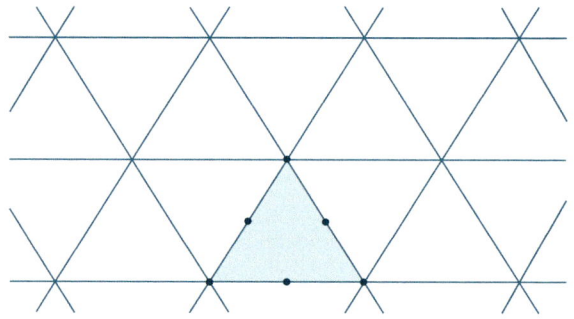

Figure 5.11 Typical triangular mesh

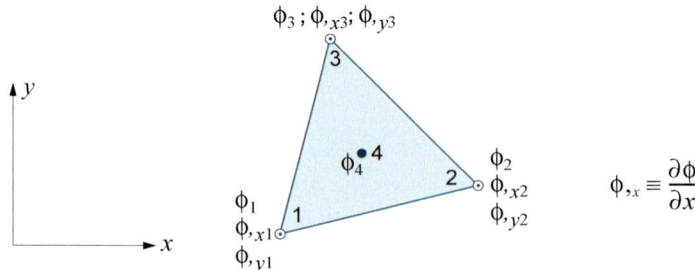

Figure 5.12 An alternative 10-*dof* triangular element

5.4 Nodal points and degrees of freedom

We have already commented on both the position of the nodal points, relative to the element geometry, and on the type of degrees of freedom **v** (*dofs*) we define at these nodal points. However, before we proceed it may be useful to take a closer look at the nodal points and their *dofs* since they are important premises for the shape functions.

We will use a triangular 2D element to exemplify problems and possibilities. In order to approximate an arbitrary state of the field variable(s), in our case displacement(s), over the problem domain, we have basically two options: we can use many simple elements or fewer, but more elaborate or complex elements. The simple elements have few *dofs* and can therefore describe only simple displacement fields within themselves. Complex elements have more *dofs* and can therefore uniquely define higher order polynomial shape functions and thus describe more complex variations of the field variable(s) within each individual element.

If we consider a problem with *one* field variable ϕ, the simplest triangular element is one with nodal points at the corners only, and with the nodal value ϕ_i at each node as its *dofs*, 3 in all as shown in Fig. 5.10a. The 3 *dofs* can describe a *linear* field uniquely, which means that the element edges remain straight also after "deformation". By including nodal points also at the mid-point of the edges, see Fig. 5.10b, we double the number of *dofs* and with 6 parameters we can uniquely define a quadratic variation over the element. We can go further and put two nodes on each edge, resulting in a total of 9 *dofs*. However, a complete cubic polynomial has 10 terms, so we would probably include an internal node (at the centre of the element area), as shown in Fig. 5.10c.

In a computational sense, side nodes are not as "effective" as corner nodes. Figure 5.11 shows a portion of a typical mesh of triangular elements, and we see that while a corner node is (on average) common to 6 elements, a side node is never common to more than 2 elements. Hence, if we, by some magic, could have moved the side nodes and their *dofs* to the corners, without degrading the element, we would, for a given element mesh, have reduced the total number of (system) *dofs*, and hence the computational effort significantly. The internal node is even less effective since it is associated with only one element. However, since it is local to the element we can eliminate it (by static condensation) at the element level and its efficiency is therefore of little concern – though it does complicate the programming slightly, as do the side nodes.

Returning to the elements of Fig. 5.10, is there any way we can replace side *dofs* by corner *dofs*? By including the nodal values of the first derivatives of the field variable ϕ, that is $\phi_{,x}$ and $\phi_{,y}$, at each corner node, we would have a 9-*dof* element with only corner nodes. If we insist on a complete cubic polynomial we will again have to include an internal *dof*, as shown in Fig. 5.12. For a thin plate bending problem, where ϕ is the transverse deflection *w*, the *dofs* of Fig. 5.12 are the most common ones. In fact, as will be shown in Chapter 14, for thin plate bending problems we will also explore elements with the second order derivatives (the curvatures) as nodal *dofs* at corner nodes.

Element analysis I - Natural coordinates and interpolation

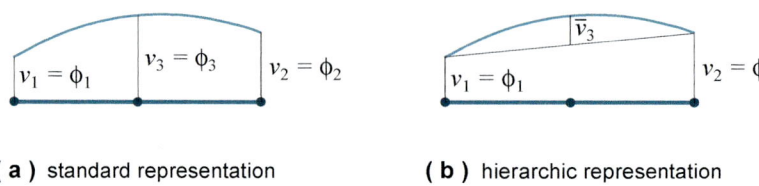

(a) standard representation   (b) hierarchic representation

Figure 5.13 Standard and hierarchic *dofs*

So far we have dealt with *dofs* which are nodal values of the field variable ϕ itself and possibly the nodal values of some of its derivatives. These are the *standard dofs*, and by far the most commonly used. Another type of *dofs* which are sometimes used, are the so-called *hierarchic* or *relative dofs*; this type is best explained by a simple example. Figure 5.13 shows a 1D element with 3 nodes and 3 *dofs*. The element is capable of representing a quadratic variation of ϕ, as shown in Fig. 5.13a with 3 standard *dofs*, and in Fig. 5.13b with two standard *dofs* and one hierarchic *dof* ($\bar{v}_3$) defined as the *departure* from the linear variation between the two end *dofs*, v_1 and v_2, that is

$$\bar{v}_3 = v_3 - \frac{1}{2}(v_1 + v_2) \tag{5-38}$$

The pros and cons of hierarchic *dofs* will be touched upon later (in Section 5.7). Before we leave this topic it should be mentioned that we will also encounter so-called *node-less dofs*. These are associated with particular displacement fields that are independent of the nodal *dofs* and they are introduced in order to alleviate certain deficiencies of the element.

5.5 The shape functions

Without loss of generality we will, in this section, consider problems with only *one* field variable which we denote by ϕ, but the problem can be:

$\phi = \phi(x)$ - one-dimensional,
$\phi = \phi(x,y)$ - two-dimensional, or
$\phi = \phi(x,y,z)$ - three-dimensional.

ϕ is a displacement quantity – for bending of a plane beam or a thin plate ϕ will be the transverse displacement w, and in the case of plane stress or plane strain ϕ will be either u (x-displacement) or v (y-displacement); hence, references will also be made to strain and stress.

The key step of the element analysis is the assumption

$$\phi \approx \hat{\phi} = \sum N_i v_i = \mathbf{N}\mathbf{v} \tag{5-39}$$

The essence of this equation is that we *assume* that the displacement (ϕ) within the element can, via the shape functions N_i, be interpolated between the nodal point degrees of freedom (the *dofs*) $\mathbf{v} = [v_1 \; v_2 \; \ldots \;]^T$. In other words, the displacement within the element, and on its "boundaries", is uniquely defined by the element *dofs* $\mathbf{v}$. Depending on the problem the element *dofs* are normally the nodal values of the displacement itself (ϕ_1, ϕ_2, ϕ_3, ...) and possibly the nodal values of its derivatives. $\mathbf{N}$, $\mathbf{v}$ and the geometric form of the element are obviously closely connected and should be treated as a whole.

We will limit our discussion to shape functions consisting of polynomial expressions, and our aim in this section is to establish the requirements which the shape functions must or should satisfy. In Section 4.5 we explored two properties, displacement *continuity* and *completeness*.

Element analysis I - Natural coordinates and interpolation

Equation (4-50): $\varepsilon = \Delta u = \Delta(\mathbf{N v}) = \Delta \mathbf{N v} = \mathbf{B v}$

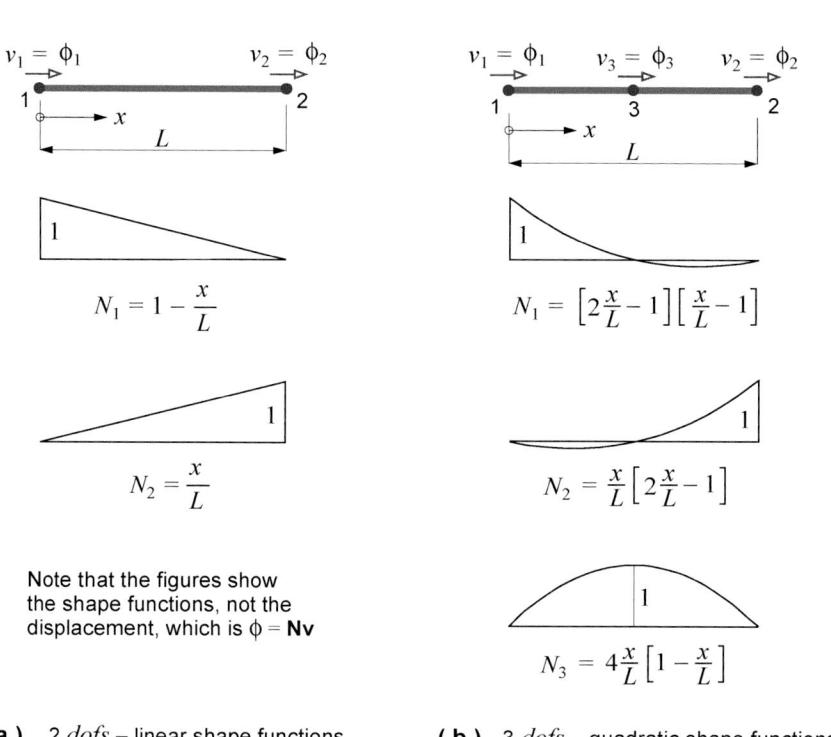

Note that the figures show the shape functions, not the displacement, which is $\phi = \mathbf{N v}$

(a) 2 *dofs* – linear shape functions

(b) 3 *dofs* – quadratic shape functions

Figure 5.14 Shape functions for 1- dimensional C^0 elements

The shape functions

The *continuity requirement* depends on the type of problem, which, in this context, is characterized by the operator Δ in the strain-displacement relation in Eq. (4-50). If the highest order of differentiation in Δ is denoted by m, then we should have continuity in the field variable ϕ itself and its derivatives up to and including order $m-1$. This characterizes a problem as a C^{m-1} problem. Figure 5.14 shows two simple, 1-dimensional elements for a C^0 problem, for which we only require continuity between elements in the variable ϕ itself; a requirement that is automatically satisfied by the fact that the elements that meet at a node share the same nodal *dof*, which is the value of ϕ at the node.

The *completeness requirement* that we formulated for the R-R process in Section 4.5 can be stated as follows:

> The shape functions must be able to describe the *rigid body* movements of the element correctly, which means that a pure rigid body movement must not cause any stress in the element, and they must be able to represent a state of *constant stress* within the element.

As the number of elements is increased the state of stress within an element approaches a constant value; hence the latter part of the requirement. For a C^0 element the completeness requirement is satisfied if the shape functions contain a complete *linear* polynomial.

For the simple 2-*dof* element in Fig. 5.14a we have two parameters, v_1 and v_2, between which we can describe uniquely a *linear* displacement variation, and the two linear shape functions, N_1 and N_2, are fairly well self-explanatory. If we include a third (internal) *dof* v_3 at the element mid-point, our three parameters, v_1, v_2 and v_3, can uniquely define a quadratic polynomial, and from the expression

$$\phi = N_1 v_1 + N_2 v_2 + N_3 v_3$$

it is fairly obvious that, for instance, N_1 must have the value 1 at node 1 and zero at nodes 2 and 3; if this was not the case the value of ϕ at node 1 would not be v_1 as it should be. We can therefore easily sketch the shape functions in Fig. 5.14b; we have anticipated events by also including their mathematical expressions.

For a C^0 element the shape functions also have to satisfy the requirement

$$\sum_i N_i = 1 \qquad (5\text{-}40)$$

at every point in the element. If this is not the case, the element will not be able to represent the rigid body movement(s) correctly. Consider for instance a rigid body translation of magnitude 1,0 in the direction of the *dofs* of the elements in Fig. 5.14. Mathematically this can, since all nodal displacements must also be equal to 1, be expressed as

$$\phi = N_1 \cdot 1 + N_2 \cdot 1 + N_3 \cdot 1 = 1{,}0 \quad \Rightarrow \quad N_1 + N_2 + N_3 = 1{,}0$$

It is left to the reader to verify that the sum of the three functions in Fig. 5.14b actually is equal to 1.

In order for a C^0 element in two dimension to satisfy completeness the shape functions must be capable of representing

For the EULER-BERNOULLI beam we have, Eq. (3-62):

$\varepsilon_x = -zw_{,xx} = -zc \quad \Rightarrow \quad$ highest order of differentiation $m = 2$

$\Rightarrow \quad C^1$ problem

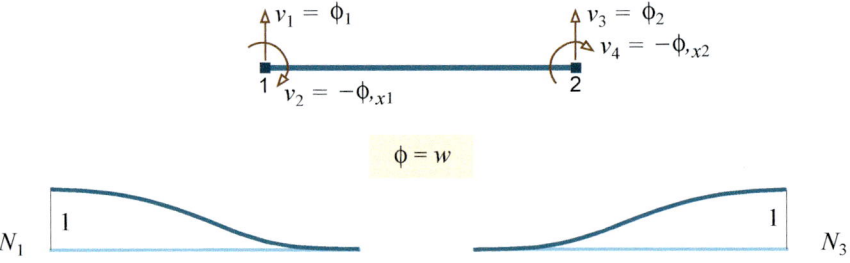

$\phi = w$

Note that in this case the shape functions coincide with the displacement

Figure 5.15 Shape functions for a 1-dimensional C^1 element

The shape functions

$$\hat{\phi}(x, y) = c_0 + c_1 x + c_2 y \qquad (5\text{-}41)$$

Thus

$$v_i = \hat{\phi}(x_i, y_i) = c_0 + c_1 x_i + c_2 y_i$$

and

$$\hat{\phi} = \sum N_i v_v = c_0 \sum N_i + c_1 \sum N_i x_i + c_2 \sum N_i y_i$$

For this expression to be the same as (5-41) we must have

$$\sum N_i = 1, \qquad \sum N_i x_i = x \quad \text{and} \quad \sum N_i y_i = y \qquad (5\text{-}42)$$

The first of these equations is the same as Eq. (5-40) and shows that this equation secures correct rigid body behaviour also in two dimensions. The last two of Eqs.(5-42) play a vital role in Section 6.1.

Next we consider a problem that requires continuity between elements not only in the variable ϕ itself, but also in its first derivative, that is a C^1 problem. A beam in bending represents such a problem, see Fig. 5.15. The element has 4 *dofs*, 2 at each end, the transverse displacement (w) and the end section rotation, which in the EULER-BERNOULLI theory is the same as the slope of the beam axis ($\approx -w_{,x}$).

The continuity requirement, which now requires continuity of both w and $w_{,x}$ is also automatically satisfied here since both these quantities are nodal *dofs* common to all elements that meet at the node.

It is again quite straightforward to argue that the shape functions shown in the figure must satisfy the requirement that any N_i must be equal to 1 "at *dof i*" and equal to zero "at all of the other element *dofs*". We see that, for instance, N_1 has zero slope at both ends as well as the value zero at end 2. We are again slightly ahead of ourselves, and so we claim without solid proof that all shape functions in Fig. 5.15 are cubic polynomials which, when combined, represent a complete cubic polynomial and thus satisfy the completeness requirement; this claim is not obvious on a simple inspection the figure, but we will come back and prove it in the next section.

Equation (5-40) does *not* apply to C^1 elements, for the simple reason that the element *dofs* are measured in different units (millimetre and radians).

We can now list the following requirements for the shape functions:

1) **Continuity** – if the order of differentiation in the operator **Δ** in the strain-displacement relation is m, the field variable(s) and its/their derivative(s) up to and including the order m-1 must be continuous along the entire "boundary" between neighbouring elements – C^{m-1} continuity.
   Compatibility and *conformity* are other words used to describe elements that satisfy the continuity requirement.

2) **Completeness** – the combination **Nv** must be capable of representing the *rigid body motions* correctly, that is, a pure rigid body motion must not produce stresses in the element, and for certain values of the nodal *dofs*

161

Element analysis I - Natural coordinates and interpolation

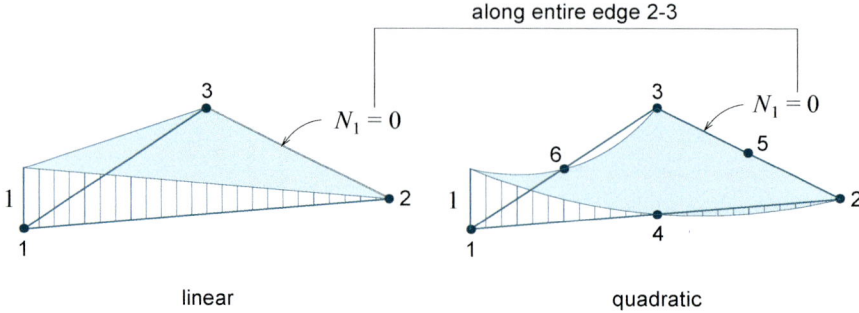

Figure 5.16 Local "support" of shape function

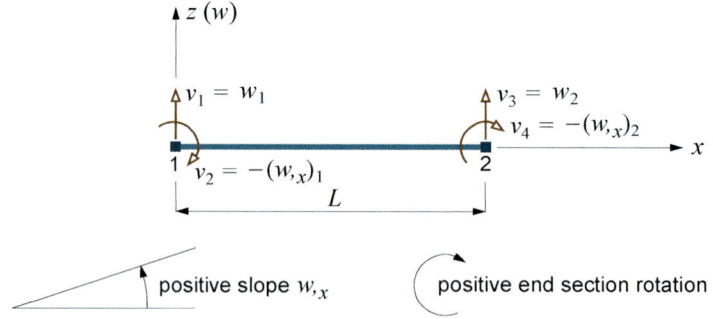

Figure 5.17 Bending of 2D beam element

Indirect interpolation – generalized displacements

(**v**) the shape functions must reproduce a state of *constant stress*. This requirement is met if the shape functions comprise complete polynomials of order *m*.

3) The *interpolation requirement* can be formulated as follows:
   - N_i must yield $v_i = 1$ and $v_j = 0$ $(j \neq i)$.
   - $N_i = 0$ "along" all edges/surfaces that do *not* contain *dof i*.
   - $\sum_i N_i = 1$.

The first point applies to all elements, the second applies only to 2- and 3-dimensional C^0 elements (see Fig. 5.16), while the last point applies to all C^0 elements.

An element that satisfies all three requirements will converge towards the correct result when the element size is decreased. Strictly speaking this statement applies to the total potential energy of the system, but for all practical purposes it applies to displacements and stresses as well. Hence the three requirements are *sufficient* for convergence, but are they also *necessary*? The last two requirements are necessary, but we shall see that the continuity requirement can be somewhat relaxed and still yield a convergent element.

We are now ready to take a closer look at how to determine mathematical expressions for the individual shape functions. To this end we can employ a *direct* or an *indirect* procedure:

- The *direct* procedure combines the use of
  - standard polynomials,
  - certain techniques and
  - experience with and insight into interpolation.
- The *indirect* approach makes use of *generalized* parameters and shape functions.

We start with the latter.

5.6 Indirect interpolation – generalized displacements

The principle is most easily demonstrated by a couple of examples, and we start with the simple 1-dimensional element of Fig. 5.15 which is reproduced in Fig. 5.17 as a beam element in bending. For the lateral displacement *w* (the field variable) we have 4 "boundary conditions", the *dofs* v_1 to v_4, which enable us to define uniquely a *cubic* polynomial. Instead of expressing *w* in terms of the *dofs*, which is not a straightforward task, we resort to some *generalized* parameters q_1 to q_4, and *assume*

$$w = q_1 + xq_2 + x^2 q_3 + x^3 q_4 = [1 \quad x \quad x^2 \quad x^3]\mathbf{q} = \mathbf{N}_q \mathbf{q} \tag{5-43}$$

The *generalized* shape functions $\mathbf{N}_q$ form a complete cubic polynomial and this expression for *w* clearly satisfies the completeness requirement. The relationship between the physical *dofs* **v** and the generalized displacements **q**, which we need not give a physical interpretation, is readily found by use of Fig. 5.17:

163

Element analysis I - Natural coordinates and interpolation

$$\mathbf{A}^{-1} = \begin{bmatrix} 1 & 0 & 0 & 0 \\ 0 & -1 & 0 & 0 \\ -\dfrac{3}{L^2} & \dfrac{2}{L} & \dfrac{3}{L^2} & \dfrac{1}{L} \\ \dfrac{2}{L^3} & -\dfrac{1}{L^2} & -\dfrac{2}{L^3} & -\dfrac{1}{L^2} \end{bmatrix}$$

$$\mathbf{N}_q = \begin{bmatrix} 1 & x & x^2 & x^3 \end{bmatrix} \longrightarrow \begin{array}{cccc} N_1 & N_2 & N_3 & N_4 \end{array}$$

$$N_1 = 1 - 3\frac{x^2}{L^2} + 2\frac{x^3}{L^3}$$

$$N_2 = -x\left(1 - 2\frac{x}{L} + \frac{x^2}{L^2}\right)$$

$$N_3 = 3\frac{x^2}{L^2} - 2\frac{x^3}{L^3}$$

$$N_4 = x\left(\frac{x}{L} - \frac{x^2}{L^2}\right)$$

Figure 5.18 Shape functions for bending of a 2D beam element

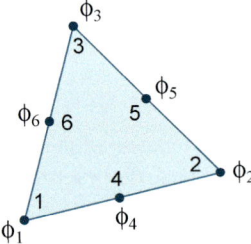

Figure 5.19 6-*dof* triangular C^0 element

Indirect interpolation – generalized displacements

$$\mathbf{v} = \begin{bmatrix} v_1 \\ v_2 \\ v_3 \\ v_4 \end{bmatrix} = \begin{bmatrix} w(0) \\ -w,_x(0) \\ w(L) \\ -w,_x(L) \end{bmatrix} = \begin{bmatrix} 1 & 0 & 0 & 0 \\ 0 & -1 & 0 & 0 \\ 1 & L & L^2 & L^3 \\ 0 & -1 & -2L & -3L^2 \end{bmatrix} \begin{bmatrix} q_1 \\ q_2 \\ q_3 \\ q_4 \end{bmatrix} = \mathbf{Aq} \quad (5\text{-}44)$$

The **A** matrix contains only constants and can therefore be inverted once and for all. Hence

$$\mathbf{q} = \mathbf{A}^{-1}\mathbf{v} \quad (5\text{-}45)$$

This particular **A** matrix is quite simple and it is not a major operation to establish

$$\mathbf{A}^{-1} = \begin{bmatrix} 1 & 0 & 0 & 0 \\ 0 & -1 & 0 & 0 \\ -\frac{3}{L^2} & \frac{2}{L} & \frac{3}{L^2} & \frac{1}{L} \\ \frac{2}{L^3} & -\frac{1}{L^2} & -\frac{2}{L^3} & -\frac{1}{L^2} \end{bmatrix} \quad (5\text{-}46)$$

The reader should verify that $\mathbf{A}\mathbf{A}^{-1} = \mathbf{I}$. If we insert Eq. (5-45) into (5-43) we get

$$w = \mathbf{N}_q \mathbf{q} = \mathbf{N}_q \mathbf{A}^{-1}\mathbf{v} = \mathbf{Nv} \quad \Rightarrow \quad \mathbf{N} = \mathbf{N}_q \mathbf{A}^{-1} \quad (5\text{-}47)$$

The computations are carried out in Figure 5.18, and we now have mathematical expressions for the shape functions in Fig. 5.15. By simply inspecting these mathematical expressions it is not obvious that they comprise a complete quadratic polynomial, but Eq. (5-43) shows that they do. In this particular case the cubic shape functions on the opposite page are of course the exact solution since they satisfy the homogeneous differential equation (2-7) of the EULER-BERNOULLI beam theory.

Next we consider the triangular plate element in Fig. 5.19. The nodal *dofs* are the 6 nodal values of the field variable ϕ. With 6 parameters we are able to define uniquely a complete quadratic polynomial in two dimensions, see Fig. 5.9. This time we will use normalized, natural (area) coordinates ζ_1, ζ_2 and ζ_3, and instead of a complete polynomial we will use the equivalent homogeneous polynomial and assume

$$\phi = [\zeta_1^2 \ \ \zeta_2^2 \ \ \zeta_3^2 \ \ \zeta_1\zeta_2 \ \ \zeta_2\zeta_3 \ \ \zeta_3\zeta_1]\mathbf{q} = \mathbf{N}_q\mathbf{q} \quad (5\text{-}48)$$

Since both ζ_2 and ζ_3 are zero at node 1 we have that

$$v_1 = \phi_1 = q_1 \quad \text{and similarly,} \quad v_2 = \phi_2 = q_2 \quad \text{and} \quad v_3 = \phi_3 = q_3$$

At node 4 we have $\zeta_1 = \zeta_2 = 1/2$ and $\zeta_3 = 0$; hence

$$v_4 = \tfrac{1}{4}q_1 + \tfrac{1}{4}q_2 + \tfrac{1}{4}q_4 \quad \text{and similarly for } v_5 \text{ and } v_6. \text{ In matrix notation:}$$

Element analysis I - Natural coordinates and interpolation

$$\mathbf{A}^{-1} = \begin{bmatrix} 1 & 0 & 0 & 0 & 0 & 0 \\ 0 & 1 & 0 & 0 & 0 & 0 \\ 0 & 0 & 1 & 0 & 0 & 0 \\ -1 & -1 & 0 & 4 & 0 & 0 \\ 0 & -1 & -1 & 0 & 4 & 0 \\ -1 & 0 & -1 & 0 & 0 & 4 \end{bmatrix}$$

$$\mathbf{N}_q = \begin{bmatrix} \zeta_1^2 & \zeta_2^2 & \zeta_3^2 & \zeta_1\zeta_2 & \zeta_2\zeta_3 & \zeta_3\zeta_1 \end{bmatrix} \longrightarrow \begin{bmatrix} N_1 & N_2 & N_3 & N_4 & N_5 & N_6 \end{bmatrix}$$

$$N_1 = \zeta_1^2 - \zeta_1\zeta_2 - \zeta_3\zeta_1 = \zeta_1(2\zeta_1 - 1)$$

$$N_2 = \zeta_2^2 - \zeta_1\zeta_2 - \zeta_2\zeta_3 = \zeta_2(2\zeta_2 - 1)$$

$$N_3 = \zeta_3^2 - \zeta_2\zeta_3 - \zeta_3\zeta_1 = \zeta_3(2\zeta_3 - 1)$$

$$N_4 = 4\zeta_1\zeta_2$$

$$N_5 = 4\zeta_2\zeta_3$$

$$N_6 = 4\zeta_3\zeta_1$$

Figure 5.20 Shape functions for a 6-*dof* triangular C^0 element

NOTE that $\quad \mathbf{A}^{-T} \equiv (\mathbf{A}^{-1})^T \equiv (\mathbf{A}^T)^{-1}$

Indirect interpolation – generalized displacements

$$v = \begin{bmatrix} v_1 \\ v_2 \\ v_3 \\ v_4 \\ v_5 \\ v_6 \end{bmatrix} = \begin{bmatrix} \phi_1 \\ \phi_2 \\ \phi_3 \\ \phi_4 \\ \phi_5 \\ \phi_6 \end{bmatrix} = \begin{bmatrix} 1 & 0 & 0 & 0 & 0 & 0 \\ 0 & 1 & 0 & 0 & 0 & 0 \\ 0 & 0 & 1 & 0 & 0 & 0 \\ \frac{1}{4} & \frac{1}{4} & 0 & \frac{1}{4} & 0 & 0 \\ 0 & \frac{1}{4} & \frac{1}{4} & 0 & \frac{1}{4} & 0 \\ \frac{1}{4} & 0 & \frac{1}{4} & 0 & 0 & \frac{1}{4} \end{bmatrix} \begin{bmatrix} q_1 \\ q_2 \\ q_3 \\ q_4 \\ q_5 \\ q_6 \end{bmatrix} = \mathbf{Aq} \quad (5\text{-}49)$$

This is a very simple **A** matrix, and its inverse can be obtained by inspection using a FALK diagram. It is shown in Fig. 5.20 together with $\mathbf{N}_q$ and the elements of **N**. If we substitute for ζ_3 in the expression for N_1 we get

$$N_1 = \zeta_1^2 - \zeta_1\zeta_2 - \zeta_1(1 - \zeta_1 - \zeta_2) = \zeta_1(2\zeta_1 - 1)$$

and, since we have complete symmetry in the area coordinates,

$$N_2 = \zeta_2(2\zeta_2 - 1) \quad \text{and} \quad N_3 = \zeta_3(2\zeta_3 - 1)$$

We will show in the next section that for this particular element we can find the shape functions more readily by direct interpolation.

Before we leave indirect interpolation and generalized displacements we should mention that the transformation of Eq. (5-45) can be "postponed". We can proceed to derive a *generalized element stiffness matrix* $\mathbf{k}_q$ based on $\mathbf{N}_q$, and then obtain **k** by *transforming* $\mathbf{k}_q$. The procedure is as follows:

$$\mathbf{u} = \mathbf{N}_q\mathbf{q} \quad \Rightarrow \quad \varepsilon = \Delta\mathbf{u} = \Delta\mathbf{N}_q\mathbf{q} = \mathbf{B}_q\mathbf{q} = \mathbf{B}_q\mathbf{A}^{-1}\mathbf{v} \quad \Rightarrow \quad \mathbf{B} = \mathbf{B}_q\mathbf{A}^{-1} \quad (5\text{-}50)$$

Using this expression for **B** in Eq. (4-54) we can express the element stiffness matrix as

$$\mathbf{k} = \mathbf{A}^{-T}\int_{V_e}\mathbf{B}_q^T\mathbf{C}\mathbf{B}_q dV \mathbf{A}^{-1} = \mathbf{A}^{-T}\mathbf{k}_q\mathbf{A}^{-1} \quad (5\text{-}51)$$

where

$$\mathbf{k}_q = \int_V \mathbf{B}_q^T\mathbf{C}\mathbf{B}_q dV \quad (5\text{-}52)$$

is the *generalized* element stiffness matrix. Similarly we can obtain the consistent element load vector by transforming a generalized load vector (by $\mathbf{A}^{-T}$). This approach will give the same end result as if we determine **N** from Eq. (5-47) and proceed from there; however, in some cases it may be computationally advantageous to first determine $\mathbf{k}_q$, since the functions in $\mathbf{N}_q$ are often very simple, and then transform as in Eq. (5-51).

Indirect interpolation is a fairly simple and general way of deriving shape functions. It can be used for both physical and natural coordinates, although care must be exercised when using it with physical coordinates, since the **A** matrix may contain exponents of length parameters which in turn may make it numerically sensitive to the units used. The procedure makes it easy to satisfy the completeness requirement.

Element analysis I - Natural coordinates and interpolation

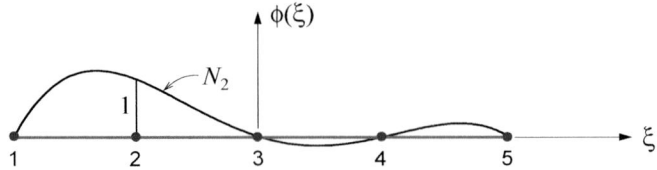

Figure 5.21 Typical shape function for a 1D C^0 element

Linear case ($n = 2$):

$$N_1 = l_1^1(\xi) = \frac{1}{2}(1 - \xi)$$

$$N_2 = l_2^1(\xi) = \frac{1}{2}(1 + \xi)$$

Quadratic case ($n = 3$):

$$N_1 = l_1^2(\xi) = -\frac{1}{2}\xi(1 - \xi)$$

$$N_2 = l_2^2(\xi) = 1 - \xi^2$$

$$N_3 = l_3^2(\xi) = \frac{1}{2}\xi(1 + \xi)$$

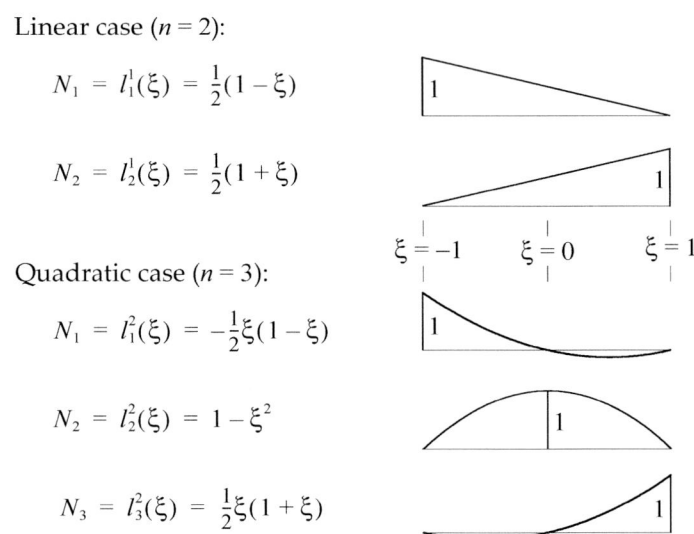

Figure 5.22 Shape functions for linear and quadratic 1D C^0 elements

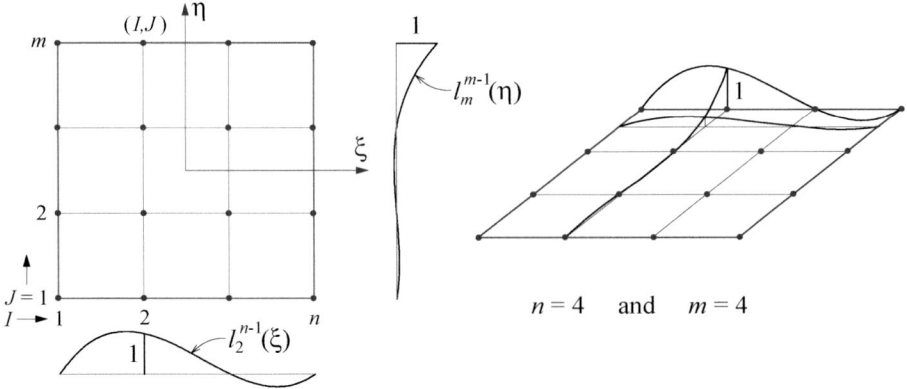

Figure 5.23 Two-dimensional LAGRANGE (square) element

The downside is the inversion of the **A** matrix which is a computationally "costly" operation when carried out numerically for each individual element in the model, as is also the transformation in Eq. (5-51). However, due to the formidable computational power we now have at our disposal this is no longer a major concern, and with efficient mathematical software it is also possible in many cases to invert **A** analytically once and for all.

5.7 Direct interpolation

It is now convenient to distinguish between C^0 and C^1 type elements, and we start with the simplest type.

C^0 **elements.** In order to simplify the discussion, and with little loss of generality, we assume that the nodal point values of the field variable ϕ are the element *dofs*, that is $v_i = \phi_i$. Hence the *dofs* have the same number as the nodal points. With reference to the basic requirements we established for the shape function N_i in Section 5.5, we seek to determine functions with the properties shown in Fig. 5.21. With words this can be expressed as:

> Given n points, fit a polynomial function of order $n-1$ through all n points such that at one of the points the function has the value one (1) and at all the other $n-1$ points the value is zero (0).

These are exactly the properties of the so-called LAGRANGE polynomials which we can express by the following general formula:

$$l_i^{n-1} = \frac{(\xi - \xi_1)...(\xi - \xi_{i-1})(\xi - \xi_{i+1})...(\xi - \xi_n)}{(\xi_i - \xi_1)...(\xi_i - \xi_{i-1})(\xi_i - \xi_{i+1})...(\xi_i - \xi_n)} \qquad n > 1 \qquad (5\text{-}53)$$

Here $n-1$ is the order of the polynomial and i is the point at which the value is 1. Figure 5.22 shows the two linear ($n = 2$) functions and the three quadratic ($n = 3$) functions. Apart from slightly different numbering these are the same functions as shown previously in Fig. 5.14.

The product of two LAGRANGE polynomials, one in each of two orthogonal directions, will produce a completely general (and indefinite) "family" of shape functions over a square/rectangle with nodal points at each intersection of the orthogonal 1-dimensional coordinate axes. With reference to Fig. 5.23 an arbitrary shape function for such an element can be expressed as

$$N_i = N_{IJ} = l_I^{n-1}(\xi) l_J^{m-1}(\eta) \qquad (5\text{-}54)$$

The number of nodal points in the two directions, n and m, can be different, but for general elements we will have symmetry in ξ and η and thus the same number of elements in both directions ($m = n$).

The two most commonly used LAGRANGE elements are the 4-node *bilinear* element and the 9-node *biquadratic* element. For the bilinear element the shape functions are, see Fig. 5.24:

$$N_1 = l_1^1(\xi) l_1^1(\eta) = \tfrac{1}{2}(1-\xi)\tfrac{1}{2}(1-\eta) = \tfrac{1}{4}(1-\xi)(1-\eta) \qquad (5\text{-}55\text{a})$$

Element analysis I - Natural coordinates and interpolation

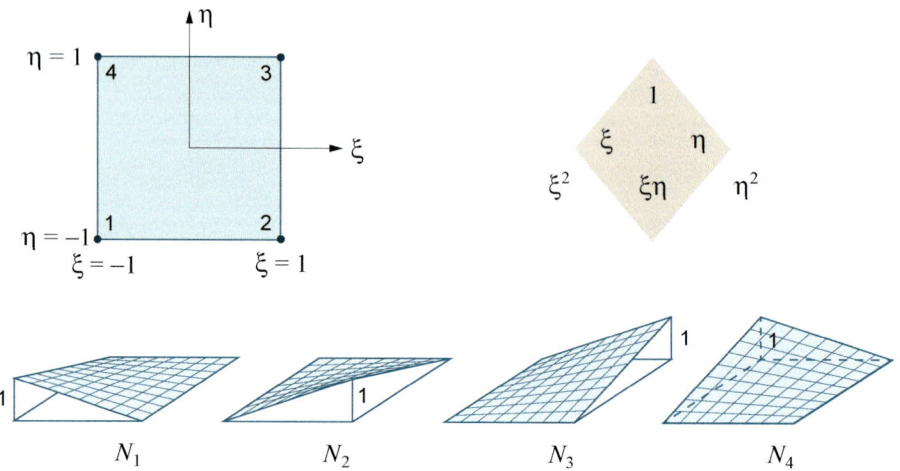

Figure 5.24 Shape functions for the bilinear 4-node LAGRANGE element

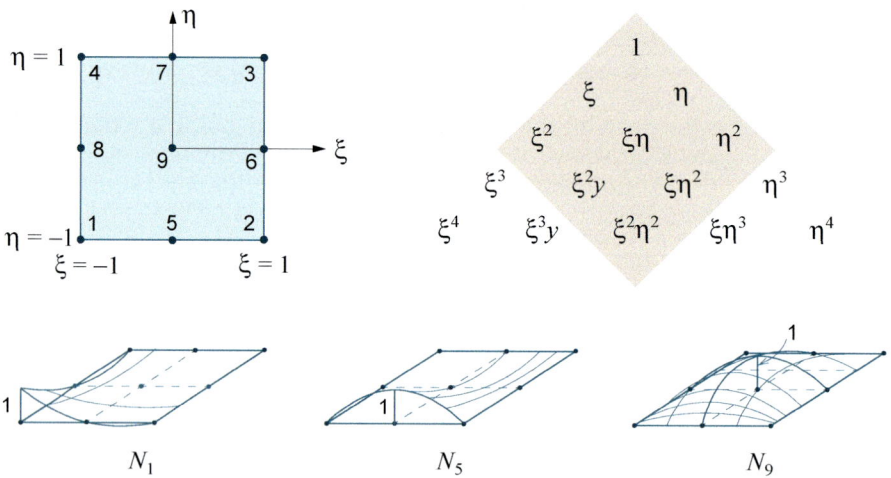

Figure 5.25 Shape functions for the bilinear 9-node LAGRANGE element

$$N_2 = l_2^1(\xi)l_1^1(\eta) = \frac{1}{4}(1+\xi)(1-\eta) \qquad (5\text{-}55\text{b})$$

$$N_3 = l_2^1(\xi)l_2^1(\eta) = \frac{1}{4}(1+\xi)(1+\eta) \qquad (5\text{-}55\text{c})$$

$$N_4 = l_1^1(\xi)l_2^1(\eta) = \frac{1}{4}(1-\xi)(1+\eta) \qquad (5\text{-}55\text{d})$$

or

$$N_i(\xi, \eta) = \frac{1}{4}(1+\xi_i\xi)(1+\eta_i\eta) \qquad i = 1, 2, 3, 4 \qquad (5\text{-}56)$$

In addition to the complete linear polynomial, the shape functions also include one quadratic term, see Fig. 5.24, which makes it bilinear.

For the biquadratic 9-node element some of the shape functions are, see Fig. 5.25:

$$N_1 = l_1^2(\xi)l_1^2(\eta) = (-\tfrac{1}{2}\xi(1-\xi))(-\tfrac{1}{2}\eta(1-\eta)) = \tfrac{1}{4}(1-\xi)(1-\eta)\xi\eta \qquad (5\text{-}57\text{a})$$

$$N_2 = l_3^2(\xi)l_1^2(\eta) = \tfrac{1}{4}(1+\xi)(1-\eta)\xi\eta \qquad (5\text{-}57\text{b})$$

$$N_5 = l_2^3(\xi)l_1^3(\eta) = -\tfrac{1}{2}(1-\xi^2)(1-\eta)\eta \qquad (5\text{-}57\text{c})$$

$$N_9 = l_2^3(\xi)l_2^3(\eta) = (1-\xi^2)(1-\eta^2) \qquad (5\text{-}57\text{d})$$

We notice that the shape function corresponding to the internal node (*dof*), the so-called "bubble" function, does not influence the field variable along the element edges (boundaries) which of course is as it should be, according to the interpolation requirement.

If we expand all the 9 shape functions into expressions of individual polynomial terms we will find that in addition to a complete quadratic polynomial, the shape functions include two cubic terms and one quartic term, see Fig. 5.25. Note that the latter three are symmetric in ξ and η.

The bicubic member of the LAGRANGE family has 16 nodal points, 4 of which are internal, see Fig. 5.23. By following the same procedure as used above for the first two members of the family, the shape functions are readily found.

Considering a mesh of many elements, we have already pointed out that the number of elements connected to or associated with a particular nodal point is a measure of the "effectiveness" of the nodal point. For a 2D element the corner nodes are therefore more effective than side nodes, and internal nodes are the least effective since they are only associated with the one element in which they reside. This lead researchers to investigate the possibility of getting rid of the internal nodes of the LAGRANGE elements, an effort that resulted in the so-called *serendipity* family of elements shown in Fig. 5.26; the name was coined by ZIENKIEWICZ and co-workers at Swansea [14]. The first element in the series is identical to the bilinear LAGRANGE element.

Element analysis I - Natural coordinates and interpolation

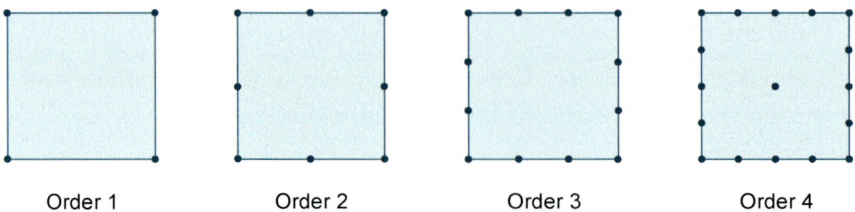

Figure 5.26 The *serendipity* family of 2D elements

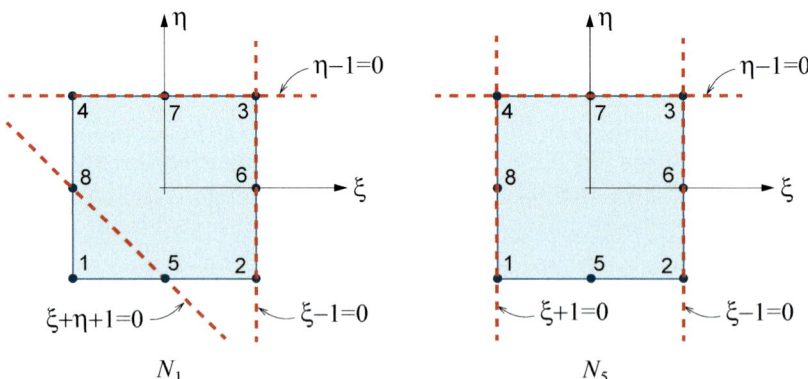

Figure 5.27 Construction of shape functions by the "method of 0-lines"

The next two elements have no internal nodes, but for the 4th and higher order elements it is not possible to satisfy the basic requirements without internal nodes. For practical purposes the most interesting of these elements is the 2nd order element.

The shape functions for the serendipity elements can be obtained by combining LAGRANGE polynomials of both similar and different order. However, they can also be constructed by a fairly simple and general method that makes use of *straight "0-lines"*. Following this method, which is applicable to both rectangular and triangular C^0 elements, a shape function can be expressed as

$$N_i = c_i l_1 l_2 ... l_p \qquad (5\text{-}58)$$

where

$$l_j = 0 \qquad j = 1, 2, ..., p \qquad (5\text{-}59)$$

is the homogeneous equation of a *straight* line, expressed as linear functions of the natural coordinates; c_i is a constant whose value enforce $N_i = 1$ at node i, and p is the polynomial order of the element, see Fig. 5.26. In the case of two-dimensional elements the method is applied as follows:

1) Choose the equations l_j as the equations of the *smallest* number of straight lines that run through *all* but node i.

2) Determine the coefficient c_i such that $N_i = 1$ at node i.

3) For each edge associated with node i, check the polynomial order of each shape function that is not equal to zero along the entire edge; if this order is n we need exactly $n+1$ nodes along this edge in order to maintain continuity between the two elements sharing the edge.

This is best explained by an example. We choose the 8-node (2nd order) serendipity element, for which we want to determine a typical corner node shape function (N_1) and a typical side node shape function (N_5), see Fig. 5.27. For N_1 we have indicated three "0-lines"; $\eta - 1 = 0$ will make $N_1 = 0$ along edge 3-4, just as $\xi - 1 = 0$ will make $N_1 = 0$ along edge 2-3. The third line, $\xi + \eta + 1 = 0$, will force N_1 to be zero at nodes 5 and 8. Hence

$$N_1 = c_1(\xi - 1)(\eta - 1)(\xi + \eta + 1)$$

For N_1 to be equal to 1 at node 1 ($\xi = \eta = -1$) we need to have

$$c_1 = 1/(-2)(-2)(-1) = -1/4$$

which gives

$$N_1 = -\tfrac{1}{4}(\xi - 1)(\eta - 1)(\xi + \eta + 1) = \tfrac{1}{4}(1 - \xi)(\eta - 1)(\xi + \eta + 1) \qquad (5\text{-}60)$$

For side node 5 we get

$$N_5 = c_5(\xi - 1)(\eta - 1)(\xi + 1)$$

and $\quad c_5 = 1/(-1)(-2)(+1) = 1/2 \quad$ which gives

$$N_5 = \tfrac{1}{2}(\xi^2 - 1)(\eta - 1) \qquad (5\text{-}61)$$

Element analysis I - Natural coordinates and interpolation

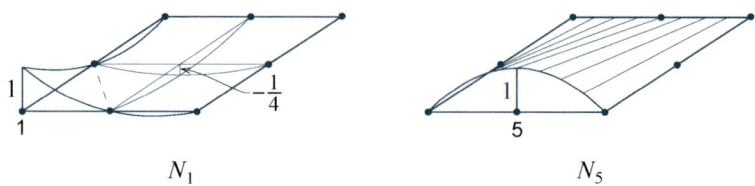

Figure 5.28 Two typical shape functions for the 8-node serendipity element

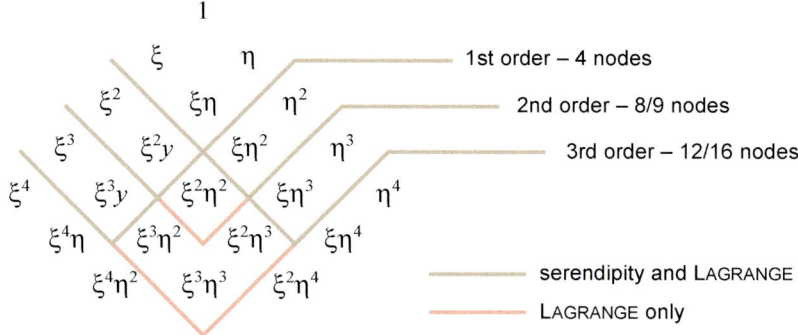

Figure 5.29 Polynomial terms included in the shape functions

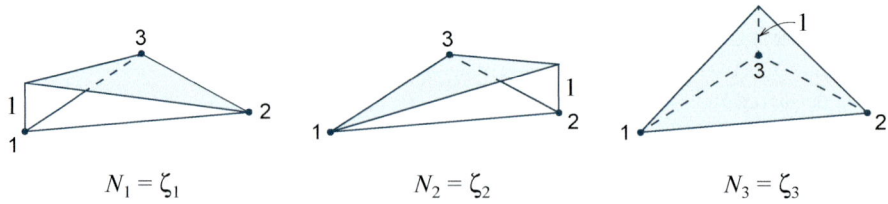

Figure 5.30 Shape functions for the linear triangle

The two shape functions are sketched in Fig. 5.28. Note the differences compared to the bilinear 9-node LAGRANGE element in Fig. 5.25. If we introduce the notation

$$\hat{\xi} = \xi_i \xi \text{ and } \hat{\eta} = \eta_i \eta \text{ where } \xi_i \text{ and } \eta_i \text{ are the coordinates of node } i,$$

then we can write all 8 shape functions for the 2nd order serendipity element as:

corner nodes: $\qquad N_i = \frac{1}{4}(1+\hat{\xi})(1+\hat{\eta})(\hat{\xi}+\hat{\eta}-1)$ (5-62a)

mid-side nodes, $\xi_i = 0$: $\qquad N_i = \frac{1}{2}(1-\xi^2)(1+\hat{\eta})$ (5-62b)

mid-side nodes, $\eta_i = 0$: $\qquad N_i = \frac{1}{2}(1+\hat{\xi})(1-\eta^2)$ (5-62c)

So what about LAGRANGE versus serendipity elements, which is the better choice? Figure 5.29 shows the polynomial terms included, for all elements of the two families, up to and including order 3. From this figure the LAGRANGE elements (of order 2 and higher) seem to have the edge, since each member of the family contains the same terms as the corresponding serendipity member plus some more. However, they do have internal nodes. For a linear analysis this is not really much of a problem, since the internal *dofs* can, as we shall see later, be eliminated or condensed out at the element level without much programming or computational effort, and with no loss of accuracy. In nonlinear analyses this may be more cumbersome, and when we move on to arbitrary, quadrilateral element shapes (*isoparametric* elements) and numerical integration, the issue becomes more subtle.

We now turn to triangular elements, and since we are still limited to C^0 elements, the interesting candidates are the three elements in Fig. 5.10. Using natural (area) coordinates, ζ_1, ζ_2 and ζ_3 (see Fig. 5.6), we see that the shape functions for the *linear triangle*, shown in Fig. 5.30, follow, more or less directly, from the definition of the area coordinates. They form a homogeneous polynomial in ζ_1, ζ_2 and ζ_3 and hence a complete linear polynomial in x and y. We see that all requirements stated in Section 5.5 are satisfied.

The next candidate is the *quadratic triangle* with nodal points not only at the corners, but also at the mid-point of each side. For a typical corner node, for instance node 1, we can use the 0-line method with two lines, $\zeta_1 = 0$ which will force the shape function to be zero along side 2-3, and $\zeta_1 - 1/2 = 0$ which will force N_1 to be zero at nodes 4 and 6, see Fig. 5.31. Hence

$$N_1 = c_1 \zeta_1 (\zeta_1 - \tfrac{1}{2}) = \zeta_1(2\zeta_1 - 1) \qquad (5\text{-}63a)$$

where c_1 has been determined so as to give $N_1 = 1$ at node 1. Similarly for the other two corner nodes:

$$N_2 = \zeta_2(2\zeta_2 - 1) \qquad (5\text{-}63b)$$

and

Element analysis I - Natural coordinates and interpolation

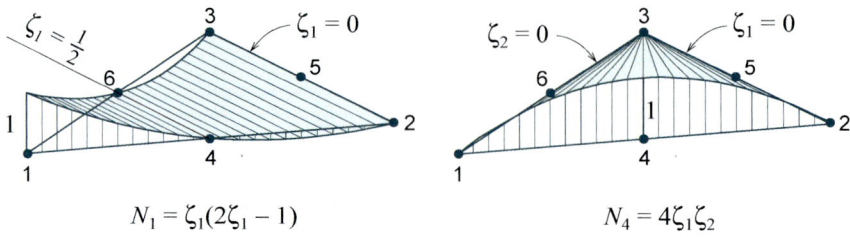

Figure 5.31 Typical shape functions for the quadratic triangle

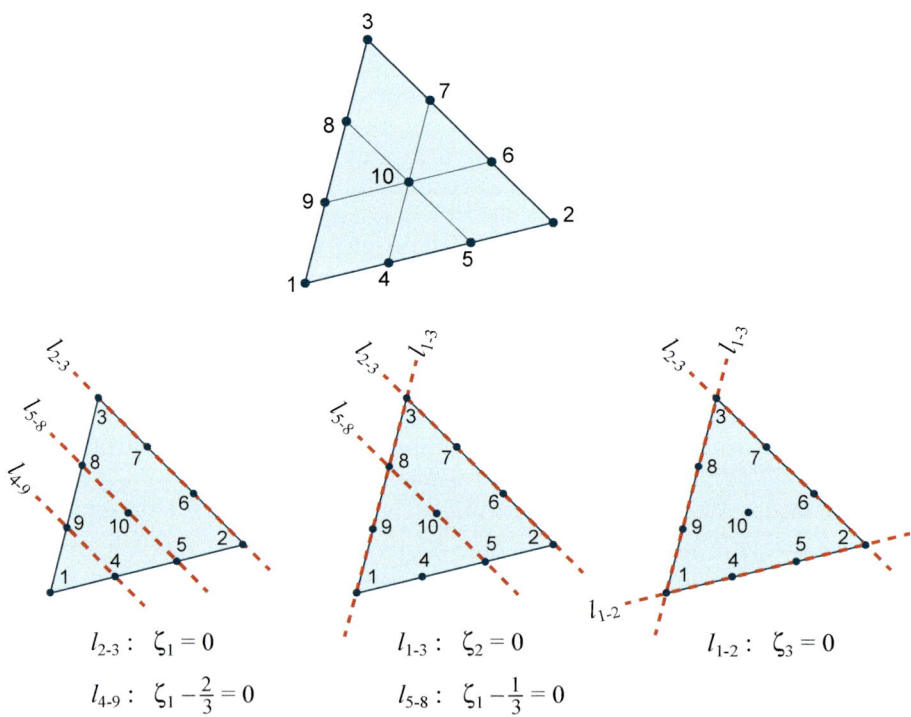

Figure 5.32 The cubic triangle

$$N_3 = \zeta_3(2\zeta_3 - 1) \qquad (5\text{-}63c)$$

The mid-side nodes are even simpler, and from Fig. 5.31 we readily find

$$N_4 = 4\zeta_1\zeta_2 \qquad (5\text{-}64a)$$

$$N_5 = 4\zeta_2\zeta_3 \qquad (5\text{-}64b)$$

$$N_6 = 4\zeta_3\zeta_1 \qquad (5\text{-}64c)$$

We see that all 6 terms of a homogeneous 2nd order polynomial in ζ_1, ζ_2 and ζ_3 are included when considering all shape functions. We leave it to the reader to verify that the sum of all six shape functions is in fact equal to 1, and from Fig. 5.31 we see that continuity of the field variable between neighbouring elements is in place. We can therefore conclude that our shape functions, which are the same as those we found by indirect interpolation, see Fig. 5.20, satisfy all formal requirements.

The next complete polynomial, the *cubic*, has ten terms. In order to match this we need to place one nodal point (number 10) inside the element and the obvious position for this node is the centre of area, that is the point with natural coordinates $\zeta_1 = \zeta_2 = \zeta_3 = 1/3$, see Fig. 5.32. The side nodes divide the sides (edges) in three equal lengths.

Typical shape functions for the cubic element are, by use of Fig. 5.32 and the method of 0-lines, found to be

$$N_1 = c_1 l_{2-3} l_{5-8} l_{4-9} = c_1 \zeta_1 (\zeta_1 - \tfrac{1}{3})(\zeta_1 - \tfrac{2}{3})$$

$$N_4 = c_4 l_{2-3} l_{1-3} l_{5-8} = c_4 \zeta_1 \zeta_2 (\zeta_1 - \tfrac{1}{3})$$

$$N_{10} = c_{10} l_{2-3} l_{1-3} l_{1-3} = c_{10} \zeta_1 \zeta_2 \zeta_3$$

$N_1 = 1$ at node 1: $\quad c_1 \cdot 1 \cdot \tfrac{2}{3} \cdot \tfrac{1}{3} = 1 \quad \Rightarrow \quad c_1 = \tfrac{9}{2}$

$N_4 = 1$ at node 4: $\quad c_4 \cdot \tfrac{2}{3} \cdot \tfrac{1}{3} \cdot \tfrac{1}{3} = 1 \quad \Rightarrow \quad c_4 = \tfrac{27}{2}$

$N_1 = 1$ at node 1: $\quad c_1 \cdot \tfrac{1}{3} \cdot \tfrac{1}{3} \cdot \tfrac{1}{3} = 1 \quad \Rightarrow \quad c_{10} = 27$

Thus

$$N_1 = \tfrac{9}{2}\zeta_1(\zeta_1 - \tfrac{1}{3})(\zeta_1 - \tfrac{2}{3}) = \tfrac{1}{2}\zeta_1(3\zeta_1 - 1)(3\zeta_1 - 2) \qquad (5\text{-}65a)$$

$$N_4 = \tfrac{27}{2}\zeta_1\zeta_2(\zeta_1 - \tfrac{1}{3}) = \tfrac{9}{2}\zeta_1\zeta_2(3\zeta_1 - 1) \qquad (5\text{-}65b)$$

$$N_{10} = 27\zeta_1\zeta_2\zeta_3 \qquad (5\text{-}65c)$$

The remaining shape functions are readily found, by cyclic permutation or by application of the same procedure as described above. It is quite straight-

Element analysis I - Natural coordinates and interpolation

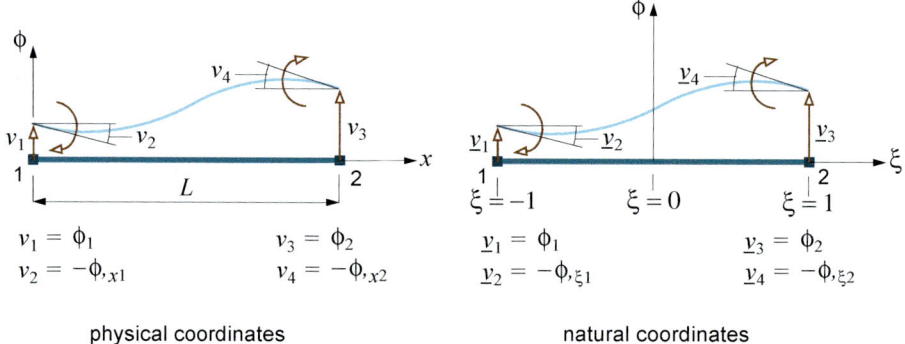

Figure 5.33 A 2D beam element – physical and natural coordinates

forward, but somewhat tedious to verify that also the cubic shape functions meet all formal requirements.

We end this discussion about shape functions for standard C^0 elements with a few words about 3D elements. The extension from 2D to 3D, that is from the rectangle to the cuboid and from the triangle to the tetrahedron, does not really present anything new. One more coordinate to keep track of, and hence more cumbersome and lengthy expressions, but otherwise much the same. Keep in mind, however, that continuity now is across surfaces, not just edges as for the 2D case; in principle the requirements are the same, but the visualization may be more demanding in 3D.

C^1 **elements.** To meet the continuity requirement for C^1 elements is far more demanding than it is for C^0 elements. For a C^1 element we must have continuity between neighbouring elements, not only in the field variable ϕ itself, but also its first derivatives. Examples are ordinary beam theory (w and $w_{,x}$) and thin plates in bending (w, $w_{,x}$ and $w_{,y}$). For the latter these continuity requirements represent quite a challenge which we will come back to in our discussion of plate bending in Chapter 14. Here we confine ourselves to the one-dimensional problem of a 2D beam element.

The problem of fitting a curve to both ordinate and slope information at given data points is known as Hermitian interpolation. We shall take the more indirect route using generalized displacements as described in Section 5.6. The point we want to make here is that care must be taken when using *natural* coordinates. Figure 5.33 shows a simple beam element and its 4 bending *dofs*, referred to both physical and natural coordinates. We have already determined the four shape functions in physical coordinates, see Figs. 5.15 and 5.18.

In natural coordinates we have

$$\phi = q_0 + q_1\xi + q_2\xi^2 + q_3\xi^3 = \underline{\mathbf{N}}_q \mathbf{q} \tag{5-66}$$

The boundary conditions are:

$$\text{for } \xi = -1: \quad \phi = v_1 \quad \text{and} \quad \phi_{,\xi} = -v_2$$
$$\text{for } \xi = +1: \quad \phi = v_3 \quad \text{and} \quad \phi_{,\xi} = -v_4$$

If we now establish the **A** matrix and its inverse, just as we did in Section 5.6, see Eqs. (5-44) and (5-46), we will, after some fairly straightforward computations, find

$$\underline{N}_1 = \tfrac{1}{4}(2 - 3\xi + \xi^3) \tag{5-67a}$$

$$\underline{N}_2 = \tfrac{1}{4}(-1 + \xi + \xi^2 - \xi^3) \tag{5-67b}$$

$$\underline{N}_3 = \tfrac{1}{4}(2 + 3\xi - \xi^3) \tag{5-67c}$$

$$\underline{N}_4 = \tfrac{1}{4}(1 + \xi - \xi^2 - \xi^3) \tag{5-67d}$$

These are the shape functions in terms of natural coordinates,

Element analysis I - Natural coordinates and interpolation

Physical coordinates :

$$N_1 = 1 - 3\frac{x^2}{L^2} + 2\frac{x^3}{L^3}, \quad N_2 = -x\left(1 - 2\frac{x}{L} + \frac{x^2}{L^2}\right), \quad N_3 = 3\frac{x^2}{L^2} - 2\frac{x^3}{L^3} \text{ and } N_4 = x\left(\frac{x}{L} - \frac{x^2}{L^2}\right)$$

Natural coordinates (modified) :

$$\underline{N}_1 = \tfrac{1}{4}(2 - 3\xi + \xi^3), \quad \underline{N}_{2,\text{mod}} = \tfrac{L}{8}(-1 + \xi + \xi^2 - \xi^3), \quad \underline{N}_3 = \tfrac{1}{4}(2 + 3\xi - \xi^3) \text{ and}$$

$$\underline{N}_{4,\text{mod}} = \tfrac{L}{8}(1 + \xi - \xi^2 - \xi^3)$$

Check :

$$\phi_{x=L/2} = \tfrac{1}{2}v_1 - \tfrac{L}{8}v_2 + \tfrac{1}{2}v_3 + \tfrac{L}{8}v_4 \qquad \phi_{\xi=0} = \tfrac{1}{2}v_1 - \tfrac{L}{8}v_2 + \tfrac{1}{2}v_3 + \tfrac{L}{8}v_4$$

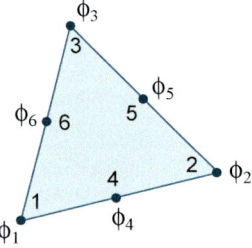

Copy of **Figure 5.19** 6-*dof* triangular C^0 element

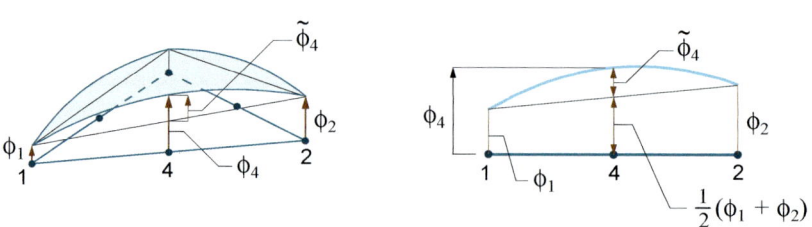

Figure 5.34 Hierarchic 6-node (quadratic) C^0 element

$$\underline{\mathbf{N}} = [\ \underline{N}_1 \quad \underline{N}_2 \quad \underline{N}_3 \quad \underline{N}_4\ ]$$

The relationship between natural and physical coordinates is, see Eq. (5.1),

$$\xi = \frac{2x}{L} - 1$$

and

$$\frac{d}{d\xi} = \frac{d}{dx}\frac{dx}{d\xi} = \frac{L}{2}\frac{d}{dx}$$

hence (5-68)

$$\underline{v}_2 = \frac{L}{2}v_2 \quad \text{and} \quad \underline{v}_4 = \frac{L}{2}v_4$$

If we now modify $\underline{\mathbf{N}}$ by multiplying the second and fourth shape function with $L/2$, we can express the field variable ($\phi = w$) as

$$\phi = \underline{\mathbf{N}}_{mod}\mathbf{v} \qquad (5\text{-}69)$$

In other words, ϕ is expressed by shape functions in natural coordinates and *dofs* in physical coordinates. This modification of the shape functions in natural coordinates is an extra operation for C^1 elements compared with C^0 elements.

Hierarchic C^0 elements (2D). We round off this section with a short discussion about hierarchic elements. We have already explained what we mean by a hierarchic *dof* for a simple 1D line element, see Fig. 5.13. Here we will extend this to 2D, and we will use the 6-node quadratic triangle as an example. For the standard version of the element shown in Fig. 5.19 we have

$$\phi = N_1\phi_1 + N_2\phi_2 + \ldots + N_6\phi_6 \qquad (5\text{-}70)$$

and the standard shape functions (N_i) are given in Eqs. (5-63) and (5-64).

In the following we will mark a hierarchic quantity by a tilde $(\sim)^1$ above its symbol. The key point of the hierarchic representation of this element is to express the value at the mid-side nodes 4, 5 and 6 as the sum of the *linear* interpolation and the *departure* from this linear variation as shown in Fig. 5.34, that is

$$\phi_4 = \tfrac{1}{2}(\phi_1 + \phi_2) + \tilde{\phi}_4 \qquad (5\text{-}71\text{a})$$

$$\phi_5 = \tfrac{1}{2}(\phi_2 + \phi_3) + \tilde{\phi}_5 \qquad (5\text{-}71\text{b})$$

$$\phi_6 = \tfrac{1}{2}(\phi_3 + \phi_1) + \tilde{\phi}_6 \qquad (5\text{-}71\text{c})$$

If we substitute Eqs. (5-71), (5-63) and (5-64) into (5-70) we get

$$\phi = \tilde{N}_1\phi_1 + \tilde{N}_2\phi_2 + \tilde{N}_3\phi_3 + \tilde{N}_4\tilde{\phi}_4 + \tilde{N}_5\tilde{\phi}_5 + \tilde{N}_6\tilde{\phi}_6 \qquad (5\text{-}72)$$

---

1. This symbol has already been used to identify a *virtual* quantity; due to a limited number of "markers" we now use it again with a different meaning; hopefully without adverse effect.

Element analysis I - Natural coordinates and interpolation

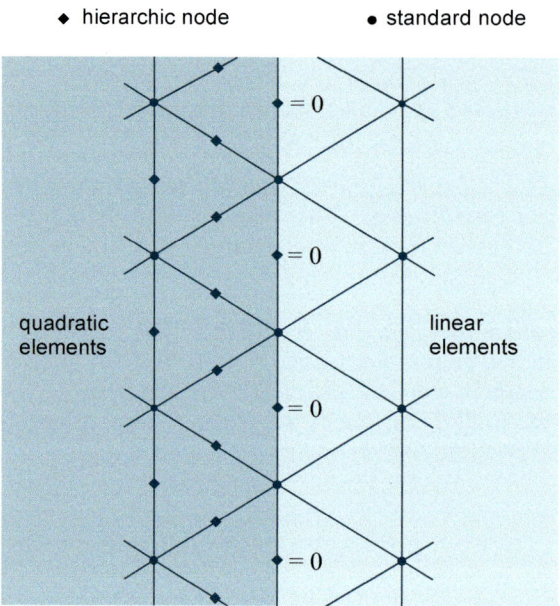

Figure 5.35 C^0 continuity between different types of elements

where

$$\tilde{N}_1 = \zeta_1, \quad \tilde{N}_2 = \zeta_2 \quad \text{and} \quad \tilde{N}_3 = \zeta_3 \quad \text{(5-73a-c)}$$

$$\tilde{N}_4 = 4\zeta_1\zeta_2 = N_4, \quad \tilde{N}_5 = 4\zeta_2\zeta_3 = N_5 \quad \text{and} \quad \tilde{N}_6 = 4\zeta_3\zeta_1 = N_6 \quad \text{(5-73d-f)}$$

We could have set this up directly since all we have done is to superimpose a quadratic mid-side variation on a linear corner node variation.

The hierarchic element is in principle exactly the same as its standard counterpart, and for the same element mesh and a consistent treatment of all loading, the two elements will give identical results provided the boundary conditions are handled in exactly the same way in both cases.

If we look at the programming, the hierarchic element has the edge. The expressions become simpler, due to simpler shape functions, and one and the same procedure (subroutine) can generate the stiffness matrix for each of a series of elements with the same geometrical shape.

Another advantage of hierarchic elements is that they enable easy transition between element meshes of different element "accuracy". Figure 5.35 shows a detail of an element mesh where simple linear triangular elements are used in one part of the problem region, and the more accurate or refined quadratic element in the other part. Full C^0 continuity is maintained along the border between the two types of elements by simply specifying that the *dof(s)* at all hierarchic nodes on the border edges are equal to *zero*. This will force the edges between linear and quadratic elements to displace as straight lines, controlled by the movements of the corner nodes only.

It should also be mentioned that we can easily obtain the element stiffness matrix **k** for the standard element from the stiffness matrix $\tilde{\mathbf{k}}$ of its hierarchic counterpart. Let

$$\mathbf{S} = \mathbf{k}\mathbf{v} \quad \text{and} \quad \tilde{\mathbf{S}} = \tilde{\mathbf{k}}\tilde{\mathbf{v}}$$

represent the two element stiffness relations. The relationship between **v** and $\tilde{\mathbf{v}}$ is

$$\tilde{\mathbf{v}} = \begin{bmatrix} \phi_1 \\ \phi_2 \\ \phi_3 \\ \tilde{\phi}_4 \\ \tilde{\phi}_5 \\ \tilde{\phi}_6 \end{bmatrix} = \begin{bmatrix} 1 & 0 & 0 & 0 & 0 & 0 \\ 0 & 1 & 0 & 0 & 0 & 0 \\ 0 & 0 & 1 & 0 & 0 & 0 \\ -\frac{1}{2} & -\frac{1}{2} & 0 & 1 & 0 & 0 \\ 0 & -\frac{1}{2} & -\frac{1}{2} & 0 & 1 & 0 \\ -\frac{1}{2} & 0 & -\frac{1}{2} & 0 & 0 & 1 \end{bmatrix} \begin{bmatrix} \phi_1 \\ \phi_2 \\ \phi_3 \\ \phi_4 \\ \phi_5 \\ \phi_6 \end{bmatrix} = \tilde{\mathbf{T}}\mathbf{v} \quad \text{(5-74)}$$

and by our transformation rule of Section 2.6 it follows that

$$\mathbf{S} = \tilde{\mathbf{T}}^T\tilde{\mathbf{S}} = \tilde{\mathbf{T}}^T\tilde{\mathbf{k}}\tilde{\mathbf{T}}\mathbf{v}$$

and

$$\mathbf{k} = \tilde{\mathbf{T}}^T\tilde{\mathbf{k}}\tilde{\mathbf{T}} \quad \text{(5-75)}$$

Blaise PASCAL (1623-1662) was a French mathematician, physicist, inventor, writer and Catholic philosopher. He was a child prodigy educated by his father, a tax collector. His earliest work was in the natural and applied sciences where he made contributions to the study of fluids, and clarified the concepts of pressure and vacuum.

At the age of 19 he started some pioneering work on calculating machines, and after three years and 50 prototypes he invented the mechanical calculator; he built 20 of these machines ("Pascaline") in the following ten years. PASCAL was a brilliant mathematician; at the age of sixteen he wrote a significant treatise on the subject of projective geometry and he later corresponded with Pierre de FERMAT on probability theory, strongly influencing the development of modern economics and social science. His "Treatise on the Arithmetical Triangle" of 1653 describes a convenient tabular presentation of binomial coefficients, now called PASCAL's triangle.

In 1646 PASCAL and his sister identified with the religious movement within Catholicism known as Jansenism. Following a mystical experience i 1654, he had his "second conversion", abandoned his scientific work, and devoted himself to philosophy and theology. He had poor health and died just a couple of months after his 39th birthday.

Joseph-Louis LAGRANGE (1736-1813) was a mathematician and astronomer, born in Turin (Italy), lived part of his life in Prussia and part in France.
LAGRANGE made professor at the Royal Artillery School in Turin at the age of 19!
He made significant contributions to all fields of analysis, to number theory, and to celestial mechanics. He succeeded EULER as director of mathematics at the Prussian Academy of Sciences in Berlin where he stayed for over twenty years. In 1787 he moved to France and became the first professor of analysis at the École Polytechnique upon its opening in 1794. LAGRANGE's treatise on analytical mechanics, written in Berlin and first published in 1788, offered the most comprehensive treatment of classical mechanics since NEWTON and formed a basis for the development of mathematical physics in the nineteenth century.

We can of course also establish the opposite transformation of Eq. (5-74) and obtain $\tilde{\mathbf{k}}$ from $\mathbf{k}$, but, from a programming point of view, that would not be a good idea.

The hierarchic version of the cubic triangle is more tricky, since we now have two choices. We can retain the node numbering and position of the standard version, see Fig. 5.36, and define all side nodes and the internal node as hierarchic and apply cubic shape functions to all of them. However, this would not enable us to go from cubic to quadratic variation along an edge since suppressing the hierarchic *dofs* would lead to a linear variation. Another possibility is shown in Fig. 5.36. By redefining the position of the side nodes, and by splitting the 7th, 8th and 9th *dofs* into *two* nodes each, as shown, we can, by applying quadratic shape functions to the mid-side nodes and cubic functions to the last 4 (including the internal node), produce all three elements, cubic, quadratic and linear, by suitable suppression of hierarchic *dofs*. The latter version is perhaps the more versatile of the two.

Hierarchic nodes/*dofs* are equally applicable to rectangular elements, but since the principle is the same we will not pursue this any further here.

As a final comment on hierarchic elements we issue a word of warning concerning the representation of loading, and mass in dynamic problems. A *consistent* approach is much to be preferred to straightforward lumping. The latter can produce erratic results if not carefully applied.

Summary of Chapter 5

We started the chapter with a short review of *types* and *shapes* of structural finite elements.

We went on to define *natural* (dimensionless) coordinates for different geometric element shapes, coordinates which are not only more convenient, but also numerically more robust than the *physical*, cartesian reference coordinates. However, the strain-displacement relationship involves differentiation with respect to physical coordinates and at the end of the element analysis we need to integrate in the physical world. This necessitates unambiguous rules for differentiation and integration, rules that emphasize the importance of the Jacobian matrix and its determinant, the Jacobian (the "scale" factor).

Next we looked at polynomials, the dominant supplier of the all important shape functions. We emphasized the importance of *complete* polynomials, and we introduced the notion of *homogeneous* polynomials in *area/volume* coordinates, and demonstrated the equivalence of complete and homogeneous polynomials of the same order.

A section on nodal points and kinematic degrees of freedom (*dofs*) is also included. Here we pointed out that the *dofs* are basically more effective if they are defined at corner nodes rather than at nodes placed on sides and surfaces. We also introduced the concept of *hierarchical dofs,* and showed how it is possible to develop "families" of elements of different order ("accuracy") that can be used in the same mesh without violating continuity requirements, simply by suppressing appropriate hierarchical *dofs*.

Element analysis I - Natural coordinates and interpolation

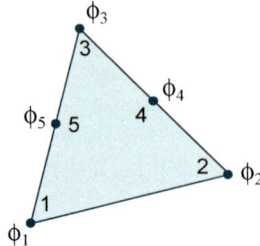

Figure 5.36 5-*dof* triangular C^0 element

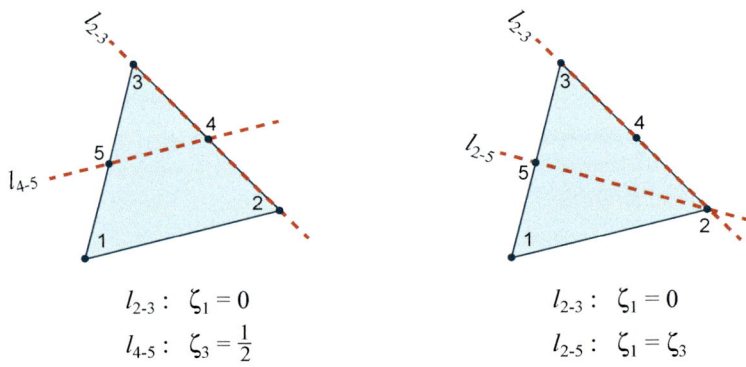

$l_{2-3}: \zeta_1 = 0$
$l_{4-5}: \zeta_3 = \frac{1}{2}$

$l_{2-3}: \zeta_1 = 0$
$l_{2-5}: \zeta_1 = \zeta_3$

Figure 5.37

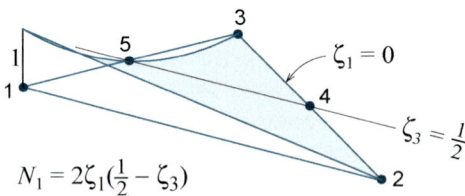

$N_1 = 2\zeta_1(\frac{1}{2} - \zeta_3)$

Figure 5.38

Example 5-2 – Shape function construction

The last and largest part of the chapter is devoted to *shape functions* and their construction. The requirements we need to impose on the shape functions, in order to be assured convergence towards the correct result when the element size is gradually decreased, are stated. Most important is the element's ability to represent *rigid body* movements correctly and to reproduce *constant stress* states, both of which are covered by the *completeness* requirement. The other basic requirement for convergence is *continuity* between neighbouring elements of displacements (C^0) and, depending on the problem, possibly also of displacement derivatives (C^1). However, it was pointed out that while satisfying both continuity and completeness will guarantee correct convergence, it is possible to derive convergent elements that do not fully satisfy the continuity requirement. Elements that do satisfy continuity are also termed *conforming* or *compatible* elements.

The two basic requirements, completeness and continuity, impose certain *interpolation* requirements that are useful guidlines when it comes to the construction of shape functions. We considered both *indirect* interpolation, making use of *generalized* displacements, and *direct* interpolation. A useful method in the latter category is the method of "0-lines".

Example 5-2 – Shape function construction

We seek the shape function N_1, in terms of area coordinates ζ_1, ζ_2 and ζ_3, for the special 5-*dof* triangular C^0 element in Fig. 5.36. Nodes 4 and 5 are placed at the midpoint of the sides 2-3 and 3-1, respectively.

Solution

We use the 0-line method, and with reference to Fig. 5.37 we see that we have two options. In both, the 0-line $\zeta_1 = 0$ must be present in order to enforce zero displacement along edge 2-3. However, in order to also make N_1 equal to zero at node 5, we have two possibilities, namely the 0-lines $l_{4\text{-}5}$ and $l_{2\text{-}5}$.

$l_{2\text{-}3}$ and $l_{4\text{-}5}$: $\quad N_1 = c_1 \zeta_1 (\zeta_3 - 0{,}5) \quad N_1 = 1$ at node 1: $\quad 1 = c_1 \cdot 1(0 - 0{,}5) \Rightarrow c_1 = -2$

$\qquad\qquad\qquad N_1 = -2\zeta_1(\zeta_3 - 0{,}5)$

$l_{2\text{-}3}$ and $l_{2\text{-}5}$: $\quad N_1 = c_1 \zeta_1 (\zeta_3 - \zeta_1) \quad N_1 = 1$ at node 1: $\quad 1 = c_1 \cdot 1(0 - 1) \Rightarrow c_1 = -1$

$\qquad\qquad\qquad N_1 = -\zeta_1(\zeta_3 - \zeta_1) = \zeta_1(\zeta_1 - \zeta_3)$

Are both of these functions admissible? Both satisfy the 0 and 1 conditions, but if we take a closer look, we see that with no side node on edge 1-2 the displacement here must vary linearly along the edge. The first function does vary linearly from 1 to 0 as it should, but the second one varies quadratically from one to zero and this function cannot be defined uniquely by only two values (one at each end). Hence the function we seek is, see Fig. 5.38,

$$N_1 = -2\zeta_1(\zeta_3 - 0{,}5) = 2\zeta_1(0{,}5 - \zeta_3)$$

Element analysis I - Natural coordinates and interpolation

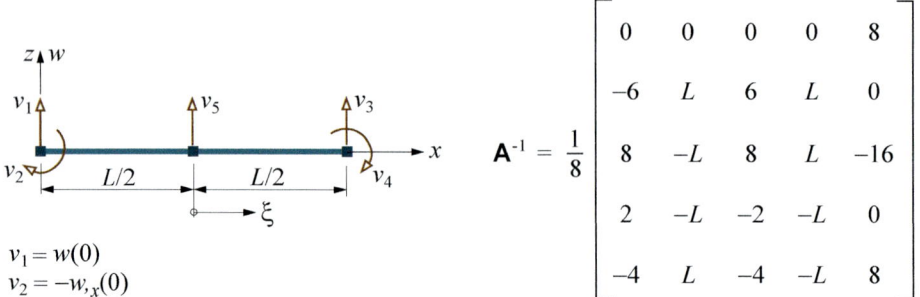

$v_1 = w(0)$
$v_2 = -w,_x(0)$

$$A^{-1} = \frac{1}{8}\begin{bmatrix} 0 & 0 & 0 & 0 & 8 \\ -6 & L & 6 & L & 0 \\ 8 & -L & 8 & L & -16 \\ 2 & -L & -2 & -L & 0 \\ -4 & L & -4 & -L & 8 \end{bmatrix}$$

Figure 5.39

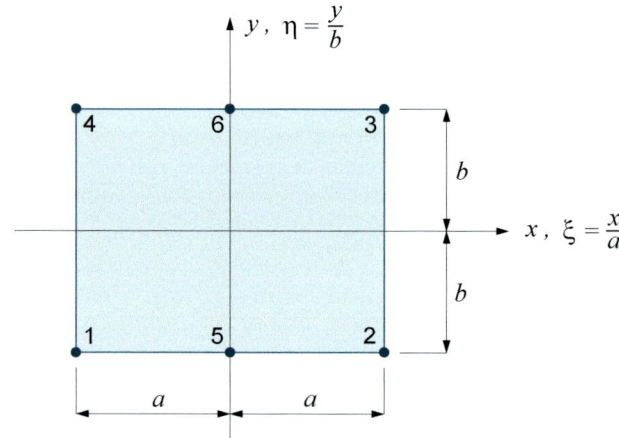

Figure 5.40

Problems

Problem 5.1

Show that the area of a triangle is equal to twice the determinant of the matrix in Eq. (5-20).

Problem 5.2

Verify that the matrix in Eq. (5-21) is the inverse of the matrix in Eq. (5-20).

Problem 5.3

Verify that the shape functions of Fig. 5.14b satisfy Eq. (5-40).

Problem 5.4

Show that $\sum N_i = 1$ is satisfied by the shape functions for the quadratic triangle of Fig 5.10b, that is by the functions defined by Eqs. (5-63) and (5-64).

Problem 5.5

Figure 5.39 shows an EULER-BERNOULLI beam element with 5 (bending) degrees of freedom. Material and cross sectional properties are constant over the element length. The element is developed by use of *generalized* displacements based on a quartic polynomial in ξ

$$w = [\,1 \quad \xi \quad \xi^2 \quad \xi^3 \quad \xi^4\,]\mathbf{q} = \mathbf{N}_q \mathbf{q}$$

a) Find $\mathbf{B}_q$ in the relation $\varepsilon_x = -z\mathbf{B}_q\mathbf{q}$
b) Determine the generalized stiffness matrix $\mathbf{k}_q$.
c) Find the matrix $\mathbf{A}$ defining $\mathbf{v}$ in terms of $\mathbf{q}$ ($\mathbf{v} = \mathbf{Aq}$).
d) The inverse of $\mathbf{A}$ is given in Fig. 5.39. Verify that your $\mathbf{A}$ from question c) is consistent with the inverse matrix in Fig. 5.39.
e) Find explicit expressions for the interpolation (shape) functions in $\mathbf{N}$ ($w = \mathbf{Nv}$).

Problem 5.6

a) The rectangular C^0 element in Fig. 5.40 has 6 degrees of freedom, the nodal values of the field variable ϕ. Determine the shape functions $\mathbf{N}$ in terms of natural coordinates ξ and η, and discuss their properties with respect to the basic requirements.

b) It is also possible to use generalized displacements $\mathbf{q}$ and corresponding shape functions $\mathbf{N}_q$. What is wrong with the following assumption

$$\mathbf{N}_q = [\,1 \quad x \quad y \quad x^2 \quad xy \quad y^2\,]\,?$$

Propose another set of generalized shape functions that will give the element the same displacement properties as those obtained in question a).

Problem 5.7

A triangular C^1 element for a problem with one field variable (ϕ) is to be based on a *complete* quartic (4th order) polynomial in x and y. The element has straight edges. Suggest nodal points (number and position) and degrees of freedom for the element and discuss the inter element continuity properties.

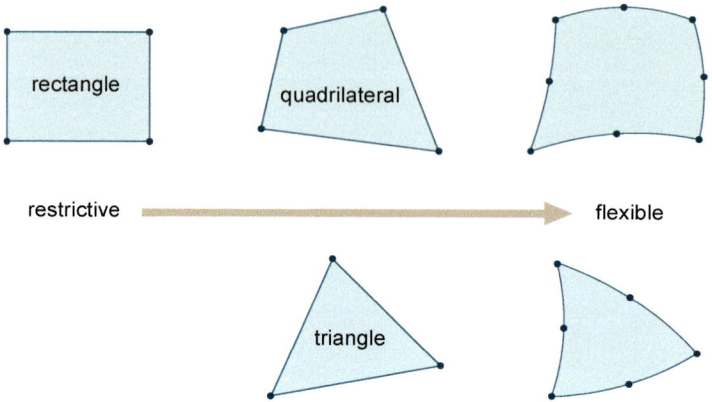

Figure 6.1 Different shapes of 2D elements

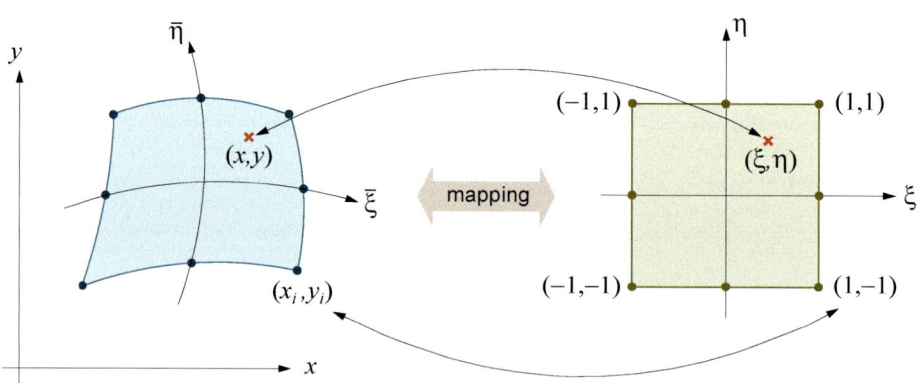

Figure 6.2 Mapping

6

Element analysis II
Mapping and numerical integration

We continue with the "tools of the trade" for displacement based finite elements, and now we turn our attention to elements with arbitrary shapes, including curved edges/surfaces. Mapping becomes essential, and as a result integration is no longer a straightforward operation, forcing us to turn to numerical integration. In this chapter we also introduce the important concept of the isoparametric formulation.

6.1 Mapping – isoparametric formulation

With the exception of the triangle and the tetrahedron, the element shapes of the previous chapter are not well suited for practical use. The rectangle, for instance, has severe restrictions when it comes to changing the element size in a mesh of elements, and also in modelling irregular boundaries. An arbitrary polygon with 4 sides, a *quadrilateral*, provides far more flexibility, almost on a par with the triangle. For the higher order elements we need far fewer elements in order to obtain a certain accuracy, and, to limit the geometric error when modelling curved boundaries, it might be useful to have an element for which one or more edges (or surfaces in 3D) can be made to curve, see Fig. 6.1.

How do we handle the necessary mathematical operations for these irregular elements? The answer is *mapping*. In the physical coordinate space (x,y), the natural coordinates $(\bar{\xi},\bar{\eta})$ for an arbitrary element are, in the general case, *curvilinear*. We would like to normalize these curvilinear coordinates such that we can perform all mathematical operations on the simplest possible geometry, as shown to the right in Fig. 6.2. In order to do that we need to establish unambiguous and reversible relations between the physical coordinates (x,y) for an arbitrary point inside the element and the corresponding coordinates (ξ,η) in the mapped system. With such a relationship in place we can express the shape functions, the derivation and integration in (normalized) natural coordinates ξ and η, and then transform established element properties (matrices) back to global, physical coordinates x and y.

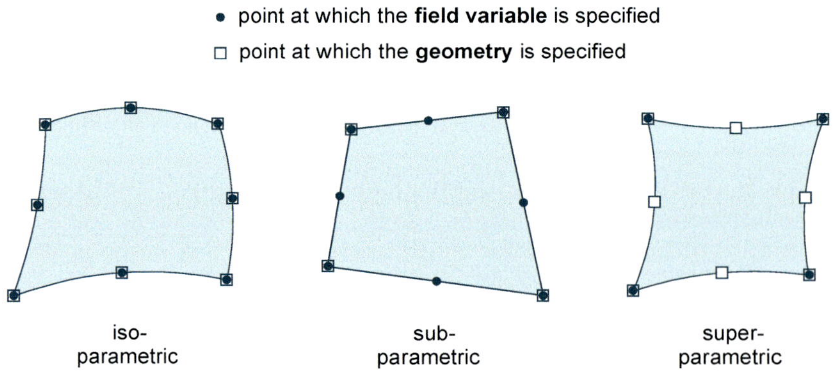

Element analysis II - Mapping and numerical integration

- point at which the **field variable** is specified
- point at which the **geometry** is specified

iso-parametric   sub-parametric   super-parametric

Figure 6.3 Different geometry mappings

Mapping – isoparametric formulation

An explicit relationship between x and y on the one hand and ξ and η on the other (or between x, y, z and ξ, η, ζ) of the kind we have between x and y and the area coordinates ζ_1, ζ_2 and ζ_3 for a triangle with straight sides, is not easily established. Fortunately we can do without such a relationship. What we need is relations of the type

$$x = f_x(\xi, \eta) \tag{6-1a}$$

$$y = f_y(\xi, \eta) \tag{6-1b}$$

where the nodal point values x_i and y_i satisfy

$$x_i = f_x(\xi_i, \eta_i) \tag{6-2a}$$

$$y_i = f_y(\xi_i, \eta_i) \tag{6-2b}$$

With the understanding that the positions of the nodal points uniquely define the geometric form of the element, it seems quite reasonable to express the element geometry as an *interpolation* between the nodal points, that is as

$$x = \sum_i N_{gi} x_i = \mathbf{N}_g \mathbf{x} \tag{6-3a}$$

$$y = \sum_i N_{gi} y_i = \mathbf{N}_g \mathbf{y} \tag{6-3b}$$

where $\mathbf{N}_g$ contains *shape functions* in terms of natural coordinates ξ and η, and $\mathbf{x}$ and $\mathbf{y}$ are vectors containing the nodal point coordinates in the physical reference system (x,y). The shape function N_{gi} has exactly the same properties as the standard C^0 shape functions described in the previous chapter. A point inside the element is uniquely defined by the physical nodal point coordinates and its natural ξ- and η-coordinates. Using the same requirements for N_{gi} as for the C^0-functions of the previous section, an edge (surface) is also uniquely defined by the nodes on that edge and only those nodes; this ensures complete geometric connection between neighbouring elements of the same type.

If we use the same interpolation for the geometry as for the field variable (f), that is if

$$\mathbf{N}_g = \mathbf{N} \tag{6-4}$$

then we have an *isoparametric* element. This approach to the mapping process was first suggested by TAIG [15] and generalized by IRONS [16,17]. If the geometry is defined by lower order shape functions than the field variable, that is by fewer points than the number of nodal points with *dofs*, then we have a *subparametric* element. In principle, but hardly in practice, we could also have a *superparametric* element, see Fig. 6.3.

Having defined the mapping, how do we proceed? Clearly there is some mathematics involved here. The procedure is perhaps best explained by an example, and the 4-node C^0 quadrilateral is a good candidate.

4-node C^0 quadrilateral. This is a typical isoparametric element, and as we shall see later it is an important plane stress/plane strain finite element. In

Element analysis II - Mapping and numerical integration

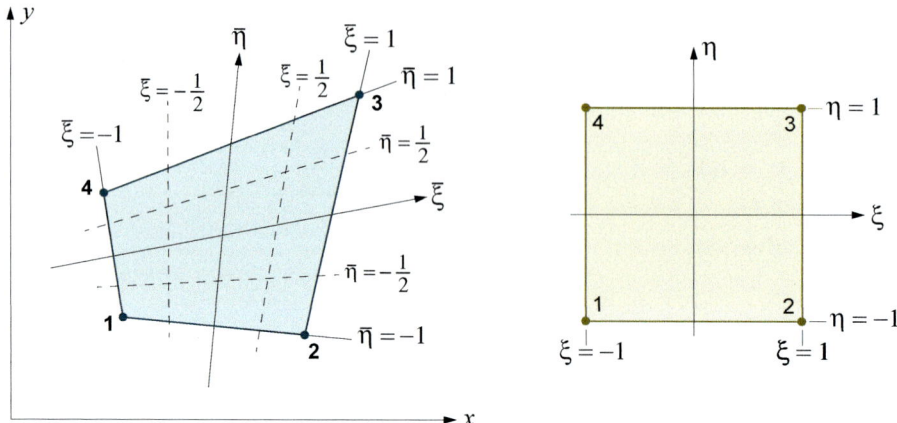

Figure 6.4 4-node quadrilateral C^0 element in physical and mapped space

order to avoid repeating much of this in a later chapter and also to enable a comprehensive description, we now consider *two* field variables, the two orthogonal displacement components u and v. The vector **v** of nodal *dofs* therefore contains the four nodal values of u and four of v. For simplicity we arrange the nodal *dofs* in two sub-vectors $\mathbf{v}_x$ and $\mathbf{v}_y$. With reference to Fig. 6.4 we have

$$\mathbf{v}^T = [\mathbf{v}_x^T \quad \mathbf{v}_y^T] = [\, u_1 \quad u_2 \quad u_3 \quad u_4 \quad v_1 \quad v_2 \quad v_3 \quad v_4] \tag{6-5}$$

$$\mathbf{x}^T = [\, x_1 \quad x_2 \quad x_3 \quad x_4 \,] \tag{6-6a}$$

$$\mathbf{y}^T = [\, y_1 \quad y_2 \quad y_3 \quad y_4 \,] \tag{6-6b}$$

The mapping, that is the relationship between physical and natural coordinates, we now express as

$$x = \sum_i N_i x_i = \mathbf{N}_0 \mathbf{x} \quad \text{and} \quad y = \sum_i N_i y_i = \mathbf{N}_0 \mathbf{y} \tag{6-7}$$

where the shape functions N_i are the bilinear functions in ξ and η that we already established in the previous chapter, see Eq. (5-56), that is

$$N_i(\xi, \eta) = \tfrac{1}{4}(1 + \xi_i \xi)(1 + \eta_i \eta) \qquad i = 1, 2, 3, 4$$

We see that with this definition $\bar{\xi}$ and $\bar{\eta}$ coincide with the medians of the quadrilateral. The point $\xi = \eta = 0$ can be regarded as the centre of the element (but in the general case it does not coincide with the centre of area). We shall return to the question of unambiguity of the mapping defined by Eq. (6-7).

The displacement components are defined by their nodal values and the same shape functions as the geometry, that is

$$\mathbf{u} = \begin{bmatrix} u \\ v \end{bmatrix} = \begin{bmatrix} \mathbf{N}_0 & 0 \\ 0 & \mathbf{N}_0 \end{bmatrix} \begin{bmatrix} \mathbf{v}_x \\ \mathbf{v}_y \end{bmatrix} = \mathbf{N}\mathbf{v} \tag{6-8}$$

The strains are defined by Eq. (4-50) which we repeat here:

$$\boldsymbol{\varepsilon} = \Delta \mathbf{u} = \Delta(\mathbf{N}\mathbf{v}) = \Delta \mathbf{N} \mathbf{v} = \mathbf{B}\mathbf{v} \tag{6-9}$$

For 2-dimensional problems, the operator matrix, given by Eq. (3-41) is

$$\Delta = \begin{bmatrix} \dfrac{\partial}{\partial x} & 0 \\ 0 & \dfrac{\partial}{\partial y} \\ \dfrac{\partial}{\partial y} & \dfrac{\partial}{\partial x} \end{bmatrix} \tag{6-10}$$

Hence we need to determine $\dfrac{\partial N_i(\xi, \eta)}{\partial x}$ and $\dfrac{\partial N_i(\xi, \eta)}{\partial y}$ which is not a trivial

195

Element analysis II - Mapping and numerical integration

Sir Isac NEWTON (1642-1727) was an English physicist, mathematician, astronomer, natural philosopher, alchemist and theologian, who has been "considered by many to be the greatest and most influential scientist who ever lived".

His monograph *Philosophiæ Naturalis Principia Mathematica*, published in 1687, provides the foundations for most of classical mechanics. Here NEWTON described universal gravitation and the three laws of motion, which dominated the scientific view of the physical universe for almost the next three centuries. The *Principia* is generally considered to be one of the most important scientific books ever written, not only for the specific physical laws the work successfully described, but also for the style of the work, which assisted in setting standards for scientific publication up to the present time.

NEWTON built the first practical reflecting telescope and developed a theory of colour. He also formulated an empirical law on cooling and studied the speed of sound. In mathematics, NEWTON shares the credit with Gottfried LEIBNIZ (1646-1716) for the development of differential and integral calculus. He also demonstrated the generalised binomial theorem, developed NEWTON's method for approximating the roots of a function, and contributed to the study of power series.

NEWTON was born after the death of his father, a prosperous farmer, and he was brought up by his maternal grandmother. He was educated at The King's School, Grantham, where he became the top-ranked student. In 1661 he was admitted to Trinity College, Cambridge, where he obtained his degree four years later. He was appointed Lucasian Professor of Mathematics at the University of Cambridge in 1669.

In 1696 NEWTON moved to London to take up the post of warden of the Royal Mint, a position he held until his death. In 1703 he was made President of the Royal Society and in 1705 he was knighted by Queen Anne.

NEWTON was also a member of the Parliament of England from 1689 to 1690 and also in 1701, but according to some accounts his only comments were to complain about a cold draught in the chamber and request that the window be closed.

NEWTON had differences of opinion with several of his contemporaries. His dispute with LEIBNIZ over priority in the development of infinitesimal calculus ended in a bitter controversy which marred the lives of both these giants. He also differed with Robert HOOKE.

NEWTON was also highly religious. He was an unorthodox Christian, and wrote more on religious matters and on occult studies than on science and mathematics.

It has been suggested that NEWTON perhaps suffered from ASPERGER's syndrome.

NEWTON died a bachelor. After his death his body was discovered to have massive amounts of mercury in it, probably as a result from his alchemical pursuits. Mercury poisoning could explain NEWTON's eccentricity in late life.

task. However, we have already touched upon this problem in Section 5.2, and the clue is to invoke the *chain* rule, that is

$$\frac{\partial N_i}{\partial x} = \frac{\partial N_i}{\partial \xi}\frac{\partial \xi}{\partial x} + \frac{\partial N_i}{\partial \eta}\frac{\partial \eta}{\partial x}$$

$$\frac{\partial N_i}{\partial y} = \frac{\partial N_i}{\partial \xi}\frac{\partial \xi}{\partial y} + \frac{\partial N_i}{\partial \eta}\frac{\partial \eta}{\partial y}$$

or, in matrix notation

$$\begin{bmatrix} \frac{\partial N_i}{\partial x} \\ \frac{\partial N_i}{\partial y} \end{bmatrix} = \begin{bmatrix} \frac{\partial \xi}{\partial x} & \frac{\partial \eta}{\partial x} \\ \frac{\partial \xi}{\partial y} & \frac{\partial \eta}{\partial y} \end{bmatrix} \begin{bmatrix} \frac{\partial N_i}{\partial \xi} \\ \frac{\partial N_i}{\partial \eta} \end{bmatrix} = \mathbf{J}^{-1} \begin{bmatrix} \frac{\partial N_i}{\partial \xi} \\ \frac{\partial N_i}{\partial \eta} \end{bmatrix} \quad (6\text{-}11)$$

where

$$\mathbf{J}^{-1} = \frac{\partial(\xi, \eta)}{\partial(x, y)} = \begin{bmatrix} \frac{\partial \xi}{\partial x} & \frac{\partial \eta}{\partial x} \\ \frac{\partial \xi}{\partial y} & \frac{\partial \eta}{\partial y} \end{bmatrix} \quad (6\text{-}12)$$

where $\mathbf{J}^{-1}$ actually is the Jacobian matrix of (ξ,η) with respect to (x,y), but in "FEM technology" it is usually termed "the inverse Jacobian matrix". It is the inverse of

$$\mathbf{J} = \frac{\partial(x, y)}{\partial(\xi, \eta)} = \begin{bmatrix} \frac{\partial x}{\partial \xi} & \frac{\partial y}{\partial \xi} \\ \frac{\partial x}{\partial \eta} & \frac{\partial y}{\partial \eta} \end{bmatrix} \quad (6\text{-}13)$$

which is the Jacobian matrix of (x,y) with respect to (ξ,η), but normally termed simply the Jacobian matrix. This is consistent with Section 5.2.

We need the inverse Jacobian matrix, $\mathbf{J}^{-1}$, and the only way to find this matrix is in fact to invert the Jacobian matrix which is not too hard to find:

$$\frac{\partial x}{\partial \xi} = \frac{\partial}{\partial \xi}\mathbf{N}_0\mathbf{x} = \sum_{i=1}^{4}\frac{\partial N_i}{\partial \xi}x_i = \frac{1}{4}\sum_{i=1}^{4}\xi_i(1+\eta_i\eta)x_i$$

$$\frac{\partial x}{\partial \eta} = \frac{\partial}{\partial \eta}\mathbf{N}_0\mathbf{x} = \sum_{i=1}^{4}\frac{\partial N_i}{\partial \eta}x_i = \frac{1}{4}\sum_{i=1}^{4}\eta_i(1+\xi_i\xi)x_i$$

Similarly for y, and the result is

$$\mathbf{J} = \begin{bmatrix} \frac{1}{4}\sum_{i=1}^{4}\xi_i(1+\eta_i\eta)x_i & \frac{1}{4}\sum_{i=1}^{4}\xi_i(1+\eta_i\eta)y_i \\ \frac{1}{4}\sum_{i=1}^{4}\eta_i(1+\xi_i\xi)x_i & \frac{1}{4}\sum_{i=1}^{4}\eta_i(1+\xi_i\xi)y_i \end{bmatrix} = \begin{bmatrix} J_{11} & J_{12} \\ J_{21} & J_{22} \end{bmatrix} \quad (6\text{-}14)$$

Element analysis II - Mapping and numerical integration

$$\mathbf{P} = \begin{vmatrix} N_{1,\xi} & N_{2,\xi} & N_{3,\xi} & N_{4,\xi} \\ N_{1,\eta} & N_{2,\eta} & N_{3,\eta} & N_{4,\eta} \end{vmatrix} \mathbf{X} = \begin{vmatrix} x_1 & y_1 \\ x_2 & y_2 \\ x_3 & y_3 \\ x_4 & y_4 \end{vmatrix} = \begin{vmatrix} J_{11} & J_{12} \\ J_{21} & J_{22} \end{vmatrix} = \mathbf{J} \quad \Rightarrow \quad J \text{ and } \mathbf{J}^{-1}$$

$$\mathbf{J}^{-1} = \frac{1}{J}\begin{vmatrix} J_{22} & -J_{12} \\ -J_{21} & J_{11} \end{vmatrix} \mathbf{P} = \begin{vmatrix} N_{1,\xi} & N_{2,\xi} & N_{3,\xi} & N_{4,\xi} \\ N_{1,\eta} & N_{2,\eta} & N_{3,\eta} & N_{4,\eta} \end{vmatrix} = \begin{vmatrix} N_{1,x} & N_{2,x} & N_{3,x} & N_{4,x} \\ N_{1,y} & N_{2,y} & N_{3,y} & N_{4,y} \end{vmatrix} = \begin{vmatrix} \mathbf{N}_{0,x} \\ \mathbf{N}_{0,y} \end{vmatrix}$$

The *Jacobian*, the determinant of **J**, is

$$J = |\mathbf{J}| = \det(\mathbf{J}) = J_{11}J_{22} - J_{12}J_{21} \qquad (6\text{-}15)$$

and

$$\mathbf{J}^{-1} = \frac{1}{J}\begin{bmatrix} J_{22} & -J_{12} \\ -J_{21} & J_{11} \end{bmatrix} \qquad (6\text{-}16)$$

Then, from Eq. (6-9),

$$\mathbf{B} = \Delta \mathbf{N} = \begin{bmatrix} \dfrac{\partial}{\partial x} & 0 \\ 0 & \dfrac{\partial}{\partial y} \\ \dfrac{\partial}{\partial y} & \dfrac{\partial}{\partial x} \end{bmatrix} \mathbf{N} = \begin{bmatrix} \mathbf{N}_{0,x} & 0 \\ 0 & \mathbf{N}_{0,y} \\ \mathbf{N}_{0,y} & \mathbf{N}_{0,x} \end{bmatrix} \qquad (\mathbf{N}_{0,x} \equiv \dfrac{\partial}{\partial x}\mathbf{N}_0) \qquad (6\text{-}17)$$

Computationally, matrix **B** may be determined as follows:

First we define two auxiliary matrices, **P** and **X** as

$$\mathbf{P} = \begin{bmatrix} \mathbf{N}_{0,\xi} \\ \mathbf{N}_{0,\eta} \end{bmatrix} \quad \text{and} \quad \mathbf{X} = [\ \mathbf{x} \ \ \mathbf{y}\ ] \quad \Rightarrow \quad \mathbf{J} = \mathbf{PX} \quad \Rightarrow \quad J \text{ and } \mathbf{J}^{-1}$$

from which we find $\begin{bmatrix} \mathbf{N}_{0,x} \\ \mathbf{N}_{0,y} \end{bmatrix} = \mathbf{J}^{-1}\mathbf{P}$ and we have the two sub-matrices that make up matrix **B**.

The element stiffness matrix, defined by Eq. (4-54), is

$$\mathbf{k} = \int_V \mathbf{B}^T \mathbf{C} \mathbf{B}\, dV = \iint_A t \mathbf{B}^T \mathbf{C} \mathbf{B}\, dA \qquad (6\text{-}18)$$

The constitutive matrix **C** is normally a constant matrix; the thickness *t* may be a function of ξ and η (expressed in terms of the corner thicknesses) and matrix **B** is, in general, a function of ξ and η, while the integration in (6-18) is taken over the physical area. We have already seen, in Section 5.2, how we can perform the integral in natural coordinates by expressing *dA* by *dξ* and *dη* as

$$dA = J\, d\xi\, d\eta \qquad (6\text{-}19)$$

This is a general mathematical relationship that can be demonstrated by the *vector* or *cross product* as shown on the next page.

For mapped (isoparametric) elements the equation for the stiffness matrix is now written as follows:

Element analysis II - Mapping and numerical integration

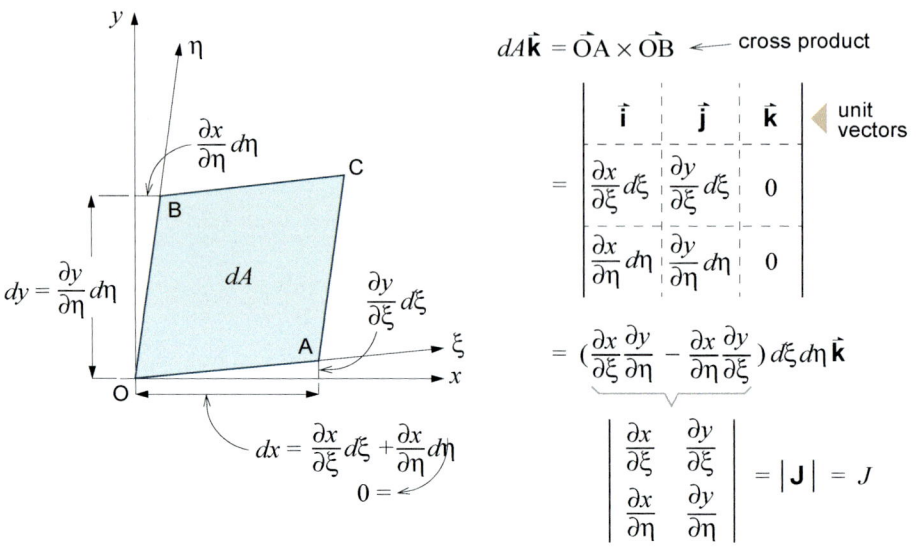

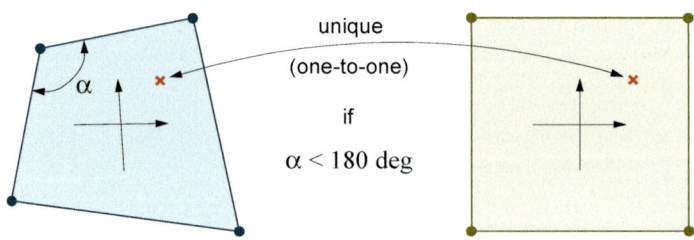

Figure 6.5 Unique mapping

Mapping – isoparametric formulation

$$\mathbf{k} = \int_{-1}^{1}\int_{-1}^{1} t(\xi, \eta)\mathbf{B}^T(\xi, \eta)\mathbf{CB}(\xi, \eta)J d\xi d\eta \qquad (6\text{-}20)$$

It should be emphasized that $\mathbf{J}^{-1}$ and hence $1/J$ is incorporated into matrix $\mathbf{B}$; this makes the expression in Eq. (6-20) sensitive to the value of J.

For an arbitrary quadrilateral element it is extremely difficult, if at all possible, to carry out the integrations in Eq. (6-20) analytically. In practice therefore, *numerical integration* is used exclusively for all mapped elements.

It appears that the Jacobian J plays an important role in the evaluation of mapped elements, so let us take a closer look at it. As already pointed out in Section 5.2, J plays the role of a *scale factor* between physical and natural coordinates, and in the general case it is a function of the natural coordinates. Since $\mathbf{J}^{-1}$ appears in the expression for matrix $\mathbf{B}$, J appears three times in each element of the integrand matrix of Eq. (6-20), once in the numerator and twice in the denominator. The upshot of this is that we have J in the denominator of all elements of the integrand matrix; hence, in general, we need to integrate *rational functions* of the natural coordinates.

For the simple 4-node quadrilateral of Fig. 6.4, J varies from point to point in the element, and according to Eq. (6-14) we can express this variation as

$$J = |\mathbf{J}| = J_0 - J_1\xi - J_2\eta - J_3\xi\eta \qquad (6\text{-}21)$$

where

$$J_3 = \frac{1}{16}[(\sum \xi_i\eta_i x_i)(\sum \eta_i\xi_i y_i) - (\sum \xi_i\eta_i y_i)(\sum \eta_i\xi_i x_i)] = 0$$

and, for the 4-node quadrilateral, we have

$$J = J_0 - J_1\xi - J_2\eta \qquad (6\text{-}22)$$

For the rectangle we have already shown, see Eq. (5-11), that J is constant. It can also be shown that it is a constant for a parallelogram. That is

for rectangles and parallelograms $J = J_0 = $ constant

How much can we "deform" a four-sided polygon from the square shape and still have a unique mapping? It can be shown that the condition for unique mapping is that the Jacobian has the same sign everywhere within the element. With the numbering scheme we have used consistently so far, in the anti-clockwise direction, that means the Jacobian determinant must be positive throughout the element. Hence the condition for unique mapping is

$$J = |\mathbf{J}| > 0 \qquad (6\text{-}23)$$

For the bilinear, 4-node quadrilateral this means that all "inside" angles must be less than 180 degrees, see Fig. 6.5. The quadrilateral element can degenerate into a triangle in two ways, see Fig. 6.6a and b. For the element in Fig. 6.6a the Jacobian is zero at node 3, and at this point stresses are not defined (infinite strains); strictly speaking, such an element can be tolerated if an open quadrature (numerical integration) rule is used (all integration points

201

Element analysis II - Mapping and numerical integration

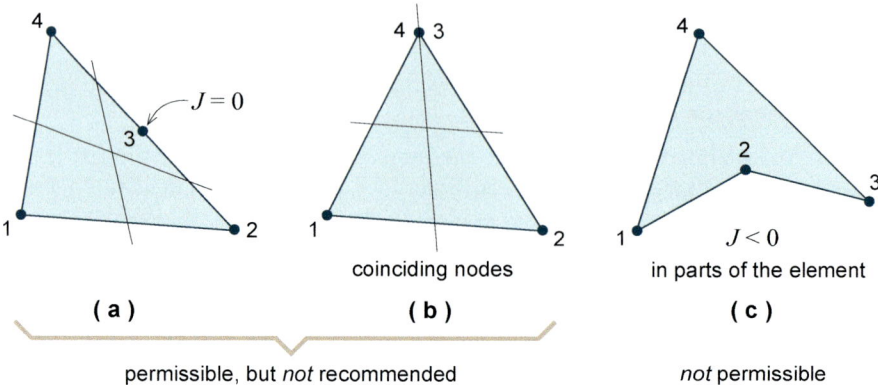

Figure 6.6 Special element shapes

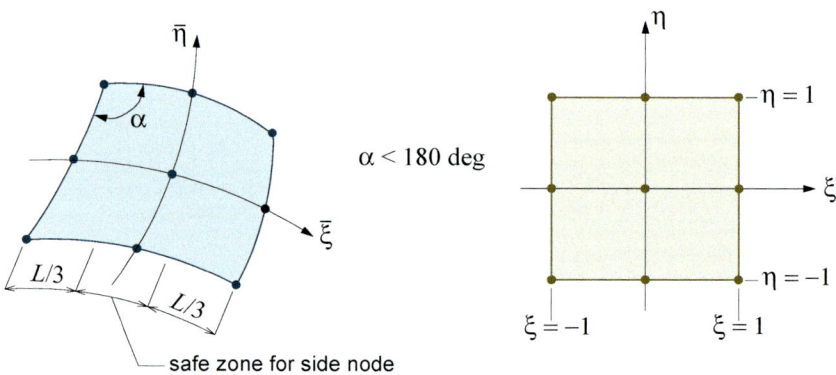

Figure 6.7 Unique mapping of quadratic C^0 element with curved edges

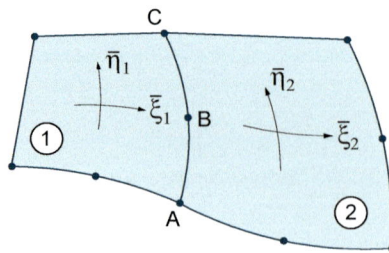

Figure 6.8 Continuity between isoparametric elements

are placed *inside* the element), and stresses must *not* be computed directly at the nodes. The same applies to the element in Fig. 6.6b; the corner points must *not* be used as integration points. Nevertheless, both these degenerated elements should be avoided. The element in Fig. 6.6c *must* be avoided since the mapping in this case is ambiguous – the Jacobian is negative in parts of the element.

Higher order C^0 quadrilaterals – curved edges. For an element that also has side nodes it is possible, by suitable positioning of these nodes, to give the element curved edges in the physical space. The physical coordinates $\bar{\xi}$ and $\bar{\eta}$ are now curvilinear coordinates that map onto the natural ξ- and η-coordinates of the "parent" square in exactly the same way as described above for the 4-node element, with $\mathbf{N}_g = \mathbf{N}$, where $\mathbf{N}$ contains the shape functions already presented for the parent elements, see Eqs. (5.57) and (5-62). This means of course that **x** and **y** must now contain the physical coordinates of all nodes, not only the corner nodes.

The question then arises as to where the side nodes should be placed in order for the mapping to be unique. It turns out that side nodes need not divide the side in equal lengths. For the 8-node serendipity and 9-node LAGRANGE elements, for instance, ZIENKIEWICZ and TAYLOR [18] claim that if the side nodes are placed within the middle third, the mapping is unique, see Fig. 6.7.

Uniqueness of mapping is one thing, performance of the element is another. For practical purposes one should always try to keep the shape of a quadrilateral element as close to the square as possible, and for a quadratic element with given corner nodes, the one with straight edges will perform better than the one with curved edges. And the more a curved edge deviates from the straight line the poorer is the performance. If one settles for straight edges, the geometry of the quadratic (8- or 9-node) quadrilateral elements may be defined by the bilinear shape functions, that is, a subparametric formulation may be used (Fig. 6.3).

What about the *validity* of the isoparametric formulation? Or, in other words, what about *continuity* and *completeness*? Along edge A-B-C in Fig. 6.8 we have

$$\bar{\xi}_1 = \text{const.}, \quad \bar{\xi}_2 = \text{const.} \quad \text{and} \quad \bar{\eta}_1 = \bar{\eta}_2$$

Furthermore, the shape functions for elements 1 and 2 along A-B-C are identical functions of the natural (parent element) coordinate η, and these functions *only* depend on parameters at nodes A, B and C. Since these parameters are the same for both elements, whether physical coordinates or displacement components, we may state:

> If the shape functions $\mathbf{N} = \mathbf{N}(\xi,\eta)$ are such that C^0 continuity is satisfied between elements in mapped coordinates ξ and η, then continuity is also satisfied, for both geometry and displacements, for the elements in the physical space.

This argument and the above statement also hold for sub- and superparametric elements.

Element analysis II - Mapping and numerical integration

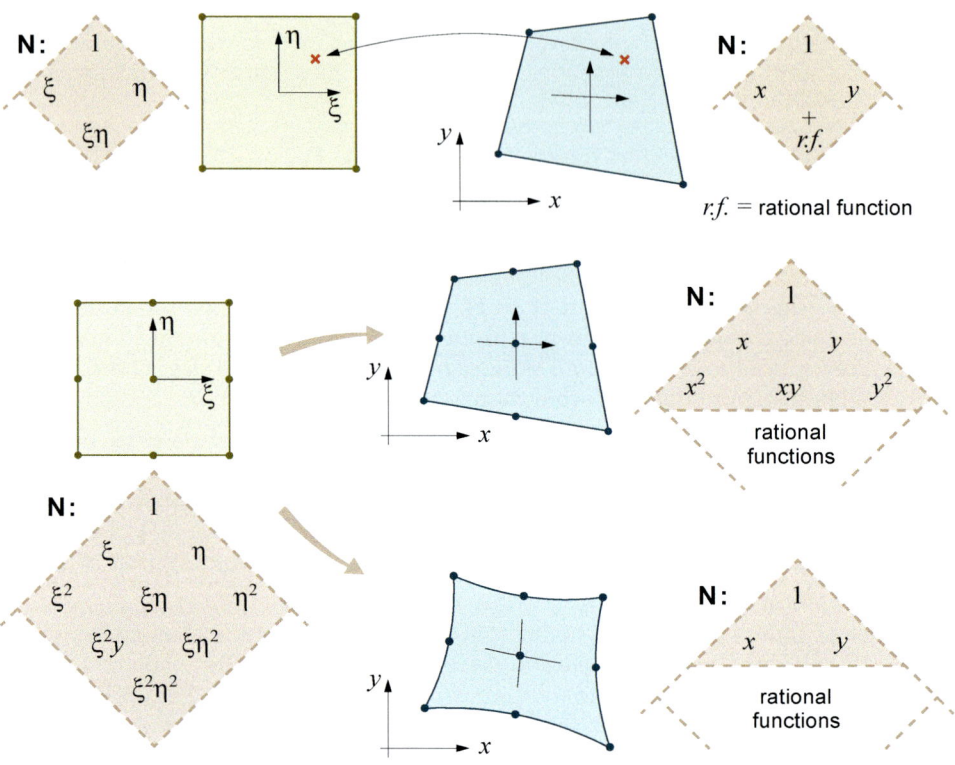

Figure 6.9 Shape functions in mapped and physical coordinates

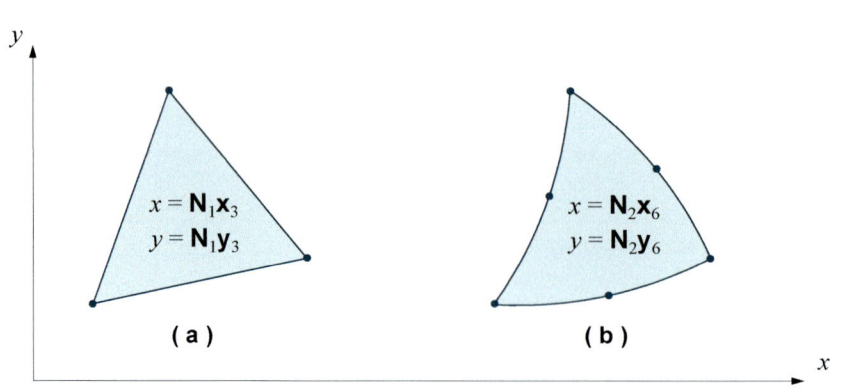

Figure 6.10 Triangular isoparametric elements

As for *completeness* – correct representation of rigid body movements and constant strain (stress) – it is necessary for the field variable ϕ (= u or v) to be able to represent a complete linear polynomial, that is,

$$\phi = c_0 + c_1 x + c_2 y$$

with nodal values

$$\phi_i = c_0 + c_1 x_i + c_2 y_i$$

If we substitute this into the assumption

$$\phi = \sum_i N_i \phi_i$$

we find

$$\phi = c_0 \sum_i N_i + c_1 \sum_i N_i x_i + c_2 \sum_i N_i y_i = c_0 \sum_i N_i + c_1 x + c_2 y$$

and the completeness requirement is satisfied if

$$\sum_i N_i = 1 \tag{6-24}$$

This is the same result as we arrived at in Eq. (5-42), and since this condition must hold for the shape functions, independent of mapping or not, we can conclude that the isoparametric formulation does not introduce any new requirements that the shape functions have to meet in order to satisfy continuity and completeness.[1]

What kind of shape functions do we have in the physical element, expressed in terms of x and y, when the starting point is shape functions assumed for the mapped (parent) element in terms of natural coordinates ξ and η? The answer indicated in Fig. 6.9 is from [19]. This figure reinforces what we have already indicated: for a 4-sided element the square shape gives the best accuracy (has the highest number of polynomial terms), and curved edges give lower accuracy than straight edges (provided the corner nodes of both elements have the same position). While curved edges will reduce the discretization error in the case of a complex (curved) geometry, compared to straight edges, it is obtained at the expense of poorer performance. This trade-off has no clear winner, but strongly curved edges should be avoided.

It should also be mentioned that higher order elements are more robust with respect to geometric distortion than lower order elements, and it is worth noting that in this respect the 9-node LAGRANGE element is significantly better than the 8-node serendipity element.

Triangular, isoparametric C^0 elements. The simple 3-node, linear triangle in Fig. 6.10, with shape functions in natural area coordinates,

$$\mathbf{N}_i = [\; \zeta_1 \;\; \zeta_2 \;\; \zeta_3 \;]$$

may be considered (and treated) as an isoparametric element. However, as shown in Section 5.2, the Jacobian for a triangle with straight edges is a

---

1. Completeness is normally *not* satisfied for superparametric elements.

Element analysis II - Mapping and numerical integration

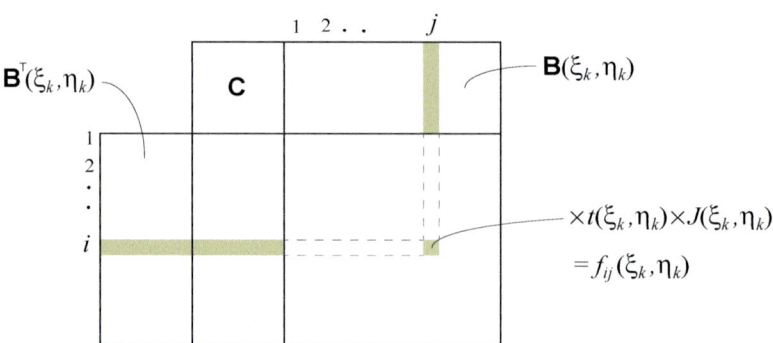

constant ($J = 2A$), see Eq. (5-28). Hence, for a triangular element with straight edges, we need not invoke mapping and the Jacobian matrix. We can express the cartesian coordinates in terms of the area coordinates, see Eq. (5-19), and vice versa, see Eq. (5-21), and thus establish expressions for the derivatives with respect to x and y of any function of the area coordinates, see Eq. (5-24).

Higher order triangular elements, with nodes on the sides, like the quadratic element in Fig. 6.10, can accommodate curved edges. For such an element the Jacobian will vary from point to point within the element, and we need to employ mapping onto a triangle with straight edges in precisely the same way as for the quadrilateral. Again this will lead to rational functions in the integrand of the stiffness matrix, see Eq. (6-20), and we will have to resort to numerical integration.

Since the procedure for a triangular C^0 element with curved edges is the same as for a C^0 quadrilateral we need not present a repetition.

6.2 Numerical integration

The elements of the stiffness matrix **k** (and the consistent load vector **S**^{0}) are integrals of the form

$$I = \int_L f(x)\,dx = \int_{-1}^{1} f(\xi)\,d\xi \qquad (6\text{-}25a)$$

$$I = \int_A f(x,y)\,dA = \begin{cases} \int_A f(\zeta_1,\zeta_2,\zeta_3)\,dA & \text{(triangle)} \\[1em] \int_{-1}^{1}\int_{-1}^{1} f(\xi,\eta)\,d\xi\,d\eta & \text{(square)} \end{cases} \qquad (6\text{-}25b)$$

$$I = \int_V f(x,y,z)\,dV = \int_{-1}^{1}\int_{-1}^{1}\int_{-1}^{1} f(\xi,(\eta,\zeta))\,d\xi\,d\eta\,d\zeta \qquad (6\text{-}25c)$$

For some elements, for instance 1-dimensional and 2-dimensional elements of rectangular or triangular shape, these integrals can be evaluated analytically. However, for many elements, and among them most isoparametric elements, this is not feasible, mainly because it is difficult if not impossible to derive analytical expressions for the elements of the integrand matrix. For these elements we therefore need to employ *numerical* integration.

With reference to Eq. (5-95) it is quite straightforward to evaluate numerically an element of the integrand matrix at, for instance, point (ξ_k,η_k):

$$\mathbf{B}^T(\xi_k,\eta_k)\mathbf{C}\mathbf{B}(\xi_k,\eta_k)t(\xi_k,\eta_k)J(\xi_k,\eta_k) = [f_{ij}(\xi_k,\eta_k)]$$

Hence

Element analysis II - Mapping and numerical integration

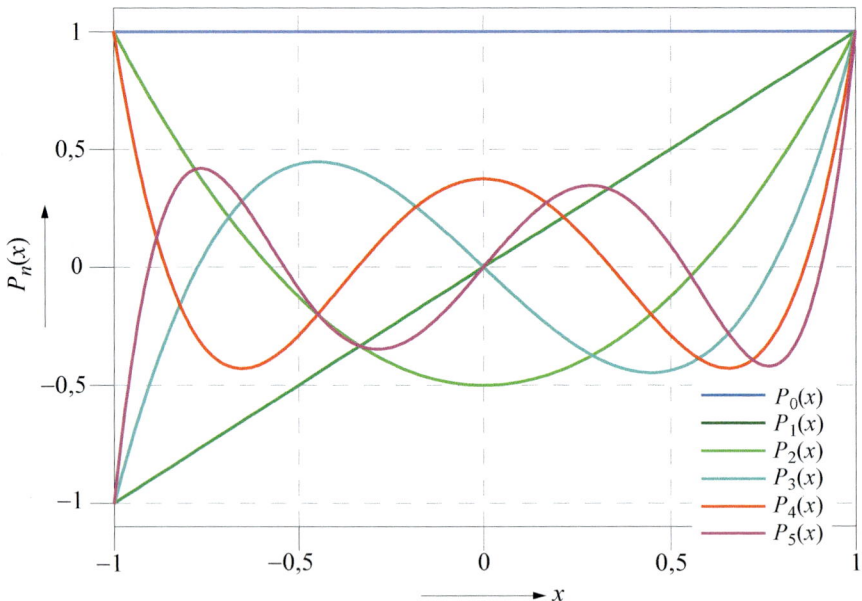

General formula (BONNET's recursion formula):

$$(n+1)P_{n+1}(x) = (2n+1)xP_n(x) - nP_{n-1}(x)$$

$P_0 = 1$
$P_1 = x$

Figure 6.11 LEGENDRE polynomials

Numerical integration

$$k_{ij} = \int_{-1}^{1}\int_{-1}^{1} f_{ij} d\xi d\eta \approx \sum_k w_k(\xi_k, \eta_k) f_{ij}(\xi_k, \eta_k) \tag{6-26}$$

where w_k is the *weight* for the particular integration scheme used; these weights depend on the number and position of *integration points k*.

One-dimensional element. For such an element we have

$$I = \int_{-1}^{1} f(\xi) d\xi \approx \sum_{k=1}^{n} w_k f(\xi_k) \tag{6-27}$$

In order to establish formulas for w_k and ξ_k we have two basically different approaches:

1) NEWTON-COTES *quadrature*[1]

   The integration points are determined *a priori* as equidistant points, including the end points. A polynomial of suitable order is passed through the function values, f_k, and integrated exactly. Since n function values defines a polynomial of order $n-1$ uniquely, three points, for instance, will integrate a 2nd order polynomial exactly. The first three formulas are:

   $n = 2$: $\quad I = f(-1) + f(1) \qquad$ trapezoidal rule

   $n = 3$: $\quad I = \frac{1}{3}[f(-1) + 4f(0) + f(1)] \qquad$ SIMPSON's 1/3 rule

   $n = 4$: $\quad I = \frac{1}{4}[f(-1) + 3f(-\frac{1}{3}) + 3f(\frac{1}{3}) + f(1)]$

   and so on. Since the method always includes the end points it is referred to as a *closed* method.

2) GAUSS *quadrature*

   If we also let the positions, ξ_k, of the k integration points, and not just the weights, w_k, be "free variables" in the process of fitting a polynomial with as high a degree as possible through the points (ξ_k, w_k), then we have $2n$ parameters at our disposal. This enables us to fit, and integrate exactly, a polynomial of degree $2n-1$ with n points.

   It can be shown that the positions of the integration points coincide with the zero-crossings of so-called LEGENDRE polynomials, see Fig. 6.11. This integration method is therefore often referred to as GAUSS-LEGENDRE quadrature.

   The proof goes as follows:

   Assume a complete polynomial in ξ of order $2n-1$,

   $$g(\xi) = a_1 + a_2\xi + a_3\xi^2 + \ldots a_{2n-1}\xi^{2n-2} + a_{2n}\xi^{2n-1} \tag{6-28}$$

   which has $2n$ terms.

   Exact (analytical) integration between -1 and 1 gives:

---

1. "Quadrature", which means "area", is a commonly used alternative to "numerical integration".

Element analysis II - Mapping and numerical integration

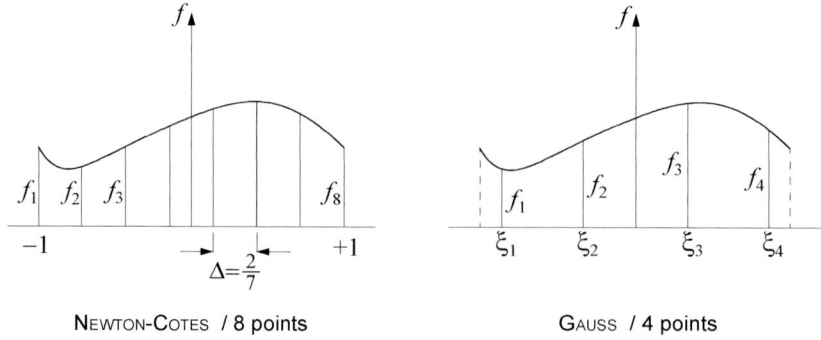

NEWTON-COTES / 8 points   GAUSS / 4 points

Both integrate a 7th order polynomial exactly

$$I = \int_{-1}^{1} g(\xi)d\xi = 2a_1 + \frac{2}{3}a_3 + \frac{2}{5}a_5 + \ldots + \frac{2}{2n-1}a_{2n-1} \qquad (6\text{-}29)$$

This expression has n terms. We use n integration points:

$$I = \int_{-1}^{1} g(\xi)d\xi \approx \sum_{k=1}^{n} w_k\, g(\xi_k) \qquad (6\text{-}30)$$

With $g(\xi)$ defined by Eq. (6-28) we obtain

$$I = a_1 \sum_{k=1}^{n} w_k + a_2 \sum_{k=1}^{n} w_k \xi_k + a_3 \sum_{k=1}^{n} w_k \xi_k^2 + \ldots + a_{2n} \sum_{k=1}^{n} w_k \xi_k^{2n-1} \qquad (6\text{-}31)$$

Comparing Eqs. (6-29) and (6-31) provides the following $2n$ conditions from which ξ_k and w_k ($k = 1, 2, \ldots, n$) can be determined:

$$\sum_{k=1}^{n} w_k = 2 \; ; \; \sum_{k=1}^{n} w_k \xi_k = 0 \; ; \; \sum_{k=1}^{n} w_k \xi_k^2 = \frac{2}{3} \; ; \ldots \; \sum_{k=1}^{n} w_k \xi_k^{2n-1} = 0 \qquad (6\text{-}32)$$

For the first three values of n we find, for a normalized region between -1 and +1:

n	ξ_k	w_k	Integrates exactly
1	0,0	2,0	1st order polynomial
2	$\pm 1/\sqrt{3}$	1,0	3rd order polynomial
3	$\pm\sqrt{0,6}$ 0,0	5/9 8/9	5th order polynomial

The position and weights are tabulated, for many values of n, in a number of books – more information in APPENDIX C.

GAUSS quadrature is an optimal integration formula, and it is *open* in the sense that the end points of the region are never included as "sampling" (that is, integration) points. In the computer, the actual numbers for the positions and the corresponding weights are of no concern (they are preprogrammed once and for all), and it is therefore quite obvious that we opt for GAUSS integration; it gives a specified accuracy for half the effort compared to NEWTON-COTES.

A few simple examples will demonstrate the "power" of the GAUSS quadrature.

Element analysis II - Mapping and numerical integration

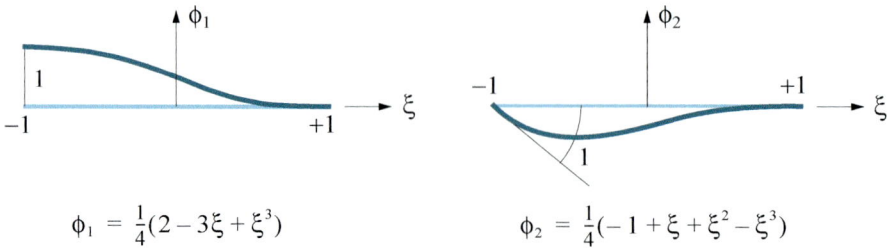

$$\phi_1 = \tfrac{1}{4}(2 - 3\xi + \xi^3)$$

$$\phi_2 = \tfrac{1}{4}(-1 + \xi + \xi^2 - \xi^3)$$

Figure 6.12 Two cubic (beam) functions in natural coordinates

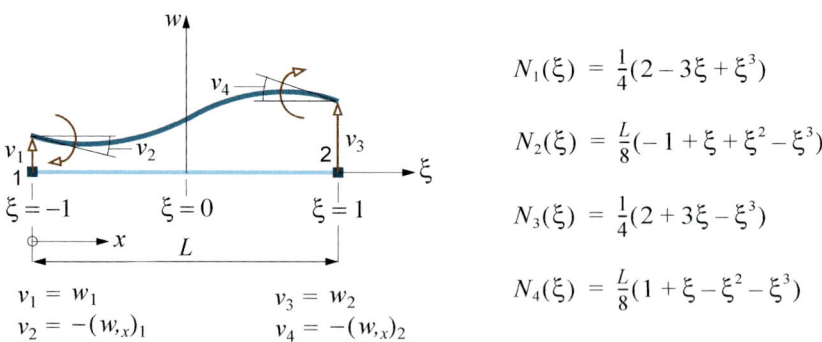

$$N_1(\xi) = \tfrac{1}{4}(2 - 3\xi + \xi^3)$$

$$N_2(\xi) = \tfrac{L}{8}(-1 + \xi + \xi^2 - \xi^3)$$

$$N_3(\xi) = \tfrac{1}{4}(2 + 3\xi - \xi^3)$$

$$N_4(\xi) = \tfrac{L}{8}(1 + \xi - \xi^2 - \xi^3)$$

$v_1 = w_1$
$v_2 = -(w_{,x})_1$
$v_3 = w_2$
$v_4 = -(w_{,x})_2$

Figure 6.13 Beam element in bending – natural coordinates

Example 6-1

Figure 6.12 shows two 3rd order "beam functions", see Eqs. (5-67). Exact, analytical integration gives,

$$I_1 = \int_{-1}^{1} \phi_1 d\xi = \left| \frac{1}{2}\xi - \frac{3}{8}\xi^2 + \frac{1}{16}\xi^4 \right|_{-1}^{1} = 1{,}0 \quad \text{and}$$

$$I_2 = \int_{-1}^{1} \phi_2 d\xi = \left| -\frac{1}{4}\xi + \frac{1}{8}\xi^2 + \frac{1}{12}\xi^3 - \frac{1}{16}\xi^4 \right|_{-1}^{1} = -\frac{1}{3}$$

	ξ_k	w_k	f_k	$w_k f_k$
ϕ_1 / $n = 1$:	0	2	1/2	1,0
ϕ_2 / $n = 1$:	0	2	−1/4	−0,5
ϕ_1 / $n = 2$:	$-1/\sqrt{3}$	1	0,8849	0,8849
	$+1/\sqrt{3}$	1	0,1159	0,1159
				1,0000
ϕ_2 / $n = 2$:	$-1/\sqrt{3}$	1	−0,2629	−0,2629
	$+1/\sqrt{3}$	1	−0,0704	−0,0704
				−0,3333

Both are exact for $n = 2$, as they should be. That ϕ_1 is exact also for $n = 1$ has to do with the function's antisymmetry about the ϕ_1-axis ($\xi = 0$).

Example 6-2 − beam element

We now consider the simple (2D) EULER-BERNOULLI beam element (in bending only) as an "ordinary" finite element for which we have, according to Eq. (4-54)

$$\mathbf{k} = \int_{V_e} \mathbf{B}^T \mathbf{C} \mathbf{B} \, dV$$

The field variable is the lateral displacement w for which we assume $w = \mathbf{Nv}$, where $\mathbf{N} = \mathbf{N}(\xi)$ contains the modified shape functions of Eq. (5-69). Note that while $\mathbf{N}$ is a function of the natural coordinate ξ, $\mathbf{v}$ contains the *physical dofs* (the rotations/slopes are the derivatives with respect to x).

Recall, from Eqs. (2-1) and (2-2),

$$\varepsilon = \varepsilon_x = \frac{du}{dx} = z\frac{d\theta}{dx} = -z\frac{d^2 w}{dx^2} = \left(-z\frac{d^2}{dx^2}\right) w \quad \text{and}$$

$$\sigma = \sigma_x = E\varepsilon_x = E\left(-z\frac{d^2}{dx^2}\right) w$$

Element analysis II - Mapping and numerical integration

$$\mathbf{B}(\xi) = \begin{bmatrix} \overset{B_1}{\tfrac{3}{2}\xi} & \overset{B_2}{\tfrac{L}{4}(1-3\xi)} & \overset{B_3}{-\tfrac{3}{2}\xi} & \overset{B_4}{-\tfrac{L}{4}(1+3\xi)} \end{bmatrix}$$

	B_1	B_2	B_3	B_4
B_1	B_1^2			
B_2				
B_3		$\mathbf{B}(\xi)^{\mathrm{T}}\mathbf{B}(\xi)$		
B_4				

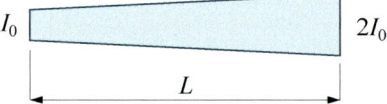

Hence $\mathbf{C} = [E]$ and $\Delta = -z\dfrac{d^2}{dx^2}$.

Recall also, see Eq. (5-4), $dx = \dfrac{L}{2}d\xi \implies \dfrac{d\xi}{dx} = \dfrac{2}{L}$ and

$$\frac{d^2}{dx^2} = \frac{d}{dx}\left(\frac{d}{dx}\right) = \frac{d}{d\xi}\left(\frac{d}{d\xi}\frac{d\xi}{dx}\right)\frac{dx}{d\xi} = \frac{4}{L^2}\frac{d^2}{d\xi^2}$$

Thus $\mathbf{B} = \Delta\mathbf{N} = -z\dfrac{4}{L^2}\dfrac{d^2}{d\xi^2}\mathbf{N}(\xi) = -z\dfrac{4}{L^2}\mathbf{B}(\xi)$ where $\mathbf{B}(\xi) = \dfrac{d^2}{d\xi^2}\mathbf{N}(\xi)$, and

$$\mathbf{k} = \int_{V_e}\mathbf{B}^T\mathbf{C}\mathbf{B}\,dV = \int_A\int_{-1}^{1} z^2\frac{4}{L^2}\mathbf{B}^T(\xi)E\frac{4}{L^2}\mathbf{B}(\xi)\,dA\underbrace{\frac{L}{2}d\xi}_{dx} = \int_{-1}^{1}\mathbf{B}^T(\xi)EI\mathbf{B}(\xi)\frac{8}{L^3}d\xi$$

The elements of $\mathbf{B}(\xi)$ are readily found to be as shown on the opposite page, from which the elements of the integrand matrix follow, for instance,

$$k_{11} = \frac{8E}{L^3}\int_{-1}^{1}\left(\frac{3}{2}\xi\right)^2 I\,d\xi$$

a) $I = $ constant and two integration points ($n = 2$) yield

$$k_{11} = \frac{8EI}{L^3}\left[\left(\frac{3}{2}\cdot 0{,}57735\right)^2 \cdot 1 \cdot 2\right] = \frac{12EI}{L^3} \quad \text{which is "exact"}$$

b) I varies linearly from I_0 at the left end to $2I_0$ at the right end, that is

$$I = \frac{I_0}{2}(3+\xi)$$

The integrand is now a cubic function of ξ, but $n = 2$ should still provide the exact result,

$$k_{11} = \frac{4EI_0}{L^3}\left[\frac{9}{4}(-0{,}57735)^2(3-0{,}57735) + \frac{9}{4}(0{,}57735)^2(3+0{,}57735)\right] = \frac{18EI_0}{L^3}$$

In terms of integration this is the exact result. However, the cubic polynomial we have assumed for w is no longer capable of representing the correct displacement since I is no longer constant. The correct answer, which can be obtained by introducing one more *dof* (the displacements w at the mid-point) and assuming a quartic polynomial for w, is found to be

$$k_{11} = \frac{17{,}24EI_0}{L^3}$$

Before we leave the one-dimensional case it should be mentioned that optimal integration formulas that *always* include the *end points* have also been developed; so-called LABOTTO quadrature. The trapezoidal rule and SIMPSON's 1/3 rule are the first two such formulas, but for $n = 4$ and higher, the intervals are no longer of equal length.

Element analysis II - Mapping and numerical integration

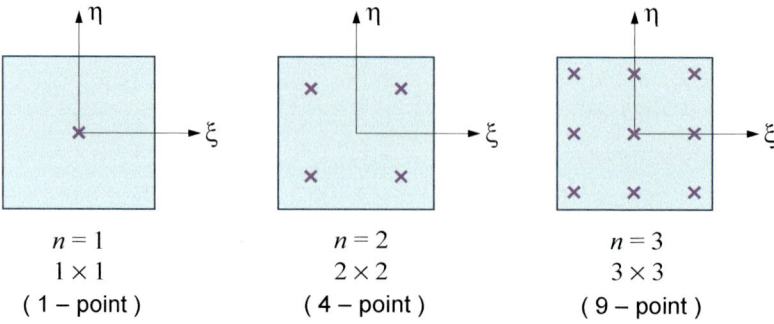

Figure 6.14 Common integration schemes for a 2 × 2 square

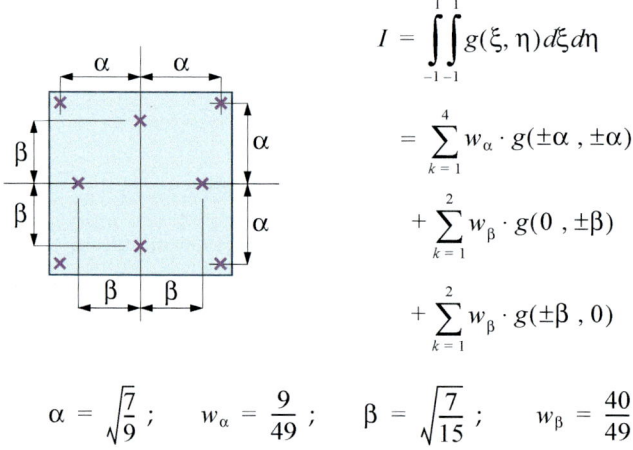

$$I = \int_{-1}^{1}\int_{-1}^{1} g(\xi, \eta)\,d\xi\,d\eta$$

$$= \sum_{k=1}^{4} w_\alpha \cdot g(\pm\alpha, \pm\alpha)$$

$$+ \sum_{k=1}^{2} w_\beta \cdot g(0, \pm\beta)$$

$$+ \sum_{k=1}^{2} w_\beta \cdot g(\pm\beta, 0)$$

$$\alpha = \sqrt{\frac{7}{9}}\,; \quad w_\alpha = \frac{9}{49}\,; \quad \beta = \sqrt{\frac{7}{15}}\,; \quad w_\beta = \frac{40}{49}$$

Figure 6.15 Optimal rule for quintic polynomials in both directions for a 2 × 2 square

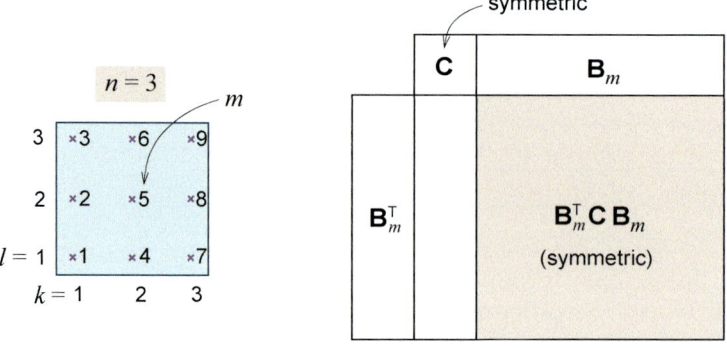

Figure 6.16 Numerical integration of the element stiffness matrix **k**

Two- and three-dimensional elements. Numerical integration over a square or cuboid is customarily carried out using the same one-dimensional rule in both or all three (orthogonal) directions (ξ, η and ζ), see Fig. 6.14. This is simple, but not necessarily optimal. The 9-point rule, often referred to as 3×3, is exact for polynomials of order 5 in each direction. However, the 8-point rule in Fig. 6.15 has the same accuracy.

It is the same for three dimensions. A 6-point rule has the same accuracy as 2×2×2 GAUSS (8 points), and a 14-point rule has the same accuracy as 3×3×3 GAUSS (27 points), see HUGHES [20].

The computation of the element stiffness matrix, for a 2D element, by GAUSS quadrature becomes, see Eq. (6-20),

$$\begin{aligned}
\mathbf{k} &= \int_{-1}^{1}\int_{-1}^{1} t(\xi,\eta)\mathbf{B}^T(\xi,\eta)\mathbf{CB}(\xi,\eta)J d\xi d\eta \\
&\approx \sum_{k=1}^{n}\sum_{l=1}^{n} \mathbf{B}^T(\xi_k,\eta_l)\mathbf{CB}(\xi_k,\eta_l)t(\xi_k,\eta_l)J(\xi_k,\eta_l)w_k w_l \\
&= \sum_{m=1}^{n\times n} \mathbf{B}^T(\xi_m,\eta_m)\mathbf{CB}(\xi_m,\eta_m)t(\xi_m,\eta_m)J(\xi_m,\eta_m)W_m
\end{aligned} \qquad (6\text{-}33)$$

where $W_m = w_k \cdot w_l$. For $n = 3$, k, l and m may take on the meaning shown in Fig. 6.16, and Eq. (6-33) suggests that each element of the matrix $\mathbf{B}^T\mathbf{CB}$ needs to be evaluated and multiplied by the product $t_m \cdot J_m \cdot W_m$ at all 9 points. This operation should be carefully executed in the computer, with due regard to symmetry.

For a *triangular* element let the field variable ϕ be a function of the *area* coordinates ζ_1, ζ_2 and ζ_3, that is $\phi = \phi(\zeta_1,\zeta_2,\zeta_3)$. The integration formula, in the general case, reads

$$\int_A \phi dA = \frac{1}{2}\sum_{k=1}^{n} \phi(\zeta_{1k},\zeta_{2k},\zeta_{3k})w_k J_k \qquad (6\text{-}34)$$

For a triangle with *straight* edges we have $J = 2A$ and the formula becomes

$$\int_A \phi dA = A\sum_{k=1}^{n} \phi(\zeta_{1k},\zeta_{2k},\zeta_{3k})w_k \qquad (6\text{-}35)$$

Formulas, both open and closed, with a varying number of integration points have been derived, some of which are quoted in APPENDIX C.

Element analysis II - Mapping and numerical integration

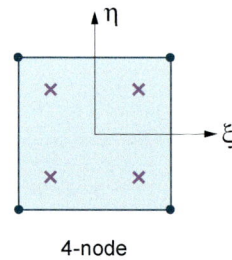

4-node

Full integration: 2×2

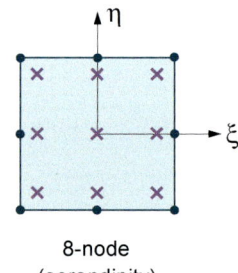

8-node
(serendipity)

Full integration: 3×3

6.3 Integration schemes

For numerically integrated elements an important question for the choice of quadrature rule is: how accurately do we need to integrate? Before we try to answer this question we introduce some important terminology, and this again requires the following clarification:

> Henceforth we will assume that for quadrilateral and hexahedral elements the *same* one-dimensional (GAUSS) quadrature rule will be used in *both/all three* (orthogonal) directions (ξ, η and ζ), even though this may not be the computationally most efficient rule.

Full integration. In order to give this notion a precise meaning we first need to define what we mean by an element with *undistorted* geometry. In mathematical terms it is an element geometry for which the *Jacobian* is constant throughout the element. For a four-sided element this holds true if it has *rectangular* or *parallelogram* shape, and if side nodes are uniformly spaced along the edges. For a triangle the Jacobian is constant if the edges are straight and side nodes are uniformly spaced along the edges.

For an element with a constant Jacobian, all elements of the integrand in the expression for **k**, see Eq. (6-20), are *polynomials*, whereas they are *rational* functions if the Jacobian varies within the element.

Now we can define *full integration* as

> the quadrature rule with the lowest order that *exactly* integrates the element stiffness matrix **k** when the *element geometry is "undistorted"*.

For a quadrilateral element for instance, this means that we must be able to integrate **k** exactly in the mapped configuration, that is for the 2×2 square in natural coordinates ξ and η. It should be noted that full integration may not, and usually will not integrate exactly in physical coordinates.

For quadrilateral C^0 elements we have for:

4-node elements with constant thickness:

> Matrix **B** contains ξ and η which means the integrand contains terms with ξ^2 and η^2.
> Full integration therefore requires a 2×2 GAUSS quadrature.

8-node (and also 9-node) elements with constant thickness:

> The highest order polynomials in matrix **B** are $\xi\eta^2$ and $\xi^2\eta$ (and also $\xi^2\eta^2$ for the 9-node element); this means the highest order polynomials in the integrand are $\xi^4\eta^2$ and $\xi^2\eta^4$ (and also $\xi^4\eta^4$ for the 9-node element).
> Full integration therefore requires a 3×3 GAUSS quadrature.

Since a 2×2 rule integrates exactly 3rd order polynomials and 3×3 integrates exactly 5th order polynomials, we "over-integrate" slightly in both cases.
We can in fact have a bilinear variation of the thickness over the element and still obtain full integration, for all three elements, with the suggested quadrature rules.

Element analysis II - Mapping and numerical integration

Johann Carl Friedrich GAUSS (1777-1855) was a German mathematician and physical scientist. He made significant contributions to many fields, including number theory, statistics, analysis, differential geometry, geodesy, geophysics, electrostatics, astronomy and optics.

GAUSS had a great influence in many fields of mathematics and science, and he is ranked as one of history's most influential mathematicians. He referred to mathematics as "the queen of sciences".

GAUSS was the son of poor working-class parents. He was a child prodigy, and made his first outstanding mathematical discoveries while still a teenager. His intellectual abilities attracted the Duke of Braunschweig who sent him to what is now Technische Universität Braunschweig and then to the University of Göttingen. He was supported by the Duke for a number of years, but he doubted the security of this arrangement, and also did not believe pure mathematics to be important enough to deserve support. Thus he sought a position in astronomy, and in 1807 was appointed Professor of Astronomy and Director of the astronomical observatory in Göttingen, a post he held for the remainder of his life.

GAUSS was an ardent perfectionist and a hard worker, but he was not a prolific writer. He refused to publish work he did not consider complete and above criticism. This was in keeping with his personal motto *pauca sed matura* ("few, but ripe"). GAUSS had six children, two daughters and four sons. He had some conflicts with his sons; he did not want any of them to enter mathematics or science for "fear of lowering the family name".

Reduced integration. This is

the quadrature rule that is *one order lower* than full integration.

The motivation for this type of integration is two-fold:

1) It is computationally "cheaper" – the integration comprises a large part of the computational effort necessary to establish the element matrices.

2) In many cases it actually gives *better* (more accurate) results – for linear problems this is probably as important as reduced "cost".
It appears that reduced integration tends to dampen the effect of higher order terms, with a "softening" effect on an otherwise too "stiff" element. This remedy falls into the category that "it is almost too good to be true", and it is. As we shall see there is a downside to this technique.

Selective reduced integration. This is a scheme whereby we

use different order quadrature rules for different contributions to the stiffness matrix.

For a plane stress problem, for instance, we may integrate the contribution from normal strain by one rule (usually full integration) and the contribution from shear strain by another rule (usually a lower order rule). We will come back to this scheme, which belongs to the "bag of tricks", in Chapter 11.

6.4 Integration and convergence

Numerical integration is an important ingredient for isoparametric elements. It has a very direct influence on the computational effort, but perhaps more important is the way the integration scheme is used to "improve" element performance. A considerable amount of research has dealt with this issue and the pros and cons are now fairly well established. Here we will try to give a brief summary; the interested reader should consult more in-depth treatments presented, for instance, by ZIENKIEWICZ and TAYLOR [18] or HUGHES [20].

Two questions are crucial:

1) What is the minimum order of the quadrature rule that will preserve convergence towards the correct result as the element size goes to zero?

2) What is the minimum order of the quadrature rule that will preserve the *rate* of convergence?

We are slightly ahead of ourselves here, since accuracy and convergence is dealt with later, in Chapter 8. Nevertheless, the arguments below can, to a large extent, stand on their own.

We need some "measure" of element size and total number of *dofs* in the finite element model, and introduce

h as a typical parameter designating the element *size*, and

n_{dof} as the total number of *dofs*.

Convergence in connection with FEM is usually associated with

Element analysis II - Mapping and numerical integration

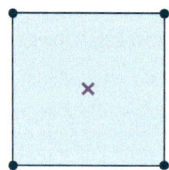

4-node quadrilateral – constant thichness t

$$tJ = t(c_0 + c_1\xi + c_2\eta)$$

the integrand is linear in ξ and η

$\Rightarrow$ a single point (1×1) quadrature rule is adequate

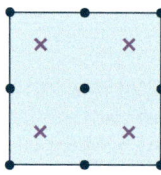

9-node (LAGRANGE) quadrilateral – constant thichness t

$$tJ = f(c,\ \xi,\ \eta,\ \xi^2,\ \xi\eta,\ \eta^2,\ \xi^3,\ \xi^2\eta^2,\ \eta^3)$$

the integrand is cubic in ξ and η

$\Rightarrow$ a 2×2 quadrature rule is adequate

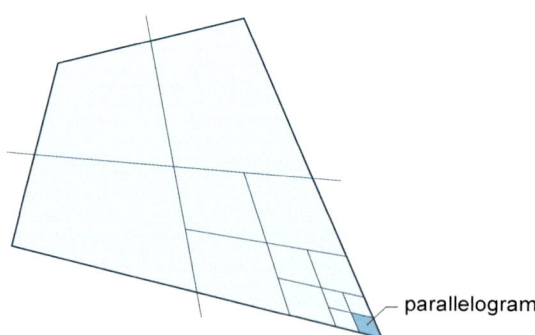

parallelogram

Figure 6.17 Uniform subdivision of a quadrilateral

$$U_e = \int_V U_0 dV \qquad (6\text{-}36)$$

the *strain energy* of the element, where $U_0 = \frac{1}{2}\sigma^T\varepsilon$ is the strain energy per unit volume. As $n_{dof} \to \infty$, $h \to 0$, and U_0 becomes constant (since the element, in order to be convergent in the first place, must be capable of reproducing a state of constant strain/stress).

In order to preserve convergence the strain energy, U, has to be integrated exactly in the limit, that is when $n_{dof} \to \infty$, when we have

$$U_e = \int_V U_0 dV = U_0 \int_V dV \qquad (6\text{-}37)$$

In other words, we need to be able to integrate the volume exactly. For a 2D quadrilateral element of thickness t this means we have to be able to evaluate the integral

$$\int_V dV = \int_{-1}^{1}\int_{-1}^{1} tJ(d\xi)d\eta \qquad (6\text{-}38)$$

exactly in order to preserve convergence, since it will guarantee correct strain energy in the limit when the strain (stress) in the element is constant.

For the 4-node, bilinear quadrilateral with constant thickness a 1×1 (1 point) GAUSS quadrature will suffice, and for the 9-node LAGRANGE element (and also for the 8-node serendipity element) a 2×2 quadrature will do the job provided the thickness is constant, see opposite page. However, as $h \to 0$, then $t \to$ constant and the element shape becomes more and more like a parallelogram (see Fig 6.17) and hence $J \to$ constant. In the limit therefore, a 1×1 (1 point) GAUSS quadrature is sufficient also for the 8- and 9-node quadrilaterals.

Convergence is vital, but also the *rate* of convergence is important. We will deal with this in more detail in Chapter 8; our concern here is that we do not want our integration scheme to lower the rate of convergence inherit in the element formulation. It can be shown, see for instance ZIENKIEWICZ and TAYLOR [18], that if the *strain energy* of the element,

$$U_e = \frac{1}{2}\mathbf{S}^T\mathbf{v} = \frac{1}{2}\mathbf{v}^T\mathbf{S} = \frac{1}{2}\mathbf{v}^T\mathbf{k}\mathbf{v} = \frac{1}{2}\mathbf{v}^T\left(\int_V \mathbf{B}^T\mathbf{C}\mathbf{B}dV\right)\mathbf{v} \qquad (6\text{-}39)$$

is integrated exactly to an order of the integrand equal to $2(p\text{-}m)$, where p is the order of the highest *complete* polynomial present in the shape functions $\mathbf{N}$ and m is the differential order of the operator Δ, then *no loss of convergence rate will occur.*

If we consider, for instance, a plane stress/plane strain problem, we have $m=1$, and for the 4-node quadrilateral $p=1$, whereas $p=2$ for the 8- and 9-node

Plane stress/plane strain:

$$\boldsymbol{\varepsilon} = \begin{bmatrix} \varepsilon_x \\ \varepsilon_y \\ \gamma_{xy} \end{bmatrix} \quad \text{and} \quad \mathbf{k} = \sum_k W_k J_k \mathbf{G}_k$$

where matrix $\mathbf{G}_k$ is:

	C (3×3) — rank 3	$\mathbf{B}_k$
$\mathbf{B}_k^T$		$\mathbf{G}_k$ (n×n) rank 3

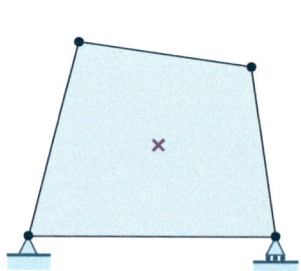

Figure 6.18 A 4-node plane stress element as a "structure"

quadrilateral. From this and the above statement we can conclude that *reduced* integration will preserve the rate of convergence for these elements.

What has been said in this section for 2D quadrilateral C^0 elements is equally applicable to 3D hexahedral C^0 elements.

6.5 Element instability – mechanisms

In some cases finite elements exhibit *instabilities* that can lead to a *singular* system stiffness matrix (**K**). These (mathematical) instabilities are matrix *defects*, usually caused by the numerical integration scheme and characterized by displacement patterns that can develop *without* accompanying forces. They go by several names, such as

- mechanisms or spurious modes,
- kinematic modes,
- singular modes,
- hourglass modes, and
- zero energy modes.

Underintegration, such as reduced integration, is the main reason why these modes occur. In order to demonstrate this we consider a single element as a "structure" with minimal support, just enough to prevent rigid body motion. Thus **r** = **v** and

$$U = U_e = \tfrac{1}{2}\mathbf{r}^T\mathbf{k}\mathbf{r} = \tfrac{1}{2}\mathbf{v}^T \int_{V_e} \mathbf{B}^T\mathbf{C}\mathbf{B}dV\mathbf{v} = \tfrac{1}{2}\int_{V_e} \boldsymbol{\varepsilon}^T\mathbf{C}\boldsymbol{\varepsilon}dV \qquad (6\text{-}40)$$

Let the problem be one of plane stress, for which **C** is a 3×3 matrix. Replacing the integral of Eq. (6-40) by a weighted sum of k individual matrices, one per integration point, the final stiffness matrix is obtained as a sum of *rank* 3 matrices, see opposite page. The $\mathbf{G}_k$ matrix is obtained as a linear combination of rows and columns of the rank 3 matrix **C**, and this process cannot raise the rank of the resulting matrix ($\mathbf{G}_k$) above 3. Hence

$$\text{rank}(\mathbf{k}) = \min\{(k\times 3 - r_b), (n_{dof} - r_b)\} \qquad (6\text{-}41)$$

where r_b is the number of rigid body motions. In other words, the matrix rank gets a "contribution" of 3 per integration point.

Example 6-3

Figure 6.18 shows a "structure" consisting of a single 4-node bilinear plane stress quadrilateral element. Rigid body movements (r_b) are prevented by suppressing (the minimum) 3 *dofs*. The total number of *dofs* is 8 (n_{dof} = 8). If we remove rows and columns corresponding to the suppressed *dofs* we are left with (m =) 5 equations (one for each unknown displacement).

Only one integration point (as indicated in Fig. 6.18) provides a stiffness matrix **k** with *rank* 3, which means the matrix has a *defect* or *rank deficiency* of $d = m - r_b = 5 - 3 = 2$. Each deficiency corresponds to a *spurious* or *zero energy mode*. If we use a 2×2 quadrature rule, the spurious modes clearly disappear.

Element analysis II - Mapping and numerical integration

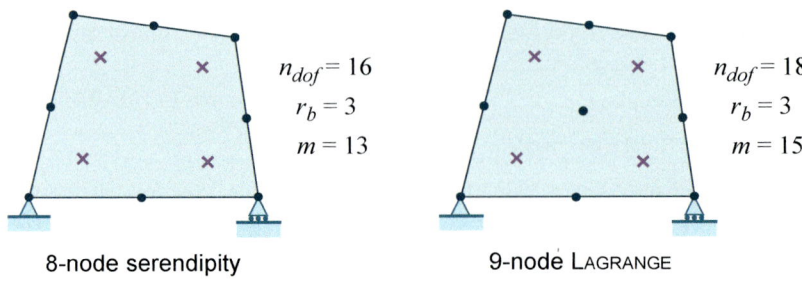

Figure 6.19 2×2 quadrature rule used for two "higher" order elements

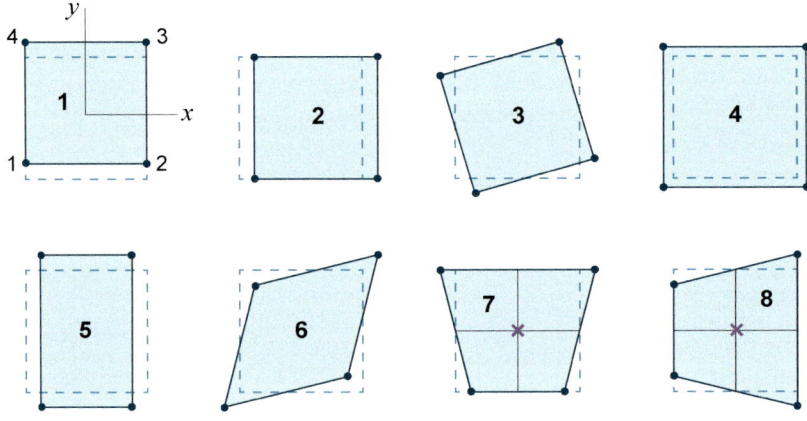

Figure 6.20 Independent displacement modes for the 4-node plane element

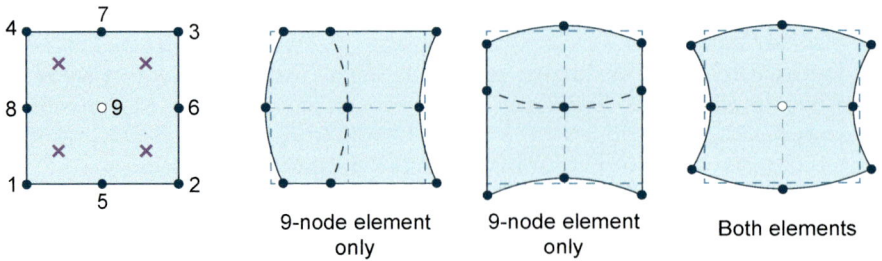

Figure 6.21 Spurious modes for 8- and 9-node elements due to reduced integration

Figure 6.19 shows the 8-node serendipity and the 9-node LAGRANGE elements as "structures" with minimal support (just enough to prevent rigid body motions). A 2×2 quadrature rule provides 12 (= 4×3) independent equations for both elements. Hence, for the

8-node element: $d = n_{dof} - r_b - k \times 3 = 16 - 3 - 4 \times 3 = 1$, whereas for the

9-node element: $d = n_{dof} - r_b - k \times 3 = 18 - 3 - 4 \times 3 = 3$.

In other words, a reduced integration scheme leaves the 8-node serendipity element with *one spurious mode* and the 9-node LAGRANGE element with *three spurious modes*.

Figure 6.20 shows the 8 independent displacement modes of the 4-node plane element, see COOK et al [21]. These are in fact the *eigenvectors* of the element stiffness matrix which we shall come back to in Section 8.5. The first three modes are the *rigid body modes* (all of which have zero eigenvalues), while the last 5 are deformation modes; it should be noted that mode 5 also includes elongation in the *x*-direction, obtained by multiplication by minus one. At the origin (centre of the element) we see that $\varepsilon_x = \varepsilon_y = \tau_{xy} = 0$ for modes 7 and 8. With only one integration point (at the origin) the integration procedure therefore produces zero strain energy (zero energy modes), and these two modes are the spurious modes or mechanisms we found in the example above.

For the 8- and 9-node elements the spurious modes, inherent in the element stiffness matrix **k** obtained by reduced (2×2) integration, are not easy to detect. However, an eigenvalue/eigenvector analysis of the element stiffness matrix (se Section 8.5) will reveal these modes which are indicated in Fig. 6.21, see COOK et al [21].

We can live with mechanisms at the element level, but we cannot accept mechanisms at the system (or structure) level as this will cause the system stiffness matrix **K** to become *singular*. In many practical situations, the boundary conditions will "block" the element mechanisms (zero energy modes) from propagating into the system mesh, see COOK et al [21] for examples. Some spurious modes are also of such a nature that they cannot develop in a mesh; the last mode in Fig. 6.21, for instance, cannot be present in two neighbouring elements for kinematic reasons. We say that this mode is *noncommunicable*.

We conclude this section by pointing out that for nonlinear problems, and also for some dynamic problems, the computational effort at the element level (due to a many iterations) can be very significant, and every possible way of reducing this effort is clearly of interest. Reducing the number of integration points has a very direct and quite noticeable effect on the computational effort, and for this reason more robust solutions to the problem of mechanisms than those indicated above, have attracted the interest of many researchers. Without going into details here – we shall come back to this issue in Section 11.2 – we mention that *stabilization* techniques, which are inexpensive, retain rigid body modes and constant strain modes, and which do not require any decisions by the *software user*, have been developed and implemented in most large, general purpose software packages.

Element analysis II - Mapping and numerical integration

Adrien-Marie Legendre (1752-1833) was a French mathematician. The watercolor caricature is apparently the only known "portrait" of Legendre; for two centuries, until the recent discovery of the error in 2009, a side-view portrait of an obscure French politician (Louise Legendre) was purported to be the far better known mathematician.

Legendre was born to a wealthy family and given an excellent education at the Collège Mazarin in Paris; he defended his thesis in physics and mathematics at the age of 18. Most of his work was brought to perfection by others: his work on roots of polynomials inspired Galois theory; Abel's work on elliptic functions was built on Legendre's; some of Gauss' work in statistics and number theory completed that of Legendre. He developed the method of least squares, which has broad application in linear regression, signal processing and curve fitting.

Legendre is known for the Legendre transformation which is used to go from the Lagrangian to the Hamiltonian formulation of classical mechanics, but he is perhaps best known as the author of *Éléments de geometrie*, which was published in 1794 and was the leading text on the topic for a long time.

Summary of Chapter 6

In this chapter we have introduced the important concept of *mapping* which enable us to handle 4-sided plane elements of arbitrary, quadrilateral shape, and similarly, general hexahedrons in 3D.

The Jacobian matrix and its determinant, the *Jacobian*, are the mathematical tools that facilitate going from physical coordinates (x,y) to natural, dimensionless coordinates (ξ,η) and back.

The *isoparametric* concept, in which the element geometry is defined in terms of the nodal point coordinates, by the *same* shape functions as those that define the displacements within the element, is a central theme of mapped elements. Requirements for unique mapping have been established, and although the technique permits elements with curved edges/faces, elements perform best when their geometric shape is as "regular" as possible, that is squares do better than rectangles and rectangles do better than quadrilaterals, and, for given corner nodes, elements with straight edges do better than those with curved edges.

Even though polynomials are used as shape functions in the natural, mapped coordinate space, the need to determine the strains calls for differentiation with respect to physical coordinates. For elements with irregular shape this leads to *rational* functions in the integrals we need to evaluate in order to determine the element stiffness matrix **k**. The only practical solution to this is *numerical integration*. For 4-sided plane elements (quadrilaterals) and 6-faced volume elements (hexahedrons), GAUSS quadrature rules are used almost exclusively. In the one-dimensional case this is the optimal method of numerical integration. For two- and three-dimensional elements we use the *same* one-dimensional rule in both or all three (orthogonal) directions; while not necessarily optimal it is simple and straightforward.

Since numerical integration is a dominant part of the computational effort of the determination of the element matrices, it is important to limit this effort without compromising accuracy or convergence. We introduced the notion of *full integration* as the minimum quadrature rule that will integrate the expression for **k** *exactly* when the element is *undistorted*, that is when it is a square in the 2D case and a cube in 3D. *Reduced integration*, which implies a degree of "under-integration", is the quadrature rule that is *one* order lower than full integration. Reduced integration will in some cases give better results than full integration, it will also preserve convergence properties inherent in the formulation, but it comes at a price. *Zero energy modes* are introduced at the element level, and if these are not contained or stabilized in the mesh of elements representing the system (problem at hand), we will end up with a *singular* system stiffness matrix **K**.

Element analysis II - Mapping and numerical integration

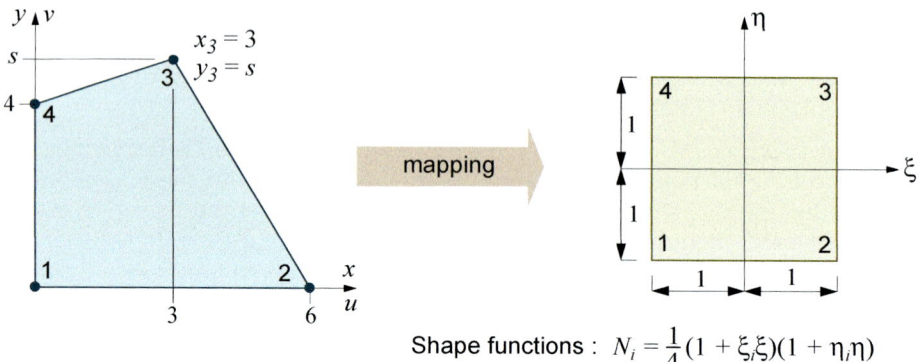

Shape functions : $N_i = \frac{1}{4}(1 + \xi_i\xi)(1 + \eta_i\eta)$

Figure 6.22 A 4-node quadrilateral C^0 element

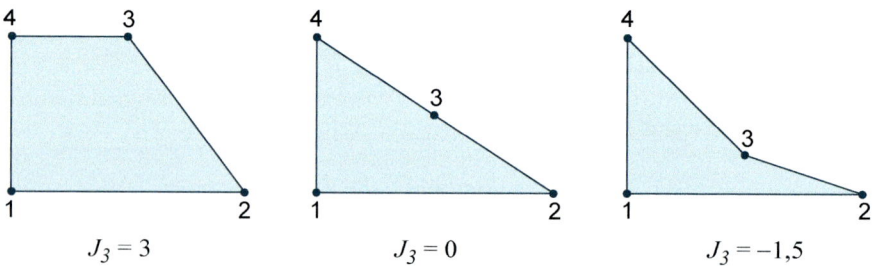

$J_3 = 3$ $\qquad\qquad$ $J_3 = 0$ $\qquad\qquad$ $J_3 = -1,5$

Example 6-4 – The Jacobian for a 4-node quadrilateral

Figure 6.22 shows a 4-node quadrilateral C^0 element. Nodal point 3 has a constant x-coordinate (= 3), but a variable y-coordinate (= s). Determine the Jacobian as a function of s.

Solution

Using Eq. (6-14) we find:

$$J_{11} = \tfrac{1}{4}[0 + 1(1-\eta)6 + 1(1+\eta)3 + 0] = \tfrac{3}{4}(3-\eta)$$

$$J_{12} = \tfrac{1}{4}[0 + 0 + 1(1+\eta)s - 1(1+\eta)4] = \tfrac{1}{4}[s(1+\eta) - 4(1+\eta)]$$

$$J_{21} = \tfrac{1}{4}[0 - 1(1+\xi)6 + 1(1+\xi)3 + 0] = -\tfrac{3}{4}(1+\xi)$$

$$J_{22} = \tfrac{1}{4}[0 + 0 + 1(1+\xi)s + 1(1-\xi)4] = \tfrac{1}{4}[s(1+\xi) + 4(1-\xi)]$$

At node 3 we have $\xi = \eta = 1$; hence, the Jacobian matrix at node 3 is:

$$\mathbf{J}_3 = \begin{bmatrix} \tfrac{3}{2} & \tfrac{s}{2} - 2 \\ -\tfrac{3}{2} & \tfrac{s}{2} \end{bmatrix}$$

The Jacobian at point 3 is the determinant of this matrix, which is:

$$J_3 = \tfrac{3}{4}s + \tfrac{3}{2}\left(\tfrac{s}{2} - 2\right) = \tfrac{3}{2}s - 3$$

For $s = 4$: $J_3 = 6 - 3 = 3$
For $s = 2$ (2-3-4 becomes a straight line): $J_3 = 3 - 3 = 0$
For $s = 1$: $J_3 = 1,5 - 3 = -1,5$

Element analysis II - Mapping and numerical integration

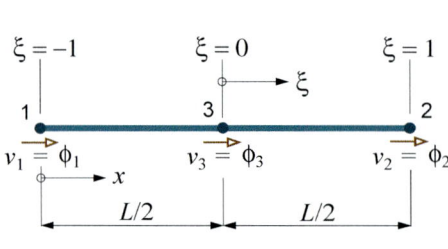

Figure 6.23

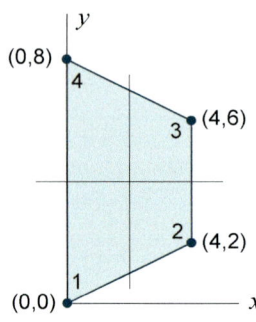

Figure 6.24

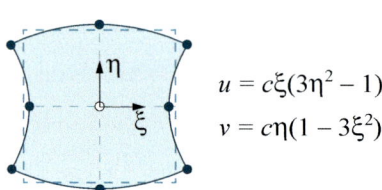

$u = c\xi(3\eta^2 - 1)$
$v = c\eta(1 - 3\xi^2)$

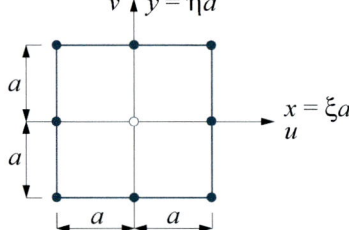

Figure 6.25

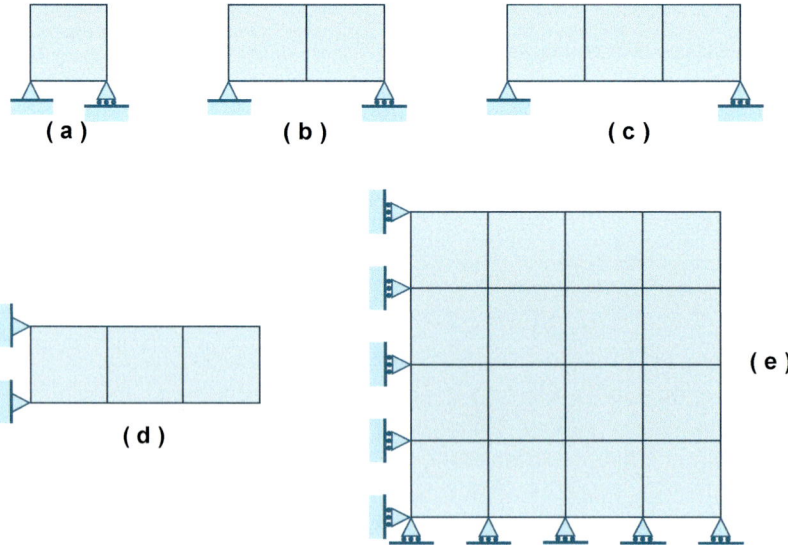

Figure 6.26

Problems

Problem 6.1

Figure 6.23 shows a 3-node (quadratic) bar element of length L. The cross section area varies linearly as $A = A_0(3+\xi)/2$, and YOUNG's modulus E is constant throughout.

a) Use an *isoparametric* formulation to determine the stiffness matrix for the element. We assume that node 3 is located at the mid-point of the element, as shown. Perform the integration analytically.

b) Establish an expression for ε_x in terms of the Jacobian J, the natural coordinate ξ and the nodal displacements **v**. Use this expression to determine how far node 3 can be moved away from the element mid-point when we require ε_x to be *finite* at node 1.

Problem 6.2

Verify that the Jacobian is a constant for a parallelogram.

Problem 6.3

Determine the Jacobian matrix for the 4-node quadrilateral element in Fig. 6.24. The numbers in parentheses are the physical (x,y) coordinates.

Problem 6.4

Verify that the expressions for u and v in Fig. 6.25 yield zero strain at the 4 integration points of a 2×2 GAUSS quadrature rule for a square element.

Problem 6.5

Evaluate the diagonal terms of the stiffness matrix for the element of Problem 6.1 by use of a two point GAUSS quadrature rule.

Problem 6.6

Figure 6.26 shows 5 plane stress "structures" modelled by 1, 2, 3, 3 and 16 quadrilateral elements, respectively. Each nodal point has two *dofs* (u and v).
The boundary conditions are as shown. Three different elements are used:
   4-node quadrilaterals, 8-node (serendipity) elements and 9-node LAGRANGE elements.

For each structure, determine if the system stiffness matrix (**K**) is *regular* (non-singular) or *singular* when

a) 4-node elements and 1×1 GAUSS quadrature is used,

b) 8-node elements and 2×2 GAUSS quadrature is used, and

c) 9-node elements and 2×2 GAUSS quadrature is used.

$$\mathbf{S}_{\varepsilon 0}^{0} = -\int_{V_e} \mathbf{B}^\mathsf{T} \mathbf{C} \boldsymbol{\varepsilon}_0 dV \qquad (4\text{-}56\text{a})$$

$$\mathbf{S}_{F}^{0} = -\int_{V_e} \mathbf{N}^\mathsf{T} \mathbf{F} dV \qquad (4\text{-}56\text{b})$$

$$\mathbf{S}_{\Phi}^{0} = -\int_{S_T} \mathbf{N}^\mathsf{T} \boldsymbol{\Phi} dS \qquad (4\text{-}56\text{c})$$

$$\boldsymbol{\sigma} = \mathbf{C}(\boldsymbol{\varepsilon} - \boldsymbol{\varepsilon}_0) \qquad (3\text{-}36)$$

7

Element analysis III
Element loads and stresses

In the previous two chapters the emphasis has been on the stiffness properties of the elements. In this last chapter on the generic element analysis we will take a closer look at the element load vector (S^0).

In a displacement formulation our primary unknown parameters, which we compute from the system stiffness relation, are displacement components. However, we are normally more interested in element stresses than displacements; computation or recovery of stresses is therefore an important part of the element analysis.

7.1 Static equivalent nodal point loads

In Section 4.7 we derived the following general element stiffness relation, cf. Eq (4-53),

$$\mathbf{S} = \mathbf{kv} + \mathbf{S}^0 \qquad (7\text{-}1)$$

where $\mathbf{kv}$ are the elastic forces due to the nodal displacements $\mathbf{v}$, and $\mathbf{S}^0$ contains the "fixed end" forces due to actions within the element (assuming the nodal displacements are zero, that is $\mathbf{v} = \mathbf{0}$). The contributions to $\mathbf{S}^0$ come from *initial strain* $\boldsymbol{\varepsilon}_0$ (and/or stress, $\boldsymbol{\sigma}_0$), *volume forces* $\mathbf{F}$, or *surface forces* or *tractions* $\boldsymbol{\Phi}$, that is

$$\mathbf{S}^0 = \mathbf{S}^0_{\varepsilon 0} + \mathbf{S}^0_F + \mathbf{S}^0_\Phi \qquad (7\text{-}2)$$

Consistent element load vector. Equations (4-56), derived by a virtual work approach, represent formulas for the computation of so-called *consistent load vectors* due to the various actions on the element. The term consistent reflects the fact that the load vectors are based on the *same* assumptions ($\mathbf{N}$ and $\mathbf{B}$) as those used for the stiffness matrix; the representation of stiffness and action is consistent. For a specific action it can easily be shown that the consistent element load vector is statically equivalent to the action itself.

In the general stress-strain expression, see Eq. (3-36), we could also have included an intial stress term. However, for linear elastic materials we can

Element analysis III - Element loads and stresses

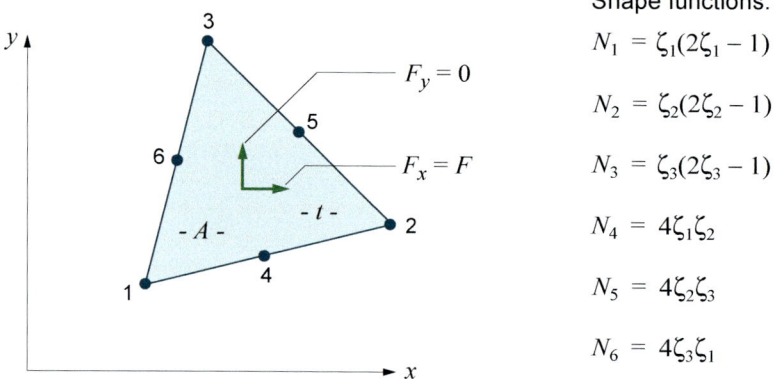

Figure 7.1 Volume forces on a quadratic triangular element

Static equivalent nodal point loads

convert initial stress to initial strain, and initial strain can therefore be used to model such effects as "temperature loading", residual stresses, shrinkage, and lack of fit. Since $\mathbf{S}_{\varepsilon 0}^0$ is statically equivalent to the action, the elements of this load vector must form a self-equilibrating set of forces, the resultant of which is zero. For actions modelled by initial strain, the consistent approach is basically the only viable method of transforming such actions into equivalent nodal forces, which is a must in the finite element displacement method.

Load lumping. For volume and surface loading we also have the option of load *lumping*. This approach, which is analogous to the approach we advocated for distributed loading on frame type models in Chapter 2, is based on the assumption that the distributed loading can be adequately represented by a series of many (small) concentrated loads acting directly in the nodal points (and in the direction of a *dof*). It is important that the resultant of the concentrated (lumped) loads is exactly the same as the resultant of the distributed loading being replaced.

The remainder of this section is basically concerned with the consistent approach, since load lumping is fairly straightforward. However, it should be pointed out that the two techniques can give results that, at the element level, are apparently quite different, as the following example will show.

Example 7-1

Figure 7.1 shows a 6-node, quadratic triangle subjected to volume forces with components F_x and F_y; for simplicity we assume $F_y = 0$, that is

$$\mathbf{F} = \begin{bmatrix} F_x \\ F_y \end{bmatrix} = \begin{bmatrix} F \\ 0 \end{bmatrix}$$

The element area is A and its constant thickness is t.

If we take the lumping approach, it seems reasonable to assign 1/6 of the total volume force FAt to the *x-dof* of each nodal point, or perhaps slightly more to the side nodes.

The consistent element nodal loads, and we consider the *x*-components only, are defined by Eq. (4-56b),

$$\mathbf{S}_{Fx}^0 = -\int_A \mathbf{N}^T F t \, dA$$

With the shape functions of Eqs. (5-63) and (5-64), which are also given in Fig. 7.1, we readily find, by use of the integration formula of Eq. (5-30),

$$\mathbf{S}_{Fx}^0 = -\frac{FAt}{3}[0 \quad 0 \quad 0 \quad 1 \quad 1 \quad 1]^T$$

That is, no forces at the corner nodes; a slightly unexpected result? At the end of the chapter we present an example for which the results are even more "strange".

For a reasonably fine mesh the results obtained with the two different load approximations will be quite similar, in spite of the difference at the element level.

Element analysis III - Element loads and stresses

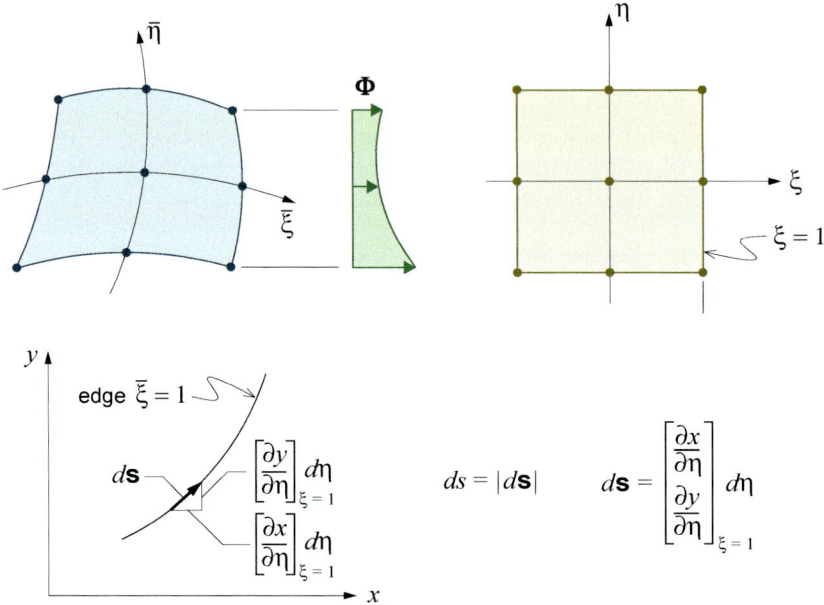

Figure 7.2 Edge loading defined in global coordinates

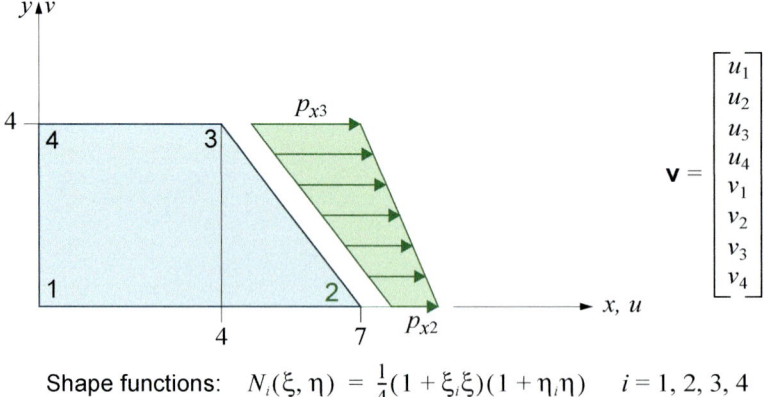

Shape functions: $N_i(\xi, \eta) = \frac{1}{4}(1 + \xi_i\xi)(1 + \eta_i\eta) \quad i = 1, 2, 3, 4$

Figure 7.3 Edge loading (in global coordinates) on a quadrilateral element

Static equivalent nodal point loads

Surface (3D) or edge (2D) loading (tractions), Eq. (4-56c), is not straightforward and will therefore be treated in more detail. For simplicity, and with little loss of generality, we limit our discussion to the plane (2D) case and write Eq. (4-56c) as

$$\mathbf{S}_\Phi^0 = -\int_S \mathbf{N}^T \mathbf{\Phi}\, ds \tag{7-3}$$

We distinguish between two cases: (1) the load is defined in global axes, and (2) the load is defined in local edge axes (which are normal and tangential to the edge, respectively).

1) Edge load defined in global axes

From Fig. 7.2,

$$ds = \sqrt{\left(\frac{\partial x}{\partial \eta}\right)^2_{\xi=1} + \left(\frac{\partial y}{\partial \eta}\right)^2_{\xi=1}}\, d\eta \tag{7-4}$$

and Eq. (7-3) can be written as

$$\mathbf{S}_\Phi^0 = -\int_{-1}^{1} \mathbf{N}^T(1,\eta)\mathbf{\Phi} \sqrt{\left(\frac{\partial x}{\partial \eta}\right)^2_{\xi=1} + \left(\frac{\partial y}{\partial \eta}\right)^2_{\xi=1}}\, d\eta \tag{7-5}$$

Example 7-2

The problem is to determine the load vector for a 4-node, 8-dof quadrilateral element with geometry and edge loading as shown in Fig. 7.3. The nodal point values of the x- and y-displacements, u_i and v_i, constitute the *dofs* of the element, and we assume that the arrangement of the elements of vector **v** is as shown in the figure. Within the element (and along its edges) the displacement components are expressed in terms of their nodal values by the shape functions of Eq. (5-56). The loading can be expressed as

$$\mathbf{\Phi} = \begin{bmatrix}\Phi_x \\ \Phi_y\end{bmatrix} = \begin{bmatrix}(N_2)_{\xi=1} & (N_3)_{\xi=1} \\ 0 & 0\end{bmatrix}\begin{bmatrix}p_{x2} \\ p_{x3}\end{bmatrix}$$

where $(N_2)_{\xi=1} = \frac{1}{2}(1-\eta)$ and $(N_3)_{\xi=1} = \frac{1}{2}(1+\eta)$. Thus

$$\left(\frac{\partial x}{\partial \eta}\right)_{\xi=1} = \sum_{i=1}^{4}\left(\frac{\partial N_i}{\partial \eta}\right)_{\xi=1} x_i = \frac{1}{4}(-2\cdot 7 + 2\cdot 4) = -\frac{3}{2}$$

$$\left(\frac{\partial y}{\partial \eta}\right)_{\xi=1} = \sum_{i=1}^{4}\left(\frac{\partial N_i}{\partial \eta}\right)_{\xi=1} y_i = \frac{1}{4}(0+8) = 2$$

since

$$\left(\frac{\partial N_i}{\partial \eta}\right)_{\xi=1} = \frac{1}{4}(1+\xi_i\cdot 1)\eta_i = \frac{1}{4}\eta_i(1+\xi_i).$$ Hence $ds = \frac{5}{2}d\eta$ and

239

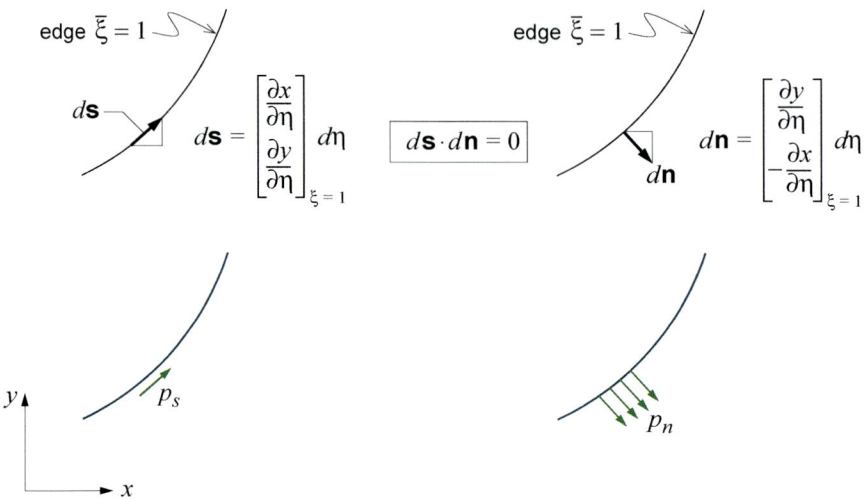

Figure 7.4 Normal and tangential edge load

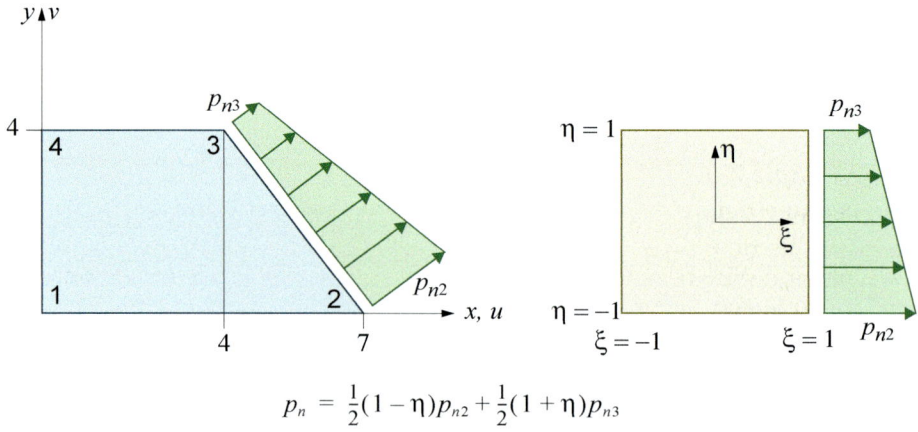

$$p_n = \tfrac{1}{2}(1-\eta)p_{n2} + \tfrac{1}{2}(1+\eta)p_{n3}$$

Figure 7.5 Normal edge loading on a quadrilateral element

$$\mathbf{S}_\Phi^0 = -\int_{-1}^{1} \mathbf{N}^T(1,\eta) \frac{1}{2} \begin{bmatrix} (1-\eta)p_{x2} + (1+\eta)p_{x3} \\ 0 \end{bmatrix} \frac{5}{2} d\eta \quad \text{where}$$

$$\mathbf{N}^T(1,\eta) = \begin{bmatrix} 0 & 0 \\ \frac{1}{2}(1-\eta) & 0 \\ \frac{1}{2}(1+\eta) & 0 \\ 0 & 0 \\ 0 & 0 \\ \vdots & \vdots \\ 0 & 0 \end{bmatrix} \begin{matrix} u \\ \\ \\ \\ v \end{matrix} \quad \Rightarrow \quad \mathbf{S}_\Phi^0 = -\frac{5}{8} \begin{bmatrix} 0 \\ \frac{8}{3}p_{x2} + \frac{4}{3}p_{x3} \\ \frac{4}{3}p_{x2} + \frac{8}{3}p_{x3} \\ 0 \\ 0 \\ \vdots \\ 0 \end{bmatrix}$$

2) *Edge load defined in local (edge) axes*

From Fig. 7.4a, for tangential load:

$$\mathbf{S}_\Phi^0 = -\int_S \mathbf{N}^T(1,\eta)(p_s d\mathbf{s}) = -\int_{-1}^{1} \mathbf{N}^T(1,\eta) p_s \begin{bmatrix} \frac{\partial x}{\partial \eta} \\ \frac{\partial y}{\partial \eta} \end{bmatrix}_{\xi=1} d\eta \quad (7\text{-}6)$$

and from Fig. 7.4b, for normal load:

$$\mathbf{S}_\Phi^0 = -\int_S \mathbf{N}^T(1,\eta)(p_n d\mathbf{n}) = -\int_{-1}^{1} \mathbf{N}^T(1,\eta) p_n \begin{bmatrix} \frac{\partial y}{\partial \eta} \\ -\frac{\partial x}{\partial \eta} \end{bmatrix}_{\xi=1} d\eta \quad (7\text{-}7)$$

Example 7-3

Figure 7.5 shows the same element as in the previous example, see Fig. 7.3, but this time the element edge 2-3 is loaded by a linearly varying normal load p_n. With the same arrangement of the element *dofs* as before, our shape function matrix now reads

$$\mathbf{N}(1,\eta) = \begin{bmatrix} 0 & \frac{1}{2}(1-\eta) & \frac{1}{2}(1+\eta) & 0 & 0 & 0 & 0 & 0 \\ 0 & 0 & 0 & 0 & 0 & \frac{1}{2}(1-\eta) & \frac{1}{2}(1+\eta) & 0 \end{bmatrix}$$

since we will also have to include the *y*-components of the nodal forces.

From the previous example we have:

Element analysis III - Element loads and stresses

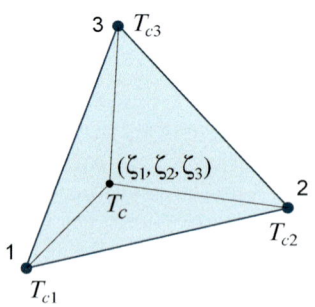

$$T_c = T_{c1}\zeta_1 + T_{c2}\zeta_2 + T_{c3}\zeta_3 = [\zeta_1 \quad \zeta_2 \quad \zeta_3]\begin{bmatrix} T_{c1} \\ T_{c2} \\ T_{c3} \end{bmatrix} = \mathbf{N}_T \mathbf{T}_c$$

Static equivalent nodal point loads

$$\begin{bmatrix} \frac{\partial y}{\partial \eta} \\ -\frac{\partial x}{\partial \eta} \end{bmatrix}_{\xi=1} = \begin{bmatrix} 2 \\ \frac{3}{2} \end{bmatrix}$$

Hence

$$\mathbf{S}_\Phi^0 = -\int_{-1}^{1} \mathbf{N}^T(1,\eta) p_n \begin{bmatrix} 2 \\ \frac{3}{2} \end{bmatrix} d\eta = -\begin{bmatrix} 0 \\ \frac{4}{3}p_{n2} + \frac{2}{3}p_{n3} \\ \frac{2}{3}p_{n2} + \frac{4}{3}p_{n3} \\ 0 \\ 0 \\ p_{n2} + \frac{1}{2}p_{n3} \\ \frac{1}{2}p_{n2} + p_{n3} \\ 0 \end{bmatrix}$$

For both of the above examples we could have found the nodal loads by fairly simple equilibrium considerations. However, we leave this for the reader, see the problems at the end of the chapter.

Thermal loading

The most common initial strain type loading is due to change of temperature. For an isotropic material subject to a temperature change T_c we have, in three and two dimensions,

$$\boldsymbol{\varepsilon}_{0T} = \begin{bmatrix} \alpha T_c \\ \alpha T_c \\ \alpha T_c \\ 0 \\ 0 \\ 0 \end{bmatrix} \quad \text{and} \quad \boldsymbol{\varepsilon}_{0T} = c \begin{bmatrix} \alpha T_c \\ \alpha T_c \\ 0 \end{bmatrix} \qquad (7\text{-}8)$$

respectively, where α is the coefficient of *thermal expansion*, and where $c = 1$ for *plane stress* and $c = 1 + \nu$ for *plane strain*. T_c is the change of temperature from a given reference temperature (for which the stresses due to temperature are assumed to be zero). Temperature change in an unconstrained structure will not cause any shear strain. For orthotropic and anisotropic materials we will normally have different values for α in the three (two) main material directions.

For simple elements the temperature change will often be taken as *one* value for each element; another common approach is to interpolated between the values at the corner nodes as shown opposite, and for a 2D case we can write

Element analysis III - Element loads and stresses

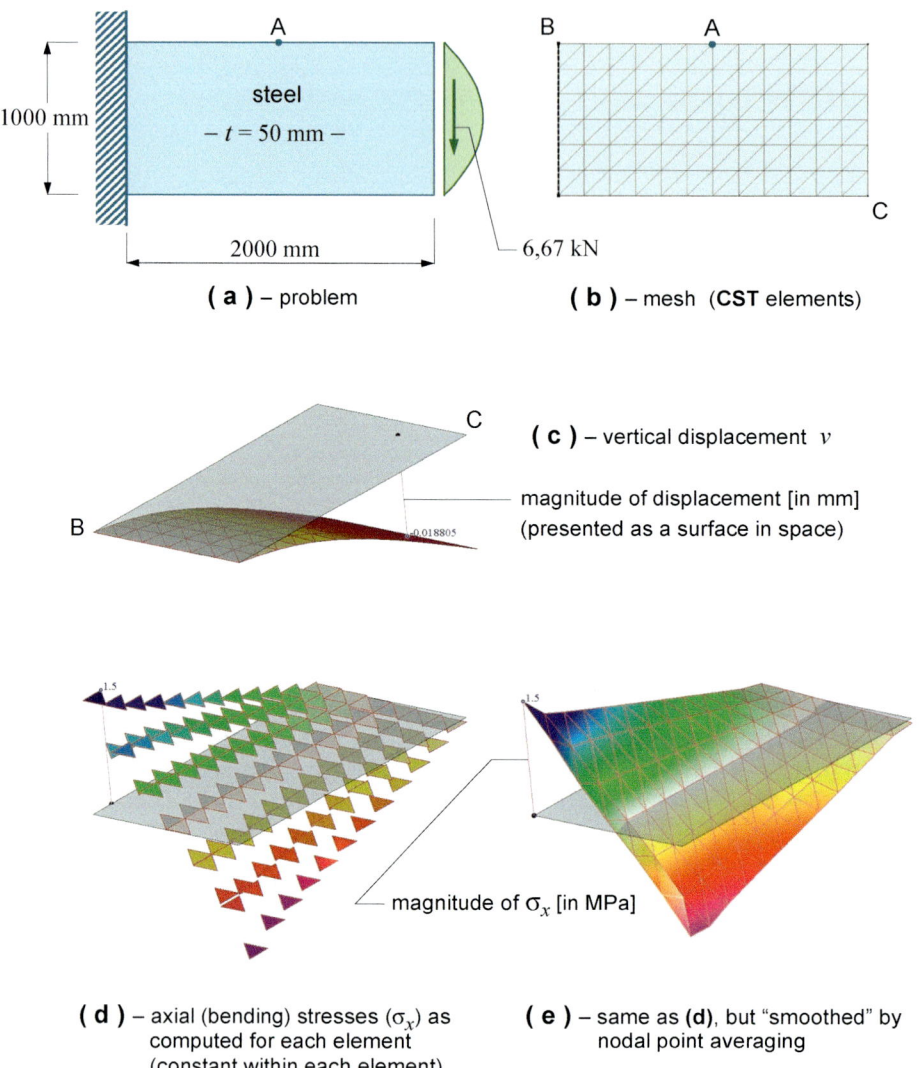

Figure 7.6 A plane stress analysis of a cantilevered plate using the simple constant strain/stress triangular element; results obtained by the **FEMplate** program

$$\boldsymbol{\varepsilon}_{0T} = \begin{bmatrix} c\alpha \\ c\alpha \\ 0 \end{bmatrix} T_c = \begin{bmatrix} c\alpha \\ c\alpha \\ 0 \end{bmatrix} \mathbf{N}_T \mathbf{T}_c \qquad (7\text{-}9)$$

and

$$\mathbf{S}_T^0 = -\int_{V_e} \mathbf{B}^T \mathbf{C} \boldsymbol{\varepsilon}_{0T} dV = -\int_A \mathbf{B}^T \mathbf{C} \begin{bmatrix} c\alpha \\ c\alpha \\ 0 \end{bmatrix} \mathbf{N}_T \mathbf{T}_c t dA \qquad (7\text{-}10)$$

In this expression the thickess, t, may also vary over the element area, and again interpolation between the thicknesses at the corner nodes is the obvious solution if it is considered necessary to account for this variation. Normally such a degree of accuracy is only called for in cases of coarse meshes, and for most practical problems such meshes should, in view of current computing devices, not be used.

For load vectors due to initial strain (temperature) and volume forces, Eqs. (7-10) and (4-56b), the computations proceed in much the same way as for the stiffness matrix. For numerically integrated elements the number of integration points can perhaps be less than for the stiffness matrix, but if in doubt use the same number; the difference in computer time is of little, if any, practical importance.

7.2 Stress recovery and stress smoothing

From the system stiffness relation $\mathbf{Kr} = \mathbf{R}$, modified with respect to the geometric or essential boundary conditions, a task we will discuss in Chapter 9, we can determine the nodal displacements, symbolically expressed as

$$\mathbf{r} = \mathbf{K}^{-1}\mathbf{R}$$

From $\mathbf{r}$ we can retrieve the nodal displacements $\mathbf{v}$ for any element, and the stress-strain relation, Eq. (3-36), and the strain-displacement relation, Eq. (4-50), enable us to compute the stresses at any point in the element:

$$\boldsymbol{\sigma} = \mathbf{C}(\boldsymbol{\varepsilon} - \boldsymbol{\varepsilon}_0) = \mathbf{CBv} - \mathbf{C}\boldsymbol{\varepsilon}_0 \qquad (7\text{-}11)$$

However, remember that matrix $\mathbf{B}$ is obtained by *differentiating* the assumed shape functions (N), and while we will usually have full (or at least partial) displacement continuity between elements, we cannot expect continuity of stresses. Figure 7.6 is an example of this. A cantilever plate is analysed by the simplest of elements, the *6-dof* constant strain triangle (**CST**), with a coarse mesh. The vertical displacement (v), presented in a rather unusual way, as a surface of triangular facets where the distance from the plate to the surface indicates the magnitude of the displacement, is continuous between elements (Fig 7.6c). The stresses, which are constant within each element due to the linear shape functions, obviously cannot be continuous between elements; this is confirmed by Fig. 7.6d. If we *smooth* the stresses, by simple nodal point averaging (discussed below), we get the continuous "stress surface" shown in Fig. 7.6e. This surface consists of triangular facets between

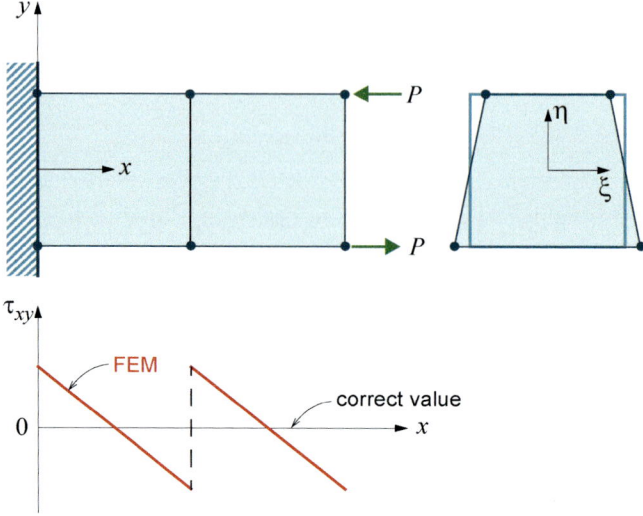

Figure 7.7 Bending of "beam" – two 4-node bilinear elements

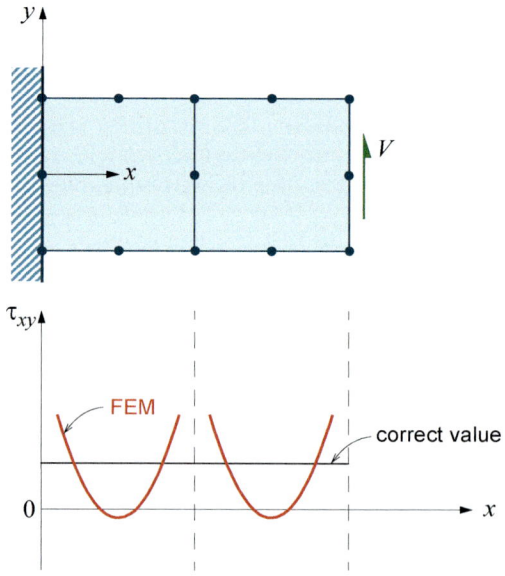

Figure 7.8 "Beam" loaded by transverse tip load – two 8-node serendipity elements

Stress recovery and stress smoothing

the "new" nodal point values (common to all elements that meet at that particular node). Keeping in mind that we are trying to determine stresses that vary linearly over the height with an element that can only represent a state of constant stress, the stress surface of Fig. 7.6e is not all that bad. At the mid-point A of the top fibre, the FEM solution gives an axial stress of 0,65 MPa whereas beam theory gives 0,8 MPa. For the tip displacement FEM gives 0,021 mm whereas beam theory, including shear deformations, gives 0,024 mm.

An alternative to Eq. (7-11) is to start with the element nodal forces defined by

$$\mathbf{S} = \mathbf{kv} + \mathbf{S}^0$$

For one-dimensional elements, bars and beams, this relationship gives us the section forces directly. In Section 4.7 we have argued that these forces, which by definition satisfy equilibrium, can be considered as *resultants* of the stresses along the element edges/surfaces also in two- and three-dimensional problems. The challenge is to convert these forces into stresses.

Stresses from computed displacements. Having determined the system *dofs* or nodal point displacements $\mathbf{r}$ we can determine the displacements $\hat{\mathbf{u}}$ within an arbitrary element, and thus the strains and stresses[1]:

$$\mathbf{r} \Rightarrow \mathbf{v} \Rightarrow \hat{\mathbf{u}} = \mathbf{Nv} \Rightarrow \hat{\boldsymbol{\varepsilon}} = \Delta\hat{\mathbf{u}} = \mathbf{Bv} \Rightarrow \hat{\boldsymbol{\sigma}} = \mathbf{C}(\hat{\boldsymbol{\varepsilon}} - \boldsymbol{\varepsilon}_0) = \mathbf{CBv} - \mathbf{C}\boldsymbol{\varepsilon}_0$$

Matrix $\mathbf{B}$ is, in general, a function of the coordinates within the element, and from the last equation above we can compute the stresses at any point in the element.

Are the stresses computed with the same accuracy at all points within the element? The answer is *no*. Figures 7.7 and 7.8, which are inspired by similar figures in COOK *et al* [21], show (qualitatively) that for a very coarse mesh, the variation within the elements of, in this case the shear stress, is far from correct. However, we also see that there are points within the elements where the shear stress is correct. Is this a coincidence or are there optimal sampling points? The answer to this question is *yes*. It turns out that for two- and three-dimensional elements, stresses are best computed at the GAUSS integration points of *one order lower than full integration*. We see that this is the case for the bilinear element in Fig. 7.7 for which a 2×2 rule represents full integration, and the shear stress is zero (as it should be) at the centre, which coincide with the single point of a 1×1 GAUSS quadrature rule.

For the 8-node element in Fig. 7.8, full integration is obtained by a 3×3 rule, and we see that the shear stresses are correct at points that seem to (and do) coincide with the points of a 2×2 rule.

This observation is credited to BARLOW [22], and the optimal sampling points for stresses are thus also known as BARLOW points. The explanation has to do with a special property of the GAUSS integration points, illustrated by Fig. 7.9: a polynomial uniquely defined by the function values at the *n* first integration points, which will be of order *n*–1, turns out to be the *least square* approximation of a polynomial of order *n*.

---

1. The "hat" signals an approximate finite element result

Element analysis III - Element loads and stresses

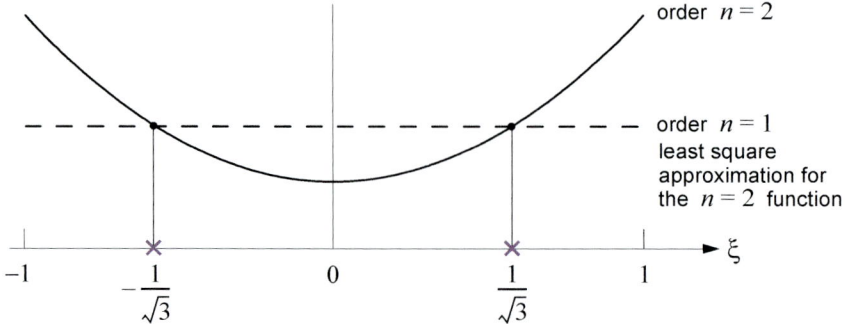

Figure 7.9 Property of Gauss integration points

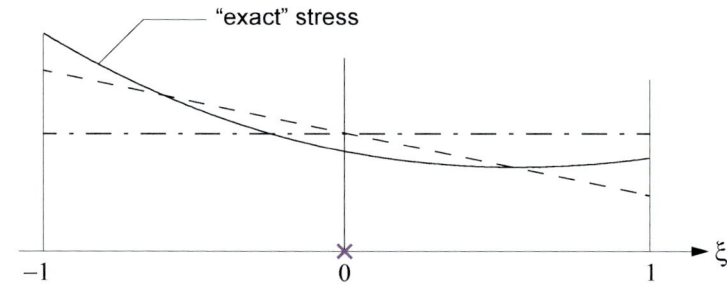

Figure 7.10 One-dimensional element with linear displacement $\Rightarrow p = 1$

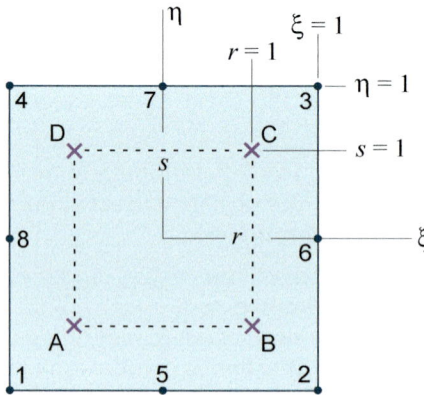

Figure 7.11 Extrapolation of "optimal" stresses

With reference to Fig. 7.10 consider:
- a FEM approximation with an element based on displacement functions containing a complete polynomial of order p, and
- select the stress(es) at the integration points of a GAUSS quadrature rule with p points;
- this gives a polynomial of order $p-1$ for the stress(es) which is a least square approximation of a p'th order polynomial for the stress(es).

For 2- and 3-dimensional stress states: p GAUSS points in each direction.

What has been said here about optimal sampling points for stresses applies only to elements with regular shape, such as rectangles and rectangular prisms. For such elements the stresses at these *superconvergent* (BARLOW) points are of the same accuracy as the displacements. For more irregular element shapes, and for elements with curved edges/surfaces, the GAUSS points are not necessarily optimal, but they are still a good (and simple) choice.

Interpolation / extrapolation. Stresses computed at the BARLOW points can be interpolated and extrapolated to other points in the element, for instance the nodal points, and these values are usually more accurate than those computed by Eq. (7-11) at the same points.

With reference to Fig. 7.11 we have at, say point C:

$$r = s = 1 \quad \text{and} \quad \xi = \eta = 1/\sqrt{3}$$

Hence, at nodal point 3:

$$r = \xi\sqrt{3} \quad \text{and} \quad s = \eta\sqrt{3}$$

A stress component σ_P (σ_x, σ_y or τ_{xy}) at an arbitrary point P, with coordinates r_P and s_P can be expressed as

$$\sigma_P = \sum_{\alpha = A}^{D} N_\alpha \sigma_\alpha$$

Where N_α ($\alpha = A, B, C$ and D) are the shape functions of a 4-node square (corresponding to the 4 integration points) in terms of coordinates r and s,

$$N_\alpha = \frac{1}{4}(1 + r_\alpha r)(1 + s_\alpha s)$$

At nodal point 1, for instance, we have $r = s = -\sqrt{3}$, and an extrapolated stress component, say σ_x, at this point reads

$$\sigma_{x1} = 1{,}866\sigma_{xA} - 0{,}5\sigma_{xB} + 0{,}134\sigma_{xC} - 0{,}5\sigma_{xD}$$

For lower order elements, such as the bilinear, 4-node quadrilateral which has only *one* BARLOW point (at the origin of the natural coordinate system), it is hardly advisable to base element nodal point stresses on the stress components at the one optimal sampling point – it would mean constant stresses throughout the element (including of course the nodal points). If we

Element analysis III - Element loads and stresses

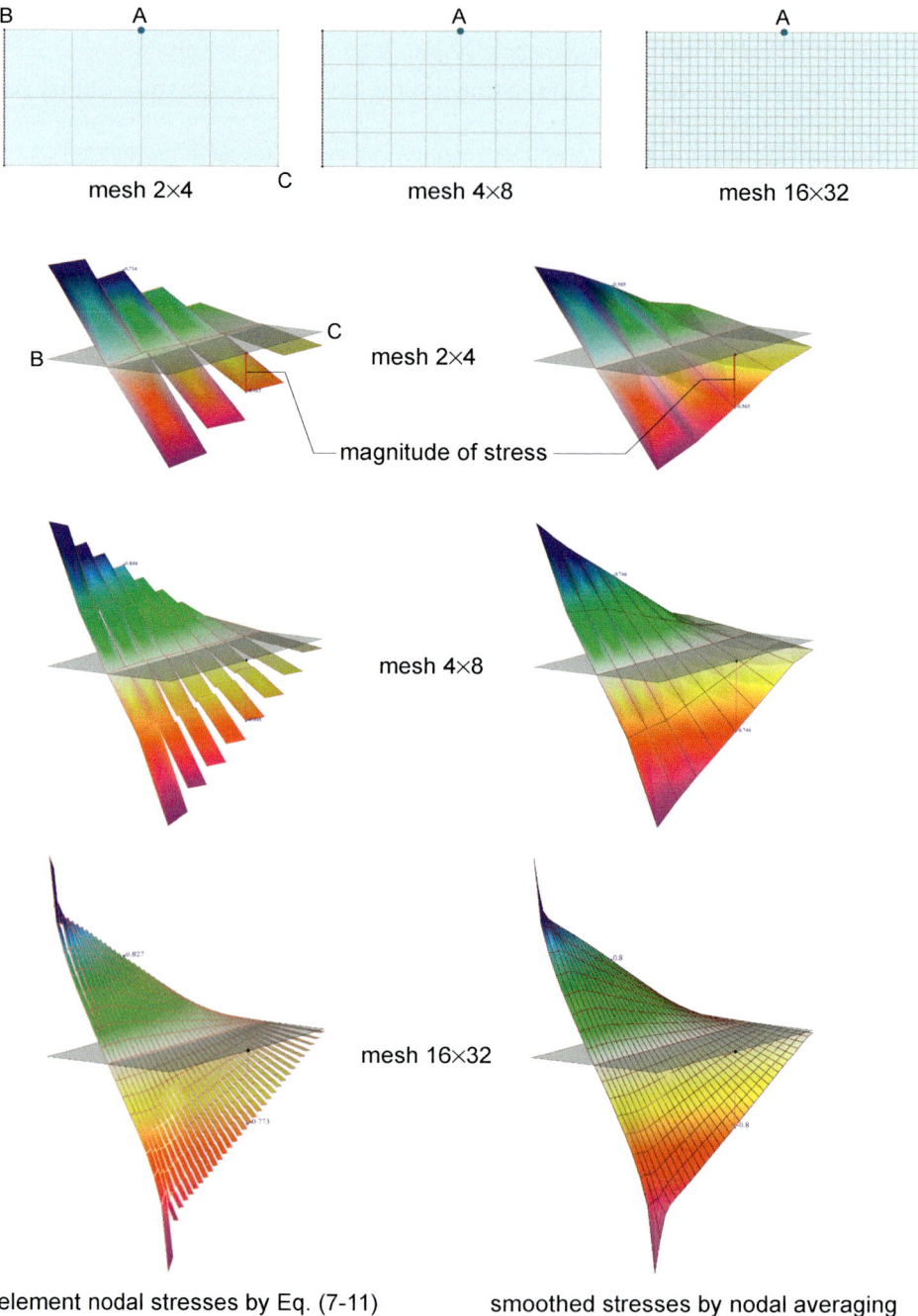

element nodal stresses by Eq. (7-11)   smoothed stresses by nodal averaging

Figure 7.12 Plane stress analysis of the plate problem in Fig. 7.6 computed by **FEMplate**, using the bilinear 4-node quadrilateral element; example of simple nodal point averaging of axial stress σ_x

look at larger regions or patches of elements, then the more accurate stresses at the BARLOW points are the natural basis for stress smoothing techniques, even for lower order elements like the 4-node quadrilateral.

Nodal point averaging. Since stresses obtained by most displacement based elements are discontinuous across element boundaries, some kind of stress smoothing is usually applied. A simple, and fairly robust technique is straightforward *nodal point averaging* also referred to as *element smoothing*:

> At any given "system node" the individual values of a particular stress component computed at this node, in all *n* elements associated with (connected to) this particular node, are added together and divided by *n*, and this "average" value is taken to be the final stress at this system node.

The technique is independent of the method used to compute the element nodal stresses; we also note that each element is given the same weight[1]. Care must be taken so as not to average across lines/surfaces where stresses are (and should be) discontinuous, such as sudden change of thickness or change of material properties.

We have already seen an example of element smoothing in Fig. 7.6. Figure 7.6d shows the "raw" element values for the axial stress component (σ_x), and in Fig. 7.6e we see a continuous stress surface (made up of triangular facets) based on nodal values obtained through nodal point averaging. Figure 7.12 shows similar results obtained for exactly the same cantilever plate problem, but this time the bilinear, 4-node quadrilateral element (the basic version) is used, for three different meshes, two very coarse meshes and one medium fine mesh. For each element the value of σ_x at the 4 (corner) nodes are computed by Eq. (7-11); these values are shown on the left-hand side of Fig. 7.12 (as triangular facets in 3D space). It should be noted that the **FEMplate** program which is described in Section 10.4, presents most results as surfaces made up of triangular facets. So also for results obtained by quadrilateral elements, each of which is divided into four triangles by also considering the value at the centroid, obtained from the four corner values.

Figure 7.12 shows that element stresses, computed directly at the nodal points by Eq. (7-11), are not too bad, even for this simple element. Furthermore, straightforward nodal point averaging also seems to work quite well. Even the coarsest mesh predicts an averaged stress at point A which is off by 40%; for the next mesh the error is about 7% while it is correct to three digits for the finest mesh (which, by today's standards, is not really a fine mesh).

Global smoothing. It is beyond the scope of this presentation to deal with this topic in any detail; an outline of a technique developed by ZIENKIEWICZ and ZHU [23] and often referred to as *"global L2 projection"* (the L2 stands for "least square") goes as follows:

Nodal values of the smoothed stress field $\boldsymbol{\sigma}^*$ are determined by a *least square fit* that minimizes the square of the difference between the smoothed stress

---

[1]. Different weighting schemes have been suggested, but simple, straightforward averaging, giving the same weight to all elements, seems to work quite well.

Element analysis III - Element loads and stresses

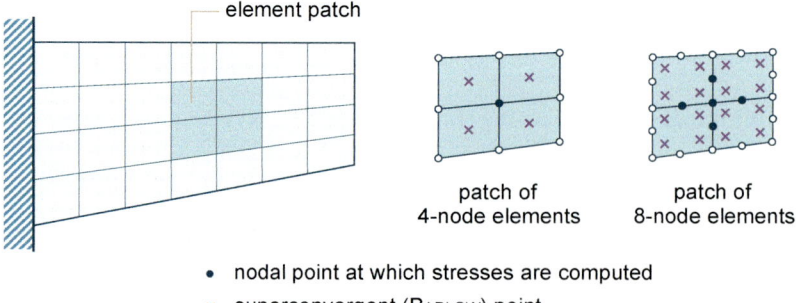

- nodal point at which stresses are computed
- × superconvergent (BARLOW) point

Figure 7.13 Example of superconvergent patch recovery (SPR)

field $\boldsymbol{\sigma}^*$ and the discontinuous stress field $\boldsymbol{\sigma}^h$ determined by the FEM analysis using Eq. (7-11). The *functional*

$$F_G = \sum_e \int_{V_e} (\boldsymbol{\sigma}^* - \boldsymbol{\sigma}^h)^2 dV \tag{7-12}$$

where e indicates that the sum is taken over the number of elements, is minimized with respect to the vector of nodal stresses, $\boldsymbol{\sigma}_n^*$, for all structure (or system) nodes. Within an element, the smoothed stress field $\boldsymbol{\sigma}^*$ is obtained from the (not yet determined) nodal stresses $\boldsymbol{\sigma}_n^*$ through the relationship

$$\boldsymbol{\sigma}^* = \mathbf{N}\boldsymbol{\sigma}_n^* \tag{7-13}$$

This is an element relation. Substitution gives

$$F_G = \sum_e \int_{V_e} (\mathbf{N}\boldsymbol{\sigma}_n^* - \boldsymbol{\sigma}^h)^T (\mathbf{N}\boldsymbol{\sigma}_n^* - \boldsymbol{\sigma}^h) dV = \sum_e \int_{V_e} ((\boldsymbol{\sigma}_n^*)^T \mathbf{N}^T - (\boldsymbol{\sigma}^h)^T)(\mathbf{N}\boldsymbol{\sigma}_n^* - \boldsymbol{\sigma}^h) dV$$

$$= \sum_e \int_{V_e} ((\boldsymbol{\sigma}_n^*)^T \mathbf{N}^T \mathbf{N} \boldsymbol{\sigma}_n^* - 2(\boldsymbol{\sigma}_n^*)^T \mathbf{N}^T \boldsymbol{\sigma}^h + (\boldsymbol{\sigma}^h)^T \boldsymbol{\sigma}^h) dV$$

and

$$\frac{\partial F_G}{\partial \boldsymbol{\sigma}_n^*} = 0 \quad \Rightarrow \quad \sum_e \int_{V_e} 2(\mathbf{N}^T \mathbf{N} \boldsymbol{\sigma}_n^* - \mathbf{N}^T \boldsymbol{\sigma}^h) dV = 0$$

$$\Rightarrow \quad \sum_e \left(\mathbf{a}_e^T \int_{V_e} \mathbf{N}^T \mathbf{N} dV \mathbf{a}_e \right) \boldsymbol{\sigma}_N^* = \sum_e \mathbf{a}_e^T \int_{V_e} \mathbf{N}^T \boldsymbol{\sigma}^h dV$$

Here we have introduced the relation $\boldsymbol{\sigma}_n^* = \mathbf{a}_e \boldsymbol{\sigma}_N^*$, where $\boldsymbol{\sigma}_N^*$ contains *all* nodal stresses, and $\mathbf{a}_e$ is a "connectivity matrix" analogous to the **a**-matrices of Eq. (2-15). We can write this as

$$\mathbf{A}\boldsymbol{\sigma}_N^* = \mathbf{b} \tag{7-14}$$

where $\quad \mathbf{A} = \sum_e \left(\mathbf{a}_e^T \int_{V_e} \mathbf{N}^T \mathbf{N} dV \mathbf{a}_e \right) \quad$ and $\quad \mathbf{b} = \sum_e \mathbf{a}_e^T \int_{V_e} \mathbf{N}^T \boldsymbol{\sigma}^h dV$

The shape functions are the same as those used for the displacements, and the integrals are evaluated numerically, using the superconvergent BARLOW points. Matrix **A** has the same sparse pattern as the stiffness matrix, and by solving Eq. (7-14), one stress component at the time, the computational effort is significantly less than the effort required for the solution of $\mathbf{Kr} = \mathbf{R}$.

The method can be modified and used to find improved stresses at nodes inside a *patch* of elements, see Fig. 7.13. Stresses at nodal points associated *only* with the elements of the patch are determined on the basis of the FEM stresses computed at the optimal sampling (BARLOW) points of the elements of the patch. This is referred to as *superconvergent patch recovery* – SPR [24,25].

Element analysis III - Element loads and stresses

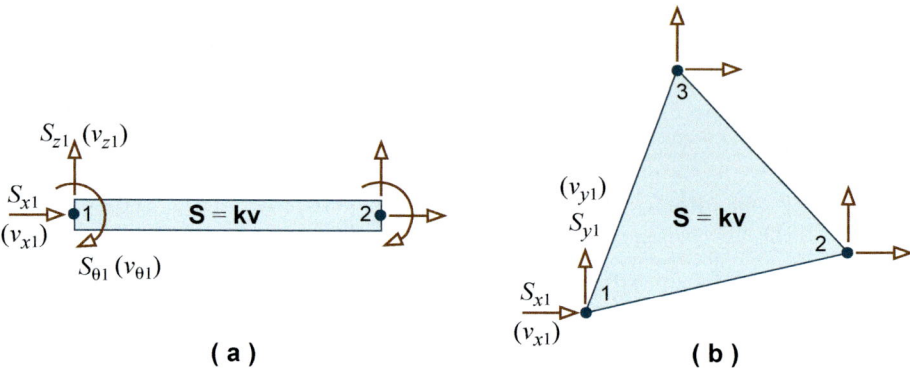

Figure 7.14 Element nodal forces (**S**) due to elastic deformations (**v**)

These improved stresses are mainly used in connection with *error estimation* schemes and *adaptive meshing*, themes we will deal with in a sketchy manner in the next chapter.

Stresses from nodal point forces. In order to simplify the discussion we limit ourselves to a situation where all loading consists of statically equivalent nodal loads. In other words, the load vector $\mathbf{S}^0$ in Eq. (4-57) is equal to zero, and the equation simplifies to

$$\mathbf{S} = \mathbf{k}\mathbf{v} \qquad (7\text{-}15)$$

Possible "fixed end" (initial) effects, resulting in a non-zero $\mathbf{S}^0$, can be dealt with as a separate correction. The sum of the element nodal forces for all elements associated with a particular node, balance the externally applied loads at this node; in other words, they satisfy equilibrium. Thus $\mathbf{S}$ ought to be a good basis for computing stresses. For the 1-dimensional beam element, Fig. 7.14a, the nodal forces are in fact the section forces of stress *resultants*, and they therefore yield the exact stresses. For 2- and 3-dimensional elements, such as the one in Fig. 7.14b, the nodal forces $\mathbf{S}$ can still be considered as resultants of the stresses along the element edges, but for such elements there is no simple relationship between $\mathbf{S}$ and $\boldsymbol{\sigma}$. Another, more "roundabout" way has to be found.

WILSON and IBRAHIMBEGOVIC [26] suggest the following approach for determining $\boldsymbol{\sigma}$ from $\mathbf{S}$:

1) Assume the stress distribution within the element to be of the form

$$\boldsymbol{\sigma} = \mathbf{P}\boldsymbol{\beta} \qquad (7\text{-}16)$$

where $\mathbf{P}$ contains assumed functions of the spatial coordinates and $\boldsymbol{\beta}$ contains some generalized stress parameters. Equation (7-16) should satisfy the *homogeneous* equilibrium equations.

Consider a two-dimensional problem, and assume a linear stress distribution within the element, that is

$$\begin{aligned}\sigma_x &= \beta_1 + \beta_2 x + \beta_3 y \\ \sigma_y &= \beta_4 + \beta_5 x + \beta_6 y \\ \tau_{xy} &= \beta_7 + \beta_8 x + \beta_9 y\end{aligned} \qquad (7\text{-}17)$$

The homogeneous equilibrium equations read, see Eqs. (3.12),

$$\sigma_{x,x} + \tau_{xy,y} = 0 \qquad (7\text{-}18\text{a})$$

$$\sigma_{x,x} + \tau_{xy,y} = 0 \qquad (7\text{-}18\text{b})$$

Equations (7-17) satisfy Eqs. (7-18) identically if

$$\beta_8 = -\beta_6 \quad \text{and} \quad \beta_9 = -\beta_2$$

Hence

Element analysis III - Element loads and stresses

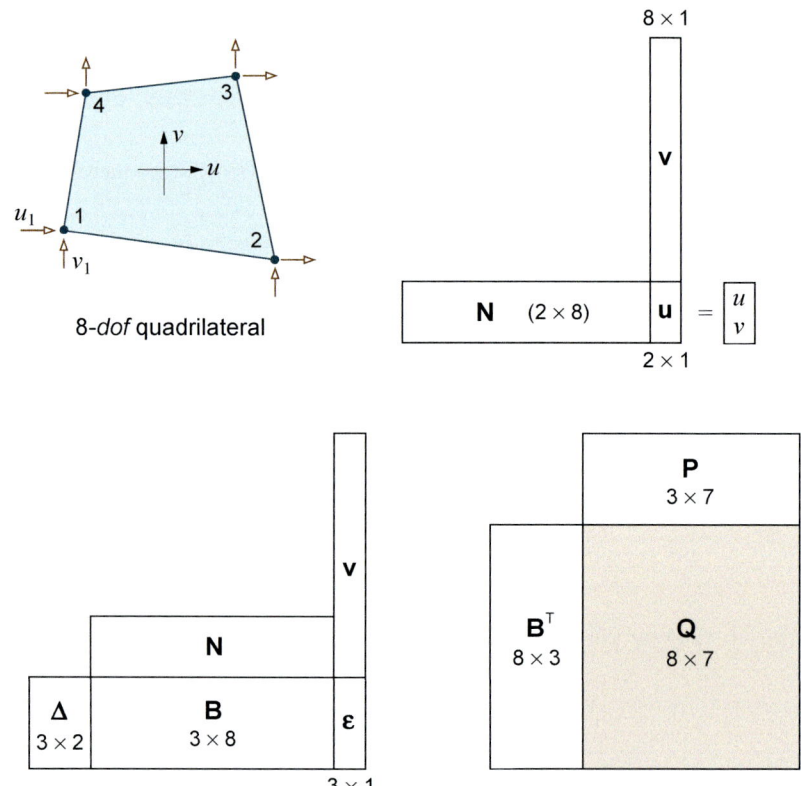

256

$$\sigma = \begin{bmatrix} \sigma_x \\ \sigma_y \\ \tau_{xy} \end{bmatrix} = \begin{bmatrix} 1 & x & y & 0 & 0 & 0 & 0 \\ 0 & 0 & 0 & 1 & x & y & 0 \\ 0 & -y & 0 & 0 & 0 & -x & 1 \end{bmatrix} \begin{bmatrix} \beta_1 \\ \beta_2 \\ \cdot \\ \cdot \\ \cdot \\ \beta_7 \end{bmatrix} = P\beta \quad (7\text{-}19)$$

2) We now require the stresses σ to be in equilibrium with the nodal forces **S**, and for this we use the principle of *virtual displacements*. Subjected to a set of virtual nodal displacements $\tilde{v}$ and corresponding kinematically compatible virtual strains

$$\tilde{\varepsilon} = \Delta\tilde{u} = \Delta N\tilde{v} = B\tilde{v} \quad (7\text{-}20)$$

the equilibrium condition is enforced by equating the "external" virtual work done by the element nodal forces to the internal virtual work done by the stresses, that is

$$\tilde{v}^T S = \int_{V_e} \tilde{\varepsilon}^T \sigma dV = \tilde{v}^T \int_{V_e} B^T P \beta dV \quad (7\text{-}21)$$

or, since this must hold for any $\tilde{v}$,

$$\int_{V_e} B^T P dV \beta = S \quad \text{or} \quad Q\beta = S \quad (7\text{-}22)$$

where we have introduced the notation

$$Q = \int_{V_e} B^T P dV \quad (7\text{-}23)$$

3) Equation (7-22) normally represents an overdetermined system of equations; for the 4-node quadrilateral plane stress element, for instance, **Q** is an 8 by 7 matrix. Such a system is usually solved by the method of least squares (see APPENDIX A), that is, by the solution of

$$Q^T Q \beta = Q^T S = Q^T k v \quad (7\text{-}24)$$

Once β is computed, the stresses may be found at any point within the element by use of Eq. (7-19).

The few results presented in [26] indicate good accuracy, but the method is not widely used. The fact that stresses due to a non-zero S^0 (from, for instance, initial strain) will have to be handled separately (by some, not obvious, scheme), is an extra complication.

In conclusion we may safely state that for most 2- and 3-dimensional displacement based elements, Eq. (7-11) is the basis for stress recovery. Also, for many practical applications of FEM in linear structural mechanics, simple nodal point averaging of stresses computed directly at the element nodal points by Eq. (7-11), provides sufficient accuracy (since very fine element meshes can be used at very low "cost"). This is not to say that the more

Element analysis III - Element loads and stresses

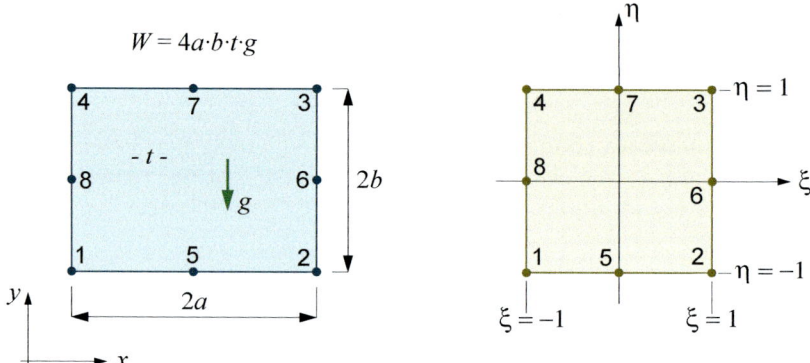

Figure 7.15 Serendipity element subjected to own weight g (volume force)

refined methods of smoothing based on (superconvergent) stresses, computed at optimal sampling points, are superfluous, but their role in some types of problems is somewhat reduced by the steady increase in computational capabilities. Examples in chapters to come will support this claim, and we shall also see that some (higher order) elements provide nodal point stresses of an accuracy on a par with that obtained for the displacements.

Example 7-4 – Consistent element load vector

For an 8-node serendipity element of constant thickness t and a weight density of g per unit volume, see Fig. 7.15, find the consistent load vector $\mathbf{S}_F^0$ due to the own weight of the element, expressed in terms of the total weight W of the element. For simplicity we let the element have rectangular shape, with side lengths $2a$ and $2b$, respectively.

Solution

The load vector we seek is defined by Eq. (4-56b),

$$\mathbf{S}_F^0 = -\int_{V_e} \mathbf{N}^T \mathbf{F} \, dV$$

where $\mathbf{N}$ is the shape function matrix and $\mathbf{F}$ contains the volume forces (per unit volume):

$$\mathbf{N} = \begin{bmatrix} N_1 & 0 & N_2 & 0 & \ldots & N_8 & 0 \\ 0 & N_1 & 0 & N_2 & \ldots & 0 & N_8 \end{bmatrix} \quad \text{and} \quad \mathbf{F} = \begin{bmatrix} 0 \\ -g \end{bmatrix}$$

Gravity is positive in negative y-direction, hence the minus sign in $\mathbf{F}$. We see that the load vector will have non-zero elements in the y-direction only, and it suffices to determine one corner node value (at node 1) and one mid-side node value (at node 5). For this we need the shape functions N_1 and N_5, both of which we have established earlier in Chapter 5, Eqs. (5-60) and (5-61), respectively:

$$N_1 = \tfrac{1}{4}(1-\xi)(\eta-1)(\xi+\eta+1) \quad \text{and} \quad N_5 = \tfrac{1}{2}(\xi^2-1)(\eta-1)$$

Recall

$$\int_{V_e} dV = t\int_{A_e} dA = t\int_{-1}^{1}\int_{-1}^{1} J\,d\xi\,d\eta = tab\int_{-1}^{1}\int_{-1}^{1} d\xi\,d\eta \quad \text{and} \quad W = 4tabg$$

Hence

$$F_{y1} = tab\int_{-1}^{1}\int_{-1}^{1} N_1(-g)\,d\xi\,d\eta = -\frac{tabg}{4}\int_{-1}^{1}\int_{-1}^{1}(\xi^2\eta+\xi\eta^2-\xi^2-\eta^2-\xi\eta+1)\,d\xi\,d\eta$$

$$= -\frac{tabg}{4}\int_{-1}^{1}\left[\frac{\xi^3}{3}\eta+\frac{\xi^2}{2}\eta^2-\frac{\xi^3}{3}-\xi\eta^2-\frac{\xi^2}{2}\eta+\xi\right]_{-1}^{1} d\eta$$

Element analysis III - Element loads and stresses

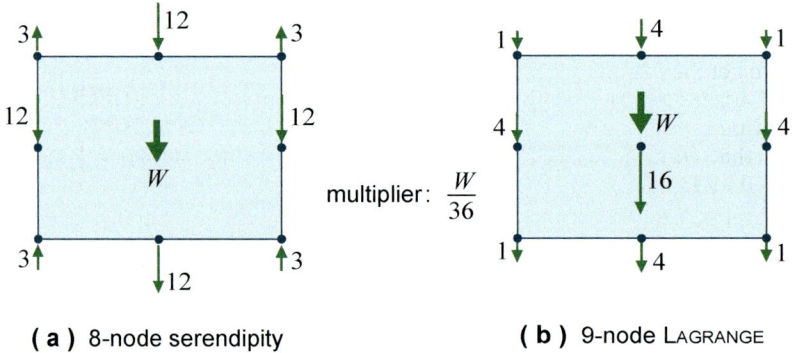

Figure 7.16 Consistent nodal loads due to own weight (W) of rectangular element

Example 7-4 – Consistent element load vector

$$F_{y1} = -\frac{tabg}{4}\int_{-1}^{1}\left(\frac{2}{3}\eta - \frac{2}{3} - 2\eta^2 + 2\right)d\eta = -\frac{tabg}{4}\left[\frac{2}{6}\eta^2 - \frac{2}{3}\eta - \frac{2}{3}\eta^3 + 2\eta\right]_{-1}^{1}$$

$$= -\frac{tabg}{4}\left[-\frac{4}{3} - \frac{4}{3} + 4\right] = -\frac{tabg}{3} = -\frac{4tabg}{12} = -\frac{W}{12}$$

Similarly for node 5:

$$F_{y5} = tab\int_{-1}^{1}\int_{-1}^{1}N_5(-g)d\xi d\eta = -\frac{tabg}{2}\int_{-1}^{1}\int_{-1}^{1}(\xi^2 - 1)(\eta - 1)d\xi d\eta$$

$$= -\frac{tabg}{2}\int_{-1}^{1}\left[\frac{\xi^3}{3}\eta - \frac{\xi^3}{3} - \xi\eta + \xi\right]_{-1}^{1}d\eta = -\frac{tabg}{2}\int_{-1}^{1}\left(\frac{4}{3}\eta - \frac{4}{3}\right)d\eta$$

$$= -\frac{tabg}{2}\left[\frac{4}{6}\eta^2 - \frac{4}{3}\eta\right]_{-1}^{1} = -\frac{tabg}{2}\left(-\frac{8}{3}\right) = \frac{4tabg}{3} = \frac{W}{3}$$

and we get the interesting result shown in Fig. 7.16a. The same procedure for the 9-node LAGRANGE element will, for a rectangle, give the result shown in Fig. 7.16b. It should be noted that the forces shown in Fig. 7.16 are in fact $-\mathbf{S}_F^0$, which are the forces that contribute to the system load vector $\mathbf{R}$ (due to the sign convention adopted by Eq. (7-1)).

For an arbitrary quadrilateral element shape the numbers will no longer be the same for all corner nodes or for all mid-side nodes, since the Jacobian J for such an element is no longer a constant, but varies from point to point within the element.

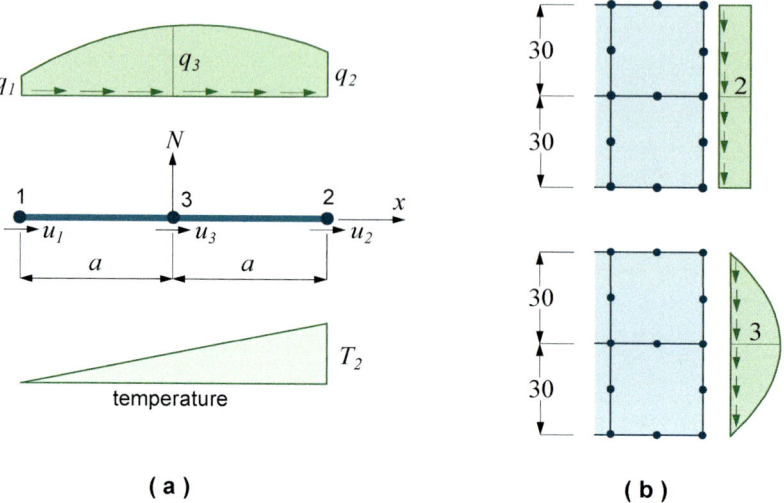

(a) (b)

Figure 7.17

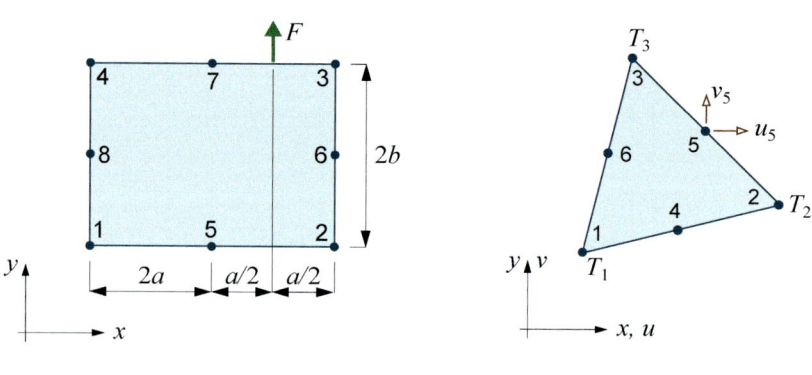

Figure 7.18 **Figure 7.19**

Problems

Problem 7.1
Verify the results shown in Fig. 7.16b for the 9-node LAGRENGE element.

Problem 7.2
Figure 7.17a shows a uniform *bar* element with three degrees of freedom (u_1, u_2, u_3) placed at end nodes (1 and 2) and a mid-side node (3). The length of the element is $2a$, and it is referred to the coordinate system shown.

a) The element is subjected to a quadratically distributed tangential traction defined by nodal values q_1, q_2, and q_3.
Determine the load vector $\mathbf{S}^0$ for this loading. Check the result for the case that the distributed loading is uniformly distributed ($q_1 = q_2 = q_3 = q$).

b) The element cross section area is A, the modulus of elasticity is E and the thermal expansion coefficient is α. A temperature change that varies linearly from 0 (at node 1) to T_2 (at node 2) is imposed. Determine the corresponding load vector $\mathbf{S}^0$. Check equilibrium.

c) Figure 7.17b shows the tip of a cantilever plate modelled by two 8-node rectangular elements over the height. Use the results from a) to determine the consistent nodal forces due to the two distributed loadings shown on the plate edge – indicate the loads on figures. Are these nodal loads statically equivalent to the distributed loads?

Problem 7.3
A concentrated force F acts on the edge of an 8-node rectangular element as shown in Fig. 7.18. Determine the three consistent nodal loads (at nodes 4, 7 and 3) due to this force, and show that these loads are statically equivalent to F.

Problem 7.4
Figure 7.19 shows a 6-node (12-*dof*) quadratic triangular element subjected to a linear *temperature field*, defined by the three corner values of the temperature differences from the "neutral state" (T_{c1}, T_{c2} and T_{c3}). The coefficient of thermal expansion is α. Consider *plane stress* and an initial strain vector $\boldsymbol{\varepsilon}_0$ defined as

$$\boldsymbol{\varepsilon}_{0T} = \begin{bmatrix} \alpha T_c \\ \alpha T_c \\ 0 \end{bmatrix}$$

and establish an expression for the load vector $\mathbf{S}_T^0$ for the element, in terms of suitable and well defined matrices.

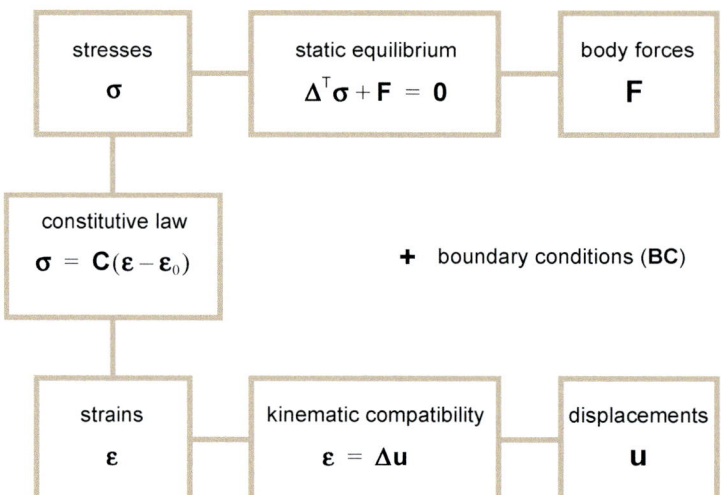

8

Accuracy and convergence

*We have already, in several of the preceding chapters, paid a visit to these topics. In this chapter we will take a closer look, but we try to avoid too much theoretical jargon. We will look at bounds in RAYLEIGH-RITZ consistent FEM models, and we shall discuss discretization errors and types of convergence. We will emphasize points of importance for the practical use of FEM, and formulate necessary and sufficient requirements for convergence to the correct answers. We shall also describe some useful element tests, in particular the all important **patch test**.*

First of all, the errors we are talking about here are the differences between the "exact" solution of the mathematical model and the numerical, finite element solution of the same mathematical model. The errors inherent in the transition from the actual, physical problem to the mathematical model of that problem are certainly not irrelevant, but they are *not* considered in this chapter. Such errors will be dealt with in later chapters.

To set the scene we recapitulate some basic properties:

For a typical structural mechanics problem, the field variables, *i.e.* displacements ($\mathbf{u}$), stresses ($\boldsymbol{\sigma}$) and strains ($\boldsymbol{\varepsilon}$), need to satisfy the *field equations* expressing the requirements of

- *equilibrium* between prescribed loads and internal stresses,
- *kinematic compatibility* between displacements and strains, and
- the *constitutive (material) law* (stress-strain relationship).

In addition the *boundary conditions* of the problem, both essential and natural, need to be satisfied. The field equations are *differential equations* representing, together with the boundary conditions, the *strong form* of the problem. The solution of the strong form (to the extent that it can be found) satisfies the field equations and the boundary conditions at every material point – *pointwise equilibrium*.

The finite element method is, as we have seen, based on the *weak form* of the problem, and it tries to satisfy equilibrium as best it can. The weak form satisfies equilibrium in an integral (average) sense rather than pointwise. Furthermore, if the finite element displacement assumption satisfies the basic requirements of continuity and completeness, two important properties are

Accuracy and convergence

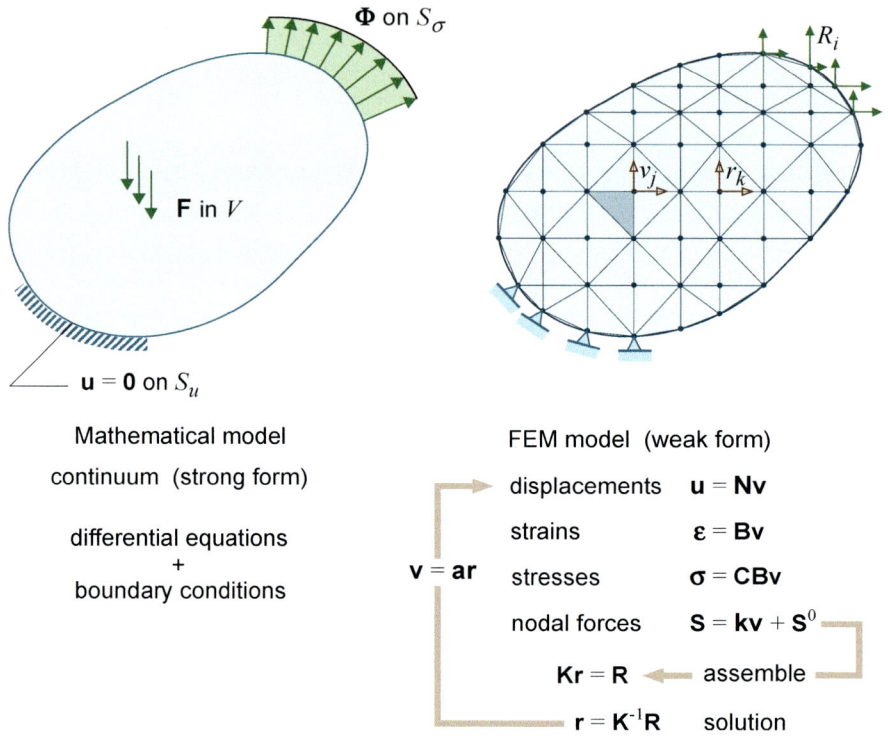

Figure 8.1 Finite element discretization

always satisfied by a finite element solution, based on the displacement method:

- *Nodal point equilibrium.*
  At any one nodal point *i* of the model, the externally applied loads **R**$_i$ (or reaction forces if the node is constrained) are in equilibrium with the sum of the element forces **S**$_j$ at that particular node, for all elements associated with node *i*.

- *Element equilibrium.*
  For any one element, the nodal forces **S** = **kv** + **S**^{0} satisfy the equilibrium conditions for that element (both force and moment equilibrium).

Both strain-displacement *compatibility* and the *material* (constitutive) *law* are satisfied (pointwise), assuming the basic continuity requirements are met.

The finite element discretization process is shown schematically in Fig. 8.1. The errors inherent in this discretization are of two types: **a)** the problem geometry may not be correctly represented, *e.g.*, a curved edge may be represented by a series of straight lines, and **b)** the displacement field is approximated by an assumed shape (**N**) in terms of discrete values (**v**) at the nodal points.

With regard to the first type of error it is clearly desirable to keep this error as small as possible, but apart from recommending that the total area (2D) or volume (3D) is modelled as accurately as possible (neither circumscribed nor inscribed, but something in between), it is not possible to specify formal requirements. It is in the nature of this type of error that it will diminish as the element size diminishes.

The second type of error, which is the error we normally associate with the notion of discretization error, depends on the type of problem we seek a solution for. It depends on the ability of the assumed displacements to approximate the true displacements within the elements and to provide the necessary continuity (C^0 or C^1) across the element boundaries.

Before we proceed we introduce the following notation:

The (mathematically) exact solution is described by letters without extra symbols, like *u*, σ and ε, whereas letters denoting finite element solutions are "marked" with the "hat" symbol (^), *e.g.*, $\hat{u}$, $\hat{\sigma}$ and $\hat{\varepsilon}$. Hence

$$e = |u - \hat{u}| \qquad (8\text{-}1)$$

is a measure of the error. Furthermore, *h* denotes (in some qualitative sense) the *size* of the elements ($h \to 0$ indicates that the element size becomes extremely small), and *p* denotes the polynomial order of the assumed shape functions.

Accuracy and convergence

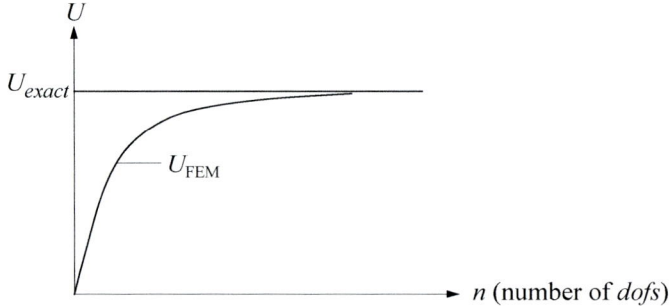

Figure 8.2 Monotonic convergence of computed strain energy

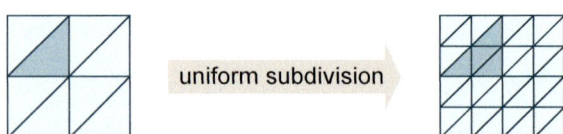

In uniform mesh refinement the coarse mesh is contained in the refined mesh

Figure 8.3 Uniform mesh refinement

8.1 Energy bounds in R-R consistent FEM solutions

A finite element displacement formulation that is consistent with the basis of the Rayleigh-Ritz method will *underestimate* the *strain energy U* of the system, that is

$$U_{exact} = \int_V \frac{1}{2}\boldsymbol{\sigma}^T\boldsymbol{\varepsilon}\,dV = \int_V \frac{1}{2}\boldsymbol{\varepsilon}^T\mathbf{C}\boldsymbol{\varepsilon}\,dV \geq \int_V \frac{1}{2}\hat{\boldsymbol{\varepsilon}}^T\mathbf{C}\hat{\boldsymbol{\varepsilon}}\,dV = \frac{1}{2}\mathbf{r}^T\mathbf{K}\mathbf{r} \qquad (8\text{-}2)$$

According to "average equilibrium" – see Eq. (4-21) – we have

$$\int_V \hat{\boldsymbol{\varepsilon}}^T\mathbf{C}(\boldsymbol{\varepsilon}-\hat{\boldsymbol{\varepsilon}})dV = \int_V \hat{\boldsymbol{\varepsilon}}^T\mathbf{C}\mathbf{e}\,dV = 0 \qquad (8\text{-}3)$$

where

$$\mathbf{e} = \boldsymbol{\varepsilon}-\hat{\boldsymbol{\varepsilon}} \qquad (8\text{-}4)$$

is the *error* in finite element strains. Hence

$$\int_V \boldsymbol{\varepsilon}^T\mathbf{C}\boldsymbol{\varepsilon}\,dV = \int_V (\mathbf{e}+\hat{\boldsymbol{\varepsilon}})^T\mathbf{C}(\mathbf{e}+\hat{\boldsymbol{\varepsilon}})dV = \int_V \mathbf{e}^T\mathbf{C}\mathbf{e}\,dV + 2\int_V \mathbf{e}^T\mathbf{C}\hat{\boldsymbol{\varepsilon}}\,dV + \int_V \hat{\boldsymbol{\varepsilon}}^T\mathbf{C}\hat{\boldsymbol{\varepsilon}}\,dV$$

In view of Eq. (8-3), the second term on the far right-hand side of this equation is equal to zero, and we can write

$$\int_V \boldsymbol{\varepsilon}^T\mathbf{C}\boldsymbol{\varepsilon}\,dV = \int_V \mathbf{e}^T\mathbf{C}\mathbf{e}\,dV + \int_V \hat{\boldsymbol{\varepsilon}}^T\mathbf{C}\hat{\boldsymbol{\varepsilon}}\,dV \qquad (8\text{-}5)$$

The first term on the right-hand side is *always* greater than or equal to zero (regardless the content of **e**), and the validity of Eq. (8-2) follows. The error decreases as the number of degrees of freedom (or elements) increases, and the computed U_{FEM} converges *monotonically* as shown in Fig. 8.2. It should be noted, however, that monotonic convergence requires mesh refinement through *uniform subdivision*, as shown in Fig. 8.3.

From Section 4.5 and the principle of minimum potential energy (PMPE) we can state that

$$\Pi_{exact} \leq \hat{\Pi}(\mathbf{r}) \qquad (8\text{-}6)$$

In other words, the finite element solution *overestimates* the *total potential energy* Π. The argument for Eq. (8-6) is that the displacement **r** is determined so that $\hat{\Pi}(\mathbf{r})$ is a minimum, but a finer mesh (more *dofs*) will produce a more accurate and *smaller* value for $\hat{\Pi}$. In view of Eq. (4-39) we can write Eq. (8-6) as

$$\Pi_{exact} \leq \underbrace{\frac{1}{2}\mathbf{r}^T\mathbf{K}\mathbf{r}}_{\mathbf{R}} - \mathbf{r}^T\mathbf{R} = -\frac{1}{2}\mathbf{r}^T\mathbf{R} \qquad (8\text{-}7)$$

and since **R** represents the loading correctly we can deduce that **r** must be too "small". Hence, *an R-R consistent finite element model is too stiff*.

Accuracy and convergence

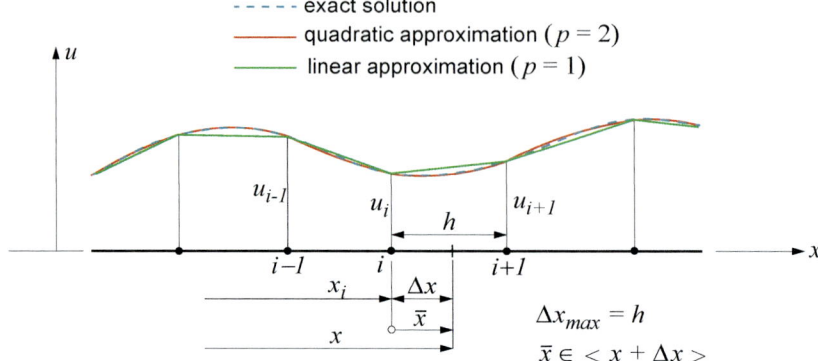

Figure 8.4 One-dimensional problem

m is the "order" of differentiation of the operator $\boldsymbol{\Delta}$ ($\boldsymbol{\varepsilon} = \boldsymbol{\Delta u}$)

It should be noted that our statements about energies of the FEM model do not provide information about individual components of **r** (except in the rare case that **R** contains only one non-zero element).

8.2 Error and rate of convergence

In the finite element method the interpolation functions (polynomials) within each element are fit to the exact solution as best they can, by satisfying, for instance, a stationary principle (PMPE) or a virtual work equation. This adaptation, within a simple one-dimensional problem, see Fig. 8.4, can be expressed as a TAYLOR series of the exact solution:

$$u(x_i+\Delta x) = u_i + \Delta x \left(\frac{du}{dx}\right)_i + \frac{\Delta x^2}{2!}\left(\frac{d^2u}{dx^2}\right)_i + \ldots + \frac{\Delta x^p}{p!}\left(\frac{d^p u}{dx^p}\right)_i + \frac{\Delta x^{p+1}}{(p+1)!}\left(\frac{d^{p+1}u}{dx^{p+1}}\right)_i + . \quad (8\text{-}8)$$

This expression represents a *complete* polynomial in Δx, and for an element with shape functions comprising a complete polynomial of order p, the terms in Eq. (8-8) up to and including the term of order p can be reproduced locally by the FEM solution. The error in u, inside the element (with length h) can therefore be expressed as

$$e_u = u - \hat{u} = \frac{\Delta x^{p+1}}{(p+1)!}\left(\frac{d^{p+1}u(\bar{x})}{dx^{p+1}}\right) + \ldots = O(h^{p+1}) \quad (8\text{-}9)$$

Here $\hat{u}$ is the FEM solution, and the notation $O(\cdot)$ denotes "of the order $(\cdot)$".

From this we conclude that the error in the displacement is proportional with h^{p+1}.

Strains and stresses are derived from the displacement (u) by differentiating u a certain number of times m. Thus, the polynomial error term of order $p+1$ corresponds to the error term of order $p-m+1$ for stress (σ) and strain (ε):

$$e_\varepsilon = \varepsilon - \hat{\varepsilon} = O(h^{p-m+1}) \quad (8\text{-}10a)$$

$$e_\sigma = \sigma - \hat{\sigma} = O(h^{p-m+1}) \quad (8\text{-}10b)$$

The strain energy U is a function of the square of the strain (or stress). Thus

$$e_U = U - \hat{U} = O(h^{2(p-m+1)}) \quad (8\text{-}11)$$

In order to reduce the error we can choose one of two methods:

1) We can stay with the same (generic) element, that is **v** is unchanged, but we increase the number of elements and thus the number of *dofs* (**r**). This is referred to as **h–convergence**.

2) We maintain the element mesh and thus the element shape, but we increase the number of element *dofs* (**v**) such as to enable a higher (complete) polynomial order p for the shape functions. With the same mesh this will increase the number of system *dofs* (**r**). This is referred to as **p–convergence**.

Clearly a combination of the two methods represents a third possibility.

Accuracy and convergence

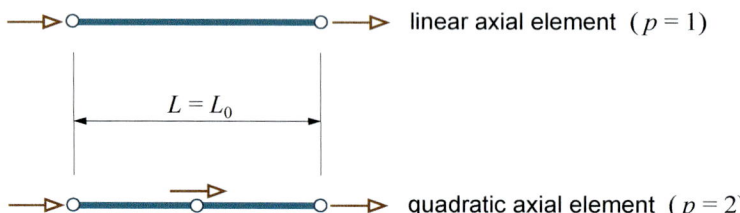

linear axial element ($p = 1$)

$L = L_0$

quadratic axial element ($p = 2$)

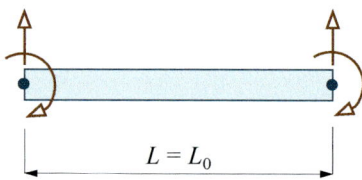

$L = L_0$

beam element – cubic interpolation ($p = 3$)

Example 8-1 – axial element in two dimensions

$$\varepsilon = \frac{du}{dx} \; ; \quad N = EA\varepsilon \; ; \quad U = \frac{1}{2}\int_0^L N\varepsilon\, dx \quad \Rightarrow \quad m = 1$$

a) Linear interpolation $\Rightarrow p = 1$

$e_u = O(h^{1+1}) = O(L^2)$ – quadratic convergence
$e_\varepsilon = e_\sigma = O(h^{1-1+1}) = O(L)$
$e_U = O(L^2)$
Halve the element length: $L = L_0 \Rightarrow L = L_0/2$

$$(e_u)_{L = L_0/2} = \frac{1}{4}(e_u)_{L = L_0}$$

$$(e_\sigma)_{L = L_0/2} = \frac{1}{2}(e_\sigma)_{L = L_0}$$

b) Quadratic interpolation $\Rightarrow p = 2$

$e_u = O(L^3); \quad e_\sigma = O(L^2); \quad e_U = O(L^4)$

$L \Rightarrow L/2: \quad (e_u)_{L = L_0/2} = \frac{1}{8}(e_u)_{L = L_0}$

Example 8-2 – beam element in two dimensions

$$\kappa = -\frac{d^2w}{dx^2} \; ; \quad M = -EI\kappa \; ; \quad U = \frac{1}{2}\int_0^L M\kappa\, dx \quad \Rightarrow \quad m = 2$$

Cubic interpolation $\Rightarrow p = 3$

$e_u = O(h^{3+1}) = O(L^4)$
$e_\sigma = O(h^{3-2+1}) = O(L^2)$ (here σ stands for M or c)
$e_U = O(L^4)$

$L \Rightarrow L/2: \quad (e_u)_{L = L_0/2} = \frac{1}{16}(e_u)_{L = L_0}$

It should be noted that for both examples the FEM results are exact if the element stiffness (AE or EI) is constant and there is no loading between the nodal points.

The arguments of this section require that we deal with a FEM solution that does in fact converge towards the correct result, that is

$$e \to 0 \quad \text{when} \quad h \to 0$$

when the displacement assumption within the element contains a complete polynomial up to and including order p. An R-R consistent FEM formulation satisfies these requirements.

Accuracy and convergence

 **Brook TAYLOR** (1685-1731) was an English mathematician.

At the age of 16 TAYLOR entered St John's College, Cambridge, where he took two degrees in law in addition to studying mathematics. One of his first mathematical achievements was a clever solution of the problem of the "centre of oscillation". However, he did not publish his solution for some time, and ended up in a dispute with Johann BERNOULLI over priority to the solution.

In 1715 TAYLOR published a work (*Methodus Incrementorum Directa et Inversa*) in which he introduced a branch of mathematics, now known as "calculus of finite differences". Among various clever applications he used it to determine the form and movement of a vibrating string. The same work contained his famous formula known as TAYLOR's *theorem,* and the TAYLOR series also stems from this work. However, the real significance of this work was not recognized until LAGRANGE almost 60 years later called it "the main foundation of differential calculus".

TAYLOR's written style was brief and sometimes even obscure; as a result some of his work went unnoticed at the time, and other ideas and findings were made available through the work of contemporary scholars.

TAYLOR was elected a fellow of the Royal Society in 1712, the same year as he served on a committee whose task it was to sort out the quarrel between NEWTON and LEIBNIZ as to who discovered *calculus*. The findings of the committee, which gave the credit to NEWTON, were cast in doubt when it came to light that NEWTON himself wrote the study's concluding remarks on LEIBNIZ.

Like many of the famous names of mathematics and physics in the 18th and 19th century, TAYLOR's studies in later life took a philosophical and religious bent.

8.3 Error estimates

The observations in the previous section give useful information about how important parameters like the element size (h) and the polynomial order (p) influence the rate of convergence. However, they do not say anything about the magnitude of the error; how good is the solution? For the practising engineer that is the important question.

For certain classes of problems and for a given element, it is possible to establish guidelines based on similar problems with known solutions. Such *a priori* estimates are often used by experienced engineers, but their formal basis is often weak, and the error is only vaguely quantified.

A posteriori error estimates are based on the numerical solution itself or on several solutions. We shall take a look at one of these methods that requires *two* different solutions. Assume that

- a certain element mesh, characterized by the element size h_1, produces the strain energy U_1, and that
- another element mesh, characterized by the element size h_2, produces the strain energy U_2.

We also assume that the "h_2-mesh" is obtained by *one* uniform subdivision of the "h_1-mesh", that is $h_2 = h_1/2$, and that we have a combination of m and p such that

$$e_U = O(h^4)$$

With these assumptions we have

$$e_{U1} = O(h_1^4)$$

$$e_{U2} = O(h_2^4) = O\left(\left[\frac{h_1}{2}\right]^4\right) = O\left(\frac{h_1^4}{16}\right) = \frac{1}{16}e_{U1}$$

Thus

$$\frac{e_{U1}}{e_{U2}} = \frac{U - U_1}{U - U_2} = 16 \quad \Rightarrow \quad U = \frac{16}{15}U_2 - \frac{1}{15}U_1 \tag{8-12}$$

and the error in the last (and best) solution is

$$e_{U2} = U - U_2 = \frac{1}{15}(U_2 - U_1) \tag{8-13}$$

This is a simple example of so-called RICHARDSON extrapolation. In addition to the assumptions made above, the example also requires that the FEM solution converges monotonically and that the solution is "well behaved" (no singularities). The technique, which requires two or more solutions, can be used for individual response parameters, but a suitable norm (which represents an "average" error) is perhaps more appropriate.

For some elements it is possible, as shown in Section 7.2, to use stresses computed at so-called super-convergent (BARLOW) points, in connection with local or global smoothing, to obtain quite good *a posteriori* error estimates

Accuracy and convergence

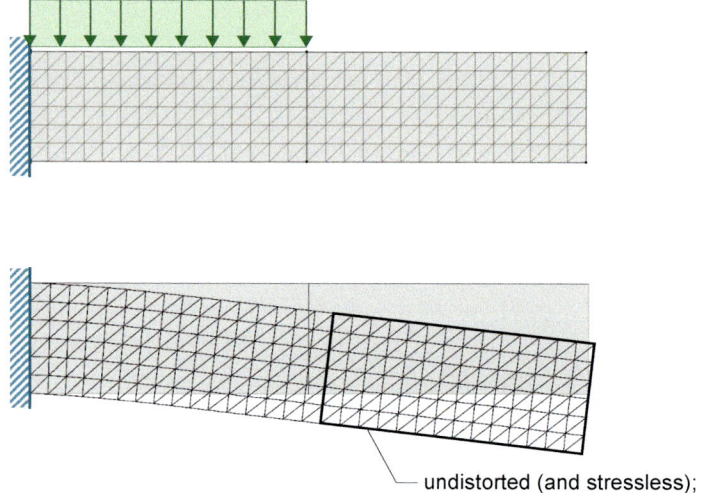

Figure 8.5 Cantilever plate – stressed and stressless regions

based on only *one* FEM solution. These estimates can be used to monitor the error level in different parts of the element mesh, and based on these estimates the mesh may (automatically) be refined in areas in order to keep the error below a certain predefined level. This process is referred to as *adaptive meshing*, and it is provided, in varying degree, by many of the commercial software packages. It is beyond the scope of this presentation to go deeper into these matters, the interested reader should consult [20], but we shall return to some practical aspects of accuracy and errors when we look at specific applications in later chapters.

8.4 Convergence criteria

In Sections 4.5 and 5.5 we have already discussed necessary and sufficient requirements for convergence of an R-R consistent FEM formulation. These requirements are concerned with *completeness* and *continuity* of the shape functions. In this section we will have a closer look, and also point out what is absolutely necessary and what can be relaxed and still provide a convergent solution.

The completeness criterion can be rewritten as two requirements dealing with *rigid body movements* and *constant strain (stress) states*, respectively:

> *Criterion 1* – The shape functions (N_i) must be capable of describing all possible *rigid body movements* correctly; this means that if $\mathbf{v}_{rbm}$ represents the nodal point displacements corresponding to any one rigid body movement of the element, then

$$\mathbf{S} = \mathbf{k}\mathbf{v}_{rbm} = \mathbf{0} \quad \text{and} \quad \mathbf{\sigma} = \mathbf{C}\mathbf{\varepsilon} = \mathbf{C}\mathbf{B}\mathbf{v}_{rbm} = \mathbf{0} \qquad (8\text{-}14)$$

This is an absolute requirement in order to avoid so-called "self-straining". Figure 8.5 shows an example of a problem for which a convergent and well-behaved element should produce zero stresses in the marked tip region. The first "column" of elements to the right of the vertical line indicating the end of the loading will probably show a low level of stresses, a spillover effect from the nodes on the said line.

As the element mesh is refined (elements are made smaller), the state of strain within each element approaches a constant value. If the nodal point *dofs* (**r**/**v**) in some portions of the mesh are compatible with a state of constant strain, it is highly desirable that the elements are capable of reproducing this state exactly. It is possible to come up with combinations of **v** and **N** that satisfy the first part (the strain approaches a constant value when the element size h approaches zero), but not the second part (for finite element size, the strain varies, not much, but it varies). Such elements will converge, but slowly. This leads to the following requirement:

> *Criterion 2* – The shape functions (N_i) must be chosen such that if **v** is compatible with a constant state of strain within the element, then the element must be capable of reproducing this state of strain,
> - preferably exactly for an arbitrary (finite) element size,
> - and definitely in the limit, as the element size goes to zero.

Accuracy and convergence

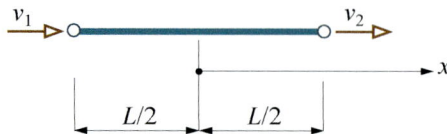

Figure 8.6 Axial element with 2 degrees of freedom

Strictly speaking it is sufficient that this criterion is satisfied in the limit, that is when $h \to 0$, but for practical use we require this criterion to be satisfied also for finite element size. In other words, this is a *necessary* requirement.

It can be argued that Criterion 2 includes Criterion 1 since a rigid body movement represents a special case of constant strain, and thus Criterion 1 also needs to be satisfied only in the limit. In view of the statement above, this is an academic proposition.

Example 8-3 – axial element in two dimensions

With reference to Fig. 8.6, we use indirect interpolation and assume

$$u = [\, x \quad x^2 \,]\mathbf{q} = \mathbf{N}_q \mathbf{q}$$

where $\mathbf{q}^T = [\, q_1 \quad q_2 \,]$ are generalized displacements. From this we find

$$(u)_{x=-L/2} = v_1 \quad \Rightarrow \quad -\frac{L}{2}q_1 + \frac{L^2}{4}q_2 = v_1$$

$$(u)_{x=L/2} = v_2 \quad \Rightarrow \quad \frac{L}{2}q_1 + \frac{L^2}{4}q_2 = v_2$$

and

$$q_2 = \frac{2}{L^2}(v_1 + v_2) \quad \text{and} \quad q_1 = \frac{2}{L}v_2 - \frac{L}{2}\cdot\frac{2}{L^2}(v_1+v_2) = -\frac{1}{L}v_1 + \frac{1}{L}v_2$$

or

$$\begin{bmatrix} q_1 \\ q_2 \end{bmatrix} = \begin{bmatrix} -\frac{1}{L} & \frac{1}{L} \\ \frac{2}{L^2} & \frac{2}{L^2} \end{bmatrix}\begin{bmatrix} v_1 \\ v_2 \end{bmatrix} = \mathbf{A}^{-1}\mathbf{v} \quad \Rightarrow \quad u = \mathbf{N}\mathbf{v} = \mathbf{N}_q \mathbf{A}^{-1}\mathbf{v}$$

Hence

$$u = N_1 v_1 + N_2 v_2 \quad \text{where} \quad N_1 = \frac{x}{L}\left(-1 + 2\frac{x}{L}\right) \quad \text{and} \quad N_2 = \frac{x}{L}\left(1 + 2\frac{x}{L}\right)$$

We see that these shape functions cannot represent the rigid body movements correctly. For instance,

$$v_1 = v_2 = v \quad \text{gives} \quad u_1 = u_2 = v \quad \text{but} \quad (u)_{x=0} = 0 \;(!)$$

However, for $v_2 = -v_1 = v$, we see that

$$\varepsilon_x = \frac{du}{dx} = \frac{2v}{L} + \frac{4x}{L^2}(v - v) = \frac{2v}{L} = \text{const.}$$

This example shows clearly that you do not sacrifice a lower order term for a higher order one – here a quadratic term has been used at the expense of the constant term.

As for *continuity*, we can rewrite the requirement we formulated in Section 4.5, in connection with a piece-wise application of the R-R method, and say

Accuracy and convergence

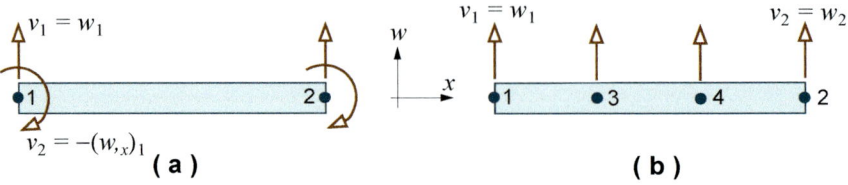

Figure 8.7 Beam elements with 4 bending degrees of freedom

that the displacement functions need to produce continuous strains within the elements and *finite strains* on the element boundaries. If the strains (ε) contain derivatives of the displacements (**u**) of order m, but not higher than m (the variational order), then we can formulate the continuity requirement as two criteria:

Criterion 3 – The element degrees of freedom (**v**) must be chosen so as to secure C^{m-1} continuity at the nodal points.

This part of the continuity requirement is generally considered to be a *necessary* requirement for convergence.

Figure 8.7 shows two choices of nodal point degrees of freedom (**v**) for a 2D beam element in bending. For this element we have $m = 2$, and we therefore need C^1 continuity at the nodes. Both elements have 4 *dofs* and can therefore describe a *cubic* displacement field uniquely, that is $p = 3$. While the *dofs* of the element in Fig. 8.7a will secure continuity of both w and its slope ($w_{,x}$), the dofs of Fig. 8.7b can only provide C^0 continuity (of w itself), the slope will be different in two neighbouring elements that meet at a common node.

Criterion 4 – The shape functions N_i must provide C^{m-1} continuity along the element boundaries.

This is a stronger requirement than Criterion 3 which is contained in this one.

Criteria 1, 2 and 4 are *sufficient* for convergence to the correct results. In many cases, however, it is difficult to satisfy all these three criteria, especially Criterion 4 for elements that need to satisfy C^1 continuity.

Fortunately it turns out that it is possible to generate convergent elements that do not fully satisfy Criterion 4, but even these elements must fully satisfy Criteria 1 and 2, that is completeness. The convergence properties of these elements are established through their performance; the next section shows how this is achieved.

8.5 Element tests

In many cases we know *a priori* that a finite element will converge to the correct solution, that is even before we have put it to the test, simply because we have satisfied all basic requirements for convergence. In other cases we may, for various reasons, have introduced modifications that are in conflict with the basic requirements, or the particular element (problem, shape and/or degrees of freedom) is such that we simply cannot satisfy all requirements. The "quality" of such elements can only be assessed by their performance. Even the elements we "know" will converge must be implemented, that is programmed for the computer, and this transition from the drawing table to the computer also needs to be verified through performance tests.

One form of performance testing is to run a series of systematic analyses on problems with known solutions. A certain amount of this type of testing is mandatory for any "new" element, but simple tests on individual elements, or a small number of elements, can also provide valuable insight and in fact provide "numerical" proof of the element's convergence properties.

A family of *norms* for an arbitrary vector **x** is defined by

$$\| \mathbf{x} \|_h = \left(\sum_{i=1}^{n} | x_i |^h \right)^{1/h}$$

$h = 2$ defines the 2-norm, usually termed the EUCLIDEAN norm. For $n = 2$ and 3 this norm expresses the *length* of *physical* vectors. This definition of vector length is generalized to any "*n*-dimensional" vector.

KRONECKER delta:
$\delta_{ii} = 1$
$\delta_{ij} = 0 \quad i \neq j$

Various tests on individual elements have been suggested. A very simple one is to check that **S** = **kv** = **0** if **v** describes a *pure rigid body movement* of the element. Another is the *single-element-test* proposed by ROBINSON [27]; this test checks the stability (no zero-energy modes), and it can also be used to check the element's robustness (with respect to aspect ratio, skewness and shape distortion). BERGAN and HANSEN [28] suggested the *individual element test* in connection with their so-called "free-formulation-approach" to the generation of finite elements. However, the two most important and useful tests, which we will take a closer look at, are the *eigenvalue test* and the *patch test*.

Eigenvalue test. This is an objective test, carried out on a single element, that can provide useful information about an element. Consider an element of a particular *kind* (problem, shape and *dofs* **v**), for which the geometrical and material properties are specified and for which we have determined the element stiffness matrix **k** numerically. The element stiffness relation is

$$\mathbf{S} = \underbrace{\mathbf{k}\mathbf{v}}_{l \times 1} \tag{8-15}$$

The number of element degrees of freedom is l. We now claim that a particular displacement vector **x** that satisfies the equation

$$\mathbf{k}\mathbf{x} = \lambda \mathbf{x} \quad \text{or} \quad (\mathbf{k} - \lambda \mathbf{I})\mathbf{x} = \mathbf{0} \tag{8-16}$$

exists. This is an *eigenvalue* problem; it has l solution pairs $(\lambda_i, \mathbf{x}_i)$, where λ_i is an *eigenvalue* and $\mathbf{x}_i$ the corresponding *eigenvector*. The eigenvector is uniquely defined save for an arbitrary *scaling* factor s, that is $s\mathbf{x}_i$ is also an eigenvector. Usually the eigenvectors are scaled such that their length is unity:

$$|\mathbf{x}_i| = \|\mathbf{x}_i\|_2 = (\mathbf{x}_i, \mathbf{x}_i)^{1/2} = (\mathbf{x}_i^T \mathbf{x}_i)^{1/2} = 1 \tag{8-17}$$

Furthermore, the eigenvectors are, or can be made, mutually *orthogonal*, that is

$$\mathbf{x}_i^T \mathbf{x}_j = \delta_{ij} \quad \text{where } \delta \text{ is the KRONECKER delta} \tag{8-18}$$

In view of Eqs. (8-16) and (8-18) the strain energy in a (scaled) eigenvector is

$$U_\mathbf{x} = \tfrac{1}{2}\mathbf{x}_i^T \mathbf{k} \mathbf{x}_i = \tfrac{1}{2}\lambda_i \mathbf{x}_i^T \mathbf{x}_i = \tfrac{1}{2}\lambda_i \tag{8-19}$$

Since the eigenvectors form a complete, orthonormal basis for the (EUCLIDIAN) vector space, *any* displacement vector **v** can be expressed as

$$\mathbf{v} = \sum_{i=1}^{l} \alpha_i \mathbf{x}_i \tag{8-20}$$

The strain energy in the element due to **v** is therefore:

$$U_\mathbf{v} = \tfrac{1}{2}\mathbf{v}^T \mathbf{k} \mathbf{v} = \tfrac{1}{2}\left(\sum_{i=1}^{l} \alpha_i \mathbf{x}_i^T\right) \mathbf{k} \sum_{j=1}^{l} \alpha_j \mathbf{x}_j = \tfrac{1}{2}\sum_{i=1}^{l} \alpha_i^2 \mathbf{x}_i^T \mathbf{k} \mathbf{x}_i = \tfrac{1}{2}\sum_{i=1}^{l} \alpha_i^2 \lambda_i \tag{8-21}$$

The tests based on the eigenvalues and corresponding eigenvectors of the element are:

Accuracy and convergence

Element kind

For a given physical problem, two elements are of the same *kind* if they have the same geometric form, the same material properties, the same nodal points and the same kinematic degrees of freedom (**v**) at the nodal points.

Element type

Two elements of the same *kind* are also of the same *type* if their element matrices are derived using exactly the same assumptions.

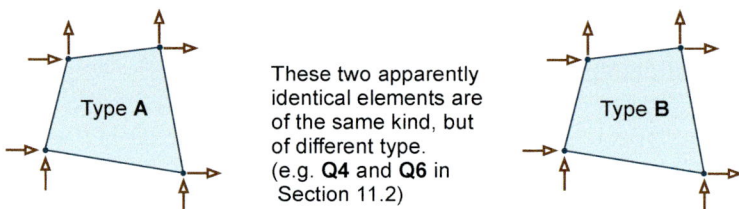

Type A — These two apparently identical elements are of the same kind, but of different type. (e.g. **Q4** and **Q6** in Section 11.2) — Type B

Figure 8.8 Element *kind* and element *type*

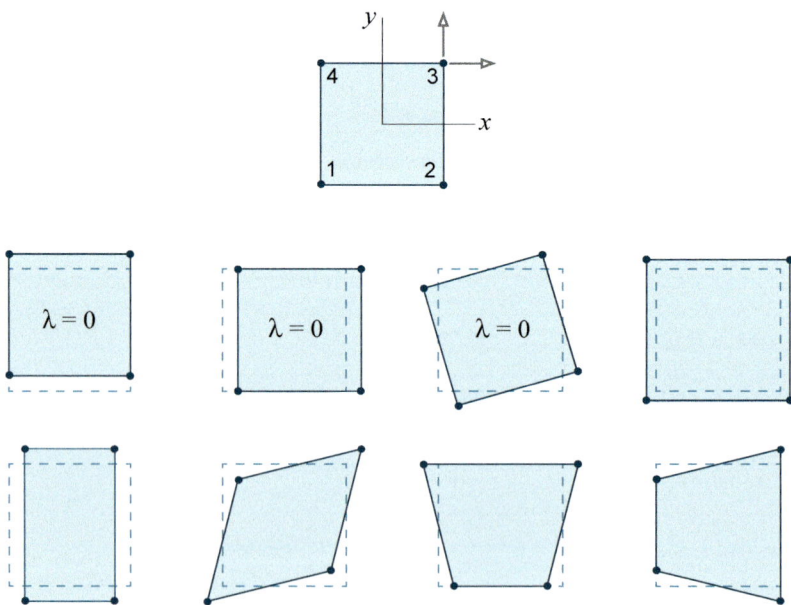

Figure 8.9 Eigenmodes of the standard 4-node (8-*dof*) quadrilateral element

- *Rigid body movements – mechanisms*
  Since the strain energy in a rigid body motion (that causes no deformation) is zero, it follows from Eq. (8-19) that **k** must have as many zero eigenvalues as the element has independent rigid body modes.
  If the number is less then clearly the element does not handle rigid body motion correctly.
  If the number of zero eigenvalues is larger than the number of rigid body modes, then the element exhibits *zero-energy-mode(s)*; in other words, it has one or more mechanisms and is consequently not stable.
- *Geometric invariance – (isotropy)*
  The eigenvalues must *not* change if the element orientation (and only the orientation) is changed. If they do change the element is not geometrically invariant, and therefore not reliable.
- *Stiffness*
  The *trace* of a square matrix is the sum of its diagonal terms, and for a symmetric matrix it can be shown that the sum of its eigenvalues is equal to its trace, that is

$$\mathrm{tr}(\mathbf{k}) = \sum_{i=1}^{l} k_{ii} = \sum_{i=1}^{l} \lambda_i$$

  With reference to Eq. (8-21) we can therefore claim that the sum of the eigenvalues is a measure of stiffness. For two apparently identical elements of the same *kind*, but of different *type* – see the definition in Fig. 8.8 – the "best" element will have the lowest stiffness, that is the lowest trace.
  This statement is strictly correct only for elements that converge monotonically from the "stiff side", as will be the case for R-R consistent displacement based elements.

An eigenvalue analysis of the stiffness matrix for the standard 4-node (8-*dof*) quadrilateral element will produce the *eigenmodes* (eigenvectors) shown in Fig. 8.9. These are the same as the displacement modes shown in Fig. 6.20. The first three modes are *rigid body* modes, and they are all associated with zero eigenvalues, and if we had used *reduced* numerical integration (1×1 GAUSS quadrature), the two modes in the lower right-hand corner of the figure would also have been associated with zero eigenvalues, indicating that, for such an integration scheme, these modes would be zero-energy-modes (mechanisms).

The patch test. The basis of this test, due to IRONS [29,30], is that an element in an arbitrary mesh that satisfies the following requirements,

- correct representation of the rigid body movements (Criterion 1),
- ability to represent constant strains within the element (Criterion 2), and
- sufficient displacement continuity at the element nodes (Criterion 3),

will converge to the correct solution, as the element size (*h*) is decreased, even if the continuity requirement (Criterion 4) is not fully satisfied.

Accuracy and convergence

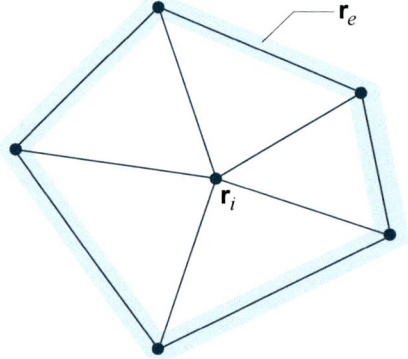

Figure 8.10 Patch of triangular elements

Basically, the test only requires the element to be able to reproduce a state of constant strain *in the limit* ($h \to 0$). However, for most (or many) element types based on assumed polynomial shape functions, the element size makes no difference, and the standard (strong) form of the test therefore requires Criterion 2 to be satisfied exactly for a patch of elements of finite size; if the element passes this form of the test, Criterion 1 about rigid body motion is also (implicitly) satisfied. If we take it for granted that Criterion 3, concerning displacement continuity at the nodes, is met, then the patch test is the "benchmark test" that any new element needs to pass, if it is in violation of the strong displacement continuity requirement.

Some doubt about the test has been raised. While the necessity of passing the test has not been questioned, it is perhaps not obvious that passing the test is also a sufficient requirement for convergence. However, there is now a common consent that *passing the patch test is both a necessary and sufficient requirement for a convergent solution*, with one reservation. This reservation has to do with zero-energy modes; a patch of elements may "block" such a mechanism from "developing" and thus mask its existence. If in doubt, run an eigenvalue test.

Although passing of the patch test is a guaranty for convergence, it says nothing about the rate of convergence, nor does it indicate what type of convergence we can expect, monotonic or oscillating; it is most likely the latter.

The test, which comes in two forms, is most conveniently explained by some examples taken from plane stress analysis.

Form 1 – *prescribed boundary displacements*

  a) Choose an *arbitrary* patch of irregular elements where at least one nodal point is surrounded by elements, see Fig. 8.10.
  b) Choose an *arbitrary* displacement pattern that can describe all rigid body movements and constant strain states over the entire element patch.
  c) Prescribe displacement values on the "outer boundary" *dofs* ($\mathbf{r}_e$) that are consistent with the chosen displacement pattern.
  d) Compute $\mathbf{r}_i$ from $\mathbf{K} \begin{bmatrix} \mathbf{r}_i \\ \mathbf{r}_e \end{bmatrix} = \mathbf{0}$ and also the strains at a number of arbitrarily placed points inside the elements.
  e) Check that the computed displacements $\mathbf{r}_i$ are in full agreement with the imposed displacement field.
  f) Check that the computed element strains are constant.

It should be emphasized that for the test to be passed, the computed values for $\mathbf{r}_e$ and $\boldsymbol{\varepsilon}$ must be exact to machine precision.

Accuracy and convergence

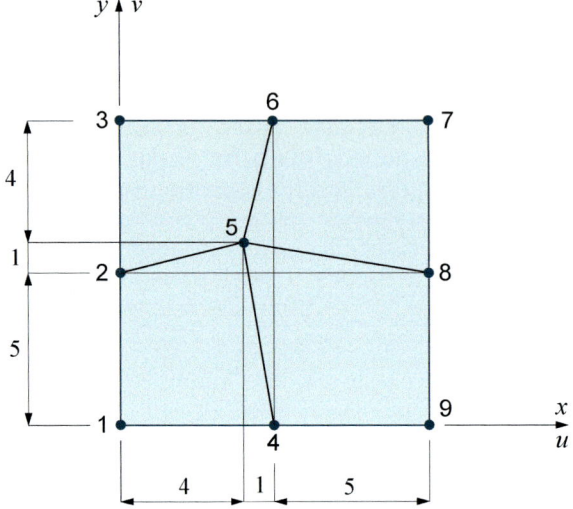

Figure 8.11 Patch test – prescribed edge displacements

Example 8-4

With reference to Fig. 8.11 we assume:

$$u = c_1 + c_2 x + c_3 y \quad \text{and} \quad v = c_4 + c_5 x + c_6 y$$

Also $\quad E = 1,0 \quad$ and $\quad \nu = 0,25$

Thus $\quad \varepsilon_x = \dfrac{du}{dx} = c_2, \quad \varepsilon_y = \dfrac{dv}{dy} = c_6 \quad$ and $\quad \gamma_{xy} = \dfrac{du}{dy} + \dfrac{dv}{dx} = c_3 + c_5$

$$\left(\sigma_x = \frac{E}{1-\nu^2}(\varepsilon_x + \nu\varepsilon_y) = \frac{16}{15}\left(c_2 + \frac{c_6}{4}\right) \right)$$

Assume that all 6 constants are equal to one (1), and impose the assumed displacement field as the sum of 6 independent cases, where each case is a simple displacement pattern:

Case	u	v
1	1	0
2	0	1
3	x	0
4	0	y
5	y	0
6	0	x

Case 3

Point	r_x	r_y
1	0	0
2	0	0
3	0	0
4	5	0
6	5	0
7	10	0
8	10	0
9	10	0

The computed results for Case 3 should be:

At point 5: $\quad r_x = 4 \quad$ and $\quad r_y = 0$

At all internal points in the elements: $\varepsilon_x = 1,0 \quad$ and $\quad \varepsilon_y = \gamma_{xy} = 0$

Form 2 – *prescribed boundary forces (tractions)*

a) As for Form 1.
b) Impose the minimum of displacement constraints at the nodes located on the edges of the patch that will prevent rigid body motion.
c) Choose a set of constant stress states, and for each set compute statically equivalent (consistent) nodal forces at the nodes on the edges of the patch.
d) Apply the computed nodal forces **R**, one set at a time, and compute the corresponding displacements **r** from **Kr** = **R**.
e) If the computed stresses anywhere in the elements are identical with the chosen state of stress (giving rise to **R**), the patch test is satisfied.
f) As an extra check, the computed nodal displacements must agree with those consistent with the constant stress state.

Accuracy and convergence

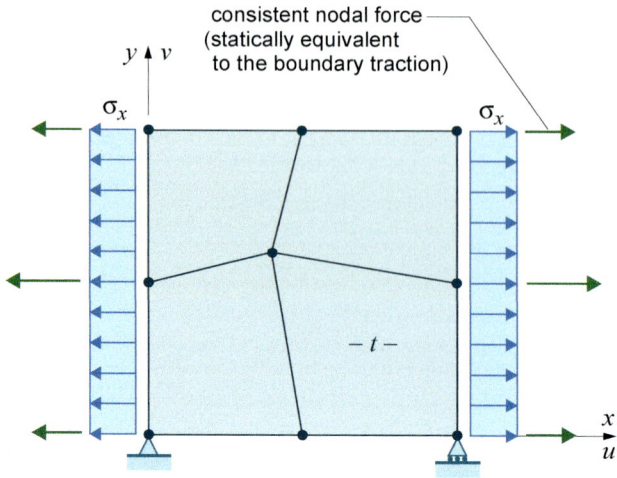

Figure 8.12 Patch test – prescribed tractions

A *non-conforming* element satisfies the *completeness* requirements, but not all *continuity* requirements.

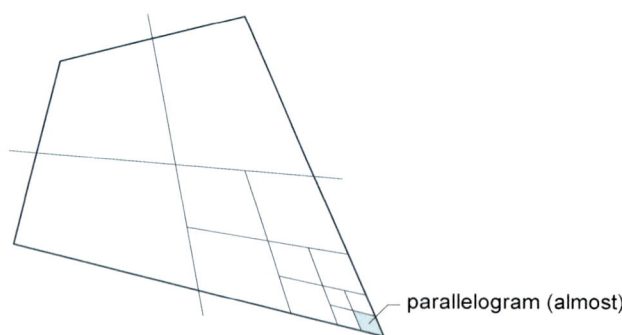

Copy of **Figure 6.17** Uniform subdivision of a quadrilateral

Example 8-5

With reference to Fig. 8.12 we assume:

$$E = 1, \quad \nu = 0{,}25 \quad \text{and} \quad t = 1$$

Also

$$u = \frac{1}{E}x \quad \text{and} \quad v = -\frac{\nu}{E}y \;\Rightarrow\; \boldsymbol{\varepsilon} = \begin{bmatrix} 1/E \\ -\nu/E \\ 0 \end{bmatrix} \;\Rightarrow\; \boldsymbol{\sigma} = \begin{bmatrix} \sigma_x \\ \sigma_y \\ \tau_{xy} \end{bmatrix} = \mathbf{C}\boldsymbol{\varepsilon} = \begin{bmatrix} 1 \\ 0 \\ 0 \end{bmatrix}$$

The nodal forces will depend on the element type.

The patch test should be carried out for more than one patch of elements, and all possible "fields" corresponding to constant strains should be tested independently to avoid that possible errors cancel out.

It is possible to design a non-conforming (or non-compatible) element such that it will (*a priori*) satisfy the patch test. An example of this will be shown in Section 11.2 (element **Q6**). To put such an element to the patch test can only prove or disprove correct implementation, that is, correct programming.

The "weak" patch test. An element that does not manage to reproduce all states of constant strain in a patch of finite size elements has not necessarily failed the patch test. If it can be shown, by repeated uniform subdivision of the element mesh, that the strains in the individual elements approach the correct, constant value, then the element satisfies the *weak patch test*. Such elements, however, will exhibit slow convergence.

An example of such an element is a special version (type) of the 4-node quadrilateral element, referred to as (unmodified) **Q6** in Section 11.2. This element can be shown to pass the patch test when it has rectangular or parallelogram form. However, as shown in Fig. 6.17, repeated uniform subdivision of an arbitrary quadrilateral will produce elements that approach parallelogram form, and thus the element will pass the weak patch test.

8.6 Exact solution at the nodal points

If the principle of *superposition* holds for a given problem, then

its *final solution* = its *particular solution* + its *complementary solution*, or

$$\mathbf{r} = \mathbf{0} + \mathbf{r}_{compl}$$

In the particular problem *all* nodal displacements are suppressed ($\mathbf{r} = \mathbf{0}$), and in the complementary problem *all* loading consists of concentrated forces acting at the nodal points and in the "directions" of the *dofs*, in other words, *all* loading is contained in **R**; all the element "fixed end forces" from the particular solution are contained in **R** in addition to concentrated forces applied at the nodes.

In the complementary problem, each individual element is governed by the problem's homogeneous differential equation; hence:

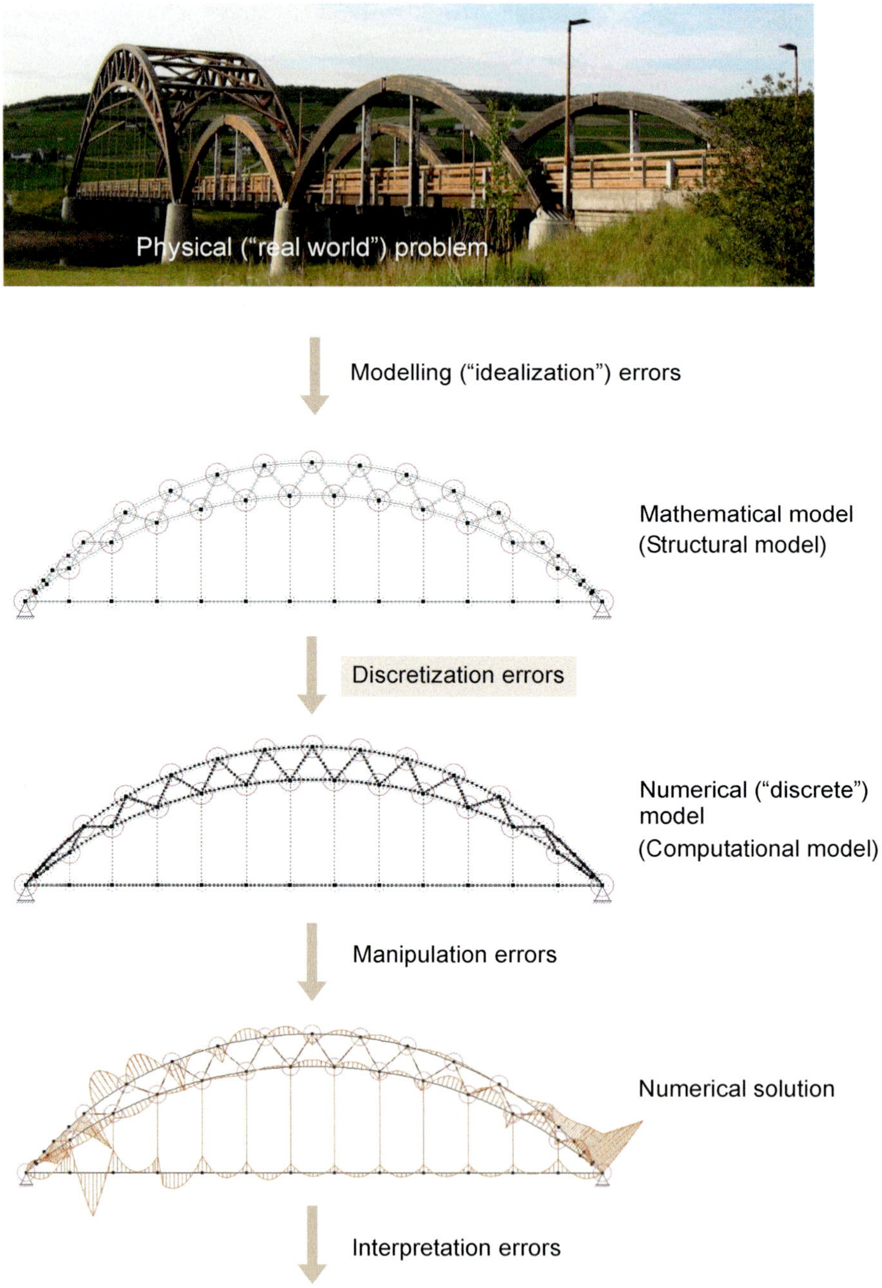

Accuracy and convergence

Figure 8.13 Classification of errors in a FEM solution of a structural problem

If we can establish shape functions, uniquely defined by the nodal displacements, that satisfy the homogeneous part of the problem's differential equation, then we will compute *exact* nodal displacements **r** (**v**) provided:
- the statically equivalent ("fixed end") nodal loads are computed correctly (consistently), and
- displacement compatibility (continuity) is satisfied over the element boundaries.

For one-dimensional *bar* and *beam* elements it is quite straightforward to establish shape functions that satisfy the corresponding homogeneous differential equations, and thus obtain solutions that are exact at the nodal points. If also the particular problem can be solved exactly, which is an easy task for bars and beams, then the total finite element based solution will be exact.

Unfortunately, for 2- and 3-dimensional problems, none of the two problems can be solved exactly, save for some very simple cases. For real engineering problems we will therefore have to settle for approximate solutions. The good news, however, is that the error of these approximate solutions can, in most cases, be made so small that they are of little practical concern. Many of the commercially available FEM software packages offer help in controlling some of the solution errors, but knowledge and experience are indispensable qualities for safe and efficient use of this powerful method.

Summary of Chapter 8

Figure 8.13 shows the types of errors that can contaminate the results of a finite element analysis. In this chapter we have concentrated on the *discretization* errors, that is the errors we make when transforming the idealized mathematical model into a discrete numerical or computational model. But as shown by the figure, these are not the only errors we need to worry about.

The *modelling phase*, that is the transition from the real world problem to a (simplified) mathematical model that captures the essential behaviour of the real structure, is a critical and difficult phase. Experience is a key factor in dealing with the many uncertainties associated with geometry, loading, material properties, boundary conditions and not least the elimination of insignificant details. We shall come back to this phase and its challenges in later chapters.

As explained in this chapter, the *discretization* errors are associated with the choice of element (degrees of freedom and choice of interpolation functions) and the finite element mesh – symbolized by p and h – and also on how we approximate curved geometry.

Manipulation errors are the errors associated with the numerical "operations" we need to perform. Choice of numerical method may play a role, but the main source for this type of errors is *truncation* and *round-off* due to the limited *precision* of the computer's representation of floating point numbers. Both this and the last type of errors, the *interpretation* errors, will be dealt with in later chapters as will *programming errors*, which are unlikely, but which can never be ruled out completely.

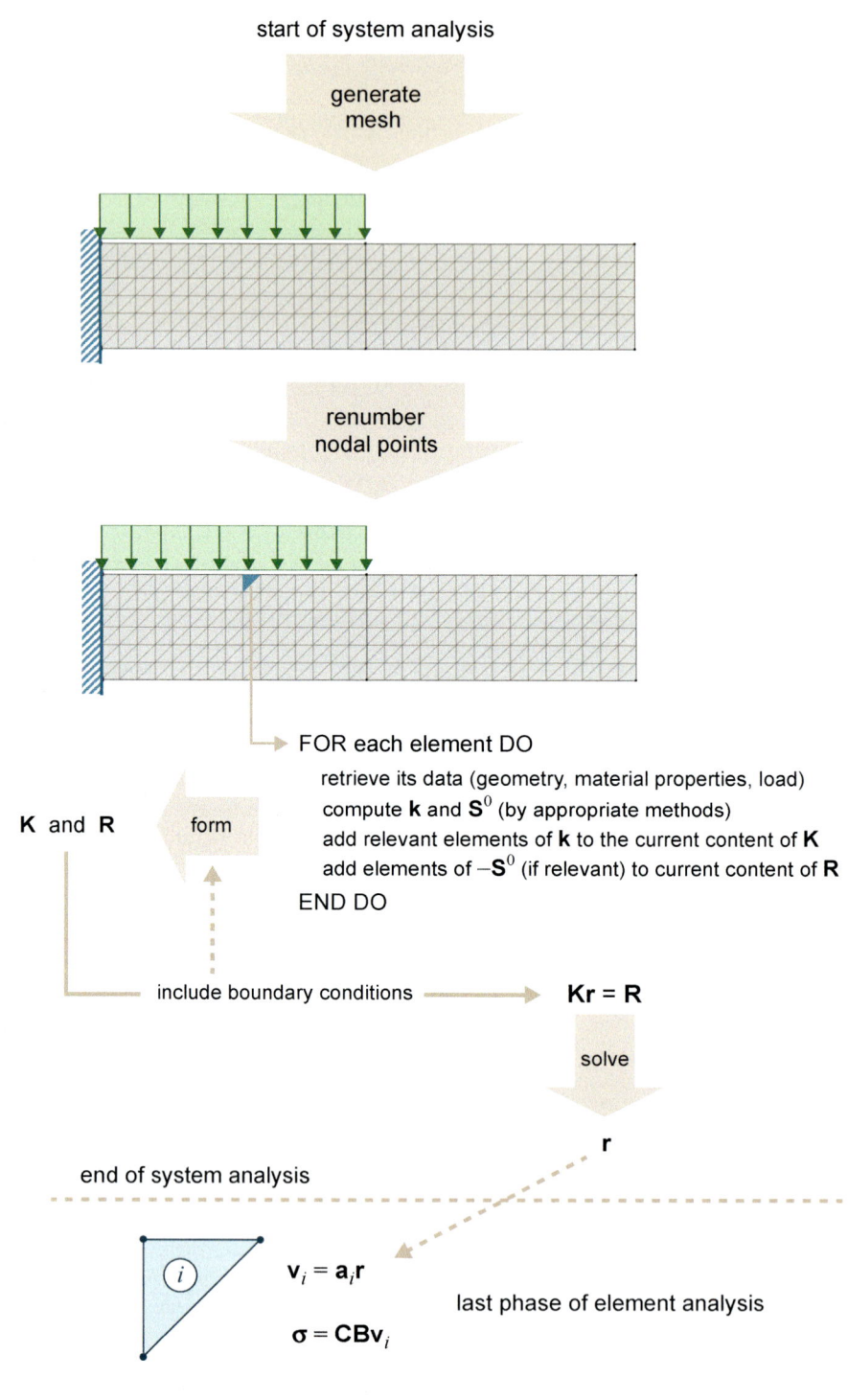

9

System analysis

*The system analysis, which is basically problem and element independent, is concerned with the assembly of **K** and **R**, from **k** and **S**^{0}, the handling of boundary conditions and the solution of the final system of linear, algebraic equations with respect to the unknown nodal point degrees of freedom (**r**). However, before the system matrices can be formed we need a finite element mesh; automatic or semi-automatic generation of such meshes is therefore, from a practical point of view, an important part of the system analysis. Storage formats for **K**, and automatic renumbering techniques are also important for solution efficiency. The partial solution technique referred to as static condensation and the related substructure analysis are also dealt with before we end the chapter with some comments on numerical problems.*

Having determined the matrices **k** and **S**^{0} of the element stiffness relation

$$\mathbf{S} = \mathbf{kv} + \mathbf{S}^0 \tag{9-1}$$

the next step is to form or build the system matrices **K** and **R** with due regard to the kinematic boundary conditions of the problem. This process results in a system of linear, algebraic equations

$$\mathbf{Kr} = \mathbf{R} \tag{9-2}$$

The solution of this system, with respect to the nodal point displacements, can be expressed mathematically as

$$\mathbf{r} = \mathbf{K}^{-1}\mathbf{R} \tag{9-3}$$

However before we can start the assembly process we need an element mesh of the problem area/volume.

9.1 Mesh generation

Since a large number of elements (many thousands) is often required, some kind of automatic or semi-automatic method for generating the mesh is absolutely essential. Mesh generation is a problem area in its own right that

System analysis

DELAUNAY triangulation

In computational geometry a DELAUNAY triangulation for a set **P** of points in a plane is a triangulation DT(**P**) such that **no** point in **P** is inside the *circumcircle* of any triangle in DT(**P**).
These triangulations, due to the Russian mathematician Boris DELAUNAY, maximize the minimum angle of all the angles of the triangles in the triangulation; thus they tend to avoid distorted triangles.

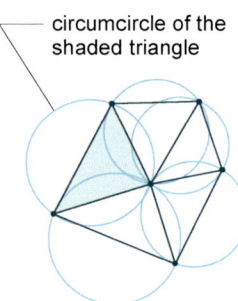

circumcircle of the shaded triangle

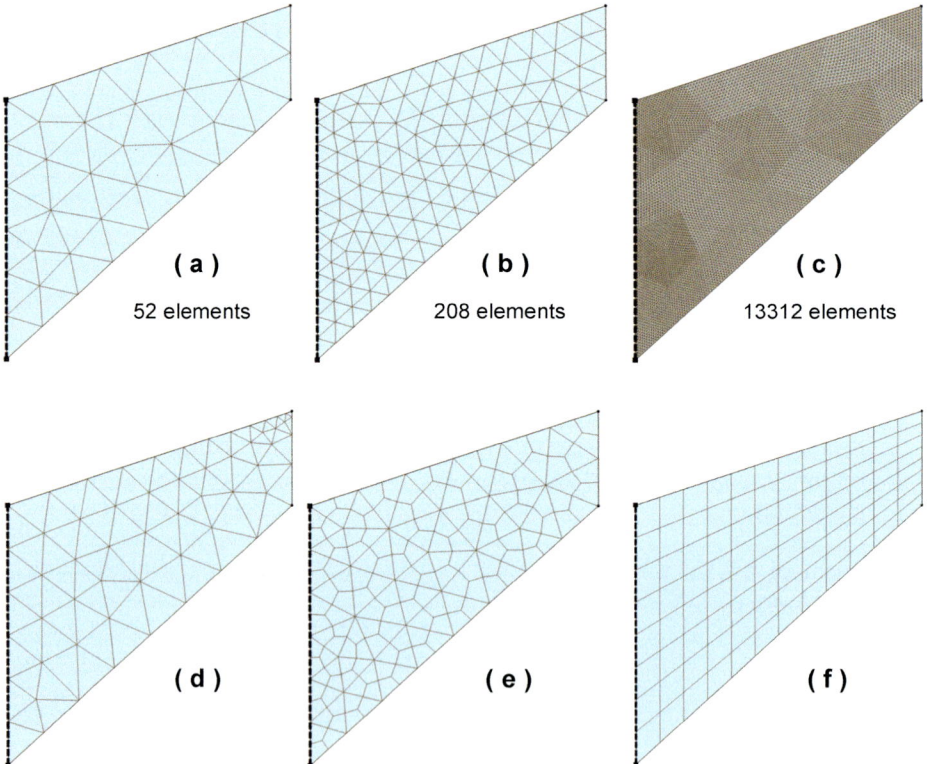

(a) 52 elements

(b) 208 elements

(c) 13312 elements

(d)

(e)

(f)

Figure 9.1 Automatic ("free") – **a** to **e** – and semi-automatic ("constrained") – **f** – meshing of a regular region

has received much attention during the last decades. The details of the various algorithms and techniques are beyond the scope of this presentation. However, some aspects of typical mesh generation offered by most, if not all, FEM based programs of some standing will be mentioned and commented upon. For simplicity we limit our discussion to two-dimensional problems; although the principles are similar in 3D, the complexity is much greater.

Most algorithms for automatic mesh generation of 2D regions make use of so-called DELAUNAY triangulation, complemented by other techniques (such as *flipping, divide and conquer* and *sweepline* algorithms). Figure 9.1a shows an *automatically* generated (coarse) mesh of triangular elements for a region with a simple and fairly regular geometry. The user's influence on this mesh is limited; she or he can obviously make it finer, usually by *uniform subdivision* as shown in Fig. 9.1b (one subdivision) and in Fig. 9.1c (four subdivisions based on the first coarse mesh), and it is often possible to ask for a finer mesh in some areas, see Fig. 9.1d.

The element shapes of the basic mesh in Fig. 9.1a are quite good (none of the elements are distorted), and their size is fairly even. Nevertheless, repeated uniform subdivision of this basic mesh reveals a certain pattern in the mesh which is vaguely visible in Fig. 9.1c (but far more obvious when viewed on a colour screen). However when zooming in on the elements we find, as was expected, that their shape is of the same quality as in the parent mesh (Fig. 9.1a), and the pattern has no practical significance for the results.

Quadrilateral meshes are often obtained from a triangular mesh. Figure 9.1e shows an example of a simple but crude mesh of quadrilateral elements based on the triangular mesh of Fig. 9.1a. Each triangle is divided into three quadrilaterals by simply drawing lines from the mid-side point of each edge to the centre of area of the triangle. The "quality" of these element shapes is not nearly as good as for the triangular parent mesh. The effect on the results is difficult to assess, but generally speaking the mesh in Fig. 9.1e is not satisfactory.

Figure 9.1f shows a *semi-automatic* mesh generated on the basis of a user specified number of elements along two adjoining sides of the quadrilateral region, assuming the numbers to be the same on opposite sides. In this particular example the sides have been divided into equal lengths, but that can easily be replaced by some other method of subdivision, for instance linear or geometric progression. This semi-automatic generation gives the user more control over the final mesh, but the process only works for simple geometries. For complex geometries it is necessary to split the total problem region into several (many) sub-regions of manageable shapes and then mesh one sub-region at a time. If the region has curved boundaries such a strategy may not be a good one, if at all possible.

Complex geometries with, for instance, holes and many sharp corners, may be too much of a challenge for even robust automatic mesh generators. Even if a solution is found it may not be a good one. In most such cases you will help the mesh generator by splitting your problem region into sub-regions, and preferably such that as many sub-regions as possible have a simple and regular shape (*e.g.* rectangles). In any case, if your program allows you to check the mesh (for distorted elements) before you continue to compute, do

297

System analysis

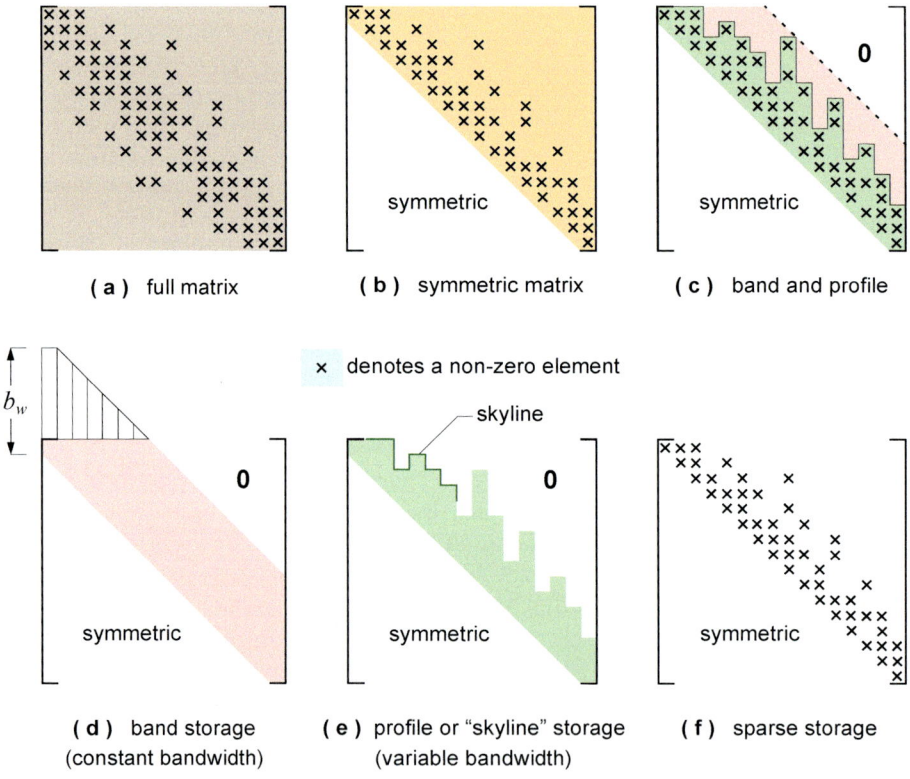

The semi-bandwidth b_w includes the element on the diagonal.

The *profile length* is the sum of all elements in the green area under the skyline, including the diagonal elements (and elements equal to zero).

Figure 9.2 Storage formats for the system stiffness matrix **K**

Storage formats and node renumbering

so, and if the program provides tools for mesh refinement/correction use them if necessary.

Having completed its task, what information will the mesh generator return to the analysis program? It must have a strategy for numbering nodal points and elements, and the minimum of information returned to the host program is:

- a table (or tables) of nodal point coordinates, and
- a *connectivity* table containing, for each element of the mesh, the global (mesh) numbers that the element nodes (1, 2,..., number of element nodes) are "connected" to.

Exactly how this information is organized will depend on the programming approach, *e.g.* object oriented or procedural, but whatever the organization is, the information must enable the host program to recreate the mesh exactly on the basis of the information it receives from the generator. The numbering strategy is normally chosen to suit the generator's needs, not those of the receiver of the information.

9.2 Storage formats and node renumbering

The solution of the system stiffness relation, Eq. (9-2), is responsible for the bulk of the computational effort in most cases of FEM analyses – 70% or more is not unusual. This is the one single operation it really pays off to optimise, and a lot of effort has been spent on this endeavour.

The solution of a system of simultaneous, linear algebraic equations, such as Eq. (9-2), can be carried out by a *direct elimination* method or by an *iterative* method. Our preferences are the direct methods, and since this operation is central to all FEM analyses, we have devoted a section (9.9) to the most used algorithms. In computer jargon we talk about the equation solver or just the *solver*; this is the code that implements the solution algorithm into the "computer commands" that will produce the solution.

Storage formats. The *storage format* of **K**, that is how the elements of **K** are organized with respect to storage in the computer memory, is very important for the efficiency of the solver. The two most important properties of **K** in this respect are its *symmetry* and its *sparseness* (**K** contains a large number of zero elements).

Figure 9.2 shows the "historical" development. At first, **K** was treated as a *full* matrix without even recognizing its symmetry, Fig. 9.2a. Next, symmetry was recognized, Fig. 9.2b, and only (a good) half of the matrix needed to be stored. By appropriate numbering of the nodes it was possible to bring all non-zero elements into a *band* along the main diagonal of **K**, Figs. 9.2c and d, which reduced both storage and the number of numerical operations during solution (fewer operations on zero elements). By utilizing the fact that the width of the band varies from row to row, or that the height of each column, measured from the diagonal element up to the last non-zero element, varies from one column to another, the *profile* or "skyline" storage format, Fig. 9.2e, was established; this proved to be a significant improvement on the band

299

System analysis

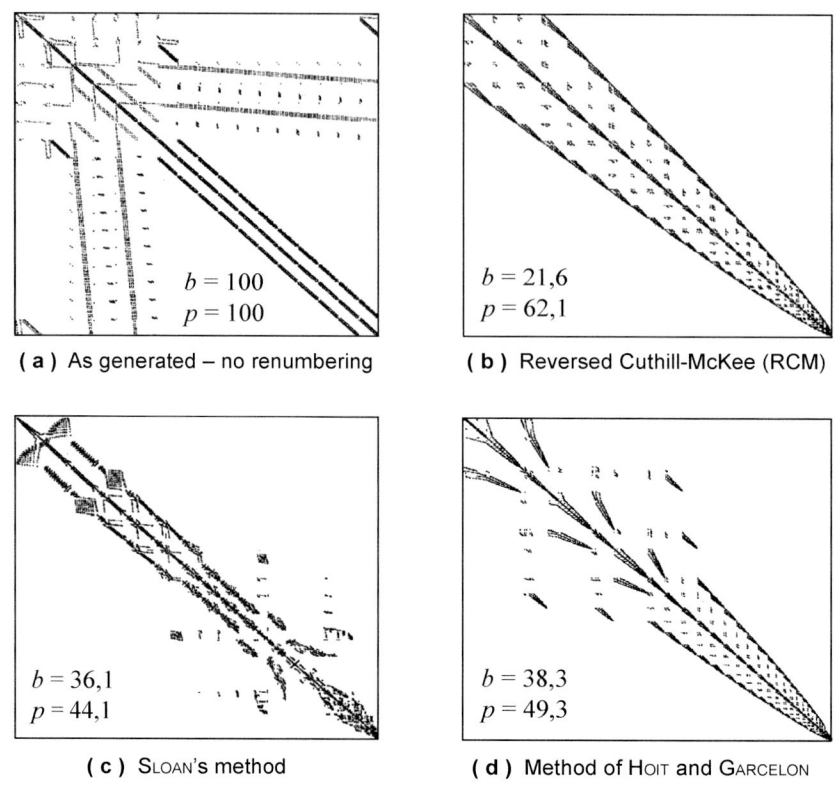

b is bandwidth and p is profile length

Figure 9.3 Stiffness matrix of a BOUSSINESQ (12×12×12) cube with 2197 nodes; original matrix (**a**) compared with three different renumbering methods, (**b**), (**c**) and (**d**)

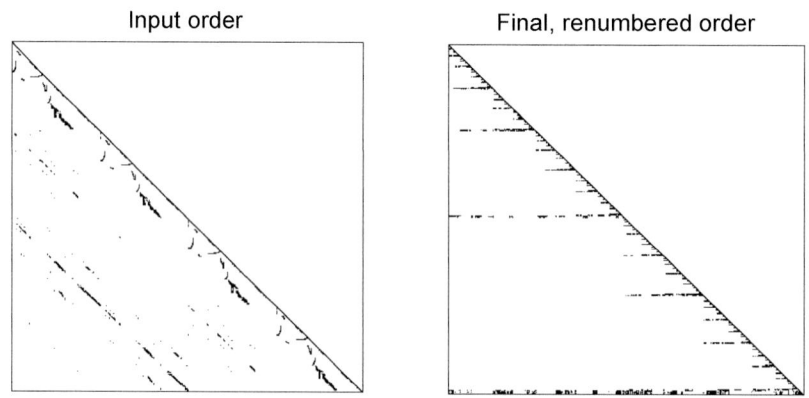

Figure 9.4 Sparse storage of **K** before and after renumbering (by incomplete nested dissection [37])

format. Better still is the *sparse* storage format. With this format only the non-zero elements of the matrix are stored, Fig. 9.2.f; in addition to the elements themselves (which are real or floating point numbers), additional (integer information) is required to uniquely define their position in the matrix. An extra complication is that some (and for large problems, many) elements of **K** that are zero after assembly will become non-zero during the solution (factorization) process.

It should be emphasized that while Fig. 9.2 uses the *upper* part to represent the symmetric matrix, the *lower* part works equally well; this is a matter of personal taste.

Renumbering schemes. The nodal point numbering resulting from an automatic or semi-automatic mesh generation is almost certainly not optimal, regardless of which storage format will be used for **K**. In terms of both storage demand and computing time it will pay handsomely to run these node numbers through a node *renumbering* procedure, the purpose of which is to *minimize* the *band width* (band format) or the *profile length* (profile format). Some mesh generators incorporate such renumbering procedures, but more often than not the renumbering of the nodal points of a generated mesh is performed as a separate operation between the mesh generation and the assembly process.

A number of renumbering schemes have been developed in parallel with the development of the finite element technique. This is a specialized area of numerical (matrix) analysis which we shall not deal with in any detail. Most renumbering algorithms rely on *graph theory* in one form or another; none of them purport to produce the best numbering scheme possible (that would in most cases require more work than solving without any renumbering). Most of the suggested procedures are sensitive to the node selected as *start node*, and sometimes a renumbered ordering may be improved by running the same procedure once more. Such improvements are in most cases small and probably not worth the extra effort.

Figure 9.3 shows how the non-zero elements of a typical stiffness matrix are placed in the matrix as it is produced by the mesh generation procedure (original ordering) and also after three different renumbering efforts (all starting from the original ordering). In each of the four patterns the bandwidth (*b*) and profile length (*p*) are given in per cent of the original ordering (which is 100%). Perhaps the best known *bandwidth* "optimizer" is the one due to CUTHILL and MCKEE [31] which, for GAUSSIAN elimination, was improved upon by GEORGE [32] who simply reversed the index numbers of the original algorithm. Use of the latter algorithm, the *reverse* CUTHILL-MCKEE (RCM) is shown in Fig. 9.3b. It reduces the bandwidth by a factor of almost five, to about 22%; this will reduce solution time by a factor of more than 20, which is a substantial reduction. The RCM algorithm has the bandwidth as its target parameter, and consequently it does not do equally well in reducing the profile length (from 100 to 62,1). The patterns in Figs. 9.3c and d are both concerned with minimizing the *profile length* (of a skyline stored matrix), and we see that they do this significantly better than RCM, with the method due to SLOAN [33] as the winner. This has proved to be both a robust and an efficient profile "minimizer". But the algorithm due to HOIT and

System analysis

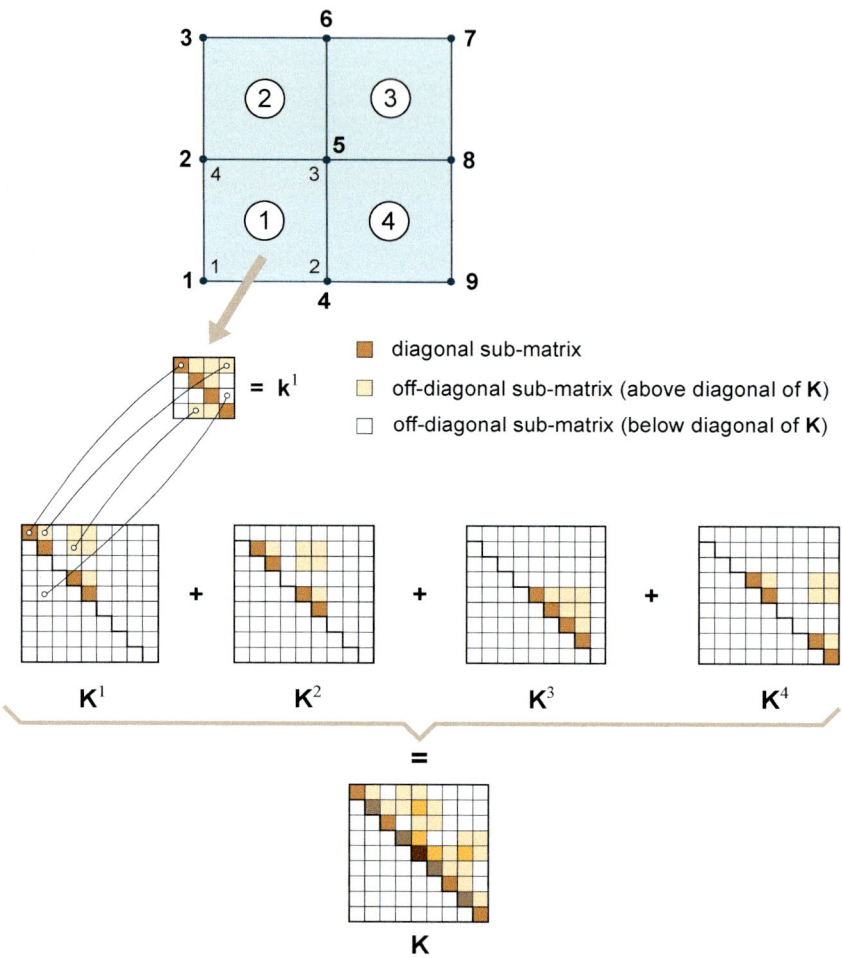

Figure 9.5 Assembly of the system stiffness matrix **K**

PVD = Principle of Virtual Displacements

GARCELON [34] also does quite well, and for another problem it might well be the better choice. It is interesting to note how differently the non-zero elements are placed in the last two figures, both of which are produced by algorithms minimizing the profile length, and yet the final results are not all that different. Note also the similarity in the lower, right-hand part of the matrix in Figs. 9.3b and d.

Figure 9.4 shows an example of renumbering in connection with a sparse storage scheme. The dimension of the matrix shown is about 210 000.

The computer effort (time) spent by all these renumbering schemes is far less than the reduction in solution time resulting from this effort. Hence, node renumbering is not just recommended, it is a must for all "ambitious" FEM programs.

It should be emphasized that this section is based on the assumption that we aim to establish **K** and **R** in full, *before* we consider the next step, the solution of the system of equations. In other words, we treat assembly and solution as two distinct and *separate* operations, which implies that we have ruled out a solution technique referred to as *frontal solution*. This technique, which we shall come back to in Section 9.9, is a form of GAUSSIAN elimination where the unknowns are eliminated *during* assembly, and where it is the *element numbering* that is important, rather than the node numbering.

9.3 Assembly of K and R

With a mesh and a good, if not necessarily optimal node number system we are ready to assemble the element stiffnesses (**k**) and loads (**S**⁰) to form the system matrices. This process is described in Section 2.4 for one-dimensional elements, that is bars and beams. It applies equally well to two- and three-dimensional elements, and for a system of m elements and concentrated nodal forces $\mathbf{R}^c$ we can write:

$$\text{Equilibrium:} \quad \mathbf{R}^c = \sum_{i=1}^{m} \mathbf{g}^i \mathbf{S}^i \qquad (9\text{-}4)$$

$$\text{Force-displacement:} \quad \mathbf{S}^i = \mathbf{k}^i \mathbf{v}^i + \mathbf{S}^{0i} \Rightarrow \mathbf{R}^c = \sum_{i=1}^{m} \mathbf{g}^i (\mathbf{k}^i \mathbf{v}^i + \mathbf{S}^{0i}) \qquad (9\text{-}5)$$

$$\text{Kinematic compatibility:} \quad \mathbf{v}^i = \mathbf{a}^i \mathbf{r} \Rightarrow \mathbf{R}^c = \sum_{i=1}^{m} (\mathbf{g}^i \mathbf{k}^i \mathbf{a}^i) \mathbf{r} + \sum_{i=1}^{m} \mathbf{g}^i \mathbf{S}^{0i} \qquad (9\text{-}6)$$

$$\text{PVD:} \quad \mathbf{g}^i = (\mathbf{a}^i)^\mathsf{T} \quad \mathbf{R}^c = \left(\sum_{i=1}^{m} (\mathbf{a}^i)^\mathsf{T} \mathbf{k}^i \mathbf{a}^i \right) \mathbf{r} + \sum_{i=1}^{m} (\mathbf{a}^i)^\mathsf{T} \mathbf{S}^{0i} = \mathbf{K} \mathbf{r} + \mathbf{R}^0 \qquad (9\text{-}7)$$

Hence

$$\mathbf{K} = \sum_{i=1}^{m} (\mathbf{a}^i)^\mathsf{T} \mathbf{k}^i \mathbf{a}^i = \sum_{i=1}^{m} (\bar{\mathbf{a}}^i)^\mathsf{T} \bar{\mathbf{k}}^i \bar{\mathbf{a}}^i = \sum_{i=1}^{m} \mathbf{K}^i \qquad (9\text{-}8)$$

The bar on top of **a** and **k** indicates that at *every* nodal point of the mesh, the element degrees of freedom (**v**) are expressed in exactly the same coordinate

System analysis

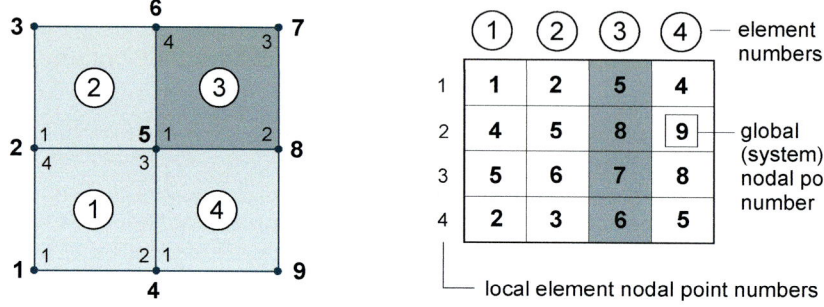

Figure 9.6 Connectivity table

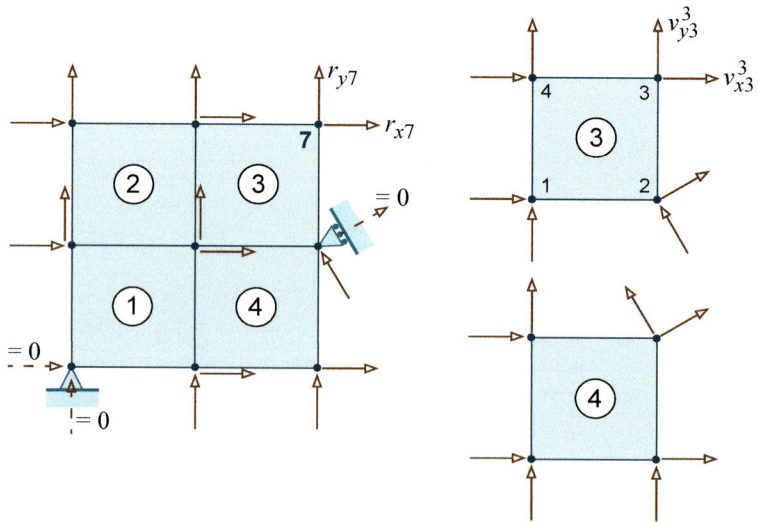

Figure 9.7 One to one correspondence of element (v) and system (r) *dofs*

system as the system degrees of freedom (**r**). Hence *all* **a**-matrices contain only 1 (a few) and 0 (many), and Eq. (9-8) is simply a *recipe* for where in **K**^{i} the elements of **k**^{i} are to be placed, see Fig. 9.5.

In Eq. (9-7) we have used the notation

$$\sum_{i=1}^{m} (\mathbf{a}^i)^T \mathbf{S}^{0i} = \mathbf{R}^0 \qquad (9\text{-}9)$$

for the "sum" of the "fixed end" forces. Equation (9-7) then becomes:

$$\mathbf{K}\mathbf{r} = \mathbf{R}^c - \mathbf{R}^0 = \mathbf{R} \qquad (9\text{-}10)$$

The system load vector (**R**) contains one part (**R**^{c}) that comes from concentrated external forces acting at the nodal points, and another part (**R**^{0}) from loading on the individual elements (which may also come from initial strain effects). The contribution from the individual elements (**S**^{0}) are "added" into **R** in exactly the same way as indicated in Fig. 9.5 for the stiffness.

While the **a**-matrix is a useful concept for a compact *mathematical* formulation, it is of course never formed in the computer. With the requirement that *every* nodal point of the mesh has a *one to one* correspondence between element degrees of freedom (**v**$_i$) and system degrees of freedom (**r**$_j$), *all* relevant information in the **a**-matrix is contained in the *connectivity* table, see Fig. 9.6. For each element in the model (mesh) this table contains the numbers of the global (system) nodes to which the local (element) nodes are attached.

The complete finite element model of a problem may include elements of different types; for each type the element nodal points must be numbered in a consistent manner, *e.g.* in a counter-clockwise sequence. The one to one correspondence between element *dofs* (v_i) and system *dofs* (r_i) at a particular node implies that both sets of *dofs* must be arranged in the same order and, for any given node its *dofs* must be referred to the *same* coordinate system at element and system level, see Fig. 9.7 (of an academic problem!)

We will return to the computer implementation of the assembly process in the next chapter.

9.4 Boundary conditions – an overview

So far we have indicated an assembly process involving *all dofs* of the system, without regard to the fact that some *dofs* are specified or *constrained.* As a minimum we need to prevent all rigid body motions of our model; for a practical problem the *support* conditions, often in the form of suppressed or fixed *dofs*, will take care of these motions (and more so). In addition to the support conditions our model may be subjected to other and more complex displacements constraints. All this information about given *dofs*, collectively referred to as *boundary conditions*, must be properly incorporated into the system stiffness relation, Eq. (9-2), *before* solution.

Definitions. We start by defining some concepts and terms.

- A *displacement constraint* or simply *constraint* defines the actual value of a particular *dof* (an element of **r**), or it defines a relationship between two or more *dofs*.

System analysis

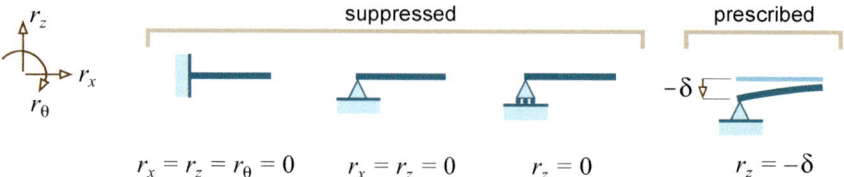

Figure 9.8 Examples of single point constraints

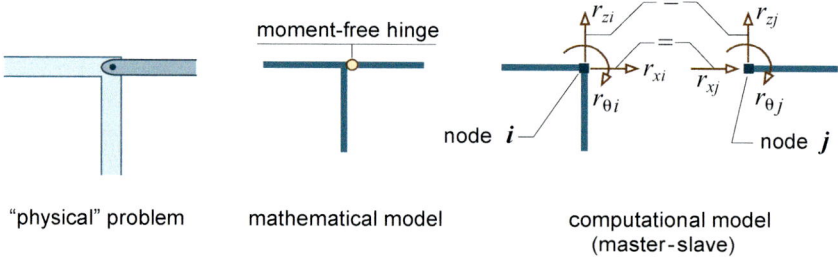

Figure 9.9 Modelling a moment release in a plane frame structure

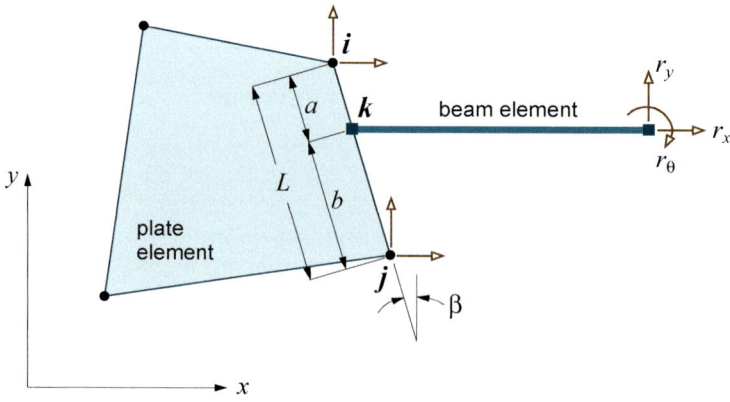

Figure 9.10 Beam – plate coupling

Boundary conditions – an overview

- A *free dof* is an element of **r** that we have *no* prior knowledge of.
- A *specified dof*, r_s, is defined by the equation

$$r_s = h_0 + h_1 r_{m1} + h_2 r_{m2} + \ldots + h_p r_{mp} \qquad (9\text{-}11)$$

where $h_0, h_1, h_2, \ldots, h_p$ are known *constants* (data), and $r_{m1}, r_{m2}, \ldots, r_{mp}$ are *free*, so-called *master dofs*. Equation (9-11) represents a general linear constraint equation, also referred to as a *multi-point constraint*, in that it involves several *dofs* which, most likely, are located at different nodal points. r_s is a *dependent dof*, often referred to as a *slave dof* (in contrast to the master *dofs*).

An important special case of Eq. (9-11) is the *single point constraint* where only one *dof* is involved, in one of two forms,

$$r_s = 0 \qquad (9\text{-}12)$$

the homogeneous form, and

$$r_s = \delta = \text{constant} \neq 0 \qquad (9\text{-}13)$$

the inhomogeneous form. The specified *dof* in Eq. (9-12) is *suppressed* (fixed and clamped are also used), while r_s in Eq. (9-13) is said to be *prescribed* (a non-zero value). Figure 9.8 shows some familiar examples of single point constraints.

The simplest and most common multi-point constraint is the case where a particular *dof* (r_s) is made equal to another, free *dof* (r_m),

$$r_s = r_m \qquad (9\text{-}14)$$

This simple constraint equation enables an elegant solution to many types of "internal" displacement *releases*, or "hinges" in engineering language. Take for instance the simple case of a beam element being joined to a rigid frame corner in such a way that the beam cannot transmit a bending moment to the frame, as shown in Fig. 9.9. The "trick" here is to include *two* independent nodal points, *i* and *j*, at the *same* geometrical point (*i* and *j* have identical coordinates). The column and the left-hand beam are both connected to node *i*, while the right-hand beam is connected to node *j*. In order to connect the right-hand beam to the rigid frame corner, with due regard to the moment release, we now introduce two multi-point constraint equations,

$$r_{xj} = r_{xi} \quad \text{and} \quad r_{zj} = r_{zi}$$

These two simple equations will ensure that points *i* and *j* will "move" together, while they have their own (uncoupled) rotations ($r_{\theta i}$ and $r_{\theta j}$) which is exactly what we want to achieve.

An example of a somewhat more complex coupling is shown in Fig. 9.10 where a standard beam element with 3×2 = 6 *dofs* is coupled to the edge of a 4-node quadrilateral (8 *dof*) plate (plane stress) element. Part of the problem here is that the beam is coupled to a point where the plate element has no nodal point. However, we do know that the element edge, for this particular

System analysis

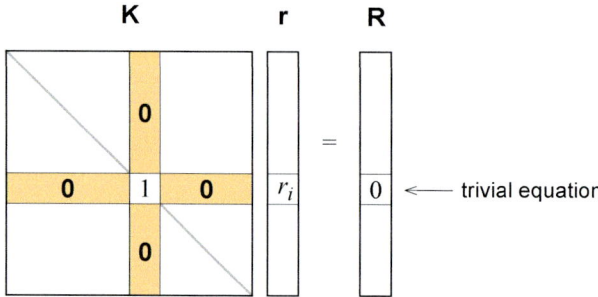

Figure 9.11 Implementing the condition $r_i = 0$

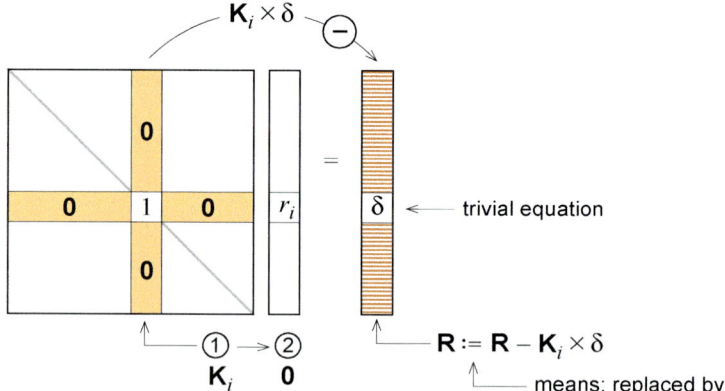

Figure 9.12 Implementing the condition $r_i = \delta$

plate element, will remain straight (between its end points) also after deformation. We define the *dofs* at node **k** as *slaves*, while the *dofs* at nodes **i** and **j** are master *dofs*. The following multi-point constraint equations, expressing the "slaves" in terms of their "masters", are readily established on the basis of "small displacement" geometry:

$$r_{xk} = \frac{b}{L}r_{xi} + \frac{a}{L}r_{xj}$$

$$r_{yk} = \frac{b}{L}r_{yi} + \frac{a}{L}r_{yj} \qquad (9\text{-}15)$$

$$r_{\theta k} = \frac{c}{L}r_{xi} + \frac{s}{L}r_{yi} - \frac{c}{L}r_{xj} - \frac{s}{L}r_{yj}$$

where the notation $c = \cos\beta$ and $s = \sin\beta$ has been used.

Implementation. The simple, but common case where all specified *dofs* are suppressed (= 0) can be implemented by simply deleting rows and columns in **K** and **R** that correspond to the suppressed components in **r**. A very simple way to do this is shown in Fig. 9.11 where, instead of deleting the row and column corresponding to suppressed *dof* r_i, a *trivial* equation ($r_i = 0$) has been introduced; this is a much simpler operation to program and execute than "packing" a matrix after deleting a row and a column.

If we also allow prescribed *dofs*, for instance $r_i = \delta$, then we need to multiply column i by δ and subtract the resulting vector from the load vector **R** before we introduce the trivial equation, see Fig. 9.12.

This rather crude implementation, which takes as its starting point that **K** and **R** are assembled *without* regard to the boundary conditions, was used in the early days of FEM. It is a simple and straightforward approach, but it has disadvantages and limitations and it has long since been replaced by far more efficient and general methods of implementing the boundary conditions into general FEM programs, such as:

- elimination of specified *dofs*,
- LAGRANGE multipliers, and
- penalty functions.

Each of these methods will be dealt with below in separate sections. But first we formalize the constraint equations into a compact matrix format.

Formal treatment. We assume that *all* displacement constraints of a given problem can be expressed as a series of equations of the form in Eq. (9-11). Collectively these equations can be written as one matrix equation,

$$\mathbf{Hr} = \mathbf{h}_0 \qquad (9\text{-}16)$$

Here **H** is a rectangular matrix – one row for each constraint equation and one column for each *dof* of the system – while $\mathbf{h}_0$ is a vector of all the c_0 coefficients. If we rearrange the *dofs* into one sub-vector containing all *free dofs* ($\mathbf{r}_f$) and one containing *all specified dofs* ($\mathbf{r}_s$), Eq. (9-16) can be written as

System analysis

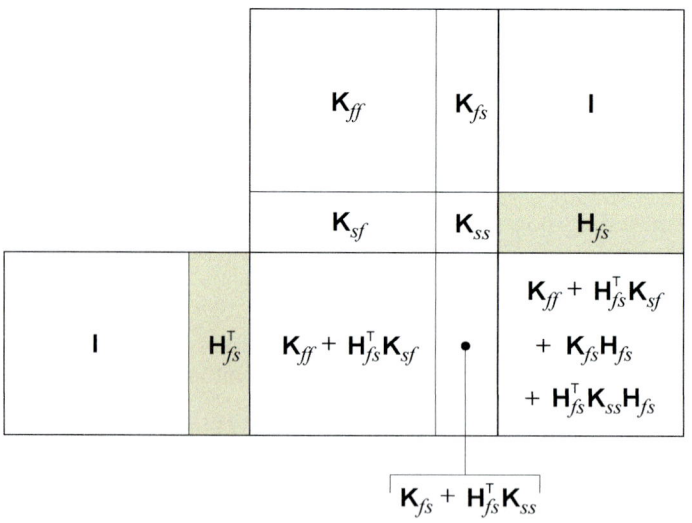

NOTE: Since $\mathbf{K}_{fs} = \mathbf{K}_{sf}^T$ then $(\mathbf{K}_{fs}\mathbf{H}_{fs})^T = \mathbf{H}_{fs}^T\mathbf{K}_{sf}$

$$\begin{bmatrix} \mathbf{H}_f & \mathbf{H}_s \end{bmatrix} \begin{bmatrix} \mathbf{r}_f \\ \mathbf{r}_s \end{bmatrix} = \mathbf{h}_0 \qquad (9\text{-}17)$$

or

$$\mathbf{r}_s = \mathbf{H}_{fs}\mathbf{r}_f + \bar{\mathbf{h}}_0 \qquad (9\text{-}18)$$

where

$$\mathbf{H}_{fs} = -\mathbf{H}_s^{-1}\mathbf{H}_f \qquad (9\text{-}19)$$

and

$$\bar{\mathbf{h}}_0 = \mathbf{H}_s^{-1}\mathbf{h}_0 \qquad (9\text{-}20)$$

Keep in mind that $\mathbf{H}$ and $\mathbf{r}_s$ have the same number of "rows". For a certain ordering of the specified *dofs* we have that $\mathbf{H}_s = \mathbf{I} = \mathbf{H}_s^{-1}$.

9.5 Boundary conditions – elimination of *dofs*

The system stiffness relation (9-2) can also be expressed in partitioned form:

$$\begin{bmatrix} \mathbf{K}_{ff} & \mathbf{K}_{fs} \\ \mathbf{K}_{sf} & \mathbf{K}_{ss} \end{bmatrix} \begin{bmatrix} \mathbf{r}_f \\ \mathbf{r}_s \end{bmatrix} = \begin{bmatrix} \mathbf{R}_f \\ \mathbf{R}_s \end{bmatrix} \qquad (9\text{-}21)$$

If we substitute for $\mathbf{r}_s$ as defined by Eq. (9-18) we get

$$\begin{bmatrix} \mathbf{K}_{ff} & \mathbf{K}_{fs} \\ \mathbf{K}_{sf} & \mathbf{K}_{ss} \end{bmatrix} \left[\begin{bmatrix} \mathbf{I} \\ \mathbf{H}_{fs} \end{bmatrix} \mathbf{r}_f + \begin{bmatrix} \mathbf{0} \\ \bar{\mathbf{h}}_0 \end{bmatrix} \right] = \begin{bmatrix} \mathbf{R}_f \\ \mathbf{R}_s \end{bmatrix}$$

or

$$\begin{bmatrix} \mathbf{K}_{ff} & \mathbf{K}_{fs} \\ \mathbf{K}_{sf} & \mathbf{K}_{ss} \end{bmatrix} \begin{bmatrix} \mathbf{I} \\ \mathbf{H}_{fs} \end{bmatrix} \mathbf{r}_f = \begin{bmatrix} \bar{\mathbf{R}}_f \\ \bar{\mathbf{R}}_s \end{bmatrix} \qquad (9\text{-}22)$$

where

$$\bar{\mathbf{R}}_f = \mathbf{R}_f - \mathbf{K}_{fs}\bar{\mathbf{h}}_0 \qquad (9\text{-}23a)$$

$$\bar{\mathbf{R}}_s = \mathbf{R}_s - \mathbf{K}_{ss}\bar{\mathbf{h}}_0 \qquad (9\text{-}23b)$$

Premultiplication of Eq. (9-22) by $\begin{bmatrix} \mathbf{I} \\ \mathbf{H}_{fs} \end{bmatrix}^T$ gives (see FALK diagram opposite):

$$[\mathbf{K}_{ff} + \mathbf{H}_{fs}^T\mathbf{K}_{sf} + \mathbf{K}_{fs}\mathbf{H}_{fs} + \mathbf{H}_{fs}^T\mathbf{K}_{ss}\mathbf{H}_{fs}]\mathbf{r}_f = \bar{\mathbf{R}}_f + \mathbf{H}_{fs}^T\bar{\mathbf{R}}_s \qquad (9\text{-}24)$$

or

$$\mathbf{K}_{ff}^*\mathbf{r}_f = \mathbf{R}_f^* \qquad (9\text{-}25)$$

If all constraint equations are of the form (9-12), that is, only suppressed *dofs*, then we have

$$\mathbf{H}_f = \mathbf{0}, \quad \mathbf{H}_s = \mathbf{I} \quad \text{and} \quad \mathbf{h}_0 = \mathbf{0}$$

and thus

$$\mathbf{H}_{fs} = \mathbf{0} \quad \text{and} \quad \bar{\mathbf{h}}_0 = \mathbf{0}$$

System analysis

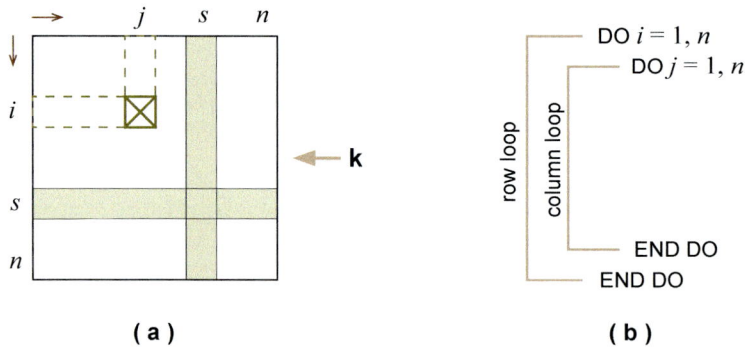

Figure 9.13 Assembly of **k** – element loops

Boundary conditions – elimination of dofs

and Eq. (9-25) becomes

$$\mathbf{K}_{ff}\mathbf{r}_f = \mathbf{R}_f \qquad (\mathbf{r}_s = \mathbf{0}) \tag{9-26}$$

We take this one step further, still with only single point constraints, but now we allow some specified *dofs* to be prescribed. Hence, $\mathbf{r}_s = \mathbf{h}_0$, which leads to

$$\mathbf{H}_f = \mathbf{0}, \quad \mathbf{H}_s = \mathbf{I} \quad \Rightarrow \quad \mathbf{H}_{fs} = \mathbf{0} \text{ and } \bar{\mathbf{h}}_0 = \mathbf{h}_0$$

This gives, see Eqs. (9-24) and (9-23a),

$$\mathbf{K}_{ff}\mathbf{r}_f = \mathbf{R}_f - \mathbf{K}_{fs}\mathbf{h}_0 \tag{9-27}$$

Thus the non-zero elements of $\mathbf{h}_0$, that is the prescribed *dofs*, end up as "load" components. In both Eqs. (9-26) and (9-27) only unknown displacements ($\mathbf{r}_f$) are present; the specified *dofs* ($\mathbf{r}_s$) have been removed from the system.

The actual implementation strategy is to account for the boundary conditions *during* the assembly process, and we only form matrices $\mathbf{K}_{ff}^*$ and $\mathbf{R}_f^*$; none of the other matrices are formed.

We assume that before the start of the assembly process, which takes place in a *loop* over all elements in the model, a *table* containing one entry for *each dof* of the model (including also the specified *dofs*) exists. For a *free dof* the table contains the equation number of that particular *dof*; this is the row/column number in $\mathbf{K}$ (actually $\mathbf{K}_{ff}^*$) for this *dof*. For the specified *dofs*, however, the table contains a special "code", *e.g.* the value 0 if it is a suppressed *dof*, a negative value if it is a prescribed *dof* and a very large number (larger than the total number of *dofs*) if it is a slave *dof* in a multi-point constraint equation. We will elaborate on this "book-keeping" aspect in the next chapter. We also assume that the element matrices ($\mathbf{k}$ and $\mathbf{S}^0$) have, if necessary, been transformed to match the system *dofs*, prior to their actual inclusion in $\mathbf{K}$ and $\mathbf{R}$; thus a one-to-one relationship between element *dofs* and the corresponding system *dofs* is guaranteed.

We limit our discussion to the stiffness; the loading is similar, but simpler. Figure 9.13 shows an element stiffness matrix $\mathbf{k}$ with n *dofs*. Two "loops" are indicated, one row-loop (with loop index i) and one column-loop (with loop index j). *Dof* number s is specified. Without going into the finer details, here is what happens during assembly of this particular element:

The "outer" (row) loop starts by assigning a value of 1 to its index i, and when the loop is completed (control reach the END DO), index i is incremented by 1, and the loop is repeated, and so on until the loop is exhausted (i becomes greater than n). For *each* value of i, the "inner" (column) loop (controlled by its index j) is executed n times ($j=1, j=2, \ldots$ and so on). If local element *dofs* i and j both correspond to *free dofs*, element k_{ij} of $\mathbf{k}$ is added to the appropriate element of $\mathbf{K}$ if this is located on or above the diagonal of $\mathbf{K}$; if not the operation is skipped. If either i or j becomes equal to s, the local element *dof* is associated with a *specified dof* and the course of action will depend on the type of constraint:

313

System analysis

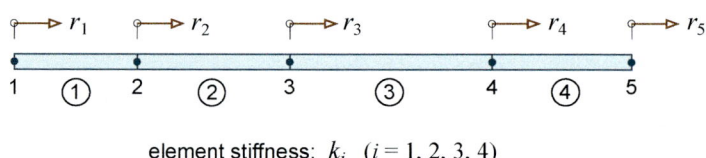

element stiffness: k_i ($i = 1, 2, 3, 4$)

Figure 9.14 A simple 1-dimensional system

Boundary conditions – elimination of dofs

- *s* is *suppressed* – no action at all, control is transferred to the end of the loop (none of the elements in row *s* and column *s* of **k** are added to **K**),
- *s* is *prescribed* – if $i = s$ no action; if $j = s$ and $i \ne s$ then k_{is} is multiplied with the prescribed value (δ) and the product is *subtracted* from the corresponding element of **R**, (column *s* of **k**, except k_{ss}, is multiplied by δ and moved to the right-hand side of the system stiffness relation); again, none of the elements in row *s* and column *s* of **k** are added to **K**),
- *s* is a "simple" *slave*, in other words $r_s = r_m$ – in this case we simply replace *s* by *m*, and add k_{sj} or k_{is} to the element of **K** corresponding to *m* and *j* or *i* and *m* (providing the corresponding element of **K** is above the diagonal; otherwise no action).

If *s* is the slave in a more complex multi-point constraint, the process is in principle the same, but the "book-keeping" becomes more involved, and we leave this for the next chapter.

The advantage of this technique is that we establish the smallest possible system of equations; the stiffness matrix emerging from this process is $\mathbf{K}^*_{ff}$ which corresponds to only "free", that is unknown, degrees of freedom. All boundary conditions are satisfied "exactly" and $\mathbf{K}^*_{ff}$ is as *stable* as the physical problem permits, it has all the positive properties, *symmetry, sparseness* and it is *positive definite*. One might argue that the method has a drawback in that the reaction forces $\mathbf{R}_s$ are "lost". However, these forces are readily recovered by a separate computation once the main problem is solved.

Example 9-1 – Incorporating a slave dof

We consider a very simple one-dimensional system with 4 elements and 5 dofs, see Fig. 9.14. The element stiffness relation is

$$\mathbf{S}^i = \begin{bmatrix} S_1^i \\ S_2^i \end{bmatrix} = \begin{bmatrix} k_i & -k_i \\ -k_i & k_i \end{bmatrix} \begin{bmatrix} v_1^i \\ v_2^i \end{bmatrix} = \mathbf{k}^i \mathbf{v}^i \tag{9-28}$$

and the system stiffness relation is readily found to be

$$\begin{bmatrix} k_1 & -k_1 & & & \\ -k_1 & k_1+k_2 & -k_2 & & \\ & -k_2 & k_2+k_3 & -k_3 & \\ & & -k_3 & k_3+k_4 & -k_4 \\ & & & -k_4 & k_4 \end{bmatrix} \begin{bmatrix} r_1 \\ r_2 \\ r_3 \\ r_4 \\ r_5 \end{bmatrix} = \begin{bmatrix} R_1 \\ R_2 \\ R_3 \\ R_4 \\ R_5 \end{bmatrix} \tag{9-29}$$

We now introduce a multi-point constraint that force nodes 2 and 4 to have the *same* displacement (as if they were connected by a rigid link), that is,

$$r_4 = r_2 \quad \text{or} \quad r_2 - r_4 = 0 \tag{9-30}$$

We want to eliminate r_4 (arbitrarily chosen to be the slave *dof*) and choose to do so via a standard transformation. To this end we need the relationship

315

System analysis

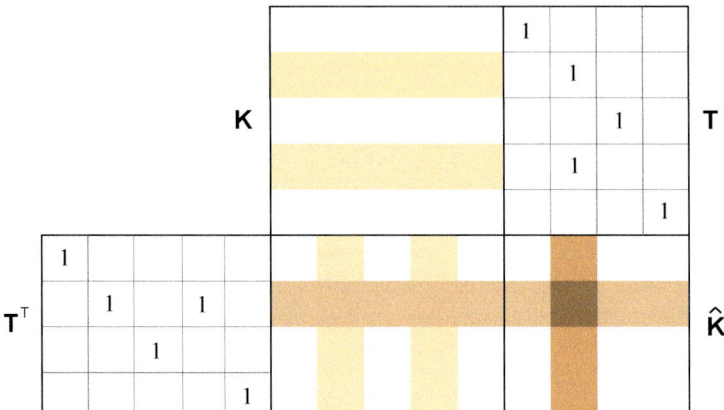

Boundary conditions – LAGRANGE multipliers

$$\mathbf{r} = \begin{bmatrix} r_1 \\ r_2 \\ r_3 \\ r_4 \\ r_5 \end{bmatrix} = \begin{bmatrix} 1 & & & \\ & 1 & & \\ & & 1 & \\ & 1 & & \\ & & & 1 \end{bmatrix} \begin{bmatrix} \hat{r}_1 \\ \hat{r}_2 \\ \hat{r}_3 \\ \hat{r}_5 \end{bmatrix} = \mathbf{T}\hat{\mathbf{r}} \qquad (9\text{-}31)$$

The stiffness matrix corresponding to $\hat{\mathbf{r}}$ is defined by

$$\hat{\mathbf{K}} = \mathbf{T}^T \mathbf{K} \mathbf{T} \qquad (9\text{-}32)$$

or, see FALK diagram opposite,

$$\hat{\mathbf{K}} = \begin{bmatrix} k_1 & -k_1 & & \\ -k_1 & \kappa & -k_2-k_3 & -k_4 \\ & -k_2-k_3 & k_2+k_3 & \\ & -k_4 & & k_4 \end{bmatrix} \quad \text{where} \quad \kappa = k_1 + k_2 + k_3 + k_4 \qquad (9\text{-}33)$$

This stiffness matrix could have been established in a more direct manner by simply adding the contributions from the individual elements directly into $\hat{\mathbf{K}}$, observing that elements of $\mathbf{k}^3$ and $\mathbf{k}^4$ that should have gone to r_4 and r_5 now need to go to $\hat{r}_2$ and $\hat{r}_4$, respectively.

9.6 Boundary conditions – LAGRANGE multipliers

In mathematical optimization the method of LAGRANGE multipliers provides a strategy for finding maxima and minima of a function subject to *constraints*. In our case the function is the *potential energy* of the system, Π, and the variables are the system *dofs* $\mathbf{r}$,

$$\Pi = \frac{1}{2}\mathbf{r}^T \mathbf{K}\mathbf{r} - \mathbf{r}^T \mathbf{R} \qquad (9\text{-}34)$$

It should be noted that here $\mathbf{r}$ comprises *all* system *dofs*, also the specified ones. We now treat the displacement constraints as additional conditions or restrictions. These are defined in Eq. (9-16) which can be written as the following homogeneous matrix equation:

$$\mathbf{H}\mathbf{r} - \mathbf{h}_0 = \mathbf{0} \qquad (9\text{-}35)$$

We multiply this equation by $\boldsymbol{\lambda}^T$, where the vector $\boldsymbol{\lambda}$ contains as many LAGRANGE multipliers as we have constraints. The result, which is a scalar, we add to the potential energy, that is,

$$\Pi = \frac{1}{2}\mathbf{r}^T \mathbf{K}\mathbf{r} - \mathbf{r}^T \mathbf{R} + \boldsymbol{\lambda}^T (\mathbf{H}\mathbf{r} - \mathbf{h}_0) \qquad (9\text{-}36)$$

The expression in parentheses is zero and we have therefore not altered Π. We make Π stationary by requiring that

System analysis

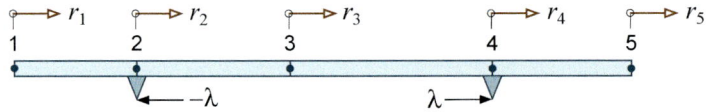

Figure 9.15 Problem of Fig. 9.14 with LAGRANGE multiplier

Boundary conditions – LAGRANGE multipliers

$$\frac{\partial \Pi}{\partial \mathbf{r}} = \mathbf{Kr} - \mathbf{R} + \mathbf{H}^T \boldsymbol{\lambda} = \mathbf{0} \qquad (9\text{-}37)$$

and

$$\frac{\partial \Pi}{\partial \boldsymbol{\lambda}} = \mathbf{Hr} - \mathbf{h}^0 = \mathbf{0} \qquad (9\text{-}38)$$

This last equation merely repeats the constraints. Combination of Eqs. (9-37) and (9-38) gives

$$\begin{bmatrix} \mathbf{K} & \mathbf{H}^T \\ \mathbf{H} & \mathbf{0} \end{bmatrix} \begin{bmatrix} \mathbf{r} \\ \boldsymbol{\lambda} \end{bmatrix} = \begin{bmatrix} \mathbf{R} \\ \mathbf{h}_o \end{bmatrix} \qquad (9\text{-}39)$$

This equation is solved with respect to both $\mathbf{r}$ and $\boldsymbol{\lambda}$. We notice that the symmetry is maintained, and if $\mathbf{K}$ is positive definite and the rows of $\mathbf{H}$ are linearly independent, then (9-39) has a unique solution.

The physical interpretation of the LAGRANGE multipliers λ_i is that they can be regarded as constraint *forces* that impose the displacement restrictions. If, for instance, Eq. (9-35) specifies only suppressed *dofs*, then the multipliers will be reaction forces.

This technique can handle very complex constraint equations, *e.g.* a slave can be coupled to other slaves as well as masters. A disadvantage of the method is that it increases the number of unknown parameters; it also increases the bandwidth of the coefficient matrix (often quite substantially) and the profile length (which is not as serious as the bandwidth increase). Furthermore, the matrix of coefficients is no longer positive definite.

Example 9-2 – LAGRANGE *multiplyer*

Same problem as in the previous section, but we now replace the rigid link, that forces node 2 and 4 to have identical displacements, by a pair of "reaction forces" $(-\lambda, \lambda)$, see Fig. 9.15:

$$\begin{bmatrix} k_1 & -k_1 & & & \\ -k_1 & k_1+k_2 & -k_2 & & \\ & -k_2 & k_2+k_3 & -k_3 & \\ & & -k_3 & k_3+k_4 & -k_4 \\ & & & -k_4 & k_4 \end{bmatrix} \begin{bmatrix} r_1 \\ r_2 \\ r_3 \\ r_4 \\ r_5 \end{bmatrix} = \begin{bmatrix} R_1 \\ R_2 - \lambda \\ R_3 \\ R_4 + \lambda \\ R_5 \end{bmatrix} \qquad (9\text{-}40)$$

Since λ is an unknown quantity we move it to the left-hand side and include it in the vector $\mathbf{r}$ of unknown displacements:

$$\begin{bmatrix} k_1 & -k_1 & & & & 0 \\ -k_1 & k_1+k_2 & -k_2 & & & 1 \\ & -k_2 & k_2+k_3 & -k_3 & & 0 \\ & & -k_3 & k_3+k_4 & -k_4 & -1 \\ & & & -k_4 & k_4 & 0 \end{bmatrix} \begin{bmatrix} r_1 \\ r_2 \\ r_3 \\ r_4 \\ r_5 \\ \lambda \end{bmatrix} = \begin{bmatrix} R_1 \\ R_2 \\ R_3 \\ R_4 \\ R_5 \end{bmatrix} \qquad (9\text{-}41)$$

319

System analysis

The term $\mathbf{r}^T\mathbf{H}^T\lfloor\alpha\rfloor\mathbf{h}_0$ of Eq. (9-46) is the sum of two terms:

$$\tfrac{1}{2}\mathbf{r}^T\mathbf{H}^T\lfloor\alpha\rfloor\mathbf{h}_0 + \tfrac{1}{2}\mathbf{h}_0^T\lfloor\alpha\rfloor\mathbf{H}\mathbf{r}$$

Both terms are *scalar* expressions, and the transpose of the second term, which is equal to the first term, evaluates to the *same* scalar.

For differentiation with respect to a vector, see APPENDIX A.

We now have 6 unknowns, but only 5 equations. By including the condition itself we obtain the following regular system of equations:

$$\begin{bmatrix} k_1 & -k_1 & & & & 0 \\ -k_1 & k_1+k_2 & -k_2 & & & 1 \\ & -k_2 & k_2+k_3 & -k_3 & & 0 \\ & & -k_3 & k_3+k_4 & -k_4 & -1 \\ & & & -k_4 & k_4 & 0 \\ 0 & 1 & 0 & -1 & 0 & 0 \end{bmatrix} \begin{bmatrix} r_1 \\ r_2 \\ r_3 \\ r_4 \\ r_5 \\ \lambda \end{bmatrix} = \begin{bmatrix} R_1 \\ R_2 \\ R_3 \\ R_4 \\ R_5 \end{bmatrix} \quad (9\text{-}42)$$

For this simple example we were able to establish the form of Eq. (9-39) by a purely physical argument. For more complex constraint equations we need to employ the mathematical argumentation.

9.7 Boundary conditions – penalty functions

If the constraint equations (9-16) are written as

$$\mathbf{t} = \mathbf{Hr} - \mathbf{h}_0 \quad (9\text{-}43)$$

then

$$\mathbf{t} = \mathbf{0} \quad (9\text{-}44)$$

implies that the constraint equations are satisfied. The system's potential energy Π can be augmented by the *penalty function*

$$\tfrac{1}{2}\mathbf{t}^T \lceil \alpha \rfloor \mathbf{t}$$

where $\lceil \alpha \rfloor$ is a diagonal matrix of penalty numbers, one number for each constraint equation. Hence

$$\Pi = \tfrac{1}{2}\mathbf{r}^T \mathbf{K}\mathbf{r} - \mathbf{r}^T \mathbf{R} + \tfrac{1}{2}\mathbf{t}^T \lceil \alpha \rfloor \mathbf{t} \quad (9\text{-}45)$$

We emphasize again that $\mathbf{r}$ contains *all dofs* of the system, also the specified ones. If $\mathbf{t} = \mathbf{0}$ the constraint equations will be satisfied and we have not changed the potential energy. If $\mathbf{t} \neq \mathbf{0}$ then the "penalty" for violating the constraint conditions increases with increasing α's.

Substituting (9-43) into (9-45) gives

$$\Pi = \tfrac{1}{2}\mathbf{r}^T \mathbf{K}\mathbf{r} - \mathbf{r}^T \mathbf{R} + \tfrac{1}{2}\mathbf{r}^T \mathbf{H}^T \lceil \alpha \rfloor \mathbf{H}\mathbf{r} - \mathbf{r}^T \mathbf{H}^T \lceil \alpha \rfloor \mathbf{h}_0 + \tfrac{1}{2}\mathbf{h}_0^T \lceil \alpha \rfloor \mathbf{h}_0 \quad (9\text{-}46)$$

The stationary condition

$$\frac{\partial \Pi}{\partial \mathbf{r}} = \mathbf{0}$$

gives

$$\mathbf{K}\mathbf{r} - \mathbf{R} + \mathbf{H}^T \lceil \alpha \rfloor \mathbf{H}\mathbf{r} - \mathbf{H}^T \lceil \alpha \rfloor \mathbf{h}_0 = \mathbf{0}$$

or

System analysis

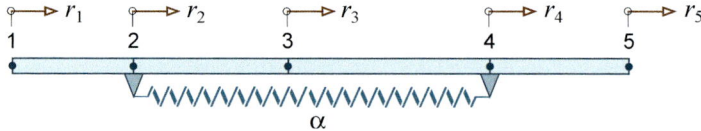

Figure 9.16 Problem of Fig. 9.14 with penalty function (spring)

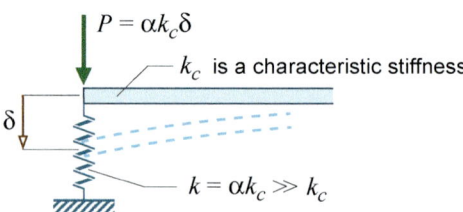

$P = \alpha k_c \delta$

k_c is a characteristic stiffness

$k = \alpha k_c \gg k_c$

Figure 9.17 Prescribed displacement implemented by a stiff spring

Boundary conditions – penalty functions

$$[\mathbf{K} + \mathbf{H}^T \lfloor \alpha \rfloor \mathbf{H}]\mathbf{r} = \mathbf{R} + \mathbf{H}^T \lfloor \alpha \rfloor \mathbf{h}_0 \qquad (9\text{-}47)$$

where $\mathbf{H}^T \lfloor \alpha \rfloor \mathbf{H}$ is the "penalty matrix". We see that $\lfloor \alpha \rfloor = \mathbf{0}$ implies that the constraint conditions are completely neglected. The more we increase the values of the penalty numbers (α_i), the better the constraint conditions will be satisfied. On the other hand, too large penalty numbers may cause numerical problems. In general this method therefore satisfies the constraint conditions approximately, and it is the user's responsibility to choose the penalty number.

Example 9-3 – Penalty function

We consider the same simple rod as in the previous two sections, see Fig. 9.14. The constraint (9-30) can be written as

$$\mathbf{H} = [\,0 \quad 1 \quad 0 \quad -1 \quad 0\,] \quad \text{and} \quad \mathbf{h}_0 = 0 \qquad (9\text{-}48)$$

We have only one penalty number α and the penalty matrix becomes

$$\mathbf{H}^T \lfloor \alpha \rfloor \mathbf{H} = \begin{bmatrix} 0 & 0 & 0 & 0 & 0 \\ 0 & \alpha & 0 & -\alpha & 0 \\ 0 & 0 & 0 & 0 & 0 \\ 0 & -\alpha & 0 & \alpha & 0 \\ 0 & 0 & 0 & 0 & 0 \end{bmatrix} \qquad (9\text{-}49)$$

When this matrix is added to the stiffness matrix $\mathbf{K}$ it has exactly the same effect as if we insert a linear spring element with stiffness α between nodes 2 and 4, se Fig. 9.16. This illustrates clearly the effect of the penalty number. For the spring to produce the desired effect, that is to force nodes 2 and 4 to have the same displacement ($r_2 = r_4$), it needs to be very stiff. In fact, the stiffer the better. However, there is a limit. For a very large penalty number the penalty matrix in (9-49) will dominate completely – the elements of $\mathbf{K}$ will become numerically "zero" in comparison – and the matrix of coefficients will have two rows/columns which are linear dependent, and this in turn means a *singular* matrix.

Most structural analysis programs offer springs as an alternative/additional means to enforce specified displacements, both *boundary springs* and simple *coupling springs*. A boundary spring connects a dof to a fixed point, whereas a coupling spring connects two (similar) *dofs* (usually defined at the same coordinates, e.g. a rotational spring simulating a semi-rigid "hinge"). The arguments for this approach are purely physical, without any reference to penalty functions, although that is basically what it is. The spring stiffnesses, which need to be specified by the user (not always an easy task), are usually chosen on the basis of other characteristic stiffnesses of the problem.

A *prescribed* displacement ($r_s = \delta$) can be modelled as a very stiff spring (with stiffness αk_c) and a very large force ($P = \alpha k_c \delta$) in the direction of the spring. For ordinary loading the spring stiffness will (for practical purposes) prevent displacement, but the large (fictitious) force P will produce the prescribed displacement δ, see Fig. 9.17.

System analysis

> **NOTE:** Consider a *homogeneous* system of equations
> $$Ax = 0$$
> where **A** is a symmetric matrix. If **A** is *regular* (non-singular), the only solution is **x** = **0**. However, if **A** is *singular* (rank deficient) we can have a solution that is different from the null vector (**0**); the number of independent elements in **x** that are different from zero is equal to the rank deficiency (the difference between the matrix dimension and its rank). See APPENDIX A

Advantages of penalty functions/springs:

- Easy to implement.
- Compared with LAGRANGE multipliers, this method does not introduce new variables; on the other hand it does not reduce the number of unknowns – all specified *dofs* are still present in the system of equations – which, compared with the elimination technique, is a disadvantage.

The disadvantage is that

- the choice of penalty numbers (spring stiffnesses) is not always easy; boundary springs do not present any problems (they can be assigned large stiffnesses without consequences), but coupling springs, like the one in Fig. 9.16, can cause numerical problems (lead to singularity or near singularity).

If we have fewer constraint equations than number of *dofs* in **r**, then **H** will have more columns than rows, and the penalty matrix $\mathbf{H}^T \lceil \alpha \rfloor \mathbf{H}$ will most certainly be *singular*. However, in some cases we may have couplings between *all dofs* in **r**, and these may cause the penalty matrix to become non-singular or regular. This is not desirable, and the reason is this:

For simplicity we assume $\lceil \alpha \rfloor = \alpha \mathbf{I}$ and $\mathbf{h}_0 = \mathbf{0}$. When α becomes very large Eq. (9-47) reduces to

$$\mathbf{H}^T \mathbf{H} \mathbf{r} = \frac{1}{\alpha} \mathbf{R} \Rightarrow 0 \qquad (9\text{-}50)$$

If $\mathbf{H}^T \mathbf{H}$ is regular **r** will tend towards **0** as α increases; the result is that the system will "lock".

Some penalty matrices that can cause locking appear "naturally" as part of the development process. An example of this is *incompressible materials*. For such materials POISSON's ratio is $\nu = 0{,}5$. A material like rubber, and materials that "flow" (plastic deformation) will have a value of ν in the neighbourhood of 0,5. For an isotropic elastic material in 3D or in *plane strain*, a value of ν equal to 0,5 is "illegal" since $(1-2\nu)$ appears in the denominator of the **C** matrix, see Eqs. (3-30) and (3-47b). It is of course possible to approximate this situation by letting ν have a value close to 0,5. However, if we do not take special precautions we will experience that the elements become stiffer and stiffer the closer POISSON's ratio becomes to 0,5; we will observe clear tendencies of *locking*. This we can explain as follows:

The shear modulus G and the bulk modulus B for an isotropic elastic material are, Eqs. (3-25) and (3-33),

$$G = \frac{E}{2(1+\nu)} \quad \text{and} \quad B = \frac{E}{3(1-2\nu)} \qquad (9\text{-}51)$$

With reference to Section 3.1 the **C** matrix in terms of G and B can be expressed as

$$\mathbf{C} = G\mathbf{C}_G + B\mathbf{C}_B \qquad (9\text{-}52)$$

where $\mathbf{C}_G$ and $\mathbf{C}_B$ are given on the opposite page.

System analysis

$$\mathbf{C}_G = \begin{bmatrix} \frac{4}{3} & -\frac{2}{3} & -\frac{2}{3} & & & \\ -\frac{2}{3} & \frac{4}{3} & -\frac{2}{3} & & \mathbf{0} & \\ -\frac{2}{3} & -\frac{2}{3} & \frac{4}{3} & & & \\ & & & 1 & & \\ & \mathbf{0} & & & 1 & \\ & & & & & 1 \end{bmatrix} \qquad \mathbf{C}_B = \begin{bmatrix} 1 & 1 & 1 & & & \\ 1 & 1 & 1 & & \mathbf{0} & \\ 1 & 1 & 1 & & & \\ & & & & & \\ & \mathbf{0} & & & \mathbf{0} & \\ & & & & & \end{bmatrix}$$

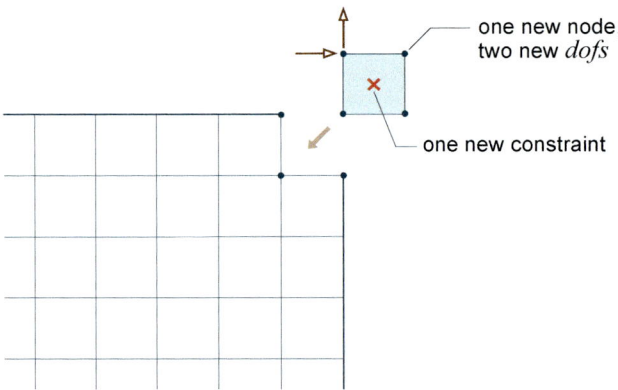

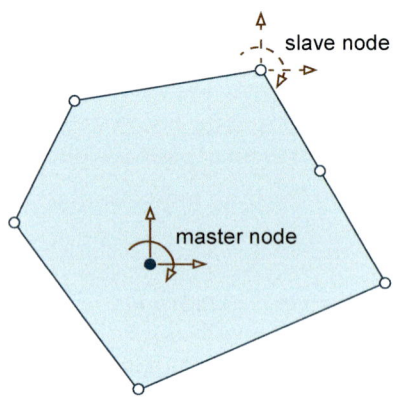

Figure 9.18 Rigid 2D element

The element stiffness matrix defined by Eq. (4-54) can now be rewritten as

$$\mathbf{k} = G \int_{V_e} \mathbf{B}^T \mathbf{C}_G \mathbf{B} dV + B \int_{V_e} \mathbf{B}^T \mathbf{C}_B \mathbf{B} dV \qquad (9\text{-}53)$$

and the system stiffness relation becomes

$$(G\mathbf{K}_G + B\mathbf{K}_B)\mathbf{r} = \mathbf{R} \qquad (9\text{-}54)$$

When ν approaches 0,5, B approaches infinity and $B\mathbf{K}_B$ will serve as a penalty matrix enforcing the incompressibility condition. The closer ν comes to 0,5 the greater the risk of numerical problems, and eventually the model will "lock", unless $\mathbf{K}_B$ is singular.

Matrix $\mathbf{C}_B$ is a *rank* 1 matrix, and in the case of numerical integration each integration point will therefore enforce *one* constraint on the system. In order for $\mathbf{K}_B$ to be singular the total sum of integration points must therefore be less than the number of equations in (9-54).

For a 4-node bilinear quadrilateral element in plane strain analysis, every new element will, on average, bring in two new equations, but only *one* constraint if we use a selective reduced integration scheme with only one (1×1) integration point for each $\mathbf{k}_B$. This will provide a singular $\mathbf{K}_B$, and we can get away with a POISSON ratio very close to 0,5.

In Chapter 14, which deals with bending of beams and plates, we shall encounter a similar locking phenomenon, so-called *shear locking*, in which the shear contribution to the element stiffness plays the role of a penalty matrix.

9.8 Boundary conditions – rigid elements

Instead of modelling something that is very stiff with elastic elements with large stiffness, it may be a better solution to represent this by completely *rigid* parts in the model. Numerically it is definitely a better solution, and the "physical" consequences of such approximations are usually negligible.

In some cases these *rigid links* may be implemented as integrated parts of the elements. For instance, beam elements with rigid "arms" at the ends are well suited for modelling eccentricities.

Rigid parts can also be modelled by completely *rigid elements*. Figure 9.18 shows an example of a two-dimensional rigid element (to be used with plane beam elements). This element may have an arbitrary geometric shape and an arbitrary number of "nodal points". The element has *three* degrees of freedom which are conveniently associated with (any) one of the element's nodal points, the *master* node; all other nodes are *slave* nodes. The rigid element has *no stiffness*. Its only purpose is to automatically generate constraint equations of the form given by Eq. (9-11).

In three dimensions, rigid elements may be rigid in only *one plane* (preferably a plane defined by two global reference axes), or it may be a rigid volume of

System analysis

COMMENT

In a paper entitled *The Top 10 Computational Methods of the 20th Century*, published in **IACM expressions** (No 11, 2001), Dan GIVOLI has FEM on top. In 2nd place he lists "Iterative Linear Algebraic Solvers", and part of his argument is the following statement. "It is well known that direct solution methods like Gaussian Elimination are effective only for small and moderately-large systems, whereas very large systems (say, of dimension larger than 10,000) must be solved iteratively."

It is interesting then to note that now, about ten years later, a direct solver based on a *sparse* storage format, solves a (square) 2D plate problem with about 370,000 unknowns in less than 13 seconds on a standard PC, without any indication of numerical degradation in the results. More than anything, this says something about the incredible development we now witness in digital computing. It has certainly changed the meaning of "a very large system". It also says something about how well *conditioned* most of our problems are.

any shape. In the latter case the element will have 6 *dofs*, and it will generate constraint equations where each slave *dof* is constrained to the 6 master *dofs*.

Whether 2 or 3D, the geometric shape of the element is irrelevant; it is defined by its nodal points and their coordinates.

9.9 Solution of Kr = R

We are now ready to solve the system stiffness relation with respect to the unknown nodal displacements **r**. This is a numerical operation, but it plays such an important role in all FEM analyses that it seems appropriate to include a brief overview. In most cases it accounts for the bulk of the computational effort.

In neutral notation our problem is to solve

$$\mathbf{A}\mathbf{x} = \mathbf{b} \tag{9-55}$$

with respect to **x**; **A** is a square, regular matrix. Two different approaches are available: the *direct* approach and the *iterative* approach. For some special problems, and for very large problems, iterative solvers are probably choice number one; see the comment on the opposite page. On the other hand, the direct solver has many advantages, the most important being its versatility. And as noted opposite, it can be used effectively for quite large problems. We therefore limit this discussion to direct solvers.

Having made this choice we then have another choice to make: a standard three step GAUSSIAN elimination procedure or the *frontal solution* technique. The latter, pioneered by IRONS [35], is particular in the sense that it "mixes" assembly and solution. Unknowns are eliminated as soon as they have received all stiffness contributions, that is, as soon as all elements contributing stiffness to a particular *dof* (or node) have been assembled. Versatility favours a procedure where assembly and solution are two distinct and separate operations. Our solution method of choice is therefore the standard three step GAUSSIAN elimination procedure:

1) *Factorize* or decompose the coefficient matrix **A** into its two *triangular* factors, **L** and **U**, that is

$$\mathbf{A} = \mathbf{L}\mathbf{U} \tag{9-56}$$

where **L** is a *lower triangular* matrix and **U** is an *upper triangular* matrix; it turns out that we can choose the diagonal elements in one of the two factors; in standard GAUSSIAN elimination it is usual to set all $L_{ii} = 1,0$.

2) *Forward substitution* (also called forward reduction) – by introducing

$$\mathbf{U}\mathbf{x} = \mathbf{z} \tag{9-57}$$

Eq. (9-55) can be written as

$$\mathbf{L}\mathbf{U}\mathbf{x} = \mathbf{L}\mathbf{z} = \mathbf{b} \quad \Rightarrow \quad \mathbf{z} = \mathbf{L}^{-1}\mathbf{b} \tag{9-58}$$

System analysis

1) Factorization

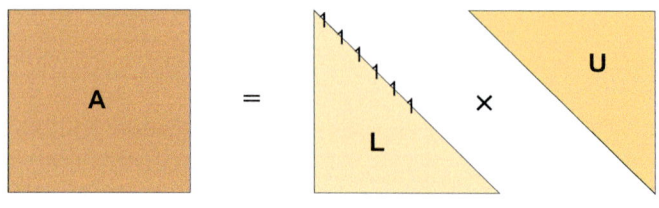

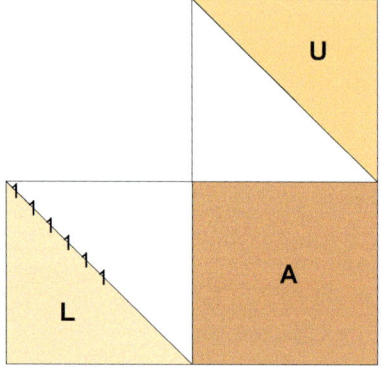

2) Forward substitution

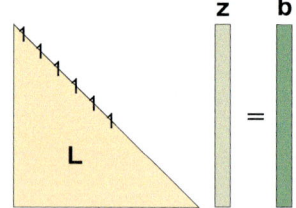

3) Backsubstitution

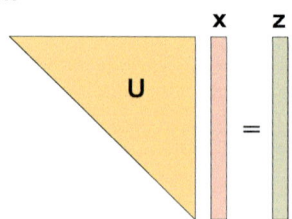

This represents a very simple system of equations, and the unknowns can be determined directly, one by one, starting from the top with z_1 as the only unknown and then proceeding down to the second equation where z_2 is now the only unknown, and so on.

3) *Backsubstitution*; with **z** being known, we find **x** as

$$\mathbf{Ux} = \mathbf{z} \quad \Rightarrow \quad \mathbf{x} = \mathbf{U}^{-1}\mathbf{z} \tag{9-59}$$

Again we have a very simple system for which we can determine the unknowns x_i, one by one, from the "bottom" up.

During factorization, which is the computationally most costly operation, the elements of **L** and **U** are stored "on top of" the corresponding elements in **A**; in other words we need no extra storage, unless we would like to "save" a copy of **A**. Note that the unit elements on the diagonal of **L** need not be stored. Similarly, the elements of **b** will first be replaced by the elements of **z** (during step 2) and then by the solution **x** (during step 3).

It should be noted that factorization is *independent* of the right-hand side **b**. For a linear problem, where **A** (= **K**) is independent of **x** (= **r**), we only need to factorize *once*, while the much "cheaper" operations in step 2 and 3 can be repeated for as many right-hand sides (load vectors) as we might wish to solve for.

Back to **K**, or $\mathbf{K}^*_{ff}$, see Eq. (9-25). It has already been pointed out that the matrix has some very important properties that the solution should take advantage of:

- *symmetry*, $\mathbf{K} = \mathbf{K}^T$
- *positive definite*, $\mathbf{x}^T\mathbf{Kx} > 0$ for all $\mathbf{x} \neq \mathbf{0}$
- *sparse* (many zero elements)

Factorization of a *symmetric* matrix can be expressed as

$$\mathbf{K} = \mathbf{LDL}^T \tag{9-60}$$

where **L**, as before, is a lower triangular matrix with unit elements on the diagonal, and **D** is a *diagonal* matrix. Comparing Eqs. (9-60) and (9-56) we find that

$$\mathbf{U} = \mathbf{DL}^T \tag{9-61}$$

for a symmetric coefficient matrix. Forward substitution is exactly as before, whereas backsubstitution will have to be slightly modified:

$$\mathbf{DL}^T\mathbf{x} = \mathbf{z} \quad \Rightarrow \quad \mathbf{x} = \mathbf{L}^{-T}(\mathbf{D}^{-1}\mathbf{z}) \tag{9-62}$$

That **K** is also *positive definite* implies that *all* elements of **D** are *positive* numbers (> 0), which means that

$$\mathbf{D} = \mathbf{D}^{1/2}\mathbf{D}^{1/2} \tag{9-63}$$

Each element in $\mathbf{D}^{1/2}$ is the square root of the corresponding element in **D**. In this case we can write (9-60) as,

System analysis

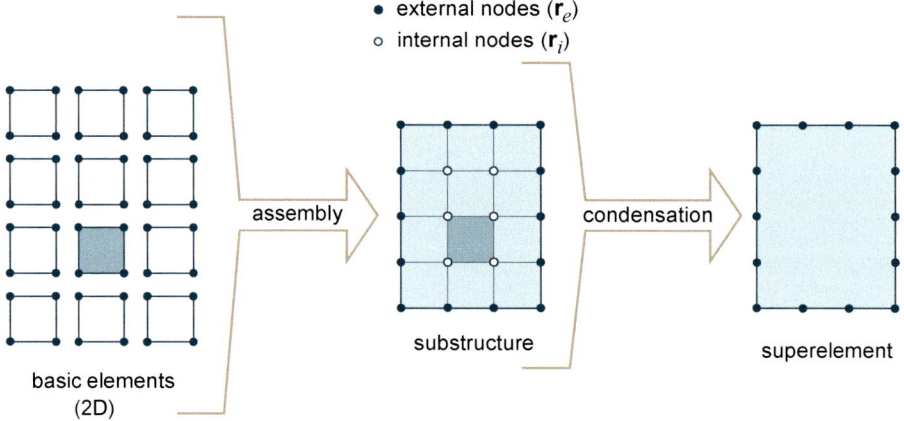

Figure 9.19 Substructure analysis

> A **basic element** is an element that cannot be broken down into smaller units; it is derived from assumed displacement (shape) functions and the application of a mechanical/mathematical principle.

$$\mathbf{K} = \mathbf{U}^T \mathbf{U} \tag{9-64}$$

This special factorization is called CHOLESKY decomposition, after the French military officer and mathematician André-Louis CHOLESKY. It has certain advantages, but the factorization of (9-60), which does not require $\mathbf{K}$ to be positive definite, and which is almost as efficient as (9-64), is more general and has a wider range of application, *e.g.* in *eigenvalue* problems. A number of solvers therefore use (9-60) instead of (9-64), and (9-60) is the preferred choice for our "number crunching work horse".

The last property of $\mathbf{K}$, its *sparseness*, was dealt with in Section 9.2.

9.10 Static condensation and substructure analysis

For very large structural problems a complete FEM model may lead to so many *dofs* that it may be convenient and even necessary to divide the model into subregions or *substructures*, which are partially solved separately. In order to explain this we consider the schematic problem of Fig. 9.19.

A series of simple *basic* elements are assembled to form a *substructure*. The degrees of freedom of the substructure are split into two groups: the *external dofs* ($\mathbf{r}_e$) and the *internal dofs* ($\mathbf{r}_i$). The internal *dofs* are associated with those nodal points that are *directly* connected *only* with elements that belong to the substructure, while the external *dofs* are also connected to elements "outside" the substructure. With this arrangement of the substructure's *dofs* we can write the substructure's stiffness relation as,

$$\begin{bmatrix} \mathbf{K}_{ii} & \mathbf{K}_{ie} \\ \mathbf{K}_{ei} & \mathbf{K}_{ee} \end{bmatrix} \begin{bmatrix} \mathbf{r}_i \\ \mathbf{r}_e \end{bmatrix} = \begin{bmatrix} \mathbf{R}_i \\ \mathbf{R}_e \end{bmatrix} \quad \text{where} \quad \mathbf{K}_{ei} = \mathbf{K}_{ie}^T \tag{9-65}$$

This matrix equation can be written as two matrix equations,

$$\mathbf{K}_{ii}\mathbf{r}_i + \mathbf{K}_{ie}\mathbf{r}_e = \mathbf{R}_i \tag{9-66a}$$

and

$$\mathbf{K}_{ie}^T \mathbf{r}_i + \mathbf{K}_{ee} \mathbf{r}_e = \mathbf{R}_e \tag{9-66b}$$

Equation (9-66a) solved with respect to $\mathbf{r}_i$ yields

$$\mathbf{r}_i = -\mathbf{K}_{ii}^{-1}\mathbf{K}_{ie}\mathbf{r}_e + \mathbf{K}_{ii}^{-1}\mathbf{R}_i \tag{9-67}$$

If we substitute this into (9-66b) and rearrange terms we obtain the following equation in terms of $\mathbf{r}_e$:

$$(\mathbf{K}_{ee} - \mathbf{K}_{ie}^T \mathbf{K}_{ii}^{-1} \mathbf{K}_{ie})\mathbf{r}_e = \mathbf{R}_e - \mathbf{K}_{ie}^T \mathbf{K}_{ii}^{-1} \mathbf{R}_i \quad \text{or} \quad \mathbf{K}_{ee}^r \mathbf{r}_e = \mathbf{R}_e^r \tag{9-68}$$

where the "reduced" matrices are defined as,

$$\mathbf{K}_{ee}^r = \mathbf{K}_{ee} - \mathbf{K}_{ie}^T \mathbf{K}_{ii}^{-1} \mathbf{K}_{ie} \tag{9-69}$$

and

$$\mathbf{R}_e^r = \mathbf{R}_e - \mathbf{K}_{ie}^T \mathbf{K}_{ii}^{-1} \mathbf{R}_i \tag{9-70}$$

System analysis

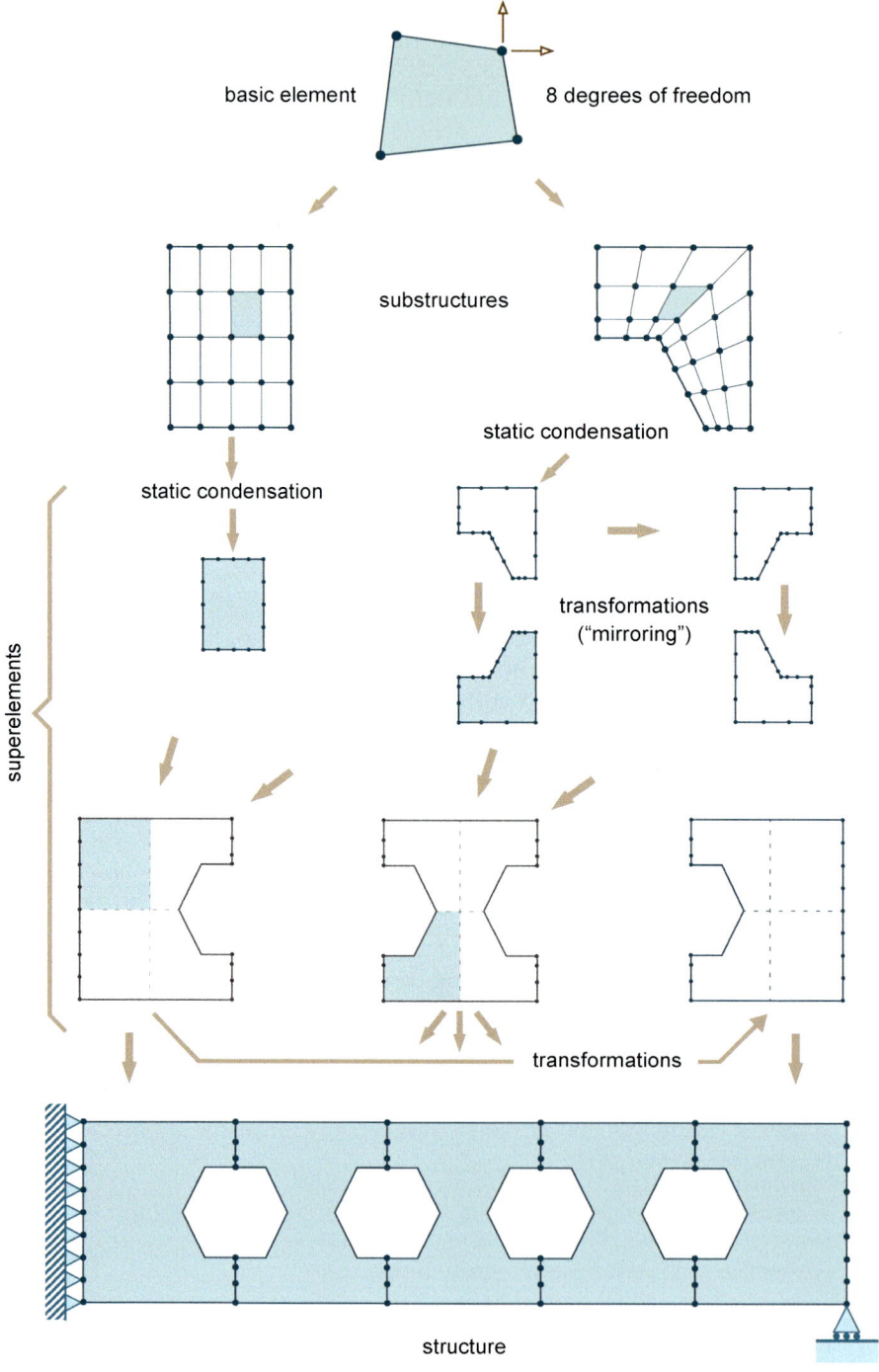

Figure 9.20 Example of substructure analysis

Static condensation and substructure analysis

If the substructure had been the entire structure, we could introduce the boundary conditions into Eq. (9-68), solve it with respect to r_e, and with r_e known we could retrieve r_i from (9-76). Such a "two-step" solution, which does *not* introduce any new approximations, will not normally pay off, since we must expect the reduced stiffness matrix K_{ee}^r to be "full" (it has lost its band or profile structure).

However, if we go back to Fig. 9.19 we see that if we eliminate, or "condense out", the internal *dofs*, this procedure generates a "composite" element, a so-called *superelement*. If we split the reduced load vector into a known part, R_e^k, and an unknown part, R_e^u, that is

$$R_e^r = R_e^k + R_e^u \qquad (9\text{-}71)$$

then we can write Eq. (9-68) as

$$S_{sup} = k_{sup} v_{sup} + S_{sup}^0 \qquad (9\text{-}72)$$

where

$$k_{sup} = K_{ee}^r = K_{ee} - K_{ie}^T K_{ii}^{-1} K_{ie} \qquad (9\text{-}73)$$

and

$$S_{sup}^0 = -R_e^k = K_{ie}^T K_{ii}^{-1} R_i - R_e \qquad (9\text{-}74)$$

are the stiffness matrix and load vector, respectively, of a superelement whose degrees of freedom are $v_{sup} = r_e$. Equation (9-71) indicates that R_e^r must be interpreted as the sum of a known part R_e^k, which comes from the loading on the substructure, and an unknown part R_e^u, that is the nodal point forces S_{sup} of the superelement.

This process, mathematically expressed by Eqs. (9-73) and (9-74), is termed *static condensation*. As already stated, it introduces *no* new approximations.

Substructure analysis and superelements are particularly useful in cases where we have several/many identical substructures, preferably with "weak" couplings (few *dofs* in r_e). The more we can repeat the use of a superelement, the greater the saving, not only in computer time, but, equally important in the amount of data preparation.

Once a superelement has been "created" it is in every respect a perfectly "normal" finite element; it has a slightly different history from the elements we have termed as "basic"; that is all. It can be used, together with other superelements and/or basic elements, to form new substructures leading to "higher level" superelements. In principle there is no limit to the number of levels. Figure 9.20 shows a schematic example of a substructure analysis. The structure itself is modelled by 5 "level 2" superelements totalling 81 unknown *dofs* (two at each node); if we had used the same number of basic elements directly, without substructure analysis, the model would have contained 967 *dofs*.

For some very large problems substructure analysis is perhaps the only practical way to obtain a fairly detailed solution. An example of this, dating back more than 20 years, is a so-called *global* analysis of a large off-shore gravity (concrete) platform of the **Condeep** type. The final structural model was made up of level 8 superelements, and this model would have comprised

System analysis

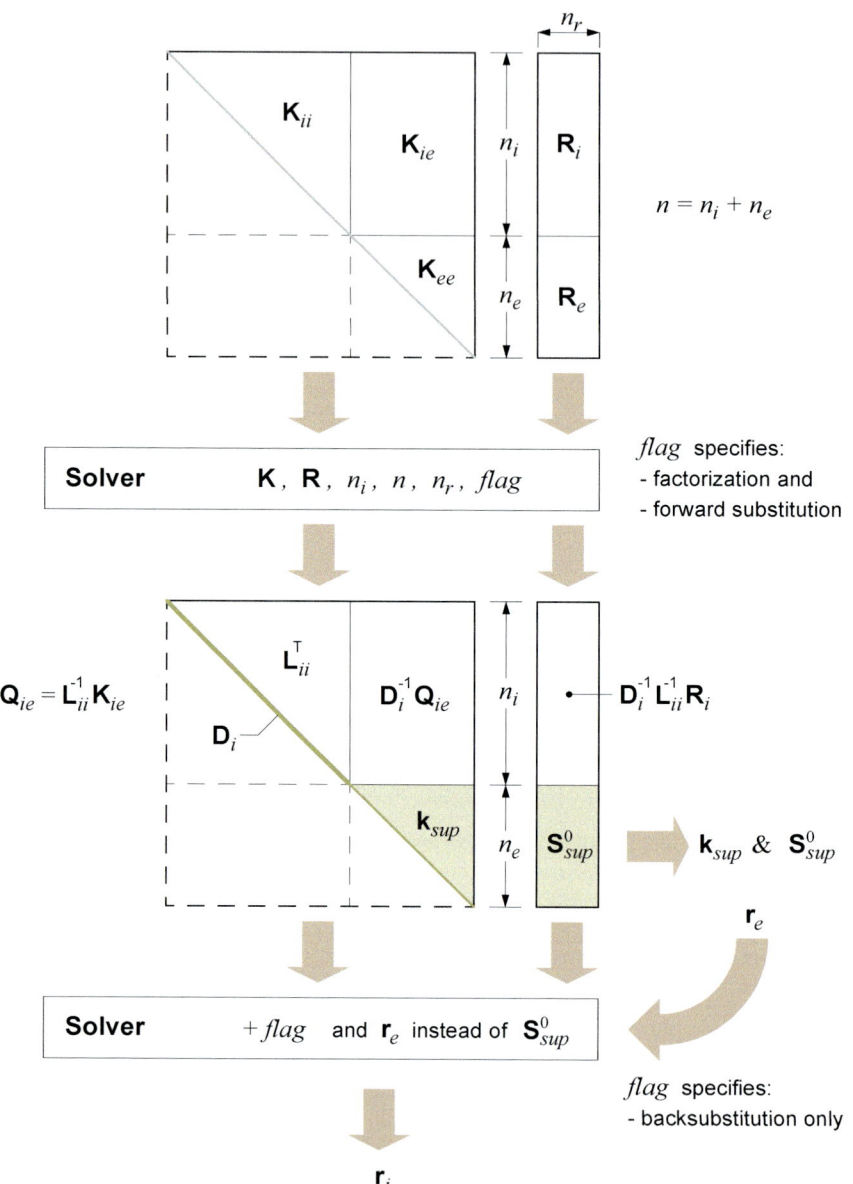

Figure 9.21 Implementation of superelement analysis

well over a million degrees of freedom if it had been analysed with just basic elements – an impossible task with the hardware of that time. Even today, the amount of data involved with these mammoth models suggests a substructure type approach.

The upside of the substructure/superelement approach literally manifests itself on the "way up", that is where one can take advantage of repeating superelements, with great savings in both storage and computing time. By the same token, the downside of the approach is experienced on the "way down" again. Superelements that, in a stiffness sense, were identical on the way up (perhaps save for a simple transformation) will, due to different responses, have to be handled independently on the way down. Hence the amount of data tends to explode on the way down, during the *retracking* phase. This represents a real "book-keeping" challenge.

Before we leave this section let us take a closer look at Eqs. (9-73) and (9-74). At first sight it looks as if the determination of $\mathbf{k}_{sup}$ and $\mathbf{S}_{sup}^0$ requires a considerable amount of work, involving both matrix inversion and matrix multiplication. However, on closer inspection this is not the case, and with a flexible equation solver that can handle any combination of the three basic steps of elimination, $\mathbf{k}_{sup}$ and $\mathbf{S}_{sup}^0$ may be determined quite efficiently. We assume a decomposition of the form defined by Eq. (9-60), that is,

$$\mathbf{K}_{ii} = \mathbf{L}_{ii}\mathbf{D}_i\mathbf{L}_{ii}^T \tag{9-75}$$

Introducing $\mathbf{Q}_{ie} = \mathbf{L}_{ii}^{-1}\mathbf{K}_{ie}$ (forward substitution) and noting that

$$\mathbf{K}_{ii}^{-1} = (\mathbf{L}_{ii}\mathbf{D}_i\mathbf{L}_{ii}^T)^{-1} = \mathbf{L}_{ii}^{-T}\mathbf{D}_i^{-1}\mathbf{L}_{ii}^{-1}$$

we may write

$$\mathbf{k}_{sup} = \mathbf{K}_{ee}^r = \mathbf{K}_{ee} - \mathbf{K}_{ie}^T\mathbf{L}_{ii}^{-T}\mathbf{D}_i^{-1}\mathbf{L}_{ii}^{-1}\mathbf{K}_{ie} = \mathbf{K}_{ee} - \mathbf{Q}_{ie}^T\mathbf{D}_i^{-1}\mathbf{Q}_{ie} \tag{9-76}$$

We now assume that our equation solver can perform any one of the three basic steps of solution; a combination of steps 1 and 2 or of steps 2 and 3, or of all three steps in one reference to the solver. The mode of operation is determined by the value of a *flag* (an integer parameter).

With reference to Fig. 9.21, we first "ask" the solver to perform a "truncated" factorization and forward substitution; truncated in the sense that it "loops" from 1 to n_i (instead of from 1 to n in the case of a full solution). On exit from this first reference $\mathbf{K}_{ee}$ and $\mathbf{R}_e$ have been replaced by $\mathbf{k}_{sup}$ and $\mathbf{S}_{sup}^0$, respectively. Also, once $\mathbf{r}_e$ (= $\mathbf{v}_{sup}$) has been determined, $\mathbf{r}_i$ can be recovered by a new reference to the solver. This time the solver is instructed (via *flag*) to perform a truncated backsubstitution (from n_i to 1). The input information to the solver, in $\mathbf{K}$ and $\mathbf{R}$, must be exactly the same information as the previous reference returned, except that $\mathbf{S}_{sup}^0$ is replaced by $\mathbf{r}_e$. The internal *dofs* ($\mathbf{r}_i$) come back to the calling program in the storage locations occupied by $\mathbf{D}_{ii}^{-1}\mathbf{L}_{ii}^{-1}\mathbf{R}_i$ on input.

Hence, $\mathbf{k}_{sup}$, $\mathbf{S}_{sup}^0$ and $\mathbf{r}_i$ are all determined quite efficiently by factorization and substitutions; no matrix inversion and no matrix multiplication.

System analysis

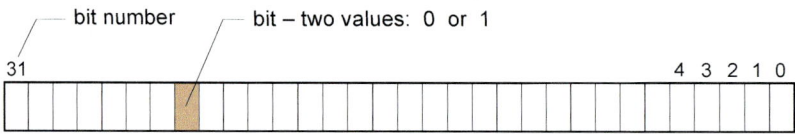

a) 32-bit "word" (addressable unit)

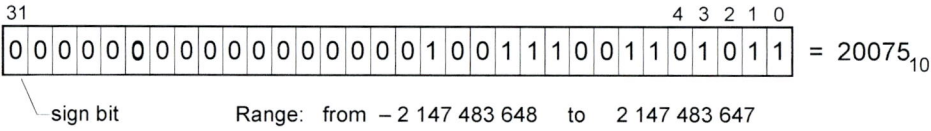

Range: from − 2 147 483 648 to 2 147 483 647

b) Binary representation of an integer number

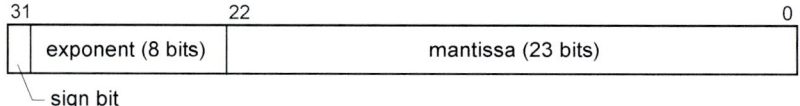

c) Representation of a floating point number (single precision)

Figure 9.22 Examples of number representation in 32-bit architecture

A *real* square matrix **A** is *normal* if $\mathbf{A}^T\mathbf{A} = \mathbf{A}\mathbf{A}^T$

9.11 Numerical issues

The system stiffness matrix **K** for a correctly formulated linear static problem, when adjusted for boundary conditions, is a *regular* (non-singular) matrix with a unique inverse. As the matrix of coefficients in a system of simultaneous, algebraic equations it will therefore provide a unique solution. This statement is basically true only if all operations during the solution process are carried out mathematically exactly. In actual computations this is not the case since *real* (or *floating point*) *numbers* are represented in the computer with limited *precision* (a limited number of significant digits). Errors due to *truncations* and/or *rounding* may therefore cause "numerical problems" during the solution process, and these problems, depending on various circumstances, may pollute the solution so as to render it useless or they may even cause it to break down completely.

We shall not dig deeply into this matter, but since it is an important part of numerical computations it warrants some comments. First of all, let us have a look at how the computer (in principle) stores data, and we distinguish between integer and real numbers. In so-called 32 bit architecture the basic, addressable unit or data *word* is 32 bits wide, see Fig. 9.22a. Each *bit* can be in one of two states, on or off, corresponding to the two "numbers" of the *binary* number system, 0 and 1. Most computers represent all numbers in binary form, and Fig. 9.22b shows how the *integer number* 20075 can be represented in a "word".

A real or *floating point number* is represented in *normalized* form as

$$r = \pm m \times b^e \tag{9-77}$$

where m is the fraction *mantissa*, b is the *base* (=10 in the decimal case, and 2 in the binary case) and e is the *exponent*. Figure 9.22c shows an example of how a real number can be stored (in binary form) in a 32-bit word. With the number of bits allocated to the exponent and mantissa we can represent floating point numbers with almost 7 significant decimal digits, and the range is from 10^{-127} to 10^{127}. This is so-called *single precision* representation, and for practical FEM computations it is insufficient. *Double precision* representation, in which *two* words are used to represent *one* floating point number, is therefore used almost exclusively in FEM computations. The entire extra word is usually allocated to the mantissa, which means the number of significant decimal digits is increased to somewhere between 16 and 17.

A measure of the "numerical stability" of a matrix is its *condition number*, which for a *normal* matrix is defined as:

$$\kappa = |\lambda_{max}/\lambda_{min}| \tag{9-78}$$

where λ_{max} and λ_{min} designate the largest and smallest *eigenvalues* of the matrix. A matrix with a low condition number is said to be *well-conditioned*, while a high condition number is an indication of an *ill-conditioned* matrix. Loosely speaking, if the condition number $\kappa(\mathbf{K}) = 10^k$, then we may loose up to k digits of accuracy, in addition to that which would be lost due rounding and truncations.

System analysis

The *sum of the eigenvalues* of a real, symmetric matrix **A** is equal to the *trace* of the matrix, that is

$$\text{tr}(\mathbf{A}) = A_{11} + A_{22} + A_{33} + \ldots + A_{nn} = \lambda_1 + \lambda_2 + \lambda_3 + \ldots + \lambda_n$$

Pivoting

In the factorization phase of GAUSSIAN elimination the *pivot* or *pivot element* is the *first* diagonal element of the "remaining" (non-factored) part of the matrix. For best numerical stability, the pivot should be as large as possible during the entire factorization. The strategy of picking the best pivot element is called *pivoting*. If the next pivot is not the element already in the correct position then rows and/or columns will have to be changed (swapped) or kept track of in some other way.

In our case symmetry must be maintained by the pivoting strategy, and also the band- or profile-structure must be considered. All in all this complicates the solution algorithm significantly, and it is a task we do not undertake unless absolutely necessary. Luckily that is not often the case.

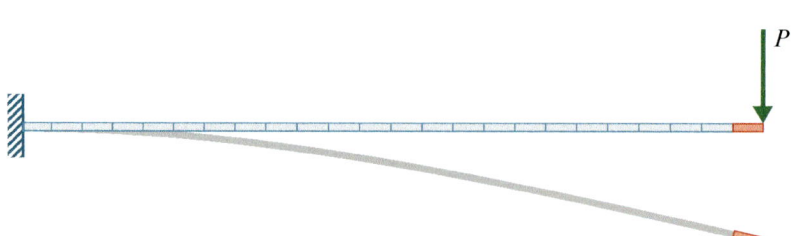

Figure 9.23 Long cantilever beam – large rigid body movement for tip element

The exact determination of κ, requiring both λ_{max} and λ_{min} to be computed, is very costly and it is seldom, if ever, undertaken unless it comes as a by-product of other computations. However for practical purposes one does not need κ to be known with great accuracy; the order of magnitude is usually sufficient. The *trace* of the matrix, tr(**K**), which is readily available, is considered a good estimate of λ_{max}. Unfortunately we do not have an equally simple and good estimate for λ_{min}. A sometimes quoted estimate for λ_{min} is the (numerically) smallest element, $|D_{ii}|_{min}$, of the diagonal matrix **D** in the decomposition of Eq. (9-60), and in [36] it is shown that while

$$\kappa_{approx} = \text{tr}(\mathbf{K})/|D_{ii}|_{min} \tag{9-79}$$

tends to overestimate the correct value, it is nevertheless near enough for practical use. The condition number, which several FEM programs will (on request) provide, is available *after* the solution and is therefore an *a posteriori* "measure" of the quality of the solution.

Perhaps a more useful test employed by many solvers is to monitor the so-called *diagonal decay*. As the solver computes the element D_{ii}, the value is compared with the corresponding element of **K**, that is K_{ii}, and if D_{ii} is much smaller than K_{ii} we may have a problem. If

$$|D_{ii}| < |\varepsilon \cdot K_{ii}| \tag{9-80}$$

where ε is a predefined small number, **K** is defined as *singular* or *near-singular*, and computations are aborted with an accompanying error message. If we express the test quantity ε as

$$\varepsilon = 10^{-r} \tag{9-81}$$

where r is a positive integer number, then as a general rule of thumb Eq. (9-80) indicates that we have "lost" roughly r significant (decimal) digits during factorization so far. In double precision a value of r in the range 8 to 10 is not uncommon. It should be noted that the test in Eq. (9-80) is more severe the larger ε is (or smaller r is), and the test is practically useless for a very small value of ε (r approaching the number of significant digits).

In general our system stiffness matrices are well conditioned, and in most cases we need not bother about *pivoting* strategies during factorization. The most common *physical* reason for ill-conditioning is a model including both very stiff and very flexible elements. Large rigid body movements of elements compared with their deformation is another indication of potential problems. Continuing uniform subdivision of the long slender cantilever beam in Fig. 9.23 will eventually cause the factorization of **K** to break down. The deformation of the tip elements will "disappear" in comparison to their rigid body motions, with apparent linear dependency as a result.

The *residual* forces represent a good indication of the quality of the computed results. At unconstrained nodes they should be zero, and at supported nodes the sum of the reaction forces in a particular direction should balance the sum of the loading in this direction.

System analysis

 (a) quadrilateral elements

 (b) triangular elements

Figure 9.24

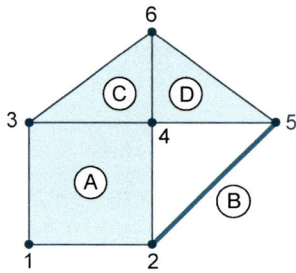

Figure 9.25

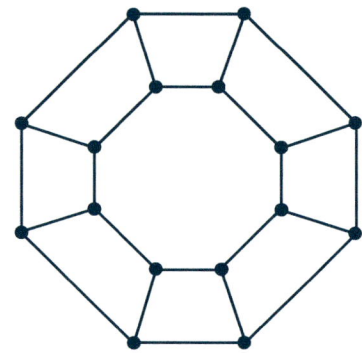

Figure 9.26

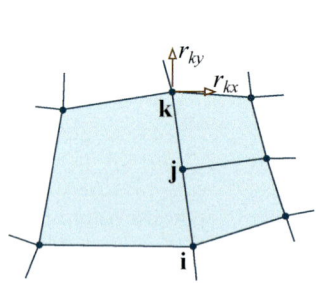

Figure 9.27

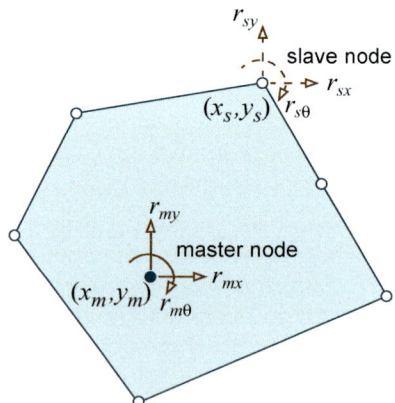

Figure 9.28

Problem 9.1

Figure 9.24 shows a strip of a 2D plate modelled by (a) quadrilateral elements and (b) triangular elements. Sketch the mesh in the zone of transition from a coarse mesh at the left end to a finer mesh at the right end; use elements with as regular shapes as possible over a transition zone of optional length.

Problem 9.2

Figure 9.25 shows a simple (and academic) assemblage of 4 finite elements, one quadrilateral, two triangles and one rod. For simplicity we assume only one degree of freedom at each of the six nodal points.

Make a 6 by 6 grid and indicate the contributions to the system stiffness matrix from each individual element by writing its letter in the appropriate "boxes" of the grid representing the stiffness matrix.

Problem 9.3

Figure 9.26 shows a structure of 16 one-dimensional (rod) elements interconnected at 16 nodal points. Assume for simplicity one *dof* at each nodal point and suggest how the nodal points should be numbered to give

a) the *smallest* semi-bandwidth b_w (see Fig. 9.2), and

b) the *smallest* profile length p_L (see Fig. 9.2).

State the corresponding numbers for b_w and p_L.

Problem 9.4

Figure 9.27 shows a patch of an element mesh where a coarse mesh of elements meets a finer mesh of the same type of elements. Nodal point *j* lies at the mid-point between *i* and *k*. Each nodal point has two *dofs*.

What do you need to do in order for this mesh to maintain displacement continuity along the element edges?

Problem 9.5

Figure 9.28 shows a *rigid* 2D element with one *master* node and several *slave* nodes. Each node, whether master or slave, has *three dofs*, two translations and one rotation.

Establish the constraint equation for each slave *dof* at a typical slave node, expressed in terms of the master *dofs* and the *x*- and *y*-coordinates of the master node and that particular slave node.

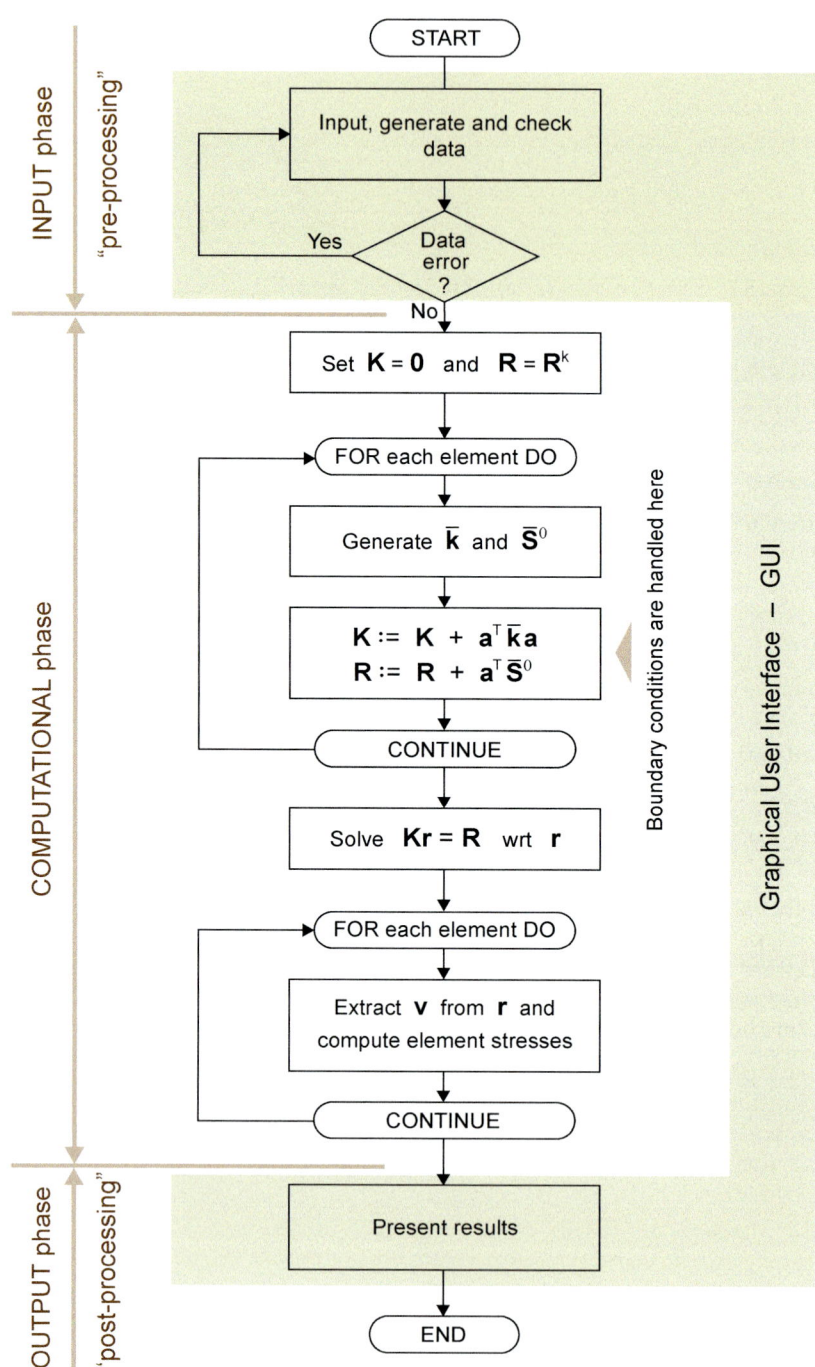

Figure 10.1 Flowchart diagram of a typical FEM analysis

10
Programming issues

Since FEM is so closely linked with the computer it seems reasonable to include some comments on how a typical finite element analysis can be programmed for execution on a digital computer. We shall only scratch the surface of this vast topic, and the approach we will comment on is but one of several approaches to the problem, and not necessarily the best. We will concentrate on the computational issues; the important part of getting the necessary information (data) into the program at the start, and the presentation of the results at the end, will not be dealt with.

At the end of the chapter we will briefly present a couple of locally developed FEM programs which we believe are fairly typical, and which are used to exemplify actual problems in several chapters of this book.

Figure 10.1 shows a simplified, graphical presentation of a typical FEM program. It is conveniently divided into three distinct phases: the input phase, the computational phase and the output phase.

In the early days (the 1960s and early 1970s) the input information was read from punched paper tape or cards, and the results were printed on line printers; boxes of cards and stacks of listing paper. Next the input data were read from prepared *files* (pre-processing) and results were written back into other files for post-processing.

As graphical devices became commonly available a more *interactive* mode of operation emerged, and today most program controlled FEM analyses are embedded in *graphical user interfaces* (GUI) which embrace both the "input" and the "output" phases, and through which the user can easily modify information, reanalyse and inspect the results in a seamless operation. For a typical, linear static analysis program the GUI, depending on its sophistication, may well account for the bulk of the total code.

However, our interest is in the "computational engine" and its development, and we shall only make brief and superfluous remarks on the GUI in the remainder of this chapter.

Programming issues

static memory allocation

1953 — John W. Backus
proposed to IBM to develop a more practical alternative to assembly language for programming their IBM 704 computer

1957 — First FORTRAN compiler

1958 — FORTRAN II

1962 — FORTRAN IV
(there was a FORTRAN III that never gained popularity)

1966 — FORTRAN 66
standardized version, by *American Standards Association*, of FORTRAN IV

1977 — FORTRAN 77
extensions to FORTRAN 66 by compiler vendors were standardized (by ANSI) – not a major revision, but important features (such as Block IF, ELSE and ELSEIF, direct-access file I/O and generic names for intrinsic functions) were made "legal".

dynamic memory allocation

1991 — Fortran 90 (ISO and ANSI standard)
a major revision that introduced such features as:
- free-form source input,
- inline comments,
- ability to operate on arrays,
- modules (instead of COMMON),
- derived/abstract data types,
- dynamic memory allocation, and
- POINTER attribute.

1995 — Fortran 95
a minor revision of Fortran 90

2003 — Fortran 2003
a major revision introducing many new features such as
- object-oriented programming support:
  e.g. type extension and inheritance, polymorphism,
- procedure pointers, and
- interoperability with C.

2010 — Fortran 2008 – a minor upgrade of Fortran 2003

Figure 10.2 A brief history of Fortran

10.1 Programming paradigms and languages

Although some primitive FEM programs were developed as early as in the late 1950s, it was not until FORTRAN IV entered the scene in 1962 that it really began.

The programming paradigm associated with Fortran is *procedural*, derived from *structured programming*. It is based upon the concept of the *procedure call*. Procedures, also known as routines, subroutines, methods or functions, simply contain a series of computational steps to be carried out. A typical feature of the procedural programming paradigm is a *main* program or a "master" procedure. Any given procedure may be called at any point during the execution of the main program (or the master procedure), including by other procedures (and even by itself if recursive calling is permitted).

Fortran and procedural programming dominated FEM programming in the 1960s, 1970s and the greater part of the 1980s. From a computer science point of view the early versions of Fortran had some serious shortcomings, but Fortran produced code that executed faster than its rivals. Considering the high price of the hardware in those days, this was a very convincing argument in favour of Fortran and it triggered a vast amount of investment in its use. To safeguard this investment standardization became a key issue for the development of Fortran, and this, probably more than anything else, can explain its longevity. Figure 10.2 shows how Fortran has evolved over the years, and it is now a language containing most features one expects to find in a "modern" programming language. A key point in the development of Fortran has been *backward compatibility*. Fortran code written back in the 1960s will still work (and it does)! Some ("purists") would say that this has kept code "alive" that should have been scrapped long ago, but in the real world this has undoubtedly been a success. The obvious downside to this strategy is the fact that the language has become large and cumbersome, and some would say that while Fortran used to be an easy language to learn and master, that is no longer the case (for Fortran 2008).

Another basically procedural programming language termed simply C, developed by RITCHIE and made available by KERNINGHAM and RITCHIE [38] in 1978, also gained followers, not least because of its close links with the popular operating system Unix. It was not until the release of C++, developed by STROUSTRUP [39] in 1983 as an enhancement to C, that we see the emergence of a new programming paradigm in FEM programming at the end of the 1980s, the *object oriented programming (OOP) paradigm*. Although object oriented programming languages had been around since DAHL and NYGAARD developed their Simula 67 [40], they were not picked up by the computational mechanics community until C++ started to gain popularity.

Fortran was developed for numerical computations (its name stems from "formula translation"), and it is in this area that it has its strength. Even the most die-hard Fortran enthusiast would not suggest Fortran as a candidate for implementation of a modern GUI. Object oriented languages like C++, and more recent languages like Java and C#, are very well suited for *event driven* programs like those making up an efficient GUI. And since they can also be used for the computational tasks – some would even say that they

Programming issues

Geometry	nodal point coordinates
	plate/shell thicknesses
	cross section properties
Material properties	elastic constants
	thermal properties
	mass density
Loading	concentrated (nodal point) loading
	distributed (element) loading
	load cases / load combinations
	initial strain (temperature)
Boundary conditions	suppressed *dofs* (= 0)
	prescribed *dofs* (= const. = 0)
	multi-point constraints (master-slave)
Element data	type information
	topology (connectivity)
Matrix data	K and R/r
Results	nodal point results
	element results

Figure 10.3 Data for a typical FEM program for linear structural analysis

also do this part of the job better than Fortran or C – we now see full fledged object oriented FEM programs emerging. However, there is still a very large amount of Fortran code out there, running FEM analyses, and even new developments use Fortran for the numerical operations on the large matrices.

A fairly recent development in computational science is *scripting*, a different programming *style* from the more traditional programming which is concerned with building (often large and monolithic) applications using Fortran, C, C++, Java or C#. Scripting means programming at a high and flexible abstraction level, not unlike computing environments such as Maple, Mathematica and Matlab. However, scripting offers more flexibility and it can also be used to "wrap" existing, well-tested and efficient, but perhaps inconvenient-to-use programs, with a modern scripting interface. *Python* is a general-purpose, high-level programming language claimed to be ideally suited for scripting. The author has no experience with scripting, but for those who might want to look into this, Ref. [41] may be useful.

The author's background started back in the 1960s and he is therefore not neutral or unbiased in the question of programming paradigms and languages. He has seen a lot of bickering about programming languages over the years, and for quite some time there was a mutual mistrust between the Fortran community and the computer scientists. This is now basically history. Today it is quite straightforward to mix different programming languages, and even paradigms, in the same program-development project, so why not use the language best suited for a particular part of the project? And best suited should not be thought of in an absolute sense. The use of available and well tested code may be a better choice than the development of new code, even if the new code could be made to run more efficiently.

The remainder of this chapter will assume Fortran, and not even the latest versions of the language. This will limit the discussion to a particular and perhaps even old-fashioned way of thinking, but the problems we shall address are general and they can easily be implemented in other programming languages.

10.2 Data structures and storage formats

Regardless of programming paradigm and programming language, the FEM program will need a considerable amount of information or data about the model, and this data will have to be organized into *data structures* that are suitable for the chosen programming environment. This is perhaps one of the most critical phases in the programming process.

Figure 10.3 gives an overview of the type of information most FEM programs need to consider. How to organize this information will depend not only on the programming paradigm and language, but also on available existing code – it is seldom such projects start from scratch. In any case, much of the information is associated with the nodal points and the elements, the natural "information carriers" in FEM programs. Without aiming for completeness we shall indicate how the information can be organized with the basic instrument offered by traditional Fortran for structuring data: the *array*. And in order to simplify communication with other programming

Programming issues

A standard **Fortran array** is a symbolic *name* which refers to a group of memory locations, all holding information of the *same* data type. Each individual memory location in an array is referenced by one or more *subscripts*, depending on the *dimension* of the array.
Example:

Mathematical notation   Fortran notation

$a_1, a_2, a_3, \ldots, a_n$ A(1), A(2), A(3), ..., A(n)

$b_{1,1}, b_{1,2}, a_{1,3}, \ldots, b_{m,n}$ B(1,1), B(1,2), B(1,3), ..., B(m,n)

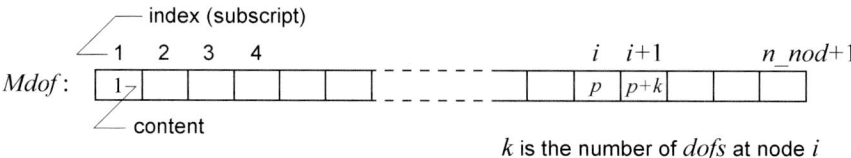

k is the number of *dofs* at node i

> **NOTE** – The following notation is adopted:
> The name of an *integer* array starts with **M**
> The name of a *real* array starts with **T**

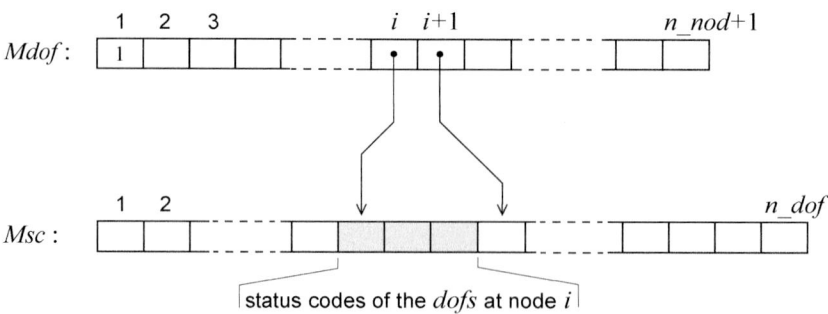

status codes of the *dofs* at node i

languages (used for instance by the GUI) we will only use simple one-dimensional arrays. Furthermore we represent *all* information in terms of two basic data *types*: *integer* numbers and *real* numbers.

To simplify our discussion, and with little loss of generality, we assume that our problem is a 2D plate problem. However our program should be able to handle different kinds and types of elements, and the nodal points can have different number of *dofs*. Within nodal points having the same *dofs*, the ordering of the *dofs* is fixed and implicitly defined (e.g. displacement w first, then slope $w_{,x}$ followed by slope $w_{,y}$).

The problem is governed by a few key *integer* variables such as

- *n_nod* - total number of nodes,
- *n_dof* - total number of *dofs*,
- *n_elm* - total number of elements
- *n_eqn* - total number of equations (dimension of $\mathbf{K} = \mathbf{K}_{ff}^*$)

The nodal points are numbered consecutively from 1 to *n_nod*, and the elements are numbered consecutively from 1 to *n_elm*.

The nodal point coordinates are stored in the *real* arrays

Tx(n_nod) where $x_i = Tx(i)$; *n_nod* is the dimension, and
Ty(n_nod) where $y_i = Ty(i)$.

The number of *dofs* at the individual nodes are stored implicitly in the *integer* array

Mdof (n_nod+1) where
Mdof (1) = 1 and
Mdof (i) = the number of *dofs* of the first i-1 nodes; in other words:
n = *Mdof (i+1)* − *Mdof (i)* is the number of *dofs* at node number *i*.

The *status* of each *dof*, e.g. free, suppressed, prescribed or slave, is recorded in the *integer* array

Msc(n_dof) where k = *Msc(i)* is the *status code* of *dof* number *i*.

The status code can, for instance, be

$k = 0$: suppressed *dof*
$k = 1$: free *dof*
$k = -indx$, where *indx* is the (*integer*) index in an array *Mcon_i* (see below) controlling the constraint equation information.

It should be noted that the entries of array *Mdof* serves as indices of array *Msc*. For instance

$$j = Msc(Mdof(i)+1)$$

is the status code of the 2nd *dof* of node number *i*.

We assume that, except for suppressed *dofs*, all specified *dofs* (prescribed and slaves) are treated as constraint equations; one constraint equation for each specified (non-suppressed) *dof*. With reference to the general constraint

Programming issues

Equation (9-11): $r_s = h_0 + h_1 r_{m1} + h_2 r_{m2} + \ldots + h_p r_{mp}$

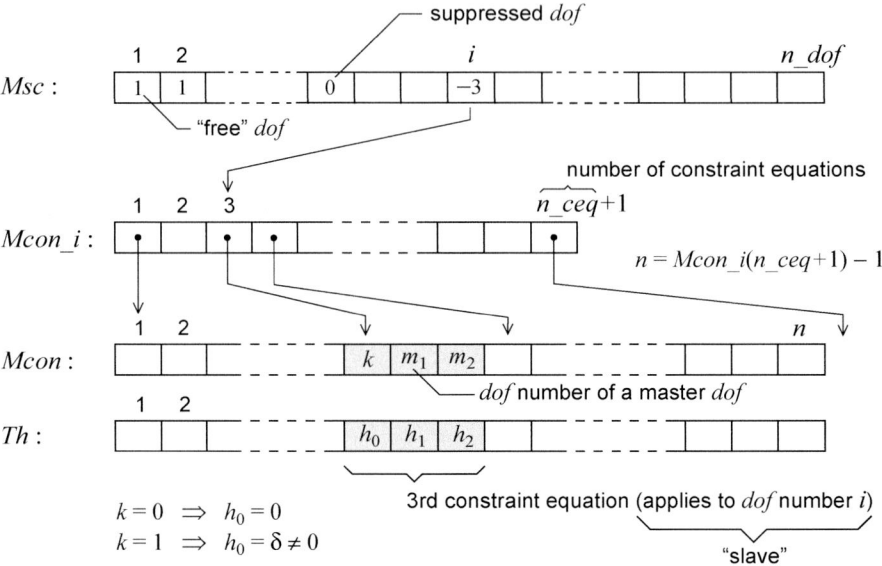

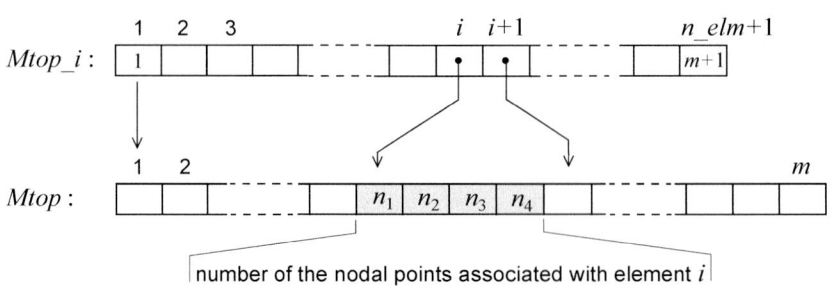

equation (9-11), a prescribed *dof* is defined by a constraint equation in which only one of the constants is different from zero, namely h_0.

The constraint equations may be recorded in three arrays, two integer arrays, *Mcon_i* and *Mcon*, and one real array, *Th*. The content of these arrays and their relationship with *Msc* are best explained by the illustration on the opposite page. We see that *Mcon_i*, *Mcon* and *Th* are only required if the problem has specified *dofs* other than suppressed *dofs*, that is if *n_ceq* > 0. If

$$Mcon\_i(i+1) - Mcon\_i(i) = 1$$

the *i*'th constraint equation defines a prescribed *dof*.

The vital element information is the "topology" or connectivity information. For maximum flexibility this information may be recorded in two *integer*, one-dimensional arrays, one array, *Mtop(m)*, containing for each element the number of the nodal points to which the element is "connected", and one index array, *Mtop_i(n_elm+1)*, "pointing" to where in *Mtop* information about a particular element starts. It is implied that the node numbers are ordered in a systematic manner (*e.g.* counter clockwise). The dimension *m* of *Mtop* is defined by the last entry of *Mtop_i*, that is (see opposite page),

$$m = Mtop\_i(n\_elm + 1) - 1$$

Other element information that needs to be available is concerned with:

> *Material properties* – this information can be organized in various ways; normally only a few material *types* are involved, and it is therefore quite common to record the parameters for the various types in a real array, and then provide an integer array with one entry for each element indicating its material type.
>
> *Plate thickness* – if the plate can have variable thickness, a simple and straightforward method may be to record the thickness at each nodal point in a real array with *n_nod* entries; however, a sudden change of thickness needs a different approach. If one-dimensional beam and/or bar elements are available objects in the modelling of the problem, it is necessary to record their cross section properties; a method similar to the one suggested for the material properties, that is a type approach, will normally work quite well.
>
> *Element types* – since the program can handle different types (and kinds) of elements, an integer array containing one entry for each element (in the form of a type number or code) is necessary.

The *loading* will need some thought. Depending on the types of loading and the combination of load patterns into load combinations, for which the program provides results, the loading may require several arrays, of both integer and real type. Following the thinking indicated above it is also fairly straightforward to design a workable data structure for the loading, the difficulty being to find a good compromise between generality and compactness. In our brief treatment we are primarily concerned with the stiffness, and we will therefore not pursue the loading any further.

The last type of information we shall look at in some detail is concerned with the storage format we choose for the system stiffness matrix. In Fig. 10.4 we

Programming issues

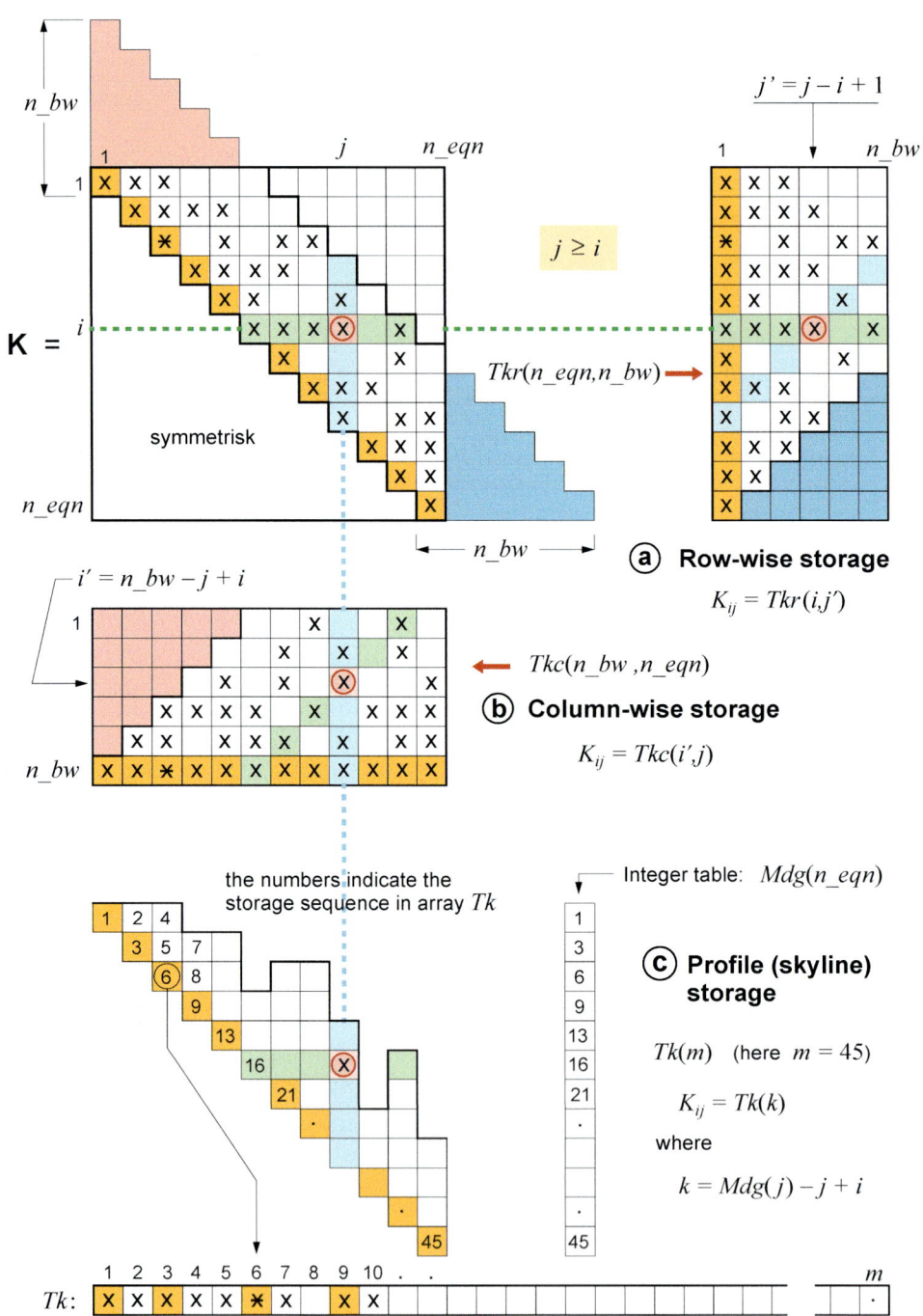

Figure 10.4 Stiffness matrix storage formats

have indicated two formats, the *band* format and the *profile* or skyline format. We choose the more efficient profile format, for which a storage scheme is indicated in the figure. The relevant part of the stiffness matrix, on and above the diagonal, is stored, column by column, in the one-dimensional real array *Tk*(*m*). In order to keep track of the elements in *Tk* we need an integer array *Mdg*(*n_eqn*) with one entry for each free (unknown) *dof* (or equation), and this entry contains the index in *Tk* of the diagonal element for that particular *dof*. Thus, the dimension of *Tk* is defined by the last entry in *Mdg*, that is, $m = Mdg(n\_eqn)$. An arbitrary element of the stiffness matrix, K_{ij}, is now easily located in *Tk* as

$$K_{ij} = Tk(k) \quad \text{where} \quad k = Mdg(j) - j + i$$

For the assembly process it is convenient to know the relationship between a particular *dof* and its position in **K**, that is, the row/column number of its diagonal position, or, in other words, its *equation* number. This information may be recorded in the array

Meqn(*n_dof*) where $j = Meqn(i)$ is

= the equation number of *dof* number *i* if this *dof* is a free *dof*,

= 0 if *dof* number *i* is suppressed, or

= −*p* if *dof* number *i* is the slave in a constraint equation;
  p is the index ("pointer") in array *Mcon_i*.

Node renumbering, if relevant, is conveniently carried out in a *pre-assembly* process. The "output" from the renumbering procedure may, for instance, be an integer array containing the "old" numbers corresponding to the new node numbers, *e.g.* entry number *i* of this array contains the old number of the node whose new number is *i*. With this array and arrays *Mdof* and *Msc*, array *Meqn* is easily determined. The nodes are inspected one by one in the "new" order, and "free" *dofs* at the node with the corresponding old number are assigned consecutive equation numbers, starting with number 1, in the sequence they are encountered. If, during this process, a zero status code is encountered in *Msc* for *dof* number *i*, *Meqn*(*i*) is also set to zero; if a negative number is encountered it is copied into *Meqn*(*i*).

Once *Meqn* has been established the temporary array with the information about the correspondence between new and old node numbers is no longer of any interest and can be used for other (scratch) purposes. It should be noted that the nodes retain their original ("old") numbers and so do the *dofs*. It is only the relationship between the *dof* numbers and the corresponding equation numbers that has changed due to the renumbering of the nodes.

In addition to checking input information for completeness and consistency, the pre-assembly process should also generate the profile definition array *Mdg*, see Fig. 10.4. This is accomplished using the information contained in

Mtop_i and *Mtop* − *Mdof* and *Meqn*, and − *Mcon_i* and *Mcon*.

This is about as much as we will say about data structures, and what we have said in this section may well be considered both cryptic (partly due to very short names) and old-fashioned. However, it makes for quite efficient

Programming issues

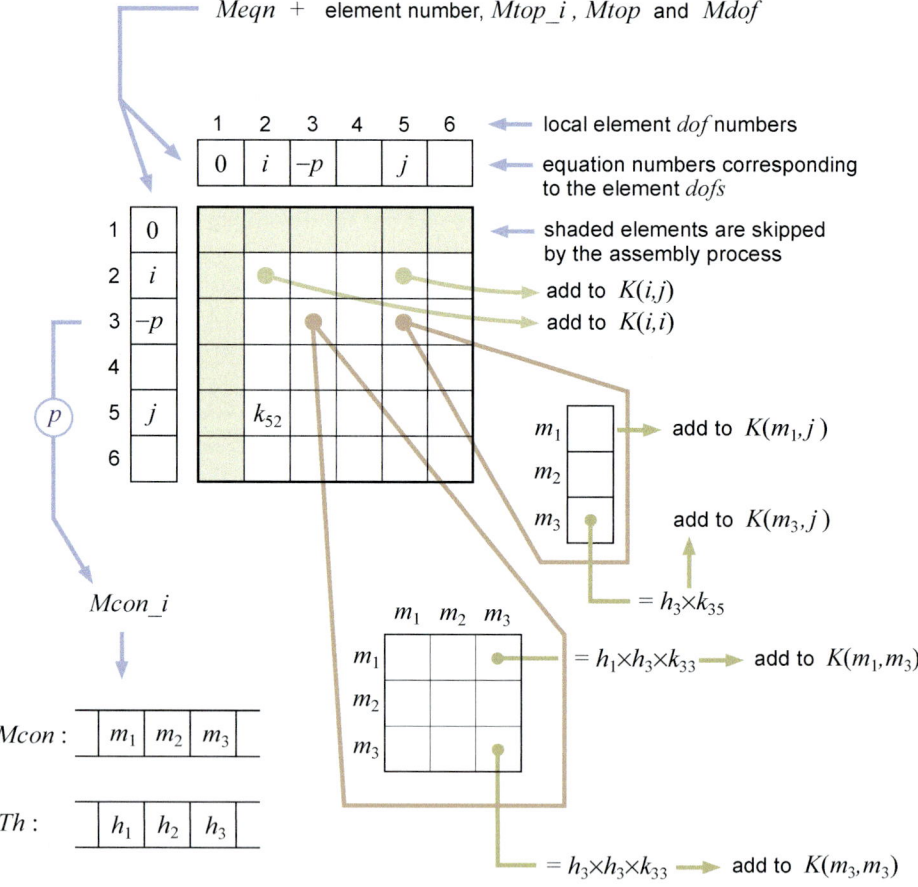

Figure 10.5 Accounting for a multi-point constraint during assembly of an element stiffness matrix

Data structures and storage formats

code, and most of the operations described can be programmed once and for all and made available as a set of general, library type subroutines.

The purpose of this section though is not to promote this type of approach to the programming of FEM analyses. Rather, we want to indicate what type of information we need to consider; how we structure this information depends on a number of factors, tradition and experience being two of them.

We conclude this section with an example of how the proposed data structure can be used to account for the boundary conditions during the assembly process.

Example 10-1 – Implementing multi-point constraints

Figure 10.5 shows, in principle, how a multipoint constraint can be accounted for while including, that is *adding*, the stiffness contribution from an arbitrary element to the system stiffness matrix **K**. The element has 6 *dofs*, and the constraint equation in question relates a *slave dof* (r_s) to three *master dofs* (r_{mi}). For simplicity we have omitted the constant term in the general constraint equation – in practice this term seldom appears together with master *dofs*; if present it usually appears alone, in which case it represents a *prescribed* displacement.

The process we describe here takes place in a loop over all elements in the finite element mesh. For this particular element we extract from the various arrays the information we need to generate the element stiffness matrix, such as nodal point coordinates, thickness and material properties. Via arrays *Mtop_i*, *Mtop* and *Mdof* we also extract from array *Meqn* the equation numbers associated with each individual element *dof*; this "table" is indicated along both the rows and the columns of the element stiffness matrix in Fig. 10.5. We see that one of the element *dofs* (the first) is suppressed at the system level (has equation number 0); we have also indicated two "free" *dofs* with equation number i and j, respectively, and one *dof*, with a negative equation number ($=-p$), is a *slave dof*.

All elements in **k** associated with the suppressed *dof* (which are shaded in the figure) are simply skipped in the process. If the equation numbers corresponding to both row and column index of **k** are positive numbers, that particular element of **k** is added to the current content of the corresponding element in **K**; for instance k_{22} is added to K_{ii} and k_{25} is added to K_{ij}, provided j is greater than i (is located above the diagonal); if not it is skipped.

But what about k_{33}, where does it end up in **K**? The value of p (which is a positive integer number) is an address as to where in *Mcon_i* we find the start address in *Mcon* and *Th* where the information about this particular slave is stored. The master *dofs* in *Mcon* are recognized by their *dof* number, and we need to invoke *Meqn* to find their equation number. Here we let m_1, m_2 and m_3 represent the equation numbers of the master *dofs* in question. If we do the mathematics here we will find that element k_{33} "expands" into a 3 by 3 symmetric matrix of elements consisting of k_{33} multiplied by two of the h constants. Six of these elements are added to the elements of **K** corresponding to the master *dofs* as shown; the remaining three elements are skipped since they would have ended up below the diagonal of **K**. The figure also indicates what happens to an element of **k**, for which only one index coincides with the slave *dof*, such as k_{35}.

Complicated you say? Perhaps so, but the logic is "clean" and general, and keep in mind, you only need to program this once (in a general, library type subroutine).

Programming issues

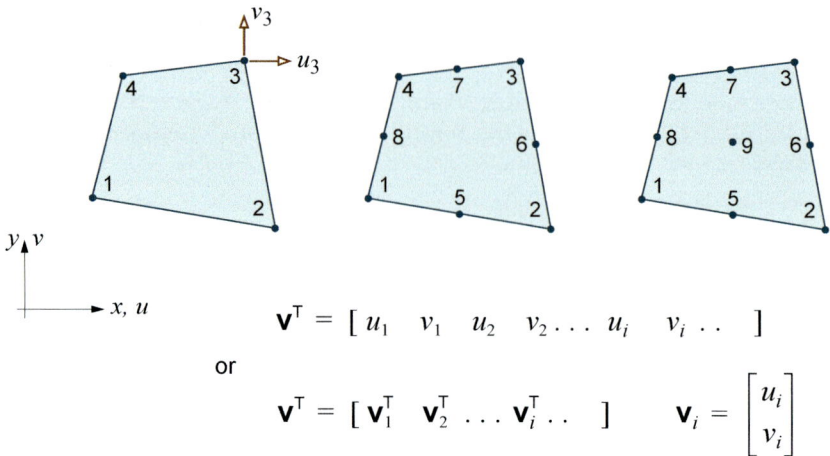

$$\mathbf{v}^\mathsf{T} = [\, u_1 \quad v_1 \quad u_2 \quad v_2 \ldots u_i \quad v_i \ldots\,]$$

or

$$\mathbf{v}^\mathsf{T} = [\, \mathbf{v}_1^\mathsf{T} \quad \mathbf{v}_2^\mathsf{T} \ldots \mathbf{v}_i^\mathsf{T} \ldots\,] \qquad \mathbf{v}_i = \begin{bmatrix} u_i \\ v_i \end{bmatrix}$$

Figure 10.6 Quadrilateral plane stress / plane strain elements

10.3 Stiffness matrix for an isoparametric element

In this section we will indicate how the element stiffness matrix **k** for a typical isoparametric element can be efficiently generated by a fairly general procedure (subroutine). In order to make this as realistic as possible we let our problem be one of plane stress or plane strain. We are getting slightly ahead of ourselves since this problem area is dealt with in detail in the next chapter. However, all necessary matrices and methods have been introduced already in previous chapters, and we only need to make a few problem-dependent assumptions, such as type and ordering of element *dofs*. One might also choose to come back to this section after having covered the next chapter.

To make the problem even more concrete, we specify the element shape to be *quadrilateral*. However, the element may have 4, 8 or 9 nodal points, numbered as shown in Fig. 10.6, the thickness may vary bilinearly between corner thicknesses and the material may be isotropic, orthotropic or anisotropic, but elastic. The latter is easily solved by letting the entire 3 by 3 *elasticity matrix* **C** be input information to the subroutine; this also solves the question of plane stress or plane strain, since the content of the **C** matrix is the only difference between the two types of problem. Each nodal point has two *dofs*, the nodal point values of the x- and y-displacements (u and v), respectively, and we assume that the stiffness matrix returned by the subroutine refers to an ordering of the element *dofs* **v** as shown in Fig. 10.6, u and v node by node.

Basic assumption (see Chapter 5):

$$\mathbf{u} = \begin{bmatrix} u \\ v \end{bmatrix} = \begin{bmatrix} N_1 & 0 & N_2 & 0 & \cdots \\ 0 & N_1 & 0 & N_2 & \cdots \end{bmatrix} \begin{bmatrix} u_1 \\ v_1 \\ u_2 \\ v_2 \\ \vdots \end{bmatrix} = \begin{bmatrix} \mathbf{N}_1 & \mathbf{N}_2 & \cdots \end{bmatrix} \begin{bmatrix} \mathbf{v}_1 \\ \mathbf{v}_2 \\ \vdots \end{bmatrix} = \mathbf{Nv} \quad (10\text{-}1)$$

where

$$\mathbf{N}_i = \mathbf{I}_2 N_i = \begin{bmatrix} N_i & 0 \\ 0 & N_i \end{bmatrix} \qquad \mathbf{I}_2 \text{ is the 2 by 2 identity matrix} \quad (10\text{-}2)$$

From Eq. (3-41) we have

$$\boldsymbol{\varepsilon} = \begin{bmatrix} \varepsilon_x \\ \varepsilon_y \\ \gamma_{xy} \end{bmatrix} = \begin{bmatrix} \frac{\partial}{\partial x} & 0 \\ 0 & \frac{\partial}{\partial y} \\ \frac{\partial}{\partial y} & \frac{\partial}{\partial x} \end{bmatrix} \begin{bmatrix} u \\ v \end{bmatrix} = \boldsymbol{\Delta}\mathbf{u} = \boldsymbol{\Delta}\begin{bmatrix} \mathbf{N}_1 & \mathbf{N}_2 & \cdots \end{bmatrix} \begin{bmatrix} \mathbf{v}_1 \\ \mathbf{v}_2 \\ \vdots \end{bmatrix} = \mathbf{Bv} \quad (10\text{-}3)$$

where

Programming issues

$$\begin{bmatrix} \dfrac{\partial N_i}{\partial x} \\ \dfrac{\partial N_i}{\partial y} \end{bmatrix} = \begin{bmatrix} \dfrac{\partial \xi}{\partial x} & \dfrac{\partial \eta}{\partial x} \\ \dfrac{\partial \xi}{\partial y} & \dfrac{\partial \eta}{\partial y} \end{bmatrix} \begin{bmatrix} \dfrac{\partial N_i}{\partial \xi} \\ \dfrac{\partial N_i}{\partial \eta} \end{bmatrix} = \mathbf{J}^{-1} \begin{bmatrix} \dfrac{\partial N_i}{\partial \xi} \\ \dfrac{\partial N_i}{\partial \eta} \end{bmatrix} \quad (6\text{-}11)$$

$$\mathbf{k} = \int_V \mathbf{B}^T \mathbf{C} \mathbf{B}\, dV = \iint_A t \mathbf{B}^T \mathbf{C} \mathbf{B}\, dA \quad (6\text{-}18)$$

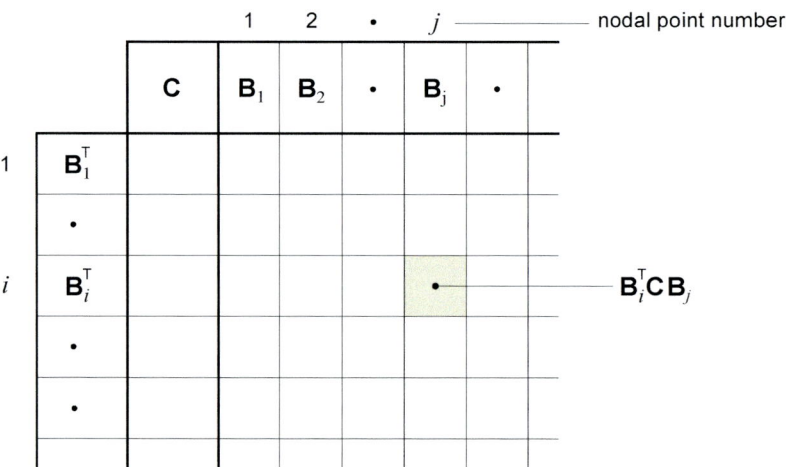

Stiffness matrix for an isoparametric element

$$\mathbf{B} = \begin{bmatrix} \mathbf{B}_1 & \mathbf{B}_2 & \cdots \end{bmatrix} \quad \text{and} \quad \mathbf{B}_i = \Delta \mathbf{N}_i = \Delta \mathbf{I}_2 N_i = \begin{bmatrix} N_{i,x} & 0 \\ 0 & N_{i,y} \\ N_{i,y} & N_{i,x} \end{bmatrix} \quad (10\text{-}4)$$

Since the shape functions N_i are defined in terms of natural coordinates ξ and η, we need to determine $N_{i,\xi}$ and $N_{i,\eta}$ before we can establish expressions for $N_{i,x}$ and $N_{i,y}$. This is all explained in Section 6.1; note however the small, but important, difference in the ordering of the nodal point *dofs* v_i. From Eq. (6-11) we have

$$\begin{bmatrix} N_{i,x} \\ N_{i,y} \end{bmatrix} = \mathbf{J}^{-1} \begin{bmatrix} N_{i,\xi} \\ N_{i,\eta} \end{bmatrix} \quad \text{where} \quad \mathbf{J} = \begin{bmatrix} x_{,\xi} & y_{,\xi} \\ x_{,\eta} & y_{,\eta} \end{bmatrix} \quad (10\text{-}5)$$

and the inverse Jacobi matrix is

$$\mathbf{J}^{-1} = \frac{1}{J} \begin{bmatrix} y_{,\eta} & -y_{,\eta} \\ -x_{,\eta} & x_{,\xi} \end{bmatrix} \quad \text{and} \quad J = x_{,\xi} y_{,\eta} - x_{,\eta} y_{,\xi} \quad (10\text{-}6)$$

Since our elements are *isoparametric* we have

$$x = \sum_i N_i x_i \quad \text{and} \quad y = \sum_i N_i y_i \quad (10\text{-}7)$$

where x_i and y_i are nodal point coordinates. Finally we need to carry out the integration of Eq. (6-18),

$$\begin{aligned}
\mathbf{k} &= \int_A t[f_{ij}(\xi,\eta)]\,dA = \int_{-1}^{1}\int_{-1}^{1} [f_{ij}(\xi,\eta)]\,tJ\,d\xi\,d\eta \\
&= \sum_{k=1}^{r}\sum_{l=1}^{r} [f_{ij}(\xi_k,\eta_l)]\,t(\xi_k,\eta_l)J(\xi_k,\eta_l)w_k w_l \qquad (10\text{-}8) \\
&= \sum_{m=1}^{r\times r} [f_{ij}(\xi_m,\eta_m)]\,t(\xi_m,\eta_m)J(\xi_m,\eta_m)w_m
\end{aligned}$$

Here n is the number of integration points in one direction, $w_m = w_k \times w_l$ is the product of the integration *weights* for the quadrature rule being used, and

$$[f_{ij}] = \mathbf{B}^T \mathbf{C} \mathbf{B}$$

From the FALK diagrams on the opposite page we see that $\mathbf{B}^T \mathbf{C} \mathbf{B}$ consists of a series of 2 by 2 submatrices whose elements we can readily establish explicit expressions for; for instance, the element in position (1,1) of $\mathbf{B}_i^T \mathbf{C} \mathbf{B}_j$ is:

$$(1,1) = C_{11}N_{i,x}N_{j,x} + C_{13}(N_{i,y}N_{j,x} + N_{i,x}N_{j,y}) + C_{33}N_{i,y}N_{j,y} \quad (10\text{-}9)$$

Similarly for the other elements.

Programming issues

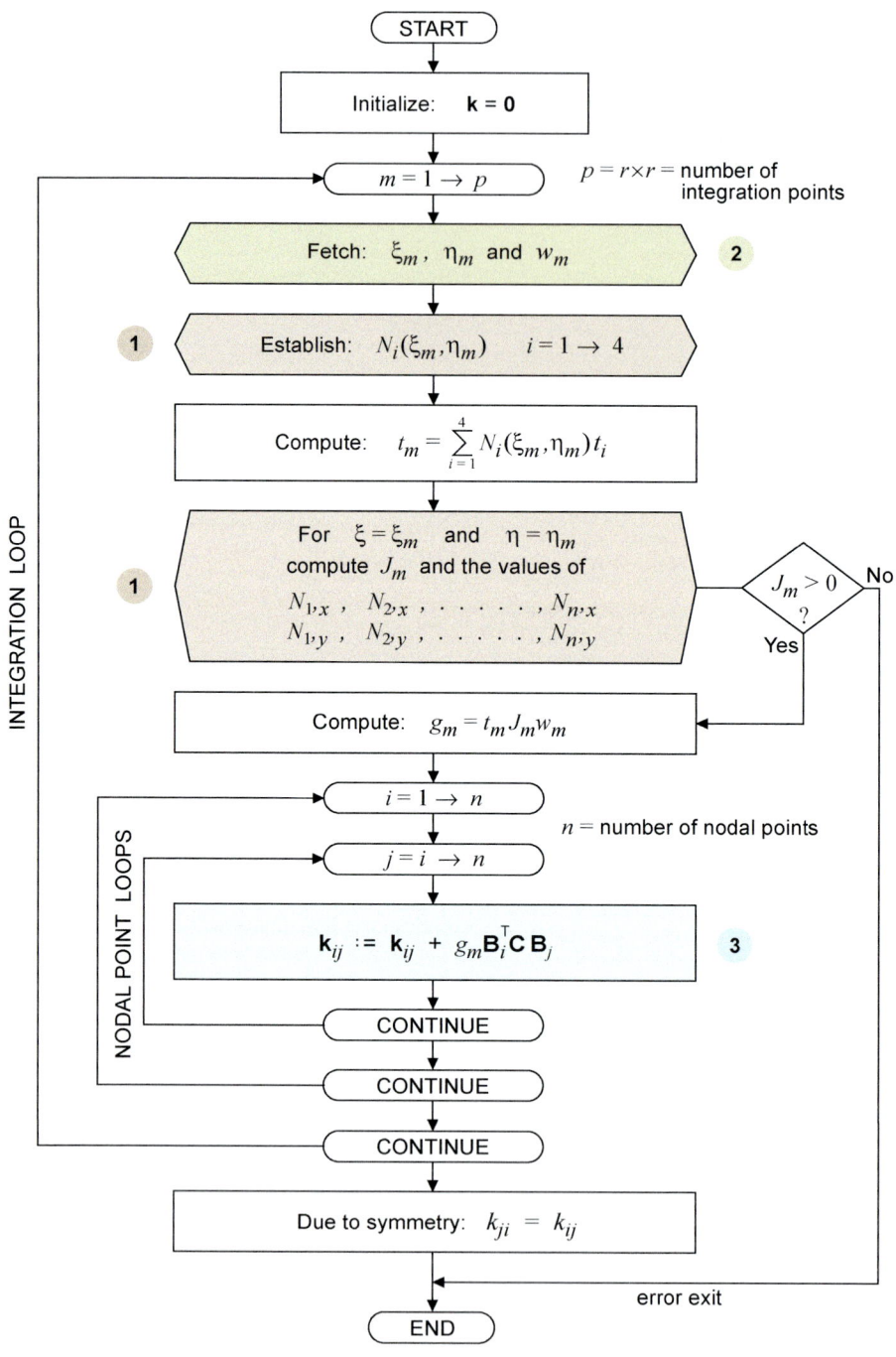

Figure 10.7 Flowchart for computation of **k** for a plane stress/ plane strain quadrilateral element with n nodal points

Figure 10.7 shows a flowchart of the procedure outlined above. It assumes a plane, isoparametric quadrilateral element with two *dofs* at each of n nodal points, where n can have values from 4 and up; the three elements of Fig. 10.6 are obvious candidates, but the procedure (subroutine) can easily be implemented to also include other, specific elements.

The thickness t is assumed to vary bilinearly between the thicknesses at the corner nodes.

A key part of the procedure, usually contained in a separate subroutine, is the *shape function routine*, indicated by the number 1 in Fig. 10.7. For a given integration point (ξ_m, η_m) this subroutine will compute the values, at this point, of each (relevant) shape function (N_i), the Jacobian (J_m), and the x- and y-derivatives of each (relevant) shape function $(N_{i,x}$ and $N_{i,y})$. In order to accomplish this, the subroutine also needs to determine $\mathbf{J}$ and $\mathbf{J}^{-1}$ as well as the derivatives of N_i with respect to ξ and η.

Figure 10.7 also indicates another subroutine, marked by the number 2 in the figure, that returns the coordinates and weights of a particular point (m) in the chosen $(r \times r)$ GAUSS quadrature rule.

The lower triangular part of $\mathbf{k}$ is determined in the two nodal point loops. The elements are determined submatrix by submatrix (in the box marked by the number 3) as the sum of the contributions from each integration point – the outer loop is the integration loop.

It should be noted that the flowchart includes a test on the value of the Jacobian. Such a test, the purpose of which is to uncover wrong or questionable input information, is important, and all numerical manipulations that involve input information should, if relevant, be tested before execution in order to avoid "illegal" or "impossible" results (such as division by zero or taking the square root of a negative number). An error or a questionable situation should be reported back to the calling program (for instance by a specific value of a designated integer variable, an "error flag"); eventually this information needs to be conveyed to the user of the program.

The programmer of any program intended for general use should keep in mind that the program should, if given "bad" input, respond with some meaningful information; and not just "die".

The necessary input information to the (main) subroutine is the number of nodal points, the nodal point coordinates, the four corner thicknesses, the six different elements of the symmetric material matrix $\mathbf{C}$, and the quadrature rule.

The interested reader may find an "old" Fortran implementation of the procedure in Fig. 10.7 in APPENDIX D.

Programming issues

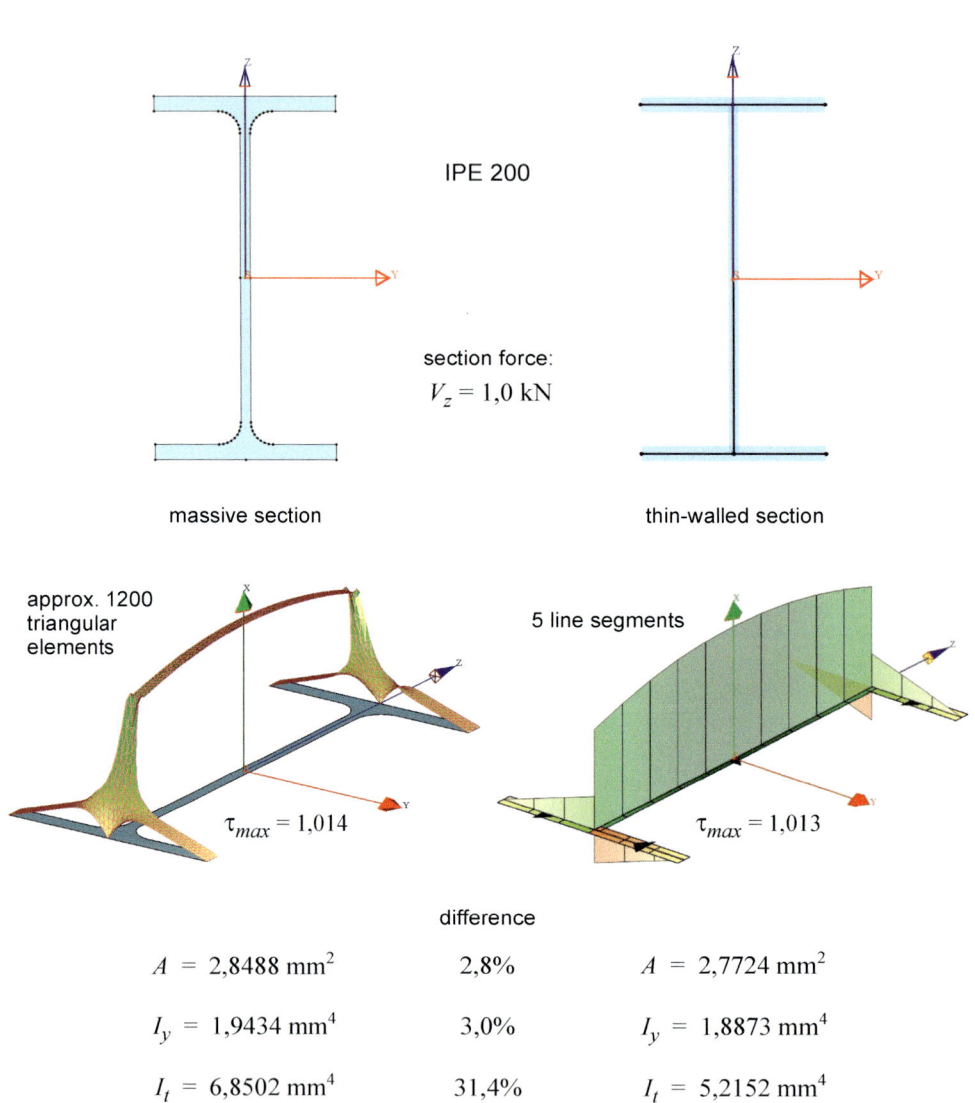

Figure 10.8 A standard steel section analyzed by CrossX

10.4 Two typical FEM programs

Application programs for structural mechanics type problems that are based on the finite element method can, broadly speaking, be divided into two categories: *general purpose programs* and *special purpose programs*. Well-known programs like ABAQUS and ANSYS fall in the first category. These are very large commercial programs aimed at solving a wide range of problems covering many fields of engineering, not only structural mechanics (although that was initially their primary target). There are a good many more such programs on the market.

In the second category we find programs aimed at solving more specific problems. These are "smaller" programs, often with a user interface that is tailored to the problem they intend to solve, and they usually solve this problem quite efficiently. The two programs described briefly below are of the special purpose type. They were developed with considerable student participation, primarily for teaching purposes, but they both have capabilities for practical problem solving. We mention these two programs for three reasons: 1) they are both good examples of programs using two programming languages, 2) we make use of both programs to exemplify FEM computations in the book, and for 3) see the note at the end of the section.

Both programs were developed around the turn of the millennium to run on the Microsoft Windows platforms at that time (Windows 98, Windows NT and Windows 2000), but they both seem to work satisfactorily also on the current Windows 7 platform. Both programs consist of a graphical user interface (GUI) developed in C++ and OpenGL, while all computations, save the mesh generation, are carried out by Fortran code.

Program CrossX. This program computes section parameters and stiffnesses of, and stress distribution on, arbitrary beam cross sections composed of one or more isotropic and linearly elastic materials. The stresses refer to any combination of specified section forces (stress resultants).

A given cross section is analysed either as a *massive* section or as a *thin-walled* section. The **massive** cross section is analysed by a finite element method, elaborated on in Chapter 16, considering only St. Venant torsion (that is, no warping restraint). A **thin-walled** cross section is made up of a series of straight line segments (thin rectangles), with or without closed cells. Both St. Venant torsion and non-uniform, warping restrained torsion are considered. The method used to analyse a thin-walled section is not a standard finite element method.

Initially the program was intended to be both a *stand-alone* cross section program and an integrated part of a general frame analysis program. However, it never came to fill the second role. As a programming project, the mesh generation and the GUI represented a greater challenge than the numerical computations, and the amount of Fortran code of the latter is small compared to the amount of C++ code used to realize the mesh generation and the GUI.

Figure 10.8 shows some results obtained by **CrossX** for a standard steel section that can adequately be analysed by both models. We see clearly the importance of the fillets, particularly for the torsion constant I_t.

Programming issues

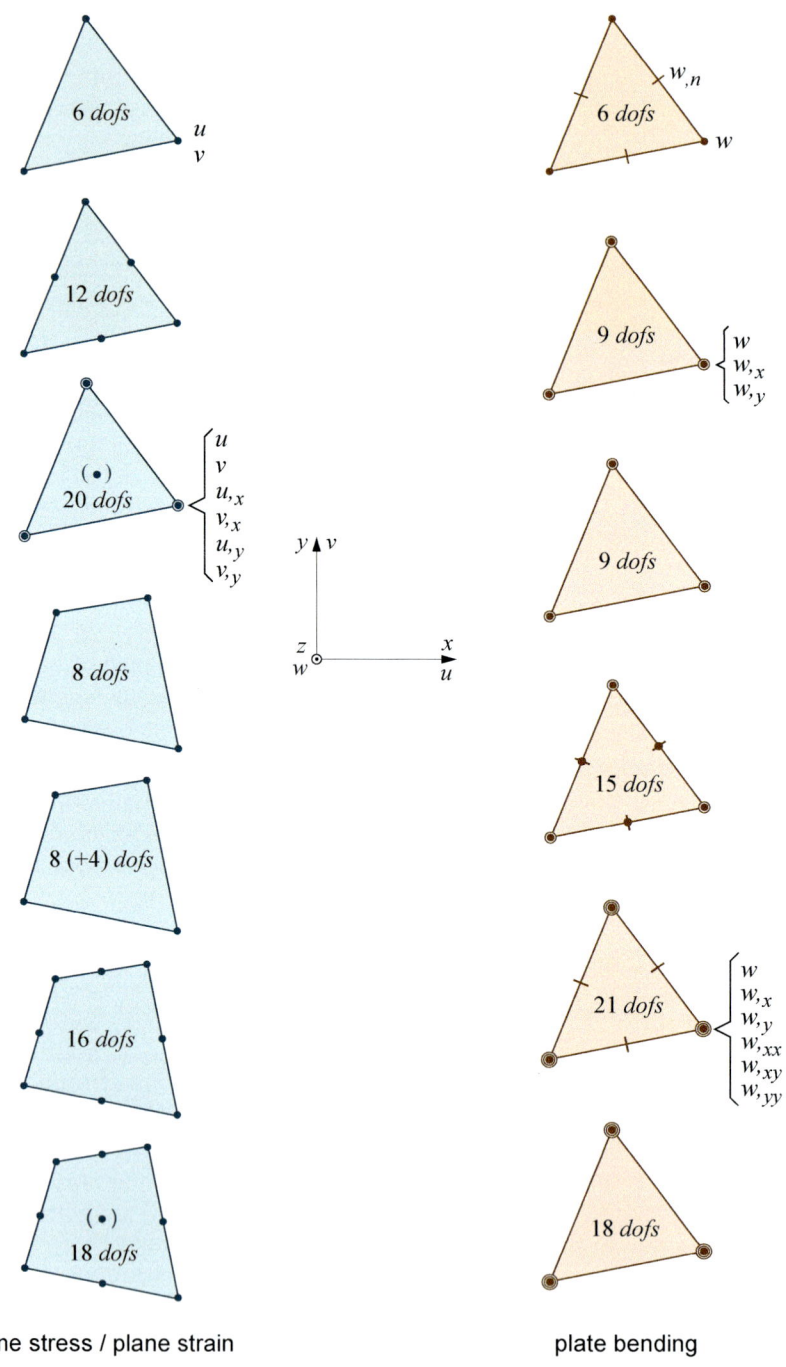

Figure 10.9 Element types in program **FEMplate**

Two typical FEM programs

Program FEMplate. This is a program for linear static analysis of 2D plate structures, developed primarily for educational purposes. It makes extensive use of the mesh generator and the graphical capabilities developed for the cross section program **CrossX**.

Both in-plane loading (plain stress and plane strain) and plate bending are considered. For in-plane analysis 7 element types – 3 triangular and 4 quadrilateral – are available, whereas for plate bending 6 element types – all triangular, and all but one based on thin plate theory – are available, see Fig. 10.9. The choice of elements included is perhaps based more on availability (as members of our local libraries) than on performance, although most of the elements are well known and well tested elements, but of rather "early" development. All but one of the elements are derived as pure displacement elements – one of the plate bending elements (with 9 *dofs*) is of the "*hybrid*" type. Most of the elements in Fig. 10.9 will be dealt with in Chapters 11 and 13.

Concentrated loading, line loading, distributed surface loading and initial strain (temperature) loading are available, and fairly general boundary conditions, including prescribed displacements, can be specified. It should be noted however that curved edges are not implemented; only elements with straight edges are available.

Examples of **FEMplate** computations are shown in Figs. 7.6 and 7.12, and more will be shown later. While **CrossX** only uses the profile storage format for the system stiffness matrix, **FEMplate** offers a choice between profile and sparse storage and corresponding equation solvers.

Although it is of little interest for the scope of this book, it should be mentioned that **FEMplate** can also perform simple plate buckling analysis for all but one of the bending elements in combination with any one of the three triangular membrane elements.

> NOTE
>
> Both **CrossX** and **FEMplate** can be downloaded by the owner of this book; see the *Software agreement* at the front of the book. More information about these programs can be found in their "User's Manual", available in PDF format from the program's HELP menu.

displacements:
$$\mathbf{u} = \begin{bmatrix} u(x,y) \\ v(x,y) \end{bmatrix}$$

strains:
$$\boldsymbol{\varepsilon} = \begin{bmatrix} \varepsilon_x \\ \varepsilon_y \\ \gamma_{xy} \end{bmatrix} = \begin{bmatrix} \frac{\partial}{\partial x} & 0 \\ 0 & \frac{\partial}{\partial y} \\ \frac{\partial}{\partial y} & \frac{\partial}{\partial x} \end{bmatrix} \begin{bmatrix} u \\ v \end{bmatrix} = \boldsymbol{\Delta}\mathbf{u} \qquad \text{Eq. (3-41)}$$

stresses (plane stress):
$$\boldsymbol{\sigma} = \begin{bmatrix} \sigma_x \\ \sigma_y \\ \tau_{xy} \end{bmatrix} = \frac{E}{1-\nu^2} \begin{bmatrix} 1 & \nu & 0 \\ \nu & 1 & 0 \\ 0 & 0 & \frac{1-\nu}{2} \end{bmatrix} \begin{bmatrix} \varepsilon_x \\ \varepsilon_y \\ \gamma_{xy} \end{bmatrix} = \mathbf{C}\boldsymbol{\varepsilon} \qquad \text{Eq. (3-43)}$$

Basic assumption: $\mathbf{u} = \mathbf{N}\mathbf{v}$ Eq. (4-1) $\Rightarrow$ $\boldsymbol{\varepsilon} = \boldsymbol{\Delta}\mathbf{N}\mathbf{v} = \mathbf{B}\mathbf{v}$ Eq. (4-50)

— element degrees of freedom (nodal point displacements)
— assumed shape functions

Element stiffness: $\mathbf{k} = \int_{V_e} \mathbf{B}^\mathrm{T} \mathbf{C} \mathbf{B} \, dV$ Eq. (4-54)

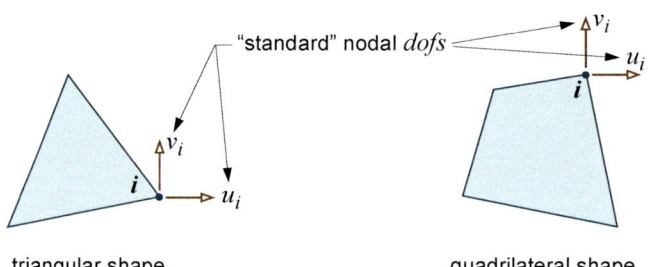

"standard" nodal *dofs*

triangular shape · quadrilateral shape

11

Plane stress and plane strain

Finally we are ready for the application of FEM to specific problems of structural mechanics, and we start with in-plane loading of flat plates, that is, problems of plane stress and plane strain. We shall look at some of the most basic and common displacement type elements of triangular and quadrilateral shape, and discuss their pros and cons, hopefully without getting lost in the details.

We shall also look at some of the "tricks" applied to improve performance and/or avoid potential problems. Singularities and the application of boundary conditions need some comment, as do efficiency and other modelling aspects. Examples of use will be shown.

For a displacement method of analysis the field variables governing the problem are the displacement components u and v in the x- and y-direction, respectively. The theoretical basis has already been covered in Sections 3.3 and 4.7, a brief summary of which is repeated on the opposite page.

From a finite element point of view the only difference between plane stress and plane strain is the elasticity matrix **C**, see Eqs. (3-43) and (3-48b) for isotropic elastic materials. Hence, by formulating the computations in terms of an arbitrary (full), but symmetric **C** matrix, our procedure (subroutine) is valid for both plane stress and plane strain, and also for different material models (isotropic, orthotropic or general anisotropic).

The *variational* order of this problem is $m = 1$, that is, the highest order of differentiation in the operator matrix Δ is 1. A RAYLEIGH-RITZ consistent finite element formulation requires $C^{m-1} = C^0$ continuity which means we only need to satisfy continuity of displacements u and v across element boundaries.

Completeness, that is correct representation of *rigid body* movements as well as the ability to represent states of constant strain (stress), requires that the assumed displacement functions (shape functions) together contain a *complete* 1st order polynomial in x and y.

The obvious geometric shapes for the elements are the *triangle* and the *quadrilateral*, and the standard element degrees of freedom (*dofs*) are the nodal point values of the displacement components u and v.

Plane stress and plane strain

		1		linear (3 terms)
	x		y	
	x^2	xy	y^2	quadratic (6 terms)
x^3	x^2y	xy^2	y^3	cubic (10 terms)

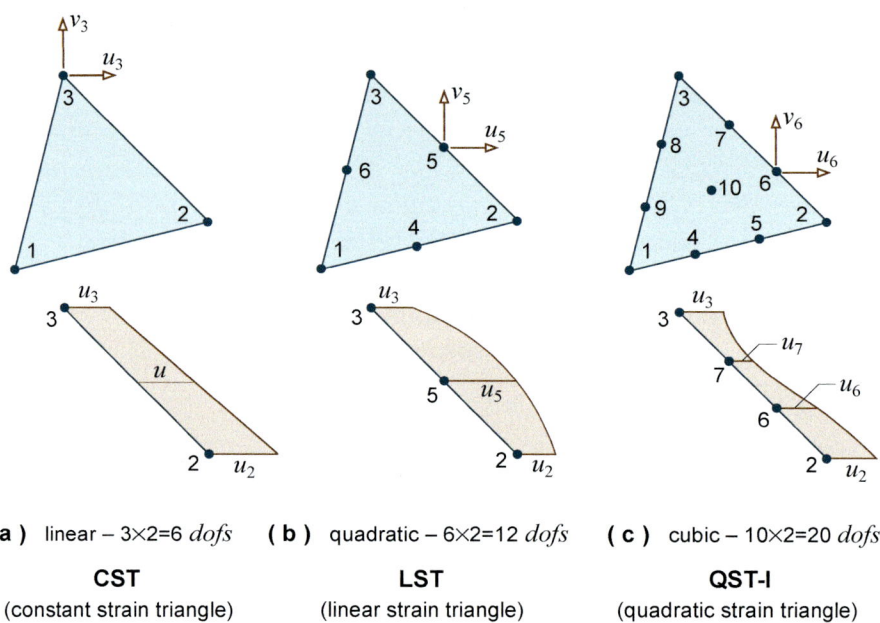

(a) linear – 3×2=6 *dofs* (b) quadratic – 6×2=12 *dofs* (c) cubic – 10×2=20 *dofs*

CST
(constant strain triangle)

LST
(linear strain triangle)

QST-I
(quadratic strain triangle)

Figure 11.1 "Standard" triangular plane stress/plane strain elements

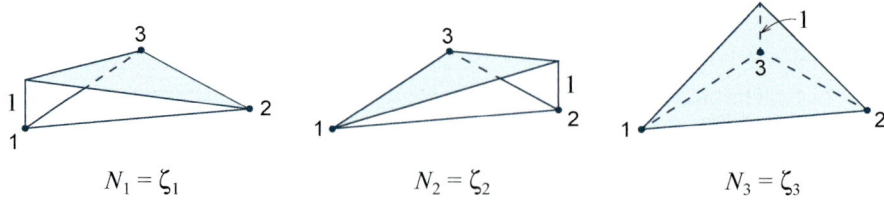

$N_1 = \zeta_1$ $N_2 = \zeta_2$ $N_3 = \zeta_3$

Copy of **Figure 5.30** Shape functions for the linear triangle

11.1 Triangular elements

We start with the triangle, the simplest and, in a modelling sense, the most versatile element shape. The triangle also has another advantage; it lends itself to the use of *complete* polynomials as shape functions, which guarantees symmetry in x and y and thus invariance with respect to coordinate transformations. Figure 11.1 shows the elements using the first three complete polynomials as shape functions for u and v, respectively.

The linear triangle. The simplest element with nodal points at the corners only and with 2×3 = 6 *dofs* assumes a displacement field for u and v described by a complete *linear* polynomial. The (physical) *dofs* (**v**) are the nodal point values of the displacement components. We can express u and v in terms of **v** using *natural* (area or triangular) coordinates, or in terms of *generalized dofs* **q** using cartesian coordinates:

$$u = [\zeta_1 \quad \zeta_2 \quad \zeta_3]\begin{bmatrix} u_1 \\ u_2 \\ u_3 \end{bmatrix} = \mathbf{N}_1 \mathbf{v}_u \quad \text{or} \quad u = [1 \quad x \quad y]\begin{bmatrix} q_{u1} \\ q_{u2} \\ q_{u3} \end{bmatrix} = \mathbf{N}_{q1}\mathbf{q}_u \quad (11\text{-}1)$$

Similarly for v. The first approach is preferable if the shape functions are readily available; the second approach, which of course can also be used in connection with natural coordinates, is basically a round about way to determine the shape functions, see Section 5.6. With cartesian coordinates, as used here, we see the polynomial terms clearly. The arrangement of the elements of **v** may depend on how the various computations are carried out, but in the end we would like to have the *dofs* ordered node by node; hence the basic assumption for the linear triangle for such an ordering would read:

$$\mathbf{u} = \begin{bmatrix} u \\ v \end{bmatrix} = \begin{bmatrix} \zeta_1 & 0 & \zeta_2 & 0 & \zeta_3 & 0 \\ 0 & \zeta_1 & 0 & \zeta_2 & 0 & \zeta_3 \end{bmatrix}\begin{bmatrix} u_1 \\ v_1 \\ u_2 \\ \vdots \\ v_3 \end{bmatrix} = \mathbf{N}\mathbf{v} \quad (11\text{-}2)$$

Similarly for the generalized shape functions of Eq. (11-1). The shape functions in Eq. (11-2) are visualized in Fig. 5.30. We see that all basic requirements for a monotonic convergent element are satisfied: the sum of the shape functions ($\mathbf{N}_1$) is equal to unity at any point within the element and along the element borders, which guarantees correct handling of the rigid body displacements; the element can definitely represent states of constant stress (in fact those are the only states of stress it can represent), and the (linear) displacement (of u and v) along an element edge is uniquely defined by *dofs* associated with this edge only, which in turn means that we have C^0 continuity between elements.

Differentiating a linear function once produces a constant, and the linear element can therefore only represent *constant strain / stress*, hence the *constant strain triangle* – **CST** for short – which is the common name for this element.

Plane stress and plane strain

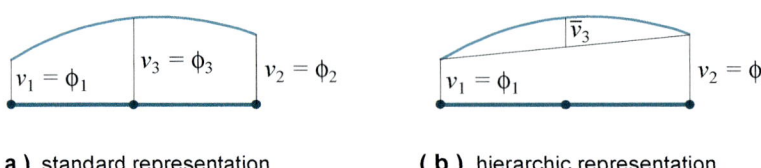

(a) standard representation   (b) hierarchic representation

Copy of **Figure 5.13** Standard and hierarchic *dofs*

Shape functions for **LST** :

$N_1 = \zeta_1^2 - \zeta_1\zeta_2 - \zeta_3\zeta_1 = \zeta_1(2\zeta_1 - 1)$ $N_4 = 4\zeta_1\zeta_2$

$N_2 = \zeta_2^2 - \zeta_1\zeta_2 - \zeta_2\zeta_3 = \zeta_2(2\zeta_2 - 1)$ $N_5 = 4\zeta_2\zeta_3$

$N_3 = \zeta_3^2 - \zeta_2\zeta_3 - \zeta_3\zeta_1 = \zeta_3(2\zeta_3 - 1)$ $N_6 = 4\zeta_3\zeta_1$

$$\mathbf{N}_1 = [\, N_1 \quad N_2 \quad N_3 \quad N_4 \quad N_5 \quad N_6 \,]$$

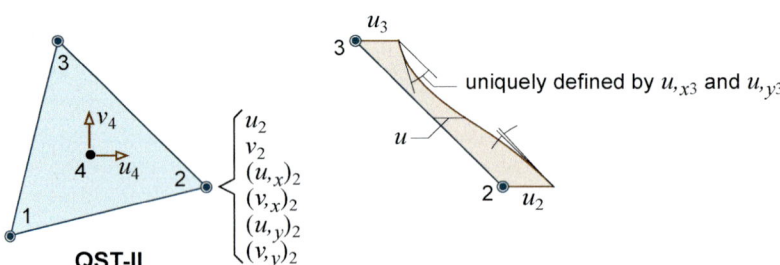

Figure 11.2 Alternative version of the quadratic strain triangle

In spite of its simplicity or perhaps because of it, this element is in many ways the perfect element, satisfying as it does all requirements with a minimum number of *dofs*. It is clearly a bit "primitive" in that it can only describe accurately constant stresses, and we will obviously need many such elements in order to obtain good approximations of rapidly changing stresses, such as the axial and shear stresses over the height of a beam in bending. Due to the formidable computational power of very affordable computers, this need not represent much of an obstacle, as we shall soon demonstrate.

The quadratic triangle. A complete quadratic polynomial in x and y contains six terms. A matching triangular element therefore has 2×6 = 12 *dofs*. In addition to corner nodes we now introduce nodal points also at the mid-point of each edge, see Fig. 11.1b. With the nodal point values of u and v as *dofs*, we see that the three values along each edge (at the two end points and the mid-point) uniquely define the quadratic displacements along the edge, and C^0 continuity is satisfied. Since the displacements vary quadratically, the strains (and stresses) vary linearly; hence **LST** – linear strain triangle.

With **LST** we have two additional options: 1) at the mid-side nodes we can define *hierarchic dofs* instead of the standard *dofs*, see Fig. 5.13, and 2) the mid-side nodes enable an isoparametric formulation with curved element edges.

In Sections 5.6 and 5.7, we have already derived the shape functions for the **LST** element, in terms of area coordinates, using both indirect and direct interpolation, see opposite page.

Like **CST**, **LST** also satisfies *a priori* all requirements (completeness and continuity) for a monotonically convergent element, and **LST** is a very good and robust element.

The cubic triangle. The next element in the "family" makes use of a complete cubic polynomial in x and y. This polynomial contains ten terms, and the element must therefore have 2×10 = 20 *dofs*. This triangular element comes in two versions. In the first version we use only standard type *dofs* (nodal point values of u and v), and in addition to corner nodes we introduce two nodal points on each element side, at the 1/3 points. This gives us 2×9 = 18 *dofs*, and we therefore need to include an internal node (at the centroid) in order to match the complete cubic displacement field for u and v, see Fig. 11.1c. With the nodal point values of u (and v) as *dofs*, we see that the four values along each edge (one at each of the two end nodes and one at each of the two side nodes) uniquely define a cubically varying displacement along the edge, and C^0 continuity is therefore satisfied. Since the displacements vary cubically, the strains (and stresses) vary quadratically; hence **QST-I** – quadratic strain triangle, version I.

The second version of the cubic triangle is shown in Fig. 11.2. Instead of side nodes, we have introduced the nodal values of the four first derivatives (with respect to x and y) of u and v as *dofs*, in addition to the standard nodal values of u and v, at the three corner nodes. Again we need the two internal *dofs* at the centroid in order to match a complete cubic displacement field for u and v respectively. If we consider an arbitrary element edge we see that each of

Plane stress and plane strain

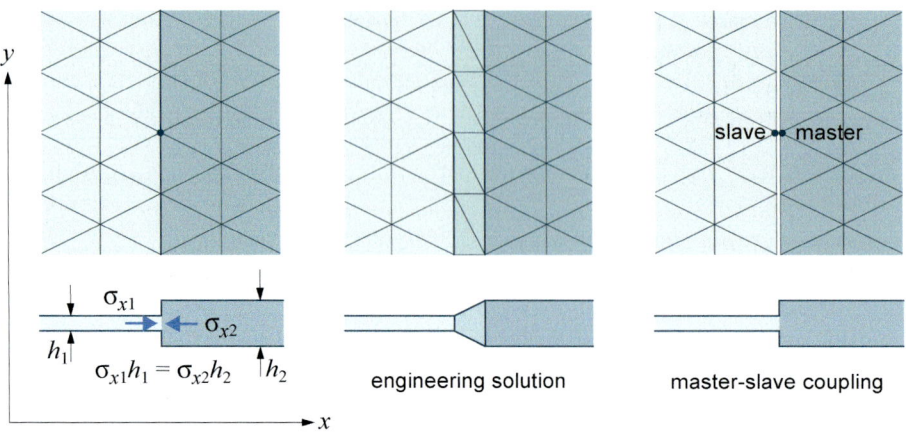

Figure 11.3 Discontinuous plate thickness – **QST II** elements

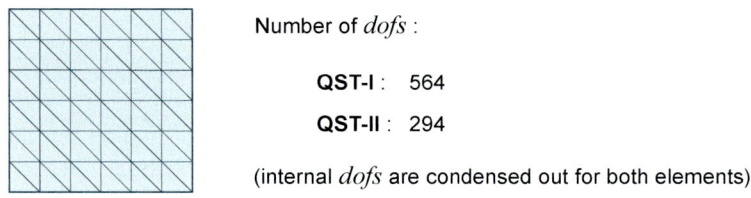

Number of *dofs* :

    QST-I :   564

    QST-II : 294

(internal *dofs* are condensed out for both elements)

Figure 11.4 Relative effectiveness of the two cubic triangles

the two displacement components has 2 "boundary conditions" at each end node, the function value itself and its "slope". Since four conditions uniquely define a cubic curve, version II of the element also satisfies C^0 continuity.

There are both advantages and disadvantages to having the derivatives as degrees of freedom. Together they define all three strain components which means that the stresses at the corner nodes are the *same* in all elements that meet at a particular node, and the accuracy of these stresses is comparable to that of the displacements. This is fine as long as we have no abrupt changes, for instance, in plate thickness. Across such a discontinuity, see Fig. 11.3, stress resultants are continuous, but stresses are not.

Figure 11.3 also indicates the engineering solution to the problem: a narrow zone of elements with varying thickness. There is also a mathematical solution: with one nodal point on each side of the discontinuity (with identical coordinates) it is possible to define one of the nodes as a *slave* of the other, the *master*, and couple the slave *dofs* to the master *dofs* in such a way that the correct continuities (of stress resultants) are maintained.

This problem of incorrect displacement continuity that may occur (if precautions are not taken) when the derivatives are included as element *dofs*, is often referred to as *over-conformity* ("too much continuity"). Another drawback of **QST-II** is that it is somewhat more difficult to assign appropriate boundary conditions to the derivatives of the displacements.

For both **QST** alternatives it is common to eliminate, by static condensation, the two internal *dofs* at the element level. Hence, for actual use both **QST-I** and **QST-II** have 18 *dofs*, and both satisfy all necessary requirements, the same as **CST** and **LST**. As we shall see, the **QST** is an excellent element.

As already pointed out in Section 5.4, corner *dofs* are more "effective" than *dofs* defined at side nodes. This favours the second alternative, **QST-II**, as Fig. 11.4 clearly indicates. As for the shape functions, they are readily found for **QST-I** by the "0-line method", see Fig 5.32. Physical shape functions (**N**) can also be found for **QST-II**, via generalized shape functions ($\mathbf{N}_q$), assumed as a 3rd order *homogeneous* polynomial in area coordinates. Alternatively, one can use the more slightly "costly" method of deriving the generalized stiffness matrix $\mathbf{k}_q$ and then transforming this matrix to the desired "physical" matrix **k** by means of the inverse of matrix **A**, see Section 5.6.

In principle one can also derive elements based on quartic and even higher order polynomials, but such elements are of little practical interest. In fact, more and cheaper computational power tends to favour the simpler elements.

Formulation for computer implementation. How do we best organize the computations involved in determining the elements of the stiffness and load matrices (**k** and $\mathbf{S}^0$) in the computer? There is no single and simple answer to this question, and the proposal outlined below for the stiffness matrix **k**, is but one of various ways it can be done. The procedure can be used for all three elements described above, but examples refer to the **LST** element. We assume *straight edges*, and therefore need not consider an isoparametric formulation.

If at all possible, we would like to derive explicit expressions for each individual element of **k** (on and above the diagonal).

Plane stress and plane strain

Equation (5-24):
$$\begin{bmatrix} \dfrac{\partial}{\partial x} \\ \dfrac{\partial}{\partial y} \end{bmatrix} = \dfrac{1}{2A} \begin{bmatrix} y_{23} & y_{31} & y_{12} \\ x_{32} & x_{13} & x_{21} \end{bmatrix} \begin{bmatrix} \dfrac{\partial}{\partial \zeta_1} \\ \dfrac{\partial}{\partial \zeta_2} \\ \dfrac{\partial}{\partial \zeta_3} \end{bmatrix}$$

Equation (5-25): $\zeta_3 = 1 - \zeta_1 - \zeta_2$

FALK diagram of Eq. (11-7)

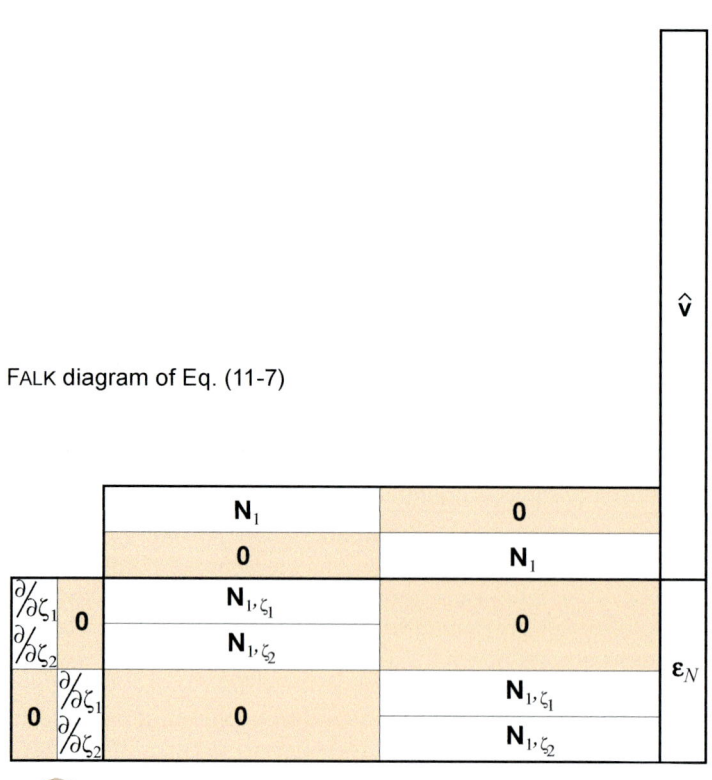

Triangular elements

Although we would like to end up with a stiffness matrix that refers to *dofs* that are ordered node by node, as in Eq. (11-2), it is convenient to start with a vector $\hat{\mathbf{v}}$ in which all *x*-displacements are listed first, node by node, and then the *y*-displacements, that is,

$$\hat{\mathbf{v}} = \begin{bmatrix} \mathbf{v}_x \\ \mathbf{v}_y \end{bmatrix} \quad \text{where} \quad \begin{matrix} \mathbf{v}_x^T = [\, u_1 \quad u_2 \quad u_3 \quad u_4 \quad u_5 \quad u_6 \,] \\ \mathbf{v}_y^T = [\, v_1 \quad v_2 \quad v_3 \quad v_4 \quad v_5 \quad v_6 \,] \end{matrix} \quad \text{for LST.} \qquad (11\text{-}3)$$

The basic assumption can then be expressed as

$$\mathbf{u} = \begin{bmatrix} u \\ v \end{bmatrix} = \begin{bmatrix} \mathbf{N}_1 & \mathbf{0} \\ \mathbf{0} & \mathbf{N}_1 \end{bmatrix} \begin{bmatrix} \mathbf{v}_x \\ \mathbf{v}_y \end{bmatrix} = \hat{\mathbf{N}}\hat{\mathbf{v}} \qquad (11\text{-}4)$$

where the shape functions $\mathbf{N}_1$, for the LST element, are

$$\mathbf{N}_1 = [\, N_1 \quad N_2 \quad N_3 \quad N_4 \quad N_5 \quad N_6 \,]$$

The strains, defined by Eq. (3-41), we now write as

$$\boldsymbol{\varepsilon} = \begin{bmatrix} \varepsilon_x \\ \varepsilon_y \\ \gamma_{xy} \end{bmatrix} = \begin{bmatrix} \dfrac{\partial}{\partial x} & 0 \\ 0 & \dfrac{\partial}{\partial y} \\ \dfrac{\partial}{\partial y} & \dfrac{\partial}{\partial x} \end{bmatrix} \begin{bmatrix} u \\ v \end{bmatrix} = \boldsymbol{\Delta}\mathbf{u} = \begin{bmatrix} u_{,x} \\ v_{,y} \\ u_{,y}+v_{,x} \end{bmatrix} = \mathbf{T} \begin{bmatrix} u_{,x} \\ u_{,y} \\ v_{,x} \\ v_{,y} \end{bmatrix} \qquad (11\text{-}5)$$

Where matrix **T** reads:

$$\mathbf{T} = \begin{bmatrix} 1 & 0 & 0 & 0 \\ 0 & 0 & 0 & 1 \\ 0 & 1 & 1 & 0 \end{bmatrix} \qquad (11\text{-}6)$$

Since our shape functions are defined in terms of area coordinates they cannot be differentiated with respect to *x* and *y* directly; we need to go via Eq. (5-24) or Eq. (5-29). Since our element has straight edges we could use Eq. (5-24), but we choose to consider only two of the area coordinates, ζ_1 and ζ_2, as independent, while the third (ζ_3) is defined by Eq. (5-25). It is now convenient to define some kind of "natural" strain,

$$\boldsymbol{\varepsilon}_N = \begin{bmatrix} u_{,\zeta_1} \\ u_{,\zeta_2} \\ v_{,\zeta_1} \\ v_{,\zeta_2} \end{bmatrix} = \begin{bmatrix} \dfrac{\partial}{\partial \zeta_1} & 0 \\ \dfrac{\partial}{\partial \zeta_2} & 0 \\ 0 & \dfrac{\partial}{\partial \zeta_1} \\ 0 & \dfrac{\partial}{\partial \zeta_2} \end{bmatrix} \begin{bmatrix} u \\ v \end{bmatrix} = \boldsymbol{\Delta}_\zeta \mathbf{u} = \boldsymbol{\Delta}_\zeta \hat{\mathbf{N}} \hat{\mathbf{v}} = \mathbf{B}_N \hat{\mathbf{v}} \qquad (11\text{-}7)$$

377

Plane stress and plane strain

Equation (5-29): $$\begin{bmatrix} \dfrac{\partial}{\partial x} \\ \dfrac{\partial}{\partial y} \end{bmatrix} = \dfrac{1}{2A} \begin{bmatrix} y_{23} & y_{31} \\ x_{32} & x_{13} \end{bmatrix} \begin{bmatrix} \dfrac{\partial}{\partial \zeta_1} \\ \dfrac{\partial}{\partial \zeta_2} \end{bmatrix} = \mathbf{J}^{-1} \begin{bmatrix} \dfrac{\partial}{\partial \zeta_1} \\ \dfrac{\partial}{\partial \zeta_2} \end{bmatrix}$$

$$x_{ij} = x_i - x_j \quad \text{and} \quad y_{ij} = y_i - y_j$$

where

$$\mathbf{B}_N = \mathbf{\Delta}_\zeta \hat{\mathbf{N}} = \mathbf{\Delta}_\zeta \begin{bmatrix} \mathbf{N}_1 & 0 \\ 0 & \mathbf{N}_1 \end{bmatrix} = \begin{bmatrix} \mathbf{B}_{N1} & 0 \\ 0 & \mathbf{B}_{N1} \end{bmatrix} \quad \text{where} \quad \mathbf{B}_{N1} = \begin{bmatrix} N_{1,\zeta_1} \\ N_{1,\zeta_2} \end{bmatrix} \quad (11\text{-}8)$$

From Eqs. (11-5) and (5-29) we have,

$$\mathbf{\varepsilon} = \mathbf{T} \begin{bmatrix} u_{,x} \\ u_{,y} \\ v_{,x} \\ v_{,y} \end{bmatrix} = \mathbf{T} \underbrace{\begin{bmatrix} \mathbf{J}^{-1} & 0 \\ 0 & \mathbf{J}^{-1} \end{bmatrix}}_{\mathbf{J}_2^{-1}} \begin{bmatrix} u_{,\zeta_1} \\ u_{,\zeta_2} \\ v_{,\zeta_1} \\ v_{,\zeta_2} \end{bmatrix} = \mathbf{T}\mathbf{J}_2^{-1}\mathbf{B}_N \hat{\mathbf{v}} = \hat{\mathbf{B}}\hat{\mathbf{v}} \quad (11\text{-}9)$$

and our standard expression for the element stiffness matrix, see Section 4.7, becomes:

$$\hat{\mathbf{k}} = \int_{V_e} \hat{\mathbf{B}}^T \mathbf{C} \hat{\mathbf{B}} dV = \int_{A_e} t \hat{\mathbf{B}}^T \mathbf{C} \hat{\mathbf{B}} dA \quad (11\text{-}10)$$

If we introduce the auxiliary matrix $\mathbf{G}$ defined as

$$\mathbf{G} = \begin{bmatrix} \mathbf{G}_1 & 0 \\ 0 & \mathbf{G}_1 \end{bmatrix} \quad \text{where} \quad \mathbf{G}_1 = \mathbf{J}^{-1}\mathbf{B}_{N1} = \mathbf{J}^{-1}\begin{bmatrix} N_{1,\zeta_1} \\ N_{1,\zeta_2} \end{bmatrix} \quad (11\text{-}11)$$

and

$$\hat{\mathbf{C}} = \mathbf{T}^T \mathbf{C} \mathbf{T} = \begin{bmatrix} \mathbf{C}_{xx} & \mathbf{C}_{xy} \\ \mathbf{C}_{yx} & \mathbf{C}_{yy} \end{bmatrix} \quad \text{where} \quad \mathbf{C}_{yx} = \mathbf{C}_{xy}^T \quad (11\text{-}12)$$

we can write Eq. (11-10) as

$$\hat{\mathbf{k}} = \int_{A_e} t \hat{\mathbf{B}}^T \mathbf{C} \hat{\mathbf{B}} dA = \int_{A_e} t \mathbf{G}^T \mathbf{T}^T \mathbf{C} \mathbf{T} \mathbf{G} dA = \int_{A_e} t \mathbf{G}^T \hat{\mathbf{C}} \mathbf{G} dA = \begin{bmatrix} \hat{\mathbf{k}}_{xx} & \hat{\mathbf{k}}_{xy} \\ \hat{\mathbf{k}}_{yx} & \hat{\mathbf{k}}_{yy} \end{bmatrix} \quad (11\text{-}13)$$

where

$$\hat{\mathbf{k}}_{xx} = \int_{A_e} t \mathbf{G}_1^T \mathbf{C}_{xx} \mathbf{G}_1 dA \quad (11\text{-}14a)$$

$$\hat{\mathbf{k}}_{yy} = \int_{A_e} t \mathbf{G}_1^T \mathbf{C}_{yy} \mathbf{G}_1 dA \quad (11\text{-}14b)$$

$$\hat{\mathbf{k}}_{xy} = \int_{A_e} t \mathbf{G}_1^T \mathbf{C}_{xy} \mathbf{G}_1 dA = \hat{\mathbf{k}}_{yx}^T \quad (11\text{-}14c)$$

The integrand of Eq. (11-14a) can, in view of Eq. (11-11), be expressed as

$$t\mathbf{G}_1^T \mathbf{C}_{xx} \mathbf{G}_1 = t\mathbf{B}_{N1}^T \mathbf{J}^{-T} \mathbf{C}_{xx} \mathbf{J}^{-1} \mathbf{B}_{N1} = t\mathbf{B}_{N1}^T \bar{\mathbf{C}}_{xx} \mathbf{B}_{N1}$$

Plane stress and plane strain

Equation (5-30): $\quad I = \int_A \zeta_1^m \zeta_2^n \zeta_3^p dA = 2A \dfrac{m!n!p!}{(m+n+p+2)!}$

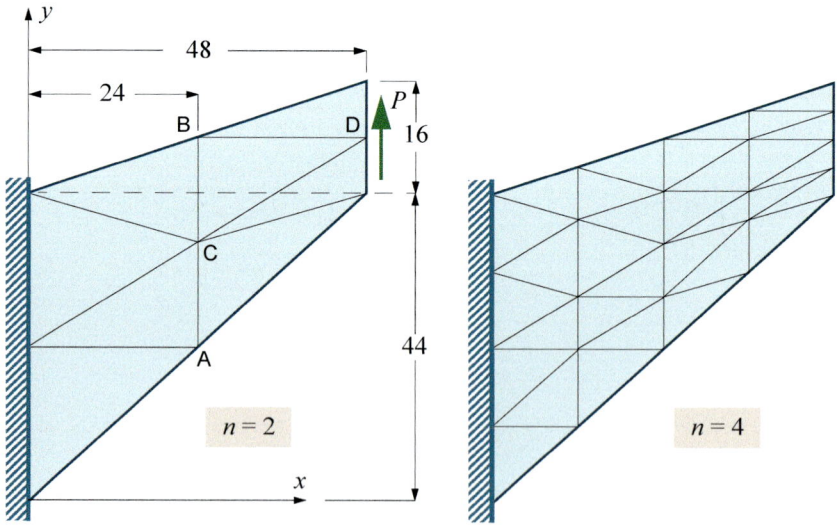

Figure 11.5 COOK's problem

where

$$\bar{\mathbf{C}}_{xx} = \mathbf{J}^{-T}\mathbf{C}_{xx}\mathbf{J}^{-1} \qquad (11\text{-}15)$$

is a 2 by 2 matrix of constants that can readily be determined explicitly, and Eq. (11-14a) becomes

$$\hat{\mathbf{k}}_{xx} = \int_{A_e} t\mathbf{B}_{N1}^T \bar{\mathbf{C}}_{xx} \mathbf{B}_{N1} dA \qquad (11\text{-}16)$$

Similarly for $\hat{\mathbf{k}}_{yy}$ and $\hat{\mathbf{k}}_{xy}$. Matrix $\mathbf{B}_{N1}$ contains simple expressions in the area coordinates, obtained through derivation of the shape functions ($\mathbf{N}_1$) with respect to ζ_1 and ζ_2, respectively. The thickness t is either constant or it varies linearly between the corner thicknesses (as simple functions of the area coordinates). In any case, if only standard *dofs* are used as for **CST**, **LST** and **QST-I**, the integrand of Eq. (11-16) is an $n\times n$ matrix, where n is the number of nodal points and each element of this matrix has the form $c \times \zeta_1^m \zeta_2^n \zeta_3^p$, where c is a constant. With the integration formula of Eq. (5-30) we can easily find explicit expressions for each element of $\hat{\mathbf{k}}_{xx}$. Easily yes, but it takes time and concentration to get it right. However, we see from Eqs. (11-14) that once we have the elements of $\hat{\mathbf{k}}_{xx}$, the elements of $\hat{\mathbf{k}}_{yy}$ and $\hat{\mathbf{k}}_{xy}$ are obtained by simply replacing the 4 constant elements of matrix $\bar{\mathbf{C}}_{xx}$ by those of $\bar{\mathbf{C}}_{yy}$ and $\bar{\mathbf{C}}_{xy}$, respectively. For maximum generality, the elasticity matrix $\mathbf{C}$, which is input information in this context, should be a *full* matrix.

It remains to get back to $\mathbf{k}$ from $\hat{\mathbf{k}}$. This is accomplished by interchanging rows and columns, and since we have determined explicit expressions for each element of $\hat{\mathbf{k}}$ (on and above the diagonal), it is merely a question of assigning correct indices to each element.

You might think that we have gone to a lot of trouble here to solve what seems to be a straightforward problem. However, once you get down to the programming details, you would probably agree that time spent in the organisation of the computations is worth while. And remember, you only do this once.

A similar line of thinking as we have used here can also be used for the simpler tasks of implementing the various load vectors ($\mathbf{S}^0$).

Example 11-1 – COOK's problem

Figure 11.5 shows a swept panel fixed along the left-hand edge and loaded by a vertical load P at the tip; the load is applied as a uniformly distributed tangential load along the right-hand (vertical) edge. The thickness of the panel is constant and the (relative) in-plane dimensions are shown in the figure; the material properties used are those of steel (with $E = 210000$ MPa and $\nu = 0{,}3$). This problem, first proposed by COOK [42], has become a frequently used "benchmark" test for in-plane (membrane) elements; it is commonly known as COOK's problem.

Plane stress and plane strain

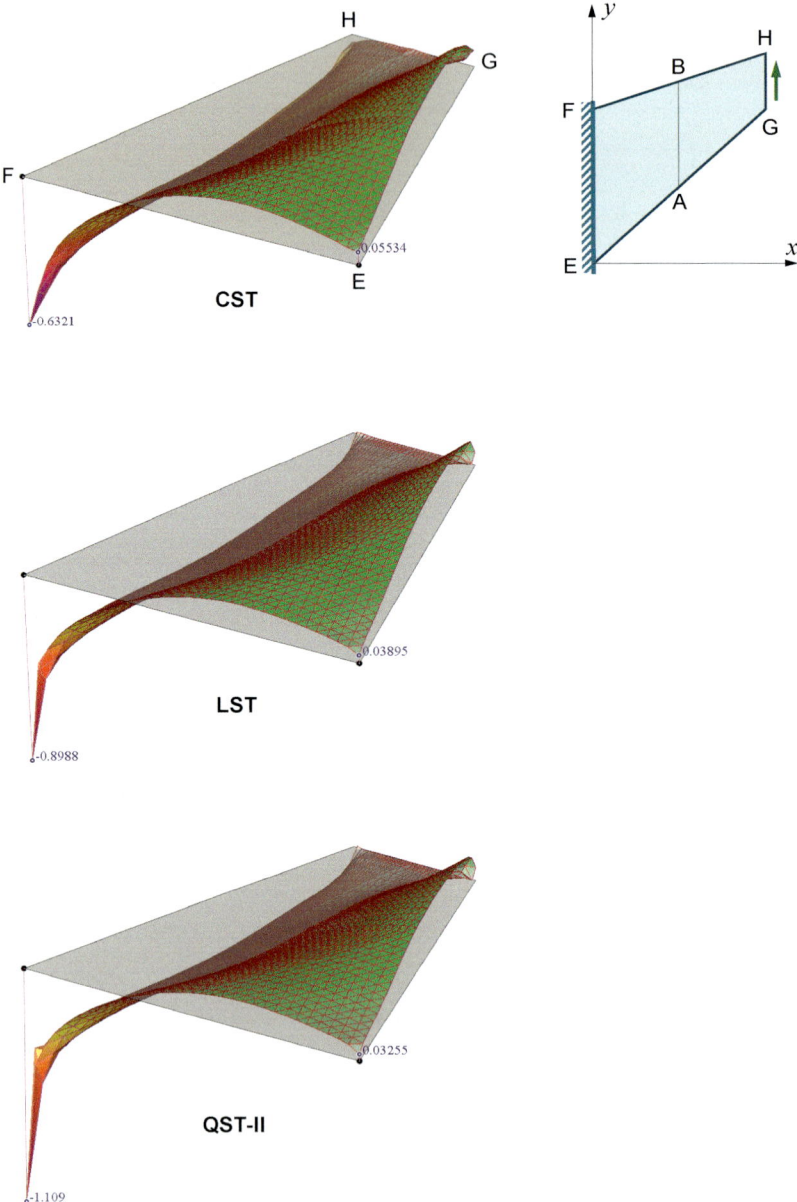

Figure 11.6 Axial stress σ_x (surfaces) for Cook's problem; analyzed by **FEMplate** for the finest mesh ($n = 32$)

Triangular elements

We have used program **FEMplate** to analyse this problem with three different elements, CST, LST and QST-II (QST-I is not implemented in **FEMplate**); 5 different meshes are used for each element. The start mesh is the one shown to the left in Fig. 11.5 ($n = 2$), the next mesh is obtained by uniform subdivision of the first ($n = 4$), and so on, up to $n = 32$.

Some results are shown in Table 11-1; vertical tip deflection at point D (v_D), axial stress σ_x at points A and B (σ_{xA} and σ_{xB}) and shear stress τ_{xy} at point C (τ_{xyC}). Each result is quoted as percentage of the correct result, taken as the result obtained with **QST-II** for $n = 32$, which is "exact" to at least 4 digits.

Table 11-1: Results, in percent of correct value, for Cook's problem

	$n =$	2	4	8	16	32
CST	v_D	49,6	75,8	91,8	97,7	99,4
	σ_{xA}	26,3	58,4	82,4	93,3	97,2
	σ_{xB}	18,0	50,7	77,8	91,1	96,4
	τ_{xyC}	81,9	93,5	96,8	98,4	99,2
LST	v_D	97,3	99,56	99,89	99,96	100,0
	σ_{xA}	105,0	101,2	100,4	100,1	100,0
	σ_{xB}	94,7	99,7	100,2	100,0	100,0
	τ_{xyC}	88,4	95,2	98,8	99,7	99,93
QST-II	v_D	98,6	99,55	99,82	99,95	100,0
	σ_{xA}	97,5	100,2	100,0	100,0	100,0
	σ_{xB}	88,9	99,4	100,0	100,0	100,0
	τ_{xyC}	119,2	102,2	100,1	100,0	100,0

When interpreting the results of Table 11-1 one should keep in mind that although the three different elements have been used for the same meshes, the actual number of unknown *dofs* is significantly different. For the finest mesh ($n = 32$), for instance, the number of unknowns is 2112, 8320 and 6402 for **CST**, **LST** and **QST-II**, respectively. In fact, **CST** with $n = 64$ has exactly the same number of unknowns (8320) as **LST** with $n = 32$. Hence, for the same computational effort, **CST** is somewhat better than the table indicates.

However, the table clearly shows the superiority of the higher order elements, and even if **LST** has, for a given mesh, the highest number of *dofs*, it provides very good results, even for the coarser meshes. It is slightly surprising that, for the same mesh, **LST** is on a par with **QST** for most results, and even sometimes slightly better. This can be due to "lucky" averaging – all stress results for **CST** and **LST** are obtained by simple nodal point averaging, whereas stress results for **QST-II** are not influenced by averaging.

Plane stress and plane strain

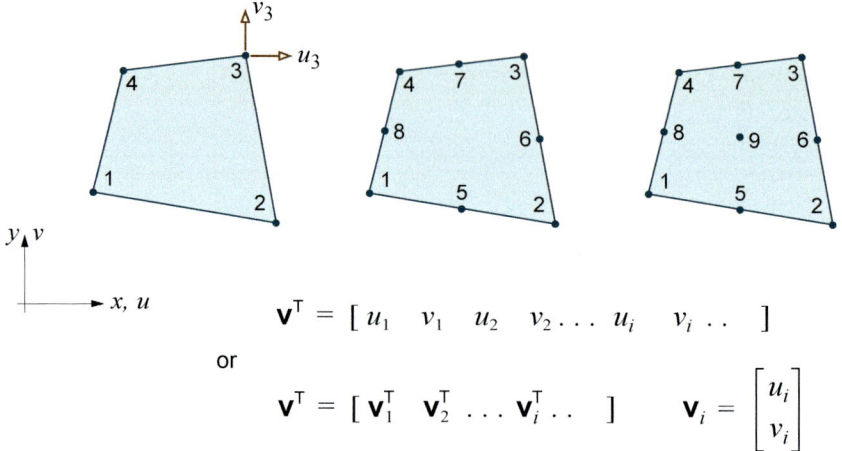

Copy of **Figure 10.6** Quadrilateral plane stress / plane strain elements

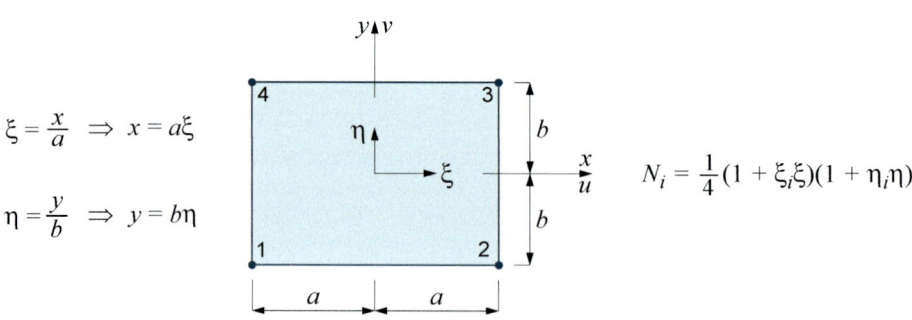

Figure 11.7 Standard or basic version of a rectangular Q4

Figure 11.6 shows the axial stress σ_x depicted as a stress "surface", for all three elements, obtained with the finest mesh. It should be noted that the scaling in the three plots of the figure is roughly, but not exactly the same. It is quite interesting to note that while point F is a singularity point (for all stress components), as expected, point E is not. For a rectangular plate both the top and the bottom points of the fixed edge are singularity points, see Fig. 7.12. For COOK's problem, not only σ_x, but also σ_y and τ_{xy} approach zero at point E, and that is perhaps not what one would expect.

Finally a note on computer times, so-called CPU times. For "small" problems like these, time measurements are very uncertain. However, for the **LST** with $n = 32$ and **CST** with $n = 64$ (8320 equations) it takes about 50[1] milliseconds to solve the system of equations, whereas it takes about 85 ms to solve the system for **QST-II** (with 6402 equations). Without making too much of a point here, this shows that solution efficiency is not just a question of counting the unknowns, the actual computer implementation of any given algorithm, including renumbering, is quite important.

We conclude this section by putting in a word for the simple **CST** element. From a practical point of view this element will provide sufficiently accurate results for most linear static problems, given a fine, but easily handled mesh.

11.2 Quadrilateral elements

The three central plane stress/plane strain elements of quadrilateral shape are those shown in Fig. 10.6, the simple 4-node element, the 8-node *serendipity* element and the 9-node LAGRANGIAN element. For simplicity we shall refer to these elements as **Q4**, **Q8** and **Q9**, respectively. All elements are shown with straight edges, in which case the side nodes are conveniently place at the mid-point. All three elements are normally formulated by use of *isoparametric* mapping, and both **Q8** and **Q9** can therefore have curved edges. The element degrees of freedom are, for all three elements, of the standard type, that is the nodal point values of the displacement components u and v.

We have already dealt quite extensively with all three elements, in a number of previous chapters, see Sections 5.7 (interpolation), 6.1 (mapping), 6.2 (numerical integration), 6.4 (integration and convergence), 6.5 (instability), 7.2 (stress recovery and smoothing) and 10.3 (programming). Since recent developments in practical use of FEM have been in favour of *simple* elements, we shall pay most attention to the 4-node element in this section, and we start with the basic version.

4-node element – the basic version. Few, if any, finite element for structural mechanics problems has received as much attention over the years as this simple element. The standard or basic version of this 8 *dof* element, shown in its rectangular form in Fig. 11.7, is based on the standard bilinear shape functions

$$N_i(\xi, \eta) = \tfrac{1}{4}(1 + \xi_i \xi)(1 + \eta_i \eta) \qquad i = 1, 2, 3, 4 \qquad (11\text{-}17)$$

---

[1]. Obtained with a sparse solver on a fairly standard PC in 2012.

Plane stress and plane strain

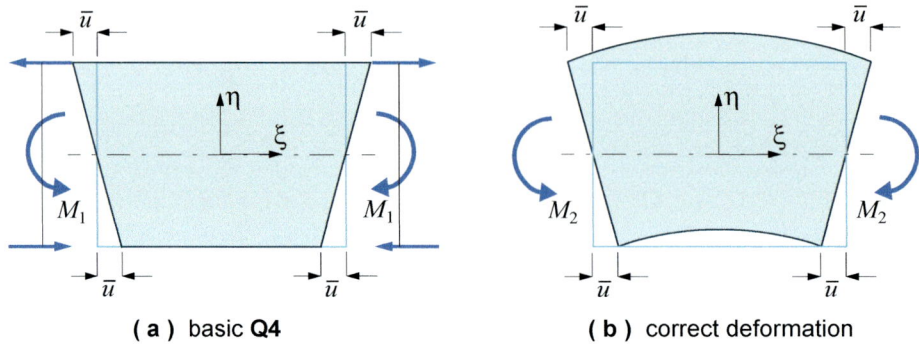

(a) basic Q4   (b) correct deformation

Figure 11.8 Pure bending

and

$$u = \sum_{i=1}^{4} N_i u_i \quad \text{and} \quad v = \sum_{i=1}^{4} N_i v_i \tag{11-18}$$

The element is R-R consistent, it satisfies both *completeness* and displacement continuity (is *conforming*), and yet, it is not without problems. One shortcoming has to do with shear deformations. A good demonstration of this is presented in COOK et al [21]; a rectangular **Q4** element is subjected to a constant moment M_1, as shown in Fig. 11.8a. For this displacement pattern we have

$$u_1 = u_3 = \bar{u}, \quad u_2 = u_4 = -\bar{u} \quad \text{and}$$

Hence, from Eqs. (11-18) and (11-17),

$$u = \xi \eta \bar{u} \quad \text{and} \quad v = 0 \tag{11-19}$$

This gives rise to the following strains

$$\varepsilon_x = \frac{\partial u}{\partial x} = \frac{1}{a}\frac{\partial u}{\partial \xi} = \eta\frac{\bar{u}}{a} \tag{11-20a}$$

$$\varepsilon_y = \frac{\partial v}{\partial y} = 0 \tag{11-20b}$$

$$\gamma_{xy} = \frac{\partial v}{\partial x} + \frac{\partial u}{\partial y} = \frac{1}{b}\frac{\partial u}{\partial \eta} = \xi\frac{\bar{u}}{b} \tag{11-20c}$$

The correct pure bending deformations, see Fig. 11.8b, are

$$\varepsilon_x = \eta\frac{\bar{u}}{a} \tag{11-21a}$$

$$\varepsilon_y = -\nu\eta\frac{\bar{u}}{a} \tag{11-21b}$$

$$\gamma_{xy} = 0 \tag{11-21c}$$

The displacement assumptions

$$u = c_1 \xi \eta \tag{11-22a}$$
$$v = c_2(1-\xi^2) + c_3(1-\eta^2) \tag{11-22b}$$

give the correct strains if

$$c_1 = \bar{u}, \quad c_2 = \frac{a\bar{u}}{2b} \quad \text{and} \quad c_3 = \nu\frac{b\bar{u}}{2a} \tag{11-23}$$

From this we conclude that the correct displacements in the case of pure bending are

$$u = \bar{u}\xi\eta \tag{11-24a}$$

$$v = \frac{a\bar{u}}{2b}(1-\xi^2) + \nu\frac{b\bar{u}}{2a}(1-\eta^2) \tag{11-24b}$$

The main problem here is the "false" shear strain, see (11-20c). While the correct deformation only contributes to the strain energy from the normal strains, the standard **Q4** element, for the same nodal displacement, will also

Plane stress and plane strain

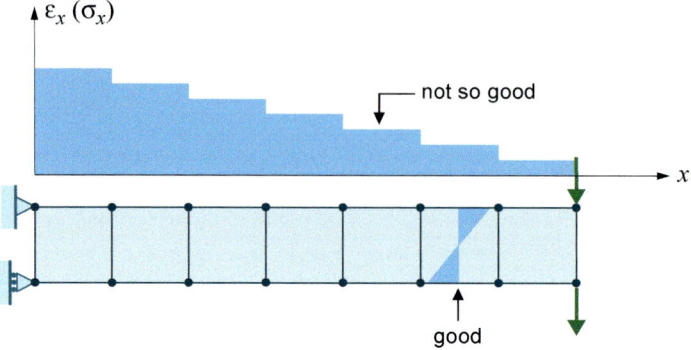

Figure 11.9 Standard **Q4** in bending

contribute to the strain energy from the false ("parasitic") shear strain. Hence, for the same nodal displacements we have

$$M_1 > M_2$$

and by computing the ratio between the strain energy for the two deformation patterns in Fig. 11.8, COOK et al [21] finds

$$\frac{M_1}{M_2} = \frac{1}{1+\nu}\left[\frac{1}{1-\nu} + \frac{1}{2}\left(\frac{a}{b}\right)^2\right] \qquad (11\text{-}25)$$

We see that the effect of the *parasitic* shear depends greatly on the ratio a/b. The larger this ratio becomes, the more energy is required to sustain the false shear strain; in the end the shear strain dominates completely and the finite element model will tend to "lock", a phenomenon usually referred to as *shear locking*.

Another shortcoming of the standard **Q4** element is indicated by Fig. 11.9. From the shape functions, Eq. (11-17), we find that within the element,

ε_x is constant in the x-direction and linear in the y-direction,

ε_y is constant in the y-direction and linear in the x-direction, and

γ_{xy} is linear in both the x-direction and in the y-direction.

A constant value of ε_x within the elements in Fig. 11.9 is clearly not advantageous, while the linear variation over the height definitely is.

The standard **Q4** element clearly does not handle bending type action all that well, and the obvious question is if this shortcoming can be overcome or alleviated? First we address the false shear strains in the case of pure bending.

For an isotropic/orthotropic material, the constitutive matrix can be expressed as

$$\mathbf{C} = \begin{bmatrix} C_{11} & C_{12} & 0 \\ C_{12} & C_{22} & 0 \\ 0 & 0 & C_{33} \end{bmatrix} = \begin{bmatrix} C_{11} & C_{12} & 0 \\ C_{12} & C_{22} & 0 \\ 0 & 0 & 0 \end{bmatrix} + \begin{bmatrix} 0 & 0 & 0 \\ 0 & 0 & 0 \\ 0 & 0 & C_{33} \end{bmatrix} = \mathbf{C}_\varepsilon + \mathbf{C}_\gamma \quad (11\text{-}26)$$

and

$$\mathbf{k} = \int_{A_e} t\mathbf{B}^\mathsf{T}\mathbf{C}\mathbf{B}\,dA = \int_{A_e} t\mathbf{B}^\mathsf{T}\mathbf{C}_\varepsilon\mathbf{B}\,dA + \int_{A_e} t\mathbf{B}^\mathsf{T}\mathbf{C}_\gamma\mathbf{B}\,dA = \mathbf{k}_\varepsilon + \mathbf{k}_\gamma \qquad (11\text{-}27)$$

If we consider pure bending, the problem is primarily associated with $\mathbf{k}_\gamma$. The shear strain in this case is correct for $\xi = 0$, see Fig 11.8a; for bending in the y-direction the shear strain is correct for $\eta = 0$. If we now modify $\mathbf{k}_\gamma$ by replacing the integrand with the constant matrix obtained by replacing matrix $\mathbf{B}(\xi,\eta)$ by $\mathbf{B}(0,0)$ we obtain an improved element that still satisfies the *patch* test. Computationally this is achieved by means of *selective* numerical integration, that is by using a 1×1 GAUSS rule for the shear contribution and a 2×2 rule for the axial contribution. In pure shear this modified element still exhibits correct shear stiffness.

Plane stress and plane strain

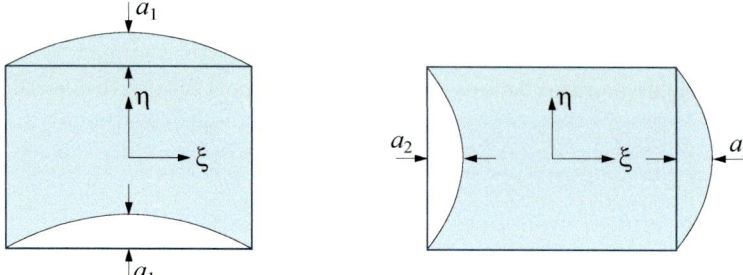

Figure 11.10 Incompatible (local) displacement modes for the 4-node quadrilateral

Quadrilateral elements

The second problem, that is the improvement of the overall bending performance of the element, needs a more fundamental "fix" which leads to a "new" element.

4-node element – incompatible version. Figure 11.8 shows that the standard version of the element "fails" due to the lack of terms $(1-\xi^2)$ and $(1-\eta^2)$ in the assumption for the v-displacement.

This observation lead WILSON et al. [43] to augment each displacement component with two *incompatible* "bending patterns", that is

$$u = \sum_{i=1}^{4} N_i u_i + (1-\xi^2)a_1 + (1-\eta^2)a_2 \tag{11-28a}$$

$$v = \sum_{i=1}^{4} N_i v_i + (1-\xi^2)a_3 + (1-\eta^2)a_4 \tag{11-28b}$$

or

$$\mathbf{u} = \begin{bmatrix} u \\ v \end{bmatrix} = \mathbf{Nv} + \mathbf{N}_a \mathbf{a} \tag{11-29}$$

where $\mathbf{N}$ contains the conforming or (compatible) bilinear shape functions, while $\mathbf{N}_a$ contains the incompatible bending patterns. The four new parameters $\mathbf{a} = [a_i]$ are "node-less", internal *dofs*; they are local to the element and therefore we no longer have C^0 continuity. Hence the element is incompatible or *non-conforming*, see Fig 11.10.

The strains now become

$$\boldsymbol{\varepsilon} = \Delta \mathbf{u} = \Delta \mathbf{Nv} + \Delta \mathbf{N}_a \mathbf{a} = \mathbf{Bv} + \mathbf{B}_a \mathbf{a} \tag{11-30}$$

Note that Eq. (11-29) is only used as the basis for the strains; for all other purposes (including the mapping) we use $\mathbf{N}$ (and forget about $\mathbf{N}_a$). Introducing the notation

$$\bar{\mathbf{B}} = [\mathbf{B} \quad \mathbf{B}_a] \quad \text{and} \quad \bar{\mathbf{v}} = \begin{bmatrix} \mathbf{v} \\ \mathbf{a} \end{bmatrix} \tag{11-31}$$

we can express the element stiffness matrix as

$$\bar{\mathbf{k}} = \int_{A_e} t \bar{\mathbf{B}}^T \mathbf{C} \bar{\mathbf{B}} dA = \begin{bmatrix} \mathbf{k} & \mathbf{E}^T \\ \mathbf{E} & \mathbf{H} \end{bmatrix} \tag{11-32}$$

where

$$\mathbf{k} = \int_{A_e} t \mathbf{B}^T \mathbf{C} \mathbf{B} dA \quad \text{(standard "compatible" stiffness)} \tag{11-33}$$

$$\mathbf{E} = \int_{A_e} t \mathbf{B}_a^T \mathbf{C} \mathbf{B} dA \tag{11-34}$$

$$\mathbf{H} = \int_{A_e} t \mathbf{B}_a^T \mathbf{C} \mathbf{B}_a dA \tag{11-35}$$

Plane stress and plane strain

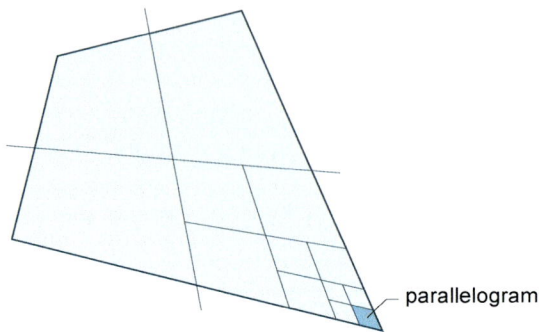

Copy of **Figure 6.17** Uniform subdivision of a quadrilateral

Since parameters **a** are local element parameters they can be eliminated at the element level, by static condensation (see Section 9.10), and a modified and incompatible 8×8 element stiffness matrix is obtained as

$$\bar{\mathbf{k}}_r = \mathbf{k} - \mathbf{E}^T \mathbf{H}^{-1} \mathbf{E} \tag{11-36}$$

For rectangular and parallelogram form this incompatible element satisfies the (strong form of the) *patch test*, and for these geometric forms it gives noticeably better results than the standard (compatible) **Q4**. Since repeated uniform subdivision of an arbitrary quadrilateral region will, in the limit, produce elements with parallelogram shape, see Figure 6.17, it can be argued that the incompatible 4-node quadrilateral element will pass the *weak* patch test. Nevertheless, for general problems of "non-rectangular shape", the incompatible 4-node quadrilateral element, in the form described above, is not an improvement on the standard, conforming **Q4** element.

So the quest continued. The strain energy of the incompatible element can be expressed as

$$U_e = \frac{1}{2}\int_V \boldsymbol{\sigma}^T \boldsymbol{\varepsilon} dV = \frac{1}{2}\int_V \boldsymbol{\sigma}^T \mathbf{B} dV \mathbf{v} + \frac{1}{2}\int_V \boldsymbol{\sigma}^T \mathbf{B}_a dV \mathbf{a} \tag{11-37}$$

In order to satisfy the patch test, the element must be able to reproduce constant strains (stresses) for all (permissible) geometries. Since the compatible part of the element meets this requirement, the incompatible element will also satisfy the requirement if we can "force" the strain energy associated with the incompatible modes to vanish identically for all constant stress states, that is by demanding that

$$\frac{1}{2}\int_V \boldsymbol{\sigma}_{const}^T \mathbf{B}_a dV \mathbf{a} = \frac{1}{2}\boldsymbol{\sigma}_{const}^T \int_V \mathbf{B}_a dV \mathbf{a} = 0 \tag{11-38}$$

Hence the requirement

$$\int_V \mathbf{B}_a dV = \mathbf{0} \tag{11-39}$$

will force the incompatible element to pass the patch test. On closer inspection one will find that Eq. (11-39) holds true *a priori* for all element shapes that possess a constant *Jacobian*, that is all elements with parallelogram shape. In order for arbitrary quadrilateral elements to satisfy (11-39), TAYLOR *et al* [44] proposed to replace the variable expressions for

$$\frac{\partial x}{\partial \xi}, \frac{\partial x}{\partial \eta}, \frac{\partial y}{\partial \xi} \text{ and } \frac{\partial y}{\partial \eta}$$

with the corresponding constant values at $\xi = \eta = 0$; these are the values that occur naturally for elements of parallelogram form.

WILSON and IBRAHIMBEGOVICZ [45] used a somewhat different approach; they added a *constant* matrix $\mathbf{B}_{ac}$ to $\mathbf{B}_a$, such that

$$\hat{\mathbf{B}}_a = \mathbf{B}_a + \mathbf{B}_{ac} \tag{11-40}$$

Plane stress and plane strain

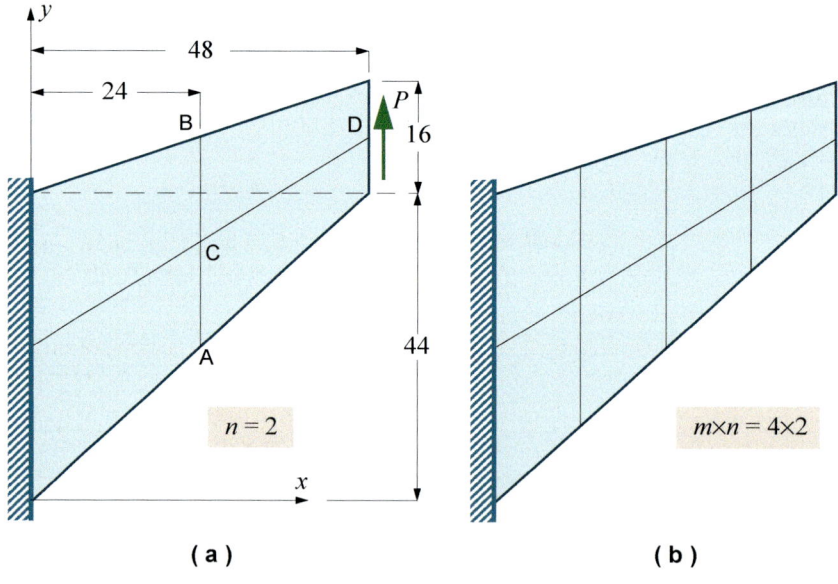

Figure 11.11 Cook's problem – quadrilateral meshes

satisfies

$$\int_V \hat{\mathbf{B}}_a dV = \int_V (\mathbf{B}_a + \mathbf{B}_{ac}) dV = 0 \qquad (11\text{-}41)$$

from which, since $\mathbf{B}_{ac}$ is constant,

$$\mathbf{B}_{ac} = -\frac{1}{V}\int_V \mathbf{B}_a dV \qquad (11\text{-}42)$$

By replacing $\mathbf{B}_a$ in Eqs. (11-34) and (11-35) by $\hat{\mathbf{B}}_a$, we have obtained a general, non-compatible 4-node quadrilateral element that satisfies the (strong) patch test; it is thus a convergent element, and it outperforms the standard, conforming bilinear element. This "double fix" – we first violate continuity and then manipulate the strain matrix (**B**) – led the mathematician Gilbert STRANG to comment that "in California two wrongs make it right" (both WILSON and TAYLOR were professors at UC Berkeley).

Both the conforming and non-conforming 4-node quadrilateral element are included in the **FEMplate** program. The basic and conforming version is termed **Q4** while the non-conforming version is termed **Q6**, which is slightly misleading since it has the same number of nodes as the basic version, that is 4. On the "outside" these two elements look the same; they are of the same *kind*, but of different *type*. It should be emphasized that neither of these two element types uses *selective* numerical integration in the **FEMplate** implementation; for **Q4** the user can choose which GAUSS rule to use, whereas **Q6** use only a 2×2 GAUSS rule, that is, full integration.

Example 11-2 – COOK's problem again

We repeat the analyses of Example 1 with the two versions of the 4-node quadrilateral element, and we start with the same basic mesh as we used for the triangular elements, see Fig 11.11a.

Table 11-2: Results, in percent of correct value, for COOK's problem

	$n =$	2	4	8	16	32
Q4	v_D	49,5	76,8	92,3	97,9	99,4
	σ_{xA}	58,5	94,3	108,3	108,0	105,0
	σ_{xB}	41,4	77,6	96,1	101,0	101,3
	τ_{xyC}	74,8	97,0	95,7	96,8	96,8
Q6	v_D	87,8	95,8	98,9	99,1	99,9
	σ_{xA}	88,5	102,3	101,9	101,2	100,7
	σ_{xB}	77,0	92,0	98,5	100,0	100,1
	τ_{xyC}	96,5	102,9	99,8	100,0	99,9

Plane stress and plane strain

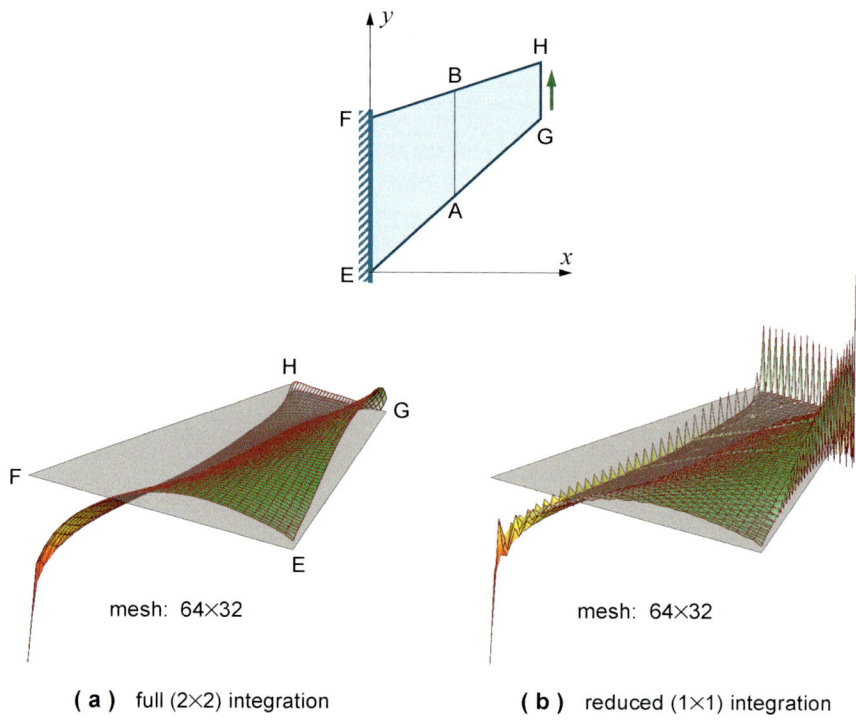

Figure 11.12 Axial stress (σ_x) for Cook's problem, analyzed by **Q4** (nodal averaging used in both cases)

The results, obtained with "full" integration (2×2 GAUSS), are shown in Table 11-2. All stress results are based on simple nodal point averaging of stresses computed at the nodal points of each individual element.

Both elements have the same number of system *dofs* as **CST** (see Table 11-1) for the same value of n. Compared with the triangular element (**CST**) the results of Table 11-2 are somewhat more erratic, but judging from this one example, the results rank **Q6** on top. There is little in it between **CST** and **Q4**, but **Q6** clearly performs better than both **Q4** and **CST**.

We emphasize that neither **Q4** nor **Q6** use selective numerical integration, and in view of what was pointed out earlier in this section (in relation to Fig. 11.8) one might suspect that parasitic shear could contaminate the results since some of the elements of the mesh are fairly elongated. In order to investigate this we have also carried out analyses based on the mesh in Fig. 11.11b, see table 11-3.

Table 11-3: Results, in percent of correct value, for COOK's problem

	$m \times n =$	4×2	8×4	16×8	32×16	64×32	64×32
Q4	v_D	69,8	88,9	96,7	99,1	99,7	100,1
	σ_{xA}	115,2	126,5	119,6	111,5	106,1	180,3
	σ_{xB}	80,1	98,8	102,9	102,6	101,6	138,7
	τ_{xyC}	42,3	99,9	99,9	99,8	99,9	98,3
Q6	v_D	98,1	99,0	99,6	99,8	100,0	
	σ_{xA}	105,1	104,3	102,7	101,5	100,8	
	σ_{xB}	93,3	99,2	100,3	100,3	100,2	
	τ_{xyC}	94,3	98,2	99,3	99,7	99,9	

The results in Table 11-3 are somewhat unexpected, particularly for **Q4**. The displacements (v_D) and the shear stresses are clearly better than those of Table 11-2, but the axial stresses are, if anything better in Table 11-2. In view of the fact that, column for column, the results of Table 11-3 are obtained with twice the number of less distorted elements than those reported in Table 11-2, one would have expected better numbers in Table 11-3. For **Q6** the results of Tables 11-2 and 11-3 are about as expected. All results in Table 11-3, except for the last column, are obtained with full (2×2) integration. For **Q4** the last column is obtained with reduced (1×1) integration.

It is quite interesting to note that we also get a solution with **Q4** when we use reduced integration (only one integration point) in spite of the fact that the element has two zero-energy modes for this integration scheme. At the system level these modes are obviously checked by the boundary conditions. Not only do we get a solution, but for both displacement and shear stress the results are also very nearly the correct ones. However, the axial stresses are not very good, a fact made very clear by Fig. 11.12. The only difference in

Plane stress and plane strain

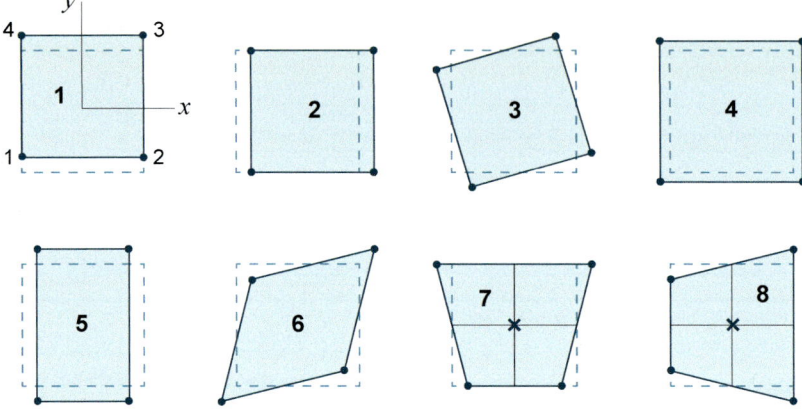

Copy of **Figure 6.20** Independent displacement modes for the 4-node plane element

the two plots is the numerical integration used for the element stiffness matrices.

Stabilization – hourglass stiffness. The presence of zero-energy ("hourglass") modes and the results of Example 2 above seem to rule out reduced integration for the 4-node quadrilateral. However, researchers were not willing to accept this verdict since the potential gain of using one-point quadrature for the 4-node bilinear element is considerable; it reduces the computational effort to about one-fourth of the full, 2×2 quadrature rule.

As already pointed out in Chapter 6 the problem is associated with the last two displacement modes of Fig. 6.20, the so-called hourglass patterns. All strain components are zero at the centre of these two modes, and with only one integration point, placed at the centre, these two modes contribute zero energy. KOSLOFF and FRAZIER [46] presented a very simple way to reintroduce the stiffness of the hourglass modes. Not only did this make it safe to use one-point integration (and thus save computer time), it also gave the element a noticeable improved bending behaviour. An outline of the technique, as described in [21], is as follows:

For simplicity we only consider the *dofs* in the x-direction,

$$\mathbf{v}_x = [\, u_1 \ u_2 \ u_3 \ u_4 \,]^\mathrm{T}$$

For modes 1 and 8 we have $\mathbf{v}_x = \mathbf{0}$; hence $\mathbf{v}_x$ can be expressed as the following linear combination

$$\mathbf{v}_x = a_2 \begin{bmatrix} 1 \\ 1 \\ 1 \\ 1 \end{bmatrix} + a_3 \begin{bmatrix} 1 \\ 1 \\ -1 \\ -1 \end{bmatrix} + a_4 \begin{bmatrix} -1 \\ 1 \\ 1 \\ -1 \end{bmatrix} + a_5 \begin{bmatrix} 1 \\ -1 \\ -1 \\ 1 \end{bmatrix} + a_6 \begin{bmatrix} -1 \\ -1 \\ 1 \\ 1 \end{bmatrix} + a_7 \begin{bmatrix} 1 \\ -1 \\ 1 \\ -1 \end{bmatrix} \tag{11-43}$$

where $a_1, a_2, \ldots, a_7$ are constants. The last term we write as $\mathbf{v}_{x7}$, that is

$$\mathbf{v}_{x7} = a_7 [\, 1 \ -1 \ 1 \ -1 \,]^\mathrm{T} \tag{11-44}$$

The *stabilization* matrix or *hourglass stiffness* matrix defined as

$$\mathbf{k}_{xhg} = \mathbf{v}_{x7}\mathbf{v}_{x7}^\mathrm{T} \tag{11-45}$$

provides mode 7 with the stiffness it lacks under one-point quadrature. The constant a_7 is determined such that a rectangular element displays the exact strain energy in a state of pure bending. Since mode 7 is orthogonal to all other modes, that is

$$\mathbf{v}_{x7}^\mathrm{T}\mathbf{v}_{xi} = 0 \quad \text{for } i = 1, 2, 3, 4, 5, 6 \text{ and } 8 \tag{11-46}$$

$\mathbf{k}_{xhg}$ will not stiffen modes other than mode 7. Similarly $\mathbf{k}_{yhg}$ is determined from mode 8.

The final step is to extend the derivation to the case of an arbitrary quadrilateral geometry. This requires the definition of the hourglass mode of Eq. (11-44) to be generalized. For details, see HUGHES [20].

Plane stress and plane strain

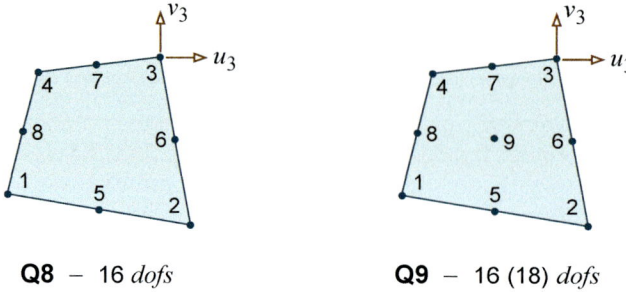

Q8 – 16 *dofs*   **Q9** – 16 (18) *dofs*

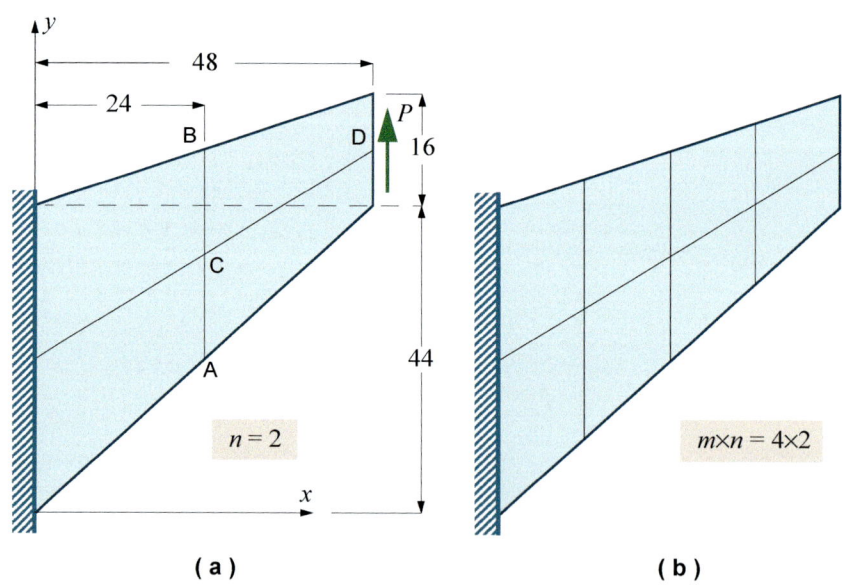

(a)   (b)

Copy of **Figure 11.11** Cook's problem – quadrilateral meshes

By adding the (total) hourglass stiffness (from modes 7 and 8) to the stiffness obtained for **Q4** using a one-point quadrature rule we obtain a 4-node quadrilateral which allegedly possesses very much the same properties as the **Q6** element, but is obtained at approximately one-fourth of the computational effort.

Higher order quadrilateral elements. We now turn to the **Q8** and **Q9** elements, both of which have been dealt with quite extensively already, see Section 5.7, Chapters 6 and 7 and Section 10.3. Here we will put them to the same test as we have the other elements. For a given mesh both elements have the same number of "system" *dofs*, since the internal *dofs* of the 9-node LAGRANGE element are eliminated, by static condensation, at the element level.

Example 3 – Cook's problem yet again

We now repeat the analyses of Examples 1 and 2 with elements **Q8** and **Q9**. We only use the mesh of Fig. 11.11a, that is, the same number of elements in both directions. The results are presented in Table 11-4, for both *full* (3×3) and *reduced* (2×2) numerical integration. All numbers in parentheses apply to reduced integration.

Table 11-4: Results, in percent of correct value, for Cook's problem

	$n =$	2	4	8	16	32
Q8	v_D	95,1 (96,9)	99,0 (99,0)	99,67 (99,67)	99,89 (99,89)	99,95 (99,95)
	σ_{xA}	120,3 (129,0)	108,8 (111,9)	102,9 (103,5)	100,7 (100,9)	100,1 (100,2)
	σ_{xB}	107,1 (108,7)	98,2 (98,9)	100,1 (103,5)	100,1 (100,1)	100,0 (100,0)
	τ_{xyC}	97,7 (93,1)	91,6 (91,5)	98,5 (98,5)	99,70 (99,69)	99,94 (99,94)
Q9	v_D	97,3 (100,7)	99,5 (100,1)	99,83 (100,1)	99,95 (100,0)	99,95 (100,0)
	σ_{xA}	121,2 (166,0)	107,3 (127,5)	102,0 (111,8)	100,5 (107,5)	100,1 (105,7)
	σ_{xB}	100,3 (161,4)	100,5 (167,5)	100,3 (152,4)	100,1 (140,4)	100,0 (131,2)
	τ_{xyC}	63,8 (40,0)	93,0 (59,9)	98,6 (83,3)	99,7 (93,8)	99,93 (97,7)

NOTE: Results in parentheses are obtained with reduced (2×2) integration.

Before we comment on the results in Table 11-4 it should be pointed out that *all* stresses are obtained by straightforward nodal point averaging of stresses computed *directly* at the corner nodes for each element.

For full (3×3) integration there is very little difference in performance between **Q8** and **Q9**. If anything **Q9** is marginally better. With reduced (2×2) integration it is a different story. The tip displacement is still better with **Q9**,

Plane stress and plane strain

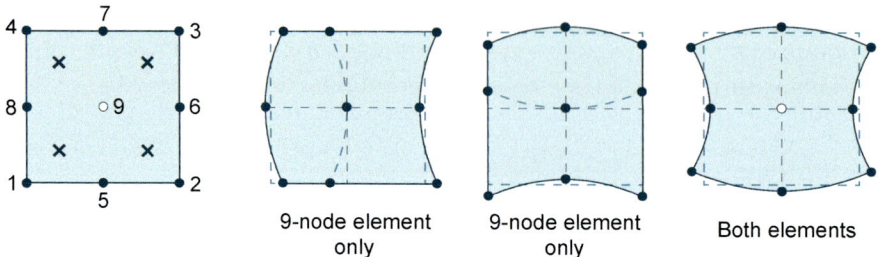

Copy of **Figure 6.21** Spurious modes for **Q8** and **Q9** elements due to reduced integration

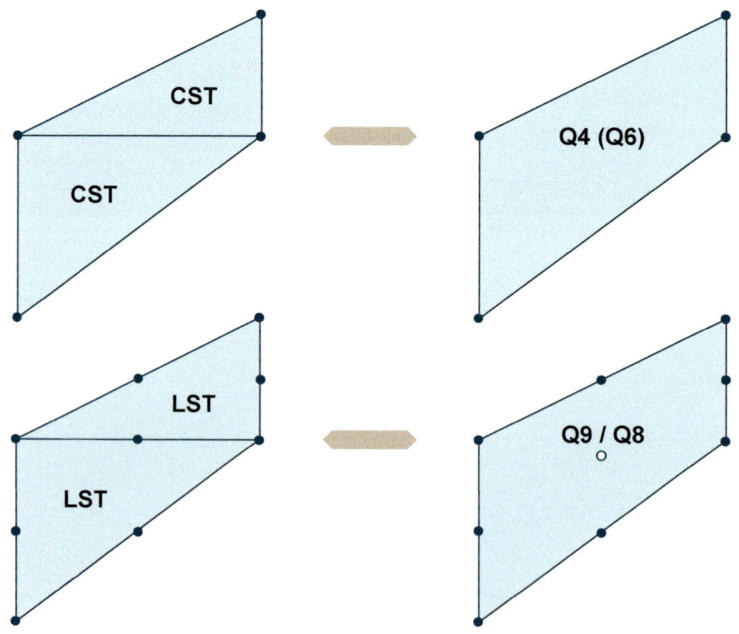

Figure 11.13 Triangular and quadrilateral counterparts

which performs extremely well, even for the coarser meshes, but we note that the displacement does not converge from the stiff side.

While stresses obtained with reduced integration are quite acceptable for **Q8**, they are remarkably poor for **Q9**. Even for the finest mesh, the axial stress σ_x at point B is off by more than 30 per cent! It is hard to see that anything we have said about this element previously can explain this poor performance. Perhaps a program bug? Possible, but not very likely. This last statement is supported by the fact that if we change strategy for the stress computation, and compute element stresses at the BARLOW points – here the integration points of the reduced (2×2) quadrature rule – and extrapolate these stresses to the nodal points before we do the averaging, then we get far better stress results. With this strategy, reduced integration for the **Q9** element gives the following σ_x stresses (in per cent of the correct value) at point B for meshes from $n = 2$ to $n = 32$:

$$95,3 \quad 101,2 \quad 100,2 \quad 100,0 \quad 100,0$$

As pointed out in Section 6.5, **Q8** and **Q9** also exhibit spurious zero energy modes if reduced (2×2) quadrature is used, see Fig. 6.21. While not as serious as for **Q4**, this represents a potential problem that is best removed if possible, especially for **Q8**. Hence in order to use reduced (2×2) integration safely, a procedure which more than halves the computational effort, stabilization by means of hourglass stiffness similar to that described for **Q4** has also been developed for **Q8** and **Q9** [20].

Regardless of which quadrature rule is used for the stiffness matrices of **Q8** and **Q9**, their stresses should always be determined at the BARLOW points, that is, at the integration points of the reduced quadrature rule.

Discussion. In dealing with particular finite elements, many textbooks and journal papers present results for very coarse meshes of simple problems with a known solution, such as the cantilever rectangular plate. While not wrong, conclusions drawn from such examples may not tell the whole story. By the same token, results from a single problem, like the one we have presented in this section, do not call for sweeping statements. Nevertheless some observations/comments can be made.

A comparison between the triangular and the quadrilateral elements reveals a noticeable difference, in approach rather than performance. If we forget about curved edges then the derivation of all three triangular elements is fairly straightforward, and all requirements are satisfied *a priori*. With exact, analytical integration we end up with robust elements, for which zero energy modes and shear locking are not an issue.

For comparison between triangular and quadrilateral elements, **CST** should be compared to **Q4** and **Q6**, whereas **LST** is the counterpart to **Q8** and **Q9**. The quadrilateral counterpart to **QST** would be the next element in the LAGRANGIAN family (with two nodes on each side and four internal nodes). When comparing results from Tables 11-1 and 11-2, **Q4**, and particularly **Q6**, perform better than **CST** for all meshes, and thus for a given number of (system) *dofs*. However for the finer meshes the difference becomes smaller, and for the finest mesh the difference between **CST** and **Q4** (without stabilization)

Plane stress and plane strain

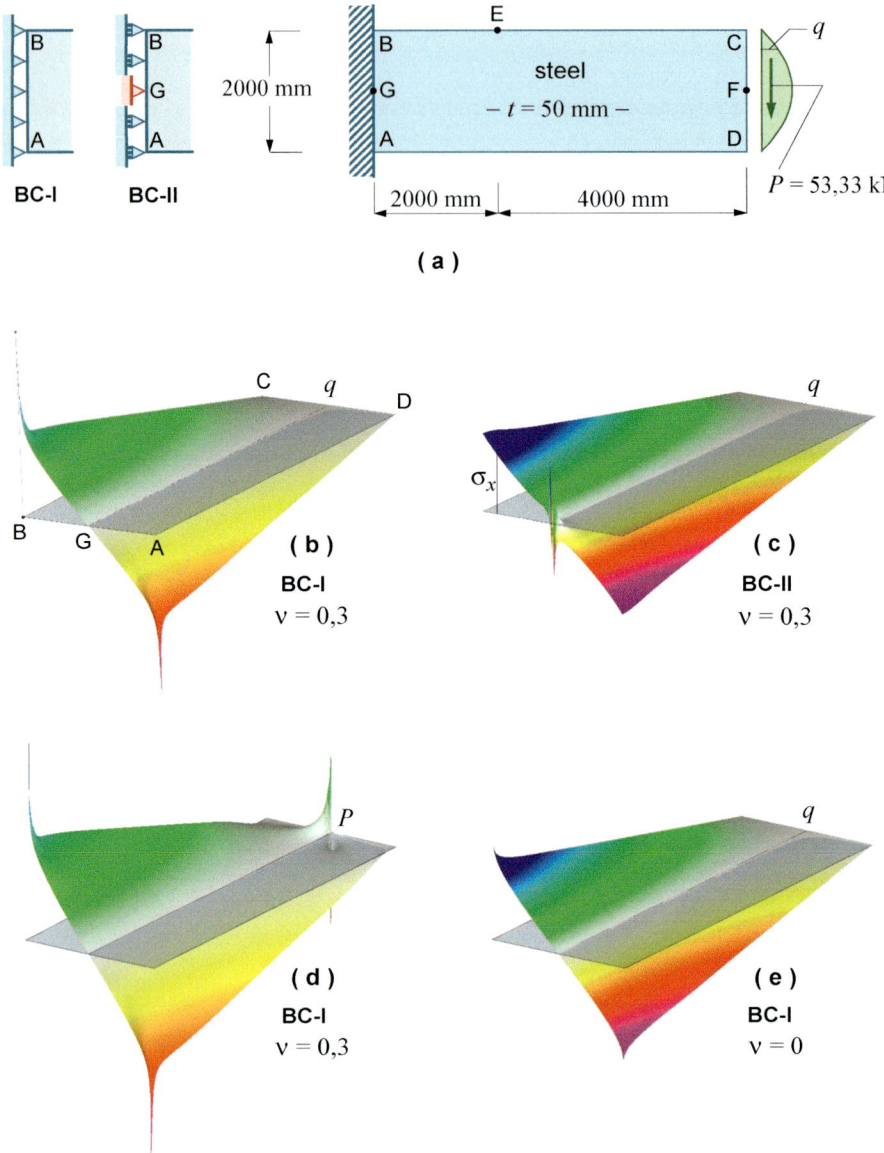

Figure 11.14 Axial stress (σ_x) computed (by **FEMplate**) for different boundary conditions, load application and Poisson's ratio

is more academic than practical. Keep in mind that none of the meshes used to produce the results presented in the tables would be considered fine by today's standards.

LST compare very well with its two quadrilateral counterparts, provided stresses are computed at the BARLOW points and extrapolated to the corner nodes for the two quadrilateral elements. If anything **LST** seems to do slightly better than the two quadrilateral elements for the coarser meshes. However, for a given mesh the **LST** model will contain more (system) *dofs*, also more than **Q9**, since its internal *dofs* are eliminated at the element level. Judging from this one example it would probably be hard to pick a winner between **LST**, **Q8** and **Q9**.

Most of the "problems" with the quadrilateral elements, which all make use of the isoparametric concept and thus numerical integration, stem from the various integration schemes. For the **Q4** element we pointed out the potential problem of shear locking tendencies in the case of pure bending, but our example, which has some bending action, did not reveal any such tendencies. Hence for both **Q4** and **Q6** full integration seems to be the natural choice. Reduced integration, which for these two elements would mean just one integration point, is not an option for **Q6**, and if applied for **Q4** it must be coupled with the augmentation of hourglass stiffnesses. Such a modified **Q4** element appears to have excellent properties, on a par with **Q6**.

While stresses should be computed at BARLOW points for **Q8** and **Q9**, no problems have been observed with stresses computed directly at the (corner) nodes of the triangular elements, or of **Q6**. How to compute stresses for **Q4** with one-point integration and hourglass stiffness is not obvious; the use of just one integration point would mean constant stresses in the element, and that seems to defeat the purpose of the "mending" exercise.

11.3 Boundary conditions and singularities

For standard type *dofs* – nodal values of the displacement components – it is, by and large, fairly straightforward to apply boundary conditions for in-plane problems of plane stress and plane strain. In many cases the actual modelling of a particular physical problem presents the more important issues to be resolved.

To give but a flavour of the type of issues and accompanying results we encounter in FEM analyses, Fig. 11.14 shows some results obtained for a cantilever plate. All results are obtained with a fine mesh of **LST** elements; 38400 elements resulting in 154560 equations should guarantee results (away from singularities) that are exact to at least 4 digits. In addition to the stress surfaces for the axial stress σ_x, the value of this stress at point E, σ_{xE}, and the tip displacement at point F, δ_F, are all recorded as typical measures of the plate behaviour. The stress surfaces are shown to roughly the same scale, but keep in mind that the "colour scale" used, always shows the maximum/minimum stress (including the singularity spikes) with the same colour.

The same tip loading is applied in two different ways, distributed as a parabola over the plate height (q) and as a concentrated load (P) at point F. Two

Plane stress and plane strain

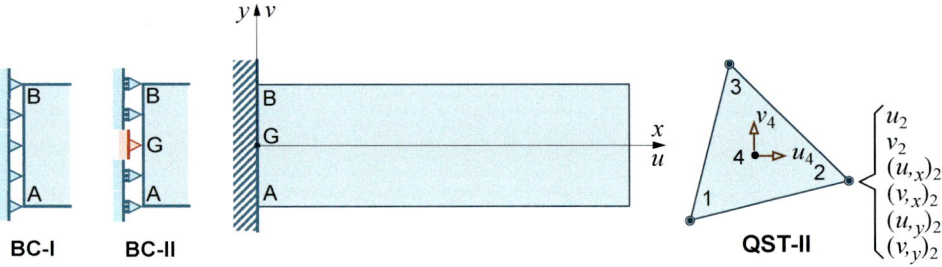

BC-I : At each nodal point on A-B: $u = v = u_{,y} = v_{,y} = 0$ while $u_{,x}$ and $v_{,x}$ are "free"

BC-II : At the nodal point at point G: $u = v = u_{,y} = v_{,y} = 0$ while $u_{,x}$ and $v_{,x}$ are "free"
At all other nodal points on A-B: $u = u_{,y} = 0$ while all other *dofs* are "free"

Figure 11.15 Boundary conditions for the **QST-II** element

different assumptions are also used for the boundary conditions at the fixed end. The most common assumption is to completely fix *all* nodes along the edge, denoted **BC-I** in the figure. Another possibility is to completely fix only *one* point and leave the other free to move in the vertical direction while preventing horizontal movement, denoted **BC-II**. With the same loading (q), Figs. 11.14b and c show the results for the axial stress σ_x for the two cases. As expected the singularities occur in different places, and for our two control parameters we find:

Fig. 11.14b: **BC-I** : $\sigma_{xE} = 6{,}3986$ Mpa   and $\delta_F = 0{,}5921$ mm
Fig. 11.14c: **BC-II** : $\sigma_{xE} = 6{,}3992$ MPa   and $\delta_F = 0{,}6117$ mm

Figure 11.14d shows the stress surface for a case which is the same as that shown in Fig. 11.14b, except for the loading which is applied as a concentrated load. The only visual difference is that we now get a singularity also at the point of load application. For the control parameters we get:

Fig. 11.14c: **BC-I** : $\sigma_{xE} = 6{,}400$ Mpa   and $\delta_F = 0{,}6064$ mm

Our last example shown in Fig. 11.14e is, save for POISSON's ratio which is now set to zero, identical to Fig. 11.14b. From this we can conclude that the singularities of Fig. 11.14b are to a large extent caused by the POISSON effect. In this case we find:

Fig. 11.14e: **BC-I** : $\sigma_{xE} = 6{,}3990$ Mpa   and $\delta_F = 0{,}5957$ mm

Apart from the singularities, the plate behaviour is hardly influenced by the two different assumptions. As for the stress singularities, the numerical stress values increase with finer meshes, but at the same time the area influenced by the singularity decreases. Figure 11.14 demonstrates ST. VENANT's principle of rapidly vanishing edge disturbances (boundary conditions and loading) quite well.

For the QST-II element the boundary conditions, for the two (physically) different approaches in Fig. 11.14, are shown in Fig. 11.15.

Problems

Problem 11.1

Use the procedure presented in Section 11.1 to determine an explicit expression for k_{11} for a **CST** element of an isotropic elastic material. The stiffness should be expressed in terms of the constat thickness t, the corner coordinates x_i and y_i ($i = 1, 2$ and 3), YOUNG's modulus E and POISSON's ratio ν.

Problem 11.2

Examine the procedure presented in Section 10.3 for the determination of the stiffness matrix for an isoparametric, quadrilateral element to see whether an intermediate arrangement of the element *dofs* into $\mathbf{v}_x$ and $\mathbf{v}_y$, as employed for the triangular elements in Section 11.1, has any merits.

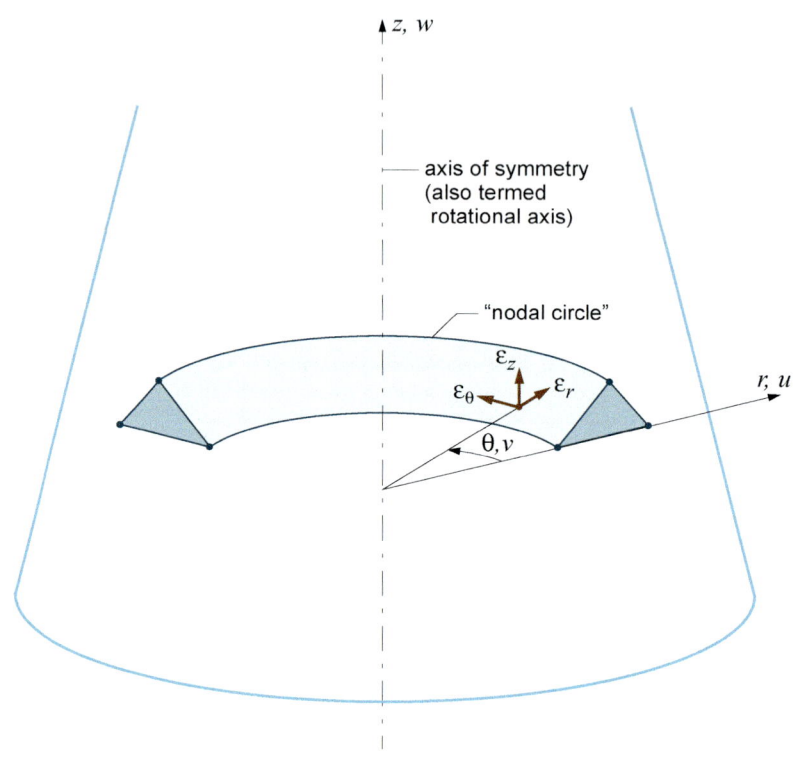

Equation (3-39):

$$\boldsymbol{\varepsilon} = \begin{bmatrix} \varepsilon_r \\ \varepsilon_\theta \\ \varepsilon_z \\ \gamma_{zr} \end{bmatrix} = \begin{bmatrix} \dfrac{\partial}{\partial r} & 0 \\ \dfrac{1}{r} & 0 \\ 0 & \dfrac{\partial}{\partial z} \\ \dfrac{\partial}{\partial z} & \dfrac{\partial}{\partial r} \end{bmatrix} \begin{bmatrix} u \\ w \end{bmatrix} = \boldsymbol{\Delta u}$$

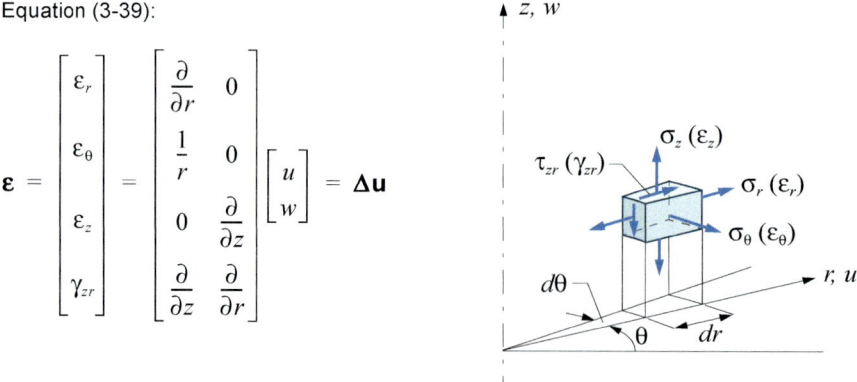

Figure 12.1 Axisymmetric body (or solid of revolution) – stress and strain

12

Axisymmetric stress analysis

In this chapter we will show that the analysis of an axisymmetric body subjected to an axisymmetric loading reduces to a two-dimensional problem. We shall also indicate that we can find the solution as the sum of solutions of two-dimensional problems even if the loading is not axisymmetric.

12.1 Axisymmetric loading

A solid of revolution is generated by revolving a plane figure about an axis in space, and such a body is most conveniently described in cylindrical coordinates r, θ and z, see Fig. 12.1. If both the structure and the loading are axisymmetric about the rotational axis (z), see Fig. 12.1, we can, as in Section 3.2, make the following assumptions:

Geometry, material properties, loading and boundary conditions are independent of the angle θ.

This implies

$$v = 0 \quad \text{and} \quad \tau_{r\theta} = \tau_{\theta z} = \gamma_{r\theta} = \gamma_{\theta z} = 0 \qquad (12\text{-}1)$$

For an arbitrary value of θ the problem is thus reduced to a 2D problem in the r-z plane. The kinematics of the problem are:

The strain-displacement relation is given by Eq. (3-39), and the stress-strain relation ($\boldsymbol{\sigma} = \mathbf{C}\boldsymbol{\varepsilon}$) is given by Eq. (3-40).

A finite element for this type of problem is a "ring" element, see Fig. 12.1, but the problem is uniquely defined by the displacements in the r-z plane (for an arbitrary value of θ), and we therefore only need to consider the 2D "cross sectional" element of the ring element.

As before we assume

$$\mathbf{u} = \begin{bmatrix} u \\ v \end{bmatrix} = \begin{bmatrix} \mathbf{N}_1 & 0 \\ 0 & \mathbf{N}_1 \end{bmatrix} \begin{bmatrix} \mathbf{v}_r \\ \mathbf{v}_z \end{bmatrix} = \hat{\mathbf{N}}\hat{\mathbf{v}} \qquad (12\text{-}2)$$

from which we find the strains as

$$\boldsymbol{\varepsilon} = \boldsymbol{\Delta}\mathbf{u} = \boldsymbol{\Delta}\hat{\mathbf{N}}\hat{\mathbf{v}} = \hat{\mathbf{B}}\hat{\mathbf{v}} \qquad (12\text{-}3)$$

Axisymmetric stress analysis

Equation (3-40) :

$$\sigma = \begin{bmatrix} \sigma_r \\ \sigma_\theta \\ \sigma_z \\ \tau_{zr} \end{bmatrix} = \frac{E}{(1+\nu)(1-2\nu)} \begin{bmatrix} 1-\nu & \nu & \nu & 0 \\ \nu & 1-\nu & \nu & 0 \\ \nu & \nu & 1-\nu & 0 \\ 0 & 0 & 0 & \frac{1-2\nu}{2} \end{bmatrix} \begin{bmatrix} \varepsilon_r \\ \varepsilon_\theta \\ \varepsilon_z \\ \gamma_{zr} \end{bmatrix}$$

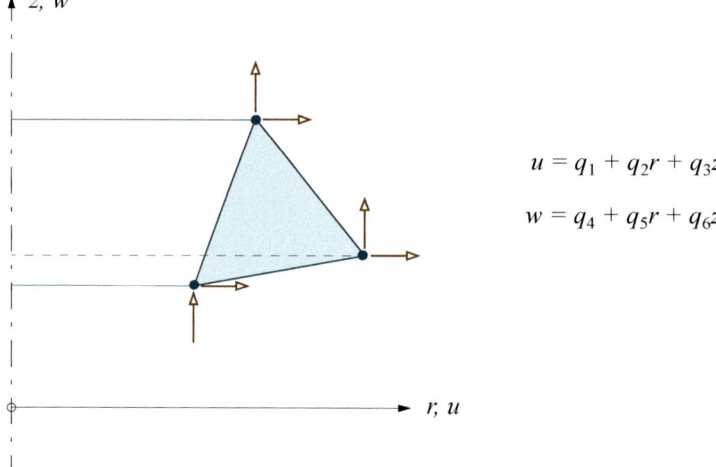

$u = q_1 + q_2 r + q_3 z$

$w = q_4 + q_5 r + q_6 z$

Figure 12.2 "CST" ring element

Axisymmetric loading

Note that the "hat" symbol (ˆ) designates that the nodal degrees of freedom ($\hat{\mathbf{v}}$) are organized in such a way that all r-displacements are listed first and then the z-displacements. The $\hat{\mathbf{B}}$ matrix now reads

$$\hat{\mathbf{B}} = \begin{bmatrix} \mathbf{N}_{1,r} & 0 \\ \dfrac{1}{r}\mathbf{N}_1 & 0 \\ 0 & \mathbf{N}_{1,z} \\ \mathbf{N}_{1,z} & \mathbf{N}_{1,r} \end{bmatrix} \qquad (12\text{-}4)$$

The second row represents the difference from the plane stress/plane strain matrix. The stiffness matrix corresponding to the particular ordering of the *dofs* becomes

$$\hat{\mathbf{k}} = \int_V \hat{\mathbf{B}}_1^T \mathbf{C} \hat{\mathbf{B}}_1 dV \quad \text{where } dV = r\,d\theta\,dA$$

and thus

$$\hat{\mathbf{k}} = \int_A \int_0^{2\pi} \hat{\mathbf{B}}_1^T \mathbf{C} \hat{\mathbf{B}}_1 r\,d\theta\,dA = 2\pi \int_A \hat{\mathbf{B}}_1^T \mathbf{C} \hat{\mathbf{B}}_1 r\,dA \qquad (12\text{-}5)$$

The elasticity matrix $\mathbf{C}$ in this equation is the 4×4 matrix defined by Eq. (3-40). Note that since r is present in the denominator of some terms of the $\hat{\mathbf{B}}$ matrix, it is also present in the denominator of some terms of the integrand.

Similarly we find

$$\mathbf{S}_F^0 = -\int_V \mathbf{N}^T \mathbf{F}\,dV = -2\pi \int_A \mathbf{N}^T \mathbf{F} r\,dA \qquad (12\text{-}6)$$

$$\mathbf{S}_\Phi^0 = -\int_S \mathbf{N}^T \Phi\,dS = -\int_S \int_0^{2\pi} \mathbf{N}^T \Phi r\,d\theta\,ds = -2\pi \int_S \mathbf{N}^T \Phi r\,ds \qquad (12\text{-}7)$$

where both $\mathbf{F}$ and Φ are axisymmetric loads. Here we have dropped the "hat" symbol, indicating that we assume the usual ordering of the nodal *dofs*. We see that the factor 2π is present in the expression for stiffness as well as the loading, and it can therefore be omitted; $\mathbf{k}$ and $\mathbf{S}^0$ are commonly expressed per radian (θ is integrated fro 0 to 1).

The only difference between plane stress/plane strain and the axisymmetric case is the additional strain and stress component in the "ring" direction, ε_θ and σ_θ. The number of nodal *dofs* and the interpolation is exactly the same.

Ring element – "CST"

With the assumptions, in terms of generalized displacements, shown in Fig. 12.2 opposite, the ring strain becomes

$$\varepsilon_\theta = \frac{u}{r} = q_1 \frac{1}{r} + q_2 + q_3 \frac{z}{r} \qquad (12\text{-}8)$$

Axisymmetric stress analysis

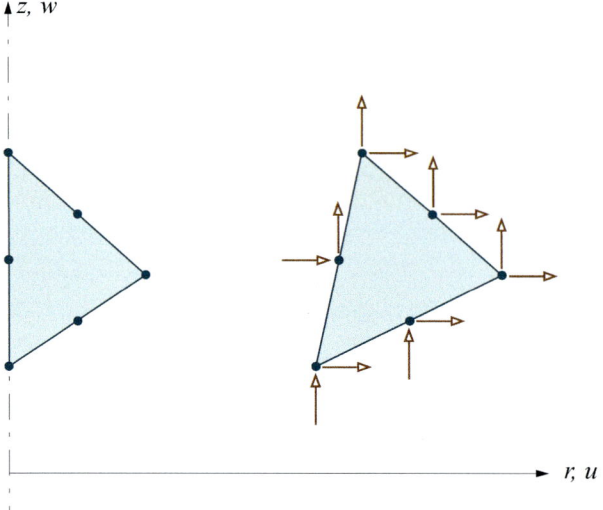

Figure 12.3 "LST" ring elements

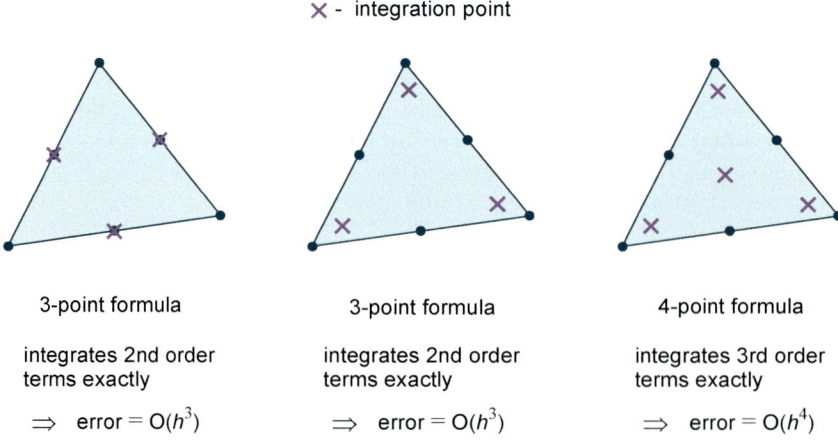

Figure 12.4 Numerical integration schemes

ε_θ is hyperbolic and it is obvious that the element cannot represent constant ring strain over the area A, particularly not for small values of r, i.e., near the axis of symmetry. The remedy is to use the value of ε_θ at the centre of the area of the element, or, in other words, use numerical integration with only *one* integration point (at the area centre). This will give, if we omit the 2π term,

$$\mathbf{k} = r_m A \mathbf{B}_m^T \mathbf{C} \mathbf{B}_m \tag{12-9}$$

where m designates the midpoint ($r_m = \frac{1}{3}(r_1 + r_2 + r_3)$). This way we also avoid the singularity problems on the axis of symmetry (where $r = 0$).

Ring element – "LST"

For this element the nodal points, the *dofs* and the interpolation functions are the same as for the corresponding plate element, see Section 11.1 and Fig. 12.3. Again the element stiffness matrix is of the form

$$\mathbf{k} = 2\pi \int_A \mathbf{B}^T \mathbf{C} \mathbf{B} r \, dA \tag{12-10}$$

and again we have the slight problem of r being present in the denominator of several terms of the integrand matrix. The computations can be accomplished both analytically and numerically.

a) Analytical evaluation

Replace $1/r$ terms in the **B** matrix by $1/r_m$ where r_m is the value of r at the element midpoint, and an analytical integration can be carried out in much the same way as in the plane (plate) case.

b) Numerical evaluation

Some possible schemes for numerical integration are shown in Fig 12.4; see APPENDIX C for more details. In order to use the first scheme, with integration points placed on the element sides, we need to avoid the problem caused by the $1/r$ term for all sides that fall on the axis of symmetry (where $r = 0$). The **B** matrix, see Eq. (12-4), must be evaluated at each integration point. When $r \to 0$ we have that $\varepsilon_\theta \to \varepsilon_r$ (ε_θ is also a radial strain at the rotational axis (at which $r = 0$); in other words

$$\frac{1}{r} N_1 \to N_{1,r}, \quad \text{when} \quad r \to 0$$

Hence: replace $\frac{1}{r} N_1$ with $N_{1,r}$ when r becomes small. It should be noted that the strains are not singular. This method is generally applicable for computing strains and stresses for points on and near the rotational axis.

Axisymmetric stress analysis

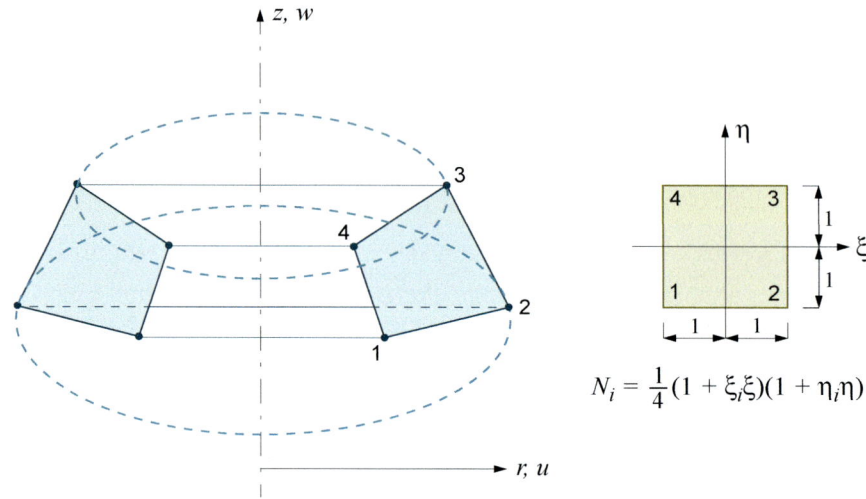

$$N_i = \tfrac{1}{4}(1 + \xi_i\xi)(1 + \eta_i\eta)$$

Figure 12.5 Isoparametric ring element

Axisymmetric loading

Isoparametric elements. Again the difference between the plane case and the axisymmetric case lies in the additional strain and stress terms (ε_θ and σ_θ). As before we assume

$$\mathbf{u} = \mathbf{Nv}$$

$$\mathbf{r} = \sum N_i r_i \quad \text{and} \quad \mathbf{z} = \sum N_i z_i$$

where r_i and z_i are the coordinates of node i. For the axisymmetric "**Q4**", the shape functions are the same as for the plane **Q4**, see Fig. 12.5, and the stiffness matrix becomes,

$$\mathbf{k} = \int_V \mathbf{B}^T \mathbf{C} \mathbf{B} dV = 2\pi \int_{-1}^{1}\int_{-1}^{1} \mathbf{B}^T(\xi,\eta)\mathbf{C}\mathbf{B}(\xi,\eta)r(\xi,\eta)J(\xi,\eta)d\xi d\eta \qquad (12\text{-}11)$$

The Jacobian $J(\xi,\eta)$ is the determinant of the Jacobi matrix defined as

$$\mathbf{J} = \frac{\partial(r,z)}{\partial(\xi,\eta)} = \begin{bmatrix} \dfrac{\partial r}{\partial \xi} & \dfrac{\partial z}{\partial \xi} \\ \dfrac{\partial r}{\partial \eta} & \dfrac{\partial z}{\partial \eta} \end{bmatrix}$$

The integrand of Eq. (12-11) will always contain rational functions and, as for the plane case, we need to use numerical integration. By using an open quadrature rule (that is all integration points are placed *inside* the element), e.g. GAUSS quadrature, we will normally not encounter any problems with the $1/r$ terms. However, since these terms vary greatly in the vicinity of the rotational axis it may be advisable (necessary) to use more integration points for elements near the z-axis (for the "core elements").

Load vectors. Axisymmetric problems often occur in machinery for which temperature can be an important "loading". As for other types of stress analysis problems we treat this kind of loading as a kind of *initial strain*, and we need to determine the statically equivalent nodal forces that we determined in Section 4.7, expressed as, see Eq. (4-56a),

$$\mathbf{S}_{\varepsilon 0}^0 = -\int_{V_e} \mathbf{B}^T \mathbf{C} \boldsymbol{\varepsilon}_0 dV$$

For an "**LST**"-type element it is reasonable to assume a linear temperature variation over the element, defined by the three corner values (T_1, T_2, T_3), that is, in terms of *area* coordinates,

$$T = \begin{bmatrix} \zeta_1 & \zeta_2 & \zeta_3 \end{bmatrix} \begin{bmatrix} T_1 \\ T_2 \\ T_3 \end{bmatrix} \qquad (12\text{-}12)$$

Hence

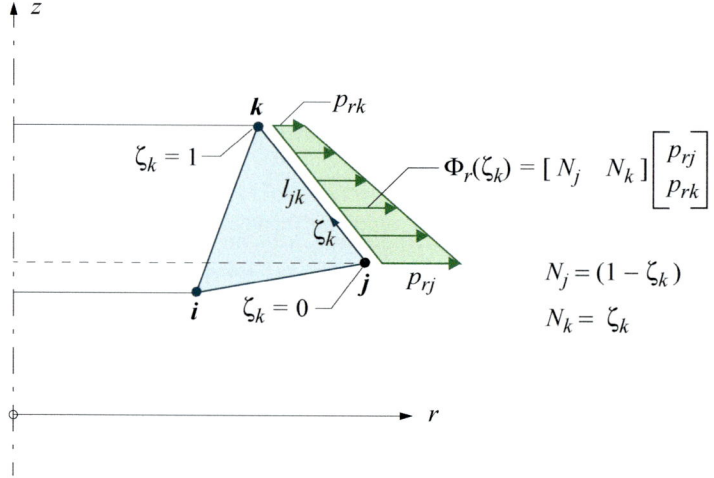

Figure 12.6 Linearly varying axisymmetric surface traction

Axisymmetric loading

$$\boldsymbol{\varepsilon}_0 = \alpha \begin{bmatrix} 1 \\ 1 \\ 1 \\ 0 \end{bmatrix} \quad T = \alpha \begin{bmatrix} \zeta_1 & \zeta_2 & \zeta_3 \\ \zeta_1 & \zeta_2 & \zeta_3 \\ \zeta_1 & \zeta_2 & \zeta_3 \\ 0 & 0 & 0 \end{bmatrix} \begin{bmatrix} T_1 \\ T_2 \\ T_3 \end{bmatrix} = \alpha \mathbf{N}_t \mathbf{t} \quad (12\text{-}13)$$

where α is the coefficient of thermal expansion. Substitution yields

$$\mathbf{S}^0_{\varepsilon 0} = -2\pi\alpha \int_{A_e} \mathbf{B}^T \mathbf{C} \mathbf{N}_t r \, dA \mathbf{t} \quad (12\text{-}14)$$

For a direct ("non-mapped") formulation we have no problems with $1/r$ (although such terms are present in the **B** matrix, the explicit r term in the integrand will cancel these terms).

Example 12-1 – "CST" element subjected to a linearly varying surface traction

The nodal forces due to a linearly varying radial surface traction Φ_r acting on the boundary l_{jk} of an axisymmetric "**CST**" triangle, as shown in Fig. 12.6, are:

$$\mathbf{S}^0_p = \begin{bmatrix} \mathbf{S}^0_{pr} \\ \mathbf{S}^0_{pz} \end{bmatrix} = \begin{bmatrix} \mathbf{S}^0_{pr} \\ 0 \end{bmatrix} = \int_S \mathbf{N}^T \Phi \, dS = 2\pi \int_{l_{jk}} \mathbf{N}^T \Phi r \, ds$$

The r-coordinate along the edge can be expressed as

$$r = N_j r_j + N_k r_k$$

and the non-zero terms of $\mathbf{S}^0_{pr}$ then becomes:

$$\mathbf{S}^0_{pr} = \begin{bmatrix} S^0_{prj} \\ S^0_{prk} \end{bmatrix} = 2\pi \int_0^1 \begin{bmatrix} 1-\zeta_k \\ \zeta_k \end{bmatrix} \begin{bmatrix} 1-\zeta_k & \zeta_k \end{bmatrix} \{(1-\zeta_k)r_j + \zeta_k r_k\} d\zeta_k \begin{bmatrix} p_{rj} \\ p_{rk} \end{bmatrix}$$

Boundary conditions. Due to the rotational symmetry the problem has only one possible *rigid body* motion, translation in the z-direction. This motion is prevented by prescribing w at *one* or more nodal circles. The radial displacement u should be suppressed (set to zero) at all nodes on the z-axis.

Convergence. Due to $1/r$, valid elements for axisymmetric solids need to satisfy a *weak patch* test.

Axisymmetric stress analysis

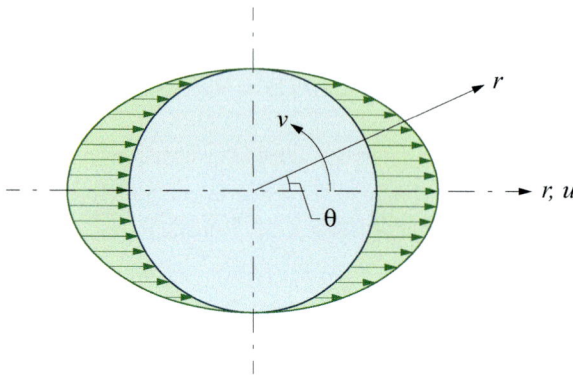

Figure 12.7 Example of loading that is not axisymmetric, but symmetric about r (at $\theta = 0$)

12.2 Non-symmetric loading

We shall not present a detailed solution to this problem, merely indicate how it can be obtained. Although the loading is no longer axisymmetric it is not completely arbitrary. We make the following assumption about the loading:

It is an arbitrary, but unique function of θ, and this function is the same for *all* values of r and z; the magnitude of the load can, however, vary both radially and vertically (along z). Such a load can be expressed as the sum of a *symmetric* part and an *antisymmetric* part; the symmetry/antisymmetry is about an r-axis at $\theta = 0$.

For simplicity we start our discussion by considering only the symmetric part of the loading, an example of which is shown in Fig. 12.7. The first step is to express the loading in terms of FOURIER series, e.g.

$$q_r = \sum_n \bar{q}_{rn}\cos(n\theta)$$

$$q_\theta = \sum_n \bar{q}_{\theta n}\sin(n\theta) \tag{12-15}$$

$$q_z = \sum_n \bar{q}_{zn}\cos(n\theta)$$

where $\bar{q}_{rn}$, $\bar{q}_{\theta n}$ and $\bar{q}_{zn}$ are amplitude values that can be functions of r and z, but not of θ. The loads in Eqs. (12-15), which can be either volume or surface forces, satisfy the symmetry property (about r at $\theta = 0$).

It can be shown that the loading of Eqs. (12-15) produces displacements

in radial direction: $\quad u = \sum_n \bar{u}_n \cos(n\theta)$

in ring direction: $\quad v = \sum_n \bar{v}_n \sin(n\theta) \tag{12-16}$

in axial direction: $\quad w = \sum_n \bar{w}_n \cos(n\theta)$

Inside each (ring) element the amplitude values ($\bar{u}_n$, $\bar{v}_n$ and $\bar{w}_n$) are interpolated between the nodal point values (of the amplitudes), that is between $\bar{\mathbf{v}}_{rn}$, $\bar{\mathbf{v}}_{\theta n}$ and $\bar{\mathbf{v}}_{zn}$, respectively,

$$u_n(r,z) = \bar{\mathbf{N}}(r,z)\bar{\mathbf{v}}_{rn}$$
$$v_n(r,z) = \bar{\mathbf{N}}(r,z)\bar{\mathbf{v}}_{\theta n} \tag{12-17}$$
$$w_n(r,z) = \bar{\mathbf{N}}(r,z)\bar{\mathbf{v}}_{zn}$$

where $\bar{\mathbf{N}}$ contains the usual interpolation functions ($= \mathbf{N}_1$ of Eq. (12-2)).

In view of Eqs. (12-16),

Axisymmetric stress analysis

Three-dimensional strain-displacement relation in cylindrical coordinates:

$$\varepsilon = \begin{bmatrix} \varepsilon_r \\ \varepsilon_\theta \\ \varepsilon_z \\ \gamma_{rz} \\ \gamma_{z\theta} \\ \gamma_{\theta r} \end{bmatrix} = \begin{bmatrix} \dfrac{\partial}{\partial r} & 0 & 0 \\ \dfrac{1}{r} & \dfrac{1}{r}\dfrac{\partial}{\partial \theta} & 0 \\ 0 & 0 & \dfrac{\partial}{\partial z} \\ \dfrac{\partial}{\partial z} & 0 & \dfrac{\partial}{\partial r} \\ 0 & \dfrac{\partial}{\partial z} & \dfrac{1}{r}\dfrac{\partial}{\partial \theta} \\ \dfrac{1}{r}\dfrac{\partial}{\partial \theta} & \left(\dfrac{\partial}{\partial r} - \dfrac{1}{r}\right) & 0 \end{bmatrix} \begin{bmatrix} u \\ v \\ w \end{bmatrix} = \Delta u$$

The corresponding elasticity matrix for an isotropic material is

$$\mathbf{C} = \dfrac{E(1-\nu)}{(1+\nu)(1-2\nu)} \begin{bmatrix} 1 & f & f & 0 & 0 & 0 \\ f & 1 & f & 0 & 0 & 0 \\ f & f & 1 & 0 & 0 & 0 \\ 0 & 0 & 0 & g & 0 & 0 \\ 0 & 0 & 0 & 0 & g & 0 \\ 0 & 0 & 0 & 0 & 0 & g \end{bmatrix}$$

where

$$f = \dfrac{\nu}{1-\nu} \quad \text{and} \quad g = \dfrac{1-2\nu}{2(1-\nu)}$$

$$\mathbf{u} = \begin{bmatrix} u \\ v \\ w \end{bmatrix} = \sum_{n=0}^{p} \begin{bmatrix} \bar{\mathbf{N}}\cos(n\theta) & 0 & 0 \\ 0 & \bar{\mathbf{N}}\sin(n\theta) & 0 \\ 0 & 0 & \bar{\mathbf{N}}\cos(n\theta) \end{bmatrix} \begin{bmatrix} \bar{\mathbf{v}}_{rn} \\ \bar{\mathbf{v}}_{\theta n} \\ \bar{\mathbf{v}}_{zn} \end{bmatrix} = \sum_{n=0}^{p} \bar{\mathbf{N}}_n \bar{\mathbf{v}}_n \quad (12\text{-}18)$$

This equation can also be written as

$$\mathbf{u} = [\bar{\mathbf{N}}_0 \ \bar{\mathbf{N}}_1 \ \ldots \ \bar{\mathbf{N}}_p] \begin{bmatrix} \bar{\mathbf{v}}_0 \\ \bar{\mathbf{v}}_1 \\ \vdots \\ \bar{\mathbf{v}}_p \end{bmatrix} \quad (12\text{-}19)$$

p is the number of FOURIER-terms deemed necessary for an adequate representation of the loading. We note that $\bar{\mathbf{N}}_n$ depends on n only through the $\cos(n\theta)$ and $\sin(n\theta)$ terms.

The problem is now three-dimensional, and we need to use the strain- and stress vectors for 3D analysis, each of which contains 6 components. The strains are expressed as

$$\boldsymbol{\varepsilon} = \sum_{n=0}^{p} \bar{\mathbf{B}}_n \bar{\mathbf{v}}_n \quad \text{where} \quad \bar{\mathbf{B}}_n = \boldsymbol{\Delta} \bar{\mathbf{N}}_n \quad (12\text{-}20)$$

and the stresses as

$$\boldsymbol{\sigma} = \mathbf{C}\boldsymbol{\varepsilon} \quad (12\text{-}21)$$

The differential operator $\boldsymbol{\Delta}$, in cylindrical coordinates, is given on the opposite page, as is the elasticity $\mathbf{C}$ matrix for an isotropic material.

The strain energy can now be expressed as

$$U_e = \frac{1}{2}\int_V \boldsymbol{\varepsilon}^T \mathbf{C} \boldsymbol{\varepsilon} \, dV = \frac{1}{2}\int_A \int_0^{2\pi} \left(\sum_{m=0}^{p} \bar{\mathbf{v}}_m^T \bar{\mathbf{B}}_m^T \right) \mathbf{C} \left(\sum_{n=0}^{p} \bar{\mathbf{B}}_n \bar{\mathbf{v}}_n \right) r \, d\theta \, dA \quad (12\text{-}22)$$

or

$$U_e = \frac{1}{2}\sum_{n=0}^{p} \left[\bar{\mathbf{v}}_n^T \int_A r \left(\int_0^{2\pi} \bar{\mathbf{B}}_n^T \mathbf{C} \bar{\mathbf{B}}_n \, d\theta \right) dA \, \bar{\mathbf{v}}_n \right] \quad (12\text{-}23)$$

since

$$\int_0^{2\pi} \sin(m\theta)\sin(n\theta)\,d\theta = 0 \quad \text{for} \quad m \neq n$$

$$\int_0^{2\pi} \cos(m\theta)\cos(n\theta)\,d\theta = 0 \quad \text{for} \quad m \neq n \quad (12\text{-}24)$$

$$\int_0^{2\pi} \sin(m\theta)\cos(n\theta)\,d\theta = 0 \quad \text{for all } m \text{ and } n$$

Jean Baptiste Joseph Fourier (1768-1830) was a French mathematician and physicist. He is best known for having initiated the investigation of Fourier series and their application to problems of vibrations and heat transfer. Fourier's law (of heat conduction) and Fourier transform (expresses a mathematical function of time as a function of frequency) are also named after him. Fourier is also recognized as being the first to discover the greenhouse effect, and indirectly he played an important role in the translation of ancient Egyptian.

Fourier, the son of a tailor, was orphaned at the age of ten. His first schooling was at a school run by a music master from Auxerre Cathedral. He proceeded in 1780 to the École Militaire of Auxerre where he soon showed an unusual talent for mathematics. In 1787 he decided to train for the priesthood, but his interest in mathematics continued, and he never took his religious vows. In 1793 a third element was added to Fourier's conflict between religion and mathematics when he became involved in politics and joined the local Revolutionary Committee. Disillusioned by the violence and terror he wanted to withdraw but found himself firmly entangled with the Revolution and unable to do so. During a very turbulent period he was arrested twice, and was lucky to be released on both occasions.

In 1794 Fourier was nominated to study at the École Normale in Paris where he was taught by Lagrange (whom he described as the first among European men of science), Laplace (whom he rated less highly) and by Monge (ingenious and very learned). In 1797 Fourier succeeded Lagrange at the École Polytechnique. He was renowned as an outstanding lecturer.

In 1798 Fourier joined Napoleon's army in its invasion of Egypt, as scientific adviser. On this expedition he was, among other things, made governor of Lower Egypt. After the French capitulated to the British in 1801 he went back to France to resume his academic post as professor at École Polytechnique. However, Napoleon decided otherwise and appointed him Prefect of the Department of Isère in Grenoble. In Grenoble Fourier happened to introduce an ink pressed copy of the Rosetta Stone, which he had brought back from Egypt, to a youngster, Champollion, who turned out to be a self-educated linguistic genius. Supported by Fourier and his elder brother, Champollion in 1822 presented a breakthrough translation of ancient Egyptian.

Fourier moved to England in 1816 but he returned to France and became the Permanent Secretary of the French Academy of Sciences in 1822. It should be mentioned that his "theory of heat" provoked a great deal of controversy and some of his work on the topic was not published in his time.

In other words,

$$U_e = \frac{1}{2}\sum_{n=0}^{p} \bar{\mathbf{v}}_n^T \mathbf{k}_n \bar{\mathbf{v}}_n \quad \text{where} \quad \mathbf{k}_n = \mathbf{k}_{nn} = \int_A r \left(\int_0^{2\pi} \mathbf{B}_n^T \mathbf{C} \mathbf{B}_n d\theta \right) dA \quad (12\text{-}25)$$

The arguments used to arrive at the stiffness can also be applied for the (consistent) loading. The FOURIER series of Eqs. (12-15) define the load amplitudes, and again we find that the cross-coupling terms between m and n vanish under the integration from 0 to 2π.

This procedure leads to $p+1$ (decoupled) relations of the type

$$\mathbf{K}_{nn} \mathbf{r}_{nn} = \mathbf{R}_{nn} \quad (12\text{-}26)$$

and the solution for the symmetric (about an r-axis) part of the loading is found as the sum of the component solutions $\mathbf{r}_{nn}$ ($n = 0, 1, 2, \ldots, p$).

If we also have an antisymmetric part of the loading we write Eqs. (12-15) as

$$q_r = \sum_n \bar{q}_{rn} \cos(n\theta) + \sum_n \underline{q}_{rn} \sin(n\theta)$$

$$q_\theta = \sum_n \bar{q}_{\theta n} \sin n\theta + \sum_n \underline{q}_{\theta n} \cos(n\theta) \quad (12\text{-}27)$$

$$q_z = \sum_n \bar{q}_{zn} \cos n\theta + \sum_n \underline{q}_{zn} \sin(n\theta)$$

where the symbols with a bar underneath represent the antisymmetric amplitudes. By writing the solution as

$$u = \sum_n \bar{u}_n \cos(n\theta) + \sum_n \underline{u}_n \sin(n\theta)$$

$$v = \sum_n \bar{v}_n \sin(n\theta) - \sum_n \underline{v}_n \cos(n\theta) \quad (12\text{-}28)$$

$$w = \sum_n \bar{w}_n \cos n\theta + \sum_n \underline{w}_n \sin(n\theta)$$

we find that the stiffness matrices for the symmetric and the antisymmetric parts are identical (due to the minus sign in the expression for v). Our task now is to solve $p+1$ systems of equations of the form given by Eq. (12-26), the only difference being that the right-hand (load) side now contains two columns, one for the symmetric loading and one for the antisymmetric loading. By the same token the solution $\mathbf{r}_{nn}$ contains two columns. Keep in mind though that the stiffness matrix $\mathbf{K}_{nn}$ needs to be established for each FOURIER term.

For a completely arbitrary loading we must resort to a full 3D analysis of the axisymmetric structure.

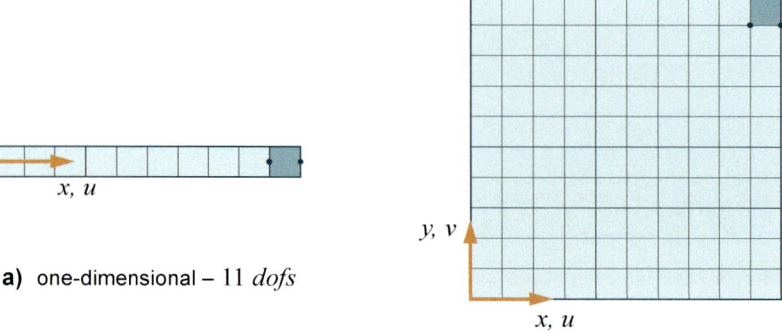

a) one-dimensional – 11 *dofs*

b) two-dimensional – 242 *dofs*

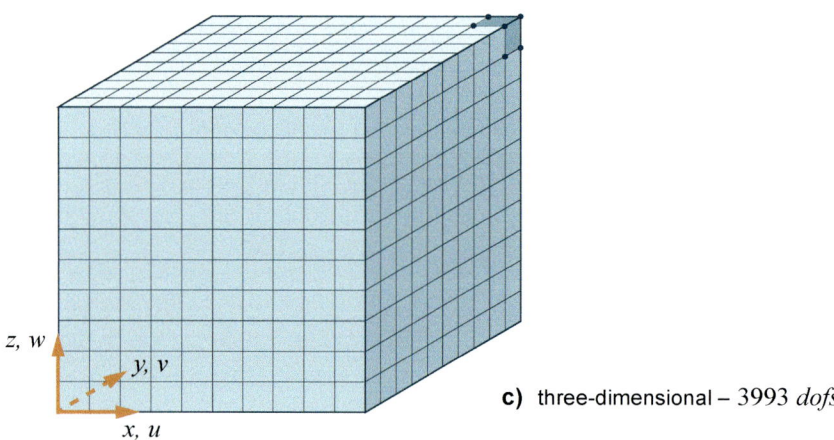

c) three-dimensional – 3993 *dofs*

Figure 13.1 Dimensionality and degrees of freedom

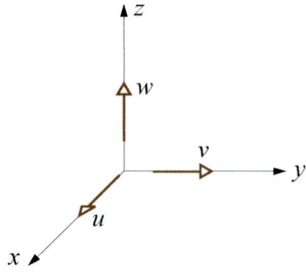

Figure 13.2 Coordinate system and field variables in 3D analysis

13

Three-dimensional stress analysis

Although not as common in structural problems as in machine type structures, 3D stress analyses are sometimes necessary in order to obtain reliable information; connections are typical candidates for such analyses. In principle three-dimensional analysis does not represent new problems, and this chapter therefore provides an overview only.

If anything, the elasticity problem as such is simpler in three dimensions than in one and two since in three dimensions we need not make any simplifying assumptions. The challenges of a three-dimensional FEM analysis, which makes use of *volume* elements, have to do with the geometrical complexity and the sheer size of such a problem. Figure 13.1 indicates how the dimensionality affects the number of degrees of freedom; in general, one more dimension increases, the number of *dofs* by more than one order of magnitude. Automatic meshing, which can also be a demanding problem in two dimensions, becomes a real challenge in three dimensions.

13.1 The basics

Theory of elasticity. General three-dimensional stress analysis is formulated with respect to a right-handed cartesian coordinate system (x,y,z), and the field variables are the displacement components in the direction of these axes, u, v and w, as shown in Fig. 13.2.

The basic equations of linear, three-dimensional theory of elasticity are given in Section 3.1, and the formulas for the element stiffness matrix (**k**) and the various consistent element load vectors (**S**^{0}), derived in Section 4.7, apply without reservation also to 3D analysis. The basic assumption now becomes

$$\mathbf{u} = \begin{bmatrix} u \\ v \\ w \end{bmatrix} = \begin{bmatrix} \mathbf{N}_1 & \mathbf{0} & \mathbf{0} \\ \mathbf{0} & \mathbf{N}_1 & \mathbf{0} \\ \mathbf{0} & \mathbf{0} & \mathbf{N}_1 \end{bmatrix} \begin{bmatrix} \mathbf{v}_x \\ \mathbf{v}_y \\ \mathbf{v}_z \end{bmatrix} = \hat{\mathbf{N}}\hat{\mathbf{v}} \qquad (13\text{-}1)$$

where the shape functions in $\mathbf{N}_1$ are now functions of all three coordinates, x, y and z. The "hat" symbol indicates that the element *dofs* are not arranged node by node, as our simple assembly procedure favours. The correct

425

Three-dimensional stress analysis

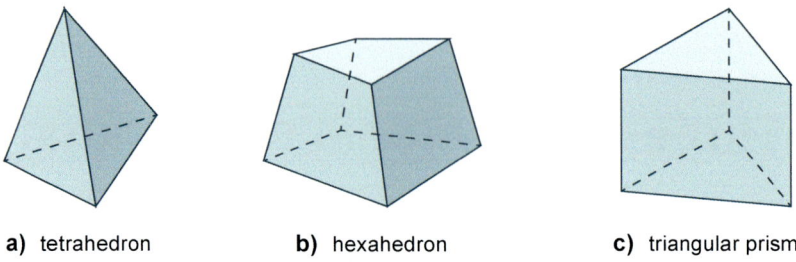

a) tetrahedron   **b)** hexahedron   **c)** triangular prism

Figure 13.3 Common element shapes for 3D analysis

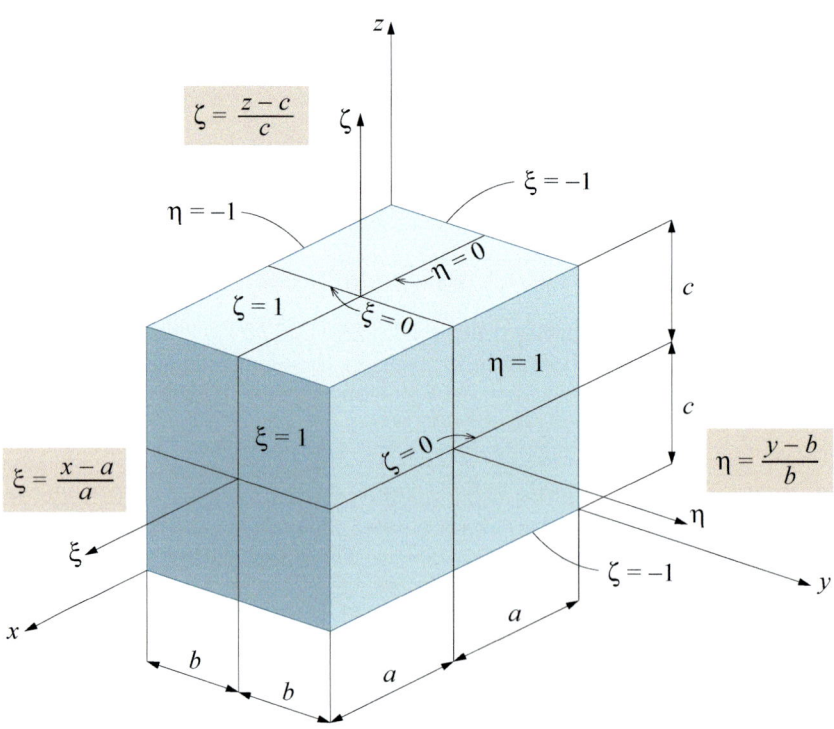

Copy of **Figure 5.5** Natural coordinates (ξ, η, ζ) for a cuboid

ordering of the element *dofs* is merely a question of rearranging rows and columns at some convenient stage of the derivation, and henceforth we therefore drop this symbol.

The strain vector **ε**, which now comprises three axial components and three shear components, is defined by Eq. (3-17), and this equation defines the **B**-matrix, that is

$$\boldsymbol{\varepsilon} = \boldsymbol{\Delta}\mathbf{u} = \boldsymbol{\Delta}\mathbf{N}\mathbf{v} = \mathbf{B}\mathbf{v} \quad \Rightarrow \quad \mathbf{B} = \boldsymbol{\Delta}\mathbf{N} \tag{13-2}$$

where

$$\boldsymbol{\Delta} = \begin{bmatrix} \frac{\partial}{\partial x} & 0 & 0 \\ 0 & \frac{\partial}{\partial y} & 0 \\ 0 & 0 & \frac{\partial}{\partial z} \\ \frac{\partial}{\partial y} & \frac{\partial}{\partial x} & 0 \\ 0 & \frac{\partial}{\partial z} & \frac{\partial}{\partial y} \\ \frac{\partial}{\partial z} & 0 & \frac{\partial}{\partial x} \end{bmatrix} \tag{13-3}$$

Finally the elasticity matrix **C**, for a linearly elastic *isotropic* material in three dimensions, is defined by Eq. (3-30), that is

$$\mathbf{C} = \begin{bmatrix} \lambda + 2G & \lambda & \lambda & 0 & 0 & 0 \\ \lambda & \lambda + 2G & \lambda & 0 & 0 & 0 \\ \lambda & \lambda & \lambda + 2G & 0 & 0 & 0 \\ 0 & 0 & 0 & G & 0 & 0 \\ 0 & 0 & 0 & 0 & G & 0 \\ 0 & 0 & 0 & 0 & 0 & G \end{bmatrix} \tag{13-4}$$

where LAMÉS *constant* is defines by Eq. (3-31), that is

$$\lambda = \frac{\nu E}{(1+\nu)(1-2\nu)}$$

Element shapes and natural coordinates. The basic element shapes are the *tetrahedron* and the *hexahedron*, see Fig. 13.3. These are the three-dimensional extensions of the two-dimensional triangle and quadrilateral, respectively. Figure 13.3 also shows a so-called wedge element in the shape of a triangular prism. Here we limit our discussion to tetrahedral and hexahedral elements.

Natural coordinates for a regular hexahedron (a cuboid), ξ, η and ζ, are, as shown in Fig. 5.5, a simple modification/extension of the natural coordinates for a rectangle (or *vice versa*).

Three-dimensional stress analysis

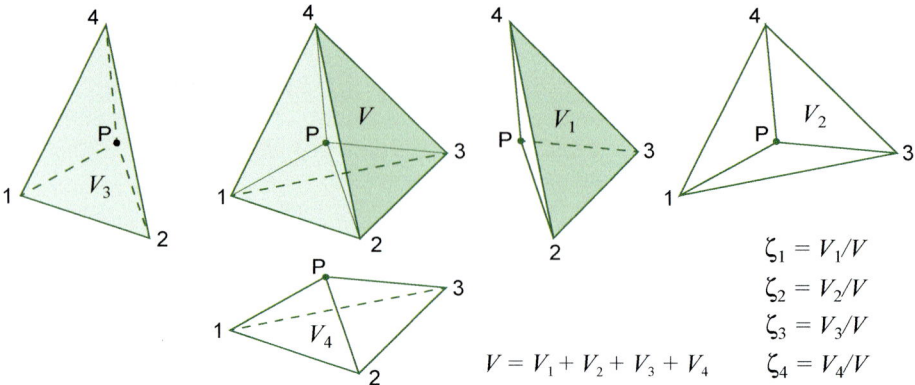

Copy of **Figure 5.8** Tetrahedron – volume coordinates

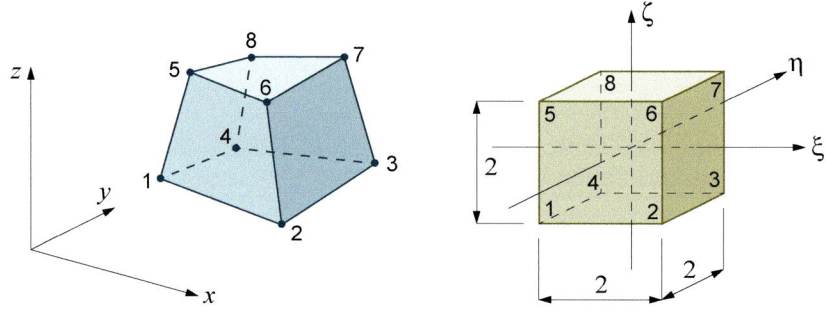

Figure 13.4 The trilinear hexahedron element (**Hex8**)

Figure 13.5 Polynomial terms of the **Hex8** element

The *volume* coordinates, see Fig. 5.8, play the same role for the tetrahedron as the area coordinates do for the plane triangle.

The order of differentiation in the strain-displacement operator Δ of Eq. (13-3) is $m = 1$; hence our problem requires $C^{m-1} = C^0$ continuity between elements, that is, the displacement components u, v and w must be continuous across element boundaries. The element boundaries are now *surfaces*, and the continuity requirement therefore means that the value of any displacement component at an arbitrary point of an element surface must be uniquely defined by shape functions associated *only* with *dofs* defined at nodal points that are in some way associated with this particular surface. Hence, shape functions associated with *dofs* at nodes *not* on the particular surface must have the value zero at all points on the surface.

13.2 Common solid elements

Hexahedral elements. The simplest hexahedral element is the 8-node *trilinear* element, often referred to as the eight-node brick element, designated Hex8, see Fig. 13.4. It is the three-dimensional counterpart of the standard quadrilateral Q4.

If formulated as an *isoparametric* element, **Hex8** can have arbitrary (irregular) shape, as indicated in Fig. 13.4. In this case both geometry and displacements are interpolated between the nodal point values, that is

$$\mathbf{x} = \begin{bmatrix} x \\ y \\ z \end{bmatrix} = \sum_{i=1}^{8} N_i(\xi, \eta, \zeta) \begin{bmatrix} x_i \\ y_i \\ z_i \end{bmatrix} = \sum_{i=1}^{8} N_i(\xi, \eta, \zeta) \mathbf{x}_i \quad (13\text{-}5)$$

and

$$\mathbf{u} = \mathbf{w} = \sum_{i=1}^{8} N_i(\xi, \eta, \zeta) \begin{bmatrix} u_i \\ v_i \\ w_i \end{bmatrix} = \sum_{i=1}^{8} N_i(\xi, \eta, \zeta) \mathbf{v}_i \quad (13\text{-}6)$$

The shape functions of **Hex8** are, analogous to **Q4**, easily found to be

$$N_i(\xi, \eta, \zeta) = \tfrac{1}{8}(1 + \xi_i \xi)(1 + \eta_i \eta)(1 + \zeta_i \zeta) \quad (13\text{-}7)$$

(x_i, y_i, z_i) and (ξ_i, η_i, ζ_i) represent the nodal coordinates in the cartesian and natural coordinate systems, respectively; $\mathbf{v}_i^T = [u_i \ v_i \ w_i]$ are the nodal *dofs* at node i.

The polynomial terms included in the shape functions for **Hex8** are shown in Fig. 13.5. We see that a complete linear polynomial is present, which secures completeness; we also have symmetry in the coordinates which guarantees an invariant element.

The Jacobian matrix in three dimensions is, analogous to the plane case,

Three-dimensional stress analysis

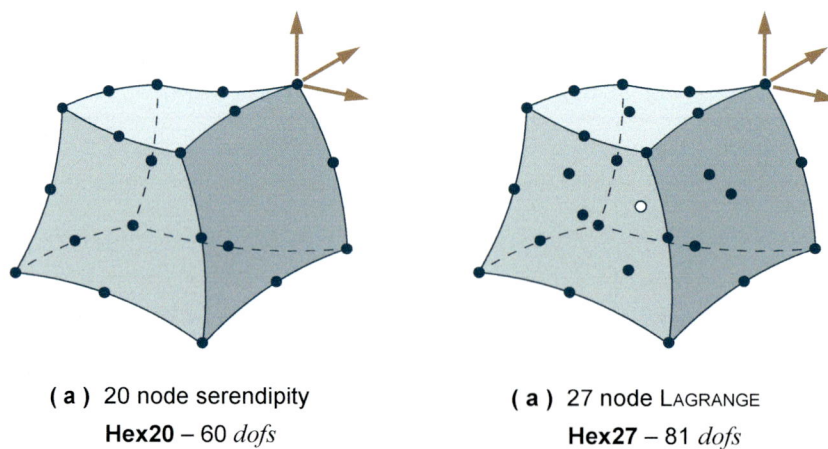

(a) 20 node serendipity
Hex20 – 60 *dofs*

(a) 27 node Lagrange
Hex27 – 81 *dofs*

Figure 13.6 Quadratic hexahedral elements

$$\mathbf{J} = \frac{\partial(x,y,z)}{\partial(\xi,\eta,\zeta)} = \begin{bmatrix} \frac{\partial x}{\partial \xi} & \frac{\partial y}{\partial \xi} & \frac{\partial z}{\partial \xi} \\ \frac{\partial x}{\partial \eta} & \frac{\partial y}{\partial \eta} & \frac{\partial z}{\partial \eta} \\ \frac{\partial x}{\partial \zeta} & \frac{\partial y}{\partial \zeta} & \frac{\partial z}{\partial \zeta} \end{bmatrix} = \sum_{i=1}^{8} \begin{bmatrix} \frac{\partial N_i}{\partial \xi} \\ \frac{\partial N_i}{\partial \eta} \\ \frac{\partial N_i}{\partial \zeta} \end{bmatrix} [x_i \ y_i \ z_i] \quad (13\text{-}8)$$

The strains are defined as derivatives of the shape functions with respect to cartesian coordinates, see Eqs. (13-2) and (13-3), and we therefore need to evaluate

$$\begin{bmatrix} \frac{\partial N_i}{\partial x} \\ \frac{\partial N_i}{\partial y} \\ \frac{\partial N_i}{\partial z} \end{bmatrix} = \mathbf{J}^{-1} \begin{bmatrix} \frac{\partial N_i}{\partial \xi} \\ \frac{\partial N_i}{\partial \eta} \\ \frac{\partial N_i}{\partial \zeta} \end{bmatrix} \quad (13\text{-}9)$$

in order to determine the elements of matrix **B**. Analogous to Eq. (6-33) we can now write the following expression for the element stiffness matrix

$$\underset{24\times 24}{\mathbf{k}} = \int_V \mathbf{B}^T \mathbf{C} \mathbf{B} \, dV = \int_{-1}^{1}\int_{-1}^{1}\int_{-1}^{1} \underset{24\times 6}{\mathbf{B}^T(\xi,\eta,\zeta)} \overset{6\times 6}{\mathbf{C}} \underset{6\times 24}{\mathbf{B}(\xi,\eta,\zeta)} J d\xi (d\eta) d\zeta \quad (13\text{-}10)$$

where we have used that $dV = J d\xi d\eta d\zeta$, J being the Jacobian (the determinant of **J**). As in the plane case **k** is evaluated by *numerical* integration, and *full* integration usually employs a 2×2×2 GAUSS quadrature rule, although this is not an optimal rule.

Hex8 has much the same shortcomings as **Q4** (noticeable mainly in connection with coarse meshes), and similar improvements ("fixes") as those applied to **Q4** can also be applied to **Hex8**. The displacement fields are augmented by nine *incompatible* modes of the type $(1-\xi^2)$, $(1-\eta^2)$ and $(1-\zeta^2)$, each multiplied by a "node-less" parameter a_i ($i = 1,....,9$), and the resulting element (**Hex11**) is then "forced" to pass the *patch* test as described in Section 11.2 for **Q6**.

Just as we did in the plane case we can also in three dimensions derive two "families" of higher order hexahedral elements, LAGRANGE and serendipity elements. The eight-node brick element is the first member in both families. The second members are shown in Fig. 13.6. Due to nodal points on the edges, and also on the surfaces in the case of the LAGRANGE element, these elements can now have curved edges/surfaces, as indicated.

Perhaps one of the most popular solid elements is the 20 node serendipity element; the additional accuracy obtained with an identical mesh of LAGRANGE (**Hex27**) elements cannot, in most cases, justify the additional computational cost. In the 2D case we can easily eliminate (by static condensation) the internal *dofs* of **Q9**, and thus make it computationally as efficient as **Q8**. In

Three-dimensional stress analysis

Shape functions for **Hex20**:

Corner nodes:

$$N_i = \tfrac{1}{8}(1+\xi_i\xi)(1+\eta_i\eta)(1+\zeta_i\zeta)(\xi_i\xi+\eta_i\eta+\zeta_i\zeta-2)$$

Typical edge nodes:

$$N_m = \tfrac{1}{4}(1-\xi^2)(1\pm\eta)(1\pm\zeta)$$

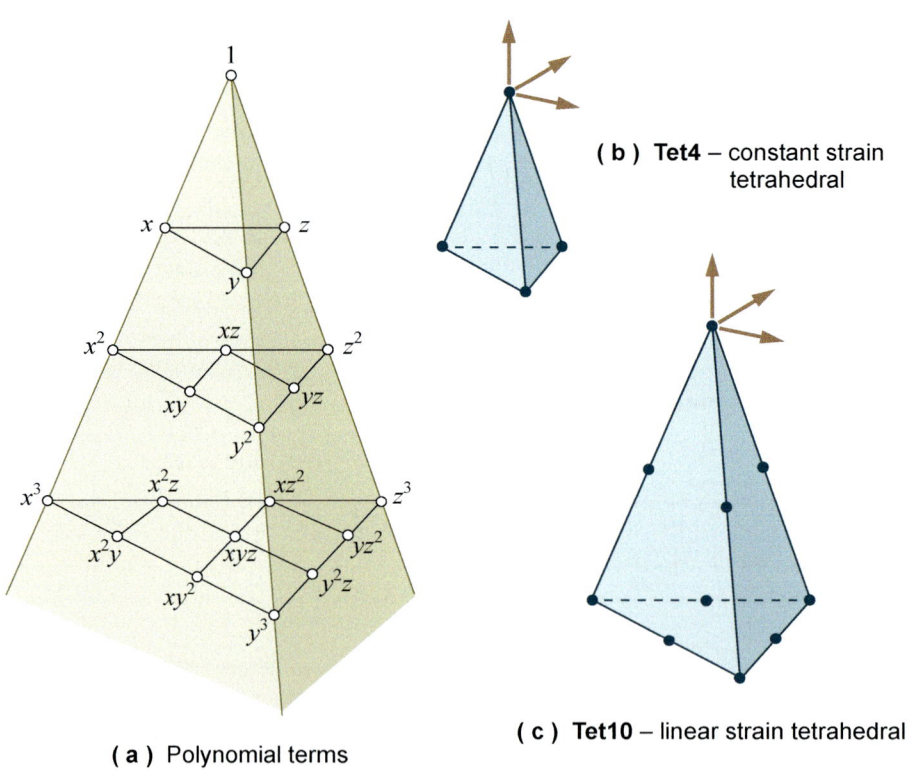

(a) Polynomial terms

(b) Tet4 – constant strain tetrahedral

(c) Tet10 – linear strain tetrahedral

Figure 13.7 Tetrahedral elements

the three-dimensional case, however, this is not possible. We can eliminate the one internal node at the element level, but we cannot eliminate the 6 mid-surface nodes, and these nodes (their *dofs* actually) are not as efficient as the other nodes since they are common to only two elements (edge nodes are common to four elements and corner nodes are common, on average, to eight elements).

Shape functions for **Hex20** are shown opposite. As for the numerical integration, a 3×3×3 GUAUSS quadrature rule is commonly used in case of *full* integration; this means 27 integration points which is not optimal. It can be shown that full integration, that is exact integration of **k** for an undistorted (cuboid shaped) element, can be obtained by a 14 point formula. *Reduced* integration uses a 2×2×2 GUAUSS quadrature rule (8 integration points); this represents an *underintegration* which leaves the element with so-called *hourglass* (or zero energy) modes. Such modes can be controlled by appropriate *stabilization* devices.

Tetrahedral elements. Analogous to the family of triangular elements in the plane case we can also in three dimensions establish a family of tetrahedral elements with much the same pros and cons as in the plane case.

The first two elements of the family, **Tet4** and **Tet10**, are shown in Fig. 13.7. We see that the simplest element (**Tet4**) makes use of a *complete* linear polynomial in cartesian coordinates, and this element can only represent a state of constant strain (or stress) – it is the counterpart of the plane **CST** element. The next member of the family, **Tet10**, makes use of a complete quadratic polynomial and is therefore capable of representing states of linear strain. From the "polynomial pyramid" in Fig.13.7a we see that the cubic member of the family has 20 nodal points, one of which is internal.

A distinct advantage of the tetrahedral elements is that the shape functions for every member of the family constitute *complete* polynomials in x, y and z, or *homogeneous* polynomials in case the of volume coordinates ζ_1, ζ_2, ζ_3 and ζ_4. However, while it takes at least two triangles to make up a quadrilateral, it takes at least *six* tetrahedral elements to make up one hexahedral element. This is a disadvantage of the Tet elements compared to the Hex elements, and the higher the order the greater is the disadvantage.

This is all we will say about three-dimensional analysis. In principle it represents nothing new compared to two-dimensional analysis; if anything the "mechanics" is simpler, but expressions become bigger and more cumbersome and in many cases the geometry is complex and difficult to visualize. The dimension of the final system stiffness relation is of course much greater than for a comparable two-dimensional problem, and even linear static problems in 3D can become very taxing in terms of computational effort. At the element level the numerical integration should be carefully examined, but perhaps more important is the efficiency of the equation solver. Given the choice, an *iterative* solver will most likely perform better than a *direct* one. Substructure analysis may also be an option for very large 3D problems.

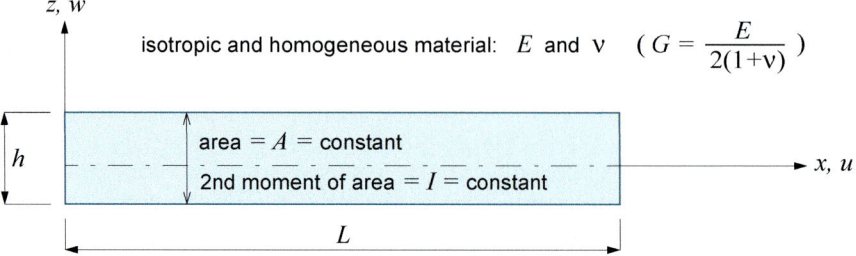

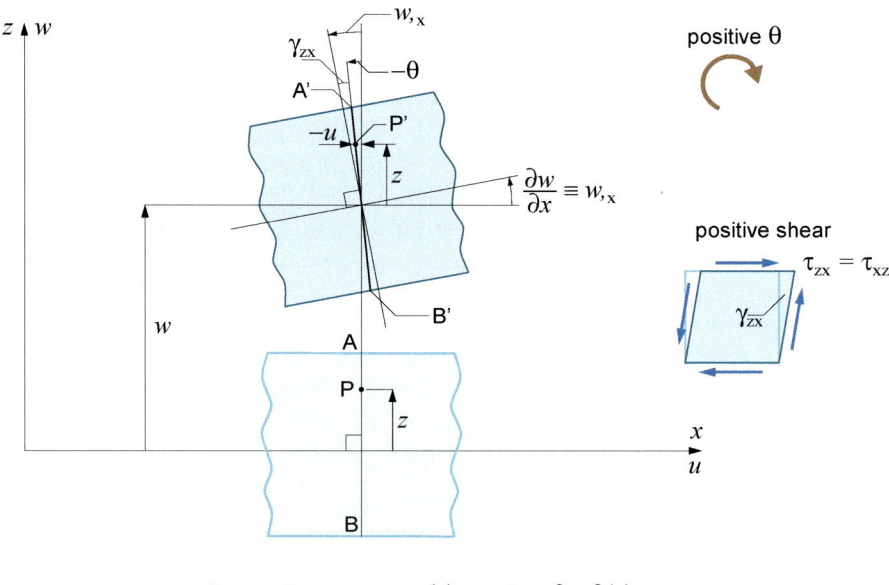

Assumptions: $w = w(x)$ and $\theta = \theta(x)$

Copy of **Figure 3.15** Bending of prismatic 2D beam

14

Bending of beams and plates

The main focus of this chapter is plate bending. However, we will start with the related but much simpler problem of a prismatic beam in bending. The purpose is two-fold: a) how to incorporate shear deformations in beams, and b) to clarify some terminology.

The emphasis is on triangular elements for classical thin plate theory, so-called KIRCHHOFF *theory, but we shall also pay a visit to elements derived by so-called* MINDLIN *theory. In a finite element context these two theories differ significantly since they lead to different continuity requirements for the elements.*

In Sections 3.4 and 3.5 we have summarized the basic equations of the *classical* theories for bending of beams and plates. We see that the variational order m (the highest order of derivative in the strain-displacement, or rather curvature-displacement, relation) is equal to 2, see Eqs. (3-62) and (3-74). Since we require C^{m-1} continuity between elements, we need to have continuity not only in the displacement (w) itself, but also in its first derivative(s), that is in the slope(s).

For the beam element this does not cause problems; by including the slope as a degree of freedom (*dof*) at the node, slope continuity is automatically achieved by requiring compatibility at the nodes (the nodal points are the only points common to beam elements). Not so for plate elements. Slope continuity at the nodes is easily obtained; the problem is slope continuity along the element edges.

The treatment below is somewhat sketchy since we assume that beam elements have been dealt with elsewhere (in a previous course), .

14.1 The two-dimensional beam problem

With reference to Section 3.4 and Fig. 3.15 we start with a slightly rewritten Eq. (3-51)

$$\theta = \phi + \gamma_{xz} = -w,_x + \gamma_{xz} \qquad (14\text{-}1)$$

Here we have introduced the slope ϕ as

Bending of beams and plates

$\kappa = 6/5$ $\quad\quad$ $\kappa = 10/9$ $\quad\quad$ $\kappa = 2$ $\quad\quad\quad\quad$ $\kappa \approx A/A_{web}$

Figure 14.2 The shear factor κ for some typical cross sections

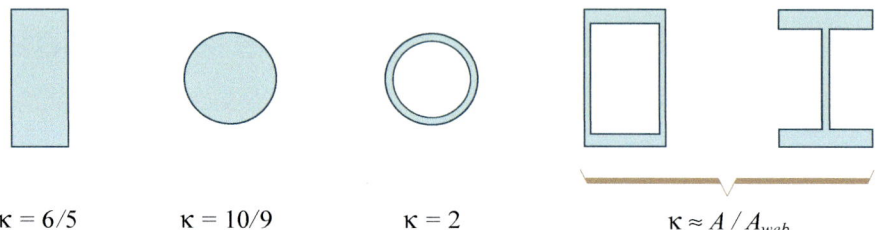

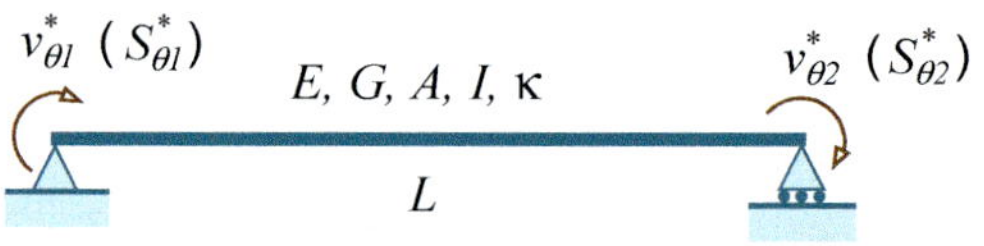

$v_{\theta 1}^*\ (S_{\theta 1}^*)$ $\quad\quad$ E, G, A, I, κ $\quad\quad$ $v_{\theta 2}^*\ (S_{\theta 2}^*)$

L

$$\mathbf{v}^* = \begin{bmatrix} v_{\theta 1}^* \\ v_{\theta 2}^* \end{bmatrix} = \frac{L}{6EI} \begin{bmatrix} \left(2+\frac{\alpha}{2}\right) & -\left(1-\frac{\alpha}{2}\right) \\ -\left(1-\frac{\alpha}{2}\right) & \left(2+\frac{\alpha}{2}\right) \end{bmatrix} \begin{bmatrix} S_{\theta 1}^* \\ S_{\theta 2}^* \end{bmatrix} = \mathbf{f}^* \mathbf{S}^*$$

$$\mathbf{k}^* = (\mathbf{f}^*)^{-1} = \frac{EI}{(1+\alpha)L} \begin{bmatrix} (4+\alpha) & (2-\alpha) \\ (2-\alpha) & (4+\alpha) \end{bmatrix}$$

Figure 14.3 Flexibility and stiffness matrix for a 2-*dof* beam element

$$\phi \approx -w_{,x} \tag{14-2}$$

It should be noted that ϕ is positive in the same direction as the rotation θ of the beam section. γ_{xz} is the shear strain, assumed to be constant over the beam height, a consequence of NAVIER's hypothesis (plane sections remain plane).

As pointed out in Section 3.4 the interpretation of Eq. (14-1) gives rise to three beam theories:

EULER-BERNOULLI beam theory. This is the common beam theory, also referred to as "technical beam theory"; it assumes *no* shear deformations, that is

$$\gamma_{xz} = \gamma = 0 \quad \Rightarrow \quad \theta = \phi = -w_{,x} \tag{14-3}$$

The basic equations of this theory are presented in Section 2.1 along with the other assumptions made. The stiffness matrix for a prismatic beam element based on this theory is easily obtained, either from a systematic use of *beam formulas* (derived from the strong form) or by a finite element type approach based on the weak form of the problem and assumed shape functions.

From the strain-displacement relationship of Eq. (2-1), that is

$$\varepsilon = -zw''$$

we conclude that this is a C^1 type problem.

TIMOSHENKO beam theory. The essence of this theory was first published in the early 1920's in connection with vibration problems, but it was not until the 1950's that TIMOSHENKO included this material in his famous structural mechanics texts. In the 3rd edition of his *Strength of Materials* book [47] he gives the following equation for the deflection w_s of a prismatic beam due to shear (in our notation):

$$\frac{dw_s}{dx} = \frac{(\tau_{xz})_{z=0}}{G} = \frac{\kappa V}{AG} \tag{14-4}$$

V is the shear force and for a rectangle his κ is 3/2. This is his expression for the shear strain γ (= γ_{xz}), and we notice that he uses the maximum shear stress. He later corrected this in recognition of the fact that the constant shear strain γ must be considered as an "average" shear strain γ_m, and it should be matched by an "average" shear stress τ_m. Hence, the shear strain should be taken as

$$\gamma = \gamma_m = \frac{\tau_m}{G} = \frac{V}{A_s G} \quad \text{where} \quad A_s = \frac{A}{\kappa} \tag{14-5}$$

is the *shear area*. The *shear deformation factor* κ depends on the shape of the cross section; it can be determined by a virtual work consideration. The value of κ for some typical beam cross sections is shown in Fig. 14.2.

The easiest way to incorporate shear deformations into the standard beam theory is to determine the *flexibility* ($\mathbf{f}^*$) matrix for the 2-*dof* beam element in Fig. 14.3, by for instance the *principle of virtual forces* ("the unit load method"), and then invert this matrix to find the 2×2 deformation stiffness matrix ($\mathbf{k}^*$). The dimensionless shear parameter α is defined as

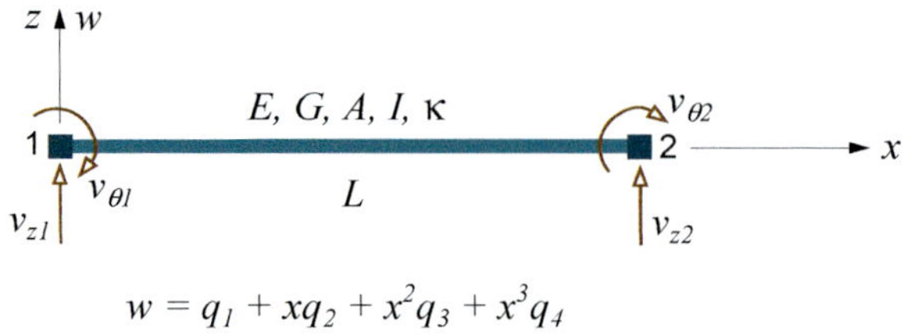

Bending of beams and plates

$$w = q_1 + xq_2 + x^2q_3 + x^3q_4$$

Figure 14.4 Assumed displacement for a 4-*dof* beam element

Equation (2-3): $M = -EIw''$

Equation (2-5): $\dfrac{dM}{dx} = M' = V$

The two-dimensional beam problem

$$\alpha = \frac{12EI}{GA_sL^2} = \frac{12\kappa EI}{GAL^2} \tag{14-6}$$

However, we can include shear deformations by a more "finite element like" approach. We consider the 4-*dof* beam element of Fig. 14.4, and use *indirect* interpolation, that is

$$w = q_1 + xq_2 + x^2q_3 + x^3q_4 = [1 \quad x \quad x^2 \quad x^3]\mathbf{q} = \mathbf{N}_q\mathbf{q} \tag{14-7}$$

In order to establish a relationship between **v** and **q** we need to take a closer look at the rotation θ. From Eq. (14-5) we have

$$\theta = \phi + \gamma = -w' + \frac{\kappa}{GA}V \tag{14-8}$$

and from Eqs. (2-5) and (2-3) we can express the shear force as

$$V = M' = -EIw''' \tag{14-9}$$

Hence

$$\theta = -w' - \frac{\kappa EI}{GA}w''' = -w' - \frac{L^2\alpha}{12}w''' \tag{14-10}$$

From the assumption (14-7) it follows that

$$w''' \equiv w_{,xxx} = [0 \quad 0 \quad 0 \quad 6]\mathbf{q} = \mathbf{N}_{q,xxx}\mathbf{q} \tag{14-11}$$

Thus

$$\begin{bmatrix} v_{z1} \\ v_{\theta 1} \\ v_{z2} \\ v_{\theta 2} \end{bmatrix} = \begin{bmatrix} w(0) \\ -w'(0) - \frac{\alpha L^2}{12}w'''(0) \\ w(L) \\ -w'(L) - \frac{\alpha L^2}{12}w'''(L) \end{bmatrix} = \begin{bmatrix} 1 & 0 & 0 & 0 \\ 0 & -1 & 0 & -\frac{\alpha L^2}{2} \\ 1 & L & L^2 & L^3 \\ 0 & -1 & -2L & -\frac{L^2(6+\alpha)}{2} \end{bmatrix} \begin{bmatrix} q_1 \\ q_2 \\ q_3 \\ q_4 \end{bmatrix} \tag{14-12}$$

or

$$\mathbf{v} = \mathbf{A}\mathbf{q}$$

We are interested in the opposite relation, and we therefore need

$$\mathbf{A}^{-1} = \frac{1}{1+\alpha}\begin{bmatrix} 1+\alpha & 0 & 0 & 0 \\ -\frac{\alpha}{L} & -\frac{2+\alpha}{2} & \frac{\alpha}{L} & \frac{\alpha}{2} \\ -\frac{3}{L^2} & \frac{4+\alpha}{2L} & \frac{3}{L^2} & \frac{2-\alpha}{2L} \\ \frac{2}{L^3} & -\frac{1}{L^2} & -\frac{2}{L^3} & -\frac{1}{L^2} \end{bmatrix} \tag{14-13}$$

Bending of beams and plates

$$\frac{d\theta}{dx} = \theta_{,x} \quad \Rightarrow \quad d\theta = \theta_{,x} dx$$

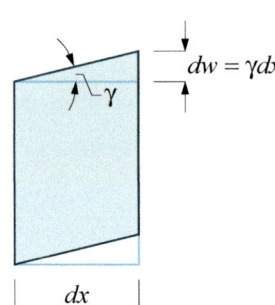

$$\mathbf{k}_q = EI \begin{bmatrix} 0 & 0 & 0 & 0 \\ 0 & 0 & 0 & 0 \\ 0 & 0 & 4L & 6L^2 \\ 0 & 0 & 6L^2 & 3L^3(4+\alpha) \end{bmatrix}$$

The reader can easily verify that $\mathbf{A}\mathbf{A}^{-1} = \mathbf{A}^{-1}\mathbf{A} = \mathbf{I}$.

We now define a set of generalized forces $\mathbf{Q}$ corresponding to $\mathbf{q}$ and seek the *generalized element stiffness matrix* $\mathbf{k}_q$ in the relationship

$$\mathbf{Q} = \mathbf{k}_q \mathbf{q} \tag{14-14}$$

To this end we employ the *principle of virtual displacements* (PVD) and assume a virtual displacement field $\tilde{w} = \mathbf{N}_q \tilde{\mathbf{q}}$ that satisfies kinematic compatibility. A tilde (~) designates a virtual quantity. Thus

$$\tilde{\mathbf{q}}^T \mathbf{Q} = \int_0^L d\tilde{\theta} M + \int_0^L d\tilde{w} V = \int_0^L \tilde{\theta}_{,x} M dx + \int_0^L \tilde{\gamma} V dx \tag{14-15}$$

The first term accounts for bending (as in E-B theory) while the second term is the contribution to the internal work from the shear force. It should be emphasized that Eq. (14-15) applies to an element with *no* loading between the nodes. From Eq. (14-10) it follows that

$$\tilde{\theta}_{,x} = -\tilde{w}_{,xx} - \frac{L^2 \alpha}{12}\tilde{w}_{,xxxx} = -\tilde{w}_{,xx} = -\mathbf{N}_{q,xx}\tilde{\mathbf{q}} = -\tilde{\mathbf{q}}^T \mathbf{N}_{q,xx}^T$$

Similarly

$$\tilde{\gamma} = -\frac{L^2 \alpha}{12}\tilde{w}_{,xxx} = -\frac{L^2 \alpha}{12}\mathbf{N}_{q,xxx}\tilde{\mathbf{q}} = -\frac{L^2 \alpha}{12}\tilde{\mathbf{q}}^T \mathbf{N}_{q,xxx}^T$$

and with $M = -EIw_{,xx} = -EI\mathbf{N}_{q,xx}\mathbf{q}$ and $V = -EIw_{,xxx} = -EI\mathbf{N}_{q,xxx}\mathbf{q}$, (14-15) can be expressed as

$$\tilde{\mathbf{q}}^T \mathbf{Q} = \int_0^L -\tilde{\mathbf{q}}^T \mathbf{N}_{q,xx}^T (-EI\mathbf{N}_{q,xx}\mathbf{q}) dx + \int_0^L \left(-\frac{L^2 \alpha}{12}\tilde{\mathbf{q}}^T \mathbf{N}_{q,xxx}^T\right)(-EI\mathbf{N}_{q,xxx}\mathbf{q}) dx$$

$$= \tilde{\mathbf{q}}^T \left[EI \int_0^L \mathbf{N}_{q,xx}^T \mathbf{N}_{q,xx} dx + \frac{\alpha L^2 EI}{12} \int_0^L \mathbf{N}_{q,xxx}^T \mathbf{N}_{q,xxx} dx \right] \mathbf{q} = \tilde{\mathbf{q}}^T \mathbf{k}_q \mathbf{q} \tag{14-16}$$

This equation must be valid for any $\tilde{\mathbf{q}}$. Hence $\mathbf{Q} = \mathbf{k}_q \mathbf{q}$ where $\mathbf{k}_q$ is a very simple matrix, see opposite page. Standard transformation of $\mathbf{k}_q$ gives the desired stiffness:

$$\mathbf{k} = \mathbf{A}^{-T}\mathbf{k}_q\mathbf{A}^{-1} = \frac{EI}{(1+\alpha)L}\begin{bmatrix} \frac{12}{L^2} & -\frac{6}{L} & -\frac{12}{L^2} & -\frac{6}{L} \\ & 4+\alpha & \frac{6}{L} & 2-\alpha \\ & & \frac{12}{L^2} & \frac{6}{L} \\ \text{sym.} & & & 4+\alpha \end{bmatrix} \tag{14-17}$$

This is precisely the same matrix as the one we obtain by transforming rigid body modes into the 2 by 2 deformation stiffness presented in Fig. 14.3. If

Bending of beams and plates

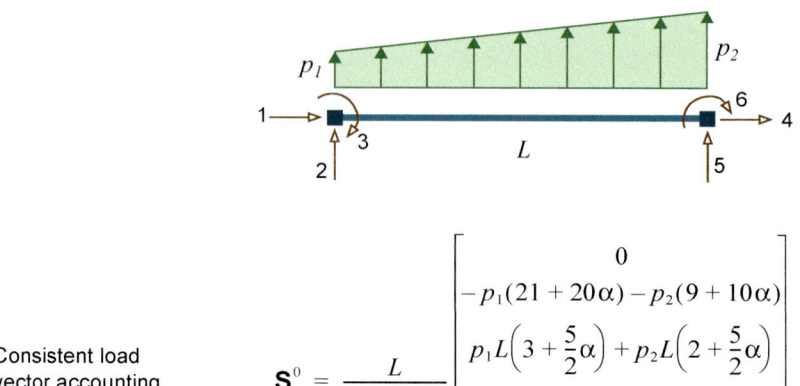

Consistent load vector accounting for shear deformation

$$\mathbf{S}^0 = \frac{L}{60(1+\alpha)} \begin{bmatrix} 0 \\ -p_1(21+20\alpha) - p_2(9+10\alpha) \\ p_1 L\left(3 + \frac{5}{2}\alpha\right) + p_2 L\left(2 + \frac{5}{2}\alpha\right) \\ 0 \\ -p_1(9+10\alpha) - p_2(21+20\alpha) \\ -p_1 L\left(2 + \frac{5}{2}\alpha\right) - p_2 L\left(3 + \frac{5}{2}\alpha\right) \end{bmatrix}$$

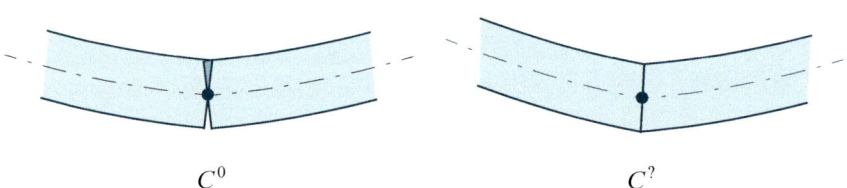

Figure 14.5 Different continuity conditions

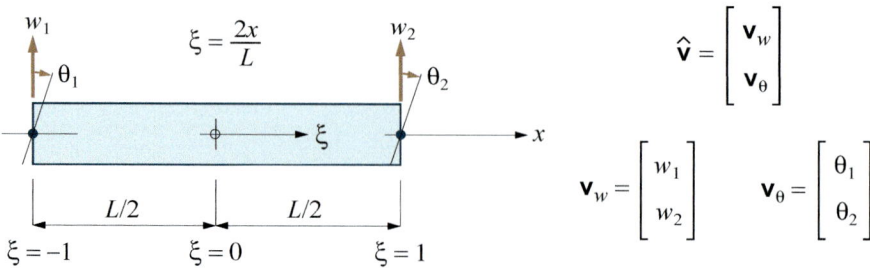

Figure 14.6 A 4-*dof* MINDLIN beam element

we neglect shear deformations, that is $\alpha = 0$, Eq. (14-17) defines the stiffness matrix for the standard, 4-*dof* EULER-BERNOULLI beam element.

The consistent load vector ($\mathbf{S}^0$) for a 6-*dof* plane TIMOSHENKO beam element due to a linearly varying transverse load is shown on the opposite page; all details have been omitted. We notice that for a uniformly distributed load ($p_1 = p_2$) the "shear factor" α cancel out.

The TIMOSHENKO beam element defined by Eq. (14-17) is a robust element that produces accurate results for beams of all practical depth to length ratios. For long slender beams it degenerates to the EULER-BERNOULLI element, as it should. An interesting, but rather academic question, can now be posed: is this a C^1 or a C^0 element? If we look at the deflection line, the TIMOSHENKO element will exhibit a "kink" between elements, as shown to the right in Fig. 14.5, and some would say that this kink is the hallmark of a C^0 element. The derivation above, however, seems to indicate that perhaps it is not quite as simple as that, and as the depth to length ratio gets smaller ($\alpha \Rightarrow 0$) the behaviour of the element relies more and more on C^1 continuity.

MINDLIN beam theory. The following beam theory is attributed to MINDLIN[1] although MINDLIN dealt with plates and not beams. However, the main idea is the same for plates and beams. To be correct one should probably call this MINDLIN-REISSNER[2] theory since both, independently of each other, put forward similar ideas.

This theory deals with shear deformations in a more direct manner. Equation (14-1) still applies, but while TIMOSHENKO theory sees this equation as a definition of θ, MINDLIN theory sees it as a definition of γ, that is,

$$\gamma = \gamma_{xz} = \theta + w_{,x}$$

and the theory assumes *independent* fields for θ and w. We exemplify this by the simple 4-*dof* element of Fig. 14.5. We only consider bending and shear since the axial deformation is decoupled and can be superimposed later.

From the expressions for the strain energies, Eqs. (3-58) and (3-59), it is evident that with independent fields for w and θ, the variational order m is equal to 1, for both U_b and U_s; hence, we only need to satisfy C^0 continuity for both fields. Thus, for the element in Fig. 14.6 we assume

$$w = [\,N_1 \quad N_2\,]\mathbf{v}_w = \hat{\mathbf{N}}\mathbf{v}_w \quad \text{and} \quad \theta = [\,N_1 \quad N_2\,]\mathbf{v}_\theta = \hat{\mathbf{N}}\mathbf{v}_\theta \quad (14\text{-}18)$$

where the shape functions are the same linear functions that we have seen a number of times now,

---

1. Raymond David MINDLIN (1906-1987) was an American scientist who made valuable contributions to many branches of applied mechanics, applied physics and engineering sciences. He spent most of his student and professional life at Columbia University in New York where he became full professor of structural engineering in 1947.
2. Eric REISSNER (1913-1996) was born and trained in Germany, but moved to the United States just before the second world war. He was a professor of applied mathematics at MIT and a professor of applied mechanics at the University of California in San Diego. He published nearly 300 papers in scientific and technical journals.

Bending of beams and plates

$\hat{\mathbf{B}}_\theta^T \hat{\mathbf{B}}_\theta$:

	0	0	$-\frac{1}{L}$	$\frac{1}{L}$
0	0	0	0	0
0	0	0	0	0
$-\frac{1}{L}$	0	0	$\frac{1}{L^2}$	$-\frac{1}{L^2}$
$\frac{1}{L}$	0	0	$-\frac{1}{L^2}$	$\frac{1}{L^2}$

$$\hat{\mathbf{k}}_b = \frac{EIL}{2}\int_{-1}^{1} \hat{\mathbf{B}}_\theta^T \hat{\mathbf{B}}_\theta \, d\xi = \frac{EI}{L}\begin{bmatrix} 0 & 0 & 0 & 0 \\ 0 & 0 & 0 & 0 \\ 0 & 0 & 1 & -1 \\ 0 & 0 & -1 & 1 \end{bmatrix}$$

$\hat{\mathbf{B}}_w^T \hat{\mathbf{B}}_w$:

	$-\frac{1}{L}$	$\frac{1}{L}$	$\frac{1}{2}(1-\xi)$	$\frac{1}{2}(1+\xi)$
$-\frac{1}{L}$	$\frac{1}{L^2}$	$-\frac{1}{L^2}$	$-\frac{1}{2L}(1-\xi)$	$-\frac{1}{2L}(1+\xi)$
$\frac{1}{L}$	$-\frac{1}{L^2}$	$\frac{1}{L^2}$	$\frac{1}{2L}(1-\xi)$	$\frac{1}{2L}(1+\xi)$
$\frac{1}{2}(1-\xi)$	$-\frac{1}{2L}(1-\xi)$	$\frac{1}{2L}(1-\xi)$	$\frac{1}{4}(1-\xi)^2$	$\frac{1}{4}(1-\xi^2)$
$\frac{1}{2}(1+\xi)$	$-\frac{1}{2L}(1+\xi)$	$\frac{1}{2L}(1+\xi)$	$\frac{1}{4}(1-\xi^2)$	$\frac{1}{4}(1+\xi)^2$

$$\hat{\mathbf{k}}_s = \frac{GA_sL}{2}\int_{-1}^{1} \hat{\mathbf{B}}_w^T \hat{\mathbf{B}}_w \, d\xi = \frac{GA_s}{L}\begin{bmatrix} 1 & -1 & -\frac{L}{2} & -\frac{L}{2} \\ -1 & 1 & \frac{L}{2} & \frac{L}{2} \\ -\frac{L}{2} & \frac{L}{2} & \frac{L^2}{3} & \frac{L^2}{6} \\ -\frac{L}{2} & \frac{L}{2} & \frac{L^2}{6} & \frac{L^2}{3} \end{bmatrix}$$

$$N_1 = \tfrac{1}{2}(1-\xi) \quad \text{and} \quad N_2 = \tfrac{1}{2}(1+\xi) \tag{14-19}$$

In order to establish the strains that appear in the strain energy expressions, Eqs. (3-58) and (3-59), we need

$$\theta_{,x} = [\,0 \quad (\hat{\mathbf{N}}_{,x})\,]\hat{\mathbf{v}} = \hat{\mathbf{B}}_\theta \hat{\mathbf{v}} \tag{14-20a}$$

and

$$\gamma = w_{,x} + \theta = [\,\hat{\mathbf{N}}_{,x} \quad \hat{\mathbf{N}}\,]\hat{\mathbf{v}} = \hat{\mathbf{B}}_w \hat{\mathbf{v}} \tag{14-20b}$$

Since

$$\frac{d}{dx} = \frac{d}{d\xi}\frac{d\xi}{dx} = \frac{2}{L}\frac{d}{d\xi}$$

we find

$$\hat{\mathbf{B}}_\theta = [\,0 \quad 0 \quad -\tfrac{1}{L} \quad \tfrac{1}{L}\,] \quad \text{and} \quad \hat{\mathbf{B}}_w = [\,-\tfrac{1}{L} \quad \tfrac{1}{L} \quad \tfrac{1}{2}(1-\xi) \quad \tfrac{1}{2}(1+\xi)\,]$$

The strain energy of the element is, according to Eqs. (3-58) and (3-59),

$$U = \frac{EI}{2}\int_{-L/2}^{L/2} \theta_{,x}^2\, dx + \frac{GA_s}{2\kappa}\int_{-L/2}^{L/2} \gamma^2\, dx = \hat{\mathbf{v}}^T\left(\frac{EI}{2}\int_{-1}^{1}\hat{\mathbf{B}}_\theta^T\hat{\mathbf{B}}_\theta\frac{L}{2}d\xi\right)\hat{\mathbf{v}} + \hat{\mathbf{v}}^T\left(\frac{GA}{2\kappa}\int_{-1}^{1}\hat{\mathbf{B}}_w^T\hat{\mathbf{B}}_w\frac{L}{2}d\xi\right)\hat{\mathbf{v}}$$

or

$$U = \tfrac{1}{2}\hat{\mathbf{v}}^T \hat{\mathbf{k}}_b \hat{\mathbf{v}} + \tfrac{1}{2}\hat{\mathbf{v}}^T \hat{\mathbf{k}}_s \hat{\mathbf{v}} = \tfrac{1}{2}\hat{\mathbf{v}}^T \hat{\mathbf{k}} \hat{\mathbf{v}} \tag{14-21}$$

where

$$\hat{\mathbf{k}} = \hat{\mathbf{k}}_b + \hat{\mathbf{k}}_s \tag{14-22}$$

The computations are shown opposite. From Eq. (14-6) it follows that $(GA_s)/L = (12EI)/(\alpha L^3)$ and rearranged with respect to $\mathbf{v}$ we can now write

$$\mathbf{k}_b = \frac{EI}{L}\begin{bmatrix} 0 & 0 & 0 & 0 \\ 0 & 1 & 0 & -1 \\ 0 & 0 & 0 & 0 \\ 0 & -1 & 0 & 1 \end{bmatrix} \tag{14-23}$$

which is the contribution to the element stiffness matrix from bending, and

$$\mathbf{k}_s = \frac{GA_s}{L}\begin{bmatrix} 1 & -\tfrac{L}{2} & -1 & -\tfrac{L}{2} \\ -\tfrac{L}{2} & \tfrac{L^2}{3} & \tfrac{L}{2} & \tfrac{L^2}{6} \\ -1 & \tfrac{L}{2} & 1 & \tfrac{L}{2} \\ -\tfrac{L}{2} & \tfrac{L^2}{6} & \tfrac{L}{2} & \tfrac{L^2}{3} \end{bmatrix} = \frac{EI}{\alpha L}\begin{bmatrix} \tfrac{12}{L^2} & -\tfrac{6}{L} & -\tfrac{12}{L^2} & -\tfrac{6}{L} \\ -\tfrac{6}{L} & 4 & \tfrac{6}{L} & 2 \\ -\tfrac{12}{L^2} & \tfrac{6}{L} & \tfrac{12}{L^2} & \tfrac{6}{L} \\ -\tfrac{6}{L} & 2 & \tfrac{6}{L} & 4 \end{bmatrix} \tag{14-24}$$

Bending of beams and plates

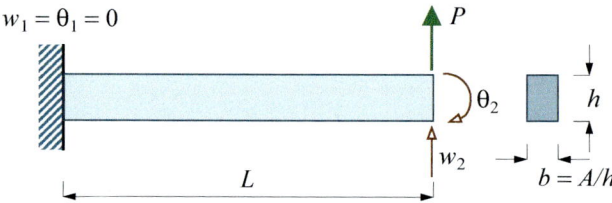

Figure 14.7 Cantilever beam subjected to a tip load P

is the contribution from shear. We notice that the rotations (θ_1 and θ_2) are coupled to the displacements (w_1 and w_2) through $\mathbf{k}_s$, but *not* through $\mathbf{k}_b$. The element stiffness matrix $\mathbf{k}$ corresponding to a nodal displacement vector $\mathbf{v}$, whose components are arranged nodal point wise, can now be written as

$$\mathbf{k} = \mathbf{k}_b + \mathbf{k}_s = \frac{EI}{\alpha L} \begin{bmatrix} \frac{12}{L^2} & -\frac{6}{L} & -\frac{12}{L^2} & -\frac{6}{L} \\ -\frac{6}{L} & 4+\alpha & \frac{6}{L} & 2-\alpha \\ -\frac{12}{L^2} & \frac{6}{L} & \frac{12}{L^2} & \frac{6}{L} \\ -\frac{6}{L} & 2-\alpha & \frac{6}{L} & 4+\alpha \end{bmatrix} \quad \text{for} \quad \mathbf{v} = \begin{bmatrix} w_1 \\ \theta_1 \\ w_2 \\ \theta_2 \end{bmatrix} \quad (14\text{-}25)$$

If we compare this to the TIMOSHENKO stiffness of Eq. (14-13) we see that the only difference is that $(1 + \alpha)$ in the denominator of the common factor is replaced by just α. A pertinent question now is, how good is this element?

Example 14-1

Let us test it in a very simple example, a cantilever beam subjected to a point load at the tip as shown in Fig. 14.7; this beam experiences both bending and shear.

We assume, for simplicity, that $A_s = A$ ($\kappa = 1{,}0$) and $G = E/2$ ($\nu = 0$). Furthermore

$$\alpha = \frac{12\kappa EI}{GAL^2} = \frac{12 \cdot 2Gbh^3}{12GAL^2} = 2\left(\frac{h}{L}\right)^2 \quad \text{and} \quad \frac{EI}{\alpha L} = \frac{Ebh^3}{12 \cdot 2\left(\frac{h}{L}\right)^2 L} = \frac{GAL}{12}$$

With one element and $w_1 = \theta_1 = 0$ at the clamped end we get the following stiffness relation

$$\frac{EI}{\alpha L} \begin{bmatrix} \frac{12}{L^2} & \frac{6}{L} \\ \frac{6}{L} & 4+\alpha \end{bmatrix} \begin{bmatrix} w_2 \\ \theta_2 \end{bmatrix} = GA \begin{bmatrix} \frac{1}{L} & \frac{1}{2} \\ \frac{1}{2} & \frac{(4+\alpha)L}{12} \end{bmatrix} \begin{bmatrix} w_2 \\ \theta_2 \end{bmatrix} = \begin{bmatrix} P \\ 0 \end{bmatrix}$$

After some straightforward manipulations we find:

$$w_2 = \frac{PL(4+\alpha)}{GA(1+\alpha)}$$

$$\text{if} \quad \frac{h}{L} \to 0 \quad \text{then} \quad \alpha \to 0 \quad \text{and} \quad w_2 \to \frac{4PL}{GA} = 4w_s$$

where $w_s = PL/GA$ is, with our simplifications, the displacement due to *shear* alone.

447

Contradictions in beam theories

EULER-BERNOULLI

The main hypothesis of this theory is that plane sections normal to the neutral axis before bending remain plane and normal also after bending. This implies *no shear deformations*, that is $\gamma = 0$. Hence $\tau = G\gamma$ should also be zero!
This make no sense, and through equilibrium considerations we determine (axis parallel) shear stresses and accept the contradiction on the basis that shear deformations are small compared to bending deformations.

MINDLIN

We consider the simple element in Fig 14.6 with independent linear displacement fields for w and θ. Note that a prime (′) denotes differentiation with respect to x.

Static considerations:

- Shear force: $V = A_s\tau = A_s G\gamma = GA_s(w' + \theta)$ $\Rightarrow$ linear variation
- Bending moment (see Eqs. (2-1) to (2-3)): $M = -EI\theta'$ $\Rightarrow$ constant
- The equilibrium condition, Eq. (2-5), that is $V = M'$ is severely violated.

Kinematic considerations:

- The approximate solution must be able to reproduce *pure bending* correctly, that is, constant curvature (c) and zero shear strain (γ)

$$c = \theta' = \text{constant} \quad \text{and} \quad \gamma = w' + \theta = 0$$

- For the given discretization (assumed displacements), using x instead of ξ:

$$w = (1 - x/L)w_1 + (x/L)w_2 \quad \Rightarrow \quad w' = (w_2 - w_1)/L$$

$$\theta = (1 - x/L)\theta_1 + (x/L)\theta_2 \quad \Rightarrow \quad \theta' = (\theta_2 - \theta_1)/L$$

- The requirement of zero shear strain requires

$$\gamma(x) = (w_2 - w_1)/L + (1 - x/L)\theta_1 + (x/L)\theta_2 = (w_2 - w_1)/L + \theta_1 + (x/L)(\theta_2 - \theta_1) = 0$$

For this to be possible γ must be independent of x; hence

$$\theta_2 - \theta_1 = 0 \quad \Rightarrow \quad c = \theta' = (\theta_2 - \theta_1)/L = 0$$

and the requirement of constant curvature is violated.

When reduced (or rather, selective) integration is used we imply that the shear strain is constant over the element, and the value is taken to be the value at the mid-point of the element:

$$\gamma(x) = \gamma = (w_2 - w_1)/L + (1 - 1/2)\theta_1 + \theta_2/2 = (w_2 - w_1)/L + (\theta_1 + \theta_2)/L$$

Kinematics: the element behaves correctly; it can describe constant curvature.

Equilibrium: Both V and M are constant; hence, equilibrium is still violated but not as severely as before.

From this we can conclude that as the depth to length ratio (h/L) decreases, the shear stiffness dominates more and more; the element becomes far too stiff and eventually "locks" due to shear, so-called *shear locking*. We also see that the factor $EI/\alpha L$ increases rapidly when $\alpha \to 0$ (**k** tends to blow up).

It should be emphasized that the h/L ratio applies to the entire beam, not to the element. Hence, subdividing the beam into more elements does not help. In short, as it stands the element is useless. However, the shortcomings can be alleviated. One-point reduced integration, suggested by HUGHES et al [48], turns out to be a significant improvement. This does not alter $\mathbf{k}_b$, but all 4 terms of the 2×2 submatrix in the lower right-hand corner of $\hat{\mathbf{k}}_s$ become $L^2/4$. Thus, with one-point reduced integration the element stiffness matrix becomes

$$\mathbf{k} = \frac{EI}{\alpha L} \begin{bmatrix} \frac{12}{L^2} & -\frac{6}{L} & -\frac{12}{L^2} & -\frac{6}{L} \\ -\frac{6}{L} & 3+\alpha & \frac{6}{L} & 3-\alpha \\ -\frac{12}{L^2} & \frac{6}{L} & \frac{12}{L^2} & \frac{6}{L} \\ -\frac{6}{L} & 3-\alpha & \frac{6}{L} & 3+\alpha \end{bmatrix} \qquad (14\text{-}26)$$

If we use this matrix in our simple Example 14-1 we find, in view of

$$\alpha = 2\left(\frac{h}{L}\right)^2 \quad \text{and} \quad GA\alpha = \frac{12EI}{L^2}, \text{ that}$$

$$w_2 = (3+\alpha)\frac{PL}{GA\alpha} = \frac{(3+\alpha)}{12}\frac{PL^3}{EI} = \frac{PL^3}{4EI} + \alpha\frac{PL^3}{12EI}$$

Hence, when $\alpha \to 0$ the displacement approaches 75% of the correct value which is $PL^3/3EI$; a considerable improvement, and locking is no longer an issue.

We can take this one step further and force the tip displacement w_2, obtained by the stiffness matrix of Eq. (14-25), to attain the correct value by replacing the shear stiffness GA_s by a modified shear stiffness GA^*. By the use of this trick, the TIMOSHENKO stiffness matrix of Eq. (14-17) is re-established! A rather clever way of achieving this correction, which MACNEAL [49] calls "residual bending flexibility", is to replace α in Eq. (14-25) by $1 + \hat{\alpha}$ and then simply remove the hat, and «voilà», all problems are solved. Or are they? What exactly is this element? What are the implications of these "manoeuvres" (or "fixes") on the displacements w and θ? Can two elements have exactly the same stiffness matrix, but different displacements? According to the derivation at the beginning of this section, the displacement w of the TIMOSHENKO element is cubic.

If we increase the number of nodal points and/or *dofs* and thus provide for higher order interpolation of w and θ, shear locking may still be a problem

Bending of beams and plates

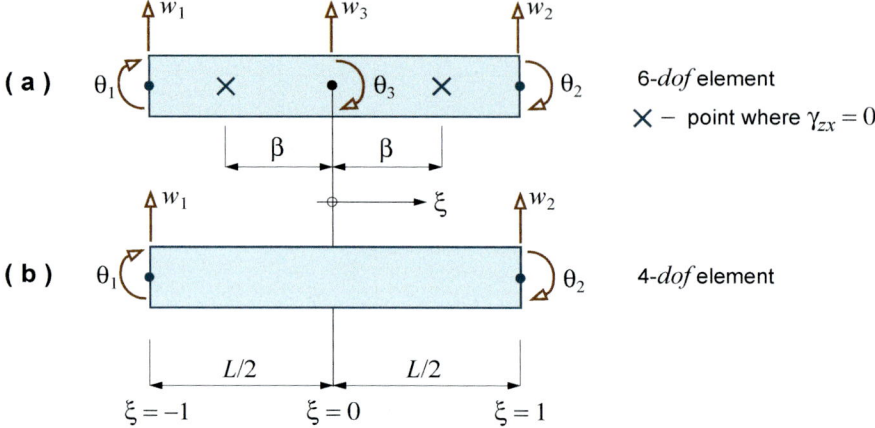

Figure 14.8 A 4-*dof* discrete KIRCHHOFF beam element

The two-dimensional beam problem

for MINDLIN type elements. Before we leave the idea of a C^0 beam element we shall mention a type of element that is based on a concept that is used with some success in plate bending elements.

A discrete KIRCHHOFF element. The basis for this approach is:

1) neglect the strain energy due to (transverse) shear deformations, and
2) use discrete (point-wise) conditions to enforce zero shear strain at specific points in the element.

The second point enforces the so-called KIRCHHOFF condition ($\gamma_{xz} = 0$); hence the name.

Figure 14.8a shows a beam element with 6 *dofs*. Eventually this element will become the 4-*dof* element of Fig. 14.8b.

Starting with the 6-*dof* element, both w and θ are interpolated by *quadratic* shape functions. At the two points with coordinates β and $-\beta$ we enforce the KIRCHHOFF condition, that is

$$\theta(-\beta) + w,_x(-\beta) = 0 \quad \text{and} \quad \theta(\beta) + w,_x(\beta) = 0$$

These two equations are used to eliminate the internal *dofs*, w_3 and θ_3. This operation leaves us with a set of modified shape functions, from which we can establish the curvature, or rather the **B** matrix. This matrix is then plugged into the strain energy equation, Eq. (3-58), from which the element stiffness matrix is obtained. Note that we only consider U_b (we neglect U_s). In MINDLIN elements the independent displacement fields for w and θ are *implicitly* coupled by the shear stiffness of the element. For discrete KIRCHHOFF elements the coupling is *explicit* at some discrete points.

So much for plane beam elements. The curious reader may find a short paper by FELIPPA [10] both interesting and amusing. He states very clearly that for the given simple geometry (straight and with constant properties) and only the four degrees of freedom ($w_1, \theta_1, w_2, \theta_2$) the TIMOSHENKO element whose stiffness matrix **k** is defined by Eq. (14-17) *cannot be improved upon*. Hence, if you only need a beam element, the TIMOSHENKO element, as defined here, is the obvious choice. It is a robust element that behaves well for all depth to length (h/L) ratios; it can only improve on the EULER-BERNOULLI element.

The reason why MINDLIN type beam elements are of interest has to do with their coupling to MINDLIN plate bending elements. The attraction of MINDLIN plate theory is of course the reduced continuity requirement (C^0) compared to KIRCHHOFF theory (C^1).

In much of the literature most elements that incorporate shear deformations by means of Eq. (14-1) seem to be termed TIMOSHENKO elements. This is unfortunate. In this author's opinion there is only one TIMOSHENKO element, just as there is only one EULER-BERNOULLI element. It may not be historically correct to associate all beam elements based on two *independent* displacement fields with MINDLIN, but at least it signals how they are derived, and what shortcomings they may possess. How the TIMOSHENKO element is derived, through sound engineering reasoning or by clever manipulations of an element of the MINDLIN type, is of less importance.

Bending of beams and plates

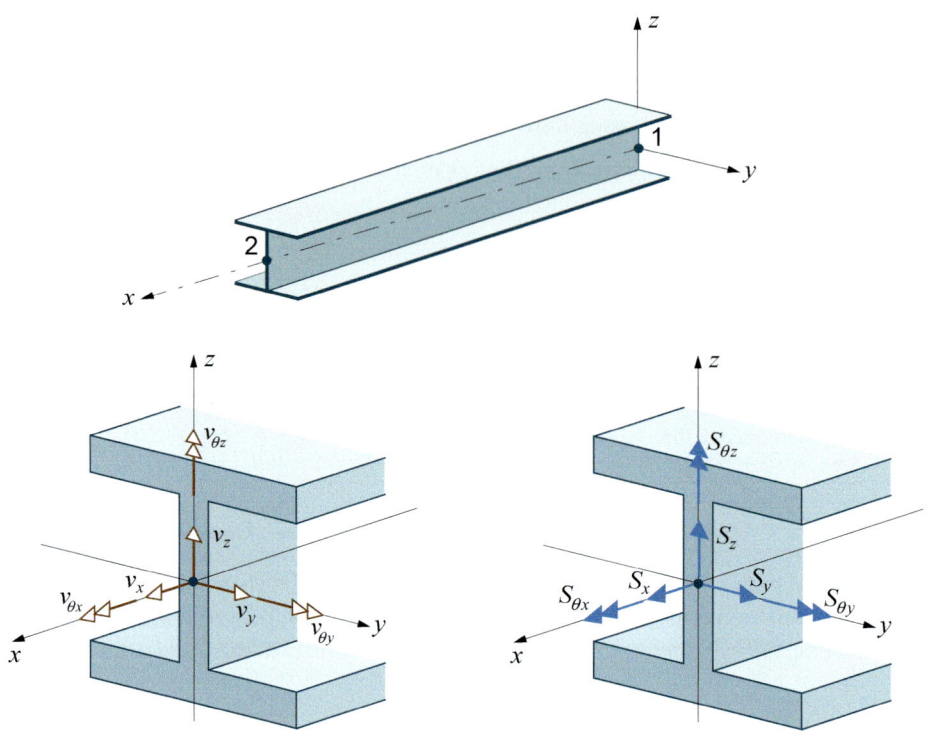

Figure 14.9 Beam element in 3D – *dofs* and corresponding forces

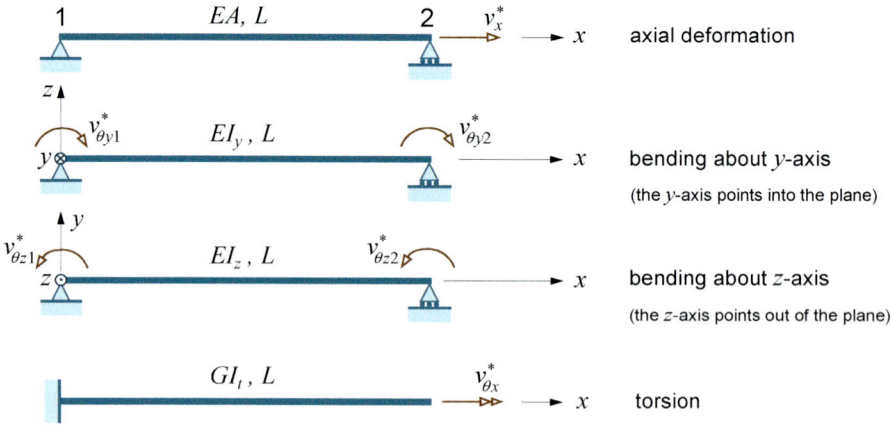

Figure 14.10 Deformation *dofs* for a beam in 3D space

14.2 Beam element in 3D space

In view of the fact that the beam element is perhaps the most important element in structural analysis, we include a brief summary of how to derive an efficient and versatile beam element in three dimensions.

We limit our presentation to a straight, prismatic element that handles *torsion* according to St. Venant's theory; in other words we assume no *warping*[1] restraints. For such an element we need the 6 degrees of freedom shown in Fig. 14.9, at each of the two nodal points, in order to describe properly the relevant deformations and rigid body displacements.

Deformation stiffness for a double symmetric cross section. If we assume for the moment a cross section that is symmetric about both principal axes (y and z), and place the two nodal points at the area centre, through which the local element x-axis runs, then we have *no* coupling between the axial force (S_x) and the bending moments ($S_{\theta y}$ and $S_{\theta z}$). For such a cross section the *shear centre* coincides with the area centre, hence there is *no* coupling between the shear forces (S_y and S_z) and the torsional moment ($S_{\theta x}$).

Figure 14.10 shows the degrees of freedom we need to uniquely describe the *deformations* of this element – all rigid body movements are prevented. The only thing that is new compared to the plane element is the *torsion*. For this particular element, torsion, which is described in detail in Chapter 16, is treated in much the same way as the axial deformation. Since we have no coupling between the four modes of deformation, the *deformation* stiffness relation for this element is very simple:

$$\begin{bmatrix} S_x^* \\ S_{\theta y1}^* \\ S_{\theta y2}^* \\ S_{\theta z1}^* \\ S_{\theta z2}^* \\ S_{\theta x}^* \end{bmatrix} = \begin{bmatrix} a & 0 & 0 & 0 & 0 & 0 \\ 0 & \frac{4+\alpha_y}{1+\alpha_y}b_y & \frac{2-\alpha_y}{1+\alpha_y}b_y & 0 & 0 & 0 \\ 0 & \frac{2-\alpha_y}{1+\alpha_y}b_y & \frac{4+\alpha_y}{1+\alpha_y}b_y & 0 & 0 & 0 \\ 0 & 0 & 0 & \frac{4+\alpha_z}{1+\alpha_z}b_z & \frac{2-\alpha_z}{1+\alpha_z}b_z & 0 \\ 0 & 0 & 0 & \frac{2-\alpha_z}{1+\alpha_z}b_z & \frac{4+\alpha_z}{1+\alpha_z}b_z & 0 \\ 0 & 0 & 0 & 0 & 0 & t \end{bmatrix} \begin{bmatrix} v_x^* \\ v_{\theta y1}^* \\ v_{\theta y2}^* \\ v_{\theta z1}^* \\ v_{\theta z2}^* \\ v_{\theta x}^* \end{bmatrix}$$

or

$$\mathbf{S}^* = \mathbf{k}^* \mathbf{v}^* \qquad (14\text{-}27)$$

The following notation has been used

---

[1]. Warping is the displacement of the points, of an originally plane cross section, in the direction of the beam axis (x) due to twisting, see Chapter 16. Such displacements are significant only for members of open thin walled sections. Warping can be included as a 7th *dof*; the corresponding force component is the so-called *bi-moment*; both these quantities are difficult to handle at the element level (in transformations), as well as at the system level (continuity and boundary conditions). This is not to say that such elements are not in use, but for a general frame analysis program St. Venant torsion is the norm.

Bending of beams and plates

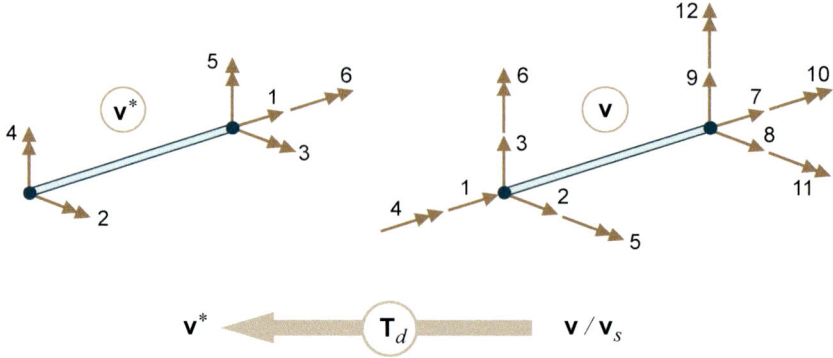

Figure 14.11 Transformation of *dofs*

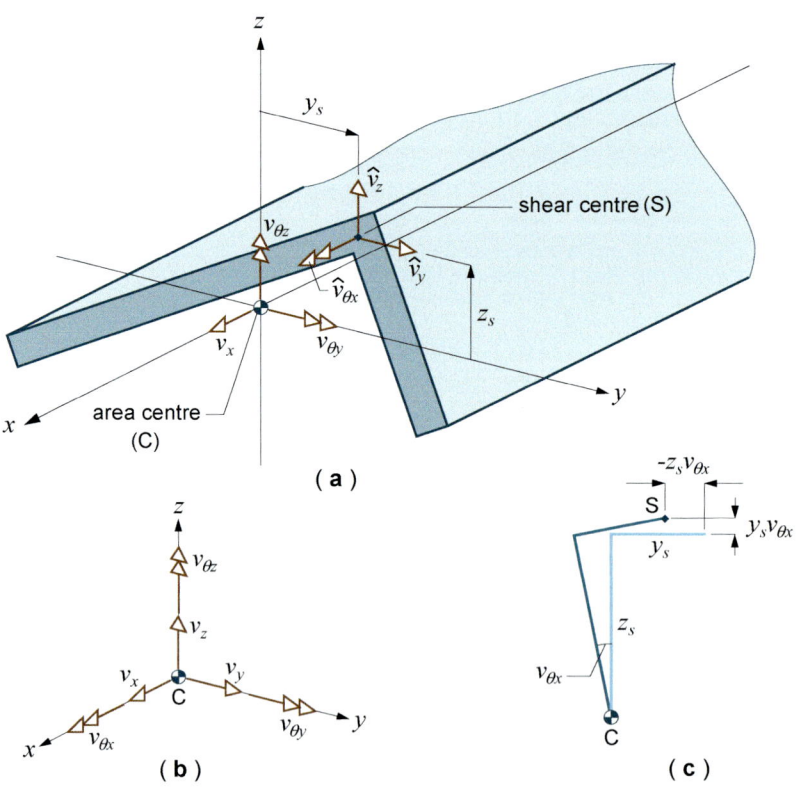

Figure 14.12 Arbitrary cross section

$$a = \frac{EA}{L}, \quad b_y = \frac{EI_y}{L}, \quad \alpha_y = \frac{12\kappa_z EI_y}{GAL^2}, \quad b_z = \frac{EI_z}{L}, \quad \alpha_z = \frac{12\kappa_y EI_z}{GAL^2}, \quad t = \frac{GI_t}{L}$$

It should be noted that while we have simple formulas for I_y and I_z, namely

$$I_y = \int_A z^2 dA \quad \text{and} \quad I_z = \int_A y^2 dA$$

this is not the case for I_t, except for circular cross sections, for which I_t is the polar moment of inertia. For other cross sections we need to use numerical solutions, see Chapter 16.

Including the rigid body modes. In order to include the rigid body modes, we need to transform from deformation *dofs* ($\mathbf{v}^*$) to the 12 *dofs* of Fig. 14.9, see also Fig. 14.11. In other words, we need the following transformation matrix $\mathbf{T}_d$:

$$\begin{bmatrix} v_x^* \\ v_{\theta y1}^* \\ v_{\theta y2}^* \\ v_{\theta z1}^* \\ v_{\theta z2}^* \\ v_{\theta x}^* \end{bmatrix} = \begin{bmatrix} -1 & 0 & 0 & 0 & 0 & 0 & 1 & 0 & 0 & 0 & 0 & 0 \\ 0 & 0 & -1/L & 0 & 1 & 0 & 0 & 0 & 1/L & 0 & 0 & 0 \\ 0 & 0 & -1/L & 0 & 0 & 0 & 0 & 0 & 1/L & 0 & 1 & 0 \\ 0 & 1/L & 0 & 0 & 0 & 1 & 0 & -1/L & 0 & 0 & 0 & 0 \\ 0 & 1/L & 0 & 0 & 0 & 0 & 0 & -1/L & 0 & 0 & 0 & 1 \\ 0 & 0 & 0 & -1 & 0 & 0 & 0 & 0 & 0 & 1 & 0 & 0 \end{bmatrix} \begin{bmatrix} v_{x1} \\ v_{y1} \\ v_{z1} \\ v_{\theta x1} \\ \cdot \\ \cdot \\ v_{\theta y2} \\ v_{\theta z2} \end{bmatrix} \quad (14\text{-}28a)$$

or

$$\mathbf{v}^* = \mathbf{T}_d \mathbf{v} \quad (14\text{-}28b)$$

Following standard transformation rules the element stiffness matrix $\mathbf{k}$ corresponding to displacements $\mathbf{v}$ is obtained as

$$\mathbf{k} = \mathbf{T}_d^\top \mathbf{k}^* \mathbf{T}_d \quad (14\text{-}29a)$$

Arbitrary cross section. For an arbitrary cross section, the area centre and the shear centre do, in general, not coincide, see Figure 14.12. We assume that axes y and z are still *principal* axes of the cross section, and that the element axis x goes through the area centre. The decoupled matrices $\mathbf{k}$ and $\mathbf{k}^*$ still apply if we place our *dofs* as shown in Fig. 14.12a, that is 3 *dofs* at the area centre (C) and 3 *dofs* at the shear centre (S). Since the stiffness matrix of Eq. (14-29a) now applies to these rather awkwardly placed *dofs* we mark it by a subscript s, that is

$$\mathbf{k}_s = \mathbf{T}_d^\top \mathbf{k}^* \mathbf{T}_d \quad (14\text{-}29b)$$

The corresponding displacement vector is

$$\mathbf{v}_s = \begin{bmatrix} \mathbf{v}_{s1} \\ \mathbf{v}_{s2} \end{bmatrix} \quad \text{where} \quad \mathbf{v}_{si}^\top = [v_x \quad \hat{v}_y \quad \hat{v}_z \quad \hat{v}_{\theta x} \quad v_{\theta y} \quad v_{\theta z}]_i, \quad (i = 1, 2)$$

For the assembly process we need all 6 *dofs* to be located at the same point,

Bending of beams and plates

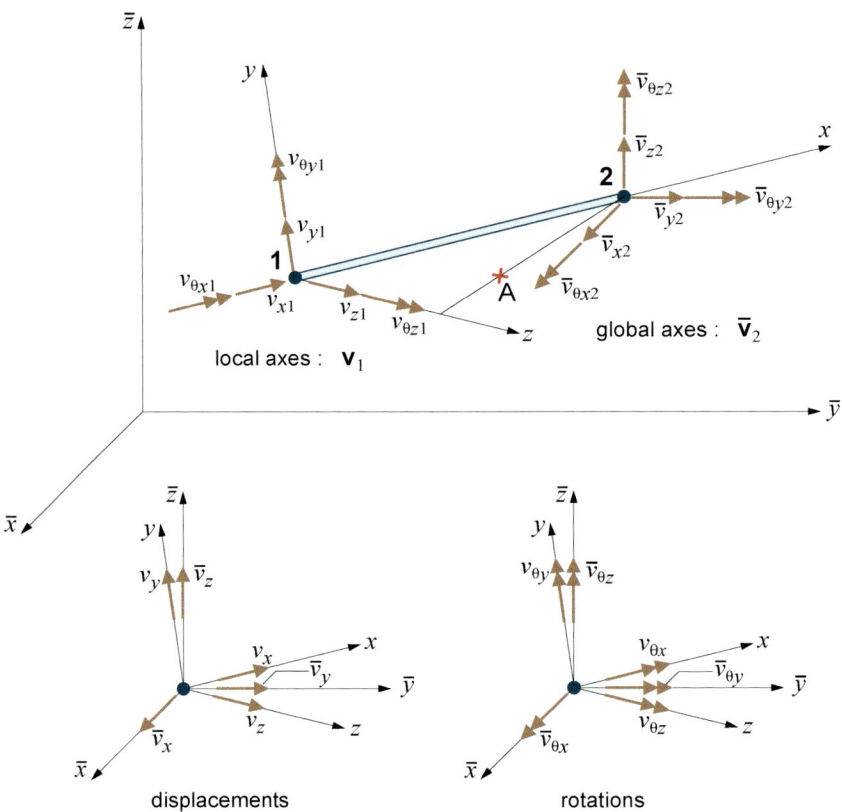

Figure 14.13 Local (x,y,z) and global $(\bar{x},\bar{y},\bar{z})$ coordinate axes

Beam element in 3D space

and the area centre is clearly the most convenient position for this point, see Fig. 14.12b. We can write the relationship between these two sets of *dofs* as

$$\mathbf{v}_s = \begin{bmatrix} \mathbf{v}_{s1} \\ \mathbf{v}_{s2} \end{bmatrix} = \begin{bmatrix} \mathbf{t}_s & \mathbf{0} \\ \mathbf{0} & \mathbf{t}_s \end{bmatrix} \begin{bmatrix} \mathbf{v}_1 \\ \mathbf{v}_2 \end{bmatrix} = \mathbf{T}_s \mathbf{v} \tag{14-30}$$

where, according to Fig. 14.11c

$$\mathbf{t}_s = \begin{bmatrix} 1 & 0 & 0 & 0 & 0 & 0 \\ 0 & 1 & 0 & -z_s & 0 & 0 \\ 0 & 0 & 1 & y_s & 0 & 0 \\ 0 & 0 & 0 & 1 & 0 & 0 \\ 0 & 0 & 0 & 0 & 1 & 0 \\ 0 & 0 & 0 & 0 & 0 & 1 \end{bmatrix} \tag{14-31}$$

y_s and z_s are the coordinates of the shear centre, and all *dofs* in $\mathbf{v}$ are placed at the element nodal points on the element axis (through the area centre of the cross section). The stiffness matrix corresponding to $\mathbf{v}$ is again obtained by standard transformation, i.e.

$$\mathbf{k} = \mathbf{T}_s^T \mathbf{k}_s \mathbf{T}_s \tag{14-32}$$

Transformation to global axes. Our *dofs* are currently referred to *local* element axes. Prior to assembly we need all element relations to be referred to a common *global* reference system ($\bar{x}, \bar{y}, \bar{z}$), see Fig. 14.13. At a given nodal point we split the 6 *dofs* into three displacements and three rotations. The displacements in the local (orthogonal) coordinates can be expressed in terms of the corresponding global components by the well-known *rotation matrix* of direction cosines, that is

$$\mathbf{v}_{di} = \begin{bmatrix} v_x \\ v_y \\ v_z \end{bmatrix}_i = \begin{bmatrix} \cos(x,\bar{x}) & \cos(x,\bar{y}) & \cos(x,\bar{z}) \\ \cos(y,\bar{x}) & \cos(y,\bar{y}) & \cos(y,\bar{z}) \\ \cos(z,\bar{x}) & \cos(z,\bar{y}) & \cos(z,\bar{z}) \end{bmatrix} \begin{bmatrix} \bar{v}_x \\ \bar{v}_y \\ \bar{v}_z \end{bmatrix}_i = \mathbf{t}_r \bar{\mathbf{v}}_{di} \tag{14-33}$$

In order to establish the rotation matrix $\mathbf{t}_r$ we need the global coordinates of at least three distinct points in the local system, for instance the nodal points 1 and 2, and a point (A) in the *xz*-plane (with a positive *z*-coordinate), see APPENDIX B.

The rotations about the three orthogonal axes can also be seen as components of a vector, a "rotation vector", and this "vector" transforms from one coordinate system to another in precisely the same way as the displacement vector $\mathbf{v}_d$, independent of the magnitude of the rotations. Hence

$$\mathbf{v} = \begin{bmatrix} \mathbf{v}_{d1} \\ \mathbf{v}_{\theta 1} \\ \mathbf{v}_{d2} \\ \mathbf{v}_{\theta 2} \end{bmatrix} = \begin{bmatrix} \mathbf{t}_r & \mathbf{0} & \mathbf{0} & \mathbf{0} \\ \mathbf{0} & \mathbf{t}_r & \mathbf{0} & \mathbf{0} \\ \mathbf{0} & \mathbf{0} & \mathbf{t}_r & \mathbf{0} \\ \mathbf{0} & \mathbf{0} & \mathbf{0} & \mathbf{t}_r \end{bmatrix} \begin{bmatrix} \bar{\mathbf{v}}_{d1} \\ \bar{\mathbf{v}}_{\theta 1} \\ \bar{\mathbf{v}}_{d2} \\ \bar{\mathbf{v}}_{\theta 2} \end{bmatrix} = \mathbf{T}_r \bar{\mathbf{v}}_e \tag{14-34}$$

Bending of beams and plates

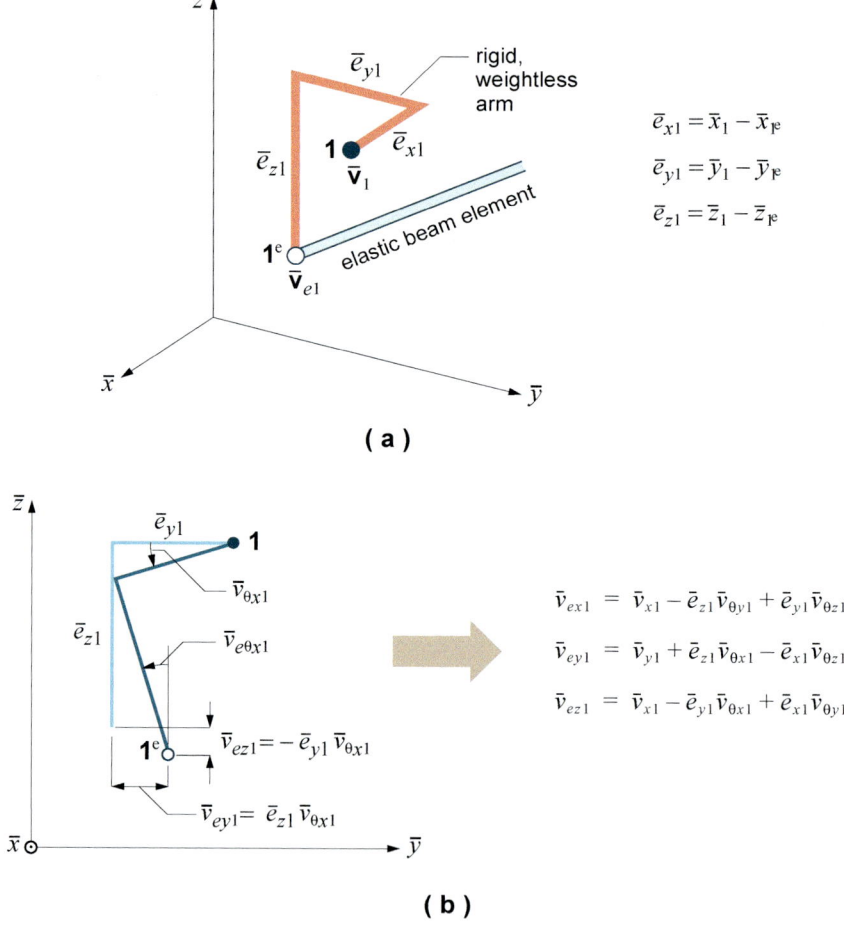

Figure 14.14 Offset, eccentrically placed, nodal point

and

$$\bar{\mathbf{k}}_e = \mathbf{T}_r^\mathsf{T} \mathbf{k} \mathbf{T}_r \tag{14-35}$$

The sub-index e on the global displacement vector ($\bar{\mathbf{v}}_e$) signifies that the nodal points are located at the ends of the flexible part of the beam element; its purpose will become more obvious later when we introduce fictitious *rigid arms* at the element ends.

Before we leave this transformation it is important to emphasize that although the rotation "vector" transforms as a vector it is not a *true* vector; it does *not obey* the rules of vector addition. For *finite* rotations we cannot add two of these "vectors" by simply adding their corresponding components; the result depends on the sequence in which we add the components. For non-linear, large displacement analysis we need to make allowances for this, which is not trivial, but for linear, small displacement theory it is quite in order to treat the rotation vector as a true vector and assume the principle of superposition to be valid.

Offset nodes – eccentricity. In order to model eccentricities it is useful to equip the element with completely rigid, but weightless, blocks that connect the element end points (1^e and 2^e) to offset nodes (1 and 2). One way this can be accomplished is to introduce three rigidly connected rigid (and weightless) *arms*, one in each global direction, as shown in Fig. 14.14a. The rigid arm is rigidly connected to the two points (1^e and 1). From purely geometrical consideration we obtain the following relationship between the displacement parameters at the two points, see Fig 14.14b (and keep in mind: small displacements):

$$\bar{\mathbf{v}}_{e1} = \begin{bmatrix} \mathbf{I} & \mathbf{t}_{e1} \\ \mathbf{0} & \mathbf{I} \end{bmatrix} \bar{\mathbf{v}}_1 \quad \text{where} \quad \mathbf{t}_{e1} = \begin{bmatrix} 0 & -\bar{e}_{z1} & \bar{e}_{y1} \\ \bar{e}_{z1} & 0 & -\bar{e}_{x1} \\ -\bar{e}_{y1} & \bar{e}_{x1} & 0 \end{bmatrix} \tag{14-36}$$

The rotations are of course the same at all points on the rigid arm. At node 2 we have, in principle, the same relationship, thus

$$\bar{\mathbf{v}}_e = \left[\begin{array}{cc|cc} \mathbf{I} & \mathbf{t}_{e1} & \mathbf{0} & \mathbf{0} \\ \mathbf{0} & \mathbf{I} & \mathbf{0} & \mathbf{0} \\ \hline \mathbf{0} & \mathbf{0} & \mathbf{I} & \mathbf{t}_{e2} \\ \mathbf{0} & \mathbf{0} & \mathbf{0} & \mathbf{I} \end{array}\right] \bar{\mathbf{v}} = \mathbf{T}_e \bar{\mathbf{v}} \tag{14-37}$$

and

$$\bar{\mathbf{k}} = \mathbf{T}_e^\mathsf{T} \bar{\mathbf{k}}_e \mathbf{T}_e \tag{14-38}$$

This is the final stiffness matrix of a general and effective 3D beam element. By use of Eqs. (14-29b), (14-32) and (14-35) we can express this matrix as

$$\bar{\mathbf{k}} = \underbrace{\mathbf{T}_e^\mathsf{T} \mathbf{T}_r^\mathsf{T} \mathbf{T}_s^\mathsf{T} \mathbf{T}_d^\mathsf{T}}_{12\times 12} \underbrace{\mathbf{k}^*}_{6\times 6} \mathbf{T}_d \mathbf{T}_s \mathbf{T}_r \mathbf{T}_e \tag{14-39}$$

This is easily programmed, by first generating the product of the four transformation matrices. An alternative route is to determine *explicit* expressions

Bending of beams and plates

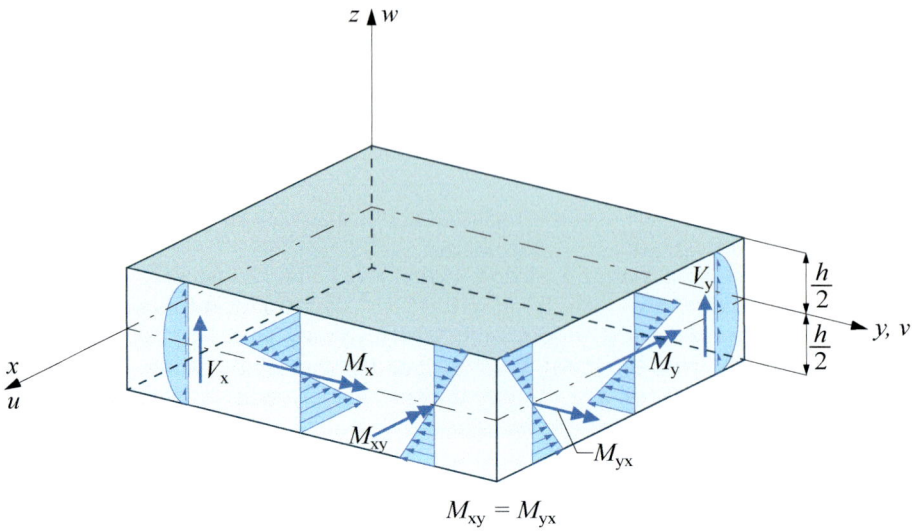

Copy of **Figure 3.17** Stresses and stress resultants – plate bending

Equation (3-74) :
$$\boldsymbol{\varepsilon} = \boldsymbol{\varepsilon}_b = \begin{bmatrix} \varepsilon_x \\ \varepsilon_y \\ \gamma_{xy} \end{bmatrix} = -z \begin{bmatrix} w_{,xx} \\ w_{,yy} \\ 2w_{,xy} \end{bmatrix} = -z\mathbf{c}_K$$

Equation (3-75) :
$$\mathbf{c}_K = \begin{bmatrix} w_{,xx} \\ w_{,yy} \\ 2w_{,xy} \end{bmatrix} = \begin{bmatrix} \dfrac{\partial^2}{\partial x^2} \\ \dfrac{\partial^2}{\partial y^2} \\ 2\dfrac{\partial^2}{\partial x \partial y} \end{bmatrix} w = \boldsymbol{\Delta}_K w$$

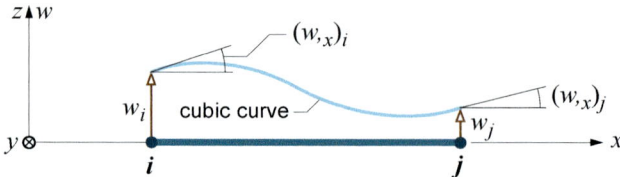

Figure 14.15 Displacement (w) along an element edge

14.3 Triangular (thin) plate bending elements

A summary of plate bending theory is presented in Section 3.5; reference coordinates and section forces are shown in Fig. 3.17. In this section we will focus on the so-called KIRCHHOFF theory for *thin* plates. This theory is analogous to the EULER-BERNOULLI beam theory in that it neglects transverse shear deformations, that is $\gamma_{yz} = \gamma_{zx} = 0$. Furthermore, we will deal mainly with pure displacement elements of triangular shape.

The only field variable in KIRCHHOFF plate theory is the lateral displacement w. The three relevant strain components are now expressed in terms of the curvature vector $\mathbf{c}_K$, see Eqs. (3-74) and (3-75); the order of differentiation of the operator $\mathbf{\Delta}_K$ is $m = 2$, and we are thus faced with a C^1 problem. In other words, we now require continuity not only in w, but also in its two first derivatives $w_{,x}$ and $w_{,y}$ along the entire common edge of two neighbouring elements.

The obvious choice of nodal parameters or *dofs*, irrespective of the element shape, is the nodal point values of the displacement w itself and the two *slopes*, $w_{,x}$ and $w_{,y}$. Figure 14.15 shows a typical element edge with nodal points at its ends. For simplicity we have placed the edge along the x-axis. For the displacement w we have four "boundary conditions" since the displacement (w) and the slope ($w_{,x}$) are known at both ends. We are therefore able to describe uniquely a *cubic* curve for w in terms of the nodal point parameters. It follows that the slope along the edge ($w_{,x}$) is also uniquely defined, but what about the slope $w_{,y}$ *normal* to the edge?

For the element to be *invariant* with respect to coordinate transformations the shape functions must exhibit symmetry in x and y. Hence if w is cubic in x it is also cubic in y. If we therefore assume that the shape functions form a complete cubic polynomial in x and y, the terms y, xy and x^2y will make the normal slope $w_{,y}$ a quadratic function of x. We need 3 boundary conditions to uniquely define a quadratic curve, but we only have two, the value of $w_{,y}$ at each end. In other words, we do not have normal slope continuity between elements; consequently we do not satisfy C^1 continuity. This is in fact an example of a general rule:

> For a KIRCHHOFF plate element it is *not* possible to determine a displacement field for w that satisfies C^1 continuity between elements by using only w and its first derivatives as nodal point parameters.

This statement holds true for any element shape and any number of nodes along an element edge. The reasoning leading up to this statement is fairly straightforward [50].

Bending of beams and plates

				1					linear (3 terms)
			x		y				
		x^2		xy		y^2			quadratic (6 terms)
	x^3		x^2y		xy^2		y^3		cubic (10 terms)
x^4		x^3y		x^2y^2		xy^3		y^4	quartic (15 terms)
x^5	x^4y	x^3y^2	x^2y^3	xy^4	y^5				quintic (21 terms)
x^6	x^5y	x^4y^2	x^3y^3	x^2y^4	xy^5	y^6			sextic (28 terms)

.
.

Copy of **Figure 5.9** Complete polynomials in two variables

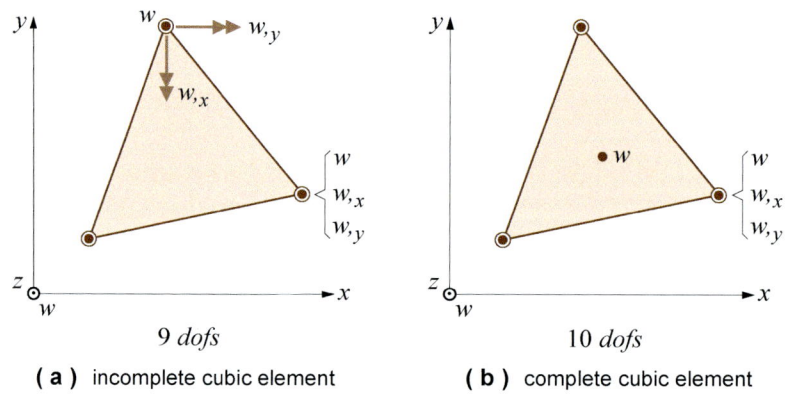

(a) incomplete cubic element — 9 dofs

(b) complete cubic element — 10 dofs

Figure 14.16 Simple triangular plate bending elements

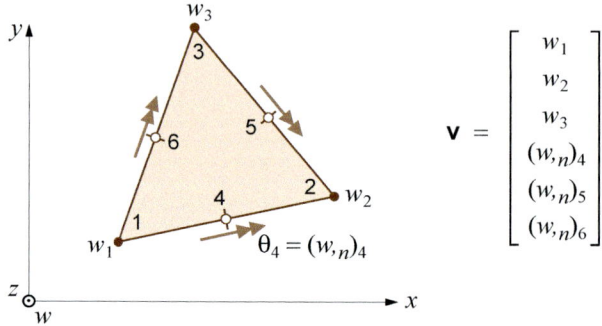

Figure 14.17 The MORLEY triangle – T6

The 1960's saw a great quest for a simple 9-*dof* triangular, thin plate bending element with the degrees of freedom shown in Fig. 14.16a. Two problems had to be overcome: C^1 continuity and an appropriate choice of polynomial terms. The latter problem was caused by the fact the 9 *dofs* do not match the 10 terms of a complete cubic polynomial, see Fig 5.9. The obvious choice was to include an internal *dof*, the lateral displacement at the centre (which could be eliminated by static condensation); however, normal slope continuity was violated and the performance of this element was poor. Attempts made at combining polynomial terms were unsuccessful.

A fully compatible 9-*dof* triangular element was derived at the expense of both internal and external constraints on the normal slope. A convergent, but overly stiff element was the result. So the decade came to an end without a satisfactory 9-*dof* element. An overview of the efforts may be found in Ref. [30].

We shall come back to some later developments that, while not confining themselves to strict displacement based ideology, turned out to be quite useful. First, however, we shall look at some other triangular displacement based elements that perform quite well, all of which make use of shape functions that constitute *complete* polynomials, or rather, *homogeneous polynomials*, since we shall be making use of triangular (or area) coordinates ($\zeta_1, \zeta_2, \zeta_3$).

So far we have been mostly concerned with continuity, but before we look at specific elements we also need to say a few words about *completeness*. Rigid body movements in plate bending are: transverse displacement and rotation about two orthogonal axes, such as $w_{,x}$ and $w_{,y}$. In order to describe these correctly (without straining) we need a complete linear polynomial to be present in the shape functions, that is, the top three terms in the triangle of Fig. 5.9. The constant strain (stress) requirement now means that the element must be capable of representing constant curvature (moment). For this to be possible we see that the shape functions must contain the three quadratic terms of the triangle.

The MORLEY triangle – T6. This 6-*dof* element, shown in Fig. 14.17 is the simplest plate bending element possible. It is based on a complete quadratic polynomial, and its *dofs* are the displacement w at each corner node, and the *normal slope* $w_{,n}$ at each mid-side nodal point. The displacement version of this element was proposed by MORLEY [51], hence the name.

Completeness is satisfied, but continuity is not at all satisfied, not even for w itself. And yet, this simple element passes the patch test and behaves, as we shall see, quite well. In spite of the element's simplicity the derivation of its stiffness matrix **k** is not trivial. Since the element is representative for this type of problem we will take the reader through the details of one way to obtain **k**; it is not the only way.

We will use area coordinates and indirect interpolation via six generalized displacement parameters **q**. A complete polynomial in x and y is equivalent to a *homogeneous* polynomial in area coordinates. Hence we assume

$$w = [\, \zeta_1^2 \quad \zeta_2^2 \quad \zeta_3^2 \quad \zeta_1\zeta_2 \quad \zeta_2\zeta_3 \quad \zeta_3\zeta_1 \,]\mathbf{q} = \mathbf{N}_q\mathbf{q} \qquad (14\text{-}40)$$

Bending of beams and plates

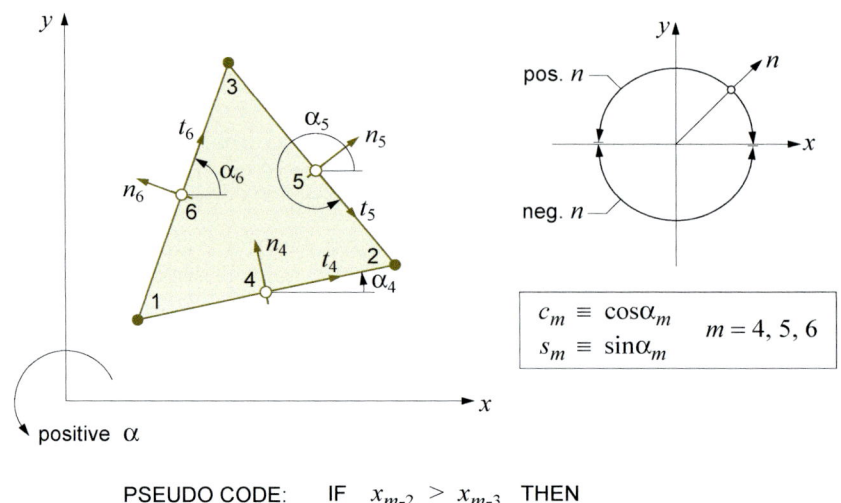

PSEUDO CODE: IF $x_{m-2} > x_{m-3}$ THEN

$$c_m = \frac{x_{m-2} - x_{m-3}}{L_m}$$

$$s_m = \frac{y_{m-2} - y_{m-3}}{L_m}$$

ELSEIF $x_{m-2} < x_{m-3}$ THEN

$$c_m = \frac{x_{m-3} - x_{m-2}}{L_m}$$

$$s_m = \frac{y_{m-3} - y_{m-2}}{L_m}$$

ELSE

$\left. \begin{array}{l} c_m = 0,0 \\ s_m = 1,0 \end{array} \right\}$ choice!

ENDIF

Figure 14.18 Unambiguous definition of the normal slope

In order to express w in terms of $\mathbf{v}$ we need to establish a relationship between $\mathbf{v}$ and $\mathbf{q}$, and then between $\mathbf{q}$ and $\mathbf{v}$. For this we need to express $w_{,n}$ in terms of derivatives with respect to ζ_1 and ζ_2; we will consider ζ_1 and ζ_2 to be our *independent* variables, while $\zeta_3 = 1 - \zeta_1 - \zeta_2$ (see Section 5.2).

First, however, we need to establish unambiguous expressions for the normal slope $w_{,n}$ along the three element edges. For this purpose a local coordinate system t and n is introduced; the t axis coincides with the element edge, and the angle between the global x-axis and the t-axis is denoted α, see Fig. 14.18. Positive direction of the n-axis is defined by Fig. 14.18, and this definition implies that α can only take values between 0 and $\pi/2$ or between $3\pi/2$ and 2π, i.e.

$$0 \le \alpha_m < \pi/2 \quad \text{or} \quad 3\pi/2 \le \alpha_m < 2\pi \quad \text{for} \quad m = 4, 5, 6 \tag{14-41}$$

We also need the relationships between x, y and t, n, that is

$$x = ct - sn \quad \text{and} \quad y = st + cn \tag{14-42}$$

$$t = cx + sy \quad \text{and} \quad n = -sx + cy \tag{14-43}$$

As in Fig. 14.18 we have used the abbreviation $c \equiv \cos\alpha$ and $s \equiv \sin\alpha$. From these equations we get

$$\frac{\partial}{\partial t} = \frac{\partial}{\partial x}\frac{\partial x}{\partial t} + \frac{\partial}{\partial y}\frac{\partial y}{\partial t} = c\frac{\partial}{\partial x} + s\frac{\partial}{\partial y} \tag{14-44a}$$

$$\frac{\partial}{\partial n} = \frac{\partial}{\partial x}\frac{\partial x}{\partial n} + \frac{\partial}{\partial y}\frac{\partial y}{\partial n} = -s\frac{\partial}{\partial x} + c\frac{\partial}{\partial y} \tag{14-44b}$$

Hence

$$w_{,n} = -sw_{,x} + cw_{,y} \tag{14-45}$$

Before we continue let us examine the general formula given by Eq. (4-54), that is

$$\mathbf{k} = \int_{V_e} \mathbf{B}^T \mathbf{C} \mathbf{B} \, dV$$

Recall Eqs. (4-50) and (4-51):

$$\boldsymbol{\varepsilon} = \Delta \mathbf{u} = \Delta(\mathbf{N}\mathbf{v}) = \Delta \mathbf{N}\mathbf{v} = \mathbf{B}\mathbf{v} \quad \Rightarrow \quad \mathbf{B} = \Delta \mathbf{N}$$

Recall also Eq. (3-74):

$$\boldsymbol{\varepsilon} = \boldsymbol{\varepsilon}_b = -z\mathbf{c}_K = -z\mathbf{B}_K\mathbf{v} \quad \Rightarrow \quad \mathbf{c}_K = \mathbf{B}_K\mathbf{v} \quad \Rightarrow \quad \mathbf{B} = -z\mathbf{B}_K$$

Therefore

$$\mathbf{k} = \int_{V_e} \mathbf{B}^T \mathbf{C} \mathbf{B} \, dV = \int_{-h/2}^{-h/2}\int_A (-z\mathbf{B}_K^T)\mathbf{C}(-z\mathbf{B}_K)\,dz\,dA = \frac{1}{12}\int_A h^3 \mathbf{B}_K^T \mathbf{C} \mathbf{B}_K \, dA \tag{14-46}$$

For constant plate thickness this can be written as

Bending of beams and plates

Equation (3-79): $\quad \mathbf{D} = \dfrac{h^3}{12}\mathbf{C}_b = D\begin{bmatrix} 1 & \nu & 0 \\ \nu & 1 & 0 \\ 0 & 0 & \dfrac{1-\nu}{2} \end{bmatrix} \quad$ and $\quad D = \dfrac{Eh^3}{12(1-\nu^2)}$

Equation (5-29): $\quad \begin{bmatrix} \dfrac{\partial}{\partial x} \\ \dfrac{\partial}{\partial y} \end{bmatrix} = \dfrac{1}{2A}\begin{bmatrix} y_{23} & y_{31} \\ x_{32} & x_{13} \end{bmatrix}\begin{bmatrix} \dfrac{\partial}{\partial \zeta_1} \\ \dfrac{\partial}{\partial \zeta_2} \end{bmatrix} = \mathbf{J}^{-1}\begin{bmatrix} \dfrac{\partial}{\partial \zeta_1} \\ \dfrac{\partial}{\partial \zeta_2} \end{bmatrix}$

$$\mathbf{A}_{21} = \begin{bmatrix} \gamma_4 & \mu_4 & 0 \\ 0 & \mu_5 & -a_5 \\ \gamma_6 & 0 & -a_6 \end{bmatrix} \qquad \mathbf{A}_{22} = \dfrac{1}{2}\begin{bmatrix} a_4 & -a_4 & -a_4 \\ \gamma_5 & -\gamma_5 & \gamma_5 \\ \mu_6 & \mu_6 & -\mu_6 \end{bmatrix}$$

where $a_m = \gamma_m + \mu_m \quad (m = 4, 5, 6)$

Triangular (thin) plate bending elements

$$\mathbf{k} = \int_A \mathbf{B}_K^T \mathbf{D} \mathbf{B}_K \, dA \quad \text{where} \quad \mathbf{D} = \frac{h^3}{12}\mathbf{C} \tag{14-47}$$

Matrix $\mathbf{D}$ is defined by Eq. (3-79).

In thin plate theory matrix $\mathbf{B}_K$ relates the KIRCHHOFF curvature vector $\mathbf{c}_K$ to the nodal displacements $\mathbf{v}$, that is

$$\mathbf{c}_K = \Delta_K w = \Delta_K \mathbf{N}_q \mathbf{q} = \Delta_K \mathbf{N}_q \mathbf{A}^{-1} \mathbf{v} \quad \Rightarrow \quad \mathbf{B}_K = \Delta_K \mathbf{N}_q \mathbf{A}^{-1} \tag{14-48}$$

where matrix $\mathbf{A}$ relates $\mathbf{v}$ to $\mathbf{q}$:

$$\mathbf{v} = \mathbf{A}\mathbf{q} \quad \Rightarrow \quad \mathbf{q} = \mathbf{A}^{-1}\mathbf{v}$$

Our first task is to determine $\mathbf{A}$, and in order to accomplish this we need to express the normal slope $w_{,n}$ in terms of derivatives with respect to area coordinates, since our shape functions are in terms of these coordinates. By use of Eq. (5-29) we get

$$(w_{,n})_m = [\, -s_m \quad c_m \,] \begin{bmatrix} w_{,x} \\ w_{,y} \end{bmatrix} = [\, -s_m \quad c_m \,] \mathbf{J}^{-1} \begin{bmatrix} w_{,\zeta_1} \\ w_{,\zeta_2} \end{bmatrix} \tag{14-49}$$

or

$$(w_{,n})_m = \gamma_m (w_{,\zeta_1})_m + \mu_m (w_{,\zeta_2})_m \tag{14-50}$$

where, in view of Eq. (5-29):

$$\gamma_m = (c_m x_{32} - s_m y_{23})/2A \tag{14-51a}$$

$$\mu_m = (c_m x_{13} - s_m y_{31})/2A \tag{14-51b}$$

where $x_{ij} = x_i - x_j$ etc, and A is the element area.

We can now establish $\mathbf{A}$:

$$\mathbf{v} = \begin{bmatrix} w_1 \\ w_2 \\ w_3 \\ (w_{,n})_4 \\ (w_{,n})_5 \\ (w_{,n})_6 \end{bmatrix} = \begin{bmatrix} \mathbf{I} & \mathbf{0} \\ \mathbf{A}_{21} & \mathbf{A}_{22} \end{bmatrix} \begin{bmatrix} q_1 \\ q_2 \\ q_3 \\ q_4 \\ q_5 \\ q_6 \end{bmatrix} = \mathbf{A}\mathbf{q} \tag{14-52}$$

The 3 by 3 submatrices, $\mathbf{A}_{21}$ and $\mathbf{A}_{22}$, are given opposite. The inverse matrix is easily obtained as

$$\mathbf{A}^{-1} = \begin{bmatrix} \mathbf{I} & \mathbf{0} \\ -\mathbf{A}_{22}^{-1}\mathbf{A}_{21} & \mathbf{A}_{22}^{-1} \end{bmatrix} \tag{14-53}$$

We have one more problem of "mixed" coordinates. In Eq. (14-46) the operator Δ_K is in terms of cartesian coordinates whereas the shape functions $\mathbf{N}_q$ are in terms of area coordinates. By repeated use of Eq. (5-29) we find

467

Bending of beams and plates

$$\mathbf{B}_q = \begin{bmatrix} 2 & 0 & 2 & 0 & 0 & -2 \\ 0 & 2 & 2 & 0 & -2 & 0 \\ 0 & 0 & 2 & 1 & -1 & -1 \end{bmatrix}$$

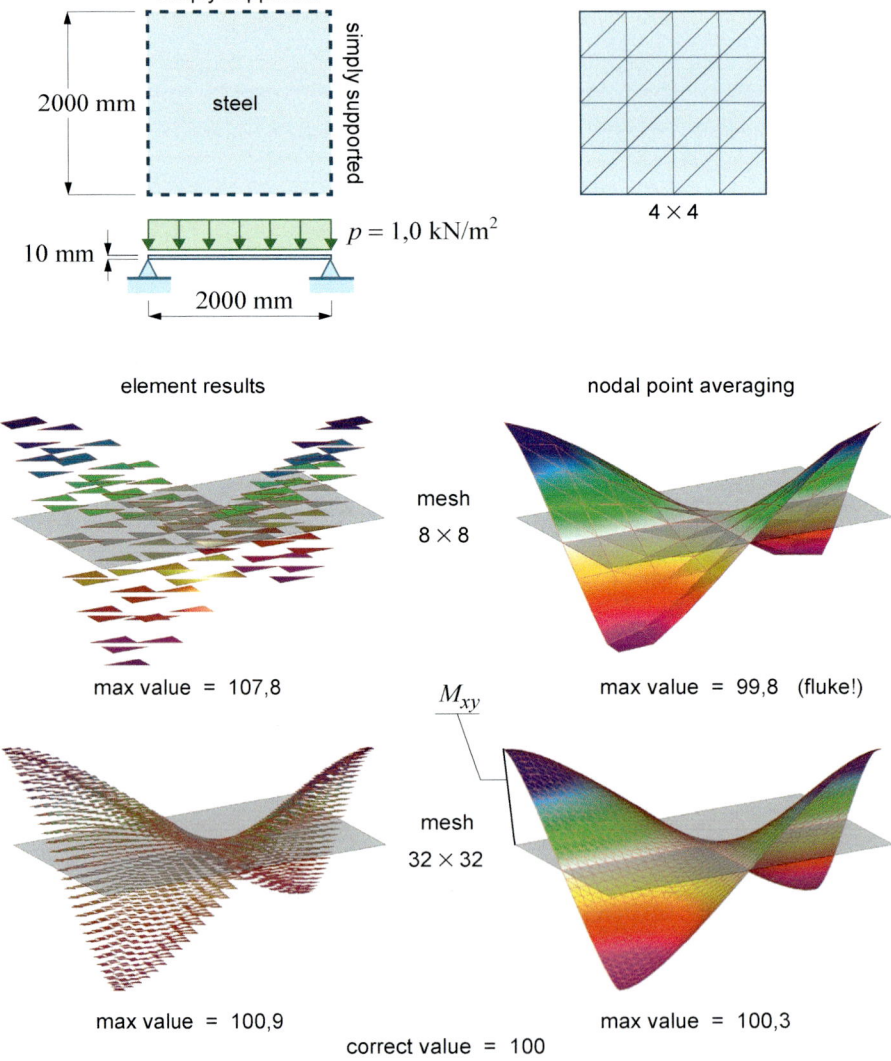

Figure 14.19 Twisting moment M_{xy} of a uniformly loaded, simply supported square plate, analysed by the **T6** (MORLEY) element

$$\begin{bmatrix} \dfrac{\partial^2}{\partial x^2} \\ \dfrac{\partial^2}{\partial y^2} \\ 2\dfrac{\partial^2}{\partial x \partial y} \end{bmatrix} = \dfrac{1}{4A^2} \begin{bmatrix} y_{23}^2 & y_{31}^2 & 2y_{31}y_{23} \\ x_{32}^2 & x_{13}^2 & 2x_{13}x_{32} \\ 2x_{32}y_{23} & 2x_{13}y_{31} & 2(x_{13}y_{23}+x_{32}y_{31}) \end{bmatrix} \begin{bmatrix} \dfrac{\partial^2}{\partial \zeta_1^2} \\ \dfrac{\partial^2}{\partial \zeta_2^2} \\ \dfrac{\partial^2}{\partial \zeta_1 \partial \zeta_2} \end{bmatrix} \qquad (14\text{-}54a)$$

or

$$\boldsymbol{\Delta}_K = \mathbf{H} \boldsymbol{\Delta}_\zeta \qquad (14\text{-}54b)$$

The $\mathbf{B}_K$ matrix of Eq. (14-46) can now be determined as

$$\mathbf{B}_K = \boldsymbol{\Delta}_K \mathbf{N}_q \mathbf{A}^{-1} = \mathbf{H} \boldsymbol{\Delta}_\zeta \mathbf{N}_q \mathbf{A}^{-1} = \mathbf{H} \mathbf{B}_q \mathbf{A}^{-1} \quad \text{where} \quad \mathbf{B}_q = \boldsymbol{\Delta}_\zeta \mathbf{N}_q \qquad (14\text{-}55)$$

The $\mathbf{B}_q$ matrix, a 3 by 6 matrix of constants, is easily found, see opposite page. If we substitute this into Eq. (14-46) we finally get

$$\mathbf{k} = \dfrac{1}{12} \int_A h^3 \mathbf{B}_K^T \mathbf{C} \mathbf{B}_K dA = \dfrac{1}{12} \int_A h^3 \mathbf{A}^{-T} \mathbf{B}_q^T \mathbf{H}^T \mathbf{C} \mathbf{H} \mathbf{B}_q \mathbf{A}^{-1} dA \qquad (14\text{-}56)$$

This looks a bit messy, but all matrices of the integrand are *constant* matrices. It is fairly straightforward to invert the 3 by 3 matrix $\mathbf{A}_{22}$ of Eq. (14-52) symbolically and thus establish explicit expressions for all elements of $\mathbf{A}^{-1}$. With a little patience it is therefore quite feasible to establish explicit expressions for each element of $\mathbf{k}$, also if the thickness varies (linearly).

A suitable element load vector $\mathbf{S}_p$ for this element, due to a distributed transverse loading $p = p(x,y)$, is obtained by straightforward *lumping* into three concentrated forces at the corner nodes.

Once the nodal point displacements ($\mathbf{v}$) have been determined (by the system analysis) the bending moments, which are constant within each element, are determined by Eq. (3-78), that is

$$\mathbf{m} = \begin{bmatrix} M_x \\ M_y \\ M_{xy} \end{bmatrix} = -\dfrac{h^3}{12} \mathbf{C} \mathbf{c}_K = -\dfrac{h^3}{12} \mathbf{C} \mathbf{B}_K \mathbf{v} = -\dfrac{h^3}{12} \mathbf{C} \mathbf{H} \mathbf{B}_q \mathbf{A}^{-1} \mathbf{v} \qquad (14\text{-}57)$$

Figure 14.19 gives an indication of what the element is capable of. A simply supported square plate, subjected to a uniformly distributed load is analysed by **FEMplate** for two different meshes, one coarse and one moderately fine. The twisting moment M_{xy}, presented as a "surface" of triangular facets, has been chosen as response parameter. The accuracy is much the same for other parameters. Results computed for the individual elements (see left-hand part of the figure) show clearly the constant value of the moment. In view of the element's simplicity and its lack of inter-element displacement continuity, the results of Fig. 14.19 are quite remarkable.

We shall come back to this example towards the end of the chapter and compare this element with some other triangular thin plate elements.

Bending of beams and plates

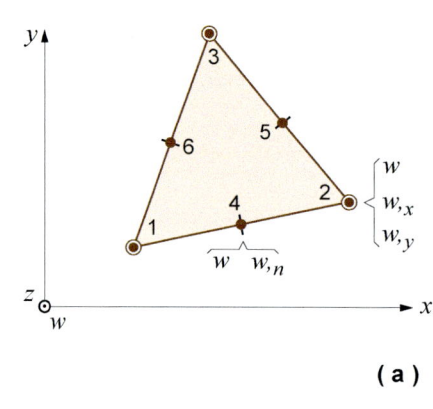

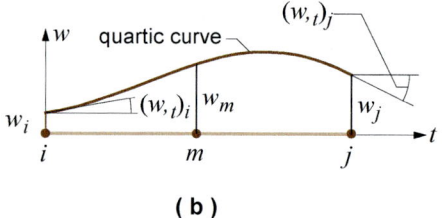

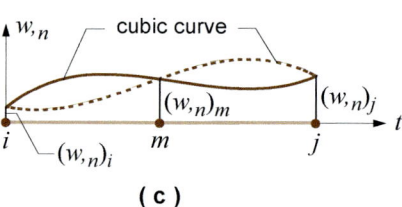

Figure 14.20 The quartic triangular **T15** element

Equation (3-82a): $V_x = M_{x,x} + M_{xy,y}$

Equation (3-82b): $V_y = M_{y,y} + M_{xy,x}$

Cubic triangle – T10. The next complete polynomial is the cubic one which has 10 terms. In order to use all 10 terms the element in Fig. 14.16b is the obvious choice; the usual three *dofs* at each corner node, and the lateral displacement w at an internal node placed at the area centre; the latter can be eliminated by static condensation.

This element, which we term **T10**, was derived by several researchers quite early (in the 1960's). It satisfies completeness but not continuity (the normal slope is not continuous between elements). The element is very flexible, and it was deemed useless. COOK et al [52] state categorically that the element does not converge. This implies that it does not pass the patch test which is somewhat surprising since its violation of the continuity requirement is not nearly as severe as that of **T6**. In any case we leave this element alone; we shall come back to some useful 9-*dof* triangular elements in the next two sections.

Quartic triangle – T15. A complete quartic polynomial has 15 terms, see Fig. 5.9. Which 15 *dofs* do we choose to match these 15 terms? One possible choice, termed **T15** [13], is shown in Fig. 14.20a. Again, completeness is not an issue; the question is continuity. For an arbitrary edge, in direction t, the displacement w varies quartically. At both end nodes we know $w_{,x}$ and $w_{,y}$ and hence both $w_{,t}$ and $w_{,n}$ are known. Together with the value of w at the mid-side node we have 5 "boundary conditions" for w, and the quartic curve is therefore uniquely defined by the nodal parameters (*dofs*). Consequently **T15** satisfies the continuity requirements for w and $w_{,s}$. However, normal slope ($w_{,n}$) continuity is not satisfied, as can be seen from Fig. 14.20c. The normal slope varies cubically along the edge and we have only three "boundary conditions" for it, the end and mid-side values. As shown in Fig. 14.20c we can pass any number of cubic curves through three points, none of which are unique. This result is no surprise as we have already stated that we cannot satisfy C^1 continuity with only w and its first derivatives as *dofs*.

The derivation of the stiffness matrix for **T15** is quite similar to that of **T6**. We now assume a 4th order homogeneous polynomial in area coordinates for the displacement,

$$w = [\,\zeta_1^4 \quad \zeta_2^4 \quad \zeta_3^4 \quad \zeta_1^3\zeta_2 \ldots \ldots \zeta_1\zeta_2^2\zeta_3 \quad \zeta_1\zeta_2\zeta_3^2\,]\mathbf{q} = \mathbf{N}_q\mathbf{q} \qquad (14\text{-}58)$$

and proceed to establish explicit expressions for the elements of the transformation matrix $\mathbf{A}^{-1}$. We will not go into the details as they will not provide new insight. With a 4th order polynomial for w it seems appropriate to let the element have a linearly varying thickness defined by the three corner thicknesses, and it is also tempting to compute KIRCHHOFF shear forces defined by Eqs. (3-82) which we can write

$$\begin{bmatrix} V_x \\ V_y \end{bmatrix} = \begin{bmatrix} \frac{\partial}{\partial x} & 0 & \frac{\partial}{\partial y} \\ 0 & \frac{\partial}{\partial y} & \frac{\partial}{\partial x} \end{bmatrix} \begin{bmatrix} M_x \\ M_y \\ M_{xy} \end{bmatrix} = \boldsymbol{\Delta}^T\mathbf{m} \qquad (14\text{-}59)$$

Substituting for $\mathbf{m}$ as defined by Eq. (14-57) we get

Bending of beams and plates

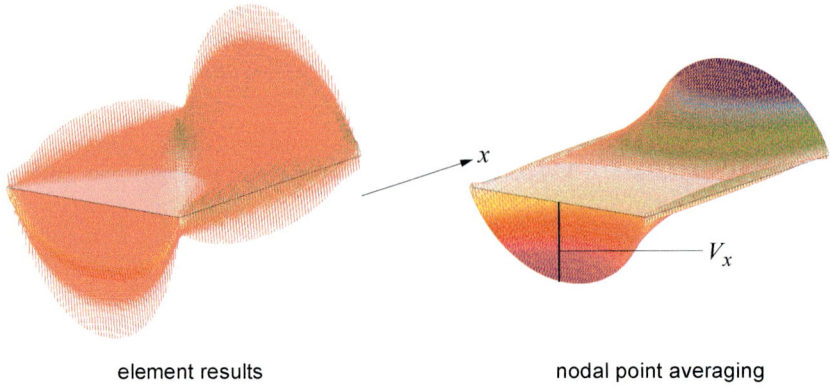

element results   nodal point averaging

Figure 14.21 KIRCHHOFF shear force V_x for the plate problem og Fig. 4.19 obtained by **FEMplate** using **T15** for a 64×64 mesh

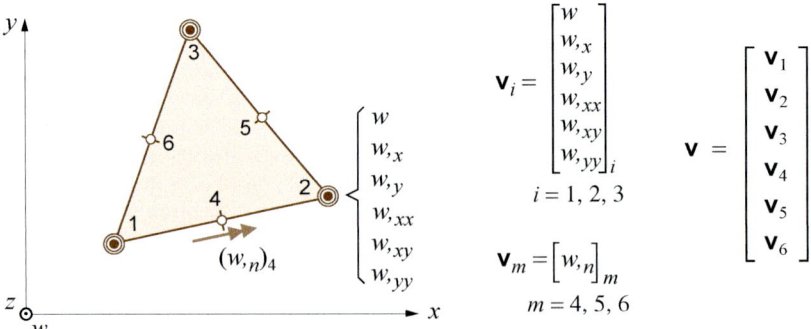

Figure 14.22 The 21-*dof* quintic triangular element – **T21**

$$\begin{bmatrix} V_x \\ V_y \end{bmatrix} = -\frac{h^3}{12}\boldsymbol{\Delta}^\mathsf{T}\mathbf{CHB}_q\mathbf{A}^{-1}\mathbf{v} \qquad (14\text{-}60)$$

Since moments vary quadratically within the element, the shear forces in **T15** vary linearly. Figure 4.21 shows how the shear force V_x, obtained by **FEMplate** for a 64×64 mesh of the square plate problem of Fig. 4.19, varies over the plate. The results computed for each element, shown on the left-hand side of the figure, are not particularly useful. The figure to the right however, obtained by simple nodal point averaging, makes far more sense.

For this element it is perhaps most appropriate to use a consistent load vector defined as, see Section 4.7 and Section 7.1,

$$\mathbf{S}^0 = -\int_A p\mathbf{N}^\mathsf{T} dA = -\int_A p\mathbf{A}^{-\mathsf{T}}\mathbf{N}_q^\mathsf{T} dA = -\mathbf{A}^{-\mathsf{T}}\int_A p\mathbf{N}_q^\mathsf{T} dA \qquad (14\text{-}61)$$

For a distributed transverse loading with linear variation over the element we have

$$p = p_1\zeta_1 + p_2\zeta_2 + p_3\zeta_3$$

where p_1, p_2 and p_3 are the load intensities at the corner nodes. The integration in Eq. (14-61) can easily be carried out analytically, using the simple formula of Eq. (5-30). As for the stiffness matrix, see Eq. (14-56), the integration for **k** can also be carried out analytically, but it is probably more convenient to do it numerically, using a 7-point formula, see APPENDIX C.

Before we leave this element it should be mentioned that the choice of *dofs* shown in Fig. 14.20 is not the only one. COOK et al [52] mention a 15-*dof* element with only one *dof* at each mid-side node, $w_{,n}$, but with 4 *dofs* at each corner node; in addition to w and its two first derivatives, the *twist* $w_{,xy}$ is included. We leave it to the reader to assess this element's continuity properties.

The quintic triangle – T21 and T18. The next complete polynomial is quintic and has 21 terms, see Fig. 5.9. A fairly obvious choice of *dofs* to match these 21 terms is shown in Fig. 14.22. At each of the three corner nodes we define the corner values of w, its two first derivatives $w_{,x}$ and $w_{,y}$ and all three second order derivatives $w_{,xx}$, $w_{,xy}$ and $w_{,yy}$ as *dofs* and in addition we define the value of the normal slope $w_{,n}$ as *dof* at each mid-side node.

This element satisfies both completeness and C^1 continuity. Completeness is obviously satisfied. As for continuity we see that we have three "boundary conditions" for w at each end of a side, the value of w and its first and second derivative with respect to the axis (*t*) along the edge. A total of 2×3 = 6 nodal point parameters, common to both elements bordering on to the edge, will uniquely define the quintically varying displacement along the edge. For the normal slope, that has a quartic variation along the edge, we need 5 parameters for its unique definition. And that is precisely what we have; two at each end, $w_{,n}$ and $w_{,nt}$, and one, $w_{,n}$, at the mid-side node.

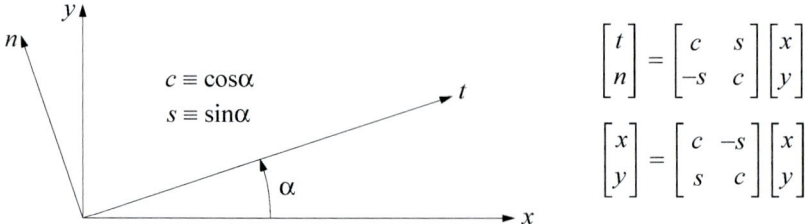

Figure 14.23 Coordinate transformation

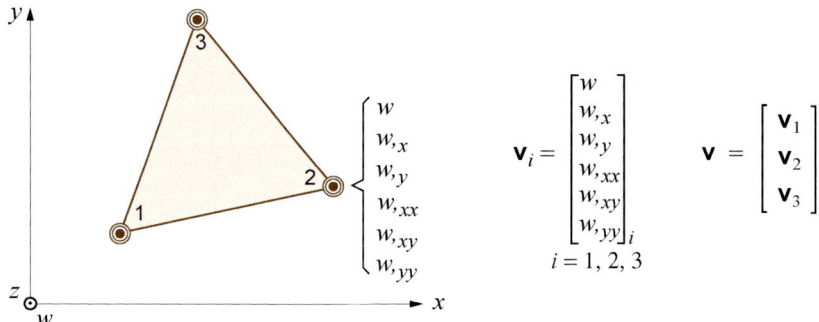

Figure 14.24 The 18-*dof* incomplete quintic triangular element – **T18**

Triangular (thin) plate bending elements

The above statements assume that if we know the first and second derivatives with respect to x and y, then we also know the same derivatives with respect to any two orthogonal axes t and n. With reference to Fig 14.23 we leave it to the reader to verify the following two transformations

$$\begin{bmatrix} \dfrac{\partial}{\partial t} \\ \dfrac{\partial}{\partial n} \end{bmatrix} = \begin{bmatrix} c & s \\ -s & c \end{bmatrix} \begin{bmatrix} \dfrac{\partial}{\partial x} \\ \dfrac{\partial}{\partial y} \end{bmatrix} \quad \text{and} \quad \begin{bmatrix} \dfrac{\partial^2}{\partial t^2} \\ \dfrac{\partial^2}{\partial t \partial n} \\ \dfrac{\partial^2}{\partial n^2} \end{bmatrix} = \begin{bmatrix} c^2 & 2sc & s^2 \\ -sc & c^2-s^2 & sc \\ s^2 & -2sc & c^2 \end{bmatrix} \begin{bmatrix} \dfrac{\partial^2}{\partial x^2} \\ \dfrac{\partial^2}{\partial x \partial y} \\ \dfrac{\partial^2}{\partial y^2} \end{bmatrix} \tag{14-62}$$

The 21-*dof* quintic element was described independently by several researchers in the late 1960's. A derivation using cartesian coordinates is reported in [13], while [12] presents derivations using both normalized cartesian coordinates and area coordinates. The element analysis can be carried out in much the same way as described for **T6**. Area coordinates are recommended and a homogeneous polynomial of order 5 is used as generalized shape functions.

The details of the derivation, which are somewhat complicated, are beyond this presentation. Instead we will take a brief look at a more convenient, but incomplete, version of this element, labelled **T18**. This element which is shown in Fig. 14.24 is obtained from **T21** by eliminating the normal slope *dof* at the mid-side nodes. In order to maintain inter-element continuity of the normal slope, the mid-side *dofs* are eliminated by imposing a *cubic* variation of $w_{,n}$ along the element edges.

It is fairly straightforward to show that for edge *i–j* along axis *t* and with mid-side node *m*, a unique cubic variation of $w_{,n}$ is obtained by the following expression

$$(w_{,n})_m = \frac{1}{2}[(w_{,n})_i + (w_{,n})_j] + \frac{L_m}{8}[(w_{,nt})_i - (w_{,nt})_j] \tag{14-63}$$

This equation and Eqs. (14-62) can now be used to establish a relation

$$\mathbf{v}_s = \mathbf{G}\mathbf{v}_c \tag{14-64}$$

where $\mathbf{v}_s$ contains the three mid-side *dofs* of **T21** and $\mathbf{v}_c$ the 18 corner *dofs*; matrix **G** impose the cubic variation. Thus

$$\mathbf{v}_{21} = \begin{bmatrix} \mathbf{v}_c \\ \mathbf{v}_s \end{bmatrix} = \begin{bmatrix} \mathbf{I}_{18} \\ \mathbf{G} \end{bmatrix} \mathbf{v}_{18} \quad (\mathbf{v}_{18} \equiv \mathbf{v}_c) \tag{14-65}$$

and

$$\mathbf{k}_{18} = \begin{bmatrix} \mathbf{I}_{18} & \mathbf{G}^T \end{bmatrix} \mathbf{k}_{21} \begin{bmatrix} \mathbf{I}_{18} \\ \mathbf{G} \end{bmatrix} \tag{14-66}$$

The consistent element load vector ($\mathbf{S}^0$) transforms through the same matrix.

Bending of beams and plates

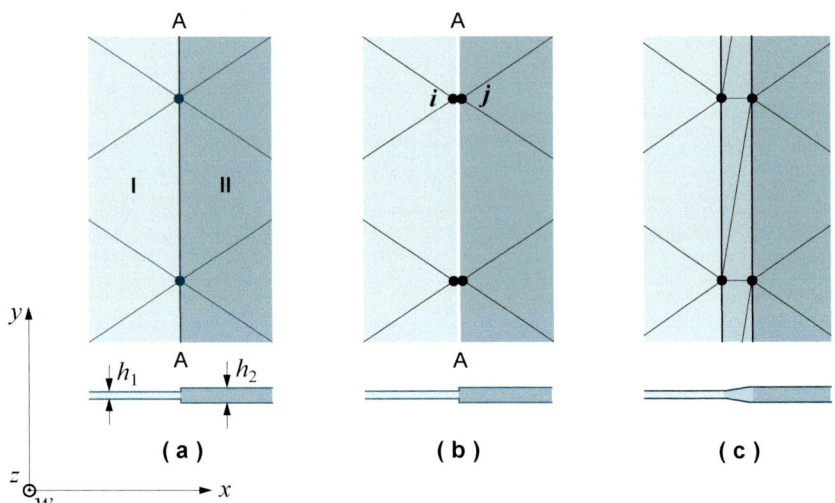

Figure 14.25 Quintic elements and change in plate thickness

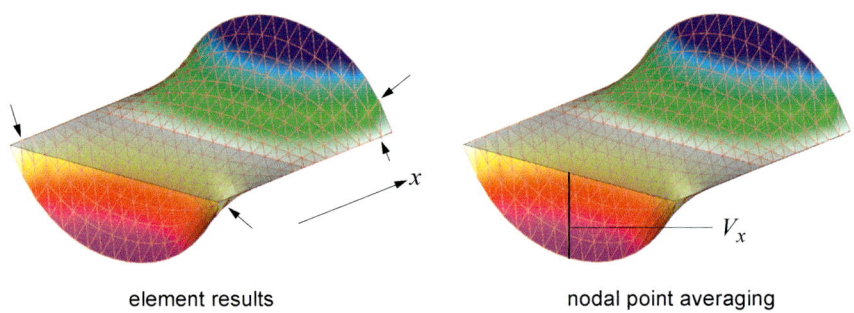

element results   nodal point averaging

Figure 14.26 KIRCHHOFF shear force V_x for the plate problem of Fig. 4.19 obtained by **FEMplate** using **T18** for a 16×16 mesh

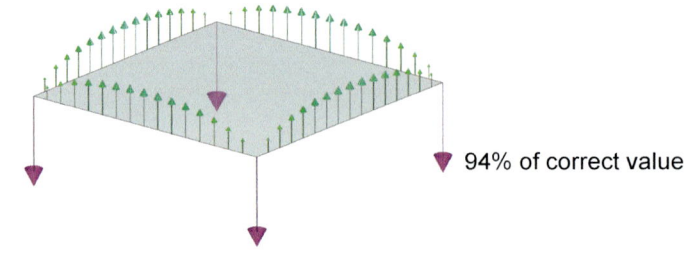

94% of correct value

Figure 14.27 Residual (reaction) forces for the plate problem of Fig. 4.19 obtained by **FEMplate** using **T18** for a 16×16 mesh

Triangular (thin) plate bending elements

An example at the end of the chapter will show that the performance of **T18** is almost indistinguishable from that of **T21**, and **T18** is more efficient since all its *dofs* are located at the corner nodes.

The inclusion of the curvatures as nodal *dofs* has both pros and cons. On the positive side is the fact that it easily enables an accurate determination of the moments at the nodal points, see Eq. (14-57); in fact, for uniform conditions the moments are the same for all elements at a common nodal point (hence no need for nodal point averaging). The drawback is that the curvatures may cause so-called *over-conformity*. Consider for instance a case of change in plate thickness as shown in Fig. 14.25a. For the nodal points on the line (A–A) of thickness discontinuity the curvature parameter $w_{,xx}$ is *not* the same on both sides of this line. Neglecting this fact will impose a continuity we do not want; hence "more conformity" than we should have.

Figures 14.25b and c indicate two ways in which we can get around this problem. The quantity that *is* continuous across line A–A is M_x (not $w_{,xx}$). We can include a double set of nodal points along line A-A, see Fig. 14.25b, and define a node on one side of the line, *e.g.* node *j*, to be *slave* of the corresponding node, *i*, on the other side. All slave *dofs*, except $(w_{,xx})_j$, are equal to the corresponding master *dof* at node *i*. By use of the moment-curvature relation, Eq. (3-78), we can establish the following expression for the remaining slave *dof*,

$$(w_{,xx})_j = c_1(w_{,xx})_i + c_2(w_{,yy})_i \tag{14-67}$$

where the two constants, c_1 and c_2, depend on the plate thicknesses and the material properties. With all slave *dofs* expressed in terms of the master *dofs* we proceed as described in Section 9.5.

Figure 14.25c shows the engineering solution to the problem. Both **T18** and **T21** allow quite large distortion of the element shape without much loss of accuracy, and it is shown in [12] that this solution will suffice in many engineering situations.

In a number of situations it can be difficult to describe correct boundary conditions for the curvature *dofs*; in a practical sense this is perhaps a more serious drawback than the problem of over-conformity.

An advantage of the quintic elements is that they predict Kirchhoff shear forces with quite good accuracy. Figure 14.26 shows the same example as the one solved with **T15** in Fig. 14.21, but here with a much coarser mesh. The comparison speaks for itself.

Figure 14.27 shows the residual forces computed for the simply supported square plate by **T18**. It should be emphasized that these are *concentrated* forces (in kN), computed at each node as the sum of the element nodal forces **S** = **kv** + **S**^{0}. For supported *dofs* the residual forces may be interpreted as reaction forces. We see that the "lift" forces at the corners are predicted with reasonable accuracy.

We shall come back to two more triangular thin plate bending elements in Sections 14.5 and 14.6, but first a short review of finite elements based on Mindlin plate bending theory.

Bending of beams and plates

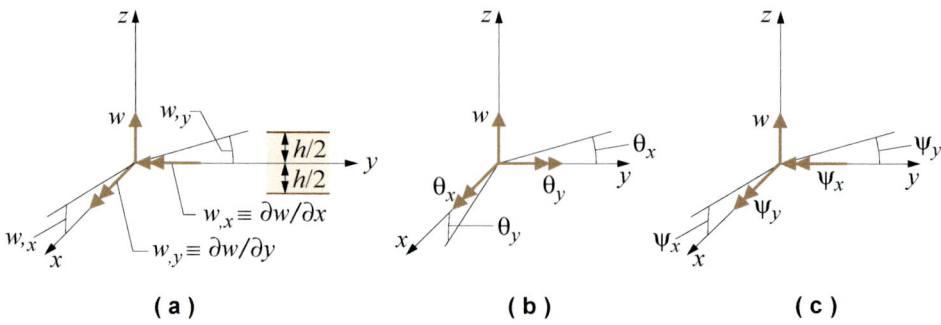

Figure 14.28 Typical plate bending *dofs*

Equations (3-70) and (3-71):

$$\boldsymbol{\varepsilon}_b = \begin{bmatrix} \varepsilon_x \\ \varepsilon_y \\ \gamma_{xy} \end{bmatrix} = -z \begin{bmatrix} -\theta_{y,x} \\ \theta_{x,y} \\ \theta_{x,x} - \theta_{y,y} \end{bmatrix} = -z\mathbf{c}_M \quad \Rightarrow \quad \mathbf{c}_M = \begin{bmatrix} -\theta_{y,x} \\ \theta_{x,y} \\ \theta_{x,x} - \theta_{y,y} \end{bmatrix}$$

$$\boldsymbol{\varepsilon}_s = \begin{bmatrix} \gamma_{yz} \\ \gamma_{zx} \end{bmatrix} = \begin{bmatrix} -\theta_x + w_{,y} \\ \theta_y + w_{,x} \end{bmatrix}$$

Equations (3-77) and (3-43):

$$\boldsymbol{\sigma}_b = \begin{bmatrix} \sigma_x \\ \sigma_y \\ \tau_{xy} \end{bmatrix} = \mathbf{C}_b \boldsymbol{\varepsilon}_b = -z\mathbf{C}_b \mathbf{c} \quad \text{and} \quad \boldsymbol{\sigma}_s = \begin{bmatrix} \tau_{yz} \\ \tau_{zx} \end{bmatrix} = \mathbf{C}_s \boldsymbol{\varepsilon}_s \quad \text{where}$$

$$\mathbf{C}_b = \frac{E}{1-\nu^2} \begin{bmatrix} 1 & \nu & 0 \\ \nu & 1 & 0 \\ 0 & 0 & \frac{1-\nu}{2} \end{bmatrix} \quad \text{and} \quad \mathbf{C}_s = G \begin{bmatrix} 1 & 0 \\ 0 & 1 \end{bmatrix}$$

14.4 MINDLIN plate bending elements

Figure 14.28 shows three different representations of the three typical nodal degrees of freedom used in plate bending analysis. The *dofs* in Fig. 14.28a are those used for thin plate (KIRCHHOFF) theory in the previous section. The rotational *dofs* are here the slopes ($w_{,x}$ and $w_{,y}$) of the middle surface of the plate. The rotational *dofs* in Fig. 14.28b, θ_x and θ_y, are now the rotations about axes x and y, respectively, of the *plate normal*. In Fig. 14.28c, the rotational *dofs*, ψ_x and ψ_y, represent basically the same rotations as θ_x and θ_y; the only difference is the notation and sign convention. Rotations ψ_x and ψ_y are preferred by some authors since they correspond better with those of Fig. 14.28a. However, we will use the *dofs* of Fig.14.28b.

The key point in MINDLIN theory is that the rotations θ_x and θ_y are *independent* of the lateral displacement w. The basic relations, as described in Section 3.5, are:

- Equations (3-70) and (3-71) define the relationship between strains (curvatures) and displacements.
- Stress-strain is defined by Eqs. (3-77) and (3-43).

As for the beam, we now *assume* independent displacement fields for all three field variables, w, θ_x and θ_y. From the expressions for U, ε_b and ε_s we see that the variational order, or equivalent, the highest order of differentiation in the strain-displacement relations, is $m = 1$ which means our problem requires C^0 continuity (for all three field variables). Hence standard isoparametric procedure applies and we assume

$$w = \sum_i N_i w_i, \quad \theta_x = \sum_i N_i \theta_{xi} \quad \text{and} \quad \theta_y = \sum_i N_i \theta_{yi} \quad (14\text{-}68)$$

or

$$\mathbf{u} = \begin{bmatrix} w \\ \theta_x \\ \theta_y \end{bmatrix} = \begin{bmatrix} N_1 & 0 & 0 & N_2 & 0 & 0 \\ 0 & N_1 & 0 & 0 & N_2 & 0 & \cdots \\ 0 & 0 & N_1 & 0 & 0 & N_2 \end{bmatrix} \begin{bmatrix} \mathbf{v}_1 \\ \mathbf{v}_2 \\ \vdots \end{bmatrix} = \mathbf{Nv} \quad (14\text{-}69)$$

where $\mathbf{v}_i^T = [\, w_i \quad \theta_{xi} \quad \theta_{yi} \,]$ are the *dofs* at element node i. The MINDLIN curvature, defined by Eq. (3-70) can now be expressed as

$$\mathbf{c}_M = \begin{bmatrix} -\theta_{y,x} \\ \theta_{x,y} \\ \theta_{x,x} - \theta_{y,y} \end{bmatrix} = \mathbf{B}_b \mathbf{v} \quad \Rightarrow \quad \varepsilon_b = -z\mathbf{c}_M = -z\mathbf{B}_b \mathbf{v} \quad (14\text{-}70)$$

where

$$\mathbf{B}_b = [\, \mathbf{B}_{b1} \quad \mathbf{B}_{b2} \quad \ldots \,] \quad \text{and} \quad \mathbf{B}_{bi} = \begin{bmatrix} 0 & 0 & -N_{i,x} \\ 0 & N_{i,y} & 0 \\ 0 & N_{i,x} & -N_{i,y} \end{bmatrix} \quad (14\text{-}71)$$

479

Bending of beams and plates

Recall from Sections 4.4 and 4.5:

Total potential energy: $\Pi = U + H$
- H — load potential
- U — strain energy

Also: $\Pi = \sum_e \Pi_e = \sum_e (\underbrace{\tfrac{1}{2}\mathbf{v}^T \mathbf{k}\mathbf{v}}_{\text{strain energy}} + \underbrace{\mathbf{v}^T \mathbf{S}^0}_{\text{load potential}})$

The strain energy of the plate is defined by Eq. (3-87):

$$U = \frac{1}{2}\int_A \mathbf{c}^T \mathbf{D}\mathbf{c}\, dA + \frac{1}{2}\int_A h\boldsymbol{\varepsilon}_s^T \mathbf{C}_s \boldsymbol{\varepsilon}_s\, dA$$

where

$$\mathbf{D} = \frac{h^3}{12}\mathbf{C}_b = \frac{Eh^3}{12(1-\nu^2)}\begin{bmatrix} 1 & \nu & 0 \\ \nu & 1 & 0 \\ 0 & 0 & \frac{1-\nu}{2} \end{bmatrix}$$

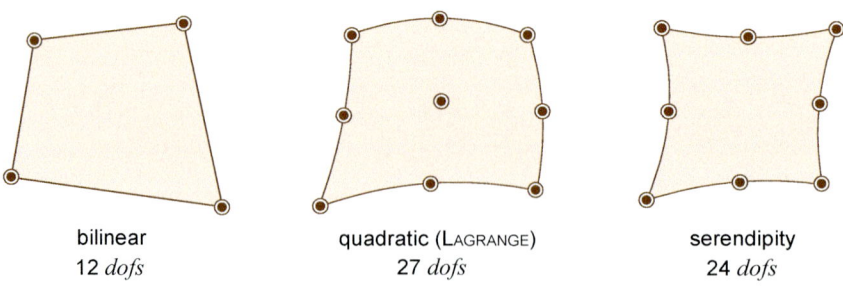

bilinear
12 *dofs*

quadratic (LAGRANGE)
27 *dofs*

serendipity
24 *dofs*

Figure 14.29 C^0 plate bending elements (MINDLIN)

MINDLIN plate bending elements

It should be noted that the curvatures depends on the rotation parameters θ_x and θ_y only. We can also express the shear strains, defined by Eq. (3-71), in terms of the nodal *dofs*,

$$\boldsymbol{\varepsilon}_s = \begin{bmatrix} \gamma_{yz} \\ \gamma_{zx} \end{bmatrix} = \begin{bmatrix} -\theta_x + w_{,y} \\ \theta_y + w_{,x} \end{bmatrix} = \begin{bmatrix} \mathbf{B}_{s1} & \mathbf{B}_{s2} & \cdots \end{bmatrix} \begin{bmatrix} \mathbf{v}_1 \\ \mathbf{v}_2 \\ \vdots \end{bmatrix} = \mathbf{B}_s \mathbf{v} \quad (14\text{-}72a)$$

where

$$\mathbf{B}_{si} = \begin{bmatrix} N_{i,y} & -N_i & 0 \\ N_{i,x} & 0 & N_i \end{bmatrix} \quad (14\text{-}72b)$$

The shape functions N_i are functions of the *natural* (mapped) coordinates ξ and η, and the derivatives with respect to physical coordinates x and y ($N_{i,x}$ and $N_{i,y}$) are found by use of the inverse Jacobian matrix, as described in Section 6.1.

From the energy relations opposite we can deduce:

$$\mathbf{k} = \int_A \mathbf{B}_b^T \mathbf{D} \mathbf{B}_b \, dA + \int_A h_s \mathbf{B}_s^T \mathbf{C}_s \mathbf{B}_s \, dA = \mathbf{k}_b + \mathbf{k}_s \quad (14\text{-}73)$$

A consistent element load vector due to a distributed transverse load $p(x,y)$ is defined as

$$\mathbf{s}_p^0 = -\int_A \mathbf{N}_b^T p \, dA \quad (14\text{-}74)$$

where $\mathbf{N}_b$ contains N_i only for the w dofs (i.e., no nodal loads corresponding to the rotational *dofs* θ_x and θ_y).

The integration in Eqs. (14-73) and (14-74) is carried out numerically in the same way as for plane stress/plane strain elements.

Once the nodal displacements have been determined, the stress resultants are found from Eq. (3-78),

$$\mathbf{m} = \begin{bmatrix} M_x \\ M_y \\ M_{xy} \end{bmatrix} = -\mathbf{D}\mathbf{c}_M = -\mathbf{D}\mathbf{B}_b \mathbf{v} \quad (14\text{-}75)$$

and from Eq. (3-80),

$$\mathbf{V} = \begin{bmatrix} V_y \\ V_x \end{bmatrix} = Gh_s \boldsymbol{\varepsilon}_s = \frac{5}{6} Gh \mathbf{B}_s \mathbf{v} \quad (14\text{-}76)$$

In the latter expression we have used the same shear factor (κ) as for beams of rectangular cross section.

Typical element shapes and nodal points, each with the three *dofs* w, θ_x and θ_y, are shown in Fig. 14.29; with respect to shape functions, they are the same as those used in plane stress/ plane stain analysis.

Bending of beams and plates

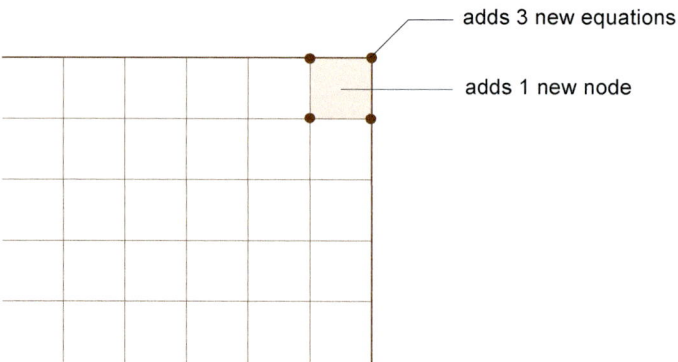

Figure 14.30 Mesh of (12-dof) bilinear quadrilateral elements

MINDLIN plate bending elements

What happens when the ratio L/h increases, that is when the plate becomes thinner? Without special precautions, the shear contribution to the strain energy, which should ideally disappear, will in fact dominate more and more as $h \to 0$, and the result is *shear locking*, just as for the MINDLIN beam. The following argument supports this statement:

According to Eq. (14-73):

$$\mathbf{Kr} = (\mathbf{K}_b + \mathbf{K}_s)\mathbf{r} = \mathbf{R} \qquad (14\text{-}77)$$

For a plate of an isotropic material and with constant thickness this can be expressed as

$$\left(\frac{Eh^3}{12(1-\nu^2)}\hat{\mathbf{K}}_b + \frac{5Gh}{6}\hat{\mathbf{K}}_s\right)\mathbf{r} = \mathbf{R}$$

If we also, for simplicity, assume that $\nu = 0$, then we can write the above as

$$\frac{Eh^3}{12}\left(\hat{\mathbf{K}}_b + \frac{5}{h^2}\hat{\mathbf{K}}_s\right)\mathbf{r} = \mathbf{R} \quad \text{or} \quad (\hat{\mathbf{K}}_b + \alpha\hat{\mathbf{K}}_s)\mathbf{r} = \frac{1}{\beta}\mathbf{R} = \hat{\mathbf{R}} \qquad (14\text{-}78)$$

where

$$\alpha = \frac{5}{h^2} \quad \text{and} \quad \beta = \frac{Eh^3}{12} \qquad (14\text{-}79)$$

Let all dimensions, except h be constant. Then

$$h \to 0 \quad \Rightarrow \quad \alpha \to \infty$$

and we see that the shear contribution to the stiffness in Eq. (14-78) will dominate more and more and we have a typical case of shear locking.

This can also be explained by the *penalty method* we presented in Section 9.7 as a way of introducing boundary constraints. The displacements **r** of a thin plate should be governed by $\mathbf{K}_b$ alone since the transverse shear deformations (γ_{yz} and γ_{zx}) are, in this case, negligible. Hence, as the "penalty parameter" α increases, matrix $\mathbf{K}_s$ (originating from U_s) ought to enforce the condition $\gamma_{yz} = \gamma_{zx} = 0$. However, as the plate thickness h gets smaller, $\mathbf{K}_s$ increases compared to $\mathbf{K}_b$, and $\mathbf{K}_s$ will act as a *penalty matrix* causing **r** to approach zero, unless $\mathbf{K}_s$ is *singular*. A singular $\mathbf{K}_s$ may enforce the condition $\gamma_{yz} = \gamma_{zx} = 0$, *without locking*.

With numerical integration of $\mathbf{k}_s$, every integration point will enforce *two restrictions* on the solution, and every new element that is added to the model will add restrictions, but also new equations. If the number of restrictions is equal to or greater than the number of equations, $\mathbf{K}_s$ will be non-singular and the solution will "lock" for thin plates.

Example 14-2 – bilinear quadrilateral element

For the mesh in Fig. 14.30 the addition of one new element will add three new equations to the system. Full, 2×2 integration of $\mathbf{k}_s$ will add 8 restrictions, and locking will occur as the plate gets thinner. Reduced 1×1 integration adds 2 restrictions and locking is avoided. But that does not necessarily mean that our problems are solved.

Bending of beams and plates

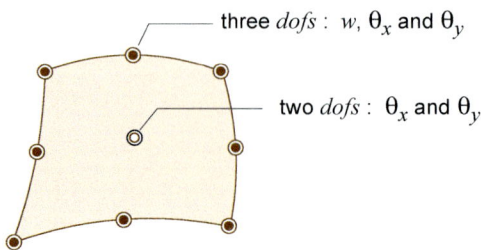

Figure 14.31 The 26 *dof* Heterosis elements

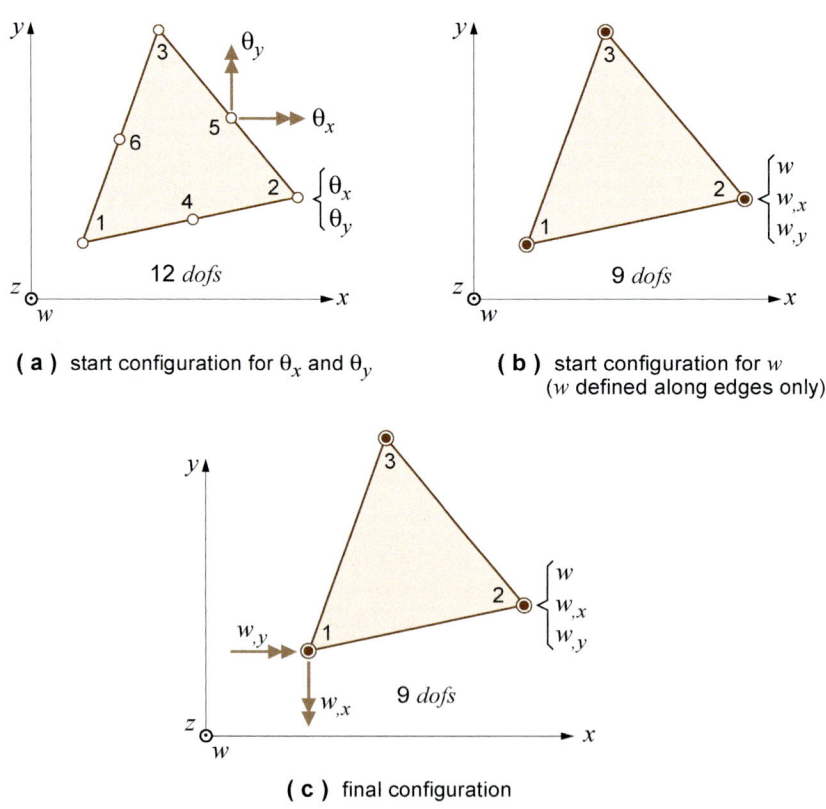

Figure 14.32 The 9-*dof* discrete KIRCHHOFF triangle (**DKT**)

Discrete KIRCHHOFF elements

 The example indicates that, as for the beam case, the solution to the shear locking problem is reduced integration. However, for plate elements this remedy may cause another problem that we have encountered before, that of *mechanisms* or *zero-energy modes*. For the bilinear quadrilateral it can be shown [21] that if we integrate both $\mathbf{k}_b$ and $\mathbf{k}_s$ with a 1-point quadrature rule the element will have 4 mechanisms. Even if we use *selective* integration with a 1-point rule for $\mathbf{k}_s$ and a 4-point rule for $\mathbf{k}_b$ we will still have 2 mechanisms. All mechanisms disappear with full integration, but then we are back to locking.

Also the other two C^0 elements in Fig. 14.29, the quadratic LAGRANGE and the serendipity elements, have problems with locking and mechanisms. There are various ways of controlling these problems, see for instance [52], but we will not pursue this any further. Instead we will take a look at a different method of deriving robust plate bending elements termed *discrete* KIRCHHOFF elements. However, before leaving the C^0 MINDLIN elements we mention that a fairly problem-free element, as far as locking and mechanisms are concerned, is the so-called *Heterosis* element [53]. This element is a "combination" of the 9 node LAGRANGE element and the 8 node serendipity element of Fig. 14.29, see Fig 14.31. The rotations θ_x and θ_y are interpolated using the 9 node LAGRANGE interpolation, while w, which is *not* defined as a *dof* at the internal node, is interpolated using the 8 node serendipity scheme. With selective integration, 3×3 for $\mathbf{k}_b$ and 2×2 for $\mathbf{k}_s$, this element has no mechanisms and it does not lock for thin plates. Yet it is not entirely satisfactory in that it only satisfies the weak patch test; the "strong" patch test is only passed by elements of rectangular or parallelogram shape [52].

14.5 Discrete KIRCHHOFF elements

The basic idea of this "family" of elements is to enforce zero transverse shear strain at specific points. We will indicate the procedure for a 9-*dof* triangular element, now known as the **DKT** element (Discrete KIRCHHOFF Triangle). This element was first published in 1969 [54].

1) The starting point is the 12-*dof* element shown in Fig. 14.32a. Each of the two field variables, θ_x and θ_y, are interpolated between the 6 nodal *dofs* by the same shape functions $\mathbf{N}_2$ as those used for the **LST** plane stress/plane strain element.

2) Next we impose restrictions on the mid-side rotations. The rotation θ_t normal to the edge (along axis *t*), see Fig. 14.33a, is "forced" to vary linearly, by simply requiring that it is equal to half the sum of the corresponding corner values, that is

$$\theta_{tk} = \frac{1}{2}(\theta_{ti} + \theta_{tj}) \qquad (14\text{-}80)$$

See Fig. 14.33 for notation. This gives us *three* coupling equations between the rotation parameters.

3) The lateral displacement w is defined *only along the element edges*, and it is assumed to vary cubically (as HERMITIAN beam functions) see Figs. 5.33 and

Bending of beams and plates

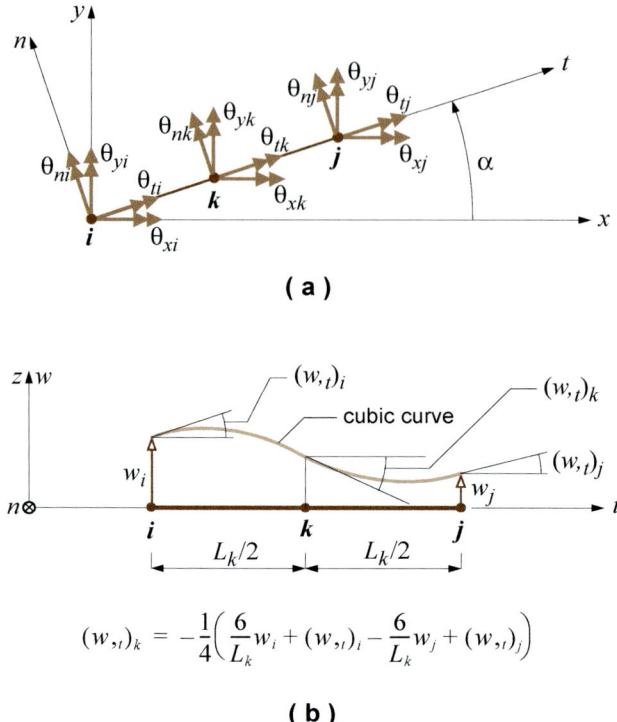

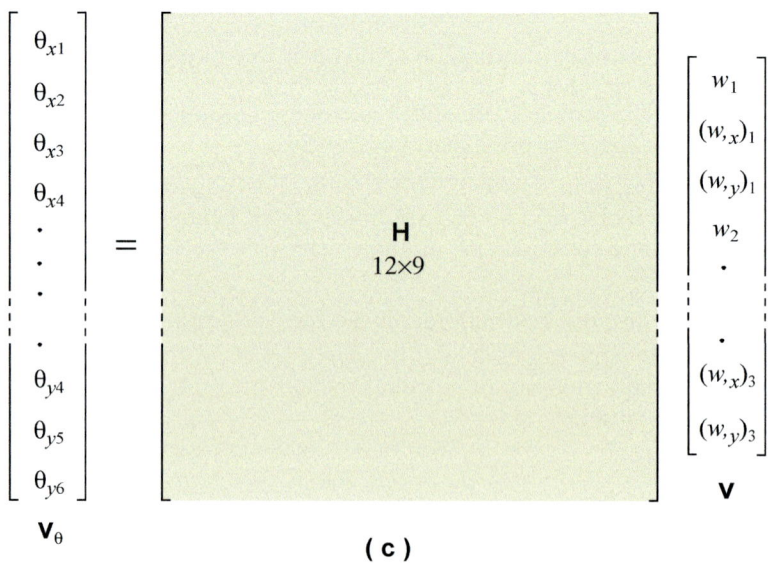

Figure 14.33 Assumptions and constraints of the **DKT** element

14.33b, between nodal parameters w, $w_{,x}$ and $w_{,y}$ at the corner nodes, see Fig. 14.32b. This introduces 9 new *dofs* and brings the total to 12+9 = 21 *dofs*.

4) Introduce KIRCHHOFF conditions (constraints) as follows:

   a) At corner nodes:

   $\gamma_{yz} = 0 \quad \Rightarrow \quad (w_{,y})_i = \theta_{xi}$

   $\gamma_{zx} = 0 \quad \Rightarrow \quad (w_{,x})_i = -\theta_{yi}$

   $i = 1, 2, 3$

   This gives 6 conditions.

   b) At mid-side nodes:

   $\gamma_{zt} = 0 \quad \Rightarrow \quad (w_{,t})_k = -\theta_{nk} \qquad k = 4, 5, 6$

   This gives 3 new conditions.

5) The 9 KIRCHHOFF constraints plus the three constraints of the type given in Eq. (14-80) give us a total of 12 constraints that enable us to eliminate 12 *dofs*, or in other words, to express the 12 rotations in Fig 14.32a in terms of the 9 final *dofs* in Fig 14.32c. In matrix notation this can be expressed as, see Fig. 14.33,

$$\mathbf{v_\theta} = \mathbf{Hv} \tag{14-81}$$

The actual implementation can be carried out in various ways. One approach is to derive the stiffness matrix $\mathbf{k_\theta}$ for the element in Fig. 14.32a and then transform this matrix to the desired one,

$$\mathbf{k} = \mathbf{H}^T \mathbf{k_\theta} \mathbf{H} \tag{14-82}$$

This matrix corresponds to the final *dof* configuration in Fig. 14.32c. Matrix $\mathbf{k_\theta}$ is obtained through standard procedures; systematic use of Figs. 14.18, 14.23 and 14.33b will lead to explicit expressions for the non-zero terms of matrix $\mathbf{H}$.

For a distributed transverse loading $p = p(x,y)$ an appropriate element load vector $\mathbf{S}^0$ is found by *lumping* the load (assumed to vary linearly between the corner values) into three statically equivalent transverse nodal forces (each corresponding to the w *dof*).

Once the nodal parameters $\mathbf{v}$ have been determined (retrieved from $\mathbf{r}$), the linearly varying moments are obtained from Eq. (3-78) which we now rewrite slightly as

$$\mathbf{m} = \begin{bmatrix} M_x \\ M_y \\ M_{xy} \end{bmatrix} = \int_{-h/2}^{h/2} \sigma_b z\, dz = -\mathbf{C}_b \int_{-h/2}^{h/2} z^2 dz \mathbf{c} = -\frac{h^3}{12}\mathbf{C}_b \mathbf{c}_M = -\mathbf{D}\mathbf{c}_M$$

By use of

$$\boldsymbol{\theta} = \begin{bmatrix} \theta_x \\ \theta_y \end{bmatrix} = \begin{bmatrix} \mathbf{N}_2 & \mathbf{0} \\ \mathbf{0} & \mathbf{N}_2 \end{bmatrix} \begin{bmatrix} \mathbf{v}_{\theta x} \\ \mathbf{v}_{\theta y} \end{bmatrix} = \mathbf{N}_\theta \mathbf{v}_\theta$$

Bending of beams and plates

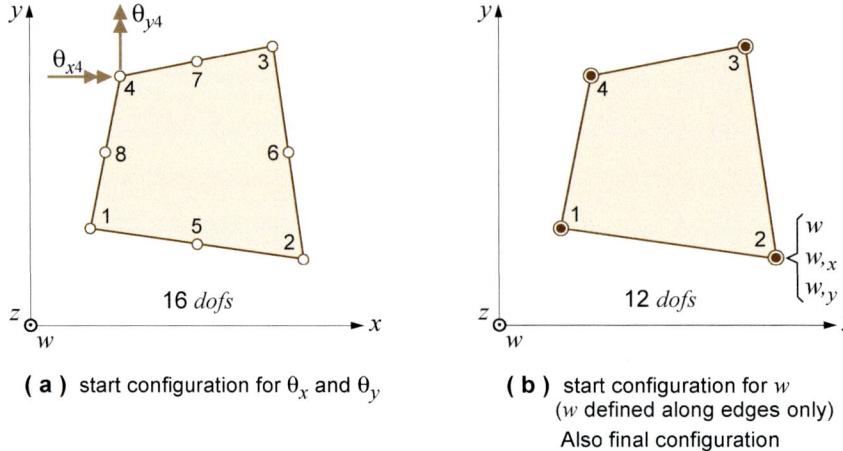

Figure 14.34 The 12-*dof* discrete Kirchhoff quadrilateral (DKQ)

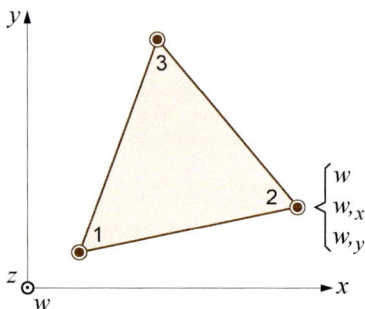

Figure 14.35 The 9-*dof* hybrid element due to Allman [55]

and Eq. (3-71) we can find

$$\mathbf{c}_M = \begin{bmatrix} -\theta_{y,x} \\ \theta_{x,y} \\ \theta_{x,x} - \theta_{y,y} \end{bmatrix} = \mathbf{B}_\theta \mathbf{v}_\theta = \mathbf{B}_\theta \mathbf{H} \mathbf{v} \quad \text{and thus} \quad \begin{bmatrix} M_x \\ M_y \\ M_{xy} \end{bmatrix} = -\mathbf{D}\mathbf{c}_M = -\mathbf{D}\mathbf{B}_\theta \mathbf{H} \mathbf{v}$$

The procedure outlined here for the **DKT** element can also be used to derive an effective quadrilateral element with the standard three *dofs* at each corner node, the 12 *dof* **DKQ** element. The starting point here will be the 8 node serendipity element with 16 rotational *dofs*, see Fig. 14.34a. Together with the 12 (standard) *dofs* in Fig. 14.34b, and a total of 16 constraints, we end up with the stiffness matrix of a robust element with the 12 *dofs* of Fig. 14.34b.

14.6 A hybrid 9-node triangular element

We include a very brief presentation of one more plate bending element. It is a *hybrid* element in the sense that it makes assumptions about moments as well as displacements. In other words this element, proposed by ALLMAN [55], is not a pure displacement element.

The element, shown in Fig. 14.35, is based on KIRCHHOFF thin plate theory, and it has the "standard" 9 *dofs*. Its derivation is somewhat lengthy and we will not get involved with the details. The *complementary strain energy* is used to derive the element *flexibility* matrix based on the following three assumptions:

1) The moments **m** are assumed to vary *linearly* over the element and are expressed in terms of 9 "moment parameters".

2) The deflection w is assumed to vary cubically along the element edges (uniquely defined by the nodal *dofs*) – w is *not* defined in the interior of the element.

3) The normal slope $w_{,n}$ varies linearly along the element edges.

The flexibility matrix is *inverted* and transformed (from moment parameters to final nodal *dofs*) to yield the element's stiffness matrix **k**.

The load vector ($\mathbf{S}^0$) due to a distributed transverse loading is obtained as for the **DKT** element.

This element is implemented in the **FEMplate** program with the name TEBA.

Bending of beams and plates

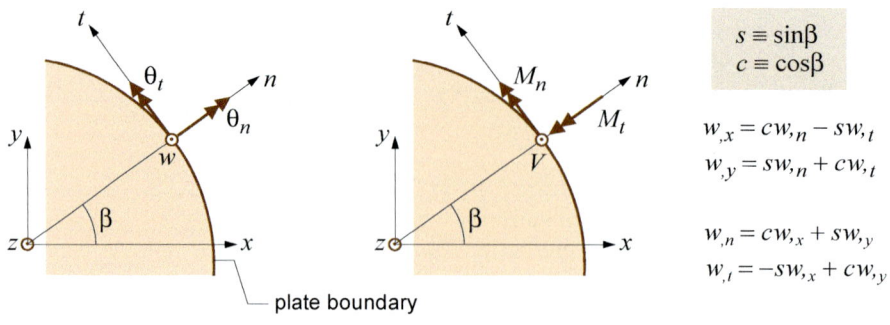

$s \equiv \sin\beta$
$c \equiv \cos\beta$

$w_{,x} = cw_{,n} - sw_{,t}$
$w_{,y} = sw_{,n} + cw_{,t}$

$w_{,n} = cw_{,x} + sw_{,y}$
$w_{,t} = -sw_{,x} + cw_{,y}$

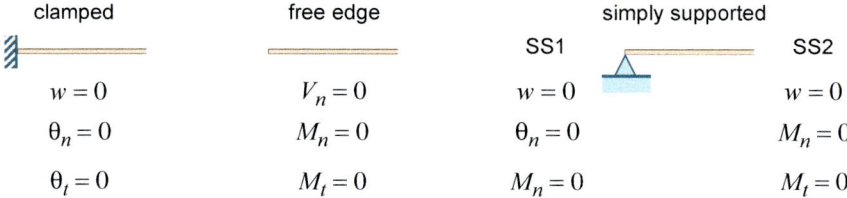

Figure 14.36 Boundary conditions at a curved plate boundary

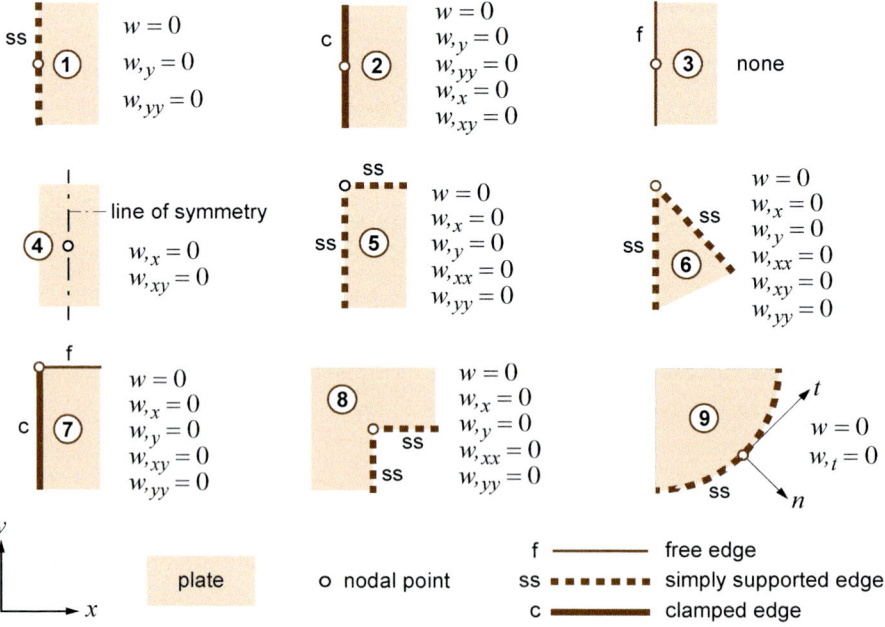

Figure 14.37 Kinematic boundary conditions for some typical thin plate cases

14.7 A note on boundary conditions

Figure 14.36 shows a nodal point at a curved edge along with the boundary conditions for three different support conditions along the edge. Both kinematic (displacements) and mechanical (forces) conditions are shown. For a simply supported boundary we have given two sets of conditions, SS1 and SS2.

For KIRCHHOFF thin plate theory we have:

$$\theta_n = w_{,t} \quad \text{and} \quad \theta_t = -w_{,n}$$

- SS1 states the boundary conditions of KIRCHHOFF theory.
- SS2 is often referred to as "finite element conditions".

When approximating a curved edge by a series of straight element edges, some authors claim that SS2 will give better results than SS1. With standard *dofs*, w, θ_x ($w_{,y}$) and θ_y ($-w_{,x}$), we can only prescribe kinematic boundary conditions; this means that according to SS2 only w should be specified at the edge nodes.

Limited testing with KIRCHHOFF elements has shown [12] that transforming the rotational *dofs* at an edge node to the *n-t* coordinate system, where t is the tangent to the curved edge at the point, and then suppressing both w and $w_{,t}$ (consistent with SS1) gives good results. However, there seems to be little in it, especially for fine meshes.

If we include the curvatures as nodal *dofs* we can also prescribe some of the mechanical boundary conditions. Figure 14.37 shows some typical cases for thin (KIRCHHOFF) plates. For case 1 we could, for an isotropic material, also specify

$$w_{,xx} = 0$$

since we know that

$$M_x = -D(w_{,xx} + \nu w_{,yy}) = 0$$

at this node. For the same reason we could specify

$$w_{,xx} + \nu w_{,yy} = 0$$

at the node in case 3. However, this complicates matters unnecessarily and we should also add that the benefits of prescribing mechanical boundary conditions are questionable. We therefore recommend that mechanical boundary conditions are *not specified*. In the case of the quintic triangles, even the specification of appropriate kinematic boundary conditions can, at times, be cumbersome.

14.8 Comparison of some plate elements

We conclude this chapter with a comparison of results obtained for an often used test problem, a square plate of uniform thickness h, made of an isotropic material and loaded by a uniformly distributed transverse load p. The elements used are:

- **T6** – the 6-*dof* MORLEY triangle.

Bending of beams and plates

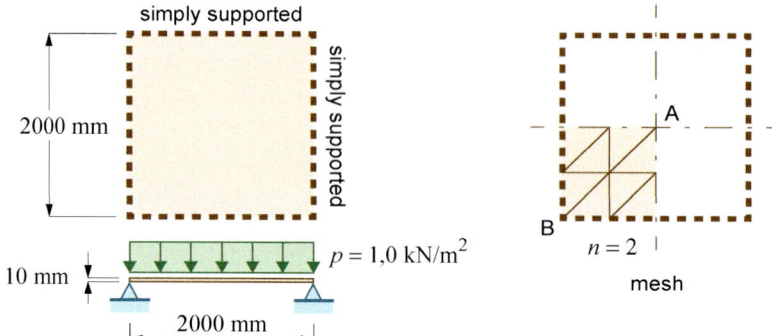

Three parameters are compared:
At mid-point A: displacement w_A and bending moment $(M_x)_A = (M_y)_A$
At corner B: twisting moment $(M_{xy})_B$

Correct values for an isptropic material with:
$E = 210\,000$ N/mm^2
and
$\nu = 0{,}3$

$w_A = 3{,}37987$ mm
$(M_x)_A = 191{,}545$ Nmm/mm
$(M_{xy})_B = 129{,}929$ Nmm/mm

Figure 14.38 Test problem – comparison of triangular plate bending elements

- **DKT** – the 9-*dof* discrete KIRCHHOFF triangle.
- TEBA – the 9-*dof* hybrid triangle due to ALLMAN.
- **T15** – the 15-*dof* incompatible quartic triangle.
- **T18** – The 18-*dof* incomplete quintic, but compatible triangle.
- **T21** – The 21-*dof* complete and compatible quintic triangle.

The problem is shown in Fig. 14.38. It is the same problem as the one used earlier, see Fig. 14.19. This time, due to the double symmetry, we only mesh one quarter of the plate, as shown in the figure; parameter n controls the fineness of the mesh.

For each mesh and element we determine the lateral displacement w, the bending moment M_x at the centre (A) of the plate and the moment of twist M_{xy} at the corner point (B). For all elements, except **T18** and **T21**, the moments (M_x and M_{xy}) are obtained by simple nodal point averaging.

The results, obtained with program **FEMplate** on a fairly standard PC (anno 2012), are presented in Table 14.1, as per cent of the correct value. The number of equations solved (n_{eq}) is also recorded for each mesh and element, and so is the computing time (in milliseconds) for the medium to fine meshes. This time is the so-called CPU time used by the Fortran kernel of the program and comprises some data preparation, assembly of **K** and **R** (which includes the evaluation of **k** and $\mathbf{S}^0$ for each element), solution of the system stiffness relation and computation of element forces.

Table 14. 1: Results obtained with 6 different triangular plate bending elements for a square, simply supported plate subjected to a uniformly distributed load.

Mesh n	Element	w_A	$(M_x)_A$	$(M_{xy})_B$	No. of eqn. n_{eq}	CPU-time [millisecs.]
1	T6	222,01	68,55	77,74	4	
	DKT	102,44	135,57	80,71	3	
	TEBA	80,73	119,40	84,80	3	
	T15	102,94	102,53	117,22	9	
	T18	99,80	92,11	102,15	10	
	T21	99,64	84,28	97,64	13	
2	T6	126,21	82,72	95,82	16	
	DKT	99,84	107,67	93,09	12	
	TEBA	95,091	106,40	95,49	12	
	T15	100,75	97,29	106,14	36	
	T18	99,999	99,793	101,18	30	
	T21	99,999	99,851	100,39	42	

Bending of beams and plates

Table 14.1: Results obtained with 6 different triangular plate bending elements for a square, simply supported plate subjected to a uniformly distributed load.

Mesh n	Element	w_A	$(M_x)_A$	$(M_{xy})_B$	No. of eqn. n_{eq}	CPU-time [millisecs.]
4	T6	106,25	95,90	100,67	64	
	DKT	100,06	102,01	98,38	48	
	TEBA	98,81	101,16	98,61	48	
	T15	100,17	99,35	102,06	144	
	T18	100,00	99,99	100,31	106	
	T21	100,00	99,97	100,10	154	
8	T6	101,54	99,07	100,75	256	
	DKT	100,03	100,46	99,62	192	
	TEBA	99,712	100,25	99,62	192	
	T15	100,04	99,85	100,65	576	
	T18	100,00	100,00	100,02	402	
	T21	100,00	100,00	100,03	594	
16	T6	100,38	99,80	100,36	1024	16
	DKT	100,01	100,10	99,92	768	16
	TEBA	99,93	100,07	99,90	768	16
	T15	100,01	99,97	100,19	2 304	31
	T18	100,00	100,00	100,02	1 570	47
	T21	100,00	100,00	100,01	2 338	62
32	T6	100,10	99,96	100,14	4 096	47
	DKT	100,00	100,02	99,99	3 072	47
	TEBA	99,98	100,01	99,97	3 072	62
	T15	100,00	99,99	100,06	9 216	109
	T18	100,00	100,00	100,00	6 210	172
	T21	100,00	100,00	100,00	9 282	234

Table 14.1: Results obtained with 6 different triangular plate bending elements for a square, simply supported plate subjected to a uniformly distributed load.

Mesh n	Element	w_A	$(M_x)_A$	$(M_{xy})_B$	No. of eqn. n_{eq}	CPU-time [millisecs.]
64	T6	100,02	99,99	100,02	16 384	109
	DKT	100,00	100,01	100,00	12 288	203
	TEBA	100,00	100,00	99,99	12 288	203
	T15	100,00	100,00	100,02	36 864	718
	T18	100,00	100,00	100,00	24 706	998
	T21	100,00	100,00	100,00	36 994	1591
128	T6	100,01	100,00	100,02	65 536	608
	DKT	100,00	100,00	100,00	49 152	1232
	TEBA	100,00	100,00	100,00	49152	1170
	T15	100,00	100,00	100,00	147 456	5335
	T18	100,00	100,00	100,00	98 562	7441
	T21	100,00	100,00	100,00	147 714	13010

We cannot make sweeping statements based on just one example, especially since this example is not all that demanding. Nevertheless, some interesting observations can be made. While the two quintic elements produce quite accurate results for a very coarse mesh ($n = 2$), the simple **T6** element requires a finer mesh, but for a very moderately fine mesh ($n = 8$), even this element produces results that are quite acceptable from a practical point of view.

Of the two higher order elements, **T18** is clearly the most attractive; it compares very well with the more "expensive" **T21**. **DKT** seems to be a good and robust element. It gives surprisingly good results for the coarsest mesh; with a total of only three free *dofs*, **DKT** produces the central deflection with an error of only 2,5% (fluke?). Apart from the coarsest mesh ($n = 1$) the hybrid element (**TEBA**) compares very well with **DKT**, not quite as good on displacements, but on the whole better on moments. The performance of the simple **T6** element is perhaps the most surprising result to come out of Table 14.1, and for those who are worried about side nodes, the computing times, particularly for the finest meshes, may come as a surprise. It is clearly not a question of just counting the number of equations. The times quoted are the actual, measured times.

While the elements show significant differences for the coarsest meshes, the differences for the finer meshes ($n > 8$) are small and of no practical consequence. For linear static problems we may therefore be well served by the simple **T6** element or the **DKT** element, both of which are robust elements

Bending of beams and plates

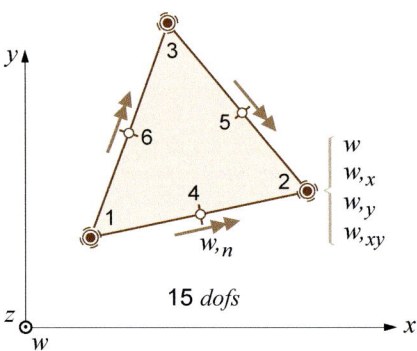

Figure 14.39 A variant of a complete quartic thin plate bending element

which, if enough are used will provide ample accuracy, and they are not likely to cause surprises. If shear forces are needed the use of a higher order KIRCHHOFF element, such as **T18**, or a well behaved MINDLIN element, might be considered. Just as the EULER-BERNOULLI beam element can be used with acceptable errors for beam members with quite a high depth to length ratio (say up to 0,15), so can KIRCHHOFF thin plate theory be used for plates with a similar kind of limit on the thickness to length (h/L) ratio.

Before we leave the plate bending problem it should be said that, for various reasons, we have presented mostly "old" elements. If we use one of the many general FEM programs that are available, we will probably find that their element libraries contain plate bending elements we have not even mentioned here. However, for linear static problems we can afford to use relatively fine meshes, and for such meshes all "proper" elements will perform satisfactorily, and probably not much better than the elements presented here.

Problems

Problem 14.1

Verify Eq. (14-53).

Problem 14.2

Derive Eq. (14-54a).

Problem 14.3

Assess the continuity properties of the thin plate element in Fig. 14.39. The element is based on a complete *quartic* displacement field; it has 4 *dofs* (as shown) at each corner node and 1 *dof* at each mid-side node.

Problem 14.4

Verify the transformations of Eqs. (14-62).

Problem 14.5

Derive Eq. (14-63).

Problem 14.6

Determine the coefficients c_1 and c_2 of Eq. (14-67).

Problem 14.7

Determine explicit expressions for all non-zero elements of the row of the transformation matrix **H** of Eq. (14-81) that correspond to element θ_{x4} of $\mathbf{v}_\theta$. In other words, express θ_{x4} (explicitly) in terms of the final *dofs* **v** of the **DKT** element.

Bending of beams and plates

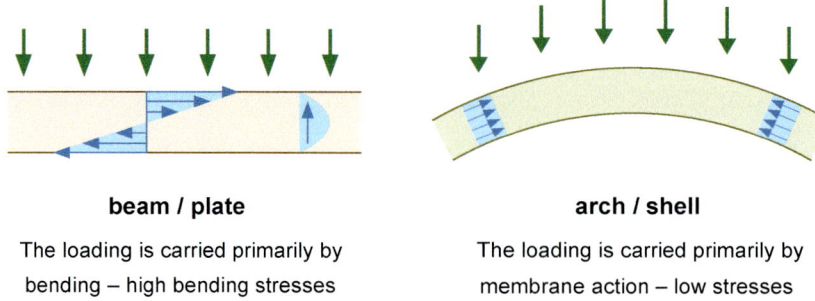

beam / plate
The loading is carried primarily by bending – high bending stresses

arch / shell
The loading is carried primarily by membrane action – low stresses

Figure 15.1 Load bearing by bending and in-plane (membrane) action

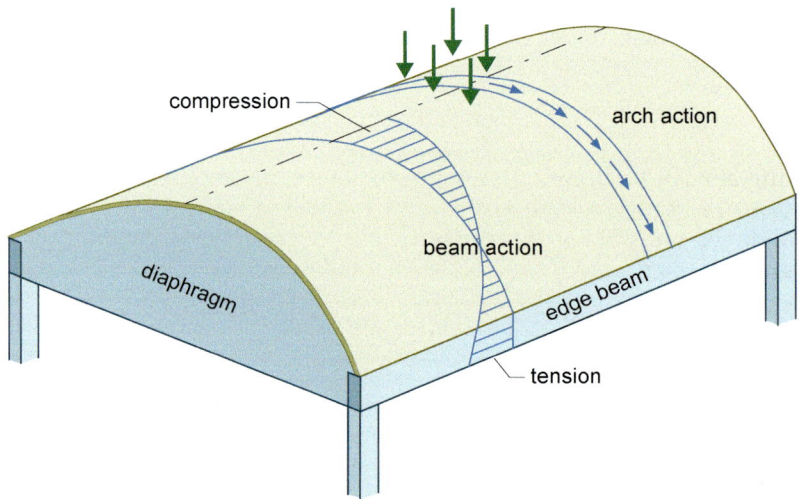

Figure 15.2 Load bearing characteristics of a cylindrical shell

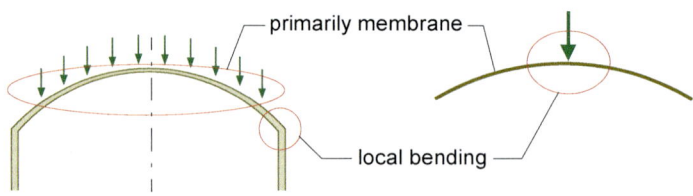

Figure 15.3 The local nature of bending stresses in shells

15

Arches and shells

This chapter will focus on the engineering approach to "curved structures", which is to approximate the curved line with a series of straight lines, and the curved surface with a series of flat facets. This approach has some important implications concerning how we apply the loading and also how fine we need or dare to mesh.

The curved structural form is in most cases a very efficient load-bearing form. Figure 15.1 shows, schematically, two very different ways of transferring a vertical loading to the end supports, by a *beam* or by an *arch*. In the case of a beam the action is bending and shear, and the stresses are normally high. For the arch, on the other hand, the loading is transferred primarily by axial or *membrane* action, and for the same loading the stress level is usually much lower than for the bending case. Figure 15.1 could equally well represent a flat *plate* (bending) and a curved *shell* (membrane).

Arches are normally circular or parabolic. Shells on the other hand come in all shapes and sizes. As load-bearing structures we normally group shell structures into three main categories:

- *Singly curved shells* – examples are cylindrical shell roofs (Fig. 15.2), cylindrical tanks and cones; the latter two also fall into the sub-class of shells of revolution. A characteristic feature of these shells is that they are *developable*; in other words, if you "slit" them lengthwise, they can be unrolled to form a *flat* sheet.
- *Doubly curved shells* – examples are spheres and hyperbolic paraboloids. These shells are *not* developable.
- *Prismatic shells* – an example is the so-called folded plate or folded roof. Although not curved, it has some of the same characteristics as the curved shell, such as the combined bending and membrane action.

Figure 15.2 shows a typical cylindrical shell roof which transfers the (vertical) loading to the end diaphragms both directly as a "beam" in the longitudinal direction and indirectly by "arch" action via two edge beams. In a well-designed shell structure, significant bending stresses are normally limited to areas in the vicinity of concentrated loading and to zones at the boundaries, see Fig. 15.3.

Arches and shells

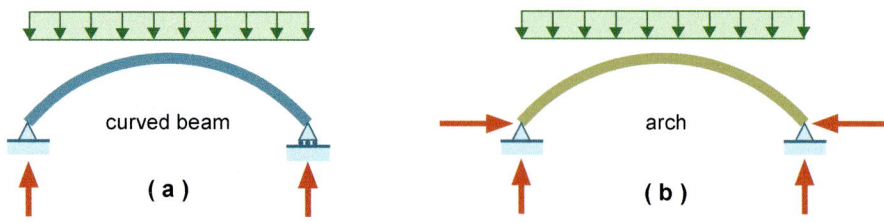

Figure 15.4 The difference between a curved beam and an arch

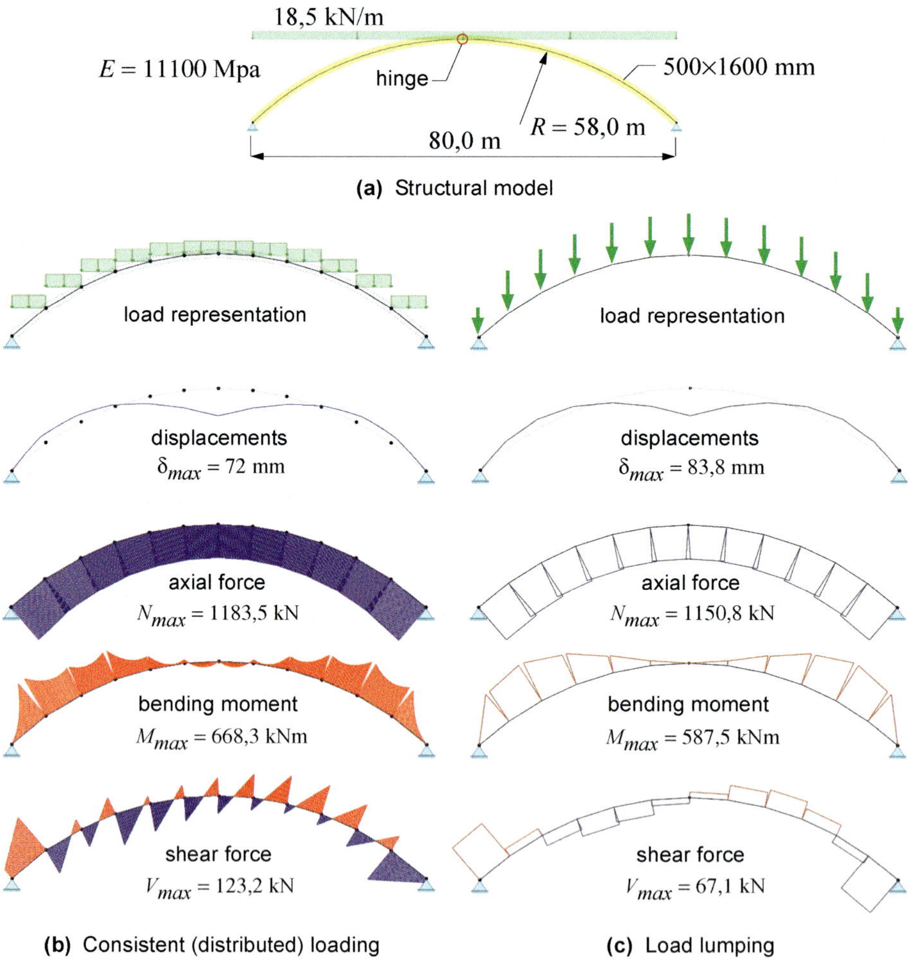

(a) Structural model

$E = 11100$ Mpa, $18{,}5$ kN/m, hinge, 500×1600 mm, $R = 58{,}0$ m, $80{,}0$ m

(b) Consistent (distributed) loading

load representation

displacements $\delta_{max} = 72$ mm

axial force $N_{max} = 1183{,}5$ kN

bending moment $M_{max} = 668{,}3$ kNm

shear force $V_{max} = 123{,}2$ kN

(c) Load lumping

load representation

displacements $\delta_{max} = 83{,}8$ mm

axial force $N_{max} = 1150{,}8$ kN

bending moment $M_{max} = 587{,}5$ kNm

shear force $V_{max} = 67{,}1$ kN

Figure 15.5 3-hinge glulam arch modelled by 12 straight beam elements

Classical shell theories lead to complex differential equations that are difficult to solve even after a series of simplifications (associated with names like VLASSOV, FLÜGGE, DONNEL and GECKELER). These theories can be viewed as modifications of plate theories where the in-plane and out-of-plane actions are combined, and most of them apply to *thin* shells for which KIRCHHOFF's hypothesis applies.

Shell theories are beyond our scope; we shall tackle this advanced problem without even as much as a reference to shell equations, but first we look at the simpler, but related problem of the arch.

15.1 Curved beams and arches

Figure 15.4 shows two different ways to support the same curved structural component. Figure 15.4a shows a curved *beam* with only vertical support reactions, whereas Fig. 15.4b shows a 2-hinge *arch*. While the beam carries a vertical loading primarily by bending and shear, the arch carries the loading to the supports primarily by axial force, and the large horizontal thrust at the supports is a characteristic feature of the arch.

The structural analysis of a curved beam is fairly straightforward. Perhaps the most important and challenging aspect of such components are found in pitch cambered glued laminated (glulam) timber beams. A "straightening" bending moment will cause tension stresses perpendicular to grain, and in view of timber's low strength for these stresses, they can cause failure. However, simple formulas are available, and if necessary a straightforward finite element plane stress analysis will provide detailed results.

We shall concentrate on the arch problem, and our preferred approach is to approximate the curved arch by a series of straight beam elements, each with constant cross section and material properties. The advantages of such an approach is that we can easily handle a varying cross section along the arch, and we can approximate any type of curvature.

Figure 15.5a shows a 3-hinge glulam arch loaded by a uniformly distributed (snow) load over the entire arch. In Fig. 15.5b the arch has been modelled by 12 straight beam elements (of equal length along the projection on the chord). The elements are rigidly connected and the loading is applied by the *consistent* load model, that is by determining S^0 for each element, and assembling these to form the load vector $R = -R^0$ (included in this load vector we will also find moments). The arch is analysed by program **fap2D** [9], a program that only offers load lumping, that is *all* distributed loading is automatically converted into statically equivalent concentrated nodal forces (no moments). In order to simulate the consistent load approach, each of the 12 straight beam elements are subdivided into 40 elements in the computational model. The computed results are shown in Fig 15.5b.

In Fig. 15.5c we have again modelled the arch using 12 straight beam elements, but there is a slight difference in that the elements are now equally long along the arch (and not along the chord). The main difference, however, is that we now lump the loading into statically equivalent vertical nodal forces as shown at the top of Fig. 15.5c. The computational results,

Arches and shells

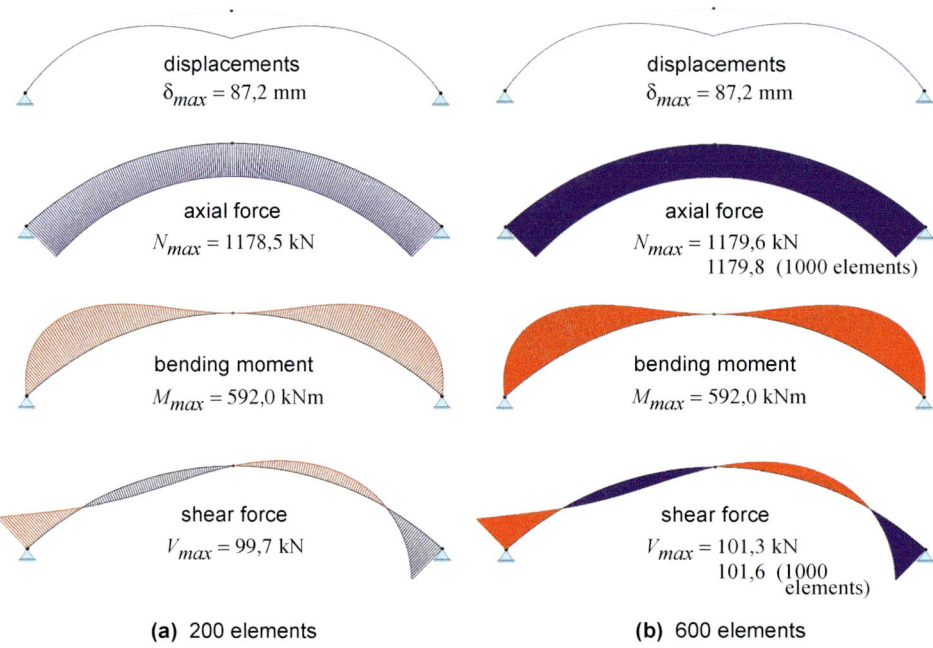

Figure 15.6 Many straight elements and load lumping for the arch of Fig 15.5

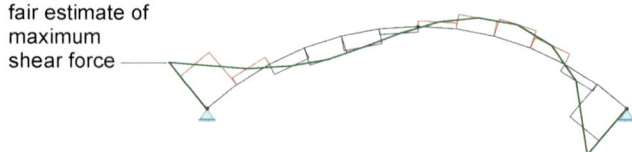

Figure 15.7 Estimating maximum shear force

which are now obtained by just 12 "computational elements" (as compared to 12×40 = 480 elements in Fig 15.5b), are shown in the figure for displacements, axial force, bending moment and shear force.

While the displacements and the axial forces are reasonably similar for the two models, the bending moments, and particularly the shear forces show a very different picture. If we take the results shown in Fig. 15.6b, obtained by 600 straight beam elements and load lumping, as the "correct" answer, we see that the shape of the diagrams on the right-hand side of Fig 15.5 make much more sense than those on the left-hand side. With only 12 elements, the step-wise constant shear forces in Fig. 15.5c are clearly very crude, but compared to the shear forces obtained by the consistent treatment of the loading for the same FEM model, the shape of the diagram is as it should be.

Keep in mind that the results in Fig. 15.5b are *correct* for the structure they are obtained for, that is a structure formed as a polygon (a series of straight beam segments rigidly interconnected), and loaded by a uniformly distributed (snow) load. But this is a *very poor model* for an arch. The picture will of course improve if we use more elements, but even with a large number of elements we will still see the "irregular" shape of both the moment diagram and particularly the shear force diagram.

The message here is loud and clear: if you approximate a curved beam or an arch by a series of rigidly connected straight beam elements you should also *lump* all distributed loading into statically equivalent nodal forces acting at the nodal points. For a reasonable number of elements this model will give you very good results.

If we compare the results shown in Figs. 15.5c and 15.6 we see that while displacements, axial forces and bending moments are, for practical purposes, determined with sufficient accuracy even for the coarsest mesh, the shear force will require more elements. We have carried out some more analyses, and the results from all analyses are as follows:

No. of elements	δ_{max} [mm]	N_{max} [kN]	M_{max} [kNm]	V_{max} [kN]
12	83,8	1150,8	587,5	67,1
24	86,3	1165,9	587,5	85,5
48	87,0	1173,1	591,5	92,5
96	87,1	1176,7	591,5	97,2
200	87,2	1178,5	592,0	99,7
600	87,2	1179,6	592,0	101,3
1000	87,2	1179,8	592,0	101,6

The slow convergence of the shear force is to be expected, but if need be, it is quite easy to estimate the "correct" result; at the middle of each element the shear force is fairly accurate, and for two consecutive elements it is straightforward to extrapolate the result at the ends of the elements, see Fig. 15.7. It is perhaps more surprising that the axial force also changes slightly, even

Arches and shells

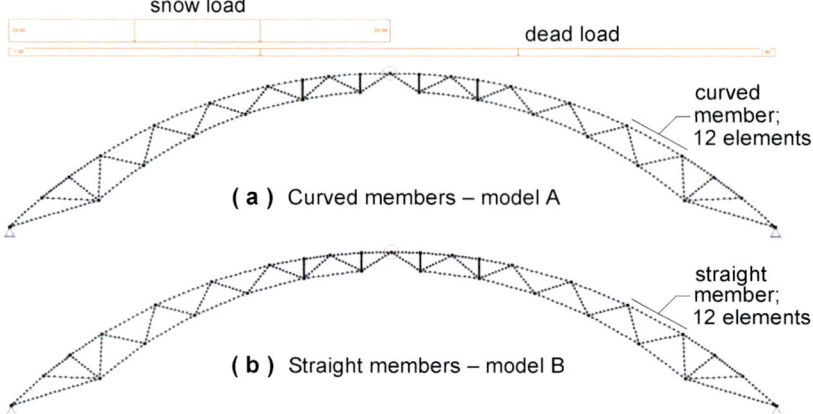

Figure 15.8 Two models for a 3-hinge steel truss arch

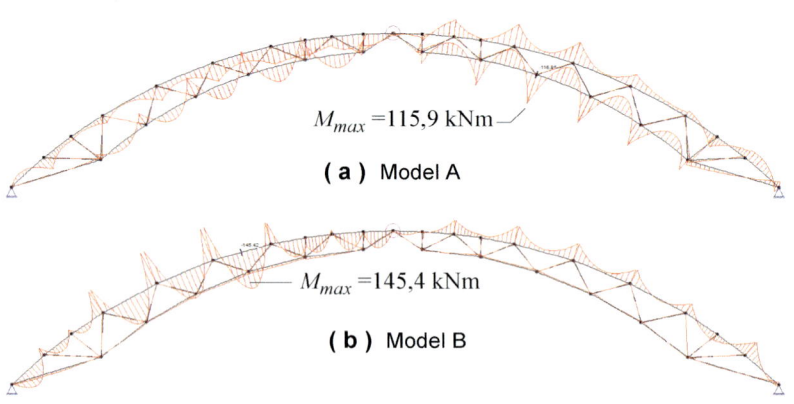

Figure 15.9 Bending moment diagrams for the two models in Fig. 15.5

from a fine to a very fine mesh. However, we are talking about an error that, for even the coarsest mesh, is about 2,5 per cent, so to the extent that this is a problem, it is an academic one.

It should be noted that a model with 1000 elements shows no sign of numerical problems, and the computing time is basically not an issue (the total time spent on the computational part of the program is about 15 milliseconds on a standard PC anno 2012). With a fairly fine mesh, say 100 to 200 elements, the geometric error we make by placing all nodal points on the centre line of the arch is also negligible.

Figure 15.8 shows a steel roof truss; it is basically a 3-hinge arch, but in the shape of a truss. In Fig. 15.8a the members between the "structure nodal points" are, in model A, represented by 12 straight beam elements following the *real curve*, whereas for model B in Fig. 15.8b the members between the truss' nodal points are assumed to be straight, and each *straight* member is modelled by 12 straight elements. By just looking at Fig. 15.8 it is difficult to see the difference between the two models; they have the same number of *dofs*.

The critical loading is the dead load of the truss itself, the uniformly distributed dead load of the roof structure and snow load on half the arch as shown in Fig 15.8. Both models are analysed for this loading, *lumped* into statically equivalent concentrated nodal forces, and while the axial forces are very similar for the two models, this is *not* the case for the bending moments, as shown in Fig. 15.9. While model B tends to overestimate the bending moments in the upper chord, this model completely neglects the moments in the lower chord. Considering the high compressive forces in the lower chord on the right-hand side of the arch, this is critical.

This example represents a *real* structure which was designed and built on the basis of the forces and moments determined by model B! It has since been strengthened. The message here is that it is vital to capture even the small deviation from the straight line that the curved member represents. The moment "arm" is small, but the axial forces are large, and the result is a very significant bending moment. This is yet another reason for modelling a curved member with a fair number of straight elements.

Before we leave the arch problem we should perhaps answer a question that has not yet been raised. Why not develop a curved beam element? The short answer is that such elements, and they have been developed, need special attention and "fixes" in order to work properly. One problem is associated with the rigid body modes; explicit inclusion of these modes is a problem for curved elements. More important is a phenomenon called *membrane locking* which makes the element overly stiff for "thin" arches with a large rise-to-span ratio. We will not examine these problems any further, the interested reader may find more information in, for instance Refs. [21] and [42]. The bottom line is that curved beam or arch elements are more trouble than they are worth, particularly in view of the excellent job a series of straight elements will do. It can be shown that a series of "ordinary" straight beam elements represents a valid model for a curved beam or an arch, and that it will converge to "exact" answers as the mesh is refined [56]. But keep in mind that even if a series of straight beam elements will provide correct section forces in an arch, such as in Fig. 15.6, the stress distribution may not be in

Arches and shells

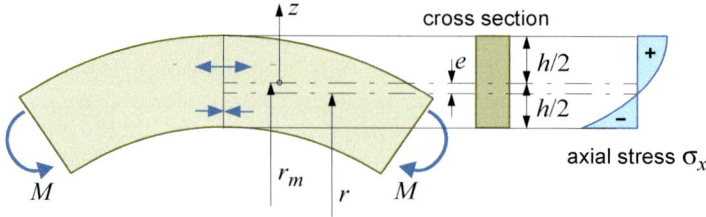

Figure 15.10 Nonlinear stress distribution in an arch due to bending

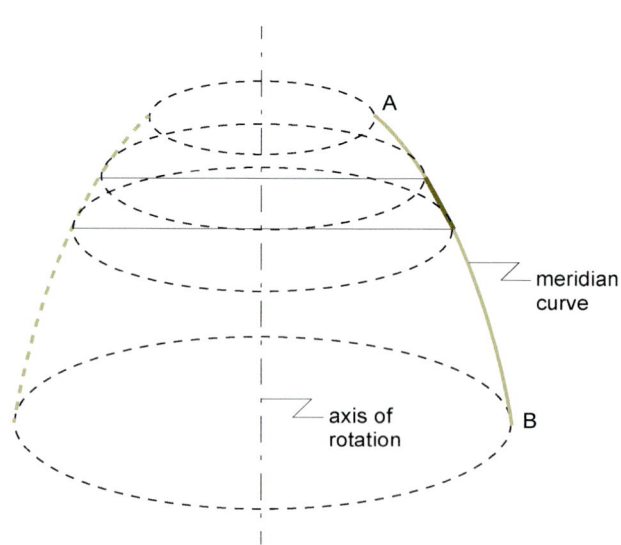

Figure 15.11 Shell of revolution

accordance with standard beam theory. Pure bending in an arch, for instance, will cause a nonlinear distribution of the axial bending stresses, as shown in Fig. 15.10. This, however, is a separate problem, and for a circular arch with a rectangular cross section it can be shown, with reference to Fig. 15.10, that the axial stress σ_x can be expressed as

$$\sigma_x = \frac{Mz}{Ae(r+z)} \tag{15-1}$$

where A is the cross sectional area and

$$e \approx \frac{h^2}{12r_m} \tag{15-2}$$

r_m is the radius of curvature of the arch centerline.

15.2 The shell problem

As already pointed out, the characteristic property of a shell structure is its combination of in-plane, *membrane*, and out-of-plane, *bending*, action. The obvious engineering solution to the modelling of an arbitrary (doubly curved) shell structure is therefore to approximate the curved surface by a series of *flat* elements that combine membrane (plane stress) and bending behaviour. This approach is particularly well suited for *thin* shells, but it can also be used for medium thick shells. Another type of shell element for medium thick to thick shells is a special, "degenerated" three-dimensional isoparametric (C^0) element. In view of the remarkable computational power now at our disposal, a third option is of course to use one or more layers of "ordinary" solid (3D) elements (for medium to thick shells).

It should also be mentioned that curved shell elements, based on classical shell theories, have been developed. However, as far as we are concerned this is a closed book; we shall only consider (briefly) the following three models:

- *Flat* elements combining membrane and bending action.
- Special, degenerated 3D isoparametric elements.
- Ordinary 3D isoparametric elements.

For general shells our preference and emphasis will be on the first option, that is the flat shell element. But first we shall briefly address a special type of shell structures.

Shells of revolution. A shell of revolution is obtained by rotating a curve (A-B) 360 degrees about an axis, see Fig. 15.11. The rotation axis becomes an axis of symmetry, and if we also assume material properties and loading to by symmetrical about this axis we have an axisymmetric problem. For the remainder of this section we assume complete axial symmetry.

For the purpose of analysis we now model the shell surface by "ring" type elements, and from what we have already seen and said about modelling of arches we approximate the meridian curve (A-B) by a series of *straight*

Arches and shells

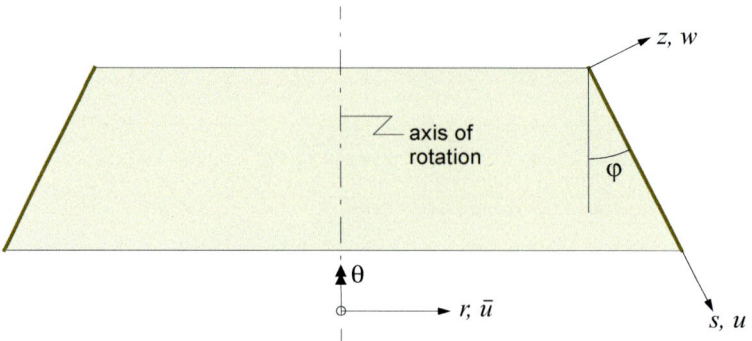

Figure 15.12 A conical shell ring element

elements. Geometry and notation for our conical "ring" element is shown in Fig. 15.12.

We need to consider two strain components, ε_s and ε_θ, that is

$$\boldsymbol{\varepsilon} = \begin{bmatrix} \varepsilon_s \\ \varepsilon_\theta \end{bmatrix} \tag{15-3}$$

The corresponding stress is

$$\boldsymbol{\sigma} = \begin{bmatrix} \sigma_s \\ \sigma_\theta \end{bmatrix} = \mathbf{C}\boldsymbol{\varepsilon} \quad \text{where} \quad \mathbf{C} = \frac{E}{1-v^2}\begin{bmatrix} 1 & v \\ v & 1 \end{bmatrix} \tag{15-4}$$

KIRCHHOFF's hypothesis about the surface normal can be expressed as

$$\boldsymbol{\varepsilon} = \boldsymbol{\varepsilon}_m + \boldsymbol{\varepsilon}_b = \boldsymbol{\varepsilon}_m - z\mathbf{c} \tag{15-5}$$

where $\boldsymbol{\varepsilon}_m$ contains the strains of the middle surface and $\mathbf{c}$ is the curvature of the middle surface. For a straight (conical) shell segment (see Fig. 15.12) the strain/curvature components are [50]:

$$\boldsymbol{\varepsilon}_m = \begin{bmatrix} \varepsilon_{s0} \\ \varepsilon_{\theta 0} \end{bmatrix} = \begin{bmatrix} \dfrac{du}{ds} \\ \dfrac{u\sin\varphi + w\cos\varphi}{r} \end{bmatrix} = \begin{bmatrix} \dfrac{d}{ds} & 0 \\ \dfrac{\sin\varphi}{r} & \dfrac{\cos\varphi}{r} \end{bmatrix}\begin{bmatrix} u \\ w \end{bmatrix} = \boldsymbol{\Delta}_m \mathbf{u} \tag{15-6}$$

$$\mathbf{c} = \begin{bmatrix} c_s \\ c_\theta \end{bmatrix} = \begin{bmatrix} \dfrac{d^2 w}{ds^2} \\ \dfrac{\sin\varphi}{r}\dfrac{dw}{ds} \end{bmatrix} = \begin{bmatrix} 0 & \dfrac{d^2}{ds^2} \\ 0 & \dfrac{\sin\varphi}{r}\dfrac{d}{ds} \end{bmatrix}\begin{bmatrix} u \\ w \end{bmatrix} = \boldsymbol{\Delta}_b \mathbf{u} \tag{15-7}$$

Subscript 0 designates the middle surface. Next we assume

$$u = \mathbf{N}_m \mathbf{v}_m \quad \text{and} \quad w = \mathbf{N}_b \mathbf{v}_b \tag{15-8}$$

or

$$\mathbf{u} = \begin{bmatrix} u \\ w \end{bmatrix} = \begin{bmatrix} \mathbf{N}_m \\ \mathbf{N}_b \end{bmatrix}\begin{bmatrix} \mathbf{v}_m \\ \mathbf{v}_b \end{bmatrix} = \mathbf{N}\mathbf{v} \tag{15-9}$$

where $\mathbf{v}_m$ are the membrane *dofs* and $\mathbf{v}_b$ the bending *dofs*. From this we find

$$\boldsymbol{\varepsilon}_m = \boldsymbol{\Delta}_m \mathbf{u} = \boldsymbol{\Delta}_m \mathbf{N}\mathbf{v} = \mathbf{B}_m \mathbf{v} \tag{15-10}$$

and

$$\boldsymbol{\varepsilon}_b = -z\mathbf{c} = -z\boldsymbol{\Delta}_b \mathbf{u} = -z\boldsymbol{\Delta}_b \mathbf{N}\mathbf{v} = -z\mathbf{B}_b \mathbf{v} \tag{15-11}$$

The strain energy of the element becomes

$$U = \frac{1}{2}\int_V \boldsymbol{\varepsilon}^T \mathbf{C}\boldsymbol{\varepsilon}\, dV = \frac{1}{2}\mathbf{v}^T \int_V (\mathbf{B}_m - z\mathbf{B}_b)^T \mathbf{C}(\mathbf{B}_m - z\mathbf{B}_b)\, dV \mathbf{v} \tag{15-12}$$

Arches and shells

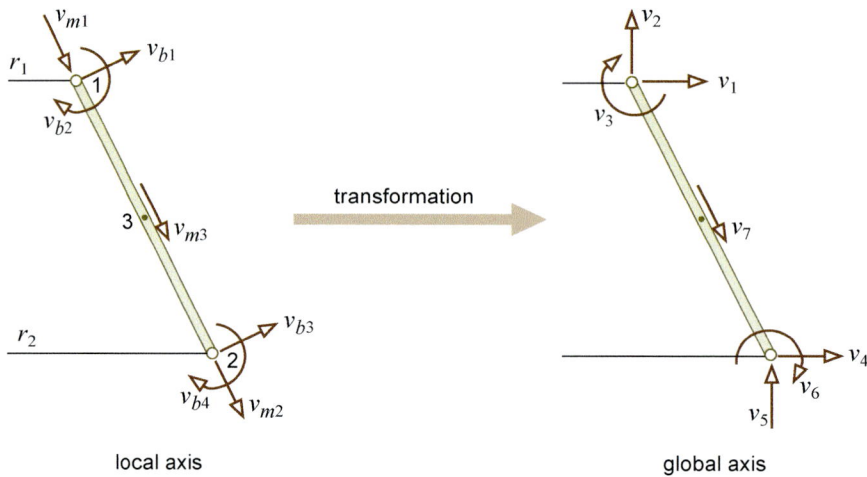

Figure 15.13 Typical conical shell ring element

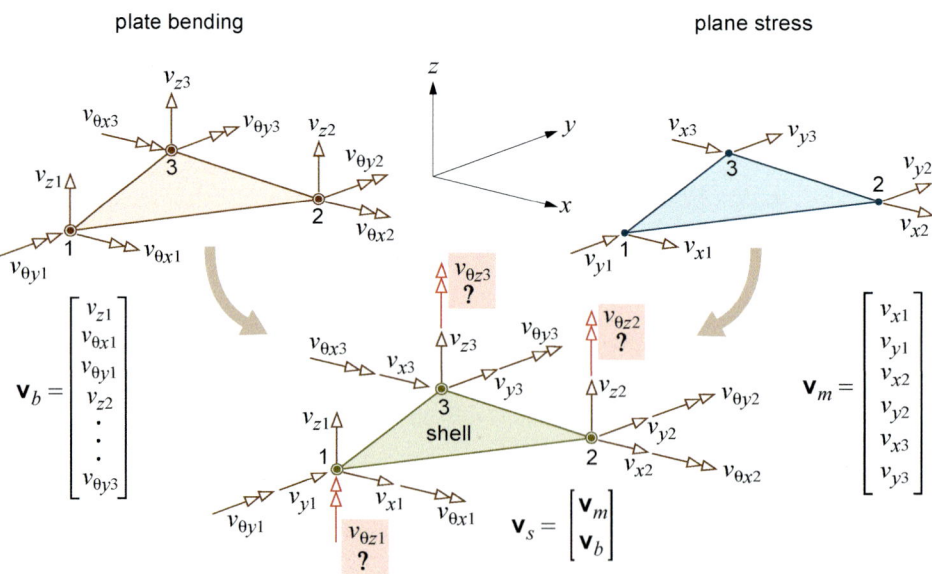

Figure 15.14 A flat, thin, triangular shell element

or

$$U = \frac{1}{2}\mathbf{v}^T\mathbf{k}_m\mathbf{v} + \frac{1}{2}\mathbf{v}^T\mathbf{k}_b\mathbf{v} = \frac{1}{2}\mathbf{v}^T(\mathbf{k}_m + \mathbf{k}_b)\mathbf{v} = \frac{1}{2}\mathbf{v}^T\mathbf{k}\mathbf{v} \qquad (15\text{-}13)$$

where, since $dV = 2\pi r\, ds\, dz$,

$$\mathbf{k}_m = 2\pi \int_0^L h \mathbf{B}_m^T \mathbf{C} \mathbf{B}_m r\, ds \qquad (15\text{-}14)$$

and

$$\mathbf{k}_b = 2\pi \int_0^L \mathbf{B}_b^T \mathbf{D} \mathbf{B}_b r\, ds \quad \text{where} \quad \mathbf{D} = \frac{h^3}{12}\mathbf{C} \qquad (15\text{-}15)$$

For a straight (conical) element the coupling terms between the membrane and the bending action vanish in the local coordinate system s,z (just as they do for the straight arch element). Since the elements of **v** in the above derivation are not ordered node by node we need to rearrange rows and columns of **k** before assembly.

Figure 15.13 shows a typical element, in local and global coordinates. Compared to an ordinary 2D beam element we have added an extra axial *dof* at the mid point, in order to facilitate a parabolic variation of the membrane strains. For this element the shape functions $\mathbf{N}_m$ are the quadratic functions of Fig. 5.14b, and the bending functions $\mathbf{N}_b$ are the standard cubic beam functions of Fig. 5.15. The transformation to global axes (before the assembly process) need not include the mid-point *dof* (v_7) since it is a *local dof*; in fact, this *dof* can be eliminated (by static condensation) at the element level, but it is probably not worth the extra programming.

The element in Fig. 15.13 is akin to the straight arch element and, as for the arch, we would recommend *lumping* the (axisymmetric) loading into concentrated nodal loads and the use of *many* elements; this will make it easy to model variable thickness and sharp bends with good accuracy.

For a medium thick to thick shell of revolution with axisymmetric loading we would recommend solid ring elements as described in Section 12.1, for instance the "LST" ring element of Fig. 12.3. It is not recommended to skimp on the number of "layers" of elements over the shell thickness.

As outlined in Section 12.2, some types of non-symmetric loading can be handled by these "ring" elements by adding together the solutions for individual FOURIER series loading terms. Alternatively we can revert to one of the element models described below; these will work for all types of loading.

Flat shell elements. The idea is illustrated in Fig. 15.14. The curved shell surface is approximated by a series of flat *facets* of triangular or quadrilateral shape. Each facet constitutes a shell element composed of a membrane part, in Fig.15.4 taken to be a **CST** element, and a bending part which in Fig. 15.14 could be a **DKT** element. At the element level there is *no* coupling between the membrane and the bending action, and, if we organize the *dofs* in a membrane part ($\mathbf{v}_m$) and a bending part ($\mathbf{v}_b$), the (local) element stiffness matrix can be expressed as

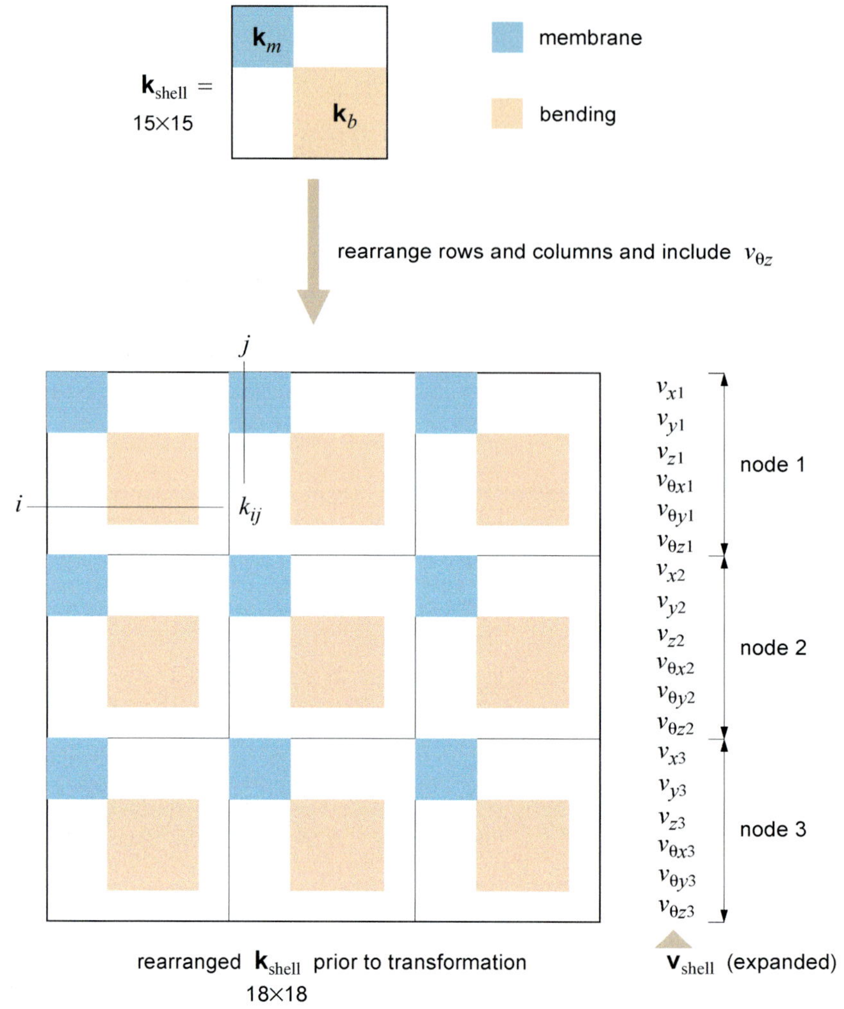

Arches and shells

Figure 15.15 Rearranging and expanding $\mathbf{k}_{shell}$ of the element in Figure 15.14

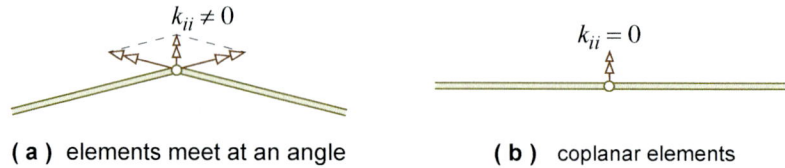

Figure 15.16 Elements at a shell node

The shell problem

$$\mathbf{k}_{shell} = \begin{bmatrix} \mathbf{k}_m & 0 \\ 0 & \mathbf{k}_b \end{bmatrix} \quad \text{where} \quad \mathbf{v}_{shell} = \begin{bmatrix} \mathbf{v}_m \\ \mathbf{v}_b \end{bmatrix} \quad (15\text{-}16)$$

Here $\mathbf{k}_m$ is the stiffness matrix for the plane stress (membrane) part, which in our example in Fig. 15.14 is the stiffness matrix for the **CST** (constant strain triangle) element, and $\mathbf{k}_b$ is the bending part, e.g. the stiffness matrix for the **DKT** (discrete KIRCHHOFF triangle) element.

For reasons that will soon be explained, we would like a shell element to have 6 *dofs* at each of the (corner) nodes, three translational *dofs* and three rotational *dofs*. However, if we take a look at the shell element in Fig. 15.14 we see that one of these *dofs*, the rotation ($v_{\theta z}$) about the z-axis, is missing. This is the so-called *drilling dof*, a special kind of membrane *dof* that we normally will not find amongst the *dofs* of plane stress elements. More about this *dof* soon.

In order to facilitate the assembly process we need to

   a) rearrange rows and columns of $\mathbf{k}_{shell}$ to make the stiffness matrix correspond with a displacement vector **v** in which the *dofs* are arranged nodal point by nodal point, see Fig. 15.15, and

   b) transform the rearranged stiffness matrix to some system reference coordinate system(s).

At a particular nodal point of the shell element in Fig. 15.14 the *dofs* are then

$$\bar{\mathbf{v}}_i = [\; \bar{v}_{xi} \quad \bar{v}_{yi} \quad \bar{v}_{zi} \quad \bar{v}_{\theta xi} \quad \bar{v}_{\theta yi} \quad \bar{v}_{\theta zi} \;]^T$$

where the bar on top indicates system reference coordinates.

So what happens if we use an element like the one shown in Fig. 15.14? If we go for 6 *dofs* per node at the system level, which is the norm, we would also include this *dof* at the nodal points at the element level, as shown in Fig. 15.15. This leaves us with an element stiffness matrix with three rows and columns that has only *zero* elements. This is not a problem if the elements attached to a particular node form an angle with each other, as shown schematically in Fig. 15.16a. In such a case the stiffness associated with the bending rotations will, when transformed, also provide stiffness to the surface normal rotation. However, if the elements that meet at a shell node are coplanar, as shown in Fig. 15.16b, we will end up with zero stiffness for the rotation about the shell normal, and our system stiffness matrix will be singular. How do we dodge this problem?

There are at least three ways in which we can avoid the singularity:

1) At each system node we define our *dofs* in and out of the *tangent plane* at the node, and we leave out the rotation about the normal to this plane, resulting in only 5 *dofs* per node. With the problem *dof* removed we no longer risk singularity of **K** (at least not from this source); the downside is that we (in principle) end up with different reference coordinates at every system node, which of course complicates the transformation from element coordinates to system coordinates. The determination of the tangent plane is another complicating factor.
A variant of this solution is to use 6 *dofs* at each node, but at all nodes

Arches and shells

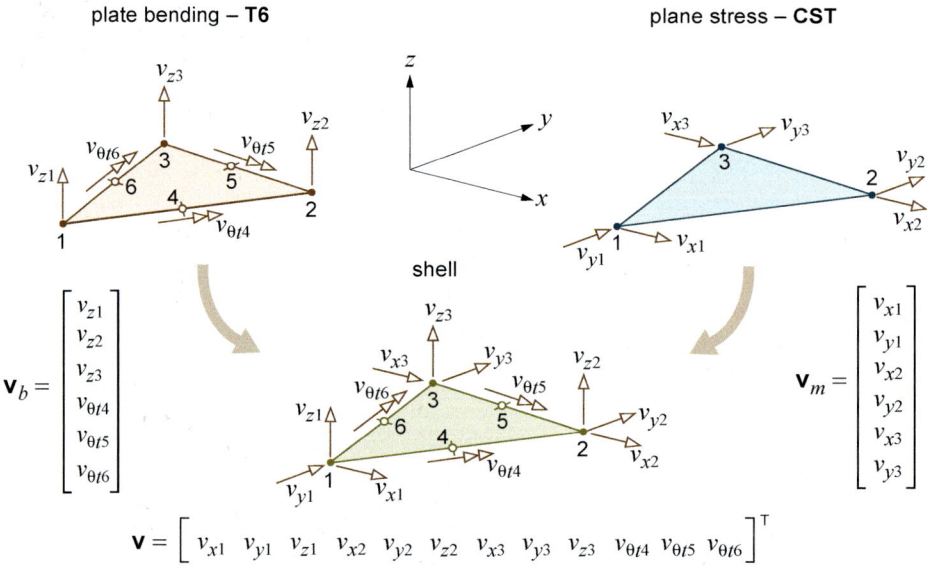

Figure 15.17 A very simple flat constant stress triangular shell element

where coplanar elements meet we transform to a coordinate system that has two axes in the plane of the elements, and we simply eliminate the *dof* representing the rotation about the normal to the plane (by specifying the *dof* equal to zero). The difficulty with this approach is to decide at which nodes the elements are coplanar, or so close to being coplanar, that we need to transform (in order to avoid numerical problems).

2) Define 6 *dofs* at each system node, and at nodes joining coplanar or near coplanar elements introduce a fictitious rotational stiffness about the surface normal. Several implementations of this idea have been suggested, see for instance COOK *et al* [21,52].

3) For the membrane part, use elements that also include the *drilling dof* at the nodes. By this method all 6 *dofs* will have stiffness at the element level and consequently also at the system level, and the singularity problem no longer exists. Several plane stress elements that also include the drilling *dof* have been developed.

The last method seems to be the obvious choice, especially since the extra drilling *dof* will also give an improved membrane action compared to the **CST** element. So that not only do we get the "missing" stiffness by this approach, we may also get some improved membrane behaviour. However, the inclusion of the drilling *dof* in the membrane element is not a straightforward operation, and to some, including this author, it is a "tricky" *dof*. Method number 1) should also be considered a viable solution, and we would recommend the latter part of this option, that is to work with a mixture of 6 and 5 *dofs* at the system nodes. We need robust methods for determining when to use 5 *dofs* and how to define the tangent plane at the system node; the rest is straightforward. This approach also lets us use all the standard plane stress elements for the membrane part (we need not worry about the drilling *dof*).

Most flat shell elements are 3-node triangles or 4-node quadrilaterals, and a wide variety of combinations of membrane and bending elements have been suggested. An extra complication for the quadrilateral element shape is that four points on a doubly curved shell surface, in general, will *not* lie in the same plane; hence some kind of an "average" plane will have to be defined for the element.

The combination shown in Fig. 15.14 is quite common. However, with these *dofs* the bending part is "better" represented than the membrane part. From what we have seen in Chapters 11 and 14, the simplest triangular elements, that is the **CST** in case of membrane action, and the MORLEY triangle, **T6**, in case of bending action, produce (from a practical point of view) excellent accuracy providing we use many elements. A combination of these two elements, both of which are *constant* stress (**CST**) and constant moments (**T6**) elements, seems to constitute a balanced and simple flat shell element, see Fig. 15.17.

This simple flat constant stress shell element seems to have been used by a number of researchers for geometrically nonlinear (large displacement) problems, but its potential for linear analysis of thin shells of arbitrary form should, in view of the formidable computer power now at our disposal, be

Arches and shells

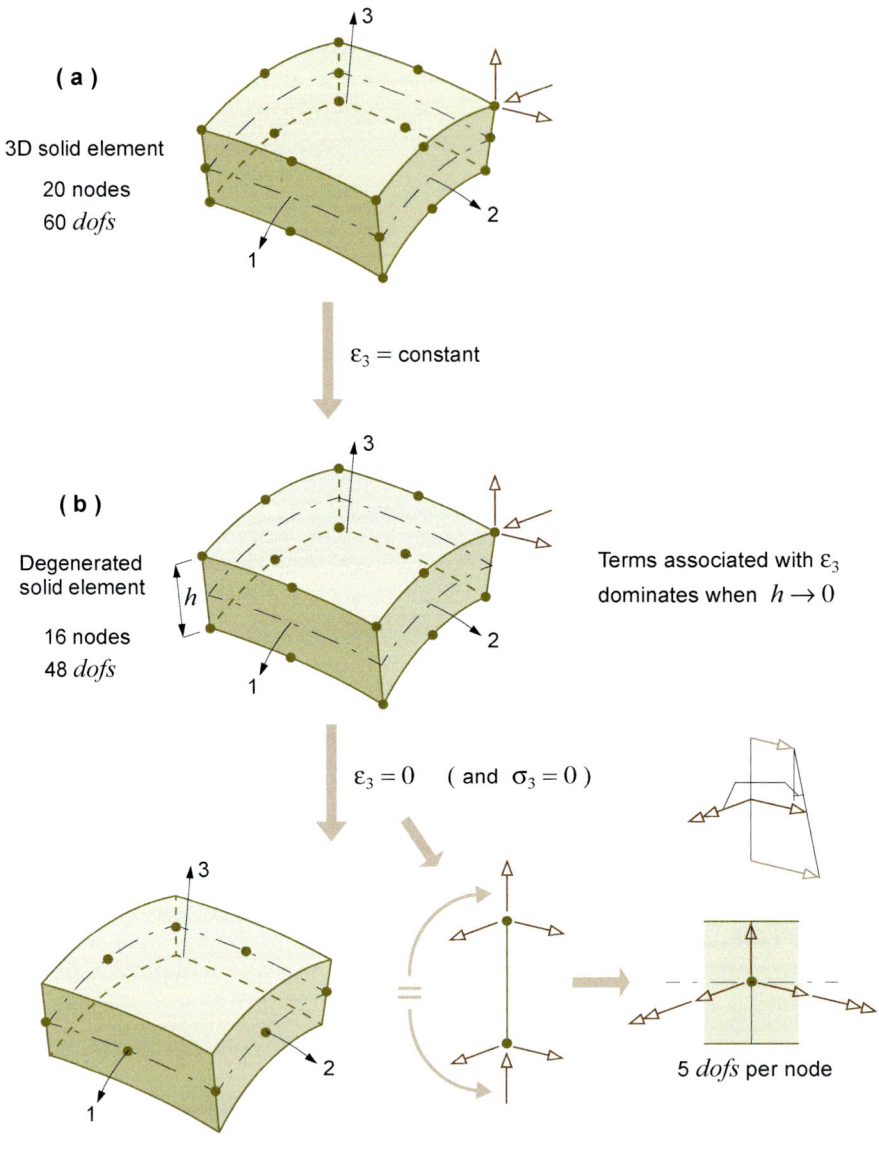

Figure 15.18 From a 3D isoparametric solid element to a thick shell element

very good. It offers an incredibly simple and apparently robust solution to a problem that is mathematically very complex. The problems with the 6th (drilling) *dof* do not exist for this element, for the simple reason that the element does not have rotational *dofs* at the corner nodes. Transformation from local element coordinates to global system coordinates is straightforward and only needs to include the 9 corner *dofs*; the mid-side rotations about the element edges need not be transformed before assembly.

Based on the results obtained with straight arch (beam) elements, we would recommend, in the case of flat shell elements, that all distributed loading on the shell is *lumped* into statically equivalent concentrated nodal forces. It is reasonable to assume that arches and shells behave similarly in this respect.

Thick shell elements. Starting with an isoparametric solid element, *e.g.* the 20-node serendipity type element shown in Fig. 15.18a, one can derive the 8-node thick shell element shown in Fig. 15.18c. Without going into the details (which can be found for instance in [56]), the reasoning is roughly as follows:

Even for a very thick shell the complete 20-node solid element with 3 nodes over the thickness of the shell is considered, or rather was considered, to provide unnecessarily many "transverse *dofs*". (The first of these elements was developed around 1970). By assuming the transverse strain (ε_3) to be *constant* over the thickness h, one can eliminate the nodes in the middle surface. The resulting 16-node element does not behave well when the thickness decreases since stiffness terms associated with ε_3 will then dominate more and more. This difficulty can be overcome by forcing nodes that are adjacent in the thickness direction to have the same displacement in the 3-direction, see Fig. 15.18c. The four remaining *dofs*, two at each of the "exterior" shell surfaces, can be moved to the middle surface as two displacements and two rotations. Thus we end up with an element with 8 nodes, all of which are located on the middle surface and have 5 *dofs*, three displacements and two rotations, a total of 40 *dofs* for the element.

Other similar elements, based on different isoparametric solid elements, have been developed. They are all of the MINDLIN type and are therefore able to represent transverse shear deformations, but care must be exercised in order to avoid problems associated with mechanisms and locking.

Solid elements in shell analysis. As already mentioned, the first elements of the type described above were developed around 1970; this was a time when computers were few and far between, their capabilities limited and their use expensive. Not so today, so why not use ordinary standard solid elements for the analysis of shells, even for thin shells? For instance, for the 20-node serendipity and the 27-node LAGRANGE elements of Fig. 13.6 and the linear strain tetrahedral element of Fig. 13.7, *one* element over the thickness would probably do a good job in many cases, but even two or three layers of such elements are well within current computational capabilities for many linear shell problems.

We conclude this chapter by suggesting that, at present, geometric description and meshing are perhaps the greatest challenges in linear, static shell analysis.

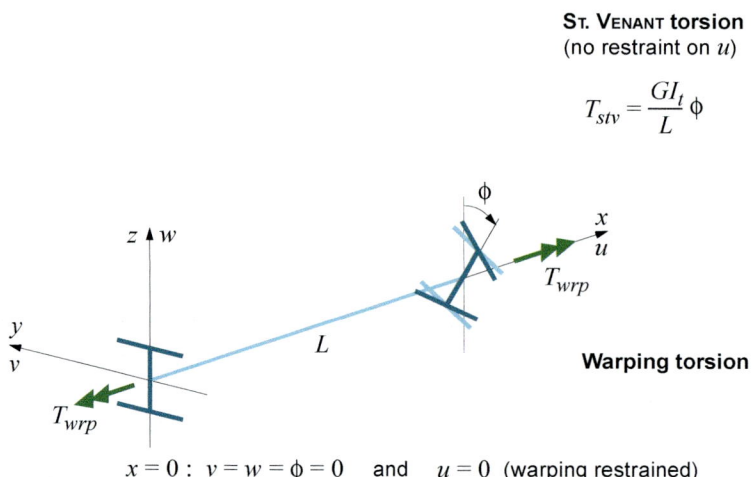

Figure 16.1 St. Venant torsion and warping torsion

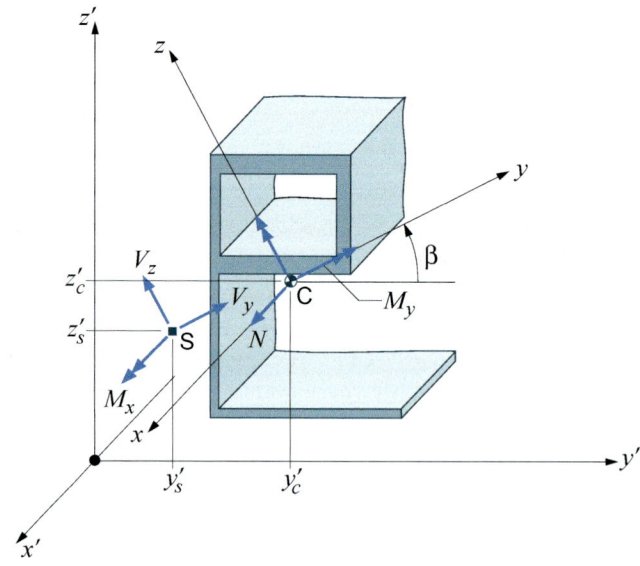

Figure 16.2 Coordinate systems and positive section forces (including moments)

16

ST. VENANT torsion

The determination of the ST. VENANT torsion constant for an arbitrary beam cross section is not a trivial task, nor is it a straightforward undertaking to find the distribution, over the cross sectional area, of shear stresses due to torsional moment and shear forces. This is, however, a problem well suited for the finite element method, as we shall see in this chapter. We start with a brief summary of the theory and proceed with a finite element formulation of the problem, and we conclude the chapter with some examples.

The *twisting* of a structural member is, depending on the shape of the cross section of the member and to some extent the boundary conditions, described by one or both of two different *torsion theories*, so-called **ST. VENANT torsion** and/or **warping torsion**. The sketches in Fig. 16.1 indicates the principal differences between the two theories.

ST. VENANT torsion assumes warping of a cross section, described by the displacement u normal to the cross section, to be unrestrained and independent of the member axis coordinate x, and the stresses resisting the *torque* (T_{stv}) are *shear stresses*. In warping torsion on the other hand, warping is restrained and the stresses resisting the torque (T_{wrp}) are *axial stresses*; this type of torsion is relevant primarily for *open thin-walled* sections.

We shall only consider ST. VENANT torsion, and our aim is to determine the *torsional constant* I_t (often denoted by J in the literature) and the shear stress distribution over the cross sectional area, for an *arbitrary* cross section. To this end we can use one of two different approaches: the one proposed by ST. VENANT himself or the one making use of a *stress function* as suggested by Ludwig PRANDTL[1]. The latter approach is the most frequently used, but we shall base our derivation on ST. VENANT's approach.

Figure 16.2 defines our coordinate systems: x',y',z' is the *reference* coordinate system where the x'-axis is parallel with the member axis and y' and z' are the axes in the cross section plane; the *member* coordinate system x, y, z is defined such that x coincide with the member axis (runs through the centre C of the cross section) while y and z are the *principal axes* of the cross section. Figure 16.2 also indicates positive section forces (including moments),

---
1. Ludwig PRANDTL (1875 – 1953) was a German scientist.

ST. VENANT torsion

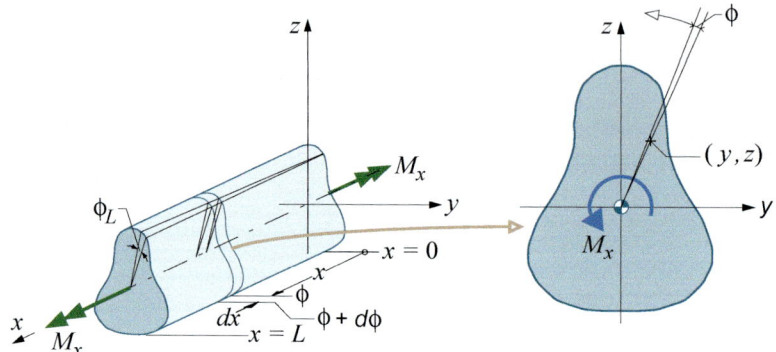

Figure 16.3 Uniform ST. VENANT torsion of prismatic bar

HOOKE's law in three dimensions:

$$\varepsilon_x = \frac{\partial u}{\partial x} = u_{,x} = \frac{1}{E}(\sigma_x - \nu(\sigma_y + \sigma_z))$$

$$\varepsilon_y = \frac{\partial v}{\partial y} = v_{,y} = \frac{1}{E}(\sigma_y - \nu(\sigma_z + \sigma_x))$$

$$\varepsilon_z = \frac{\partial w}{\partial z} = w_{,z} = \frac{1}{E}(\sigma_z - \nu(\sigma_x + \sigma_y))$$

$$\gamma_{xy} = \frac{\partial u}{\partial y} + \frac{\partial v}{\partial x} = u_{,y} + v_{,x} = \frac{1}{G}\tau_{xy}$$

$$\gamma_{yz} = \frac{\partial v}{\partial z} + \frac{\partial w}{\partial y} = v_{,z} + w_{,y} = \frac{1}{G}\tau_{yz}$$

$$\gamma_{zx} = \frac{\partial u}{\partial z} + \frac{\partial w}{\partial x} = u_{,z} + w_{,x} = \frac{1}{G}\tau_{zx}$$

referred to principal axes. A positive axial force (N) is a tensile force, and the shear forces (V_y and V_z) are positive in the direction of positive axes. The bending moments (M_y and M_z) and the torsional moment (M_x) are shown as double arrows, indicating the rotation of a right-hand screw in the arrow's direction. It should be noted that V_y, V_z and M_x are all assumed to act at the *shear centre* S.

16.1 ST. VENANT torsion – theoretical basis

Figure 16.3 shows a prismatic bar, of length L and with a "massive" cross section, subjected to a constant torsional moment or *torque*, M_x. Relative to each other the two end sections rotate an angle ϕ_L about the x-axis. Assuming the section at $x = 0$ to be fixed, the entire rotation takes place at $x = L$, as shown in the figure. At an arbitrary section ($x = x$) the *angle of twist* is ϕ, and at $x+dx$ it is $\phi + d\phi$. We denote the ratio $d\phi/dx$ by θ and refer to it as the *rate of twist*:

$$\theta = \frac{d\phi}{dx} \tag{16-1}$$

Assuming that

$$\theta = \text{constant} \tag{16-2}$$

we have

$$\phi = \theta x \tag{16-3}$$

In his *semi-inverse* method ST. VENANT made the following displacement assumptions:

$$u = \theta \psi(y, z) \tag{16-4a}$$

$$v = -\theta x z \tag{16-4b}$$

$$w = \theta x y \tag{16-4c}$$

$\psi(y, z)$ is the (as yet unknown) *warping function*, and the displacement component u, which describes the warping, is called the warping displacement. The displacement assumptions imply that the warping is the same from section to section (is independent of x) and that each cross section rotates (an angle $\phi = \theta x$) about the x-axis with *no distortion* of its shape ("rigid body" rotation).

The relationship between strain and displacements in 3D space is defined by Eq. (3-17); combined with HOOKE's law for a linear elastic, *isotropic* material as defined by Eq. (3-29), we get the relationships shown on the opposite page. From these it follows that, for the assumptions of Eqs. (16-4), we have

$$\varepsilon_x = \varepsilon_y = \varepsilon_z = 0 \tag{16-5}$$

and hence

$$\sigma_x = \sigma_y = \sigma_z = 0 \tag{16-6}$$

Furthermore

ST. VENANT torsion

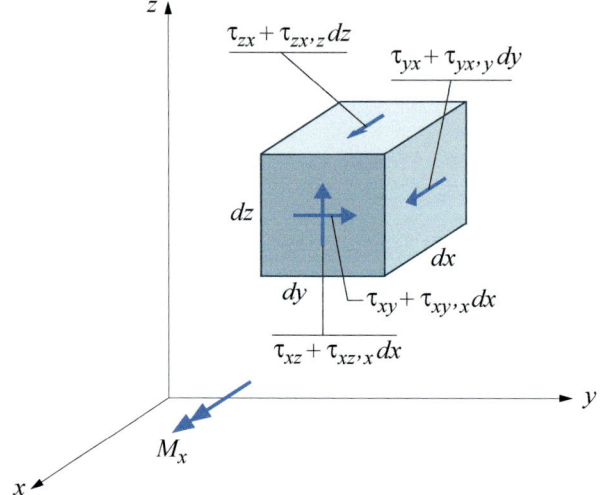

Figure 16.4 St. Venant torsion – non-zero stress components

ST. VENANT torsion – theoretical basis

$$\gamma_{xy} = \theta\frac{\partial \psi}{\partial y} - \theta z = \frac{1}{G}\tau_{xy} \qquad (16\text{-}7a)$$

$$\gamma_{yz} = -\theta x + \theta x = \frac{1}{G}\tau_{yz} \qquad (16\text{-}7b)$$

$$\gamma_{zx} = \theta\frac{\partial \psi}{\partial z} + \theta y = \frac{1}{G}\tau_{zx} \qquad (16\text{-}7c)$$

Hence, from (16-7b),

$$\tau_{yz} = 0 \qquad (16\text{-}8)$$

Figure 16.4 shows the non-zero stress components. Disregarding volume forces, force equilibrium of a small volume element gives us the following three equations

$$\tau_{yx,y}dxdydz + \tau_{zx,z}dxdydz = 0 \quad \Rightarrow \quad \tau_{yx,y} + \tau_{zx,z} = 0 \qquad (16\text{-}9a)$$

$$\tau_{xy,x}dxdydz = 0 \quad \Rightarrow \quad \tau_{xy,x} = 0 \qquad (16\text{-}9b)$$

$$\tau_{xz,x}dxdydz = 0 \quad \Rightarrow \quad \tau_{xz,x} = 0 \qquad (16\text{-}9c)$$

whereas moment equilibrium gives us the well known relations

$$\tau_{xz} = \tau_{zx} \quad \text{and} \quad \tau_{xy} = \tau_{yx} \qquad (16\text{-}10)$$

From Eqs. (16-7a) and (16-7c) we have

$$\tau_{xy} = G\theta(\psi_{,y} - z) \qquad (16\text{-}11a)$$

$$\tau_{xz} = G\theta(\psi_{,z} + y) \qquad (16\text{-}11b)$$

We see that (16-11a) and (16-11b) satisfy (16-9b) and (16-9c), respectively. Substitution of Eqs. (16-11a) and (16-11b) into (16-9a) finally gives the governing equation for the warping function:

$$G\theta(\psi_{,yy} + \psi_{,zz}) = 0$$

or

$$\psi_{,yy} + \psi_{,zz} = \nabla^2\psi(y, z) = 0 \qquad (16\text{-}12)$$

which is LAPLACE's equation in two dimensions.

An alternative representation is obtained by expressing the two non-zero shear stress components by PRANDTL's *stress function* $\Phi(y, z)$ as

$$\tau_{xy} = \frac{\partial}{\partial z}\Phi(y, z) = \Phi_{,z} \qquad (16\text{-}13a)$$

$$\tau_{xz} = -\frac{\partial}{\partial y}\Phi(y, z) = -\Phi_{,y} \qquad (16\text{-}13b)$$

ST. VENANT torsion

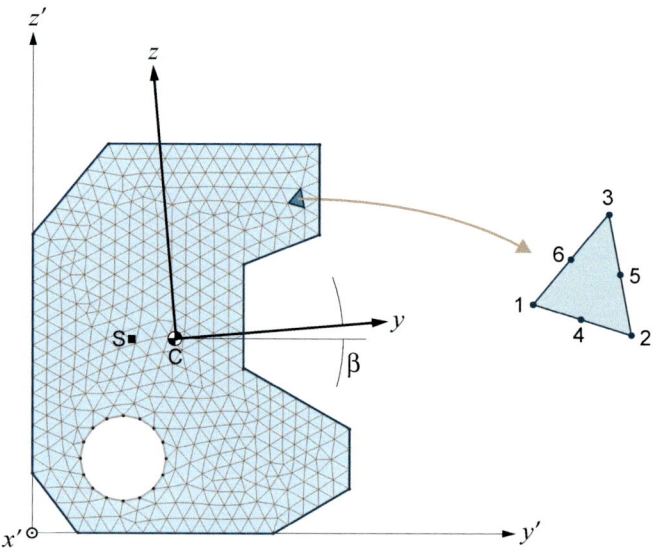

Figure 16.5 Typical finite element mesh of 6-node triangular elements

These assumptions satisfy the equilibrium equations (16-9a to c) identically, provided Φ has continuous second derivatives. Substituting Eqs. (16-13a) and (16-13b) into (16-11a) and (16-11b), respectively, give

$$\Phi_{,z} = G\theta(\psi_{,y} - z)$$

and
$$-\Phi_{,y} = G\theta(\psi_{,z} + y)$$

Eliminating the warping function ψ (by derivation) gives

$$\Phi_{,zz} + \Phi_{,yy} = -2G\theta \quad (16\text{-}14\text{a})$$

or
$$\nabla^2 \Phi = -2G\theta \quad (16\text{-}14\text{b})$$

which is a two-dimensional POISSON equation.

For arbitrary cross sections, ST. VENANT's torsion problem can only be solved by numerical methods, and a number of *finite element* solutions are reported, using either the warping function or the stress function as primary variable. As already pointed out we shall proceed with the warping function as our primary variable.

16.2 Finite element torsion analysis

Figure 16.5 shows a massive cross section with a hole, meshed automatically with triangular finite elements. The basic element we shall use is a 6-node triangle (similar to **LST**) with straight sides. The side nodes are located at mid-point, and the local node numbering system is (as used in previous chapters) shown in the figure. With this element all curved edges must be approximated by piece-wise straight lines.

The section is referred to a reference system y' and z' (the x'-axis is parallel with the beam member axis). Determination of the area centre, point **C** in Fig. 16.5, and the orientation (angle β) of the principal axes, y and z, are straightforward. The location of the shear centre **S** is, however, a more subtle problem that we shall return to briefly in the next section. It should be noted that while the material properties are constant within a particular element, they may, in principle, vary from element to element. The centre of area of the total cross section (**C**) is thus based on the "axial stiffness" ($E_i A_i$) of each individual element, rather than the true area (A_i).

We now change the reference frame to the principal axes (of which y always is the major or "strong" axis). Hence, from now on all reference to coordinates assume principal axes. The procedure described below follows [57] which in turn is based on a derivation described by MEEK [58].

The cross section in Fig. 16.5 is assumed to rotate as a *rigid* body about the shear centre **S** (also called the centre of twist) according to the displacement assumptions of Eqs. (16-4). It follows from Eq. (16-4a) that the warping u at any point in the section is proportional to θ, and it is therefore sufficient to determine u for a given value of θ. Since the most convenient value is unity we proceed, for the time being, with θ = 1, for which (16-4a) becomes

ST. VENANT torsion

Shape functions for the quadratic triangle:

$$N_1 = \zeta_1^2 - \zeta_1\zeta_2 - \zeta_3\zeta_1 = \zeta_1(2\zeta_1 - 1) \qquad N_4 = 4\zeta_1\zeta_2$$

$$N_2 = \zeta_2^2 - \zeta_1\zeta_2 - \zeta_2\zeta_3 = \zeta_2(2\zeta_2 - 1) \qquad N_5 = 4\zeta_2\zeta_3$$

$$N_3 = \zeta_3^2 - \zeta_2\zeta_3 - \zeta_3\zeta_1 = \zeta_3(2\zeta_3 - 1) \qquad N_6 = 4\zeta_3\zeta_1$$

$$\mathbf{N} = [\,N_1 \quad N_2 \quad N_3 \quad N_4 \quad N_5 \quad N_6\,]$$

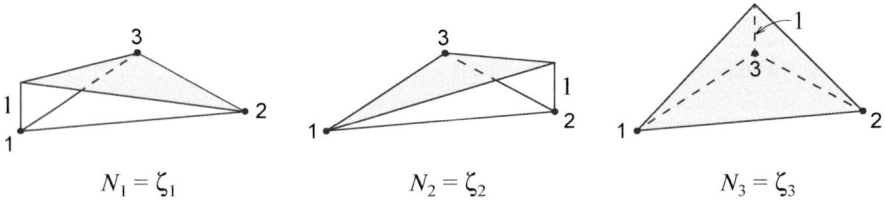

Copy of **Figure 5.30** Shape functions for the linear triangle

$$u = \psi(y, z) \tag{16-15}$$

and we continue to determine the displacement u for $\theta = 1$. The element *dofs* are the nodal point values of u or rather ψ, and as usual our first step is to assume

$$\psi = u = \mathbf{Nv} \tag{16-16}$$

where **N** contains the *shape functions* and **v** contains the element *dofs*, that is

$$\mathbf{v}^T = [\psi_1 \ \psi_2 \ \psi_3 \ \psi_4 \ \psi_5 \ \psi_6] \tag{16-17}$$

The shape functions form a second order homogeneous polynomial (which is equivalent to a complete quadratic polynomial) and they are the same as those used for the **LST** element in Chapter 11, see opposite page.

The first derivatives, in terms of **v**, becomes

$$\begin{bmatrix} u_{,y} \\ u_{,z} \end{bmatrix} = \begin{bmatrix} \mathbf{N}_{,y} \\ \mathbf{N}_{,z} \end{bmatrix} \mathbf{v} \tag{16-18}$$

From Eqs. (16-7a) and (16-7c) we find the shear strains (remember $\theta = 1$ and thus $u = \psi$),

$$\boldsymbol{\varepsilon}_\gamma = \begin{bmatrix} \gamma_{xy} \\ \gamma_{xz} \end{bmatrix} = \begin{bmatrix} \mathbf{N}_{,y} \\ \mathbf{N}_{,z} \end{bmatrix} \mathbf{v} + \begin{bmatrix} 0 & -1 \\ 1 & 0 \end{bmatrix} \begin{bmatrix} y \\ z \end{bmatrix} \tag{16-19}$$

The coordinates y and z can be expressed in terms of the corner coordinates **y** ($\mathbf{y}^T = [y_1 \ y_2 \ y_3]$) and **z** by linear interpolation functions $\mathbf{N}_1$ as

$$y = \mathbf{N}_1 \mathbf{y} \quad \text{and} \quad z = \mathbf{N}_1 \mathbf{z} \tag{16-20}$$

$\mathbf{N}_1$ contains the simple shape functions of Fig. 5.30. Hence

$$\boldsymbol{\varepsilon}_\gamma = \mathbf{Bv} + \boldsymbol{\varepsilon}_{\gamma 0} \tag{16-21}$$

where, see Eq. (5-29),

$$\mathbf{B} = \begin{bmatrix} \mathbf{N}_{,y} \\ \mathbf{N}_{,z} \end{bmatrix} = \mathbf{J}^{-1} \begin{bmatrix} \mathbf{N}_{,\zeta_1} \\ \mathbf{N}_{,\zeta_2} \end{bmatrix} \tag{16-22}$$

For the inverse Jacobian matrix of Eq. (5-29) x should be replaced by y and y by z. Furthermore

$$\boldsymbol{\varepsilon}_{\gamma 0} = \begin{bmatrix} 0 & -\mathbf{N}_1 \\ \mathbf{N}_1 & 0 \end{bmatrix} \begin{bmatrix} \mathbf{y} \\ \mathbf{z} \end{bmatrix} \tag{16-23}$$

represents the "initial strains" which will produce the "load" vector. The shear stresses (for $\theta = 1$) follow from Eqs. (16-7):

 **Adhémar Jean Claude BARRÉ de SAINT-VENANT** (1797-1886) was a French mechanician and mathematician; outside France he is known simply as ST. VENANT.

ST. VENANT entered the École Polytechnique at the age of 16 after taking the competitive examinations. He showed outstanding abilities and became the first in his class. But political events in 1814, a year after he entered the school, forced him to leave (in disgrace); all the students were mobilised to defend Paris, but the 17-year-old ST. VENANT refused to take part, saying: "My conscience forbids me to fight for an usurper ... ". He was banned for life from the prestigious school, and it was not till eight years later, after having worked as an assistant in the power industry, that he was permitted to enter the École des Pontes et Chaussées. For two years he bore the protests of the other students who did not want to have anything to do with him; however, he worked hard and graduated from the school as the first of his class. He then worked for the Service des Pontes et Chaussées until 1848 and later as *Professeur du génie rural* at the Agricultural Institute in Versailles, where he was involved with typical civil engineering work, always trying to improve the miserable living conditions in the countryside. He also gave lectures on strength of materials at the École des Pontes et Chaussées, and in 1868 ST. VENANT was elected a member of the Academy of Sciences and was the authority in mechanics at that institution to the end of his life.

ST. VENANT worked mainly on mechanics, elasticity, hydrostatics and hydrodynamics. One of his most known works, published in 1843, gave the correct derivation of the NAVIER-STOKES equations. In the 1850s he derived solutions for the torsion of bars with non-circular cross sections; he was the first to recognize that for such sections plane sections do *not* remain plane, the torsion produces *warping*, a discovery that led to his famous semi-inverse method. He also expanded the practical bending theory of NAVIER. In a presentation in 1860 ST. VENANT formulated the compatibility conditions of elastic theory for the first time and hence completed the set of equations: equilibrium conditions, material equations and kinematical relationships and compatibility conditions.

In an elementary discussion of pure bending of beams ST. VENANT formulated a principle which now carries his name; in popular terms ST. VENANT's principle says: Boundary disturbances (due to concentrated loads and/or displacement constraints) are local and will normally vanish a short distance from the point of "application".

ST. VENANT received more recognition abroad than in France. In a book by MOIGNO, in which ST. VENANT presented the history of the equations of elasticity in a most comprehensive way, the editor observes in the preface: "Fatally belittled in France of which he is the purest mathematical glory, M. de SAINT-VENANT enjoys a reputation in foreign countries which we dare to call grandiose".

$$\boldsymbol{\tau} = \begin{bmatrix} \tau_{xy} \\ \tau_{xz} \end{bmatrix} = \begin{bmatrix} G & 0 \\ 0 & G \end{bmatrix} \begin{bmatrix} \gamma_{xy} \\ \gamma_{xz} \end{bmatrix} = \mathbf{C}_\gamma \boldsymbol{\varepsilon}_\gamma = \mathbf{C}_\gamma (\mathbf{Bv} + \boldsymbol{\varepsilon}_{\gamma 0}) \qquad (16\text{-}24)$$

Defining a set of nodal point "forces" **S** corresponding to **v**, a straightforward (virtual work) approach results in the following element equation

$$\mathbf{S} = \mathbf{kv} + \mathbf{S}^0 \qquad (16\text{-}25)$$

where

$$\mathbf{k} = \int_A \mathbf{B}^T \mathbf{C}_\gamma \mathbf{B} \, dA \qquad (16\text{-}26)$$

is the element stiffness matrix, for which explicit expressions for the individual elements are readily obtained, and

$$\mathbf{S}^0 = \int_A \mathbf{B}^T \mathbf{C}_\gamma \boldsymbol{\varepsilon}_{\gamma 0} \, dA \qquad (16\text{-}27)$$

is the element "load" vector. Introducing the displacement vector for the complete assemblage (that is, the system displacements)

$$\mathbf{r}^T = [\psi_1 \ \psi_2 \ \psi_3 \ \psi_4 \ \ldots \ldots \ \psi_n]$$

standard assembly procedure gives the final equation

$$\mathbf{Kr} = \mathbf{R} \qquad (16\text{-}28)$$

This equation is solved subject to the *boundary conditions* which depend on symmetry properties:

- *No symmetry* – the entire cross section needs to be modelled. Since we are only interested in the relative displacements, it is sufficient to fix (set to zero) any one nodal displacement (*dof*).
- *Simple symmetry* – only half the cross section needs to be modelled. Since the warping is zero along the symmetry line, the displacement ($\psi = r_i$) is set to zero (fixed) at all nodes (*i*) on the symmetry line.
- *Double symmetry* – only on quarter of the cross section needs to be modelled, and the displacement is set to zero at all nodes on the symmetry lines.

Shear stress distribution and torsional stiffness. We have now determined, at each nodal point, $\psi = u(y, z)$ for $\theta = 1$. In order to determine the shear stresses and the torsional stiffness it is necessary to determine θ. The rate of twist depends on the torque M_x, which may be expressed in terms of the shear stresses as

$$M_x = \int_A (\tau_{xz} y - \tau_{xy} z) \, dA = \int_A [z \ y] \begin{bmatrix} -\tau_{xy} \\ \tau_{xz} \end{bmatrix} dA \qquad (16\text{-}29)$$

ST. VENANT torsion

$$\begin{bmatrix} z & y \end{bmatrix} \begin{bmatrix} -\tau_{xy} \\ \tau_{xz} \end{bmatrix} = \begin{bmatrix} -z & y \end{bmatrix} \begin{bmatrix} \tau_{xy} \\ \tau_{xz} \end{bmatrix}$$

$$\begin{bmatrix} -z & y \end{bmatrix} = \begin{bmatrix} -\mathbf{N}_1 \mathbf{z} & \mathbf{N}_1 \mathbf{y} \end{bmatrix} = \begin{bmatrix} -\mathbf{z}^T \mathbf{N}_1^T & \mathbf{y}^T \mathbf{N}_1^T \end{bmatrix} = \begin{bmatrix} \mathbf{y}^T & \mathbf{z}^T \end{bmatrix} \begin{bmatrix} 0 & \mathbf{N}_1^T \\ -\mathbf{N}_1^T & 0 \end{bmatrix}$$

scalar

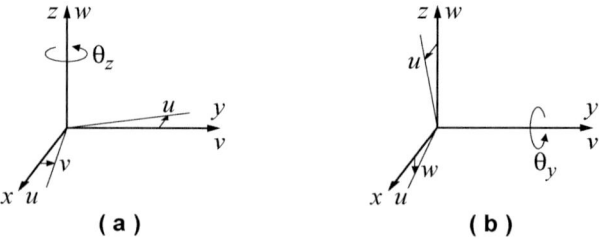

Figure 16.6 Rotations about principal axes

Finite element torsion analysis

For an arbitrary element, the shear stresses may, in view of Eqs. (16-11) be written as

$$\begin{bmatrix} \tau_{xy} \\ \tau_{xz} \end{bmatrix} = G\theta \left(\begin{bmatrix} \psi_{,y} \\ \psi_{,z} \end{bmatrix} + \begin{bmatrix} -z \\ y \end{bmatrix} \right) = G\theta \left(\mathbf{Bv} + \begin{bmatrix} 0 & -\mathbf{N}_1 \\ \mathbf{N}_1 & 0 \end{bmatrix} \begin{bmatrix} y \\ z \end{bmatrix} \right) \qquad (16\text{-}30)$$

Hence the contribution to M_x from this particular element becomes

$$\delta M_x = G\theta \int_{A_e} [\mathbf{y}^T \; \mathbf{z}^T] \begin{bmatrix} 0 & \mathbf{N}_1^T \\ -\mathbf{N}_1^T & 0 \end{bmatrix} \left(\mathbf{Bv} + \begin{bmatrix} 0 & -\mathbf{N}_1 \\ \mathbf{N}_1 & 0 \end{bmatrix} \begin{bmatrix} y \\ z \end{bmatrix} \right) dA \qquad (16\text{-}31)$$

At this stage the nodal values $\mathbf{v}$ of the warping function ψ are known, and the integral can be evaluated. By definition

$$\delta M_x = \theta(\delta GI_t) \qquad (16\text{-}32)$$

Hence

$$\delta GI_t = G \int_{A_e} [\mathbf{y}^T \; \mathbf{z}^T] \begin{bmatrix} 0 & \mathbf{N}_1^T \\ -\mathbf{N}_1^T & 0 \end{bmatrix} \left(\mathbf{Bv} + \begin{bmatrix} 0 & -\mathbf{N}_1 \\ \mathbf{N}_1 & 0 \end{bmatrix} \begin{bmatrix} y \\ z \end{bmatrix} \right) dA \qquad (16\text{-}33)$$

is the contribution to the torsional stiffness GI_t from one element. Thus

$$GI_t = \sum \delta GI_t = \frac{M_x}{\theta} \quad \Rightarrow \quad \theta = \frac{M_x}{GI_t} \qquad (16\text{-}34)$$

With θ known, Eq. (16-30) gives the shear stresses in an element.

Position of shear centre. It remains to find the position of the shear centre. MEEK [58] describes a procedure based on the following reasoning:

Equations (16-4b and c) assume the "in-plane" displacements to be zero at the origin, that is at the centre of area (C). However, this is only the case if the centre of area coincides with the shear centre (S). In general, S and C do not coincide, and if the coordinates of S, relative to C, are denoted by y_S and z_S, Eqs. (16-4b and c) should be rewritten as

$$v = -\theta x(z - z_s) \quad \text{and} \quad w = \theta x(y - y_s) \qquad (16\text{-}35)$$

indicating that v and w are zero at the shear centre, as they should be (by definition).

We now examine the expressions

$$v = \theta x z_s \quad \text{and} \quad w = -\theta x y_s$$

in view of Fig. 16.6:

$$v = \theta_z x \quad \Rightarrow \quad \theta_z = \theta z_s$$
$$w = -\theta_y x \quad \Rightarrow \quad \theta_y = -\theta y_s$$

531

ST. VENANT torsion

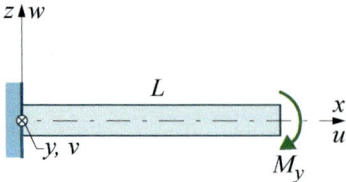

Figure 16.7 Cantilever subjected to a constant bending moment about y-axis

These rotations contribute to the warping, and the final warping displacement, with respect to principal axes (with origin at C), can now be written as

$$u(y, z) = \theta\psi(y, z) + \theta y_s z - \theta z_s y \tag{16-36}$$

It should be noted that the two additional terms will not affect the shear strains (defined by Eqs. (16-7)), nor the shear stresses.

In order to determine y_s we consider a cantilever beam subjected to a constant bending moment M_y which will produce an axial stress of the form

$$\sigma_x = E(y, z)cz$$

We subject the beam to a *virtual* twist $\tilde{\theta}$, about the shear centre, which is kinematically compatible with a virtual warping $\tilde{u}$ defined as

$$\tilde{u} = \tilde{\theta}(\psi(y, z) + y_s z - z_s y)$$

The principle of virtual displacements gives

$$M_y \cdot 0 = \int_V \sigma_x \tilde{u}\, dV = \tilde{\theta} L c \int_A E(y, z) z (\psi(y, z) + y_s z - z_s y)) dA = 0$$

Since y and z are principal axes we can now write

$$y_s = -\frac{\int_A E(y, z) z \psi(y, z) dA}{EI_y} \tag{16-37}$$

or, in terms of finite element notation,

$$y_s = -\frac{\sum E\mathbf{z}^T \int_{A_e} \mathbf{N}_1^T \mathbf{N} dA \mathbf{v}}{EI_y} \tag{16-38}$$

Similarly we find, by considering a constant moment about the z-axis,

$$z_s = \frac{\sum E\mathbf{y}^T \int_{A_e} \mathbf{N}_1^T \mathbf{N} dA \mathbf{v}}{EI_z} \tag{16-39}$$

16.3 Finite element shear analysis of prismatic beam

If we consider the cross section of Fig. 16.5, torsion is not the only problem. Finding the shear stress distribution due to shear forces V_y and/or V_z is not a trivial task, nor is it straightforward to determine the *shear deformation* factors (κ_y and κ_z) that define the so-called shear areas ($A_{sy} = A/\kappa_y$ and $A_{sz} = A/\kappa_z$) which in turn define the "average" or *mean* shear deformations ($\gamma_y^m = V_y/A_{sy}$ and $\gamma_z^m = V_z/A_{sz}$).

A detailed exposition of this analysis is beyond the scope of this presentation. However, since these tasks are solved by the **CrossX** program, we shall include a brief outline of how it is accomplished.

ST. VENANT torsion

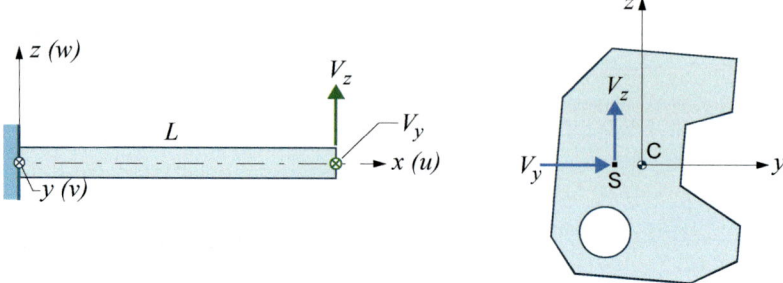

Figure 16.8 Cantilever, prismatic beam subjected to constant shear

Theoretical approach. The **CrossX** program uses a procedure described by MASON and HERRMANN [59]; they assume that the relationships between stress and stress resultants and between curvature changes and stress resultants for the simple case of a cantilever, prismatic beam subjected to an end load, approximately hold for all loading systems.

Figure 16.8 shows a prismatic cantilever with an arbitrary ("massive") cross section subjected to end loads V_y and V_z. The following assumptions are made:

- the cross section of the beam is uniform over its entire length,
- the x-axis is straight and coincides with the centroid of the cross section,
- the material is linearly elastic, continuous, homogeneous and isotropic,
- the body forces are small compared to the stresses and can be neglected,
- stresses $\sigma_y = \sigma_z = \tau_{yz} = \tau_{zy} = 0$, and
- the normal stress σ_x is distributed in the same manner as in the case of pure bending.

Furthermore the end loads, V_y and V_z, are assumed to be applied at the shear centre (**S**), which eliminates torsion, and the y- and z-axes are assumed to be *principal* axes. The latter assumption, which is not made in Ref. [59], simplifies expressions greatly without loss of generality (since we can always determine the orientation of the principal axes by a straightforward, auxiliary analysis).

Based on the above assumptions Ref. [59] quotes analytical solutions to the equations of three-dimensional linear elasticity theory for our cantilever problem. Using these solutions and the *principle of minimum potential energy,* MASON and HERRMANN [59] formulate a finite element procedure from which the shear centre position (y_s and z_s), the shear deformation factors (κ_y and κ_z) as well as the shear stress distributions are determined. The details of the derivation, which is somewhat lengthy, may be found in [59].

The upshot is that we end up with the same "stiffness" matrix as for the torsion problem, but the "load" vectors, of which there are now two (one for each of the two shear forces), are new. If we do not have symmetry, or do not take advantage of existing symmetry, the boundary conditions are the same for both the torsion case and the two shear cases, and the total problem can be solved with *one* factorization and *three* forward and backward substitutions.

However, if symmetry exists and is taken advantage of, we need to establish and solve Eq. (16-28) *three* times, each for one right-hand side. This because of different boundary conditions.

It should be noted that the position of the shear centre determined by the shear analysis coincides with the position determined by the torsion analysis only when POISSON's ratio is set to zero for the shear analysis. The shear centre position determined by torsion analysis is *not* affected by POISSON's ratio. From this we conclude that the shear centre position for a cross section

ST. VENANT torsion

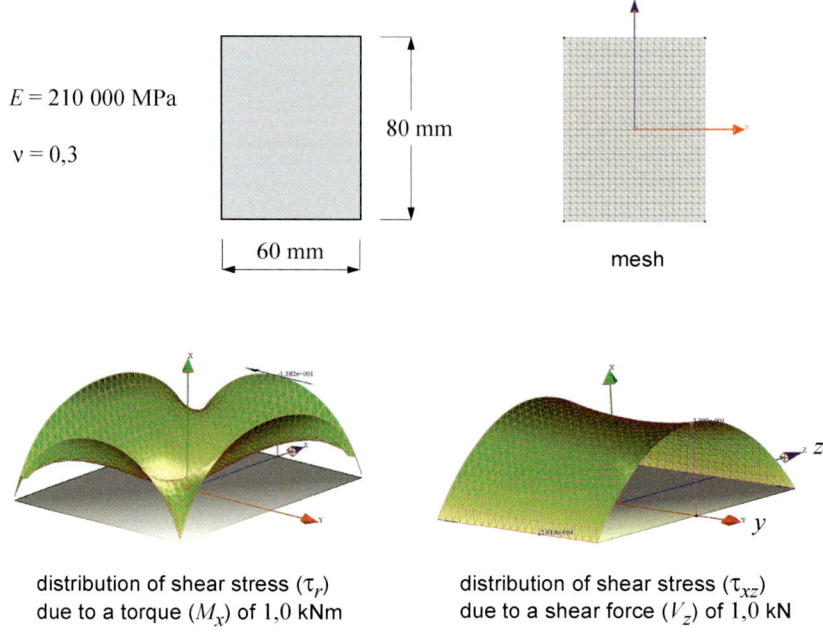

Figure 16.9 Massive rectangular (steel) section analyzed by CrossX

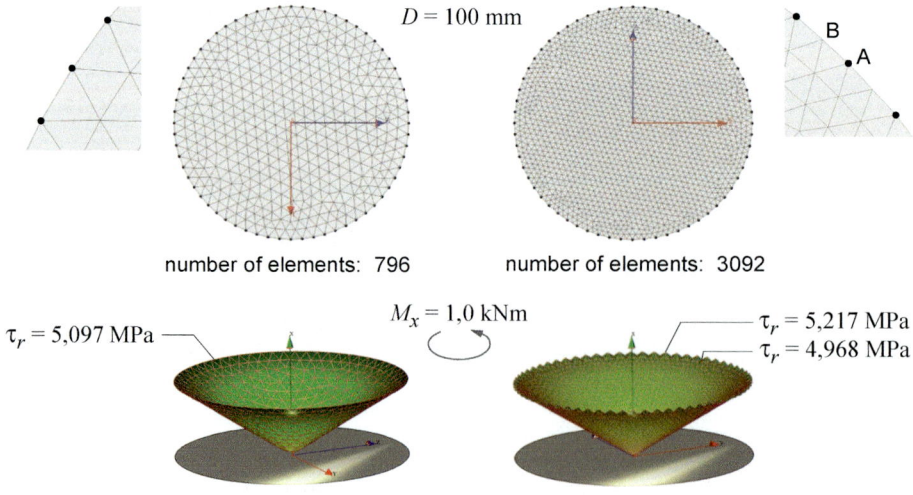

Figure 16.10 Shear stress (τ_r) distribution due to a unit torque (M_x)

composed of material(s) with arbitrary value(s) of POISSON's ratio should perhaps be determined by shear analysis, although the difference is moderate. Both options are available in **CrossX**.

16.4 Numerical examples

We have already, in Section 10.4, seen results obtained by the finite element method described in this chapter, see Fig. 10.8. Here we include a few more examples, and while the power of the method is most apparent in connection with geometrically complex sections, we shall concentrate on quite simple, almost trivial sections.

Rectangular, massive section. Figure 16.9 shows a simple, rectangular cross section (made of steel). For the indicated mesh the figure shows two distributions of shear stresses, one caused by a unit *torque* ($M_x = 1{,}0$ kNm) as the only non-zero section force, and the other caused by a unit *shear force* ($V_z = 1{,}0$ kN) in the direction of the principal axis z.

For the torsion case we find the maximum shear stress to be 15,45 MPa; this result is correct to all four digits (demonstrated by repeating the analysis using a finer mesh). For the torsion constant we find $I_t = 3{,}1183 \cdot 10^6$ mm^4 which is correct to all five digits.

The distribution of the shear stress (τ_{xz}), caused by a shear force $V_z = 1{,}0$ kN, is not exactly as we were thought in our first course in strength of materials. The parabola in the z-direction is as it "should be", but we might not have been told about the parabolic variation in the y-direction. The FEM analysis gives a maximum shear stress (at $y = \pm b/2$) of 0,3392 MPa (the correct value is 0,3391), whereas the standard formula $\tau_{max} = 3V/2A$ gives a value of 0,3125 MPa. The discrepancy here is caused by the POISSON effect. If we set ν to zero, the FEM analysis shows a distribution with no variation in the y-direction, and the value for τ_{max} is exactly that given by our well-known formula.

Another effect of POISSON's ratio is found in the values of the two shear deformation factors, κ_y and κ_z. For a rectangular massive cross section these are normally taken to be 1,2 (= 6/5); however, for a material with ν = 0,3 we find them to be 1,177. Again, if we set ν = 0, the FEM analysis gives exactly 1,2.

Massive circular section. Figure 16.10 shows the trivial case of a massive circular section with diameter 100 mm, analysed by two different meshes. In both cases the circle is approximated by 64 equal straight line segments in such a way as to represent the area of the circle as well as possible. For this section the torsional constant is equal to the polar moment of inertia, that is $I_t = I_p = \pi D^4/32 = 9{,}8175 \cdot 10^6$ mm^4; both meshes gives this value (correct to 5 digits). The exact maximum shear stress (τ_r) due to a given torque (M_x) is also easily found; for $M_x = 1{,}0$ kNm it is 5,093 MPa.

We see that the coarser mesh finds the maximum shear stress with very good accuracy, and the stress distribution is nice and smooth. For the finer mesh, however, the FEM analysis produces a "jagged" shear stress distribution along the circumference. This is quite typical for finite element analyses when curved boundaries are approximated by straight line segments. The

ST. VENANT torsion

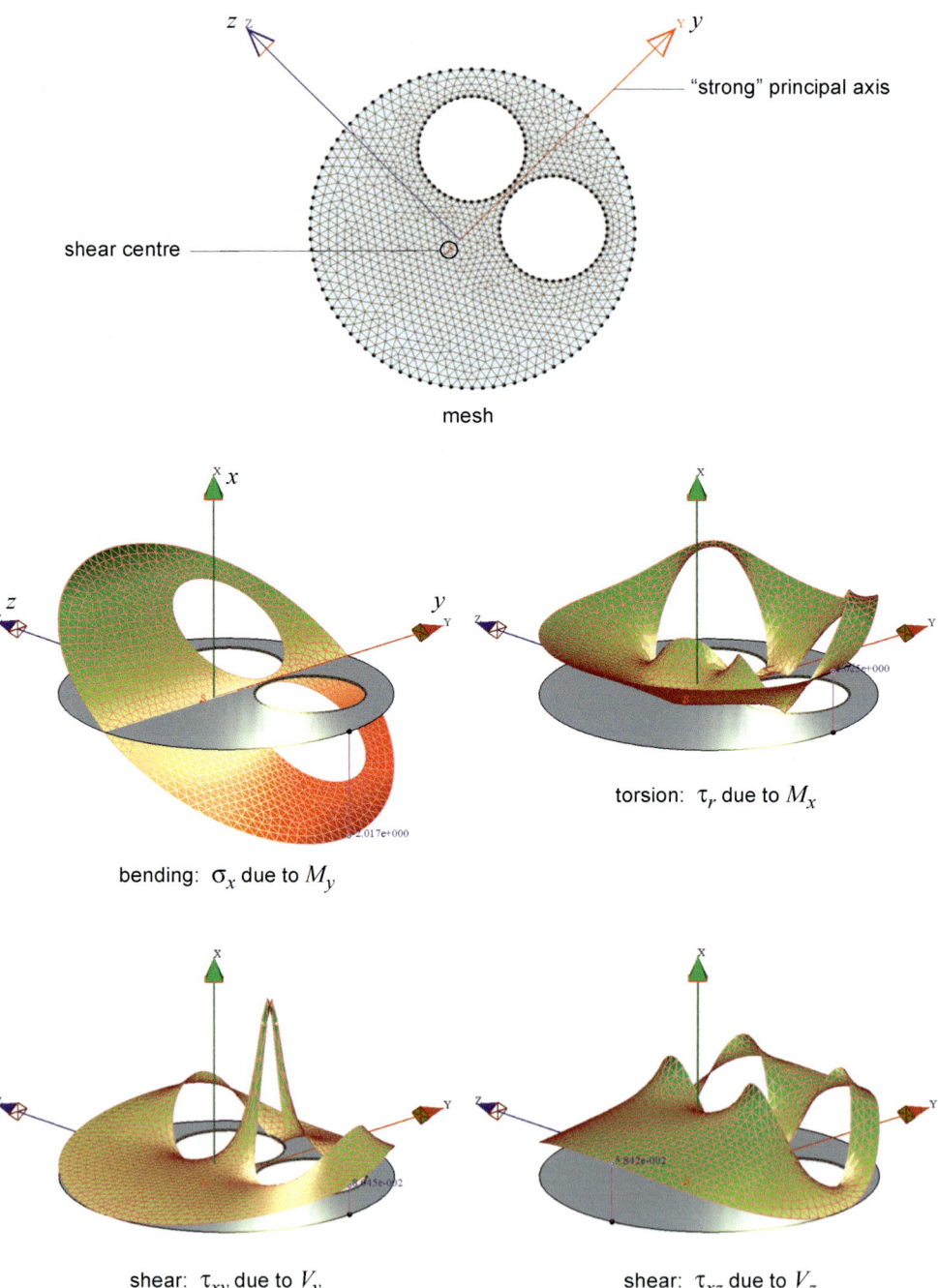

Figure 16.11 Stress distributions for a more complex cross section

difference between the two meshes of Fig. 16.10 is that the coarser mesh use only *one* element to represent the straight line segment, while the finer mesh uses two elements (see the enlarged mesh patches). Hence the finer mesh has two "types" of nodes, A and B, on the circumference, and the stresses computed at these two types of nodes are different; the highest stresses are computed at nodes of type B. Note however that the average of the stresses at points A and B is practically exact. The coarse mesh has only one "type" (A) of nodes, and if the mesh had been completely symmetrical (in a circular sense) the exact same stress would have been determined at each of theses nodes; as it is there is a slight difference (in the fourth digit).

The lesson here is that it is better to use more straight line segments to represent the circle and a mesh with one element side per line segment, than a very fine mesh on a "coarser" representation of the circle itself.

Figure 16.11 shows some stress distributions of a somewhat more complex cross section, a circular section with two (equal) "holes". For such a section a numerical solution comes into its own and is probably the only viable (or practical) solution. Even the axial stresses due to a bending moment (about a principal axis) is a fair challenge for a "manual analytical" solution for such a section.

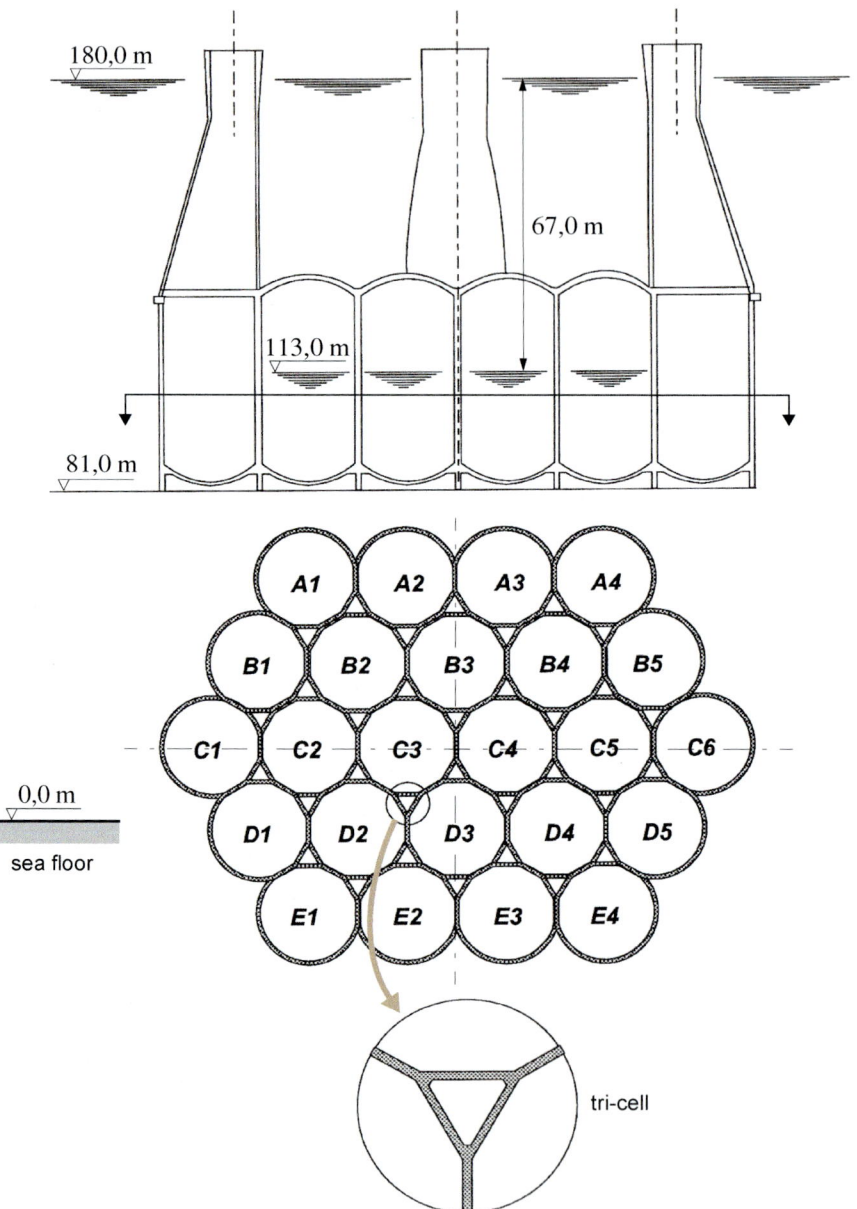

Figure 17.1 The Sleipner A platform – water levels at failure; the figure (slightly modified) is from Ref. [60]

17

Practical use of FEM

The finite element method, implemented in a well-designed program, is a powerful tool. But as most tools, the method can, if not properly used, be an accessory to disaster, as we shall see an example of in this chapter. The main focus of the chapter, however, is the positive potential of the method, and we shall try to come up with some guidelines for its safe and efficient use; not an easy task as there are no hard and fast rules. A major difficulty is that, almost regardless of how we model our problem, the program will provide an answer, most likely presented in attractive colours and with numerical results with an apparently high accuracy (if we simply judge it by the number of digits). Is this a viable result, or merely a pretty picture? And if viable, how good are the results?

17.1 The Sleipner accident

Very early in the morning on the 23rd of August 1991, the reinforced concrete substructure of the Sleipner A platform sprung a leak during a controlled ballasting test outside Stavanger in southwest Norway, and, about eighteen and a half minutes after the first noticeable rumble, it ended up as a heap of crushed concrete and twisted reinforcing bars on the sea floor 180 meters below sea level. The impact registered 3 on the Richter scale at seismological stations in Norway [60].

Fourteen people were on board when the accident happened; they were all rescued and none suffered more than wet feet. That was the good news; the bad news was that some two billion Norwegian kroner were lost (in today's money about 700 million US dollars), and the event jeopardized a very big gas delivery contract.

The gravity base structure (GBS) of the Sleipner A platform is shown in Fig. 17.1, with water levels as they were at the time of the accident. The structure is of the **Condeep** type, its base (caisson) consists of 24 cells with a total base area of 16 000 m^2. Four of the cells, B3, D3, C1 and C6, continued upwards as support and utility shafts for the platform deck. Triangular spaces between the main cells are denoted tri-cells. The total concrete volume was about 75 000 m^3 and the amount of steel was about 19 000 tons.

Practical use of FEM

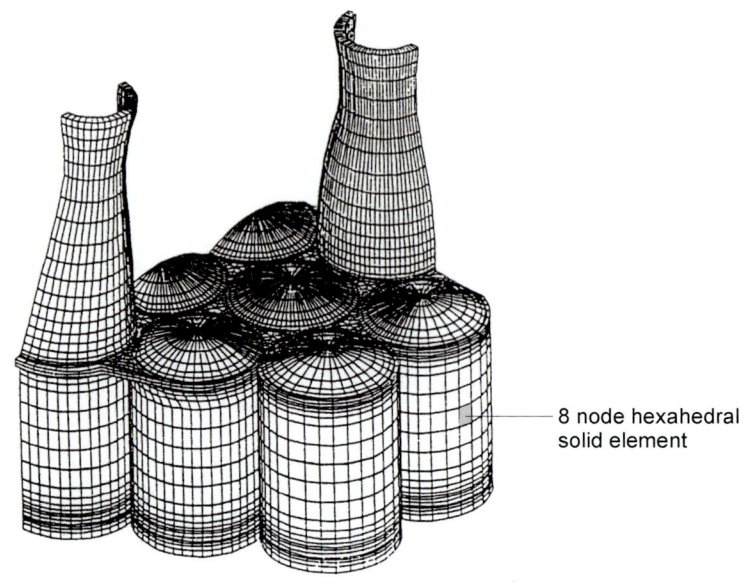

Figure 17.2 Finite element mesh of a quarter of the Sleipner A GBS

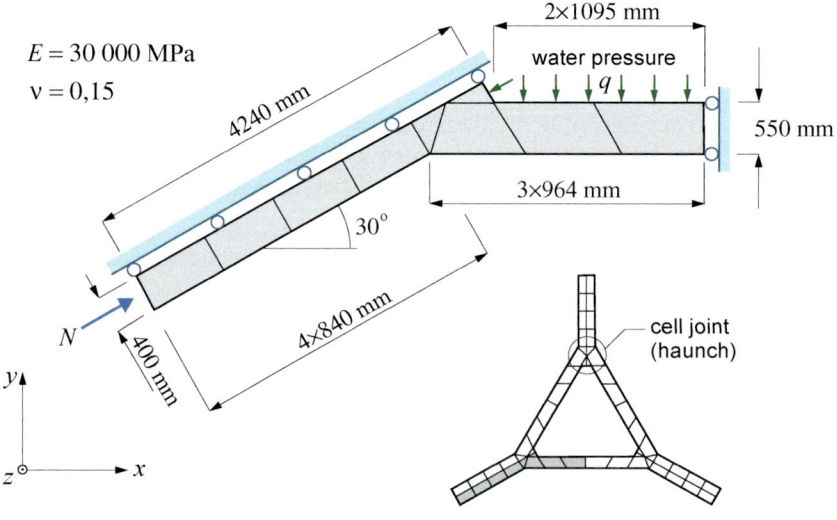

Figure 17.3 Mesh generated for tri-cell – planar model

Because of the gas delivery commitments it was paramount to establish quickly what had gone wrong, and immediately following the accident, on the very same day, the operator of the Sleipner Field and the designer-cum-contractor of the GBS each established independent investigating teams to determine the probable cause of the disaster.

The investigating teams were able to piece together a likely sinking scenario based on observations made on board the platform, from the moment the first rumble or bang from the utility shaft D3 was heard until the platform disappeared and hit the sea floor. Careful examination of the design revealed only one weak area, the walls of the tri-cells and their connection to the main cell walls.

A 500 mm diameter hole through the concrete slab covering the top of the tri-cells meant that the tri-cells were filled with water during the ballasting test. The maximum pressure on the tri-cell walls at the time the accident started was therefore caused by a 67 m water head, see Fig. 17.1. Both investigating teams came to the same conclusion: *the failure was caused by rupture of a tri-cell wall*.

The failure of the tri-cell wall was a shear failure caused in part by an unfortunate FEM analysis that underestimated the shear forces, and in part by inadequate reinforcement. Our main concern is the FEM analysis. Figure 17.2 shows an overview of the finite element mesh of the so-called global FEM analysis, performed by the program system MSC/NASTRAN (version 65 C). A quarter of the GBS, which is symmetric about two orthogonal axes, was modelled by 8 node hexahedron elements (CHEXA-8). This element is the 3D version of the modified and incompatible 4 node quadrilateral plate element described in Section 11.2 (and labelled **Q6**).

A two-dimensional picture of the mesh used in the area of failure is shown in Fig. 17.3 (where symmetry is used to the utmost). When one looks at this mesh today, a couple of questions arise; why such a coarse mesh (with such a "simple" element) and why the unnecessary distortion of the elements in the tri-cell wall? With regard to the number of elements, one must keep in mind that this was 1990-1991 – even supercomputers at that time were no match for a PC of today – and the element shape was generated by the program. Normally distorted elements were adjusted manually, but, for some unknown reason, not in this case.

However, the major error was made in the interpretation of the results. The shear force over the wall thickness was estimated based on the shear stress computed at the element centre. Thus the FEM analysis produced the shear force at four sections of the tri-cell wall (two at each half). These values are indicated by the circles in Fig. 17.4a. While not perfect, these four values are not too bad (considering the two questions raised above). The real problem was the parabolic extrapolation of the computed values used to obtain the values at the ends of the wall, see the solid line in Fig. 17.4a; compared with the correct shear forces, namely those obtained by simple beam theory, from which a maximum shear force of $qL/2$ is found, we see that the shear forces at the ends of the tri-cell walls were underestimated by some 40 to 45 per cent.

Practical use of FEM

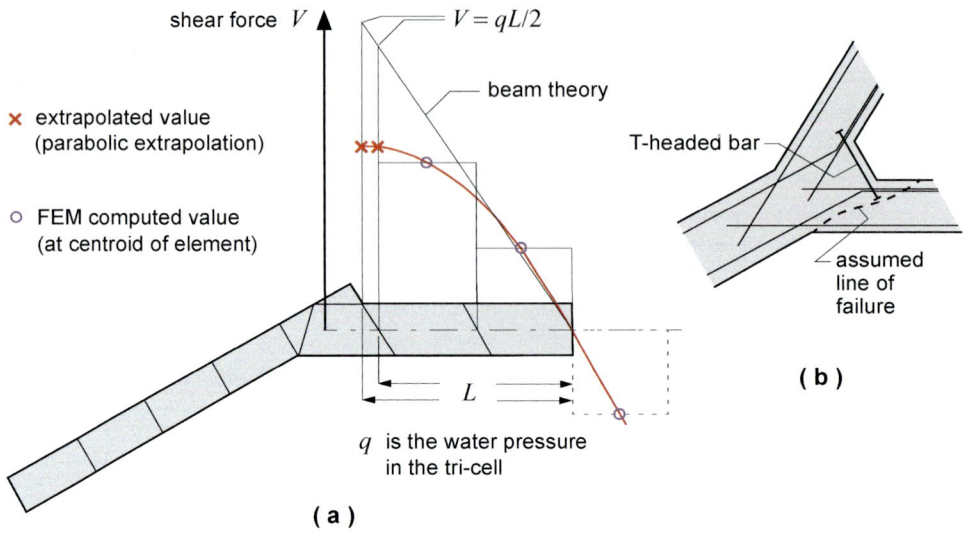

Figure 17.4 Computed shear force in tri-cell wall (**a**), and horizontal reinforcement at corner of tri-cell (**b**)

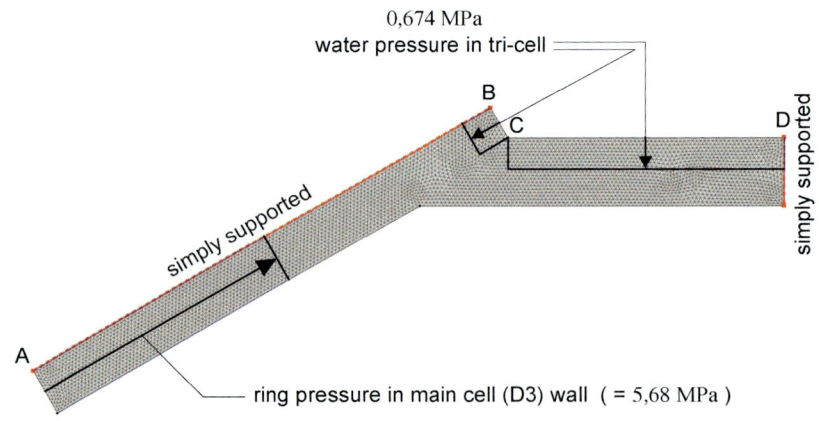

Figure 17.5 A two-dimensional (**FEMplate**) model of the critical detail; a relatively fine mesh of approximately 7300 **LST** elements

544

Figure 17.4a also shows that if a linear (least square fit) extrapolation had been used, the critical shear force would not have been underestimated by much more than 10 per cent. Axial stresses due to bending were also underestimated, but not nearly as badly as the shear stresses, and the levels of the axial stresses were not considered to be critical. Figures 17.3 and 17.4 are adaptations of similar figures in Ref. [60].

While the unfortunate finite element analyses – using distorted elements and above all a poor post processing of the results – contributed to the failure, this alone could certainly not explain it. The load situation at the time of failure was about 90 per cent of the design load and the design safety margin should easily have accommodated shear forces 30 (= 40 − 10) per cent higher than the design shear force. The real culprit was inadequate shear reinforcement.

The design aspect, including the reinforcement, is beyond the scope of this book. However, in very simple terms, the sketch in Fig. 17.4b indicates the problem. The T-headed bar, a reinforcing bar on to which a steel plate was fastened at each end, was *too short*, and there were no stirrups in the haunches of the cell joints. The figure also indicates where it is assumed that the wall failed. More information about the accident can be found in Refs. [60 and 61].

In the aftermath of the accident a fair number of analyses and also some full scale experiments of critical details were carried out and all results confirmed that the failure would have had to have been a shear failure at the support of the tri-cell wall. The economic issues in the wake of the accident were settled out-of-court, and it is interesting to note that the designer-cum-builder of the failed structure was asked to rebuild a new GBS to the same specifications as the first one. Time was of the essence and it should also be kept in mind that a huge and expensive deck superstructure had been made ready for mating with the GBS when the accident occurred. The new GBS, considerably better reinforced, was finished in record time and was ready for mating with the deck structure on May 1st, 1993. It has served the operator of the Sleipner field well for nearly 20 years.

For interest we have carried out a two-dimensional *plane strain* analysis, with program **FEMplate**, using a relatively fine mesh of **LST** elements, see Fig. 17.5. The loading is the assumed failure loading – the ring pressure in the main cell wall is due to the external water pressure on the GBS (which is approximately constant at a given horizontal plane).

The distribution of the three main stress components are shown in Fig. 17.6. Even disregarding the singularity at point C we see that the stress situation in the cell joint area is complex, and we find quite large values of all three stress components in this area. This analysis also confirms, not surprisingly, that ordinary beam theory predicts the shear stress distribution in the tri-cell wall very well.

So how could the Sleipner GBS accident happen? A good many people have asked that question. The designer-cum-builder was renown for its quality assurance system and the independent controls were conducted by well reputed bodies. And yet, obvious mistakes slipped through. However, looking

Practical use of FEM

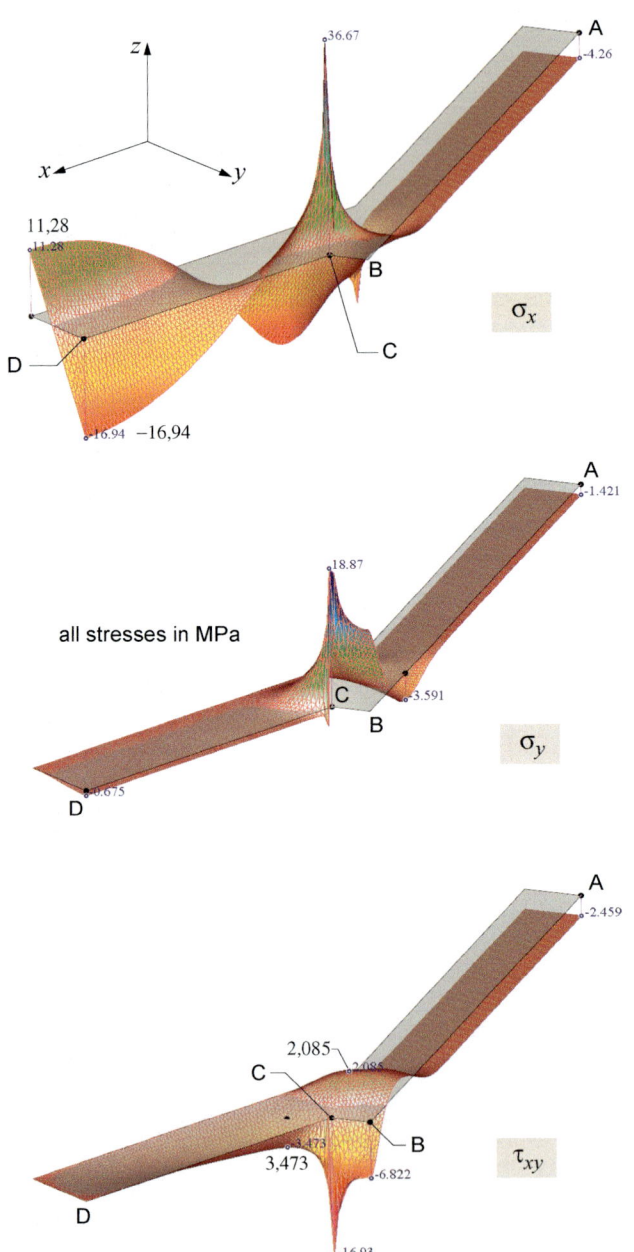

Figure 17.6 Stress distribution in the critical area of the tri-cell wall of the Sleipner A platform (**FEMplate**)

at the accident today, some 20 years later, it is only fair to point out some mitigating circumstances.

The establishment of a finite element model of that size and complexity in 1990 was a major undertaking that employed a large number of people for quite some time. Each individual focused on his or her part of the problem and it was difficult to see the bigger picture; remember that at that time computer graphics were in their infancy. Another important aspect of the modelling was the loading, and to many this was their main focus. A very large number of load cases had to be correctly modelled and combined into relevant load combinations, and although special programs had been developed to perform design checks for all these combinations at all possible critical sections, the amount of information was formidable. It was difficult to know which results out of this mass of results should be looked at more closely?

Nevertheless, these mistakes should have been picked up, and most likely would have been, if experienced engineers had been more involved in the design process. After the accident it was said that "Sleipner was a case of over-analysis and under-design". Perhaps we also see an element of complacency here? After all, this was number 12 in a series of Condeep platforms installed in the North Sea, and all the previous 11 platforms of this design had been successful. And the Sleipner GBS was, in comparison, a "simple" case; the water depth on site was only 82 m and it was therefore the first of its type to be placed in a water depth less than 100 m.

What are the "FEM-lessons" to be learnt here? Perhaps the most important are not to loose sight of the bigger picture and to exercise engineering judgement. We shall refer more specifically to this example in the next section. We conclude this section by mentioning that the Sleipner accident also made us at the university take a good look at our teaching, and some adjustments were made in our basic finite element course; more emphasis was put on the practical use of the method.

17.2 Advice and guidelines

The finite element method is a very powerful tool, but only in the hands of knowledgeable users. It is an approximate method which in itself requires insight, and while it can be made almost foolproof for very simple problems, most engineering problems need to be "prepared" before they can be analysed by a FEM based program. And equally important is the assessment of the results produced by the analysis.

Before we continue it is important to make it absolutely clear that whatever program you use to analyse your problem, *you and you alone are responsible for the results obtained and the use you make of these results in the design process*. Even in the unlikely situation that erroneous results are produced by a program error or by unfortunate or unclear program documentation, the small print you agreed to when you installed the program will most likely keep the program vendor in the clear.

Know your program and your own limitations. There are quite a number of FEM based programs on the market, from the large and powerful, *general*

Practical use of FEM

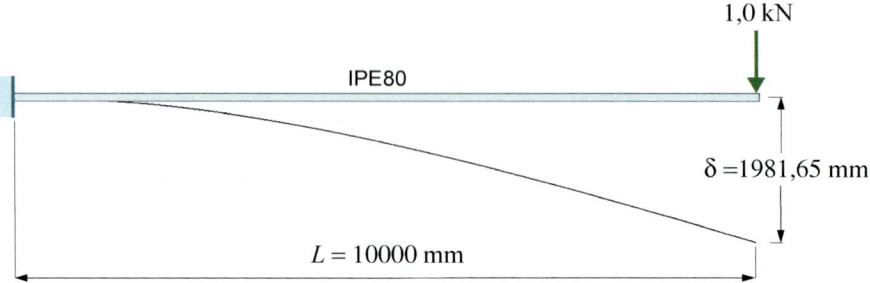

The cantilever is divided into n equally long beam elements

n	δ [mm]	n	δ [mm]
100	1981,65	100	1981,65
500	1981,65	500	1981,65
1000	1981,65	1000	1981,64
1500	1981,02	1500	1980,68
1600	1981,85	1700	1981,70
1700	1981,64	2000	1981,70
1800	1981,09	2500	1978,73
1900	1982,70	3000	1968,78
2000	1982,80	4000	1982,12
2100	1984,23	5000	1974,01
2150	1981,52	6000	1858,78
2154	1976,60	7000	1982,88
2155	"singular"	10000	1987,07

Figure 17.7 Tip deflection of a "soft" cantilever beam for different numbers of elements and different node numbering

Advice and guidelines

purpose program suites, such as **ABAQUS, ANSYS** and the like, to versatile frame type programs and easy to use *special purpose* programs, *e.g.* a simple cross section program like **CrossX** [57]. Whichever of these programs you have at your disposal as a fresh structural engineer just out of university, it will pay handsomely for you to invest time in learning how to use them effectively.

Familiarise yourself with the programs by using them on simple problems with known solutions. Check out the different elements that are available and how they respond to different element meshes. How do the programs handle different types of loading and different types of boundary conditions, such as displacement releases? In short, what are the possibilities and, equally important, what are the limitations? The large general purpose programs will have many features that we have not touched upon in preceding chapters; be careful when you use unfamiliar features, and if in doubt seek help and advice from more experienced colleagues.

It is also advisable to "push" the program, and learn how it responds to erroneous or questionable input. For instance, how fine can you mesh before numerical inaccuracy starts to creep into the solution? The cantilever beam in Fig. 17.7 may serve as an example. How many (ordinary) beam elements of equal length can we divide this rather "soft" beam into before we run into numerical difficulties? The problem is a 10 m long IPE80 steel beam subjected to a concentrated tip load of 1 kN; it should be noted that the displacement shown in the figure is drawn to scale. The "exact" beam theory tip displacement (to 6 digits) is 1981,65 mm.

We have analysed this problem with a program, **fap2D** [9], that uses a direct skyline solver, and all floating point arithmetic is carried out in *double precision* (about 15 significant digits). The solver checks for *diagonal decay* (see Section 9.11), and the parameter ε in Eq. (9-80) is set to 10^{-10}. If, during factorization, the diagonal decay test is not satisfied, the program terminates with the message: singular or near-singular stiffness matrix. Figure 17.7 has two columns of results (for the tip displacement δ); the one to the left is obtained for a numbering of the nodes (and the equations) starting with the clamped node and ending with the tip node, while the column to the right is associated with the opposite numbering, the tip *dofs* are the *first* equations.

When the tip *dofs* are numbered *last* we see that the result of the diagonal decay test is that **K** is "numerically singular" when the number of elements exceeds 2154. However, when we reverse the numbering such that the tip *dofs* are the *first* equations, the stiffness matrix passes the decay test even for 10 000 elements! In other words, the diagonal decay test is not entirely "objective". Considering how the test is carried out during factorization of the stiffness matrix, this is not all that surprising. For the numbering to the left in Fig. 17.7, the test fails at the very end of the factorization process, when the inaccuracies have had time to develop; the physical reason for numerical problems here is the large rigid body displacements at the tip compared to the *deformation* of the elements in this region. If the factorization process starts with the "critical" tip *dofs* and moves towards the more "stable" *dofs* at the clamped end, the numerical "contamination" will be far less. COOK *et al* [52] have a similar example (in Section 9.4).

Practical use of FEM

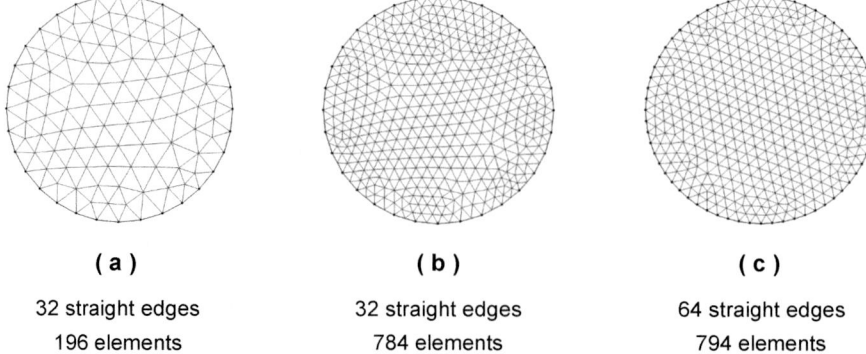

(a)
32 straight edges
196 elements

(b)
32 straight edges
784 elements

(c)
64 straight edges
794 elements

Figure 17.8 Representation of a circular area by elements with straight sides

A careful examination of the results shown in Fig. 17.7 reveals some strange effects that are not easy to explain. Consider for instance the noticeable drop in accuracy, for both directions of numbering, for $n = 1500$; in both cases the tip deflection is determined with much greater accuracy for $n = 1700$. And in the column to the right we see that the accuracy is greater for $n = 7000$ than for $n = 5000$, and far greater for $n = 10\,000$ than for $n = 6000$!

Regardless of the findings in this example, our experience with the diagonal decay test is good, and we believe it is a fairly robust and useful test.

From structure to FEM model. This is perhaps the most critical phase of any finite element analysis. If the problem at hand is large or complex, it may pay to start with a simplified (*e.g.* frame-type) model with the purpose of getting a "feel" for the structure and uncovering critical areas. This exercise also forces the engineer to capture the main load-bearing features of the structure, an insight that may be very useful when it comes to interpreting and checking results computed by a comprehensive model composed of 2D and/or 3D finite elements.

Whatever the real physical problem, the first step will always be to make some simplifications and idealisations, with regard not only to geometry, but also to material properties and the loading. This important phase requires both engineering judgement and "mechanical insight", both qualities that need experience in addition to education and training. Depending on the complexity of the problem, the inexperienced engineer should therefore have his or her solution of this phase checked by or at least discussed with an experienced colleague before proceeding to the FEM modelling.

Choice of element and meshing

Having established the mathematical model of the problem, the next step is usually the meshing. A well-designed FEM program will offer a choice of elements and automatic mesh generation. As to which element to choose, it is tempting to say a simple one, together with a fine mesh, but circumstances may suggest a different choice. In any case it is important to check the mesh and make sure that the mesh density reflects the assumed stress gradients (finer mesh in areas of expected large stress gradients) and, perhaps more important, to check that the element shapes are acceptable (not too distorted). As a user you can probably influence the mesh in various ways, for instance by establishing the problem domain as a collection of subregions of simple geometrical shapes and thus "helping" the mesh generator produce "well-shaped" elements; you can probably also specify the areas where you would like to have a denser mesh. Another possibility is that your program may provide *adaptive* meshing combined with *error estimation*, which means that the program will automatically refine the mesh in required areas; such features should be used with care.

Figure 17.8 shows two different representations of a curved boundary by straight line segments. In Fig. 17.8a the circle has been approximated by 32 straight and equally long line segments, and the automatically generated mesh has *one* element edge per line segment. The mesh in Fig. 17.8b is obtained by uniform subdivision of the mesh in Fig. 17.8a, and we see that each line segment representing the circumference of the circle now has a node

Practical use of FEM

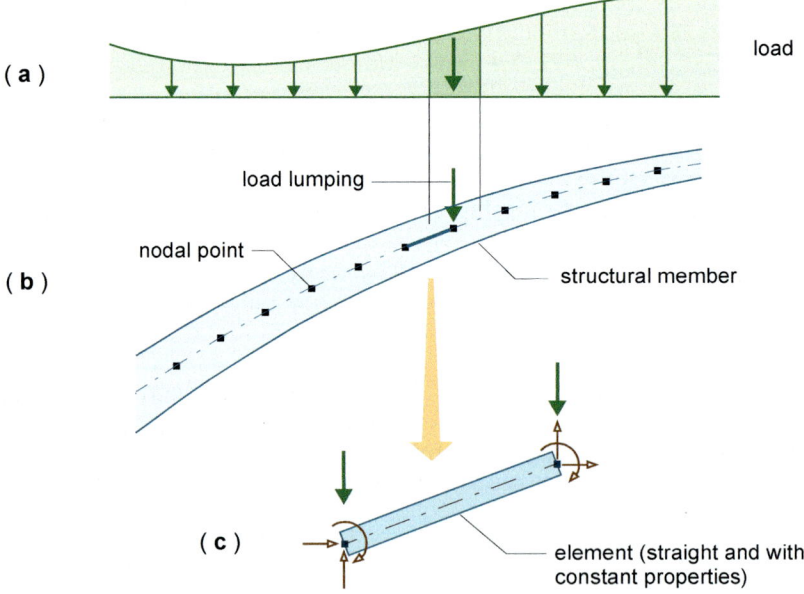

Figure 17.9 Load lumping

also at its midpoint; two colinear element edges make up the line segment. A similar mesh with almost the same number of elements is shown in Fig. 17.8c; this mesh is obtained by automatic meshing of a circle approximated by 64 equal and straight line segments. This last mesh is better than the one in Fig. 17.8b. Not only does it approximate the circle better, we also see that the mesh as such (in terms of element shapes) is better, and perhaps most important, the mesh in Fig. 17.8c does *not* give the irregular stress patterns along the boundary that the mesh in Fig. 17.8b gives; for an example of this see Fig. 16.10. In fact the much coarser mesh of Fig. 17.8a will give a much better "looking" stress picture and, depending on the element type, will not be much less accurate than that in Fig. 17.8b (both use the same number of straight line segments to approximate the circle).

Would isoparametric elements with curved edges have been a better choice for the circle in Fig. 17.8? It is hard to say; the problem of irregular stresses along the boundary, exemplified by Fig. 16.10, would disappear, but curved element edges, in general, do impair the element behaviour, and if such elements are used their edges should not deviate much from the straight line.

Before we leave problems associated with curves attention should also be paid to the arch problem described in Section 15.1, and to the steel roof truss in Fig. 15.8. Both these cases are essentially concerned with proper meshing, even if they deal with one-dimensional structural members.

For linear static problems it is tempting to recommend the following meshing strategy: Generate a rather coarse mesh of elements with acceptable shapes, perform an analysis and make a note of the results, with special attention to areas of "steepish" gradients. Refine the mesh, preferably by uniform subdivision, and repeat the analysis. Check the results and if necessary repeat the process, several times if required. Current computational capabilities favour this "sledgehammer" approach.

The Sleipner accident described in the previous section is an example where a trained eye would have picked up the unfortunate (and unnecessary) shape of the elements in the tri-cell walls, see Fig. 17.3. This could have reduced the error significantly, and possibly prevented the accident.

Loading

The definition of appropriate load cases and load combinations is perhaps the most important part of the load handling. Our concern, however, is to transform a given load cases into concentrated forces acting at the nodal points, and to make sure we include the correct amount of loading.

In a finite element context, the handling of concentrated external forces is straightforward. For distributed loading we have, in principle, two options: the *consistent* approach and the *lumping* approach; the latter is exemplified by Fig. 17.9. Your program may or may not give you these options – the rule seems to be that most programs implement the consistent approach for all but the simplest element types. If you use fine meshes there is not much in it, regardless of the type of element you use; but if you use lumping of distributed loading for higher order elements, with nodal points on element edges/surfaces, the lumping should only result in forces at the corner nodes. This is particularly important if hierarchic *dofs* are present. For higher order

Practical use of FEM

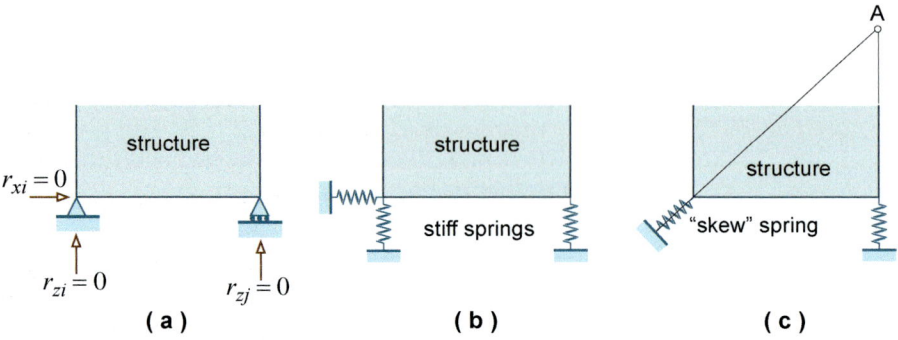

Figure 17.10 *External* boundary conditions

554

elements and relatively coarse meshes the consistent approach is the obvious solution in most cases, but we have also seen that for some problems, *e.g.* the arch modelled by straight beam elements, load lumping, as shown in Fig. 17.9 and also described in Section 15.1, is the answer.

Boundary conditions

In a *displacement* based finite element analysis the boundary conditions are almost exclusively defined in terms of specified *kinematic* (geometric) information about the structure. For some higher order elements, such as the **T18** plate bending element (Section 14.3) it is also possible to specify *mechanical* boundary conditions, but, in general, that is not recommended.

Schematic pictures of typical *external* boundary conditions are shown in Fig. 17.10. The most common way to account for the support of the structure is to suppress or *fix* (set to zero) the appropriate *dofs* at the supported nodal points as shown in Fig 17.10a. In order for the system stiffness matrix (**K**) to be *regular* we need to specify enough *dofs* to prevent, or perhaps more precisely to limit, the *rigid body* motions of the structure. If your program responds with a message indicating a *singular* **K**, rigid body motion(s) may well have caused this singularity. Remember that you also need to prevent rigid body motion in an "unloaded" direction; even if all loading on the structure in Fig. 17.10a is vertical, you still need to specify $r_{xi} = 0$.

Figure 17.10b indicates an alternative to the suppressed *dofs* of Fig. 17.10a; that is, to insert (stiff) boundary springs in the directions you wish to prevent motions. By adjusting the spring stiffnesses, through some iterative procedure, or by use of (large) artificial loads (se Fig. 9.17), boundary springs can also accommodate specified, non-zero displacements at specified nodal points (such as uneven settlements of supports).

Be aware of "skew" springs, that is springs that are *not* parallel with the axes defining the direction of the *dofs*. Such a spring is shown in Fig. 17.10c. This spring *cannot* be made to have the same effect as the two (orthogonal) springs at the same point in Fig. 17.10b. If this point only experiences a vertical displacement the horizontal spring in Fig. 17.10b remains inactive (no force in this spring). On the other hand, a vertical displacement of the same point in Fig. 17.10c will also produce a horizontal force component at the point. Furthermore, if the two springs in Fig. 10c are the only means of support for the structure, we are faced with a singular problem; these two springs cannot prevent rigid body rotation about point A (in small displacement, linear theory).

Boundary springs are normally stiff, in order to prevent a particular displacement. How stiff, can sometimes be a problem. If the spring "direction" coincides with the direction of a specific *dof*, its stiffness will only influence a diagonal term of the stiffness matrix **K**, and you can safely assign it a large stiffness. A linear boundary spring that does not coincide with the direction of a *dof* will offer resistance to more than one *dof* and its stiffness will also contribute to off-diagonal (coupling) terms in **K**. If they are very large, these off-diagonal terms may cause numerical problems. A simple way to avoid

Practical use of FEM

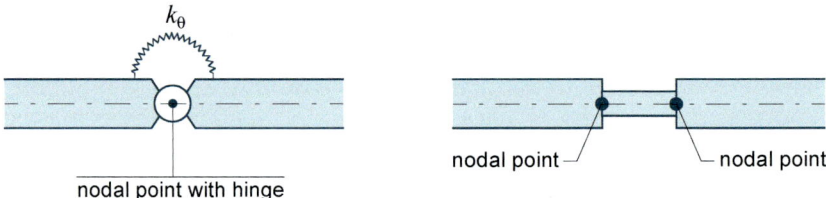

Figure 17.11 Different models of a semi-rigid joint

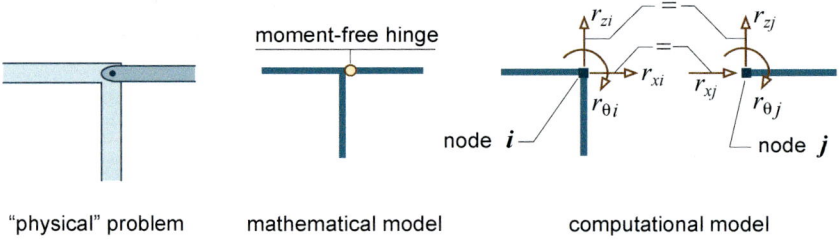

Copy of **Figure 9.9** Modelling a moment release in a plane frame structure

such difficulties is to transform the *dofs* at the particular node such that one *dof* coincide with the "skew" spring.

Be careful with *soft* springs, whether boundary springs or coupling springs. In combination with an otherwise stiff structure such springs may cause numerical problems of a nature similar to those described for the cantilever of Fig. 17.7. In general, if you mix soft and stiff elements you must be on the lookout for numerical problems (ill-conditioning). Sometimes it is better to replace a very stiff element by a completely *rigid* element and thus eliminate one or more *dofs* as *slaves*.

Depending on how your program implements boundary conditions – whether by elimination of (explicitly or implicitly) specified *dofs*, penalty functions (*e.g.* springs) or LAGRANGE multipliers – the real challenge is to model the so-called *internal* conditions, often defined by some kind of constraint equation in which a particular (*slave*) *dof* is defined as a linear combination of other (*master*) *dofs*. In structural engineering, *semi-rigid joints* represent this kind of problem. A simple version of such a joint is shown in Fig. 17.11.

The standard solution to the left in Fig. 17.11 is realized in two steps; first the moment-free hinge is inserted, implemented for instance by the master-slave approach of Fig. 9.9, and secondly a rotational spring is coupled to the two independent rotational *dofs* at the joint. The engineer's problem is to assign a suitable stiffness (k_θ) to the spring – not always an easy task. If your program does not have the facilities necessary to accomplish this solution you may have to improvise, for instance as suggested to the right in Fig. 17.11. The semi-rigid joint is replaced by a short element with a lower bending stiffness; the fictitious element is rigidly connected to the main members at both ends. This approach may be tuned to give practically the same solution as the hinge with a spring approach, but it may be as difficult to "size" this type of joint as it is to pick the "right" spring stiffness.

Interpretation and assessment of the results. Program errors cannot be completely ruled out, but it is seldom you hear about such errors these days, and here we take the position that the program will analyse correctly the problem we prepare. It then remains to check and interpret the results.

The first question we ask ourselves is: *are the results reasonable* in view of the applied loading? A well designed program will offer a number of features that facilitate this phase, the most important being good graphical presentation of the results. Checklist:

- √ Inspect the *displacements*, both form and size; keep in mind that most programs will scale the displacements so as to make the numerically largest displacement equal to, say 10 mm on the screen – it may be very small or very large, so check the numbers. Are the displacements compatible with linear (small displacement) theory?
- √ If your problem is symmetric, make sure your results are also symmetric.
- √ Check stress plots or section force diagrams for unreasonable results.

Practical use of FEM

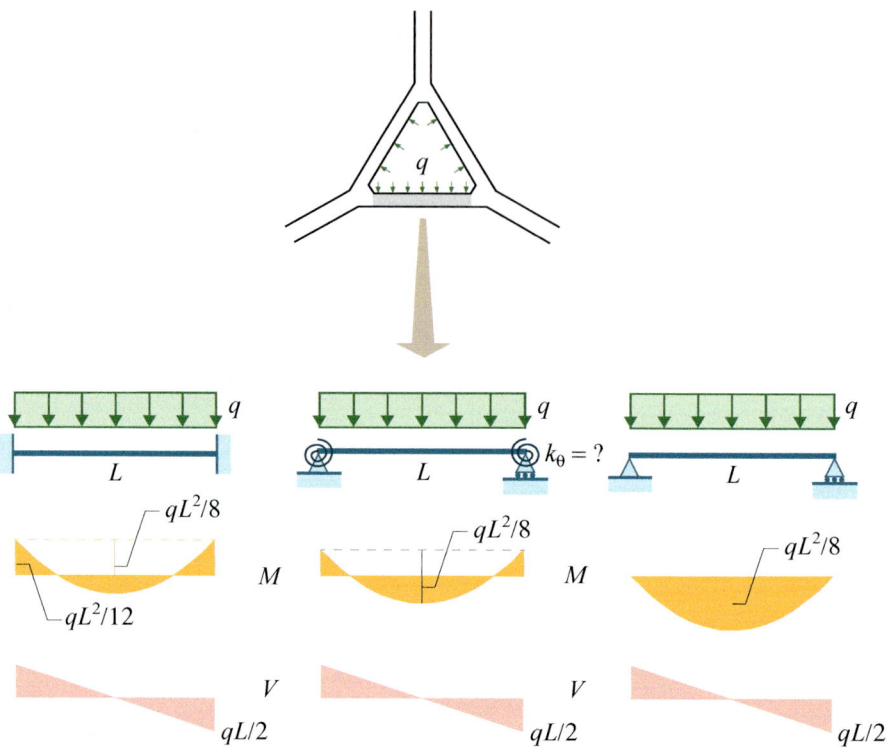

Figure 17.12 Shear force in the tri-cell wall of the Sleipner A platform; a simple equilibrium check

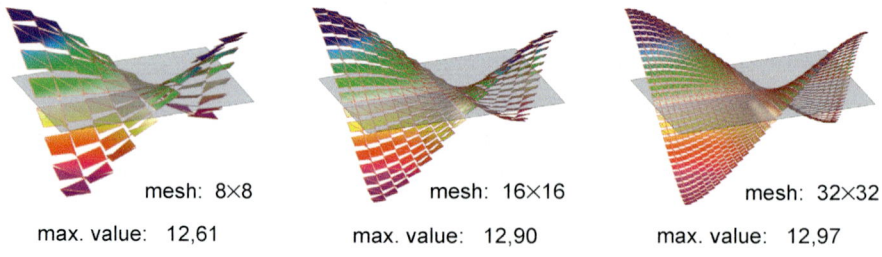

Figure 17.13 Moment of twist (M_{xy}) for a simply supported square plate with uniform loading, analysed by DKT elements; un-smoothed results

√ If possible, check if the load resultants are reasonable; some programs will sum up the loading in the directions of the reference axes and present these resultants. It is not unusual to miss some of the loading. If all loading is due to temperature changes, all load resultants should be zero.

√ Check the condition number if it is available and/or the residual forces in order to rule out numerical problems. The residual forces corresponding to unsupported *dofs* should be zero (or very close to zero); for supported *dofs* they are reaction forces.

Next, check *equilibrium,* both global and local.

A well designed program should not only sum up and return the external load resultants in the global (reference) directions, it should do the same for the resultants of the computed support reactions. If that is the case the global equilibrium check is merely a check on the numerical accuracy and on the amount of loading. If these parameters are not provided, the global check will require more work, but in most cases it should be carried out.

Local equilibrium checks should be undertaken in areas where you might suspect that the computed stresses may not be all that accurate, or in details where it is easy to check. Figure 17.12 which is related to the FEM analysis of the Sleipner platform in the previous section, is a good example. The tri-cell wall is basically a simple beam problem; the only uncertainty is associated with the degree of rigidity at the supports (*i.e.*, the stiffness of the rotational spring). The bending moment diagram has to be somewhere between the two extreme cases: simply supported and completely clamped. It is most likely to be closer to the clamped case; the careful engineer would probably make his best guess and then increase both the end and the midpoint moments somewhat. However, the shear force is *not* influenced by this at all; regardless of the end support, the maximum shear force is $qL/2$ (it is the same parabola for all the cases). As pointed out in the previous section, the Sleipner failure was a shear failure in the tri-cell wall (that explanation was never questioned), and it is therefore fair to say that this simple check could well have prevented the accident. $qL/2$ gives about 40-45 per cent higher shear force than predicted by the FEM computations, and this would most likely have made someone take a closer look at the reinforcement. The FEM analysis of the critical detail reported in Fig. 17.6 confirms the shear force diagram of Fig. 17.12, and the maximum value $qL/2$. (You will have to take the author's word for this, since the view of τ_{xy} presented in Fig. 17.6 does not support this statement).

Figure 17.6 is a good example of how one can get a better picture of the stress distribution in a critical area by using a separate model of just this area; use symmetry if possible and take the "loading" on the "cut-out sections" from the main analysis. In this particular case a 2D model suffices even though the problem is 3D.

How *good* are the results?

Having established that the results are reasonable and satisfy the equilibrium checks made, accuracy may be the next issue. Some programs will provide error estimates (usually related to energy measures). If your program lets

Practical use of FEM

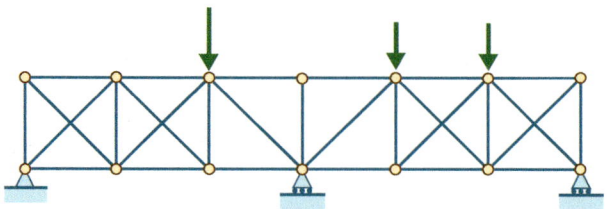

Figure 17.14

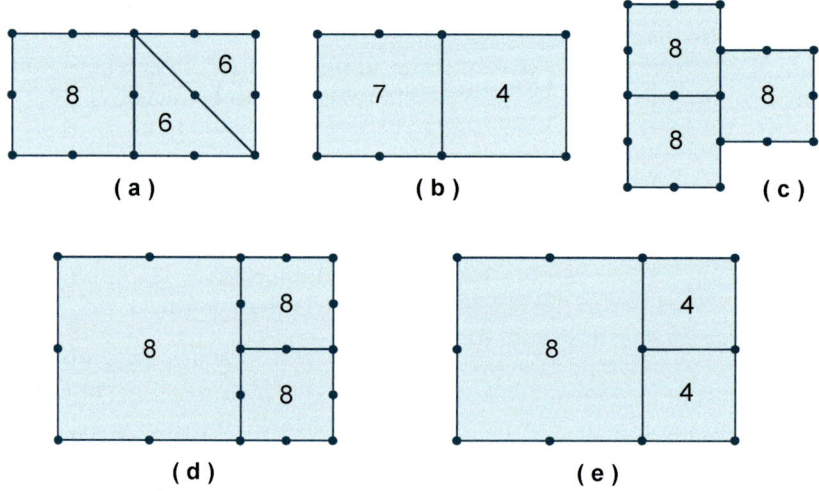

Figure 17.15

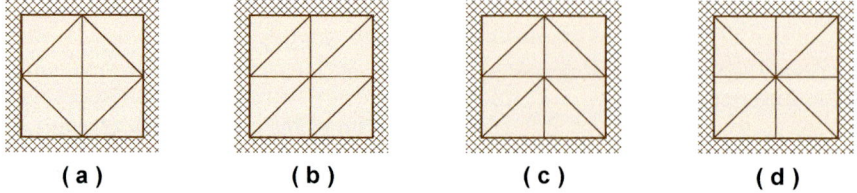

Figure 17.16

you see basic element results that have not been smoothed in any way, such as the ones in Fig. 17.13, the gap between neighbouring elements is an indication of the quality of the results, but not much more than that.

If we are talking about linear static analysis, and we are, all programs should have the capability of producing a uniform subdivision of any element mesh. For many engineering problems, comparing results from two consecutive meshes (the finer obtained by uniform subdivision of the coarser) will give a pretty good idea about the level of accuracy.

Alternative solutions and/or *independent control* should always be an option if there is any doubt about the results.

We round off this chapter with a word of caution. Both hardware and FEM software are constantly becoming better (faster) and more and more tailored to the needs of the user. This is good, but there is the danger that excellent tools, which is what we are talking about here, can mask the real problems and lower the concentration of the user. So make sure that it is you who is in command, not the program, and do not hesitate to ask for help and advice if you are uncertain; as we have seen, a mistake can have far reaching consequences.

Problems

Problem 17.1

Figure 7.14 shows a perfect truss structure, that is a moment-free structure (only axial forces). The only program at your disposal is a 2D *frame* program with *beam elements only*.
How would you use this program to obtain an engineering solution to the truss problem?
Is it possible to obtain the exact truss solution with this frame program, and if so, how?

Problem 17.2

Figure 17.15 shows five patches of plane stress elements. We assume that each of the elements involved is based on shape functions that satisfy *all* requirements for compatibility between themselves (inter-element continuity requirements).

Determine, for each patch, if you have displacement continuity, and if not, is it possible to obtain continuity by imposing extra constraints?

Problem 17.3

Figure 17.16 shows four element meshes for a *clamped*, square plate subjected to transverse loading. The plate bending element used is the 9 *dof* **DKT** element.

For each mesh, identify elements that will be *inactive*, in the sense that they do not experience any deformation, regardless of the external loading.

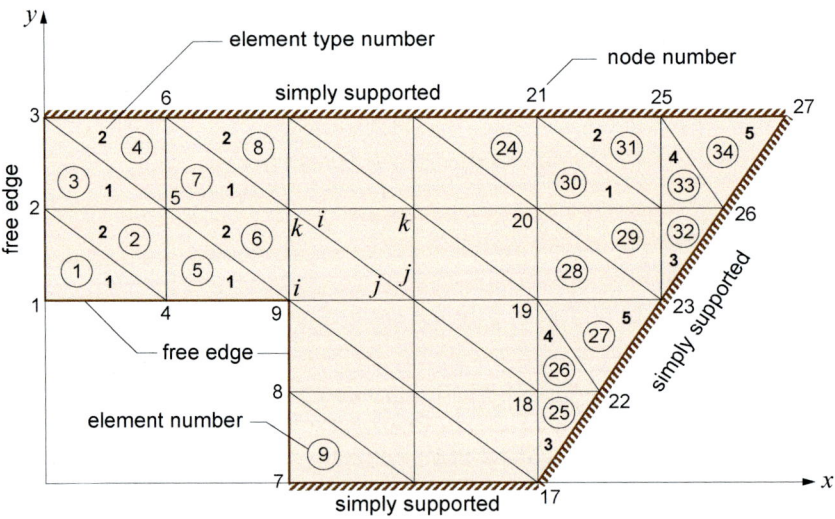

$i \rightarrow j \rightarrow k$ are local node numbers (counter clockwise)

Figure 18.1 FEM analysis anno 1968 – plate bending with element **T18**

18

Personal comments

A textbook on a theme such as structural mechanics should ideally present state-of-the-art material in an objective and unbiased manner. I have tried to do this, but I suspect that my background and personal preferences may have influenced some of my statements and recommendations. After half a century of working in various facets of the "FEM game" I have pondered and formed opinions on questions I believe are important. While my opinions may not be of general interest, I believe the questions are. Since this chapter is not an authoritative account, I feel it is most appropriate to present it as personal comments.

In order to understand the present and to plan for the future it is useful to know the past.

18.1 The past

I wrote my master thesis in the autumn of 1962 at the same time as the first computer for general use was installed at my university in Trondheim, a Danish contraption (GIER) with the following specifications: 5 KB of primary storage (1024 words of 42 bits each), 60 KB of mass storage and a performance of about 0,1 MIPS; not very impressive by today's standards, but quite a marvel in those days. The programming language was Algol. My professor managed to get me into a programming course arranged for faculty members, and I wrote a small Algol program as part of my thesis. I still have the text book with the visionary title (translated from Danish) "DASK Algol for school and home".

Three years later a Univac 1107 mainframe was installed, and our computer centre director proclaimed that with this "machine" we had enough computational capability to serve all of Northern Europe! But not for long; in 1969 we installed its big brother, the Univac 1108. Both these computers came with both Fortran and Algol compilers; I converted to Fortran during a one year stay at UC Berkeley in the mid 1960's, and that has been my programming language ever since.

Figure 18.1 shows a finite element model of a plate bending problem solved by a state-of-the-art program anno 1968. Mesh generation in those days was limited to simple geometric shapes (such as parallelograms); for an irregular

Personal comments

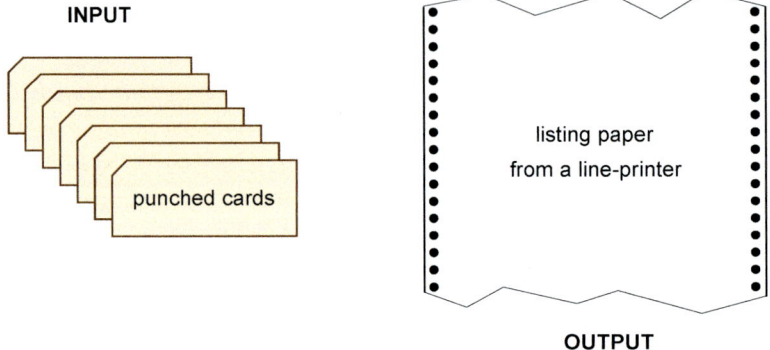

Figure 18.2 INPUT / OUTPUT media in the late 1960's

geometry like the one in Fig. 18.1 we had to specify the *x*- and *y*-coordinates as well as the boundary conditions of *each* nodal point, and for *each* element we had to specify

- the material type,
- the global node numbers of each (local) corner node, and
- the element type number.

If the plate thickness and the loading were constant over the plate, only those two values needed to be specified, otherwise one value would have to be given for each node. The element *type* concept used here needs a comment; two elements were of the same type only if they had *identical* stiffness matrices; in 1968 computer time was "expensive" and we tried to save whenever possible. The generation of the element stiffness matrix **k** for the **T18** element (which involved the inversion of an 18×18 matrix) was not a trivial operation, so if two or more elements had an identical **k** we would take advantage of this, even if this meant more book-keeping and more input information.

All input was punched on cards and all results were presented as tables printed on listing paper by a line-printer, see Fig. 18.2.

A problem similar to the one in Fig. 18.1 with the following characteristics:
- number of elements: 200
- number of equations: 726

took about 570 seconds of CPU-time to solve on a Univac 1107. The solver was a standard direct *band* solver, and the bandwidth was a function of how the user numbered the nodes (no automatic node renumbering).

My intention here is not to give a historical account, but rather to present some background information on the important question of how we should teach structural mechanics in general, and basic finite element methods in particular, to structural engineers-to-be, in this digital age? In this perspective, the development of both computer hardware and software is, in my opinion, important.

By the mid 1960's a course on *matrix theory of structures* (MTS) was offered to our students majoring in structural engineering, and a *finite element* (FEM) course followed not long after. The basic mechanics courses were not influenced by this, but some of the more advanced courses (on continuum mechanics and shell theories) were reduced in order to make room for the new, numerical based courses. Throughout the 1970's and most of the 1980's all students who took the MTS course would be expected to do some programming (mostly Fortran) as part of the course and likewise for the FEM course.

During most of the 1970's the mainframe computer was our "calculating machine". The Univac machines were special in two respects,

1) the addressable unit, the "computer word", was 36 bits long, which gave it a reasonable address space and fairly adequate accuracy even in single precision, and

2) it had a fast random access peripheral storage device, the magnetic drum.

Personal comments

Moore's law (1965): *Processing speed doubles every 18 months.*

Moore actually predicted that the number of transistors or integrated circuits placed on a certain area would double every two years – the interpretation we now attach to the law is due to Intel executive David House who predicted the period for doubling in chip *performance* (which is a combination of the effect of more transistors and their speed).

Kryder's law (2005): *Storage capacity doubles every 12 months.*

While not stated as explicitly as this by Mark Kryder, this is the popular interpretation.

However, the amount of primary storage (RAM) was limited, and in order to negotiate larger problems we were forced to work out-of-core (two levels of memory); this increased the book-keeping quite considerably. Towards the end of the decade, work stations and mini-computers appeared on the market, at first as 16-bit machines, but soon with 32-bit architecture. This decentralized computing on campus, but it did not really change the way we went about our programming, even though the *virtual memory* concept did make a difference. By the beginning of the 1980's 32-bit architecture was fairly standard, and more and more (and soon all) numerical computations were carried out in double precision (using *two* 32-bit words to store a floating point variable).

Processing speed and storage capacity improved steadily during the period from 1970 to 1990; however, even if the improvements followed the laws of MOORE and KRYDER, we did not think of them as dramatic. In the 1980's we also saw an emerging graphical capability.

Our MTS course and basic FEM course matured during the 1970's and 1980's; new developments and better explanations were incorporated, but up to the end of this period the main focus of these courses was still on the implementation of methods for computer solutions. In fact a new course on the *programming of structural computations*, based on the MTS course, was introduced in the early 1980's; it attracted a fair number of students and by the end of the decade about half our civil engineering students attended this course.

The proliferation of the PC in the early 1990's changed the digital landscape dramatically. This affordable "calculating machine" had been around for some time, but it was not until then that the use of it really took off, and it soon sparked a discussion at the university about our focus regarding information and communication technology. Shortly afterwards our syllabus was given a major shake up. New subjects and courses with reshuffled content were introduced and at the end of this process we had to acknowledge a significant reduction of the time allocated to basic mechanics courses.

The majority of our faculty members held the opinion that it was now time for our students to concentrate on the *use* of existing software; programming was no longer considered essential. I was in the minority; that is not to say that I saw a need for all engineering students to become programming geeks, but I believed then and still believe that all engineering students should take a course in programming, if for no other reason than that it provides a good grounding in logical thinking. By the end of the millennium very few civil and mechanical engineering students took computer programming courses, and our own (Fortran based) course on structural computations petered out and came to an end, in its original form, in the early years of the new millennium.

A couple of years without students who had at least some programming skills made it apparent that this was not a satisfactory situation, and a new program of study, which included a fair number of computer science courses, was introduced for a modest number of engineering students. After two years, these students are redeployed to the traditional engineering programs, and we now have the situation that most engineering disciplines

Personal comments

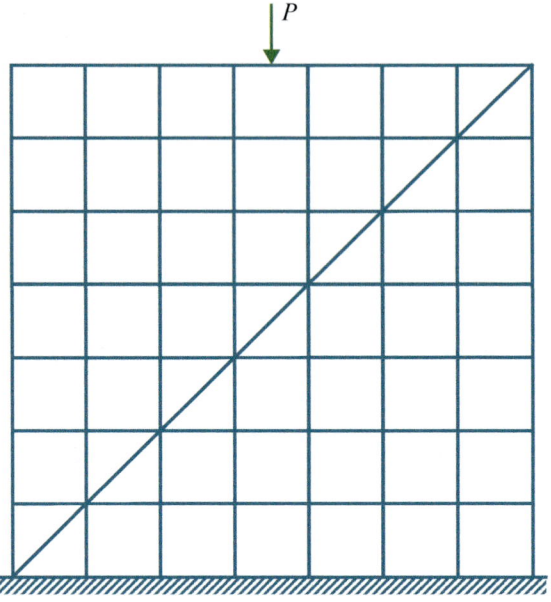

Problem characteristics:

- each of the 112 beam and column members is divided into *1000* beam elements which makes for a computational model with

  - 111 952 nodal points,
  - 112 000 beam elements, and
  - 335 832 unknown nodal *dofs*

Figure 18.3 A 2D test frame

have a small group of students with quite a good background in computer program development. In many ways this is not a bad compromise, but it means that our basic MTS and FEM courses, which we still offer, struggle to find a balance that satisfies both the majority who will mainly use FEM based programs and the few who will also develop and program the methods.

18.2 The present

When you double something that is large, but not very large, it becomes larger, but not dramatically so. However, when you double something that is already very large, and keep doubling the result, it soon becomes very, very large. This is exactly what has happened in digital computing during the last five years or so. In the early 1990's we were told that MOORE's law would not apply for much longer, and certainly not into the new millennium. Oh boy, were they wrong.

As an example of what we are talking about here, we have analysed the plane frame in Fig. 18.3. A linear static analysis for one load case was carried out by program **fap2D** [9], which uses a direct "skyline" solver. The numerical computations (performed by a Fortran subroutine) comprised the following operations:

- preparing, generating and checking integer "control" information, including node renumbering,
- generating all 112 000 element stiffness matrices and "adding" them into the system stiffness matrix (assembly),
- solving the system stiffness relation with respect to the 335 832 unknown *dofs*, and
- computing the end section forces (*M*, *V* and *N*) at both ends of each of the 112 000 elements.

On a dual core (i5-254 M CPU – 2,60 GHz) *laptop* with 8 GB of RAM and an SSD disk of 160 GB, purchased in mid-2012 for about USD 1500, the Fortran subroutine performed *all* the above operations in about 420 milliseconds; that is less than *half a second*! And the results were correct (the same as those we obtained for a model with a total of about 1000 *dofs*). The time was obtained by reading the CPU-clock on entry to, and exit from, the subroutine.

For someone with my background – remember GIER of 1962 – this is mind blowing, and I believe that this incredible computational capability, matched by equally impressive graphics, will have a profound effect not only on the practical use of FEM, but also on how we should teach mechanics in general, and finite element analysis in particular.

We still need to teach our engineering students basic equilibrium and strength of materials. They should of course be able to identify whether or not a structure is stable, statically determinate or not, and in principle they should know how to solve a statically indeterminate structure. But we do not have the time, nor is there the need, to teach them the "hand calculation methods" our generation used when solving such problems. For instance, we gave up teaching the *moment distribution method* (the "CROSS method")

Personal comments

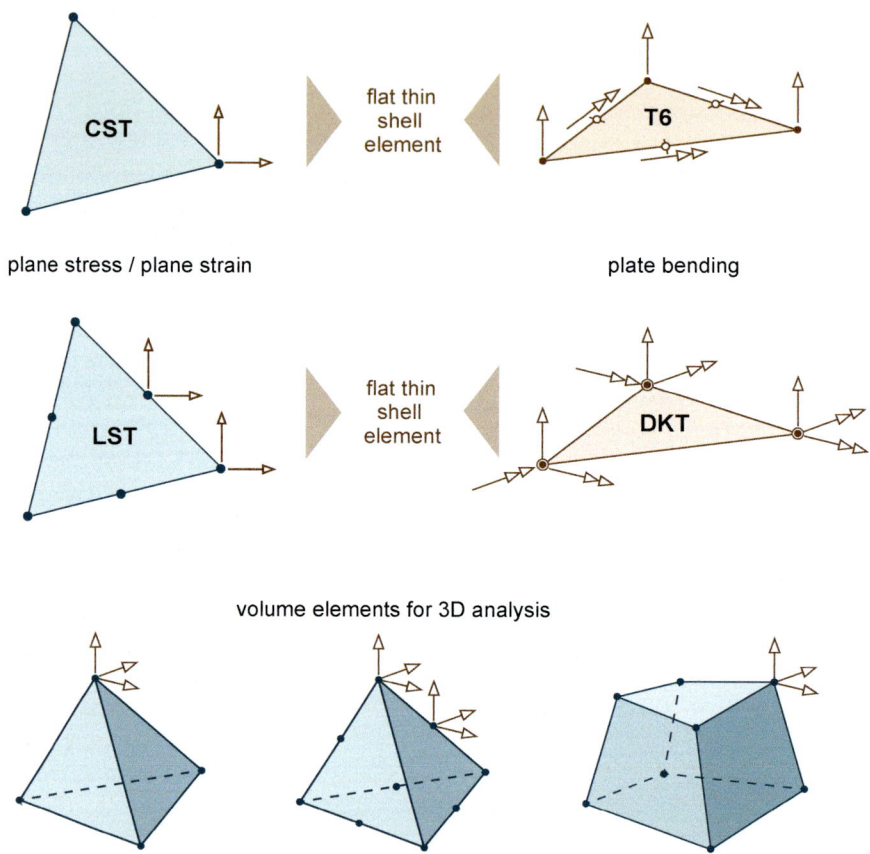

Figure 18.4 "Favourite" elements

more than 15 years ago. The engineers we are now educating will never solve anything more complicated than simple, statically determinate structures by "manual methods". For anything slightly more complicated they will use a computer program, and we should train them to use such a tool properly. In my opinion all users of such tools should know the principles of what is going on "behind the scenes", but not all the details. Furthermore we should instil a critical mindset and teach them how to check the results they obtain with tools that become more and more sophisticated.

If we assume that we have already covered basic equilibrium, strength of materials and some basic 2 and 3D elasticity in previous courses, and that we also recognize that we need to cater for students with different backgrounds and aspirations, then how should we structure the basic MTS and FEM courses? Our approach up to now has been to give a fairly thorough course in *matrix theory of structures* (MTS) [7] which still contains material aimed at computer implementation of the methods, and then a *finite element* (FEM) course that covers most of the "theoretical" material in this book. At present both courses attract a large number of students, and I believe a fair number of them struggle. It should be mentioned that we also offer a course on non-linear finite element analysis in the final year.

Would it perhaps be better to structure the material of the two basic courses differently? For instance, the first course might cover the basics of 1-, 2- and 3-dimensional elements, with emphasis on "simple", but robust, elements well suited for practical analysis. The governing principle should be many simple elements and load lumping, and material necessary for computer implementation of the method should give way to practical applications.

In view of current computational capabilities I would pick the elements shown in Fig. 18.4 as my "favourite" elements for linear problems. For 2D elements my preference is *triangular* elements, and although I can see the need for a hexahedral element for 3D problems, the *tetrahedron* can certainly do the job. The advantages of triangles and tetrahedrons are:

- *complete* polynomials for the shape functions (in most cases),
- *no* need for numerical integration (provided straight edges and plane surfaces),
- flat triangular facets can approximate a doubly curved surface, and
- most automatic mesh generators use triangulation.

Furthermore, we do not need to worry about mechanisms (zero energy modes) and these elements are not "fixed" in any way.

If the first basic course only included the elements of Fig. 18.4, minus the hexahedron, in addition to simple bars and beams, we would remove many "tripwires" for the students. These include isoparametric elements, numerical integration (full, reduced and selective), mechanisms (hourglass and zero energy modes), not to mention the "fixes" that are required for some of the other popular elements. The second course could then cover these and other theoretical topics at a detail level that would also serve the needs of program development. Some of the students who currently take both courses might then decide that such a "renovated" first course would suffice.

Personal comments

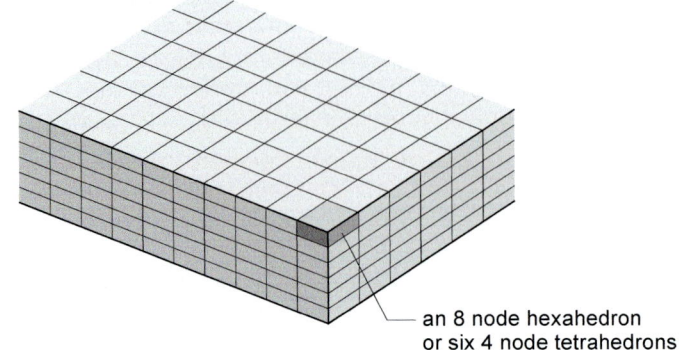

A plate (bending) corner

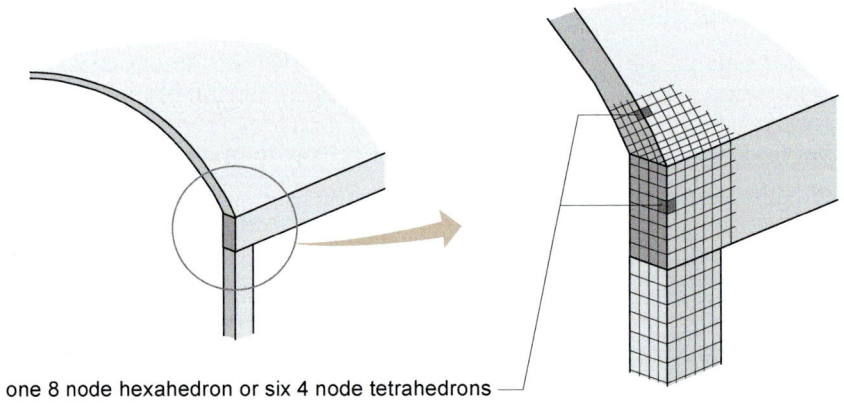

A shell-beam-post structure

Figure 18.5 Unusual but (computationally) quite feasible use of 3D elements

If this is the path I believe we should follow, why have I not written a textbook more in line with this thinking? The simple answer is that the textbook for our first course already exists. And I am sure there are valid objections to these ideas that need to be considered carefully; major changes in the content of university courses are not to be taken lightly.

Nevertheless, I have tried in a small way to "move" my presentation in the direction indicated above. I am convinced that tomorrow's textbooks on these themes must reflect the incredible computational capability we now have at our disposal much better than is currently the case.

18.3 The future

My crystal ball gets hazy very quickly when it comes to digital computing. The only thing I am fairly sure of is that tomorrow's computers will be a great deal faster than those of today, and they will have more and faster memory, both primary (RAM) and secondary (disk). I have great faith in the "hardware guys"; they have delivered and will continue to do so. There is clearly a limit as to how small a transistor can be made, but those who pretend to know about these things seem to think that MOORE's law will most likely apply until 2020, and Seagate (one of the major disk manufacturers) claim that they will be able to produce a 60 TB 3,5 inch hard drive by 2016. And on the horizon lurks quantum computing (whatever that might be).

What about software? With 64-bit architecture, which is already here, we can address an almost unlimited amount of memory, and we can easily afford to allocate more bits to our floating point variables. What if the standard *integer* and *real* "words" become 64 bits long and 128 bits are allocated to a *double precision* variable? All our present software would run practically without changes, and with such a (ridiculously) high precision we could run very, very large problems without worrying about numerical problems.

Whatever happens I feel quite confident that the "brute force" technique using many simple elements and load lumping will become the backbone of FEM software for practical use, and it will come as no surprise if more and more problems are analysed with solid (3D) elements; semi-thick plates and shells are already analysed by 3D elements, as indicated in Fig. 18.5. After all, three-dimensional elasticity is the only "pure" theory; with no simplifying hypotheses everything is fairly straightforward. Mesh generation and the piecing together of many regions, each with a very large number of (3D?) elements, will be a challenge, and much work on the post-processing of the incredible amount of stress data will be required.

With an increasing number of engineering students world wide seeking higher degrees, I am sure that new (clever) elements will be developed and some old ones will probably be reinvented. However, I am also sure that the way forward for effective and safe structural analysis will have to concentrate on other issues than the development of new elements. As to which issues, I have indicated a few, but my limited imagination suggests I should leave this question to those who will participate actively in this process.

19 References

[1] J.H. ARGYRIS: "Energy Theorems and Structural Analysis", *Aircraft Eng.*, Vol. **26**, 347-356 (Oct.) and 383-387 (Nov.), 1954.

[2] M.J. TURNER, R.W. CLOUGH, H.C. MARTIN and L. TOPP: "Stiffness and Deflection Analysis of Complex Structures", *J. Aero. Sci.*, Vol. **23**, No. 9, 805-823, Sept.1956.

[3] R.W. CLOUGH: "The Finite Element Method in Plane Stress Analysis", *Proc. 2nd ASCE Conf. on Electronic Computation*, Pittsburgh, Pa, Sept. 1960.

[4] G. STRANG and G.J. FIX: *An Analysis of the Finite Element Method*, Prentice-Hall, Englewood Cliffs, N.J., 1973.

[5] K.-E. KURRER: *The History of the Theory of Structures*, Ernst & Sohn, Berlin, 2008.

[6] S.P. TIMOSHENKO: *History of Strength of Materials*, McGraw-Hill, New York, 1953.

[7] K. BELL: *Matrisestatikk – statiske beregninger av rammekonstruksjoner*, (in Norwegian), Tapir Akademiske Forlag, Trondheim, 2011.

[8] H. CROSS: "Analysis of Continuous Frames by Distributing Fixed-End Moments", *Proc. of the ASCE*, 919-928, 1930.

[9] K. BELL: "**fap2D** – a Windows-based program for static and dynamic analysis of 2D frame type structures – User's Manual", preliminary version, Department of structural engineering, NTNU, Trondheim, 2011.

[10] C.A. FELIPPA: "The Amusing History of Shear Flexible Beam Elements", Report CU-CAS-05-1, University of Colorado, Boulder, 2005.
Also published in IACM Expressions No. 17, January 2005.

[11] F. HARTMANN: *The Mathematical Foundation of Structural Mechanics*, Springer, Berlin, 1985.

[12] K. BELL: "On the quintic triangular plate bending element", Report No. 72-2, Div. of Structural Mechanics, NTH, Trondheim, 1972.

[13] K. BELL: "Analysis of Thin Plates in Bending Using Triangular Finite Elements", Div. of Structural Mechanics, NTH, Trondheim, 1968.

[14] O.C. ZIENKIEWICZ *et al.*: "Iso-Parametric and Associated Element Families for Two- and Three-Dimensional Analysis", Chapter 13 of *Finite*

References

Element Methods in Stress Analysis (editors I. HOLAND and K. BELL), Tapir, Tech. Univ. of Norway, Trondheim, 1969.

[15] I.C. TAIG: Structural analysis by the matrix displacement method, Engl. Electric Aviation Report No. S017, 1961.

[16] B.M. IRONS: "Numerical integration applied to finite element methods", Conf. Use of Digital Computers in Struct. Eng., Univ. of Newcastle, UK, 1966.

[17] B.M. IRONS: "Engineering application of numerical integration in stiffness method", JAIAA, Vol. **14**, 2035-7, 1966.

[18] O.C. ZIENKIEWICZ and R.L. TAYLOR: The Finite Element Method, 4th edition, Volume 1 – Basic Formulation and Linear Problems, McGraw-Hill Book Company (UK), 1989.

[19] K.-J. BATHE: Finite Element Procedures, Prentice Hall, 1996.

[20] T.J.R. HUGHES: The Finite Element Method – Linear Static and Dynamic Finite Element Analysis, Prentice-Hall, 1987.

[21] R.D. COOK, D.S. MALKUS and M.E. PLESHA: Concepts and Applications of Finite Element Analysis, 3rd edition, John Wiley & Sons Inc., 1989.

[22] J. BARLOW: "Optimal Stress Locations in Finite Element Models", Int. J. Num. Meth. Eng., Vol. **10**, 243-251, 1976.

[23] O.C. ZIENKIEWICZ and J.Z. ZHU: "A simple error estimator and adaptive procedure for practical engineering analysis", Int. J. Num. Meth. Eng., Vol. **24**, 337-357, 1987.

[24] O.C. ZIENKIEWICZ and J.Z. ZHU: "The superconvergent patch recovery and a posteriori error estimates. Part 1: The recovery technique", Int. J. Numer. Meth. Eng., Vol. **33**, 1331-1364, 1992.

[25] O.C. ZIENKIEWICZ and J.Z. ZHU: "The superconvergent patch recovery and a posteriori error estimates. Part 2: Error estimates and adaptivity", Int. J. Numer. Meth. Eng., Vol. **33**, 1365-1382, 1992.

[26] E.L. WILSON and A. IBRAHIMBEGOVIC: "Use of incompatible modes for the calculation of element stiffnesses or stresses", Finite Elements in Analysis and Design, Vol. **7**, 229-241, 1990.

[27] J. ROBINSON: "A single element test", Computer Meth. in Appl. Mech. and Eng., Vol. **7**, Issue 2, 191-200, 1976.

[28] P.G. BERGAN and L. HANSEN: "A New Approach for Deriving "Good" Element Stiffness Matrices", MAFELAP 75 – The Mathematics of Finite Elements and Applications, Brunel University, 1975.

[29] B.M. IRONS: "Numerical integration applied to finite element methods", Proc. Conf. on Use of Digital Computers in Struct. Eng., University of Newcastle, 1966.

[30] G.P. BAZELEY, Y.K. CHEUNG, B.M. IRONS and O.C. ZIENKIEWICZ: "Triangular elements in bending – conforming and non-conforming solutions", Proc. 1st Conf. Matrix Methods in Struct. Mech., volume AFFDL-TR-66-80, 547-576, Wright Patterson Air Force Base, Ohio, Oct. 1966.

[31] E. CUTHILL and J. MCKEE: "Reducing the bandwidth of sparse symmetric matrices", Proc. 24th Nat. Conf. ACM, 157-172, 1969.

[32] J.A. GEORGE and J.W-H. LIU: *Computer Solution of Large Sparse Positive Definite Systems*, Prentice-Hall, 1981.

[33] S.W. SLOAN: "A Fortran Program for Profile and Wavefront Reduction", Int. J. Num. Meth. Eng., Vol. **28**, 2651-2679, 1989.

[34] M.I. HOIT and J.H. GARCELON: "An Updated Profile Front Minimization Algorithm", *J. of Computers and Structures*, Vol. **33**, No. 3, 903-914, 1989.

[35] B.M. IRONS: "A frontal solution program for finite element analysis", *Int. J. Num. Meth. Eng.*, Vol. **2**, 5-32, 1970.

[36] K. BELL: *Eigensolvers for structural problems – some algorithms for symmetric eigenproblems and their merits*, Delft University Press, 1998.

[37] A.C. DAMHAUG: "Sparse Solution of Finite-Element Equations – Part 1 Theory", Technical Report No. 97-2003, Det Norske Veritas Software, October, 2003.

[38] B.W. KERNINGHAM and D.M. RITCHIE: *The C Programming Language*, 1st edition, Prentice Hall, 1976.

[39] B. STROUSTRUP: *The Design and Evolution of C++*, Addison-Wesley, 1994.

[40] O.-J. DAHL, B. MYHRHAUG and K. NYGAARD: "Simula 67 Common Base Language", Norwegian Computing Center, 1968.

[41] H.P. LANGTANGEN: *Python Scripting for Computational Science*, 3rd ed., Springer Verlag, 2008.

[42] R.D. COOK: "Improved Two-dimensional Finite Element", *J. of Struct. Div. ASCE*, Vol. **100**, ST6, 1851-1863, 1974.

[43] E.L. WILSON, R.L. TAYLOR, W.P. DOHERTY and J. GHABOUSSI: "Incompatible displacement models", *Num. and Comp. Meth. in Struct. Mech.* (edited by S.T. FENVES), 43-57, Academic Press, 1973.

[44] R.L. TAYLOR, P.J. BERESFORD and E.L. WILSON: "A non-conforming element for stress analysis", *Int. J. Numer. Meth. Eng.*, Vol. **10**, 1211-1219, 1976.

[45] A. IBRAHIMBEGOVIC and E.L WILSON: "A modified method of incompatible modes", *Comm. Numer. Meth. Eng.*, Vol. **7**, 187-194, 1991.

[46] D. KOSLOFF and G.A. FRASIER: "Treatment of hour glass patterns in low order finite element codes", *Int. J. Numer. Meth. Geomech.*, Vol. **2**, 57-72, 1978.

[47] S.P. TIMOSHENKO: *Strength of Materials – PART I Elementary Theory and problems*, 3rd ed., Van Nostrand Company, Princeton, New Jersey, 1955.

[48] T.J.R. HUGHES, R.L. TAYLOR and W. KANOKNUKULCHAI: "A simple and efficient element for plate bending", *Int.J. Numer. Meth. Engrg.*, Vol. **11**, 1529-1543, 1977.

[49] R.H. MACNEAL: "A simple quadrilateral shell element", *Comp. & Struc.*, Vol. **8**, 175-183, 1978.

[50] A. FISKVATN: *Elementmetoden*, textbook in Norwegian, Tapir 1984, Trondheim, Norway.

[51] L.S.D. MORLEY: "The constant-moment plate-bending element", *Journal of Strain Analysis*, Vol. **6**, 20-24, 1971.

References

[52] R.D. Cook, D.S. Malkus, M.E. Plesha and R.J. Witt:
Concepts and Applications of Finite Element Analysis, 4th edition, John Wiley & Sons Inc., 2002.

[53] T.J.R. Hughes and M: Cohen: "The 'Heterosis' Family of Plate Finite Elements", *Proceedings of the ASCE Electronic Computations Conference*, St. Louis, Missouri, August 6-8, 1979.

[54] J.A. Stricklin, W. Haisler, P. Tisdale and R. Gunderson: "A Rapidly Converging Triangular Plate Element", *AIAA Journal*, Vol. **7**, No. 1, 180-181, 1969.

[55] D.J. Allman: "Triangular finite elements for plate bending with constant and linearly varying bending moments", *IUTAM* Symposium on High Speed Computing of Elastic Structures, Liège, 1970.

[56] F. Kikuchi and K. Tanizawa: "Accuracy and Locking-Free Property of the Beam Element Approximation for Arch Problems", *Computers & Structures*, Vol. **19**, No. 1-2, 103-110, 1984.

[57] K. Bell, O.V. Bleie and L. Wollebæk: "**CroosX** – A Windows-based program for computation of parameters for, and stress distribution on, arbitrary beam cross sections – User's Manual",
Report No. R-13-00, Department of structural engineering, Norwegian University of Science and Technology, Trondheim, 2000.

[58] J.L. Meek: *Computer Methods in Structural Analysis*, E &FN SPON (Chapman & Hall), 1991.

[59] W.E. Mason and L.R. Herrman: "Elastic Shear Analyis of General Prismatic Beams", *Journal of the Eng. Mech. Div.*, ASCE, Vol. **94**, 965-983, August 1968.

[60] I. Holand: "The Sleipner accident", chapter in *From Finite Elements to the Troll Platform – Ivar Holand 70th Anniversary*, edited by K. Bell, Department of Structural Engineering, The Norwegian Institute of Technology, Trondheim, 1994.

[61] B. Jakobsen and F. Rosendahl: "The Sleipner Platform Accident", *Struct. Engineering International*, No. 3, 1994, 190-193.

Appendix

Matrix algebra

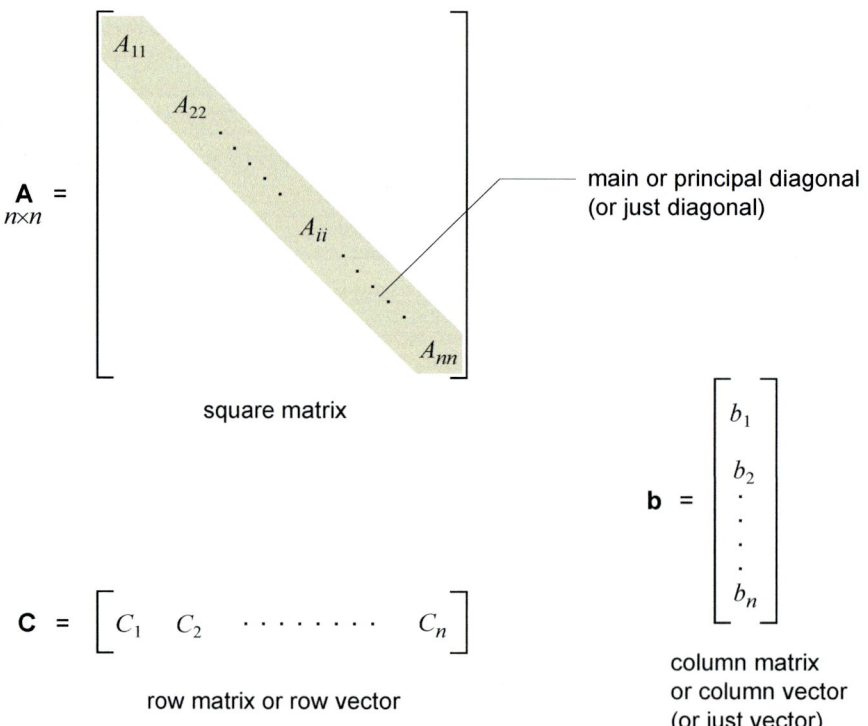

Matrix algebra

A certain familiarity with matrices and matrix manipulations is almost a prerequisite for getting on the "inside" of FEM. In this appendix we summarize some definitions, properties and rules concerning matrices, vectors and determinants, without much proof.

A.1 Definitions

A *matrix* is a rectangular array of numbers or quantities. The individual quantities, A_{11}, A_{12}, ... , A_{mn}, are the *elements* of the matrix, usually real or complex numbers. The elements are ordered in *rows* and *columns*, and a matrix with m rows and n columns has *dimension* $m \times n$.

Matrices are often denoted by bold-faced letters (**A**, **B**, **k**, ..), a notation used consistently in this book. A typical element of a matrix **A** is denoted A_{ij}, where the following rule always applies: the first index (i) denotes the *row* number while the second index (j) denotes the *column* number of the element's position in the matrix. Another common way of denoting a matrix is to put the typical matrix element in square brackets, *e.g.* $[A_{ij}]$.

A matrix with the same number of rows and columns, that is $m = n$, is a *square* matrix, of order n; also called an $n \times n$ matrix. The diagonal from the upper left-hand corner to the lower right-hand corner of a square matrix is its main or *principal diagonal*; the other diagonal is called the bidiagonal. Square matrices play an important role in FEM.

A matrix with only one row is termed a *row matrix* or *row vector*. Similarly, a matrix with only one column is a *column matrix* or a *column vector*, or simply a *vector*.

Matrix notation varies from author to author. Some use upper case bold-faced letters for matrices and lower case bold-faced letters for vectors. You will also find that some use special brackets to denote matrices ([ ]), column vectors ({ }) and row vectors (< >). We have chosen not to distinguish between matrices and vectors; both are denoted by bold-faced letters (upper or lower case), and their elements are denoted by the same symbol as the matrix or vector symbol, in *italic* (ordinary, not bold print). When brackets are considered necessary, which they sometimes are, we will only use square

A   **581**

Matrix algebra

$$\mathbf{A} = \begin{bmatrix} 2 & -8 & 1 \\ 2 & 0 & 3 \end{bmatrix} \qquad \mathbf{A}^\mathsf{T} = \begin{bmatrix} 2 & 2 \\ -8 & 0 \\ 1 & 3 \end{bmatrix}$$

$$\mathbf{I} = (\mathbf{I}_5) = \begin{bmatrix} 1 & 0 & 0 & 0 & 0 \\ 0 & 1 & 0 & 0 & 0 \\ 0 & 0 & 1 & 0 & 0 \\ 0 & 0 & 0 & 1 & 0 \\ 0 & 0 & 0 & 0 & 1 \end{bmatrix} = \begin{bmatrix} \delta_{ij} \end{bmatrix}$$

$$\mathbf{D} = \begin{bmatrix} D_{11} & & & & \\ & D_{22} & & \mathbf{0} & \\ & & \ddots & & \\ & \mathbf{0} & & \ddots & \\ & & & & D_{nn} \end{bmatrix} = \begin{bmatrix} D_{ii} \end{bmatrix} \qquad D_{ij} = 0 \text{ for } i \neq j$$

$$\mathbf{D} = \begin{bmatrix} 2 & 0 & 0 \\ 0 & 4 & 0 \\ 0 & 0 & 6 \end{bmatrix} = \begin{bmatrix} 2 & 4 & 6 \end{bmatrix}$$

Definitions

brackets. A vector symbol, *e.g.* **v**, will *always* mean a column vector. For a row vector we will use the superscript T on the vector symbol, indicating the *transpose* of the column vector; hence, if **C** is a (column) vector, then **C**^{T} is a row vector with the same elements as **C**.

The *transpose* of a rectangular $m{\times}n$ matrix **A** = $[A_{ij}]$ is, by definition, another rectangular $n{\times}m$ matrix denoted **A**^{T} whose rows are the columns of **A** and whose columns are the rows of **A**; in other words, we transpose a matrix by interchanging its rows and columns,

$$\mathbf{A}^T = [A_{ji}] = \begin{bmatrix} A_{11} & A_{21} & . & . & A_{m1} \\ A_{12} & A_{22} & . & . & A_{m2} \\ . & . & . & . & . \\ . & . & . & . & . \\ A_{1n} & A_{2n} & . & . & A_{mn} \end{bmatrix} \qquad (C\text{-}1)$$

It should be noted that we have generalized the term vector to mean something more than a directional line segment in 3D space. Our vectors are "*n*-dimensional"; vector spaces play an important role in *linear algebra*, but we manage here without getting involved with these concepts. If you have a problem with the *n*-dimensional "space", think of the vector as a (simple) matrix.

A *scalar* is a single number or variable denoted by ordinary print.

As already mentioned a matrix is usually a rectangular array of numbers whose mutual relationship makes it convenient to treat them as a whole. However, the elements of a matrix may also be mathematical expressions, such as trigonometric functions, algebraic expressions, derivatives, integrals or even other matrices.

Two matrices are *equal* if, and only if, each element of one of the matrices is equal to the corresponding element of the other matrix, that is

$$\mathbf{A} = \mathbf{B} \quad \text{if} \quad A_{ij} = B_{ij} \quad \text{for all } i \text{ and } j \qquad (C\text{-}2)$$

Before we proceed further we define some special, square matrices.

The null or zero matrix

This is a matrix (not necessarily square) for which *all* elements are identically equal to zero; we denote it by the symbol **0**. If it is necessary to indicate the dimension $n{\times}n$ of a square zero matrix we will write $\mathbf{0}_n$.

The identity (or unit) matrix

This is a matrix for which all diagonal elements are equal to 1 (unity) while all other elements are zero, that is

$$\mathbf{I} = [\delta_{ij}] \qquad (C\text{-}3)$$

where δ_{ij} is the Kronecker delta whose properties are

$$\delta_{ii} = 1 \quad \text{for all } i, \text{ and} \quad \delta_{ij} = 0 \quad \text{for } i \neq j \qquad (C\text{-}4)$$

Matrix algebra

$$\begin{bmatrix} 3 & 7 & -2 \\ 7 & 4 & 0 \\ -2 & 0 & 8 \end{bmatrix}$$

symmetric

$$\begin{bmatrix} 0 & 2 & -3 \\ -2 & 0 & 5 \\ 3 & -5 & 0 \end{bmatrix}$$

skew-symmetric

× denotes an element different from zero

symmetric

band matrix (constant band width), band width: b_w

profile or skyline matrix (variable band width)

tridiagonal matrix

$$\mathbf{L} = \begin{bmatrix} L_{11} & & & & 0 \\ L_{21} & L_{22} & & & \\ \vdots & \vdots & \ddots & & \\ \vdots & \vdots & & \ddots & \\ L_{n1} & L_{n2} & \cdots & & L_{nn} \end{bmatrix}$$

lower triangular matrix

$$\mathbf{U} = \begin{bmatrix} U_{11} & U_{12} & \cdots & U_{1n} \\ & U_{22} & \cdots & U_{2n} \\ & & \ddots & \vdots \\ 0 & & & U_{nn} \end{bmatrix}$$

upper triangular matrix

Definitions

An identity matrix of dimension $n \times n$ is denoted $\mathbf{I}_n$ or just $\mathbf{I}$.

The diagonal matrix

This matrix, often denoted by $\mathbf{D}$, has *non-zero* elements *only* on the diagonal; special brackets are sometimes used to indicate a diagonal matrix.
If all diagonal elements are *equal* ($= s$) the matrix is a **scalar matrix** $\mathbf{S}$.

Symmetric and skew-symmetric matrices

A square matrix is *symmetric* if it is equal to its own transpose,

$$\mathbf{A} = \mathbf{A}^\mathsf{T} \tag{C-5}$$

which implies that $A_{ij} = A_{ji}$ for all i and j. We see that the symmetry is about the (principal) diagonal; hence the notion of matrix symmetry only applies to square matrices.

A square matrix $\mathbf{H}$ is *skew-symmetric* if

$$\mathbf{H}^\mathsf{T} = -\mathbf{H} \quad \text{that is} \quad H_{ij} = -H_{ji} \text{ for all } i \text{ and } j \tag{C-6}$$

It follows from this definition that all diagonal elements of a skew-symmetric matrix must be zero.

An arbitrary square matrix $\mathbf{A}$ can always be expressed as the sum of a symmetric and a skew-symmetric matrix:

$$\mathbf{A} = \underbrace{\tfrac{1}{2}(\mathbf{A} + \mathbf{A}^\mathsf{T})}_{\text{symmetric}} + \underbrace{\tfrac{1}{2}(\mathbf{A} - \mathbf{A}^\mathsf{T})}_{\text{skew-symmetric}} \tag{C-7}$$

Band and profile matrices

A square matrix whose non-zero elements are *all* contained within a band along the diagonal is called a *band matrix*. The most important FEM matrix, the system stiffness matrix $\mathbf{K}$, has this property, in addition to being symmetric. By taking advantage of the band structure we can save both storage space and computer time when such matrices are involved in numerical operations. Even greater savings are obtained if we also take advantage of the variable band width, in which case we are talking about a *profile* or "skyline" matrix.

Better still is the so-called *sparse* storage format which only stores the non-zero elements, along with information about their position in the matrix. However, the utilization of this in numerical operations requires a fair amount of extra "book-keeping".

A special form of band matrix is the *tridiagonal* matrix; this matrix has non-zero elements only on the diagonal and on the sub- and super-diagonals. This matrix plays an important role in some *eigenvalue* algorithms.

Triangular matrices

A square matrix that has only zero elements above or below the diagonal is called a *triangular* matrix. Such a matrix is either *lower triangular* or *upper tri-*

Matrix algebra

$$\begin{bmatrix} A_{11} & A_{12} & \cdots\cdots\cdots & A_{1n} \\ A_{21} & A_{22} & \cdots\cdots\cdots & A_{2n} \\ \vdots & \vdots & & \vdots \\ A_{m1} & A_{m2} & \cdots\cdots\cdots & A_{mn} \end{bmatrix} \begin{bmatrix} x_1 \\ x_2 \\ \vdots \\ x_n \end{bmatrix} = \begin{bmatrix} b_1 \\ b_2 \\ \vdots \\ b_m \end{bmatrix}$$

angular, depending on where the zero elements are located. We often use **L** and **U** to denote lower and upper triangular matrices, respectively.

Triangular matrices play an important role in direct solution of systems of linear equations.

A.2 Addition and multiplication

Addition and subtraction

Addition of two matrices **A** and **B** is only possible if the two matrices have the same dimension. The sum is a new matrix **C**, with the same dimension, and its elements are found as the sum of corresponding elements in **A** and **B**, that is

$$\mathbf{C} = \mathbf{A} + \mathbf{B} \quad \text{means that} \quad C_{ij} = A_{ij} + B_{ij} \tag{C-8}$$

for all *i* and *j*. Similarly

$$\mathbf{G} = \mathbf{A} - \mathbf{B} \quad \text{means that} \quad G_{ij} = A_{ij} - B_{ij} \tag{C-9}$$

for all *i* and *j*. Obviously

$$\mathbf{A} + \mathbf{B} = \mathbf{B} + \mathbf{A} \quad \text{(commutative)} \tag{C-10}$$

and

$$\mathbf{A} + (\mathbf{B} + \mathbf{C}) = (\mathbf{A} + \mathbf{B}) + \mathbf{C} \quad \text{(associative)} \tag{C-11}$$

Here we can replace one or more + signs with − signs. Also

$$\mathbf{A} + \mathbf{0} = \mathbf{A} \quad \text{and} \quad \mathbf{A} + (-\mathbf{A}) = \mathbf{0} \tag{C-12}$$

Multiplication by a scalar constant

Every element of the matrix is multiplied by the constant, that is,

$$\mathbf{E} = s\mathbf{A} = s[A_{ij}] = [sA_{ij}] \quad \text{or} \quad E_{ij} = sA_{ij} \quad \text{for all } i \text{ and } j \tag{C-13}$$

It follows that

$$(s_1 + s_2)\mathbf{A} = s_1\mathbf{A} + s_2\mathbf{A}, \quad s(\mathbf{A} + \mathbf{B}) = s\mathbf{A} + s\mathbf{B} \quad \text{and} \quad s_1(s_2\mathbf{A}) = (s_1 s_2)\mathbf{A}$$

Hence also the *distributive* law applies to matrix addition and subtraction.

Matrix multiplication

Solution of simultaneous, linear equations and linear transformations are central operations in linear algebra. It is therefore natural to define matrix multiplication, which is the "backbone" of matrix algebra, with due regard to these problems. The matrix equation

$$\mathbf{Ax} = \mathbf{b} \tag{C-14}$$

where **A** is a matrix of coefficients, **x** is a vector of unknown parameters and **b** a vector of known numbers (the right-hand side), symbolizes a system of linear equations.

Equation (A-14) implies multiplication of **A** and **x** that needs to be defined in such a way that an arbitrary element of **b** is defined by

Matrix algebra

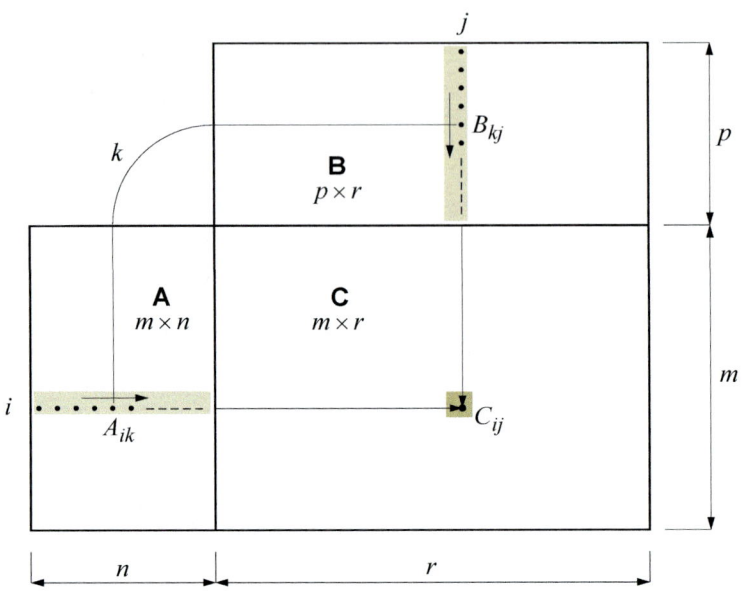

FALK-diagram for the matrix multiplication **C = AB**

Fortran 90 code for the same multiplication:

```
DO  i = 1,m                                  ! row-loop
    DO  j = 1,n                              ! column-loop
        C(i,j) = 0.0
        DO  k = 1,p                          ! product-sum-loop
            C(i,j) = C(i,j) + A(i,k) * B(k,j)
        END DO
    END DO
END DO
```

Addition and multiplication

$$b_i = \sum_{k=1}^{n} A_{ik}x_k \qquad \text{for all } i\ (=1, 2, \ldots, m) \tag{C-15}$$

This we can generalize to read as follows for the multiplication of two matrices **A**, with dimension $m \times n$, and **B**, with dimension $p \times r$:

In order for the matrix product **AB** (in this sequence) to be defined, matrix **A** must have the same number of columns as **B** has rows, and for $n = p$ the product **AB** is an $m \times r$ matrix **C** whose general element is defined by the product sum

$$C_{ij} = A_{i1}B_{1j} + A_{i2}B_{2j} + \ldots = \sum_{k=1}^{n} A_{ik}B_{kj} \tag{C-16}$$

The FALK diagram to the left is a very useful and illustrative way to demonstrate matrix multiplication. In this diagram we also see very clearly the requirement the two matrices need to satisfy in order for them to *conform* for multiplication, namely $n = p$.

Matrix multiplication is a demanding computational operation. If we define one multiplication of two scalars and one succeeding addition as one *operation*, the number of operations is $m \times r \times n$ for the matrix multiplication on the opposite page; multiplying two 100×100 matrices requires one million operations.

The computer is extremely efficient when it comes to repeating a series of simple operations in so-called *loops*. This is one of the reasons why matrices and matrix formulations are so popular in connection with program driven numerical computations. Just about all basic matrix algebra operations can be formulated as loops that are easy to program. The example opposite indicates how the code of a typical programming language (here Fortran 90) may look for straightforward matrix multiplication.

Matrix multiplication is, except in some very special cases, *not commutative*, that is

$$\mathbf{AB} \neq \mathbf{BA} \tag{C-17}$$

If **A** and **B** are not square matrices, and **AB** is defined, **BA** is normally not even defined, see example on opposite page. We can also have **AB** = **0** even if both **A** and **B** are non-zero matrices.

Example

		2	-4			1	2
		-1	2			3	6
1	2	0	0	2	-4	-10	-20
3	6	0	0	-1	2	5	10

　　　　AB = 0　　　　　　　　　　**BA ≠ AB**

Matrix algebra

$$A = \begin{bmatrix} 1 & 0 \\ 2 & 0 \end{bmatrix} \qquad B = \begin{bmatrix} 2 & 3 \\ -4 & 1 \end{bmatrix} \qquad C = \begin{bmatrix} 2 & 3 \\ 1 & 5 \end{bmatrix}$$

$$AB = AC = \begin{bmatrix} 2 & 3 \\ 4 & 6 \end{bmatrix}$$

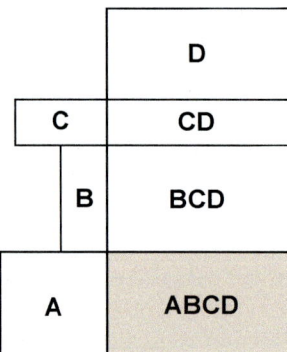

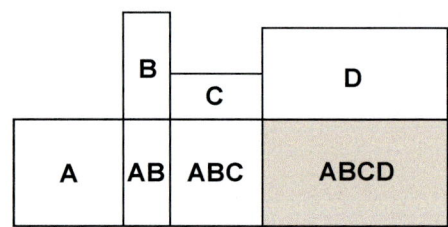

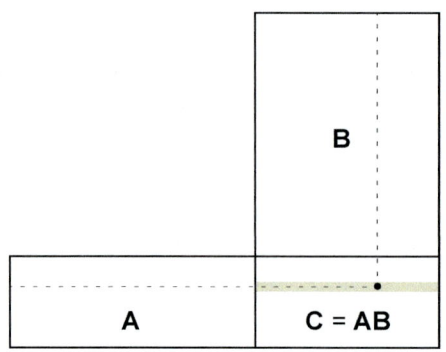

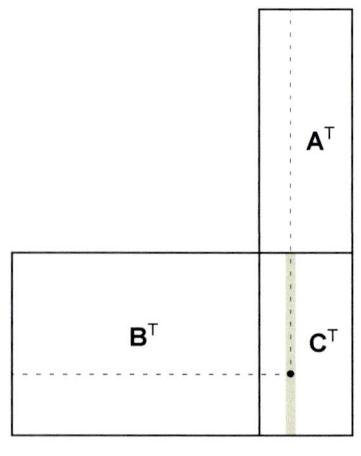

$C^T = B^T A^T$

Addition and multiplication

Matrix multiplication obeys both the *distributive* and *associative* laws, that is

$$(AB)C = A(BC) \qquad (C\text{-}18)$$

$$A(B+C) = AB + AC \qquad (C\text{-}19)$$

$$s(AB) = (sA)B = A(sB) \qquad (C\text{-}20)$$

For square matrices the following notation is often used

$$AA = A^2, \quad AAA = A^3 \quad \text{etc}$$

It should be noted that

$$AB = AC$$

does not necessarily mean that $B = C$. For that to be the case matrix A must be square and have certain properties (it must be non-singular); more about this later.

In the multiplication AB we say that B is *premultiplied* by A while A is *postmultiplied* by B.

The FALK diagram is particularly useful in connection with matrix products of more than two matrices, for instance

$$P = ABCD$$

If we start from "behind", that is with matrix D, we get the diagram to the left on the opposite page. In the diagram to the right, on the other hand, we start with matrix A. The final result is the same, but the number of operations required and the storage space needed may be quite different, depending on the sequence in which the multiplications are carried out.

The *transpose* of the product of two matrices is the same as the product of the transposed matrices *in reversed order*, that is

$$(AB)^T = B^T A^T \qquad (C\text{-}21)$$

That this is so is perhaps most easily seen by the FALK diagram opposite. Equation (A-21) can be generalized to the following important rule:

$$(ABC...S)^T = S^T...C^T B^T A^T \qquad (C\text{-}22)$$

Vector products

The *scalar* or *inner product*, also called the *dot product* of two (column) vectors u and v is defined as

$$s = u^T v = v^T u = \sum_{k=1}^{n} u_k v_k \qquad (C\text{-}23)$$

It is assumed that both vectors have the same dimension (n). The inner product is a very simple matrix product, and the multiplication of two general matrices A and B can be viewed as a series of inner vector products. If we consider the $m \times n$ matrix A as a collection of m row vectors a_i^T of dimension n,

Matrix algebra

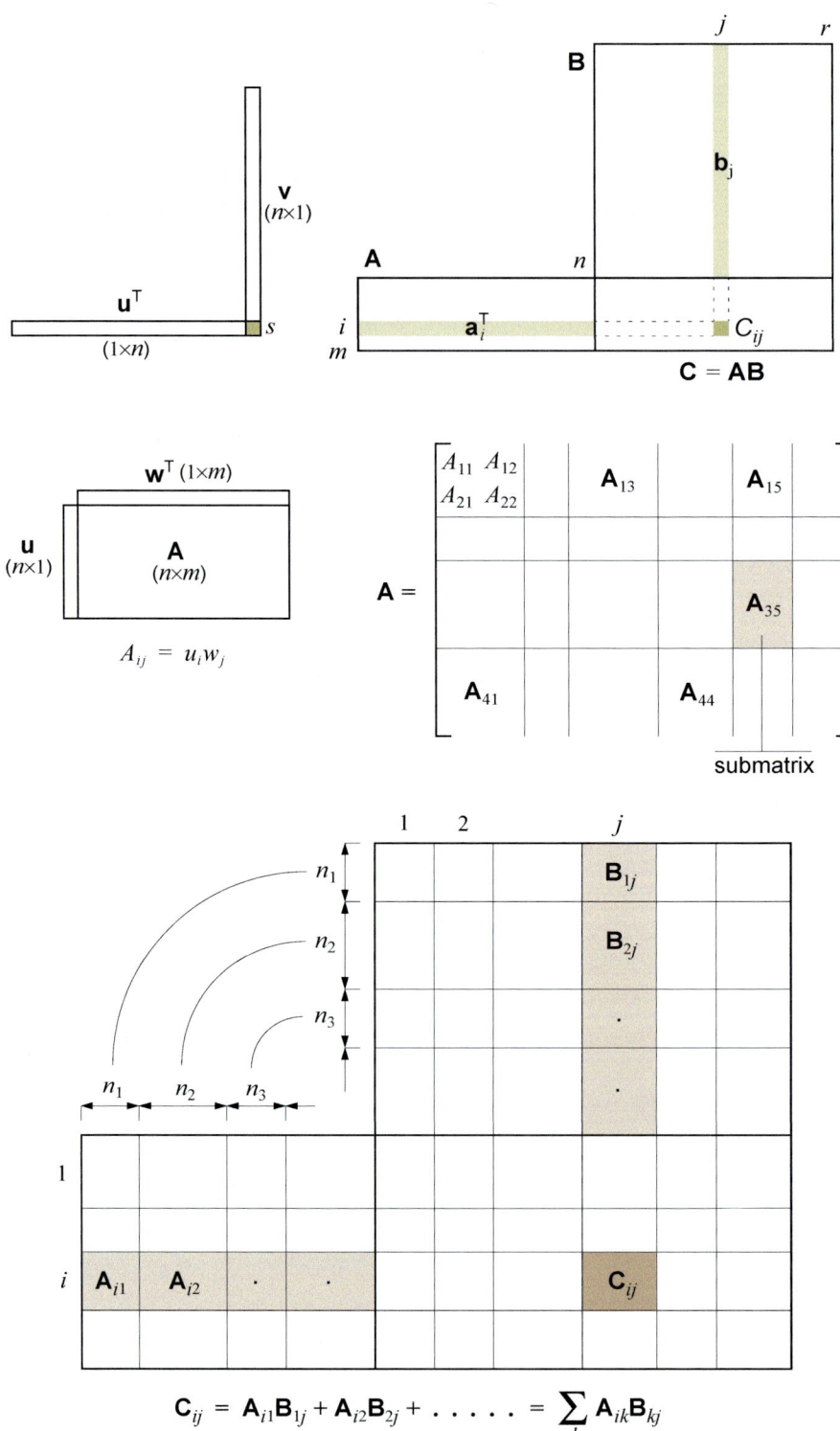

and the $n \times r$ matrix **B** as a collection of r column vectors $\mathbf{b}_j$ of dimension n, then the general element of the matrix $\mathbf{C} = \mathbf{AB}$ can be expressed $C_{ij} = \mathbf{a}_i^T \mathbf{b}_j$

The *outer product* of two column vectors **u** and **w** is a matrix

$$\mathbf{A} = \mathbf{u}\mathbf{w}^T \tag{C-24}$$

is also called the *dyadic product* of **u** and **w**. This again is a special matrix product, see opposite page. **u** and **w** may have different dimensions, but in general we have

$$\mathbf{u}\mathbf{w}^T \neq \mathbf{w}\mathbf{u}^T$$

even if the vectors have the same dimension. On the other hand

$$\mathbf{u}\mathbf{w}^T = (\mathbf{w}\mathbf{u}^T)^T$$

The *vector* or *cross* product, which applies to "physical" vectors in 3D space, will be dealt with in APPENDIX B.

A.3 Matrix partitioning – submatrices

It is sometimes convenient to partition a matrix by vertical and horizontal lines between its columns and rows, respectively. The rectangular "blocks" between these lines are called *submatrices*. A matrix **A** is thus made up of submatrices $\mathbf{A}_{ij}$ in which the indices have the same meaning as for scalar elements, that is, i is the number of the submatrix row and j the number of the submatrix column.

The partitioning is arbitrary and defined by whatever purpose it is meant to serve; however, in order for the basic matrix operations, such as addition and multiplication, also to apply when the "matrix elements" are submatrices, we need to make sure that the submatrices *conform* to the operation. This is particularly important for multiplication of two partitioned matrices, **A** and **B**, as shown on the opposite page.

A.4 Determinants

The *determinant* is a scalar property of a *square* matrix that, both directly and indirectly, characterizes the matrix.

The determinant of an $n \times n$ matrix **A** is commonly denoted as:

$$\det(\mathbf{A}) = \det\mathbf{A} = |\mathbf{A}| = \begin{vmatrix} A_{11} & A_{12} & \cdots & A_{1n} \\ \cdot & & & \\ & \cdot & & \\ A_{n1} & \cdot & \cdot & \cdot \end{vmatrix} \tag{C-25}$$

For $n = 2$, the basic case, the determinant is defined as:

Matrix algebra

> **CRAMER's *rule***
>
> Consider a system of n linear equations in matrix notation:
>
> $$\mathbf{Ax} = \mathbf{b}$$
>
> where the $n \times n$ matrix $\mathbf{A}$ has a *non-zero* determinant $\det(\mathbf{A})$.
>
> CRAMER's rule now states that this system of equations has a unique solution, and the individual values for the unknowns are given by
>
> $$x_i = \frac{\det(\mathbf{A}_i)}{\det(\mathbf{A})}$$
>
> where $\mathbf{A}_i$ is the matrix formed by replacing the ith column of $\mathbf{A}$ by vector $\mathbf{b}$.

$$\det(\mathbf{A}) = \begin{vmatrix} A_{11} & A_{12} \\ A_{21} & A_{22} \end{vmatrix} = A_{11}A_{22} - A_{12}A_{21} \qquad \text{(C-26)}$$

This definition can be extended to an arbitrary value of n by introducing the concepts of minor and cofactor.

The *minor* M_{ij} of $\det(\mathbf{A})$ is the determinant we are left with after we have deleted row number i and column number j of $|\mathbf{A}|$. The minor is itself a determinant, but of one order lower than the determinant it is associated with.

The *cofactor* C_{ij} of element A_{ij} of $|\mathbf{A}|$ is defined as

$$C_{ij} = (-1)^{i+j} M_{ij} \qquad \text{(C-27)}$$

The determinant of an arbitrary ($n > 2$) matrix $\mathbf{A}$ can now be defined as

$$|\mathbf{A}| = \sum_{k=1}^{n} A_{ik} C_{ik} \quad \text{developed by line } i \qquad \text{(C-28)}$$

or

$$|\mathbf{A}| = \sum_{k=1}^{n} A_{kj} C_{kj} \quad \text{developed by column } j \qquad \text{(C-29)}$$

These general formulas are *recursive*, C_{ik} and C_{kj} are themselves determinants, but of one order lower than $|\mathbf{A}|$, and we can work our way "down" until we are left with many determinants of order 2. However, computing the determinant by these formulas is extremely time consuming since the number of operations is $n!$ (n factorial).

Example

For a computer that can perform 1 billion (1 000 000 000) operations in a second it will take about half a second to compute the determinant of a 12×12 matrix by the recursive formulas above. However, already for $n = 20$ it will take about 77 years (!).

From this it follows that if we need to determine the determinant explicitly – which is not very often – we cannot use the development formula (A-28) or (A-29); we need to use other (indirect) methods, one of which is mentioned below. For the same reason, straightforward use of CRAMER's rule (see opposite page) is not a very useful method for solving systems of linear equations.

Without proof we list the following useful properties of determinants:

1) The value of the determinant does not change if rows and columns are interchanged, that is,

$$\det(\mathbf{A}) = \det(\mathbf{A}^T) \qquad \text{(C-30)}$$

2) If any two rows (or columns) of a determinant are interchanged the determinant changes sign, but its absolute value remains unchanged.

3) If *all* elements of a row (or column) of $\det(\mathbf{A})$ are multiplied by a scalar c, the value of the determinant is $c \times \det(\mathbf{A})$.

Matrix algebra

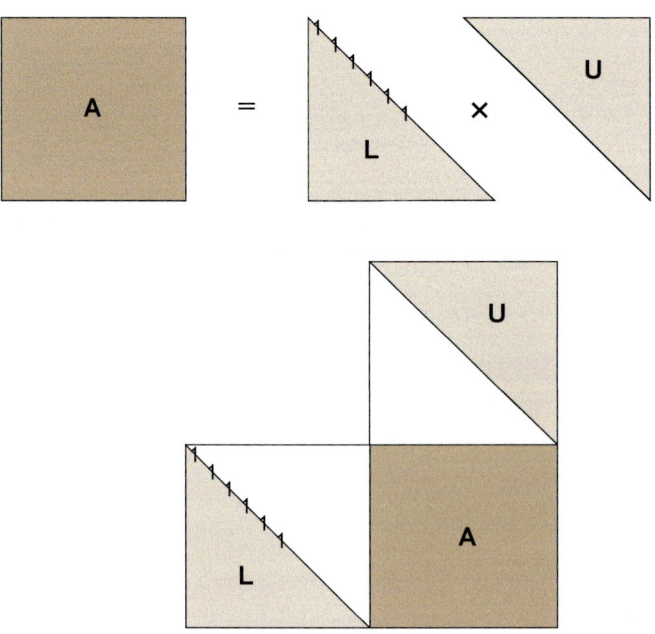

Determinants

4) The determinant does not change its value if a constant multiple of a row (column) is added to another row (column).

5) If *all* elements of a row (column) are equal to zero, the determinant is also equal to zero.

6) If corresponding elements of two rows (or two columns) of a determinant are *proportional*, the value of the determinant is zero.

7) The determinant of a product of matrices is equal to the product of the determinants of the individual factors:

$$\det(\mathbf{AB}) = \det(\mathbf{A})\det(\mathbf{B}) \qquad \text{(C-31)}$$

A and **B** must both be *square*.

Some of these properties can be used to simplify the computation of the determinant of a matrix **A**. Triangular matrices are central here. Consider for instance a lower triangular matrix **L**. If we develop the determinant of **L** by its first row and use property 5 we have:

$$\det(\mathbf{L}) = L_{11} \begin{vmatrix} L_{22} & 0 & 0 & . & . \\ L_{32} & L_{33} & 0 & . & . \\ . & . & . & . & . \\ . & . & . & . & . \\ L_{n2} & L_{n3} & . & . & L_{nn} \end{vmatrix} = L_{11} L_{22} \begin{vmatrix} L_{33} & 0 & . & . \\ L_{43} & L_{44} & 0 & . \\ . & . & . & . \\ L_{n3} & L_{n4} & . & L_{nn} \end{vmatrix} \quad \text{etc}$$

Hence

$$\det(\mathbf{L}) = L_{11} L_{22} \ldots L_{nn} \qquad \text{(C-32)}$$

The same applies to an upper triangular matrix U, that is,

$$\det(\mathbf{U}) = U_{11} U_{22} \ldots U_{nn} \qquad \text{(C-33)}$$

If the determinant of an $n \times n$ matrix **A** is *different* from zero, the matrix can *always* be decomposed (or factorized) and expressed as the product of a lower and an upper triangular matrix, that is,

$$\mathbf{A} = \mathbf{LU}$$

Consequently

$$\det(\mathbf{A}) = \det(\mathbf{L})\det(\mathbf{U}) = L_{11} L_{22} \ldots L_{nn} U_{11} U_{22} \ldots U_{nn} \qquad \text{(C-34)}$$

It is fairly straightforward to determine **L** and **U**, and (A-34) can be further simplified by the fact that we can choose the diagonal elements of one of the triangular matrices, for instance $L_{ii} = 1$ (or $U_{ii} = 1$) for all i.

Matrix algebra

$$\mathbf{a}_1 = \begin{bmatrix} a_{11} \\ a_{21} \\ a_{31} \\ \cdot \\ \cdot \\ \cdot \\ \cdot \\ a_{m1} \end{bmatrix} \quad \mathbf{a}_2 = \begin{bmatrix} a_{12} \\ a_{22} \\ a_{32} \\ \cdot \\ \cdot \\ \cdot \\ \cdot \\ a_{m2} \end{bmatrix} \quad \cdots \quad \mathbf{a}_i = \begin{bmatrix} a_{1i} \\ a_{2i} \\ a_{3i} \\ \cdot \\ \cdot \\ \cdot \\ \cdot \\ a_{mi} \end{bmatrix} \quad \cdots \quad \mathbf{a}_n = \begin{bmatrix} a_{1n} \\ a_{2n} \\ a_{3n} \\ \cdot \\ \cdot \\ \cdot \\ \cdot \\ a_{mn} \end{bmatrix}$$

$$\mathbf{A} = \begin{bmatrix} \mathbf{a}_1 & \mathbf{a}_2 & \cdots & \mathbf{a}_i & \cdots & \mathbf{a}_n \end{bmatrix}$$

A.5 Linear dependencies – rank

We say that n vectors $\mathbf{a}_1, \mathbf{a}_2, \ldots, \mathbf{a}_n$, of the same dimension m, are **linear dependent** if

$$c_1\mathbf{a}_1 + c_2\mathbf{a}_2 + \ldots\ldots + c_n\mathbf{a}_n = \mathbf{0} \qquad \text{(C-35)}$$

and not all coefficients $c_1, c_2, \ldots, c_n$ are equal to zero. If (A-35) can only be satisfied if *all* coefficients (c_i) are equal to zero, the vector system is said to be **linear independent**. The largest number of *linear independent* vectors that can be picked from the system of n vectors is defined as the **rank** of the system and denoted r. Obviously

$$0 \leq r \leq n$$

and $r = 0$ means that all vectors are equal to $\mathbf{0}$.

It can also be shown that the rank cannot be greater than m, the dimension of the individual vectors; hence

$$0 \leq r \leq \min\{n, m\} \qquad \text{(C-36)}$$

An $m \times n$ matrix $\mathbf{A}$ can be viewed as a system of n vectors, that is

$$\mathbf{A} = [\mathbf{a}_1 \ \mathbf{a}_2 \ \ldots \ \mathbf{a}_n]$$

where $\mathbf{a}_j$ is column number j of the matrix. Rank is therefore also a property of a matrix; it is denoted rank($\mathbf{A}$) or just r. From what is said above it follows that

$$0 \leq \text{rank}(\mathbf{A}) \leq \min\{m, n\} \qquad \text{(C-37)}$$

rank($\mathbf{A}$) = $\mathbf{0}$ is possible only if $\mathbf{A} = \mathbf{0}$.

The rank defined above for $\mathbf{A}$ is the *column rank* for the matrix, and it is the largest number of linearly independent columns of $\mathbf{A}$. We can also define a *row rank* for $\mathbf{A}$ which is the largest number of linearly independent rows of $\mathbf{A}$. A third definition of the rank of an $m \times n$ matrix $\mathbf{A}$ is its *determinant rank* defined as follows: The rank of $\mathbf{A}$ is equal to the dimension of the largest square submatrix in $\mathbf{A}$ that has a *determinant* different from zero. It can be shown that all three ranks are the same, that is,

rank($\mathbf{A}$) = column rank($\mathbf{A}$) = row rank($\mathbf{A}$) = determinant rank($\mathbf{A}$)

A *square $n \times n$* matrix $\mathbf{A}$ with rank equal to n, that is $r = \text{rank}(\mathbf{A}) = n$, which is the same as saying that det($\mathbf{A}$) $\neq 0$, is said to be *regular* or *non-singular*. If, on the other hand, det($\mathbf{A}$) = 0, than $\mathbf{A}$ is said to be *singular*. The **defect** d of a square matrix is the difference between its dimension and rank, that is,

$$d = n - r \qquad \text{(C-38)}$$

A matrix whose defect is greater than zero is said to be *rank deficient*.

The concepts of linear dependencies, rank and defect are closely related to solution of systems of linear equations and matrix inversion.

Matrix algebra

$$\mathbf{Ax} = \mathbf{b}$$

$$\begin{pmatrix} x_1 \\ x_2 \\ \cdot \\ \cdot \\ \cdot \\ x_i \\ \cdot \\ \cdot \\ x_n \end{pmatrix}$$

$\mathbf{a}_1 \quad \mathbf{a}_2 \quad \cdots \quad \mathbf{a}_i \quad \cdots \quad \mathbf{a}_n \quad \mathbf{b}$

A.6 Linear systems of equations

In general, a system of simultaneous, linear equations can comprise m equations in n unknowns. We will assume that $m = n$, that is, we have the same number of equations as we have unknowns. Hence, the matrix of coefficients is *square*, and we denote this $n \times n$ matrix by **A** while the unknown quantities $x_1, x_2, \ldots, x_n$ are the elements of vector **x**.

The *homogeneous* system of equations

$$\mathbf{Ax} = \mathbf{0} \tag{C-39}$$

has *non-trivial* solutions, that is solutions for which $\mathbf{x} \neq \mathbf{0}$, if and only if the columns of **A** (i.e. $\mathbf{a}_1, \mathbf{a}_2, \ldots, \mathbf{a}_n$) are linear dependent. We see this by expanding **Ax**:

$$\mathbf{Ax} = \mathbf{a}_1 x_1 + \mathbf{a}_2 x_2 + \ldots + \mathbf{a}_n x_n$$

and comparing this with the definition of linear dependency, see Eq. (A-35). The condition for a non-trivial solution of (A-39) is therefore that

$$\text{rank}(\mathbf{A}) < n \quad \text{or} \quad \det(\mathbf{A}) = 0$$

and it can be shown that we have $d = n - \text{rank}(\mathbf{A})$ linearly independent solution vectors.

If the right-hand side $\mathbf{b} \neq \mathbf{0}$ we have an *inhomogeneous* system of equations:

$$\mathbf{Ax} = \mathbf{b} \tag{C-40}$$

where $\mathbf{b}^T = [b_1 \; b_2 \; \ldots \; b_n]$; the elements $b_1, b_2, \ldots, b_n$ represent *known* quantities that are not all zero. Equation (A-40) can also be written as

$$\mathbf{a}_1 x_1 + \mathbf{a}_2 x_2 + \ldots + \mathbf{a}_n x_n = \mathbf{b} \tag{C-41}$$

that is, we seek a linear combination of the columns of **A** that is equal to **b**. It follows from the above that for (A-40) to have a solution we must have

$$\text{rank}(\mathbf{A}) = n \quad \text{or} \quad \det(\mathbf{A}) \neq 0$$

or **A** must be *regular* (*non-singular*). It can be shown that this is also a necessary and sufficient condition for a *unique* solution.

A.7 Matrix inversion

The *inverse* of an $n \times n$ matrix **A** is another $n \times n$ matrix, denoted $\mathbf{A}^{-1}$, that satisfies the condition

$$\mathbf{AA}^{-1} = \mathbf{A}^{-1}\mathbf{A} = \mathbf{I} \tag{C-42}$$

If **A** has an inverse, which it does not necessarily have, then $\mathbf{A}^{-1}$ is *unique*. The following argument supports this. Assume that both **B** and **C** are the inverse of **A**; then

$$\mathbf{AB} = \mathbf{I} \quad \text{and} \quad \mathbf{CA} = \mathbf{I}$$

It follows that

Matrix algebra

It can be shown that

$$\mathbf{A}^{-1} = \frac{1}{\det(\mathbf{A})} \begin{bmatrix} C_{11} & C_{21} & . & C_{n1} \\ C_{12} & C_{22} & . & . \\ . & . & . & . \\ C_{1n} & C_{2n} & . & C_{nn} \end{bmatrix} = \frac{\text{adj}(\mathbf{A})}{\det(\mathbf{A})}$$

where the *adjoint* of matrix A is defined in terms of its cofactors as

$$\text{adj}(\mathbf{A}) = [C_{ij}]^T$$

Consider

$$\mathbf{A}\mathbf{x} = \mathbf{b}$$

The solution can be expressed as

$$\mathbf{A}^{-1}\mathbf{A}\mathbf{x} = \mathbf{A}^{-1}\mathbf{b} \quad \Rightarrow \quad \mathbf{I}\mathbf{x} = \mathbf{x} = \mathbf{A}^{-1}\mathbf{b}$$

Matrix inversion

$$B = IB = (CA)B = C(AB) = CI = C$$

The inverse of a matrix can also be expressed in terms of its *adjoint* matrix and its determinant as shown opposite. This expression can be used to establish the following general formula for the inverse of a 2×2 matrix:

$$\mathbf{A} = \begin{bmatrix} A_{11} & A_{12} \\ A_{21} & A_{22} \end{bmatrix} \quad \Rightarrow \quad \mathbf{A}^{-1} = \frac{1}{A_{11}A_{22} - A_{12}A_{21}} \begin{bmatrix} A_{22} & -A_{12} \\ -A_{21} & A_{11} \end{bmatrix} \quad \text{(C-43)}$$

For larger matrices however this approach, which again requires a number of operations proportional with the factorial of the matrix dimension, does not work in practice.

If we require the inverse matrix explicitly we need to use a different approach, and in practice this almost inevitably involves solution of a system of linear equations, as indicated by the following reasoning:

Denote the columns of $\mathbf{A}^{-1}$ by $\mathbf{z}_1, \mathbf{z}_2, \ldots, \mathbf{z}_n$, that is

$$\mathbf{A}^{-1} = [\mathbf{z}_1 \ \mathbf{z}_2 \ \ldots \ \mathbf{z}_n]$$

Then

$$\mathbf{A}\mathbf{A}^{-1} = [\mathbf{A}\mathbf{z}_1 \ \mathbf{A}\mathbf{z}_2 \ \ldots \ \mathbf{A}\mathbf{z}_n] = \mathbf{I} = [\mathbf{e}_1 \ \mathbf{e}_2 \ \ldots \ \mathbf{e}_n]$$

where the vector $\mathbf{e}_i$ has only zero elements except for element number i which is equal to 1 (one). The matrices $\mathbf{A}\mathbf{A}^{-1}$ and $\mathbf{I}$ must be identical column by column; hence

$$\mathbf{A}\mathbf{z}_i = \mathbf{e}_i$$

$\mathbf{A}^{-1}$ is therefore obtained by solving a system of equations, for which $\mathbf{A}$ is the matrix of coefficients, for n (simple) right-hand sides.

Inversion can of course also be used to solve systems of equations, as indicated opposite. However, in view of the above reasoning this is not at all efficient. Hence, while

$$\mathbf{x} = \mathbf{A}^{-1}\mathbf{b} \quad \text{(C-44)}$$

is a mathematically convenient way to express the solution vector, it should be emphasized that we will, in practice, *never* invert the matrix of coefficients in connection with equation solving (unless, in the unlikely event, we need to find the inverse matrix for other reasons).

By using the properties of determinants we see that

$$\det(\mathbf{A}\mathbf{A}^{-1}) = \det(\mathbf{I}) = 1$$

or

$$\det(\mathbf{A})\det(\mathbf{A}^{-1}) = 1 \quad \Rightarrow \quad \det(\mathbf{A}^{-1}) = \frac{1}{\det(\mathbf{A})} \quad \text{(C-45)}$$

We leave it to the reader to verify the following properties of the inverse

$$(\mathbf{A}^{-1})^{-1} = \mathbf{A} \quad \text{(C-46)}$$

Matrix algebra

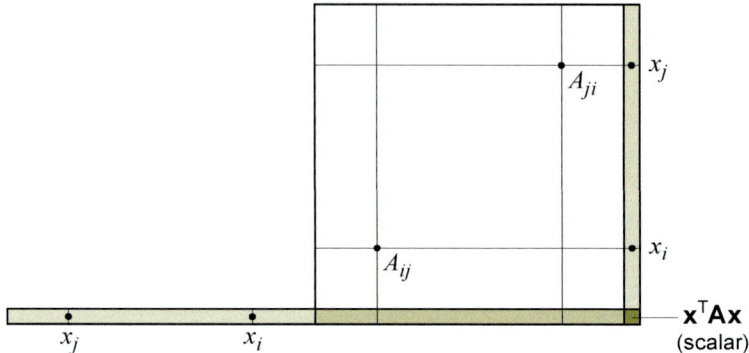

$$(\mathbf{A}^{-1})^{\mathsf{T}} = (\mathbf{A}^{\mathsf{T}})^{-1} = \mathbf{A}^{-\mathsf{T}} \tag{C-47}$$

$$(\mathbf{AB})^{-1} = \mathbf{B}^{-1}\mathbf{A}^{-1} \tag{C-48}$$

The last term of (A-47) indicates usual notation.

A square matrix **A** is *orthogonal* if its inverse is equal to its transpose, that is if

$$\mathbf{A}^{-1} = \mathbf{A}^{\mathsf{T}} \quad \Rightarrow \quad \mathbf{A}^{-1}\mathbf{A}^{\mathsf{T}} = \mathbf{A}^{\mathsf{T}}\mathbf{A}^{-1} = \mathbf{I} \tag{C-49}$$

If we consider **A** to be a collection of vectors $(\mathbf{a}_i)$ then (A-49) means that $\mathbf{a}_i^{\mathsf{T}}\mathbf{a}_j = \delta_{ij}$. Hence the vectors $\mathbf{a}_i$ and $\mathbf{a}_j$ $(i \neq j)$ are orthogonal, that is, they are normal to each other in n-dimensional space.

If matrix **A** satisfies the condition

$$\mathbf{A}^{\mathsf{T}}\mathbf{A} = \mathbf{A}\mathbf{A}^{\mathsf{T}} \tag{C-50}$$

it is said to be *normal*. A real, symmetric matrix is normal.

A.8 Quadratic forms and definiteness

The *scalar* quantity $\mathbf{x}^{\mathsf{T}}\mathbf{A}\mathbf{x}$, where **A** is a *square* matrix and **x** is an arbitrary vector of the same dimension as **A**, is a so-called *quadratic form*. It follows from the FALK diagram opposite, that the quadratic form of a matrix is the same as that of its transpose, that is,

$$\mathbf{x}^{\mathsf{T}}\mathbf{A}\mathbf{x} = \mathbf{x}^{\mathsf{T}}\mathbf{A}^{\mathsf{T}}\mathbf{x} \tag{C-51}$$

From Eq. (A-6) we see that the quadratic form of a skew-symmetric matrix is zero, and since an arbitrary square matrix can be expressed as the sum of a symmetric component and a skew-symmetric component, see Eq. (A-7), it follows that the quadratic form of the arbitrary matrix is equal to the quadratic form of its symmetric component. If

$$\mathbf{x}^{\mathsf{T}}\mathbf{A}\mathbf{x} > 0 \quad \text{for all} \quad \mathbf{x} \neq \mathbf{0} \tag{C-52}$$

matrix **A** is *positive definite*. This is an important property in connection with several numerical operations.

It follows from the definition (A-52) that the determinant of a positive definite matrix must be different from zero. To show this we start by assuming that (A-52) is also valid for the case that $\det(\mathbf{A}) = 0$. However, as explained in connection with Eq. (A-39), $\det(\mathbf{A}) = 0$ implies that the homogeneous system of equations $\mathbf{A}\mathbf{x} = \mathbf{0}$ has a non-trivial solution, and this is not consistent with (A-52). Hence

$$\det(\mathbf{A}) \neq 0 \quad \text{if} \quad \mathbf{A} \text{ is positive definite} \tag{C-53}$$

The opposite is *not* the case, that is, we cannot say that **A** is positive definite simply because $\det(\mathbf{A}) \neq 0$.

A matrix *cannot* be positive definite unless *all* its diagonal elements are *positive*. We see that this must be so by choosing a special, but permissible **x**:

Matrix algebra

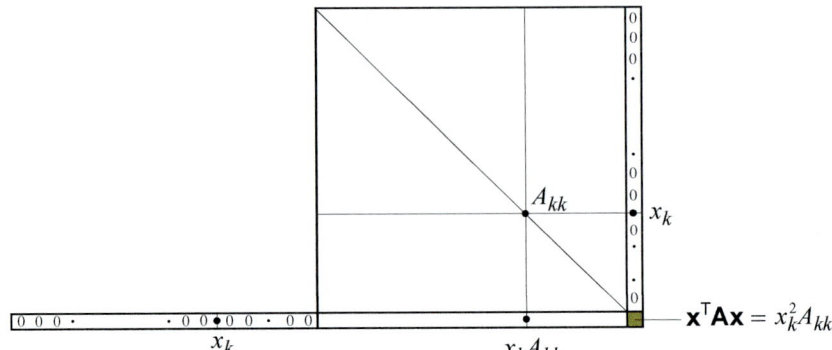

$$\mathbf{x}^T = [0 \ 0 \ 0 \ \ldots \ 0 \ x_k \ 0 \ \ldots \ 0]$$

With this **x**-vector the quadratic form becomes

$$\mathbf{x}^T \mathbf{A} \mathbf{x} = A_{kk} x_k^2$$

For this form to be positive, A_{kk} must be positive, and the claim is verified. A quadratic matrix **A** is *positive semi-definite* if

$$\mathbf{x}^T \mathbf{A} \mathbf{x} \geq 0 \quad \text{for all} \quad \mathbf{x} \neq \mathbf{0} \tag{C-54}$$

and it is *indefinite* if

$$(\mathbf{x}^T \mathbf{A} \mathbf{x})(\mathbf{y}^T \mathbf{A} \mathbf{y}) < 0 \quad \text{for some} \quad \mathbf{x} \text{ and } \mathbf{y} \tag{C-55}$$

A.9 The eigenvalue problem

We consider two *real* and *symmetric* $n \times n$ matrices **A** and **B**. The equation

$$(\mathbf{A} - \lambda \mathbf{B})\mathbf{q} = \mathbf{0} \quad \text{or} \quad \mathbf{A}\mathbf{q} = \lambda \mathbf{B}\mathbf{q} \tag{C-56}$$

defines the *general, symmetric eigenvalue problem*. If **B** is the *unit* matrix, that is $\mathbf{B} = \mathbf{I}$, Eq. (A-56) becomes

$$(\mathbf{A} - \lambda \mathbf{I})\mathbf{q} = \mathbf{0} \quad \text{or} \quad \mathbf{A}\mathbf{q} = \lambda \mathbf{q} \tag{C-57}$$

which defines the *special* (standard), *symmetric eigenvalue problem*. In these equations the scalar λ represents the *eigenvalue* of the problem, while **q** is the corresponding *eigenvector*.

The special eigenvalue problem (A-57) has n non-trivial solutions the ith of which is defined by the eigen-pair $(\lambda_i, \mathbf{q}_i)$ that satisfies

$$\mathbf{A}\mathbf{q}_i = \lambda_i \mathbf{q}_i \tag{C-58}$$

We normally order the eigenvalues such that

$$\lambda_1 \leq \lambda_2 \leq \ldots \leq \lambda_n$$

The eigenvectors are unique save for a scale factor; if $\mathbf{q}_i$ is an eigenvector, then $s\mathbf{q}_i$, where s is a scalar, is also an eigenvector. Hence, the eigenvectors are usually *normalized*, for instance such that their "length" is equal to unity, that is

$$|\mathbf{q}_i| = \sqrt{\|\mathbf{q}_i\|} = \sqrt{\mathbf{q}_i^T \mathbf{q}_i} = 1 \tag{C-59}$$

The scalar $\|\mathbf{q}_i\| = \mathbf{q}_i^T \mathbf{q}_i$ is the EUCLIDEAN norm or 2-*norm* of the vector $\mathbf{q}_i$.

For equation (A-57) to have non-trivial solutions we need to have

$$\det(\mathbf{A} - \lambda \mathbf{I}) = 0 \tag{C-60}$$

We can develop this determinant into an algebraic *polynomial* in λ of order n,

$$p(\lambda) = \lambda^n + \alpha_1 \lambda^{n-1} + \alpha_2 \lambda^{n-2} + \ldots + \alpha_{n-1}\lambda + \alpha_n = 0 \tag{C-61}$$

Matrix algebra

Properties of eigenvalues and eigenvectors of symmetric matrices
- All eigenvalues of a symmetric matrix are *real*.
- All eigenvalues of a *positive definite* matrix are positive (> 0).
- The sum of the eigenvalues of a matrix is equal to the *trace* of the matrix, i.e.,
$$\mathrm{tr}(\mathbf{A}) = A_{11} + A_{22} + .. + A_{nn} = \lambda_1 + \lambda_2 + .. + \lambda_n$$
- The product of the eigenvalues of a matrix is equal to the determinant of the matrix, i.e.,
$$\det(\mathbf{A}) = \lambda_1 \lambda_2 \lambda_3 \ldots \lambda_n$$
- Since the determinant of a *triangular* matrix is simply the product of the *diagonal* elements, it follows that the eigenvalues of a triangular matrix, and hence also of a diagonal matrix, are equal to the diagonal elements.
- All eigenvectors corresponding to distinct eigenvalues are mutually *orthogonal*. Eigenvectors corresponding to multiple eigenvalues can be made orthogonal to each other.

$p(\lambda)$ is the *characteristic polynomial* of matrix **A**. This polynomial has n roots which are the eigenvalues of **A**; the roots are not necessarily distinct which means that two or more eigenvalues may coincide.

Let
$$\Lambda = \lceil \lambda_1 \; \lambda_2 \ldots \lambda_i \ldots \lambda_n \rfloor \qquad \text{(C-62)}$$
be a diagonal matrix of all eigenvalues, and
$$\mathbf{Q} = [\, \mathbf{q}_1 \; \mathbf{q}_2 \ldots \mathbf{q}_i \ldots \mathbf{q}_n \,] \qquad \text{(C-63)}$$
be a square matrix whose columns are the corresponding eigenvectors. We can now write (A-57) as
$$\mathbf{AQ} = \mathbf{Q}\Lambda \qquad \text{(C-64)}$$
and (A-56) as
$$\mathbf{AQ} = \mathbf{BQ}\Lambda \qquad \text{(C-65)}$$
Some important properties of eigenvalues and eigenvectors of *real, symmetric* matrices are listed opposite.

A.10 Differentiation

Let **x** and **y** be vectors of dimensions n and m, respectively,
$$\mathbf{x} = [x_i] = \begin{bmatrix} x_1 \\ x_2 \\ \vdots \\ x_n \end{bmatrix} \qquad \mathbf{y} = [y_i] = \begin{bmatrix} y_1 \\ y_2 \\ \vdots \\ y_m \end{bmatrix} \qquad \text{(C-66)}$$

The component y_i can be a function of all x_j; this can be expressed by saying that **y** is a function of **x**, or
$$\mathbf{y} = \mathbf{y}(\mathbf{x}) \qquad \text{(C-67)}$$
If $n = 1$, **x** reduces to a scalar we denote by x, and if $m = 1$, **y** reduces to a scalar denoted y.

We now define the derivative of a scalar with respect to a vector as
$$\frac{\partial y}{\partial \mathbf{x}} \stackrel{\text{def}}{=} \left[\frac{\partial y}{\partial x_j}\right] = \begin{bmatrix} \dfrac{\partial y}{\partial x_1} \\ \dfrac{\partial y}{\partial x_2} \\ \vdots \\ \dfrac{\partial y}{\partial x_n} \end{bmatrix} \qquad \text{(C-68)}$$

and the derivative of a vector with respect to a scalar as

Matrix algebra

$y = \mathbf{x}^T\mathbf{A}\mathbf{x}$

$$\frac{\partial y}{\partial x_i} = [0\ 0\ 0\ \ldots\ 1\ \ldots\ 0]\mathbf{A}\mathbf{x} + \mathbf{x}^T\mathbf{A}\begin{bmatrix}0\\0\\0\\\vdots\\1\\\vdots\\0\end{bmatrix}$$

Since both terms are scalars they do not change by being transposed; hence

$$\frac{\partial y}{\partial x_i} = [0\ 0\ 0\ \ldots\ 1\ \ldots\ 0][\mathbf{A} + \mathbf{A}^T]\mathbf{x}$$

and

$$\frac{\partial y}{\partial \mathbf{x}} = \begin{bmatrix}\frac{\partial y}{\partial x_1}\\\frac{\partial y}{\partial x_2}\\\vdots\end{bmatrix} = \begin{bmatrix}1 & & & \\ & 1 & & \\ & & 1 & \\ & & & \ddots \\ & & & & 1\end{bmatrix}[\mathbf{A} + \mathbf{A}^T]\mathbf{x} = \mathbf{A}\mathbf{x} + \mathbf{A}^T\mathbf{x}$$

Summary
Derivatives of a vector

y	$\frac{\partial y}{\partial \mathbf{x}}$
$\mathbf{A}\mathbf{x}$	$\mathbf{A}^T$
$\mathbf{x}^T\mathbf{A}$	$\mathbf{A}$
$\mathbf{x}^T\mathbf{x}$	$2\mathbf{x}$
$\mathbf{x}^T\mathbf{A}\mathbf{x}$	$\mathbf{A}\mathbf{x} + \mathbf{A}^T\mathbf{x}$

The elements of **A** are independent of **x**

Differentiation

$$\frac{\partial \mathbf{y}}{\partial x} \stackrel{\text{def}}{=} \left[\frac{\partial y_i}{\partial x}\right]^T = \left[\frac{\partial y_1}{\partial x} \quad \frac{\partial y_2}{\partial x} \quad \cdots \quad \frac{\partial y_m}{\partial x}\right] \qquad \text{(C-69)}$$

The derivative of a vector with respect to a vector follows naturally from these two definitions:

$$\frac{\partial \mathbf{y}}{\partial \mathbf{x}} = \begin{bmatrix} \frac{\partial y_1}{\partial x_1} & \frac{\partial y_2}{\partial x_1} & \cdots & \frac{\partial y_m}{\partial x_1} \\ \frac{\partial y_1}{\partial x_2} & \frac{\partial y_2}{\partial x_2} & \cdots & \frac{\partial y_m}{\partial x_2} \\ \vdots & \vdots & \ddots & \vdots \\ \frac{\partial y_1}{\partial x_n} & \frac{\partial y_2}{\partial x_n} & \cdots & \frac{\partial y_m}{\partial x_n} \end{bmatrix} \qquad \text{(C-70)}$$

From this it follows that if $\mathbf{A}$ is a matrix of constants and

$$\mathbf{y} = \mathbf{A}\mathbf{x} \quad \text{then} \quad \frac{\partial \mathbf{y}}{\partial \mathbf{x}} = \begin{bmatrix} A_{11} & A_{21} & \cdots & A_{n1} \\ A_{12} & A_{22} & \cdots & A_{n3} \\ \vdots & \vdots & \ddots & \vdots \\ A_{1n} & A_{2n} & \cdots & A_{nn} \end{bmatrix} = \mathbf{A}^T \qquad \text{(C-71)}$$

and if

$$\mathbf{y} = \mathbf{x}^T \mathbf{A} \quad \text{then} \quad \frac{\partial \mathbf{y}}{\partial \mathbf{x}} = \mathbf{A} \qquad \text{(C-72)}$$

Furthermore, if

$$y = y = \mathbf{x}^T \mathbf{A} \mathbf{x} \quad \text{then} \quad \frac{\partial y}{\partial \mathbf{x}} = \mathbf{A}\mathbf{x} + \mathbf{A}^T \mathbf{x} \qquad \text{(C-73)}$$

This result follows from what has been shown above; an alternative explanation is given opposite. If $\mathbf{A}$ in Eq. (A-73) is *symmetric* then

$$\frac{\partial y}{\partial \mathbf{x}} = 2\mathbf{A}\mathbf{x} \qquad \text{(C-74)}$$

If we continue the differentiation process in Eq. (A-73) we find that

$$\frac{\partial^2 y}{\partial \mathbf{x}^2} = \frac{\partial}{\partial \mathbf{x}}\left(\frac{\partial y}{\partial \mathbf{x}}\right) = \mathbf{A}^T + \mathbf{A} \qquad \text{(C-75)}$$

and if $\mathbf{A}$ is symmetric

$$\frac{\partial^2 y}{\partial \mathbf{x}^2} = 2\mathbf{A} \qquad \text{(C-76)}$$

Matrix algebra

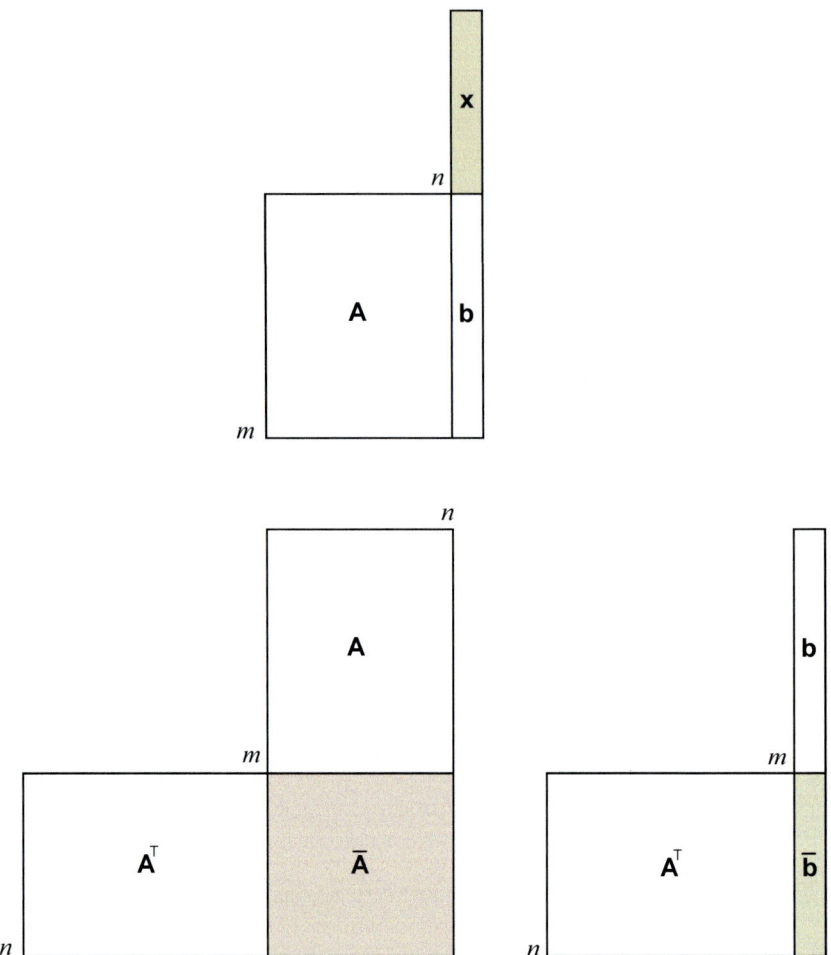

A.11 Least square approximation

Consider the following system of linear equations

$$\underset{m\times n}{\mathbf{A}} \underset{n\times 1}{\mathbf{x}} = \underset{m\times 1}{\mathbf{b}} \qquad m > n \qquad \text{(C-77)}$$

Here we have more equations than unknowns. Premultiplication by $\mathbf{A}^T$ yields

$$\underset{n\times n}{\bar{\mathbf{A}}} \underset{}{\mathbf{x}}^{n\times 1} = \underset{n\times 1}{\bar{\mathbf{b}}} \qquad \text{(C-78)}$$

where

$$\bar{\mathbf{A}} = \mathbf{A}^T \mathbf{A} = \bar{\mathbf{A}}^T \qquad \text{(C-79)}$$

is a symmetric (usually full) $n \times n$ matrix, and

$$\bar{\mathbf{b}} = \mathbf{A}^T \mathbf{b} \qquad \text{(C-80)}$$

Equation (A-78) is solved by standard methods to give

$$\mathbf{x} = \bar{\mathbf{A}}^{-1} \bar{\mathbf{b}} \qquad \text{(C-81)}$$

The *error* $\mathbf{e}$ of this solution is

$$\mathbf{e} = \mathbf{A}\mathbf{x} - \mathbf{b} \qquad \text{(C-82)}$$

We will now show that

$$e = \sum e_i^2 = \mathbf{e}^T \mathbf{e} \qquad \text{(C-83)}$$

is a *minimum* for $\mathbf{x}$ defined by Eq. (A-81), which means that the $\mathbf{x}$ we find by solving Eq. (A-78) is a *least square approximation* to Eq. (A-77).

From (A-82):

$$\mathbf{e}^T = \mathbf{x}^T \mathbf{A}^T - \mathbf{b}^T$$

Hence

$$e = \mathbf{e}^T \mathbf{e} = \mathbf{x}^T \mathbf{A}^T \mathbf{A} \mathbf{x} - \mathbf{x}^T \mathbf{A}^T \mathbf{b} - \mathbf{b}^T \mathbf{A} \mathbf{x} + \mathbf{b}^T \mathbf{b}$$

Here $\mathbf{b}^T \mathbf{A} \mathbf{x} = \mathbf{x}^T \mathbf{A}^T \mathbf{b}$ (a scalar is its own transpose); therefore

$$e = \mathbf{e}^T \mathbf{e} = \mathbf{x}^T \mathbf{A}^T \mathbf{A} \mathbf{x} - 2 \mathbf{x}^T \mathbf{A}^T \mathbf{b} + \mathbf{b}^T \mathbf{b} \qquad \text{(C-84)}$$

We seek the solution $\mathbf{x}$ that will give e a minimum; for this to happen we must have

$$\frac{\partial e}{\partial x_i} = 0 \quad \text{for } i = 1, 2, \ldots, n \quad \text{or} \quad \frac{\partial e}{\partial \mathbf{x}} = 0 \qquad \text{(C-85)}$$

or

$$\frac{\partial e}{\partial \mathbf{x}} = \mathbf{A}^T \mathbf{A} \mathbf{x} + (\mathbf{A}^T \mathbf{A})^T \mathbf{x} - 2 \mathbf{A}^T \mathbf{b} = 0 \quad \text{which means}$$

$$\mathbf{A}^T \mathbf{A} \mathbf{x} = \mathbf{A}^T \mathbf{b} \quad \text{QED} \qquad \text{(C-86)}$$

Matrix algebra

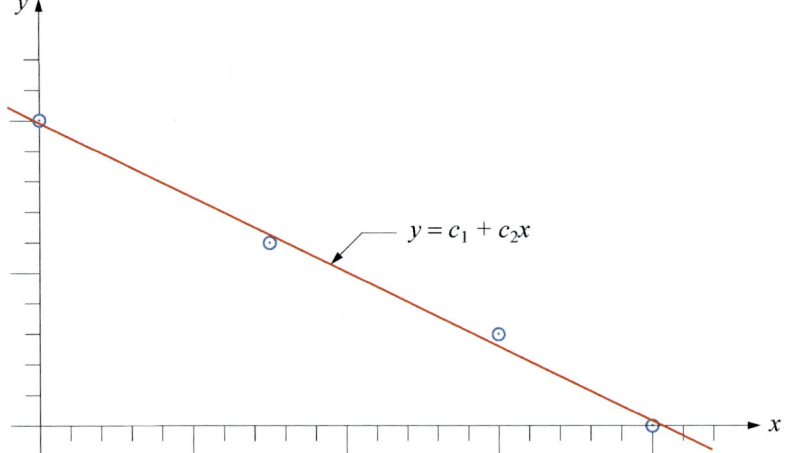

$y = c_1 + c_2 x$

Example – curve fitting

Four points are given in the x-y plane:

$$(0\,;1{,}0),\ (0{,}75\,;0{,}6),\ (1{,}5\,;0{,}3)\ \text{and}\ (2{,}0\,;0)$$

and we want to fit a linear curve,

$$y = c_1 + c_2 x$$

to these points. The error at each of the given points is

$$e_i = c_1 + c_2 x_i - y_i$$

or

$$\begin{bmatrix} e_1 \\ e_2 \\ e_3 \\ e_4 \end{bmatrix} = \underbrace{\begin{bmatrix} 1 & 0 \\ 1 & 0{,}75 \\ 1 & 1{,}5 \\ 1 & 2 \end{bmatrix}}_{\mathbf{A}} \begin{bmatrix} c_1 \\ c_2 \end{bmatrix} - \underbrace{\begin{bmatrix} 1{,}0 \\ 0{,}6 \\ 0{,}3 \\ 0 \end{bmatrix}}_{\mathbf{b}}$$

Thus

$$\mathbf{A}^T\mathbf{A}\mathbf{c} = \bar{\mathbf{A}}\mathbf{c} = \mathbf{A}^T\mathbf{b} = \bar{\mathbf{b}}$$

or

$$\begin{bmatrix} 4{,}0 & 4{,}25 \\ 4{,}25 & 6{,}81 \end{bmatrix}\begin{bmatrix} c_1 \\ c_2 \end{bmatrix} = \begin{bmatrix} 1{,}9 \\ 0{,}9 \end{bmatrix} \quad\Rightarrow\quad \begin{bmatrix} c_1 \\ c_2 \end{bmatrix} = \begin{bmatrix} 0{,}992 \\ 0{,}483 \end{bmatrix}$$

and

$$\mathbf{y} = \mathbf{A}\mathbf{c} = \begin{bmatrix} 0{,}992 \\ 0{,}635 \\ 0{,}27 \\ 0{,}018 \end{bmatrix}$$

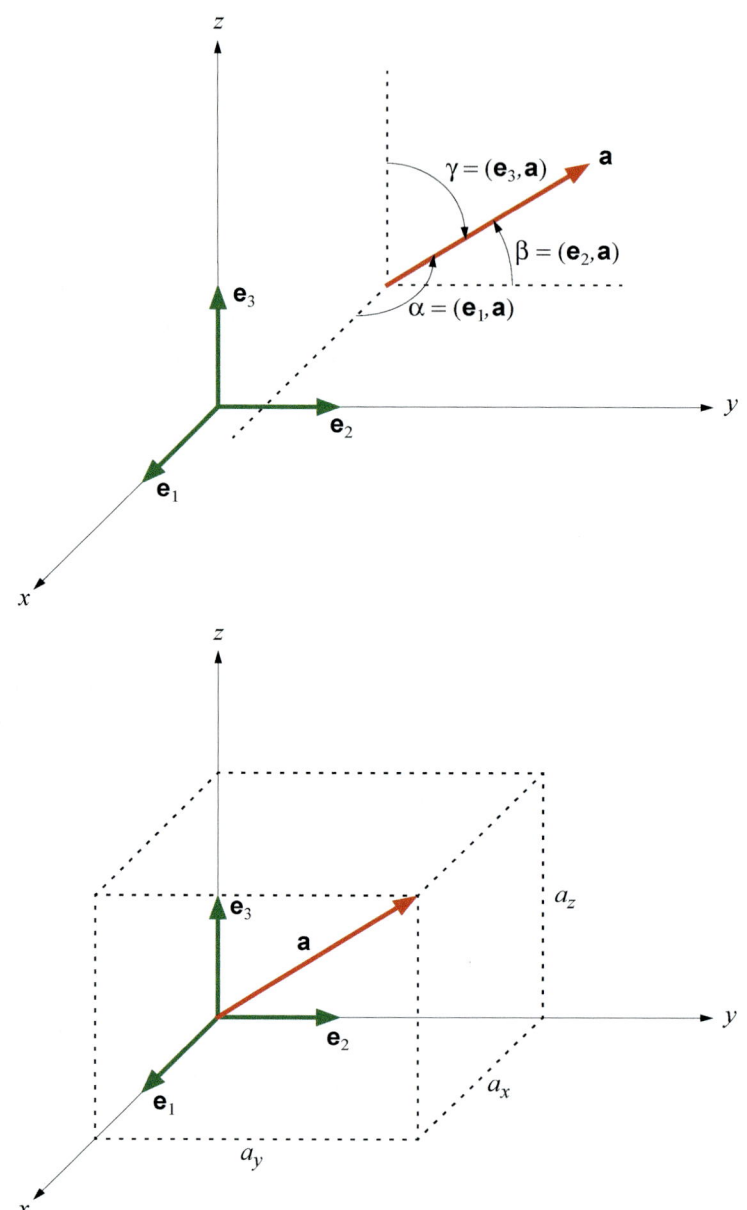

Coordinate transformation

In this appendix we summarize some vector algebra with the purpose of deriving coordinate transformation matrices. Our presentation, which is limited to physical vectors in 3D space, concludes with a Fortran subroutine that computes the rotation matrix, given the global coordinates of three points associated with the rotated (local) coordinate system.

Direction cosines

We consider an arbitrary vector **a** referred to a right-handed, cartesian coordinate system x, y, z whose *base vectors* are $\mathbf{e}_1$, $\mathbf{e}_2$ and $\mathbf{e}_3$, respectively. The base vectors are all *unit* vectors, that is,

$$|\mathbf{e}_1| = |\mathbf{e}_2| = |\mathbf{e}_3| = 1 \tag{C-1}$$

where $|\mathbf{e}_i|$ designates the magnitude or length of $\mathbf{e}_i$. The angles $\alpha = (\mathbf{e}_1, \mathbf{a})$, $\beta = (\mathbf{e}_2, \mathbf{a})$ and $\gamma = (\mathbf{e}_3, \mathbf{a})$ define the *direction* of the vector **a**, and the cosines of these angles are the *direction cosines* of the vector, e.g.

$$\cos\alpha = \cos(\mathbf{e}_1, \mathbf{a}) = \cos(\mathbf{a}, \mathbf{e}_1) \tag{C-2}$$

Since the angle between $\mathbf{e}_1$ and **a** is the same as the angle between **a** and $\mathbf{e}_1$ we introduce the notation

$$c_{1a} \equiv \cos(\mathbf{e}_1, \mathbf{a}) = \cos(\mathbf{a}, \mathbf{e}_1) \equiv c_{a1} \tag{C-3}$$

We can write the component of **a** along the x-axis, for instance, as

$\mathbf{a}_x = a_x \mathbf{e}_1$ where a_x is the *scalar component* of **a** along the x-axis.

The *scalar* or *dot product* of **a** and $\mathbf{e}_1$ is, by definition,

$$\mathbf{a} \cdot \mathbf{e}_1 = a_x = |\mathbf{a}| \cdot |\mathbf{e}_1| \cos(\mathbf{e}_1, \mathbf{a}) = |\mathbf{a}| c_{1a} \tag{C-4}$$

From this it follows that

$$c_{1a} = \frac{a_x}{|\mathbf{a}|} \qquad c_{2a} = \frac{a_y}{|\mathbf{a}|} \qquad c_{3a} = \frac{a_z}{|\mathbf{a}|} \tag{C-5}$$

According to the Pythagorean theorem:

$$|\mathbf{a}|^2 = a_x^2 + a_y^2 + a_z^2 = |\mathbf{a}|^2 c_{1a}^2 + |\mathbf{a}|^2 c_{2a}^2 + |\mathbf{a}|^2 c_{3a}^2$$

Coordinate transformation

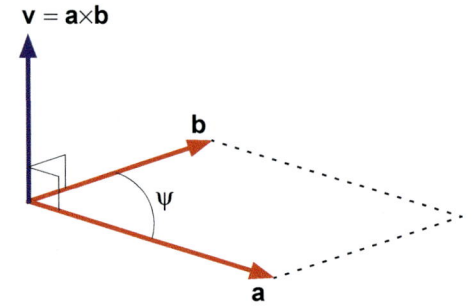

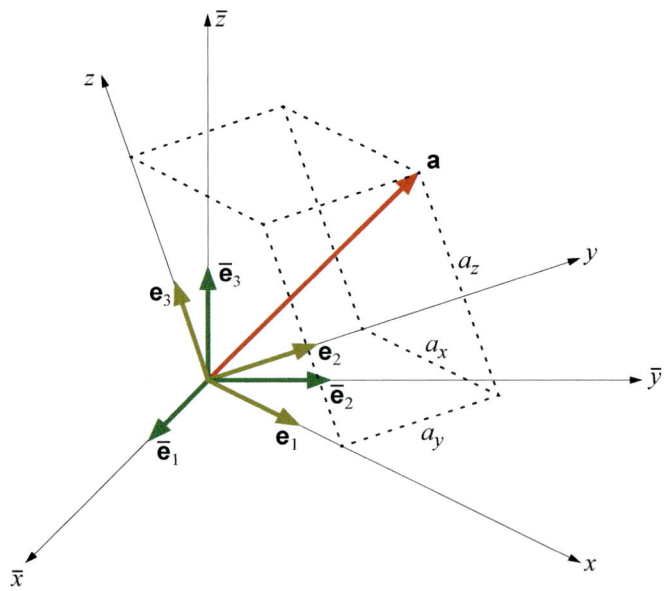

or
$$c_{1a}^2 + c_{2a}^2 + c_{3a}^2 = 1 \tag{C-6}$$

Furthermore, by use of the law of vector addition (*parallelogram law*), we can write

$$\mathbf{a} = a_x\mathbf{e}_1 + a_y\mathbf{e}_2 + a_z\mathbf{e}_3 \tag{C-7}$$

and

$$\mathbf{e} = c_{a1}\mathbf{e}_1 + c_{a2}\mathbf{e}_2 + c_{a3}\mathbf{e}_3 \tag{C-8}$$

where $\mathbf{e}$ is a unit vector along $\mathbf{a}$ and c_{ai} is the cosine of the angle between $\mathbf{a}$ and x ($i = 1$), y ($i = 2$) and z ($i = 3$).

The *vector* or *cross product* of two vectors $\mathbf{a}$ and $\mathbf{b}$ is a vector $\mathbf{v}$, expressed as

$$\mathbf{v} = \mathbf{a} \times \mathbf{b} \tag{C-9}$$

which is *normal* to the plane through $\mathbf{a}$ and $\mathbf{b}$, and such that $\mathbf{a}$, $\mathbf{b}$ and $\mathbf{v}$, in this order, form a *right-handed* system. The length of the vector $\mathbf{v}$ is equal to the area of the parallelogram that has $\mathbf{a}$ and $\mathbf{b}$ as the non-parallel sides, that is

$$|\mathbf{v}| = |\mathbf{a}||\mathbf{b}|\sin\psi \tag{C-10}$$

where ψ is the angle between $\mathbf{a}$ and $\mathbf{b}$. It can be shown (see a standard textbook on the subject) that the cross product can be expressed in terms of the scalar vector components and the base vectors of a cartesian right-handed coordinate system as

$$\mathbf{v} = \mathbf{a} \times \mathbf{b} = (a_y b_z - a_z b_y)\mathbf{e}_1 + (a_z b_x - a_x b_z)\mathbf{e}_2 + (a_x b_y - a_y b_x)\mathbf{e}_3 \tag{C-11}$$

It should be noted that

$$\mathbf{a} \times \mathbf{b} = -\mathbf{b} \times \mathbf{a} \tag{C-12}$$

and

$$\mathbf{a} \times (\mathbf{b} + \mathbf{c}) = (\mathbf{a} \times \mathbf{b}) + (\mathbf{a} \times \mathbf{c}) \tag{C-13}$$

Coordinate transformation

Consider a vector $\mathbf{a}$ referred to two right-handed cartesian coordinate systems, a "local" system x, y, z and a "global" system $\bar{x}$, $\bar{y}$, $\bar{z}$, with base vectors $\mathbf{e}_1$, $\mathbf{e}_2$, $\mathbf{e}_3$ and $\bar{\mathbf{e}}_1$, $\bar{\mathbf{e}}_2$, $\bar{\mathbf{e}}_3$, respectively. For both coordinate systems the origin is at the start point of vector $\mathbf{a}$. According to Eq. (B-7)

$$\mathbf{a} = a_x\mathbf{e}_1 + a_y\mathbf{e}_2 + a_z\mathbf{e}_3 = \bar{a}_x\bar{\mathbf{e}}_1 + \bar{a}_y\bar{\mathbf{e}}_2 + \bar{a}_z\bar{\mathbf{e}}_3 \tag{C-14}$$

The scalar component of $\mathbf{a}$ along the local x-axis, that is a_x, can be expressed in terms of the scalar components along the global axes by scalar multiplication of (B-14) by $\mathbf{e}_1$, i.e.,

$$\mathbf{a} \cdot \mathbf{e}_1 = a_x = \bar{a}_x\bar{\mathbf{e}}_1 \cdot \mathbf{e}_1 + \bar{a}_y\bar{\mathbf{e}}_2 \cdot \mathbf{e}_1 + \bar{a}_z\bar{\mathbf{e}}_3 \cdot \mathbf{e}_1 \tag{C-15}$$

From Eq. (B-4) we have

$$\bar{\mathbf{e}}_j \cdot \mathbf{e}_i = (1)(1)\cos(\bar{\mathbf{e}}_j, \mathbf{e}_i) = \bar{c}_{ji} \tag{B-16a}$$

Coordinate transformation

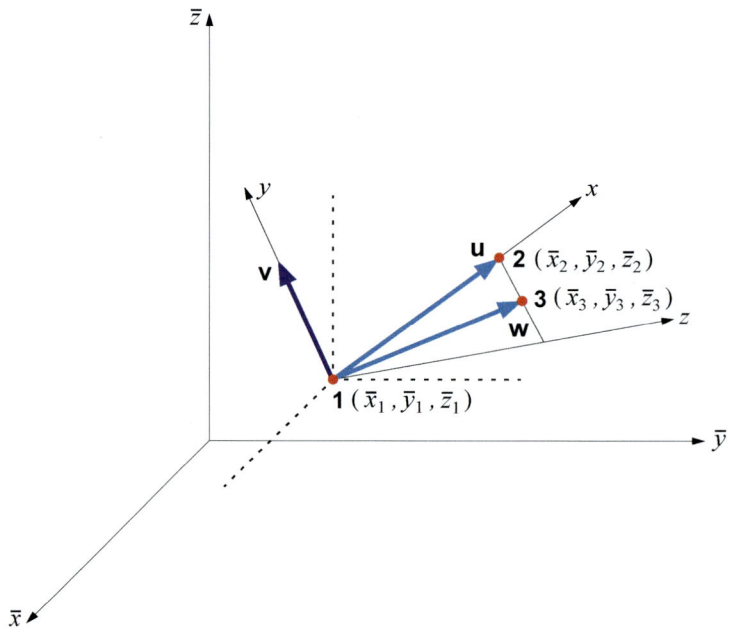

$$\mathbf{e}_i \cdot \bar{\mathbf{e}}_j = (1)(1)\cos(\mathbf{e}_i, \bar{\mathbf{e}}_j) = c_{ij} \tag{B-16b}$$

Also

$$c_{ij} = \bar{c}_{ji} \qquad (\mathbf{e}_i \cdot \bar{\mathbf{e}}_j = \bar{\mathbf{e}}_j \cdot \mathbf{e}_i) \tag{C-17}$$

and Eq. (B-15) can be written as

$$a_x = c_{11}\bar{a}_x + c_{12}\bar{a}_y + c_{13}\bar{a}_z$$

By applying the same reasoning for $\mathbf{e}_2$ and $\mathbf{e}_3$ we can establish the following transformation between the two coordinate systems:

$$\begin{bmatrix} a_x \\ a_y \\ a_z \end{bmatrix} = \begin{bmatrix} c_{11} & c_{12} & c_{13} \\ c_{21} & c_{22} & c_{23} \\ c_{31} & c_{32} & c_{33} \end{bmatrix} \begin{bmatrix} \bar{a}_x \\ \bar{a}_y \\ \bar{a}_z \end{bmatrix} \qquad \text{or} \qquad \mathbf{a} = \mathbf{c}\bar{\mathbf{a}} \tag{C-18}$$

where

$$c_{ij} = \cos(\mathbf{e}_i, \bar{\mathbf{e}}_j) = \cos(\bar{\mathbf{e}}_j, \mathbf{e}_i) = \bar{c}_{ji} \tag{C-19}$$

The *rotation* matrix $\mathbf{c}$ of Eq. (B-18) is identical with the matrix $\mathbf{t}_r$ in Eq. (14-34). If we repeat the above procedure, but the other way around, we find

$$\mathbf{a} \cdot \bar{\mathbf{e}}_1 = \bar{a}_x = \bar{c}_{11}a_x + \bar{c}_{12}a_y + \bar{c}_{13}a_z \qquad \text{or} \qquad \bar{\mathbf{a}} = \bar{\mathbf{c}}\mathbf{a} \tag{C-20}$$

From (B-17) we have

$$\bar{\mathbf{c}} = \mathbf{c}^T \tag{C-21}$$

It also follows from (B-18) and (B-20) that

$$\bar{\mathbf{c}} = \mathbf{c}^{-1} \tag{C-22}$$

Hence

$$\mathbf{c}^{-1} = \mathbf{c}^T \qquad \Rightarrow \qquad \mathbf{c}\mathbf{c}^T = \mathbf{I} \tag{C-23}$$

and the rotation matrix is *orthogonal*.

Our goal is to establish expressions (formulas) for the elements (c_{ij}) of the rotation matrix, based on the "global" coordinates ($\bar{x}_i, \bar{y}_i, \bar{z}_i$) of three characteristic points, 1, 2 and 3, of the "local" coordinate system (x, y, z). Point 1 is the origin of the local system, point 2 is located on the positive x-axis, while point number 3 lies in the xz-plane, and its z-coordinate is positive.

According to the parallelogram law, see Eq. (B-8), the relationship between the base vectors of the two coordinate systems can be expressed as:

$$\mathbf{e}_i = c_{i1}\bar{\mathbf{e}}_1 + c_{i2}\bar{\mathbf{e}}_2 + c_{i3}\bar{\mathbf{e}}_3 \tag{C-24}$$

and
$$i = 1, 2, 3$$

$$\bar{\mathbf{e}}_i = \bar{c}_{i1}\mathbf{e}_1 + \bar{c}_{i2}\mathbf{e}_2 + \bar{c}_{i3}\mathbf{e}_3 \tag{C-25}$$

Local x-axis: the vector $\mathbf{u}$ from point 1 to point 2 can be expressed as

$$\mathbf{u} = (\bar{x}_2 - \bar{x}_1)\bar{\mathbf{e}}_1 + (\bar{y}_2 - \bar{y}_1)\bar{\mathbf{e}}_2 + (\bar{z}_2 - \bar{z}_1)\bar{\mathbf{e}}_3$$

and

Coordinate transformation

```
      SUBROUTINE Dircos (X,Y,Z,C)
!
!======================================================
!
! This subroutine determines the rotation matrix C, relating the three
! components of an arbitrary (physical) vector V in a local coordinate
! system, Vl, to the vector's three components in the global coor-
! dinate system, Vg,
!                 Vl = C*Vg
! The local system is defined by three points:
!  1 - the origin
!  2 - a point on the positive local x-axis
!  3 - a point in the xz-plane having a positive z-coordinate
! Arrays X, Y and Z contains the global coordinates of the three points
! NOTE: Both coordinate systems are right-handed.
!
! Programmed by :  K.Bell
! Date/version  :  10.02.1995 / 1.0
!
!======================================================
!
      IMPLICIT     NONE
      INTEGER, PARAMETER  :: DP = KIND(1.0D0)
!
      REAL(DP), INTENT(IN)  :: X(3),Y(3),Z(3)
      REAL(DP), INTENT(OUT) :: C(3,3)
!
!                  local variables
      REAL(DP)           :: cl,cx,cy,cz
!_____
!
! — The direction cosines of the local x-axis
!
      cx = X(2)-X(1)
      cy = Y(2)-Y(1)
      cz = Z(2)-Z(1)
      cl = SQRT(cx*cx + cy*cy + cz*cz)
!
      C(1,1) = cx/cl
      C(1,2) = cy/cl
      C(1,3) = cz/cl
!
! — The direction cosines of the local y-axis
!
      cx = (Y(3)-Y(1))*(Z(2)-Z(1)) - (Z(3)-Z(1))*(Y(2)-Y(1))
      cy = (Z(3)-Z(1))*(X(2)-X(1)) - (X(3)-X(1))*(Z(2)-Z(1))
      cz = (X(3)-X(1))*(Y(2)-Y(1)) - (Y(3)-Y(1))*(X(2)-X(1))
      cl = SQRT(cx*cx + cy*cy + cz*cz)
!
      C(2,1) = cx/cl
      C(2,2) = cy/cl
      C(2,3) = cz/cl
!
! — The direction cosines of the local z-axis
!
      C(3,1) = C(1,2)*C(2,3) - C(1,3)*C(2,2)
      C(3,2) = C(1,3)*C(2,1) - C(1,1)*C(2,3)
      C(3,3) = C(1,1)*C(2,2) - C(1,2)*C(2,1)
!
      RETURN
      END
```

$$\mathbf{e}_1 = \frac{\mathbf{u}}{|\mathbf{u}|} = \frac{\bar{x}_2-\bar{x}_1}{|\mathbf{u}|}\bar{\mathbf{e}}_1 + \frac{\bar{y}_2-\bar{y}_1}{|\mathbf{u}|}\bar{\mathbf{e}}_2 + \frac{\bar{z}_2-\bar{z}_1}{|\mathbf{u}|}\bar{\mathbf{e}}_3$$

If we compare this expression for $\mathbf{e}_1$ with (B-24) we find:

$$c_{11} = \frac{\bar{x}_2-\bar{x}_1}{|\mathbf{u}|} \qquad c_{12} = \frac{\bar{y}_2-\bar{y}_1}{|\mathbf{u}|} \qquad c_{13} = \frac{\bar{z}_2-\bar{z}_1}{|\mathbf{u}|} \qquad \text{(C-26)}$$

where

$$|\mathbf{u}| = \sqrt{(\bar{x}_2-\bar{x}_1)^2 + (\bar{y}_2-\bar{y}_1)^2 + (\bar{z}_2-\bar{z}_1)^2} \qquad \text{(C-27)}$$

Local y-axis: the vector $\mathbf{w}$ from point 1 to point 3 can be expressed as

$$\mathbf{w} = (\bar{x}_3-\bar{x}_1)\bar{\mathbf{e}}_1 + (\bar{y}_3-\bar{y}_1)\bar{\mathbf{e}}_2 + (\bar{z}_3-\bar{z}_1)\bar{\mathbf{e}}_3$$

The cross product of vectors $\mathbf{w}$ and $\mathbf{u}$ is a vector $\mathbf{v}$ in the direction of positive y-axis; according to Eq. (B-11):

$$\mathbf{v} = \mathbf{w} \times \mathbf{u} = \bar{v}_x\bar{\mathbf{e}}_1 + \bar{v}_y\bar{\mathbf{e}}_2 + \bar{v}_x\bar{\mathbf{e}}_3$$

where

$$\bar{v}_x = (\bar{y}_3-\bar{y}_1)(\bar{z}_2-\bar{z}_1) - (\bar{z}_3-\bar{z}_1)(\bar{y}_2-\bar{y}_1)$$
$$\bar{v}_y = (\bar{z}_3-\bar{z}_1)(\bar{x}_2-\bar{x}_1) - (\bar{x}_3-\bar{x}_1)(\bar{z}_2-\bar{z}_1)$$
$$\bar{v}_z = (\bar{x}_3-\bar{x}_1)(\bar{y}_2-\bar{y}_1) - (\bar{y}_3-\bar{y}_1)(\bar{x}_2-\bar{x}_1)$$

Since $\mathbf{e}_2 = \mathbf{v}/|\mathbf{v}|$ it follows from the above expression for $\mathbf{v}$ and Eq. (B-24) that

$$c_{21} = \frac{\bar{v}_x}{|\mathbf{v}|} \qquad c_{22} = \frac{\bar{v}_y}{|\mathbf{v}|} \qquad c_{23} = \frac{\bar{v}_z}{|\mathbf{v}|} \qquad \text{(C-28)}$$

where

$$|\mathbf{v}| = \sqrt{\bar{v}_x^2 + \bar{v}_y^2 + \bar{v}_x^2} \qquad \text{(C-29)}$$

Local z-axis: the base vector along the z-axis is, by definition,

$$\mathbf{e}_3 = \mathbf{e}_1 \times \mathbf{e}_2$$

or, by use of Eq. (B-24),

$$\mathbf{e}_3 = (c_{11}\bar{\mathbf{e}}_1 + c_{12}\bar{\mathbf{e}}_2 + c_{13}\bar{\mathbf{e}}_3) \times (c_{21}\bar{\mathbf{e}}_1 + c_{22}\bar{\mathbf{e}}_2 + c_{23}\bar{\mathbf{e}}_3)$$
$$= c_{11}c_{22}\bar{\mathbf{e}}_3 - c_{11}c_{23}\bar{\mathbf{e}}_2 - c_{12}c_{21}\bar{\mathbf{e}}_3 + c_{12}c_{23}\bar{\mathbf{e}}_1 + c_{13}c_{21}\bar{\mathbf{e}}_2 - c_{13}c_{22}\bar{\mathbf{e}}_1$$
$$= (c_{12}c_{23} - c_{13}c_{22})\bar{\mathbf{e}}_1 + (c_{13}c_{21} - c_{11}c_{23})\bar{\mathbf{e}}_2 + (c_{11}c_{22} - c_{12}c_{21})\bar{\mathbf{e}}_3$$

Hence

$$c_{31} = c_{12}c_{23} - c_{13}c_{22}, \quad c_{32} = c_{13}c_{21} - c_{11}c_{23} \quad \text{and} \quad c_{33} = c_{11}c_{22} - c_{12}c_{21} \quad \text{(C-30)}$$

and we have explicit expressions for all elements of the rotation matrix $\mathbf{c}$.

A straightforward Fortran 90 code, in the form of a subroutine, is shown opposite.

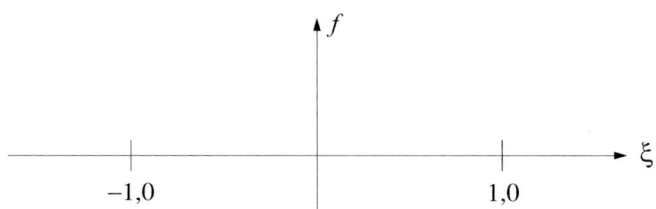

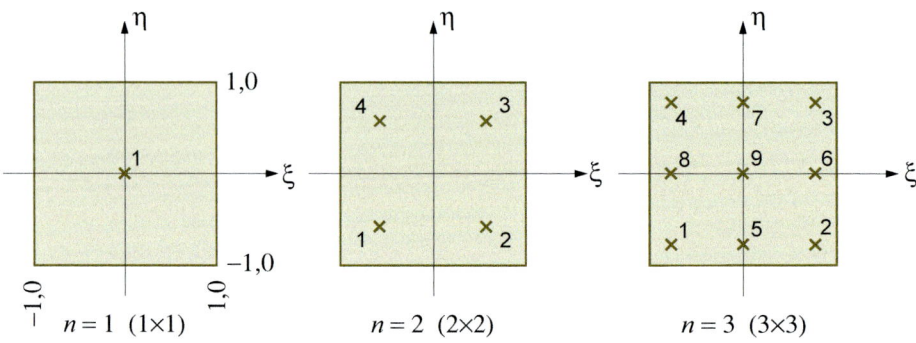

For the 3×3 scheme we have:

m	ξ_m	η_m	w_m
1	$-\sqrt{0{,}6}$	$-\sqrt{0{,}6}$	25/81
2	$\sqrt{0{,}6}$	$-\sqrt{0{,}6}$	25/81
3	$\sqrt{0{,}6}$	$\sqrt{0{,}6}$	25/81
4	$-\sqrt{0{,}6}$	$\sqrt{0{,}6}$	25/81
5	0	$-\sqrt{0{,}6}$	40/81
6	$\sqrt{0{,}6}$	0	40/81
7	0	$\sqrt{0{,}6}$	40/81
8	$-\sqrt{0{,}6}$	0	40/81
9	0	0	64/81

C

Numerical integration

Here we summarize some useful numerical integration rules for quadrilaterals and triangles – natural coordinates are assumed.

C.1 Quadrilaterals

Numerical integration of a function $f(\xi,\eta)$ over a 2×2 square, although not necessarily optimal, is usually accomplished by use of 1-dimensional GAUSSIAN quadrature in each of the two orthogonal directions (ξ and η). In one dimension we have

$$I = \int_{-1}^{1} f(\xi)\,d\xi \approx \sum_{k=1}^{n} w_k f(\xi_k)$$

As shown in Section 6.2 the abscissas and weights for the three first values of n are:

n	ξ_k	w_k	Integrates exactly
1	0,0	2,0	1st order polynomial
2	$\pm 1/\sqrt{3}$	1,0	3rd order polynomial
3	$\pm\sqrt{0,6}$ 0,0	5/9 8/9	5th order polynomial

For the 2×2 square then,

$$I = \int_{-1}^{1}\int_{-1}^{1} f(\xi,\eta)\,d\xi\,d\eta \approx \sum_{k=1}^{n}\sum_{l=1}^{n} w_k w_l f(\xi_k,\eta_l) = \sum_{m=1}^{n\times n} w_m f(\xi_m,\eta_m) \quad \text{where } w_m = w_k \cdot w_l$$

The abscissas, ξ_m and η_m, and weights w_m for a 3×3 GAUSS integration over a 2×2 square are shown opposite. The numbering of the 9 integration points is arbitrary and a matter of taste; that shown to the left is different from that shown in Fig. 6.16, but it coincides with the numbering used in the next appendix. In finite element analysis 3×3 GAUSS is about the highest order of integration used for 2D quadrilaterals; however, the extention to a higher

Numerical integration

p	Point k	ζ_{1k}	ζ_{2k}	ζ_{3k}	Weight w_k
1	1	1/3	1/3	1/3	1,0
3	1	1/3	1/3	1/3	−0,5625
	2	3/5	1/5	1/5	0,520833333333333
	3	1/5	3/5	1/5	0,520833333333333
	4	1/5	1/5	3/5	0,520833333333333
5	1	1/3	1/3	1/3	0,225
	2	α_1	β_1	β_1	0,125939180544827
	3	β_1	α_1	β_1	0,125939180544827
	4	β_1	β_1	α_1	0,125939180544827
	5	α_2	β_2	β_2	0,132394152788506
	6	β_2	α_2	β_2	0,132394152788506
	7	β_2	β_2	α_2	0,132394152788506

$\alpha_1 = 0,797426985353087 \qquad \beta_1 = 0,101286507323456$

$\alpha_2 = 0,059715871789770 \qquad \beta_2 = 0,470142064105115$

p is the degree of the complete polynomial integrated exactly

order scheme is straightforward, and abscissas and corresponding weights are easily found in textbooks or on the internet.

Extention to three dimensions and the 2×2×2 cube is straightforward.

C.2 Triangles

For triangular finite elements we have recommended the use of natural coordinates in the form of *triangular* or *area* coordinates ζ_1, ζ_2 and ζ_3. Numerical integration for triangles will therefore be of the form

$$\int_A f(\zeta_1, \zeta_2, \zeta_3) dA \approx \sum_{k=1}^{n} w_k f(\zeta_{1k}, \zeta_{2k}, \zeta_{3k})$$

Various schemes have been developed, and those shown opposite are amongst the most popular. These and other formulas have been reported by several researchers – those opposite are some of the rules due to DUNAVANT[1]. They are all *open* rules in the sense that all integration points are *inside* the triangle; this is an advantage for certain types of problems, *e.g.*, axisymmetric problems. The 7-point formula opposite works very well for the quintic plate bending elements **T21** and **T18**, see Ref. [12].

Similar formulas can be found (most easily on the internet) for tetrahedrons using volume coordinates.

---

1. D.A. DUNAVANT: "High Degree Efficient Symmetrical Gaussian Quadrature Rules for the Triangle", *Int. J. Num. Meth. Eng.*, **21**, 1129-1148 (1985).

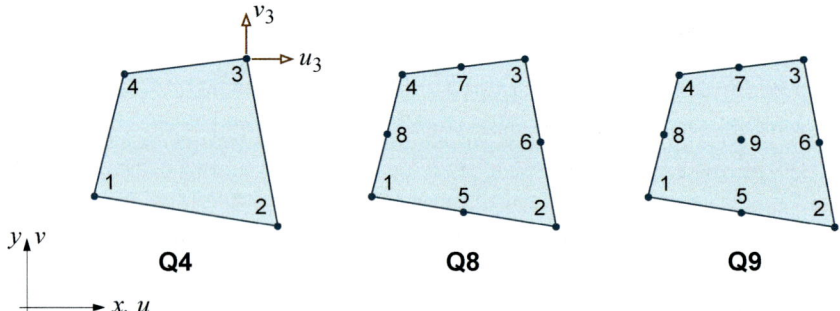

D

Source code

This appendix contains Fortran (77) source code for subroutines computing the element stiffness matrices and other relevant matrices for a family of quadrilateral elements for plane stress and plane strain analysis.

D.1 Introduction

The suite of subroutines contains 4 "public" subroutines,

MPQ61 – generates the element stiffness matrix (**k**),
MPQ62 – generates the element stress matrix (= **CB**(ξ_i, η_i), see Eq. (7-8)),
MPQ63 – generates a consistent element load vector ($\mathbf{S}^0_\Phi / \mathbf{S}^0_F$), and
MPQ64 – generates a consistent element load vector ($\mathbf{S}^0_{\varepsilon 0}$)

for quadrilateral elements with from 4 to 9 nodal points, including the "standard" elements **Q4**, **Q8** and **Q9** (see opposite). Non-corner nodes may be "standard" or "hierarchical", and node 9 may be eliminated (by static condensation) or retained. The integration rule is optional (1×1, 2×2 or 3×3).

In addition to the "public" subroutines, the suite comprises two "private" subroutines,

SHPQ49 – determines the Jacobian, the values of the shape functions and their x and y derivatives at any point within the element.
GAUSQ2 – determines the natural coordinates and the corresponding weights for 1×1, 2×2 or 3×3 GAUSS integration over a 2 by 2 square.

The subroutines, which are believed to be self contained (in terms of documentation), are coded in straightforward Fortran 77. No claim is made as to the efficiency of the subroutines, but they have behaved well in the **FEMplate** program.

It should be noted that the ("doubly fixed") **Q6** element is *not* included.

Source code

D.2 Subroutine MPQ61

```
      SUBROUTINE MPQ61 (C,XG,YG,TH,MNS,NDIM,NIP,IFLAG,SM,S9,IERR)
C
C ***********************************************************************
C
C  S A M  library routine :  MPQ61           Group 6 / Public
C
C  T A S K :  Determine the stiffness matrix, SM, for an isoparametric
C             plane stress / plane strain quadrilateral element with
C             from 4 to 9 nodes.
C             Both Lagrangian and Serendipity elements are available,
C             with "standard" or "hierarchical" non-corner nodes.
C             The type of element is uniquely defined by the "node
C             status" array MNS.  A value of 0, 1 or -1 (see below)
C             must be specified for all 9 nodes, located as shown,
C             in all cases.
C  4 - 7 - 3
C  !   !        MNS(I) = 1 - indicates a standard or ordinary node
C  8   9   6    MNS(I) = 0 - indicates a passive (non-existent) node
C  !   !        MNS(I) =-1 - indicates a hierarchical node (I > 4)
C  1 - 5 - 2        I = 1,2,....,9
C
C             Geometric nodal coordinates are stored in XG and YG,
C             respectively.  Only standard nodes need to be included,
C             since the geometric transformation (that is the Jacobian
C             matrix and its determinant) is based on standard nodes
C             only.  Hence, if node I is hierarchical or passive, XG(I)
C             and YG(I) are irrelevant (but present!).
C             For a (Lagrangian) element with an active node 9, this
C             interior node is automatically eliminated, by static con-
C             densation, if it is defined as a hierarchical node.  If
C             it is defined as a standard node (MNS(9)=1) it is, how-
C             ever, retained.
C             The dimension of SM, that is NDIM, is 2 times the number
C             of retained (active) nodes.
C             For an element with  MNS(9)=-1  the array S9(2,18)
C             contains the submartices, Kie and Kii-1, of the element
C             stiffness matrix associated with the "internal" dofs.
C             (Kii-1 is contained in the last two columns).  These
C             matrices are necessary for generating reduced, consistent
C             load vectors.
C             Integration is accomplished by Gauss quadrature, using
C             NIP (=1, 4 or 9) integration points.
C
C             The element thickness is interpolated (bilinearly)
C             between the four corner values stored in array TH(4).
C
C             All six independent members of the elasticity matrix are
C             stored in array C(6):
C
C             C(1)=C11, C(2)=C22, C(3)=C33, C(4)=C12, C(5)=C13, C(6)=C23
C
C             If IFLAG < 0  only  relevant elements of S9(2,18) are
C                 determined.
C
C
C  ROUTINES CALLED/REFERENCED :  GAUSQ2 and SHPQ49  (SAM-6)
C
C  PROGRAMMED BY :  Kolbein Bell
C  DATE/VERSION :  15.03.1990 / 1.0
C
C ***********************************************************************
C
      IMPLICIT    NONE
C
```

cont.

cont.

```
      INTEGER       IERR,IFLAG,NDIM,NIP,   MNS(9)
      DOUBLE PRECISION  C(6),SM(NDIM,NDIM),S9(2,9),TH(4),XG(*),YG(*)
C
C                          ! local variables
C
      INTEGER       I,II,J,JJ,N,NH,NN,   LA(9)
      DOUBLE PRECISION  C11,C12,C13,C22,C23,C33,CB11,CB12,CB21,CB22,
     +           CB31,CB32,DJ,X1,ZERO,
     +           SHP(3,9),XN(9),YN(9),WG(9)
C
      PARAMETER       ( ZERO = 0.0D0 )
C ──────────────────────────────
      IF (IFLAG.LT.0 .AND. MNS(9).GE.0) THEN
        IERR =-3
        GO TO 100
      ENDIF
C
C — initialize
      IERR = 0
      IF (IFLAG .GE. 0) THEN
        DO 10 J=1,NDIM
          DO 5 I=J,NDIM
            SM(I,J) = ZERO
    5     CONTINUE
   10   CONTINUE
      ENDIF
      IF (MNS(9) .LT. 0) THEN
        DO 20 J=1,18
          DO 15 I=1,2
            S9(I,J) = ZERO
   15     CONTINUE
   20   CONTINUE
      ENDIF
      DO 25 I=1,9
        IF (I .LT. 5) THEN
          LA(I) = 1
          MNS(I) = 1
        ELSE
          LA(I) = 0
        ENDIF
   25 CONTINUE
C                          ! highest node number
C                              of a retained node
      NH = 4
      DO 30 I=4,8
        IF (MNS(I) .NE. 0)  NH = I
   30 CONTINUE
      IF (MNS(9) .GT. 0) NH = 9
C                          ! integr. parameters
C
      IF (NIP.EQ.1 .OR. NIP.EQ.4 .OR. NIP.EQ.9) THEN
        NN = NIP
      ELSEIF (MNS(9) .EQ. 0) THEN
        NN = 4
      ELSE
        NN = 9
      ENDIF
      CALL GAUSQ2 (NN,XN,YN,WG)
C ──────────────────────────────
C  Form lower triangular part of stiffness - submatrix by submatrix
C ──────────────────────────────
      DO 50 N=1,NN
C                          ! thickness

        CALL SHPQ49 (XN(N),YN(N),XG,YG,LA,-1,DJ,SHP)
```

cont.

Source code

cont.

```
      X1 = ZERO
      DO 35 I=1,4
        X1 = X1 + SHP(3,I)*TH(I)
   35 CONTINUE
C
C —— shape functions and their x and y derivatives - Jacobian
C
      CALL SHPQ49 (XN(N),YN(N),XG,YG,MNS,1,DJ,SHP)
      IF (DJ .LE. ZERO) THEN
        IERR =-1
        GO TO 100
      ENDIF
C
      X1 = X1*WG(N)*DJ
      C11 = X1*C(1)
      C12 = X1*C(4)
      C13 = X1*C(5)
      C22 = X1*C(2)
      C23 = X1*C(6)
      C33 = X1*C(3)
C
      J = 1
      DO 45 JJ=1,9
        IF (MNS(JJ) .NE. 0) THEN
          CB11 = C11*SHP(1,JJ) + C13*SHP(2,JJ)
          CB12 = C12*SHP(2,JJ) + C13*SHP(1,JJ)
          CB21 = C12*SHP(1,JJ) + C23*SHP(2,JJ)
          CB22 = C22*SHP(2,JJ) + C23*SHP(1,JJ)
          CB31 = C13*SHP(1,JJ) + C33*SHP(2,JJ)
          CB32 = C23*SHP(2,JJ) + C33*SHP(1,JJ)
C
          IF (IFLAG.GE.0 .AND. JJ.LE.NH) THEN
            I = J
            DO 40 II=JJ,NH
              IF (MNS(II) .NE. 0) THEN
                SM(I  ,J  ) = SM(I  ,J  ) + SHP(1,II)*CB11
     +                      + SHP(2,II)*CB31
                SM(I  ,J+1) = SM(I  ,J+1) + SHP(1,II)*CB12
     +                      + SHP(2,II)*CB32
                SM(I+1,J  ) = SM(I+1,J  ) + SHP(2,II)*CB21
     +                      + SHP(1,II)*CB31
                SM(I+1,J+1) = SM(I+1,J+1) + SHP(2,II)*CB22
     +                      + SHP(1,II)*CB32
                I = I+2
              ENDIF
   40       CONTINUE
          ENDIF
          IF (MNS(9) .LT. 0) THEN
            S9(1,J  ) = S9(1,J  ) + SHP(1,9)*CB11 + SHP(2,9)*CB31
            S9(1,J+1) = S9(1,J+1) + SHP(1,9)*CB12 + SHP(2,9)*CB32
            S9(2,J  ) = S9(2,J  ) + SHP(2,9)*CB21 + SHP(1,9)*CB31
            S9(2,J+1) = S9(2,J+1) + SHP(2,9)*CB22 + SHP(1,9)*CB32
          ENDIF
          J = J+2
        ENDIF
   45 CONTINUE
   50 CONTINUE
C
      IF (MNS(9) .LT. 0) THEN
C
C — form kii-1 in last two columns of S9
C
        J = J-2
        DJ = S9(1,J)*S9(2,J+1) - S9(1,J+1)*S9(2,J)
```

cont.

Subroutine MPQ61

```
      cont.
            IF (DJ .LE. ZERO) THEN
              IERR =-2
              GO TO 100
            ENDIF
            C11 = S9(2,J+1) / DJ
            C12 =-S9(1,J+1) / DJ
            C22 = S9(1,J) / DJ
            S9(1,17) = C11
            S9(1,18) = C12
            S9(2,17) = C12
            S9(2,18) = C22
C
            IF (IFLAG .GE. 0) THEN
C
C —— eliminate internal dofs by static condensation
C
            J = 1
            DO 70 JJ=1,8
              IF (MNS(JJ) .NE. 0) THEN
                CB11 = C11*S9(1,J  ) + C12*S9(2,J  )
                CB12 = C11*S9(1,J+1) + C12*S9(2,J+1)
                CB21 = C12*S9(1,J  ) + C22*S9(2,J  )
                CB22 = C12*S9(1,J+1) + C22*S9(2,J+1)
                I = J
                DO 60 II=JJ,8
                  IF (MNS(II) .NE. 0) THEN
                    SM(I  ,J  ) = SM(I  ,J  ) - S9(1,I  )*CB11
     +                                        - S9(2,I  )*CB21
                    SM(I  ,J+1) = SM(I  ,J+1) - S9(1,I  )*CB12
     +                                        - S9(2,I  )*CB22
                    SM(I+1,J  ) = SM(I+1,J  ) - S9(1,I+1)*CB11
     +                                        - S9(2,I+1)*CB21
                    SM(I+1,J+1) = SM(I+1,J+1) - S9(1,I+1)*CB12
     +                                        - S9(2,I+1)*CB22
                    I = I+2
                  ENDIF
60              CONTINUE
                J = J+2
              ENDIF
70          CONTINUE
          ENDIF
        ENDIF
C
        IF (IFLAG .GE. 0) THEN
C
C —— upper part of stiffness matrix by symmetry
C
          DO 80 I=1,NDIM-1
            DO 75 J=I+1,NDIM
              SM(I,J) = SM(J,I)
75          CONTINUE
80        CONTINUE
        ENDIF
C
 100    RETURN
        END
```

Source code

D.3 Subroutine MPQ62

```
      SUBROUTINE MPQ62 (SM,B,C,XG,YG,XI,ETA,MNS,NDIM,S9,IERR)
C
C**********************************************************************
C
C   S A M  library routine : MPQ62           Group 6 / Public
C
C   T A S K : Determine the stress matrix (SM) and the stress
C             correctiona matrix (B), at an arbitrary point (with
C             natural coordinates XI and ETA) within an isoparametric
C             plane stress/plane strain quadrilateral element with from
C             4 to 9 nodes.
C
C             The type of element is uniquely defined by the "node
C             status" array MNS. A value of 0, 1 or -1 (see below)
C             must be specified for all 9 nodes, located as shown,
C             in all cases.
C   4 - 7 - 3
C   !   !        MNS(I) = 1 - indicates a standard or ordinary node
C   8   9   6    MNS(I) = 0 - indicates a passive (non-existent) node
C   !   !        MNS(I) =-1 - indicates a hierarchical node (I > 4).
C   1 - 5 - 2    I = 1,2,....,9
C
C             Geometric nodal coordinates are stored in XG and YG,
C             respectively.  Only standard nodes need to be included,
C             since the geometric transformation (that is the Jacobian
C             matrix and its determinant) is based on standard nodes
C             only.  Hence, if node I is hierarchical or passive, XG(I)
C             and YG(I) are irrelevant (but present!).
C             For a (Lagrangian) element with an active node 9, this
C             interior node is automatically eliminated, by static con-
C             densation, if it is defined as a hierarchical node.  If
C             it is defined as a standard node (MNS(9)=1) it is, how-
C             ever, retained.
C             The dimension parameter NDIM is 2 times the number of
C             retained (active) nodes.
C             For an element with MNS(9)=-1, the array S9(2,18)
C             contains the submartices, Kie and Kii-1, of the element
C             stiffness matrix associated with the "internal" dofs.
C             (Kii-1 is contained in the last two columns).  These
C             matrices are necessary for determining the corrections
C             due to internal dofs.
C
C
C   ROUTINES CALLED/REFERENCED : SHPQ49   (SAM-6)
C
C   PROGRAMMED BY :  Kolbein Bell
C   DATE/VERSION  :  13.05.2001 / 1.0
C
C**********************************************************************
C
      IMPLICIT      NONE
C
      INTEGER       IERR,NDIM,   MNS(9)
C
      DOUBLE PRECISION ETA,XI
      DOUBLE PRECISION B(3,*),C(6),SM(3,NDIM),S9(2,*),XG(*),YG(*)
C
C                           ! local variables
C
      INTEGER       I,II,J
C
      DOUBLE PRECISION AUX,DJ,ZERO,   SHP(3,9)
C
```

cont.

cont.

```
      PARAMETER    ( ZERO = 0.0D0 )
C
C ─────────────────────────────────────────────
C
      IERR = 0
C
C — Shape functions and their x and y derivatives
C
      CALL SHPQ49 (XI,ETA,XG,YG,MNS,1,DJ,SHP)
      IF (DJ .LE. ZERO) THEN
        IERR =-1
        GO TO 100
      ENDIF
C
C — Stress matrix
C
      I  = 1
      DO 20 II=1,8
        IF (MNS(II) .NE. 0) THEN
          SM(1,I  ) = C(1)*SHP(1,II) + C(5)*SHP(2,II)
          SM(2,I  ) = C(4)*SHP(1,II) + C(6)*SHP(2,II)
          SM(3,I  ) = C(5)*SHP(1,II) + C(3)*SHP(2,II)
          SM(1,I+1) = C(5)*SHP(1,II) + C(4)*SHP(2,II)
          SM(2,I+1) = C(6)*SHP(1,II) + C(2)*SHP(2,II)
          SM(3,I+1) = C(3)*SHP(1,II) + C(6)*SHP(2,II)
          I = I+2
        ENDIF
  20  CONTINUE
      IF (MNS(9) .GT. 0) THEN
        SM(1,17) = C(1)*SHP(1,9) + C(5)*SHP(2,9)
        SM(2,17) = C(4)*SHP(1,9) + C(6)*SHP(2,9)
        SM(3,17) = C(5)*SHP(1,9) + C(3)*SHP(2,9)
        SM(1,18) = C(5)*SHP(1,9) + C(4)*SHP(2,9)
        SM(2,18) = C(6)*SHP(1,9) + C(2)*SHP(2,9)
        SM(3,18) = C(3)*SHP(1,9) + C(6)*SHP(2,9)
      ELSEIF (MNS(9) .LT. 0) THEN
        B(1,1) = C(1)*SHP(1,9) + C(5)*SHP(2,9)
        B(2,1) = C(4)*SHP(1,9) + C(6)*SHP(2,9)
        B(3,1) = C(5)*SHP(1,9) + C(3)*SHP(2,9)
        B(1,2) = C(5)*SHP(1,9) + C(4)*SHP(2,9)
        B(2,2) = C(6)*SHP(1,9) + C(2)*SHP(2,9)
        B(3,2) = C(3)*SHP(1,9) + C(6)*SHP(2,9)
      ENDIF
  50  CONTINUE
C
      IF (MNS(9) .LT. 0) THEN
C
C ── Determine the stress correction matrix B (internal loads)
C
      AUX    = B(1,1)*S9(1,17) + B(1,2)*S9(2,17)
      B(1,2) = B(1,1)*S9(1,18) + B(1,2)*S9(2,18)
      B(1,1) = AUX
      AUX    = B(2,1)*S9(1,17) + B(2,2)*S9(2,17)
      B(2,2) = B(2,1)*S9(1,18) + B(2,2)*S9(2,18)
      B(2,1) = AUX
      AUX    = B(3,1)*S9(1,17) + B(3,2)*S9(2,17)
      B(3,2) = B(3,1)*S9(1,18) + B(3,2)*S9(2,18)
      B(3,1) = AUX
C
C ── Modify the stress matrix
C
      I  = 1
```

cont.

Source code

cont.

```
      DO 40 II=1,8
        IF (MNS(II) .NE. 0) THEN
          J = I+1
          SM(1,I) = SM(1,I) - B(1,1)*S9(1,I) - B(1,2)*S9(2,I)
          SM(2,I) = SM(2,I) - B(2,1)*S9(1,I) - B(2,2)*S9(2,I)
          SM(3,I) = SM(3,I) - B(3,1)*S9(1,I) - B(3,2)*S9(2,I)
          SM(1,J) = SM(1,J) - B(1,1)*S9(1,J) - B(1,2)*S9(2,J)
          SM(2,J) = SM(2,J) - B(2,1)*S9(1,J) - B(2,2)*S9(2,J)
          SM(3,J) = SM(3,J) - B(3,1)*S9(1,J) - B(3,2)*S9(2,J)
          I = I+2
        ENDIF
 40   CONTINUE
      ENDIF
C
 100  RETURN
      END
```

D.4 Subroutine MPQ63

```
      SUBROUTINE MPQ63 (F,XG,YG,TH,MNS,NDIM,NIP,IFLAG,PI,PJ,S9,IERR)
C
C ***********************************************************************
C
C   S A M  library routine : MPQ63           Group 6 / Public
C
C   T A S K : Determine a kinematically consistent load vector, F, for
C             an isoparametric plane stress / plane strain quadri-
C             lateral element with from 4 to 9 nodes, due to
C             - constant volume forces (IFLAG = 0), or
C             - linearly varying edge loading (IFLAG .NE. 0).
C             Only "standard" non-corner side-nodes are considered,
C             and edge loading assumes a straight edge.
C
C             The type of element is uniquely defined by the "node
C             status" array MNS.  A value of 0, 1 or -1 (see below)
C             must be specified for all 9 nodes, located as shown,
C             in all cases.
C   4 - 7 - 3
C   |   |       MNS(I) = 1 - indicates a standard or ordinary node
C   8   9   6   MNS(I) = 0 - indicates a passive (non-existent) node
C   |   |       MNS(I) =-1 - indicates a hierarchical node (I > 4).
C   1 - 5 - 2       I = 1,2,....,9
C
C             Geometric nodal coordinates are stored in XG and YG,
C             respectively.  Only standard nodes need to be included,
C             since the geometric transformation (that is the Jacobian
C             matrix and its determinant) is based on standard nodes
C             only.  Hence, if node I is hierarchical or passive, XG(I)
C             and YG(I) are irrelevant (but present!).
C             For a (Lagrangian) element with an active node 9, this
C             interior node is automatically eliminated, by static con-
C             densation, if it is defined as a hierarchical node.  If
C             it is defined as a standard node (MNS(9)=1) it is, how-
C             ever, retained.
C             The dimension of F, that is NDIM, is 2 times the number
C             of retained (active) nodes.
C             For an element with   MNS(9)=-1  the array S9(2,18)
C             contains the submartices, Kie and Kii-1, of the element
C             stiffness matrix associated with the "internal" dofs.
C             (Kii-1 is contained in the last two columns).  These
C             matrices are necessary for generating reduced, consistent
C             load vectors.
C             Integration is accomplished by Gauss quadrature, using
C             NIP (=1, 4 or 9) integration points.
C
C             The element thickness is interpolated (bilinearly)
C             between the four corner values stored in array TH(4).
C
C
C   ROUTINES CALLED/REFERENCED : GAUSQ2 and SHPQ49  (SAM-6)
C
C   PROGRAMMED BY :  Kolbein Bell
C   DATE/VERSION :  13.05.2001 / 1.0
C
C ***********************************************************************
C
      IMPLICIT    NONE
C
      INTEGER     IERR,IFLAG,NDIM,NIP,   MNS(9)
C
      DOUBLE PRECISION  F(NDIM),PI(2),PJ(2),S9(2,*),TH(4),XG(*),YG(*)
C
```

cont.

Source code

cont.

```fortran
C                         ! local variables
C
      INTEGER      I,II,J,K,N,NH,NN,    IPERM(4),LA(9)
      DOUBLE PRECISION C11,C12,C22,CB11,CB12,CB21,CB22,DJ,FIX,FIY,
     +         C,S,SL,  P1(2),P2(2),
     +         SHP(3,9),XN(9),YN(9),WG(9),X1,ZERO
C
      PARAMETER    ( ZERO = 0.0D0 )
C
      DATA      IPERM(1),IPERM(2),IPERM(3),IPERM(4) / 2,3,4,1 /
C————————————————————————————
C
C — initialize
C
      IERR = 0
C
      DO 10 I=1,NDIM
        F(I) = ZERO
   10 CONTINUE
C
      IF (IFLAG .EQ. 0) THEN
C
C————————————————————————————
C     LOAD VECTOR DUE TO VOLUME FORCES
C————————————————————————————
C
      DO 20 I=1,9
        IF (I .LT. 5) THEN
          LA(I) = 1
          MNS(I) = 1
        ELSE
          LA(I) = 0
        ENDIF
   20 CONTINUE
C
      IF (MNS(9) .LT. 0) THEN
        FIX = ZERO
        FIY = ZERO
      ENDIF
C                         ! highest node number
C                           of a retained node
      NH = 4
      DO 30 I=4,8
        IF (MNS(I) .NE. 0)  NH = I
   30 CONTINUE
      IF (MNS(9) .GT. 0) NH = 9
C                         ! integr. parameters
C
      IF (NIP.EQ.1 .OR. NIP.EQ.4 .OR. NIP.EQ.9) THEN
        NN = NIP
      ELSEIF (MNS(9) .EQ. 0) THEN
        NN = 4
      ELSE
        NN = 9
      ENDIF
      CALL GAUSQ2 (NN,XN,YN,WG)
C
C —— Form load vector by numerical integration
C
```

cont.

cont.

```
      DO 50 N=1,NN
C                         ! thickness
C
      CALL SHPQ49 (XN(N),YN(N),XG,YG,LA,-1,DJ,SHP)
      X1 = ZERO
      DO 35 I=1,4
        X1 = X1 + SHP(3,I)*TH(I)
  35  CONTINUE
C
C ——— shape functions and their x and y derivatives - Jacobian
C
      CALL SHPQ49 (XN(N),YN(N),XG,YG,MNS,1,DJ,SHP)
      IF (DJ .LE. ZERO) THEN
        IERR =-1
        GO TO 100
      ENDIF
C
      X1  = X1*WG(N)*DJ
      I   = 1
      DO 40 II=1,8
        IF (MNS(II) .NE. 0) THEN
          F(I  ) = F(I  ) + PI(1)*X1*SHP(3,II)
          F(I+1) = F(I+1) + PI(2)*X1*SHP(3,II)
          I = I+2
        ENDIF
  40  CONTINUE
      IF (MNS(9) .GT. 0) THEN
        F(I  ) = F(I  ) + PI(1)*X1*SHP(3,9)
        F(I+1) = F(I+1) + PI(2)*X1*SHP(3,9)
      ELSEIF (MNS(9) .LT. 0) THEN
        FIX   = FIX + PI(1)*X1*SHP(3,9)
        FIY   = FIY + PI(2)*X1*SHP(3,9)
      ENDIF
  50  CONTINUE
C
      IF (MNS(9) .LT. 0) THEN
C
C ——— eliminate internal dofs by static condensation
C
      C11 = S9(1,17)
      C12 = S9(1,18)
      C22 = S9(2,18)
      I = 1
      DO 60 II=1,8
        IF (MNS(II) .NE. 0) THEN
          CB11 = C11*S9(1,I  ) + C12*S9(2,I  )
          CB12 = C11*S9(1,I+1) + C12*S9(2,I+1)
          CB21 = C12*S9(1,I  ) + C22*S9(2,I  )
          CB22 = C12*S9(1,I+1) + C22*S9(2,I+1)
          F(I  ) = F(I  ) - CB11*FIX - CB21*FIY
          F(I+1) = F(I+1) - CB12*FIX - CB22*FIY
          I = I+2
        ENDIF
  60  CONTINUE
      PI(1) = FIX
      PI(2) = FIY
      ENDIF
C
      ELSE
C
C————————————————————————
C     LOAD VECTOR DUE TO LINEARLY VARYING EDGE LOAD
C————————————————————————
C
```

cont.

Source code

cont.

```fortran
      IF (IFLAG .GT. 0) THEN
        I = IFLAG
        J = IPERM(I)
C
        SL = SQRT((XG(J)-XG(I))**2 + (YG(J)-YG(I))**2)
        S = (YG(J)-YG(I))/SL
        C = (XG(J)-XG(I))/SL
C
        P1(1) = PI(1)*C - PI(2)*S
        P1(2) = PI(1)*S + PI(2)*C
        P2(1) = PJ(1)*C - PJ(2)*S
        P2(2) = PJ(1)*S + PJ(2)*C
C
      ELSE
C
        I = ABS(IFLAG)
        J = IPERM(I)
        SL = SQRT((XG(J)-XG(I))**2 + (YG(J)-YG(I))**2)
        P1(1) = PI(1)
        P1(2) = PI(2)
        P2(1) = PJ(1)
        P2(2) = PJ(2)
      ENDIF
C
      K = I+4
C
      IF (MNS(K) .GT. 0) THEN
C                         Mid-side node
        F(2*I-1) = SL*P1(1)/6.0D0
        F(2*I)   = SL*P1(2)/6.0D0
        F(2*J-1) = SL*P2(1)/6.0D0
        F(2*J)   = SL*P2(2)/6.0D0
        F(2*K-1) = SL*(P1(1)+P2(1))/3.0D0
        F(2*K)   = SL*(P1(2)+P2(2))/3.0D0
      ELSE
C                         No mid-side node
        F(2*I-1) = SL*(P1(1)+P2(1))/4.0D0
        F(2*J-1) = SL*(P1(1)+P2(1))/4.0D0
        F(2*I)   = SL*(P1(2)+P2(2))/4.0D0
        F(2*J)   = SL*(P1(2)+P2(2))/4.0D0
      ENDIF
C
      ENDIF
C
  100 RETURN
      END
```

D.5 Subroutine MPQ64

```
      SUBROUTINE MPQ64 (F,FI,C,XG,YG,TH,MNS,NDIM,NIP,EC,S9,IERR)
C
C ***********************************************************************
C
C   S A M  library routine : MPQ64         Group 6 / Public
C
C   T A S K :  Determine a kinematically consistent load vector, F, for
C              an isoparametric plane stress / plane strain quadri-
C              lateral element with from 4 to 9 nodes, due to a bi-
C              linear varying initial strain (E0).
C
C          The type of element is uniquely defined by the "node
C          status" array MNS. A value of 0, 1 or -1 (see below)
C          must be specified for all 9 nodes, located as shown,
C          in all cases.
C   4 - 7 - 3
C   |   |        MNS(I) = 1 - indicates a standard or ordinary node
C   8   9   6    MNS(I) = 0 - indicates a passive (non-existent) node
C   |   |        MNS(I) =-1 - indicates a hierarchical node (I > 4).
C   1 - 5 - 2        I = 1,2,....,9
C
C          Geometric nodal coordinates are stored in XG and YG,
C          respectively. Only standard nodes need to be included,
C          since the geometric transformation (that is the Jacobian
C          matrix and its determinant) is based on standard nodes
C          only. Hence, if node I is hierarchical or passive, XG(I)
C          and YG(I) are irrelevant (but present!).
C          For a (Lagrangian) element with an active node 9, this
C          interior node is automatically eliminated, by static con-
C          densation, if it is defined as a hierarchical node. If
C          it is defined as a standard node (MNS(9)=1) it is, how-
C          ever, retained.
C          The dimension of F, that is NDIM, is 2 times the number
C          of retained (active) nodes.
C          For an element with MNS(9)=-1, the array S9(2,18)
C          contains the submartices, Kie and Kii-1, of the element
C          stiffness matrix associated with the "internal" dofs.
C          (Kii-1 is contained in the last two columns). These
C          matrices are necessary for generating a reduced, con-
C          sistent load vector.
C          Integration is accomplished by Gauss quadrature, using
C          NIP (=1, 4 or 9) integration points.
C
C          The element thickness is interpolated (bilinearly)
C          between the four corner values stored in array TH(4).
C
C
C   ROUTINES CALLED/REFERENCED :  GAUSQ2 and SHPQ49  (SAM-6)
C
C   PROGRAMMED BY :  Kolbein Bell
C   DATE/VERSION  :  13.05.2001 / 1.0
C
C ***********************************************************************
C
      IMPLICIT      NONE
C
      INTEGER       IERR,IFLAG,NDIM,NIP,  MNS(9)
      DOUBLE PRECISION  C(6),EC(3,4),F(NDIM),FI(*)
      DOUBLE PRECISION  S9(2,*),TH(4),XG(*),YG(*)
C
C                       ! local variables
C
      INTEGER       I,II,J,K,N,NH,NN,   LA(9)
```

cont.

cont.

```
      DOUBLE PRECISION  C11,C12,C22,CB11,CB12,CB21,CB22,DJ,FIX,FIY,
     +           E0(3),SHP(3,9),XN(9),YN(9),WG(9),X1,ZERO
C
      PARAMETER       ( ZERO = 0.0D0 )
C
C_____
C
C — initialize
C
      IERR = 0
C
      DO 10 I=1,NDIM
        F(I) = ZERO
  10  CONTINUE
C
      DO 20 I=1,9
        IF (I .LT. 5) THEN
          LA(I) = 1
          MNS(I) = 1
        ELSE
          LA(I) = 0
        ENDIF
  20  CONTINUE
C
      IF (MNS(9) .LT. 0) THEN
        FIX = ZERO
        FIY = ZERO
      ENDIF
C                         ! highest node number
C                              of a retained node
      NH = 4
      DO 30 I=4,8
        IF (MNS(I) .NE. 0)  NH = I
  30  CONTINUE
      IF (MNS(9) .GT. 0) NH = 9
C                         ! integr. parameters
C
      IF (NIP.EQ.1 .OR. NIP.EQ.4 .OR. NIP.EQ.9) THEN
        NN = NIP
      ELSEIF (MNS(9) .EQ. 0) THEN
        NN = 4
      ELSE
        NN = 9
      ENDIF
      CALL GAUSQ2 (NN,XN,YN,WG)
C
C — Form load vector by numerical integration
C
      DO 50 N=1,NN
C                         ! thickness & init.str.
C
        CALL SHPQ49 (XN(N),YN(N),XG,YG,LA,-1,DJ,SHP)
        X1   = ZERO
        E0(1) = ZERO
        E0(2) = ZERO
        E0(3) = ZERO
        DO 35 I=1,4
          X1 = X1 + SHP(3,I)*TH(I)
          E0(1) = E0(1) + SHP(3,I)*EC(1,I)
          E0(2) = E0(2) + SHP(3,I)*EC(2,I)
          E0(3) = E0(3) + SHP(3,I)*EC(3,I)
  35    CONTINUE
C
```

cont.

cont.

```
C —— shape functions and their x and y derivatives - Jacobian
C
      CALL SHPQ49 (XN(N),YN(N),XG,YG,MNS,1,DJ,SHP)
      IF (DJ .LE. ZERO) THEN
        IERR =-1
        GO TO 100
      ENDIF
C
      E0(1) = X1*WG(N)*DJ*E0(1)
      E0(2) = X1*WG(N)*DJ*E0(2)
      E0(3) = X1*WG(N)*DJ*E0(3)
      C11  = C(1)*E0(1) + C(4)*E0(2) + C(5)*E0(3)
      C12  = C(4)*E0(1) + C(2)*E0(2) + C(6)*E0(3)
      C22  = C(5)*E0(1) + C(6)*E0(2) + C(3)*E0(3)
      I = 1
      DO 40 II=1,8
        IF (MNS(II) .NE. 0) THEN
          F(I  ) = F(I  ) + SHP(1,II)*C11 + SHP(2,II)*C22
          F(I+1) = F(I+1) + SHP(2,II)*C12 + SHP(1,II)*C22
          I = I+2
        ENDIF
   40 CONTINUE
      IF (MNS(9) .GT. 0) THEN
        F(I  ) = F(I  ) + SHP(1,9)*C11 + SHP(2,9)*C22
        F(I+1) = F(I+1) + SHP(2,9)*C12 + SHP(1,9)*C22
      ELSEIF (MNS(9) .LT. 0) THEN
        FIX  = FIX + SHP(1,9)*C11 + SHP(2,9)*C22
        FIY  = FIY + SHP(2,9)*C12 + SHP(1,9)*C22
      ENDIF
   50 CONTINUE
C
      IF (MNS(9) .LT. 0) THEN
C
C —— eliminate internal dofs by static condensation
C
        C11 = S9(1,17)
        C12 = S9(1,18)
        C22 = S9(2,18)
        I = 1
        DO 60 II=1,8
          IF (MNS(II) .NE. 0) THEN
            CB11 = C11*S9(1,I  ) + C12*S9(2,I  )
            CB12 = C11*S9(1,I+1) + C12*S9(2,I+1)
            CB21 = C12*S9(1,I  ) + C22*S9(2,I  )
            CB22 = C12*S9(1,I+1) + C22*S9(2,I+1)
            F(I  ) = F(I  ) - CB11*FIX - CB21*FIY
            F(I+1) = F(I+1) - CB12*FIX - CB22*FIY
            I = I+2
          ENDIF
   60   CONTINUE
        FI(1) = FIX
        FI(2) = FIY
      ENDIF
C
  100 RETURN
      END
```

Source code

D.6 Subroutine SHPQ49

```
      SUBROUTINE SHPQ49 (XI,ETA,XG,YG,MNS,IFLAG,DJ,SHP)
C
C  **********************************************************************
C
C  S A M  library routine : SHPQ49          Group 6 / Private
C
C  T A S K :  Compute the Jacobian (DJ) and the values of the shapeS
C             functions, SHP(3,I), and its x- and y-derivatives,
C             SHP(1,I) and SHP(2,I), at "natural" coordinates XI and
C             ETA inside a quadrilateral with from 4 to 9 nodes,
C             located as shown.
C
C  4 - 7 - 3     The node "status" is recorded in array MNS(9):
C  !       !       MNS(I) = 1 - indicates ordinary node
C  8   9   6       MNS(I) = 0 - indicates a passive (non-existent) node
C  !       !       MNS(I) =-1 - indicates a hierarchical node (I > 4).
C  1 - 5 - 2
C             Geometric nodal coordinates are stored in XG and YG,
C             respectively.
C             Functions of both Lagrangian (MNS(9).NE.0) and
C             Serendipity (MNS(9)=0, and at least one side node) type
C             are considered, and side nodes may be "ordinary" or so-
C             called "hierarchical".
C             If IFLAG < 0 only the shape functions (and their natural
C                   coordinate derivatives) are computed.
C             If IFLAG = 0 above + Jacobian are computed.
C             If IFLAG > 0 shape functions and their geometric (x and
C                   y) coordinate derivatives are computed, as
C                   is the Jacobian.
C
C             NOTE :  Only standard (non-hierarchical) nodes are used
C                in the coordinate mapping (that is in computing
C                the Jacobian matrix and its determinant).
C
C  ROUTINES CALLED/REFERENCED :  None
C
C  PROGRAMMED BY :  Kolbein Bell
C  DATE/VERSION :   13.03.1990 / 1.0
C
C  **********************************************************************
C
      IMPLICIT         NONE
C
      INTEGER          IFLAG,  MNS(9)
      DOUBLE PRECISION DJ,ETA,XI,  SHP(3,*),XG(*),YG(*)
C
C                      ! local variables
      INTEGER          I,J,K,NH
      DOUBLE PRECISION AJ11,AJ12,AJ21,AJ22,FOUR,HALF,ONE,S2,T2,ZERO
      DOUBLE PRECISION S(4),T(4)
C
      SAVE      S,T
C
      PARAMETER        (ZERO=0.0D0,HALF=0.5D0,ONE=1.0D0,FOUR=4.0D0)
C
      DATA     S /-0.5D0, 0.5D0,0.5D0,-0.5D0/
      DATA     T /-0.5D0,-0.5D0,0.5D0, 0.5D0/
C
C  _____
C                      ! highest node number
C                         of a retained node
      NH = 4
```

cont.

cont.

```
      DO 5 I=5,8
        IF (MNS(I) .NE. 0)  NH=I
    5 CONTINUE
      IF (MNS(9) .GT. 0)    NH=9
C
C — form shape functions and their natural coordinate derivatives for
C   4-node quadrilateral
C
      DO 10 I=1,4
        SHP(3,I) = (HALF + S(I)*XI)*(HALF + T(I)*ETA)
        SHP(1,I) = S(I)*(HALF + T(I)*ETA)
        SHP(2,I) = T(I)*(HALF + S(I)*XI)
   10 CONTINUE
C
      IF (NH .GT. 4) THEN
C
        DO 20 J=5,NH
          DO 15 I=1,3
            SHP(I,J) = ZERO
   15     CONTINUE
   20   CONTINUE
C
C —— add quadratic functions as specified by MNS
C
        S2 = HALF*(ONE - XI*XI)
        T2 = HALF*(ONE - ETA*ETA)
C                             ! side nodes
C                               (Serendipity)
        IF (MNS(5) .NE. 0) THEN
          SHP(3,5) = S2*(ONE-ETA)
          SHP(1,5) =-XI*(ONE-ETA)
          SHP(2,5) =-S2
        ENDIF
        IF (MNS(6) .NE. 0) THEN
          SHP(3,6) = T2*(ONE+XI)
          SHP(1,6) = T2
          SHP(2,6) =-ETA*(ONE+XI)
        ENDIF
        IF (MNS(7) .NE. 0) THEN
          SHP(3,7) = S2*(ONE+ETA)
          SHP(1,7) =-XI*(ONE+ETA)
          SHP(2,7) = S2
        ENDIF
        IF (MNS(8) .NE. 0) THEN
          SHP(3,8) = T2*(ONE-XI)
          SHP(1,8) =-T2
          SHP(2,8) =-ETA*(ONE-XI)
        ENDIF
C
        IF (MNS(9) .NE. 0) THEN
C                             ! interior node
C                               (Lagrangian)
          SHP(3,9) = FOUR*S2*T2
          SHP(1,9) =-FOUR*XI*T2
          SHP(2,9) =-FOUR*ETA*S2
C
        IF (MNS(9) .GT. 0) THEN
C
C —— correct edge nodes for presence of standard interior node
C
          DO 40 I=1,3
            DO 30 J=1,4
              SHP(I,J) = SHP(I,J) - SHP(I,9)/FOUR
   30       CONTINUE
```

cont.

Source code

cont.

```fortran
          DO 35 J=5,8
            IF (MNS(J).NE.0)  SHP(I,J)=SHP(I,J)-HALF*SHP(I,9)
 35       CONTINUE
 40     CONTINUE
      ENDIF
    ENDIF
C
C —— correct corner nodes for presence of standard side nodes
C
    K = 8
    DO 60 J=1,4
      IF (MNS(K) .EQ. 1) THEN
        DO 50 I=1,3
          SHP(I,J) = SHP(I,J) - HALF*SHP(I,K)
 50     CONTINUE
      ENDIF
      K = J+4
      IF (MNS(K) .EQ. 1) THEN
        DO 55 I=1,3
          SHP(I,J) = SHP(I,J) - HALF*SHP(I,K)
 55     CONTINUE
      ENDIF
 60  CONTINUE
C
    ENDIF
C
    IF (IFLAG .GE. 0) THEN
C
C —— determine the Jacobian matrix and its determinant
C
      AJ11 = ZERO
      AJ12 = ZERO
      AJ21 = ZERO
      AJ22 = ZERO
      DO 70 J=1,NH
        IF (MNS(J) .GT. 0) THEN
          AJ11 = AJ11 + XG(J)*SHP(1,J)
          AJ12 = AJ12 + YG(J)*SHP(1,J)
          AJ21 = AJ21 + XG(J)*SHP(2,J)
          AJ22 = AJ22 + YG(J)*SHP(2,J)
        ENDIF
 70   CONTINUE
      DJ = AJ11*AJ22 - AJ21*AJ12
C
      IF (IFLAG .GT. 0) THEN
C
C ——— transform to x and y derivatives
C
        S2 = DJ
        IF (S2 .LE. ZERO)  S2 = ONE
        DO 80 J=1,9
          IF (MNS(J) .NE. 0) THEN
            T2    = (AJ22*SHP(1,J) - AJ12*SHP(2,J)) / S2
            SHP(2,J) = (-AJ21*SHP(1,J) + AJ11*SHP(2,J)) / S2
            SHP(1,J) = T2
          ENDIF
 80     CONTINUE
      ENDIF
    ENDIF
C
    RETURN
    END
```

D.7 Subroutine GAUSQ2

```
      SUBROUTINE GAUSQ2 (NIP,XN,YN,W)
C
C **********************************************************************
C
C   S A M  library routine : GAUSQ2         Group 6 / Private
C
C   T A S K : To return natural coordinates and (resulting) weights
C             for 1-, 2- or 3-point Gauss quadrature in two dimensions.
C             The points are numbered as follows in arrays XN, YN and
C             W:
C                          4      3         4   7   3
C
C                1                           8   9   6
C
C                          1      2         1   5   2
C
C             1-point rule    2-point rule    3-point rule
C               NIP = 1         NIP = 4         NIP = 9
C
C
C
C   ROUTINES CALLED/REFERENCED :  SQRT   (Fortran library)
C
C   PROGRAMMED BY :  Kolbein Bell
C   DATE/VERSION :   14.03.1990 / 1.0
C
C **********************************************************************
C
      IMPLICIT      NONE
C
      INTEGER       NIP
      DOUBLE PRECISION  W(*),XN(*),YN(*)
C                       ! local variables
C
      DOUBLE PRECISION  FIVE,FOUR,G,ONE,THREE,W1,W2,W3,ZERO
C
      PARAMETER (ZERO=0.0D0,ONE=1.0D0,THREE=3.0D0,FOUR=4.0D0,FIVE=5.0D0)
      PARAMETER (W1=25.0D0/81.0D0, W2=40.0D0/81.0D0, W3=64.0D0/81.0D0)
C ----------------------------------------------------------------------
C
      IF (NIP .EQ. 1) THEN
C
         XN(1) = ZERO
         YN(1) = ZERO
         W(1)  = FOUR
C
      ELSEIF (NIP .EQ. 4) THEN
C
         G     = ONE/SQRT(THREE)
         XN(1) =-G
         XN(2) = G
         XN(3) = G
         XN(4) =-G
         YN(1) =-G
         YN(2) =-G
         YN(3) = G
         YN(4) = G
         W(1)  = ONE
         W(2)  = ONE
         W(3)  = ONE
         W(4)  = ONE
C
      ELSEIF (NIP .EQ. 9) THEN
C
```

cont.

Source code

cont.

```
      G    = SQRT(THREE/FIVE)
      XN(1) =-G
      XN(2) = G
      XN(3) = G
      XN(4) =-G
      XN(5) = ZERO
      XN(6) = G
      XN(7) = ZERO
      XN(8) =-G
      XN(9) = ZERO
      YN(1) =-G
      YN(2) =-G
      YN(3) = G
      YN(4) = G
      YN(5) =-G
      YN(6) = ZERO
      YN(7) = G
      YN(8) = ZERO
      YN(9) = ZERO
      W(1)  = W1
      W(2)  = W1
      W(3)  = W1
      W(4)  = W1
      W(5)  = W2
      W(6)  = W2
      W(7)  = W2
      W(8)  = W2
      W(9)  = W3
      ENDIF
C
      RETURN
      END
```

Index

A
ABAQUS 549
adaptive meshing 277
Algol 563
ANSYS 549
arch 501
    load modelling 503
    rise-to-span ratio 505
    stress distribution 507
area coordinates
    definition 145
    differentiation 147
    integration 151
array (Fortran) 351
assembly process 117, 303
axisymmetric
    load 409
    problem 409
    strain 413

B
backsubstitution 331
backward compatibility 347
bandwidth 299
Barlow points 247, 403
basic assumption 95, 125
basic element 139, 333
beam curvature 79
beam theory
    Euler-Bernoulli 17, 79, 437
    Mindlin 81, 443
    Timoshenko 79, 437
bending action (shell) 499, 507
BERNOULLI, J. 24
BETTI, E. 38
bi-moment 453
bit 339
boundary conditions 305, 407, 555
    axial symmetry 417
    elimination of dofs 311
    essential 101
    implementation 309
    kinematic 491
    Kirchhoff theory 491
    Lagrange multipliers 317
    mechanical 491
    natural 101
    penalty functions 321
    rigid elements 327
boundary spring 323
bulk modulus 69

C
calculus of variation 99, 100
CAUCHY, A.L. 58
Cauchy's equation 57
chain rule 140, 197
Cholesky decomposition 333
circumcircle 296
coefficient of thermal expansion 71
cofactor 595
compatibility matrix 27
complementary problem 37
complementary strain energy 489
completeness 111, 369
completeness requirement 159
compliance matrix 67
computational model 21
computer word 565
Condeep platform 335, 541
condition number 339
connectivity information 353
connectivity matrix 117
connectivity table 299, 305
consistent load vector 235
constraint
    equation 307, 309, 353
    multi-point 307
    single point 307
continuity 369
    C0 115
continuity requirement 159
contragredient transformation 41
convergence
    axial symmetry 417
    completeness criterion 277
    continuity requirement 281
    necessary requirements 163
    rate of 271

sufficient requirements 163
Cook's problem 381, 395, 401
coordinates
    area 145
    curvilinear 191
    global 21
    local 17, 21
    natural 139, 143
    right-handed cartesian 15
    transformation 39
    triangular 145
    unit triangle 145
    volume 151, 429
coplanar shell elements
    drilling dof 515
    fictitious stiffness 515
    tangent plane 513
coupling spring 323
Cramer's rule 594
Cross method 569
cross section
    massive 365, 521
    principal axes 519, 525
    shear centre 525, 531
    thin-walled 365
CrossX (program) 365
CST - constant strain triangle 371
cubic triagle (plate bending) 471
cuboid 143
curvature
    Kirchhoff 461, 467
    Mindlin 479
curved beam 501
cyclic permutation 149

D

data structures 349
data word 339
deformation stiffness 453
degrees of freedom 155
    dependent 307
    drilling 513
    free 307
    hierarchic 157
    master 307
    node-less 157, 391
    prescribed 307, 351
    rotational 15
    slave 307, 351
    specified 307

    standard 157
    status 351
    suppressed 307, 351
    translational 15
Delaunay triangulation 296
determinant
    definition 595
    properties of 595
diagonal decay 341, 549
diaphragm 499
differentiation wrt a vector 609
dilatation 69
direct stiffness method 33
direction cosine 55, 617
discrete Kirchhoff
    beam element 451
    plate bending element 485
    triangle - DKT 485
displacement
    deformation 23
    rigid body 23
displacement constraint 305
displacement method 29, 33
displacement release 307
displacement vector 21
divergence theorem 57
drilling dof 513, 515
dyadic product 593

E

eccentricity 459
eigenvalue problem 607
eigenvalue properties 608
eigenvalue test 283
elasticity matrix 67
element
    curved edges 203
    equilibrium 267
    flexibility matrix 23
    hexahedral 429
    higher order quadrilateral 203
    isoparametric triangles 205
    kind 285, 395
    load vector 37
    quadrilateral 195
    rigid 327
    ring 409
    stiffness matrix 23, 117, 127
    tetrahedral 433
    type 285, 395, 565

element instability 225
element load vector 117
   consistent 127
element Q4
   basic version 385
   incompatible version 391
element Q6 395
element Q8 (serendipity) 401
element Q9 (Lagrange) 401
energy bounds 269
engineering approach 1, 5
equation solvers
   direct 329
   frontal 329
   iterative 329
equilibrium
   average 109, 269
   element 267
   neutral 103
   nodal point 267
   stable 103
   unstable 103
equilibrium equations 59
equilibrium matrix 25
error 267
   a posteriori estimates 275
   a priori estimates 275
   discretization 267, 293
   interpretation 293
   manipulation 293
   modelling 293
   programming 293
   strain 271
   strain energy 271
   stress 271
EULER, L. 26
Euler-Bernoulli beam theory 17, 79, 437
Euler-Lagrange equation 99
exponent (real number) 339

F

factorization 329
Falk diagram 167, 589
fap2D (program) 47
FEMplate 367
flat shell element 515
flexibility matrix 489
   beam 437
flexural rigidity of a plate 85
force method 29

force-displacement 23
form
   operator 97
   residual 99
   strong 97
   weak 101
Fortran 347
   source code 629
forward substitution 329
FOURIER, J.B.J. 422
framework method 5
functional 99, 101
fundamental requirements 21
   kinematic compatibility 27, 265
   material law 23, 265
   static equilibrium 25, 265

G

GALERKIN, B.G. 120
Galerkin's method 121
Gauss quadrature 363
GAUSS, J.C.F. 220
Gaussian elimination 329
Gauss-Legendre quadrature 209
general purpose program 365
generalized
   displacements 163
   element stiffness matrix 167, 441
geometric distortion 205
geometric invariance 285
geometrical imperfection 49
global L2 projection 251
graphical user interface 345

H

h-convergence 271
Heterosis element 485
hexahedral element 429
hexahedron 143
hierarchic degrees of freedom 157
hierarchic elements 181
hinge 307
HOOKE, R. 66
Hooke's law 67
hourglass modes 225
hourglass stiffness 405
hybrid element 489

I

identity matrix 583
image 97
incompatible modes 391, 431
influence coefficient 25
information carrier 349
initial strain 243
integration by parts 98
interpolation
    direct - 0-lines 173
    direct - C0 elements 169
    Hermitian 179
    indirect 163, 463
    indirect - natural coordinates 179
interpolation function 95
inverse of a matrix 601
isoparametric element 193
isoparametric formulation 195
    completeness 205
    continuity 203

J

JACOBI, C.G. 140
Jacobian 141, 199
    determinant 141
    for parallelogram 201
    scale factor 141
Jacobian matrix 143, 197, 431

K

kinematic compatibility 27, 63
kinematic degrees of freedom 15, 137
kinematic modes 225
Kirchhoff plate theory 83, 461
Kirchhoff shear force 477
KIRCHHOFF, G.R. 84
Kirchhoff's hypothesis 83
Kronecker delta 283, 583
Kryder's law 567

L

Labotto quadrature 215
Lagrange elements 169
Lagrange multipliers 317
Lagrange ploynomials 169
LAGRANGE, J.L. 184
LAMÉ, G.L. 74
Lamé's constant 69
lateral contraction 65
lattice analogy 5

LDLT factorization 331
least square approximation 613
LEGENDRE, A.M. 228
linear dependency 599
linear dependent 323
linear theory 17
linear triangle 175
load lumping 37, 237
load vector 21
    axial symmetry 415
    consistent 235
    edge loading 239, 241
    initial strain 237, 245
    volume forces 237
loading 553
    Fourier series 419
locking 325
    membrane 505
    shear 389, 449, 483
LST - linear strain triangle 373

M

macro elements 139
mantissa (real number) 339
Maple 349, 461
mapping 191, 195
    unique 201
material
    anisotropic 75
    incompressible 69, 325
    isotropic 65
    orthotropic 75
Mathematica 349, 461
mathematical approach 3, 5
Matlab 349
matrix
    adjoint 602
    band 35, 585
    characteristic polynomial 609
    condition number 339
    defect 225, 599
    definition 581
    diagonal 585
    differentiation 609
    dimension 581
    element stiffness 127
    flexibility 437
    hourglass stiffness 399
    ill-conditioned 339
    indefinite 607

inverse 601
lower triangular 585
multiplication 587
normal 605
notation 581
operator 59
orthogonal 39, 605
partitioning 593
positive definite 35, 331, 605
positive semi-definite 607
profile 585
rank 225, 599
rank deficient 599
regular 35, 599
rotation 39
shape function 115
singular 323, 599
skew-symmetric 585
sparse 35, 585
stabilization 399
symmetric 585
trace 285, 341, 608
transpose 583
tridiagonal 585
upper trianguler 585
MAXWELL, J.C. 38
mechanisms 225
membrane action (shell) 499, 507
membrane locking 505
mesh generation 297
    automatic 297
    semi-automatic 297
method
    collocation 121
    Galerkin 121
    least square 121
    Rayleigh-Ritz 103
    weighted residual 121
method of 0-lines 173
Mindlin beam theory 81, 443
Mindlin plate theory 479
mini-computers 567
minor 595
modulus of elasticity 65
moment distribution 5, 47
Moore's law 567
Morley triangle 463
multi-point constraint 307, 357

N

NASTRAN 543
NAVIER, C.L. 28
Navier's hypothesis 19, 77
NEWTON, I. 196
Newton-Cotes
    quadrature 209
nodal point 137, 155
    equilibrium 267
nodal point averaging 251
nodal point offset 459
node renumbering 301
    Cuthill-McKee 301
    graph theory 301
    RCM 301
    Sloan 301
non-conforming element 391
norm
    Euclidean 282
normal slope 465, 471
number
    binary 339
    floating point 339
    integer 339
    real 339
    representation 339
numerical integration 207
    beam element 213
    convergence 221
    full 219, 431
    Gauss quadrature 209
    integration point 209
    Newton-Cotes quadrature 209
    quadrilaterals 625
    reduced 221
    ring elements 413
    selective 389, 395
    selective reduced 221
    trapezoidal rule 209
    triangles 627
    triangular elements 217
    weight 209
numerical precision 339

O

offset nodes 459
operator form 97
operator matrix 59, 195
over-conformity 115, 375, 477

P

parallelogram law 619
particular problem 37
PASCAL, B. 184
Pascal's triangle 153
patch test 285, 393
    prescribed displacements 287
    prescribed tractions 289
    strong form 287
    weak form 291
p-convergence 271
penalty functions 321
penalty matrix 323, 483
penalty method 483
pivot element 340
pivoting 340
plane strain 75
plane stress 73
plate bending 81
plate bending element
    DKT 485
    Heterosis 485
    T10 471
    T15 471
    T18 475
    T21 473
    T6 (Morley) 463
plate theory
    Kirchhoff (thin plate) 83
    Mindlin 479
POISSON, S.D. 66
Poisson's ratio 65
polynomials
    complete 153
    Hermitian 485
    homogeneous 153
    Lagrange 169
    Legendre 209
post-processing 345
potential energy 103, 269
pre-assembly 355
precision
    double 339, 549
    single 339
pre-processing 345
principal axes 453
principle of
    minimum potential energy 103
    superposition 17
    virtual displacements 29
program
    CrossX 365
    fap2D 47
    FEMplate 367
    general purpose 365, 549
    special purpose 365, 549
programming
    event driven 347
    object oriented 347
    paradigms 347
    procedural 347
    structured 347
    style 349
programming language
    C 347
    C# 347
    C++ 347
    Fortran 347
    Java 347
    Python 349
    Simula 67 347
properties of determinants 595
properties of K 331

Q

quadratic form 605
quadratic strain triangle 373
quadratic triangle 175
quadratic triangle (plate bending) 463
quadrature
    Gauss 209
    Labotto 215
    Newton-Cotes 209
quadrilateral element 195
quantum computing 573
quartic triangle (plate bending) 471
quintic triangle (plate bending) 473

R

rank 599
rank deficiency 25, 225
rank deficient 324
rational functions 201, 207
RAYLEIGH, Lord 120
Rayleigh-Ritz method 103
    accuracy 111
    convergence 111
    equilibrium 109
    piece-wise assumption 113

problems 111
  similarity with FEM 113
  sub-domain 113
residual 97
residual bending flexibility 449
residual forces 341, 477
results
  interpretation 557
retracking 337
Richardson extrapolation 275
rigid body modes 227, 285
rigid elements 327
ring element 409
RITZ, W. 120
rotation matrix 457, 621
rounding errors 339

S

SAINT-VENANT, A.J.C.B. de 528
scalar 583
scale factor 201
scripting 349
self-straining 277
semi-rigid joints 557
serendipity elements 173
shape function 95, 115, 157
  compatibility 161
  completeness 163
  conformity 161
  continuity 161
  interpolation requirement 163
  matrix 115
  routine 363
shear
  area 77, 437
  average strain 437
  average stress 437
  centre 453, 455, 531
  deformation factor 77, 437, 533
  false strain 389
  locking 389, 449, 483
  modulus 67
  parameter 439
  parasitic strain 389
shell
  conical ring element 509
  doubly curved 499
  FEM models 507
  flat elements 511
  of revolution 507

prismatic 499
  singly curved 499
  thick elements 517
  volume elements 517
sign convention 17
singular modes 225
singularities 405
Sleipner accident 541
special elements 139
special purpose program 365
SPR (superconv. patch recovery) 253
spring
  boundary 323
  coupling 323
spurious modes 225
St.Venant torsion 365, 453, 519
stabilization 227, 399
static condensation 139, 335
storage format
  band 299, 355
  profile 299, 355
  skyline 299
  sparse 301
strain
  energy 63, 87, 223, 269
  error 269
  initial 69, 243
  lateral 65
  normal 61
  shear 61
  vector 63
  volumetric 69
strain-displacement matrix 117, 125
stress
  hydrostatic 67
  matrix 55
  mean 67
  vector 55
stress recovery 245
  Barlow points 247
  extrapolation 249
  from nodal forces 255
  interpolation 249
  optimal sampling points 247
stress smoothing 245
  global L2 projection 251
  nodal point averaging 251
  SPR 253
strong form 97
submatrix 593

655

subparametric element 193
subparametric formulation 203
substructure analysis 139, 333
superconvergent (stress) points 249
superelement 139, 335
superparametric element 193
surface forces 57
system
    assembly 33, 303
    stiffness matrix 33
system of equations
    homogeneous 601
    inhomogeneous 601

T

TAYLOR, B. 274
tests
    eigenvalue test 283
    individual element test 283
    patch test 285
    single-element-test 283
tetrahedral elements 433
tetrahedron 151
theorem
    Betti-Maxwell 35
    reciprocity 35
thermal expansion 243
thermal loading 243
Timoshenko beam theory 79, 437
TIMOSHENKO, S.P. 80
topology 353
torque 521
torsion
    constant 531
    St.Venant 519
    stiffness 531
    stress function 519, 523
    warping 519
traction 53
transpose (of a matrix) 583
transpose of products 591
trial function 103, 111
triangle
    cubic 373
    linear 175, 371
    quadratic 175, 373
triangular matrices 585
tri-cell 541
truncation errors 339

twist
    angle of 521
    rate of 521
types of elements 139

U

undistorted geometry 219
uniform subdivision 269, 297
unit matrix 583
Univac 563
Unix 347

V

variation of a function 99
variational calculus 101
variational order 369
vector
    cross product 619
    definition 581
    inner product 591
    outer product 593
    scalar product 617
virtual memory 567
volume coordinates 151, 429
volume forces 57

W

warping 453
warping function 521
warping torsion 519
weak form 101
weight function 121
weighted residual methods 121

Y

YOUNG, T. 66

Z

zero-energy modes 225, 485